EMBRYONIC STEM CELLS – BASIC BIOLOGY TO BIOENGINEERING

Edited by **Michael S. Kallos**

Embryonic Stem Cells – Basic Biology to Bioengineering
Edited by Michael S. Kallos

Published by InTech

Janeza Trdine 9, 51000 Rijeka, Croatia

Publishing Process Manager Romina Krebel

Technical Editor Teodora Smiljanic

Cover Designer InTech Design Team

Additional hard copies can be obtained from orders@intechopen.com

Embryonic Stem Cells – Basic Biology to Bioengineering, Edited by Michael S. Kallos
p. cm.
ISBN 978-953-307-278-4

For EU product safety concerns:
IN TECH d.o.o., Prolaz Marije Krucifikse Kozulić 3, 51000 Rijeka, Croatia,
info@intechopen.com or visit our website at intechopen.com.

Meet the editor

Dr. Michael S. Kallos, PhD, PEng is an Associate Professor of Chemical Engineering in the Schulich School of Engineering, and Associate Director of the Pharmaceutical Production Research Facility (PPRF), both at the University of Calgary. He is also the Director of the Biomedical Engineering Graduate Program and the Associate Director of the Center for Bioengineering Research and Education (CBRE), both at the University of Calgary. His research interests include stem cell bioprocess design, bioreactors, tissue engineering and mass transfer. He was awarded the Alberta Science and Technology (ASTech) "Leader of Tomorrow" Award in 2002, a Faculty of Engineering Service Award in 2005, and was named one of Calgary's Top 40 Under 40 in 2009. He has also won numerous Teaching Awards from his peers and students, including "Professor of the Year" and "Outstanding Excellence in Teaching".

Contents

Preface

The isolation and culture of human embryonic stem cells by Thomson in the late 1990s has accelerated a paradigm shift in medicine that was started much earlier by Till and McCulloch in the early 1960s with the discovery of the first stem cells in mice. The burgeoning field of regenerative medicine will ultimately transform modern human health care from a molecule-based focus, which serves to alleviate symptoms, to a cell and tissue based focus which has the promise of actually restoring function. Although the potential is enormous, the road is long and there are certainly many milestones along the way. This book, *Embryonic Stem Cells - Basic Biology to Bioengineering* and its companion, *Embryonic Stem Cells - Differentiation and Pluripotent Alternatives*, serve as a snapshot of many of the activities currently underway on a number of different fronts.

This book is divided into five parts and provides a foundation upon which future therapies and uses of embryonic stem cells can be built.

Part 1: Challenges and Possibilities - From New Cell Lines to Alternative Uses of Cryopreserved Embryos

Chapters 1-3 offer a broad overview of some of the challenges in bringing embryonic stem cell based medicine to the clinic, as well as a case study of the derivation of new embryonic stem cell lines, and an alternative to the use of cryopreserved embryos.

Part 2: Methods, Tools and Technologies for Embryonic Stem Cell Culture, Manipulation and Clinical Application

Chapters 4-10 present a wide variety of tools and technologies ranging from large-scale bioreactors to scaled-down bioreactor arrays and synthetic surfaces that can be used for embryonic stem cell culture. In addition, methods for introducing foreign genes into embryonic stem cells and controlling gene expression are described. Lastly, the use of imaging is presented as a tool to measure pluripotency and early differentiation.

Part 3: Applications of Embryonic Stem Cells in Research and Development

Chapters 11-13 present methods to generate chimeric mice for use in research, and in addition, describe the use of embryonic stem cells in toxicological studies and the use of teratomas derived from embryonic stem cells as models for early development, disease, and tumorigenesis.

Part 4: Pluripotency and Molecular Biology of Embryonic Stem Cells

Chapters 14-20 describe our understanding of pluripotency as well as some of the key molecules involved in regulating not only pluripotency but cancer and early embryonic tissues.

Part 5: Lessons from Development

Chapters 21-23 examine the knowledge we have gained from studying embryonic germ cells and pluripotent gametogenic stem cells of asexually reproducing invertebrates.

In the book *Embryonic Stem Cells - Differentiation and Pluripotent Alternatives*, the story continues with a sample of some of the studies currently under way to derive neural, cardiac, endothelial, hepatic and osteogenic lineages. In addition, induced pluripotent stem cells are introduced and other unique sources of pluripotent stem cells are explored.

I would like to thank all of the authors for their valuable contributions. I would also like to thank Megan Hunt who provided me with much needed assistance and acted as a sounding board for early chapter selection, and the staff at InTech, particularly Romina Krebel who answered all of my questions and kept me on track during the entire process.

Calgary, Alberta, Canada, July 2011

Michael S. Kallos
Pharmaceutical Production Research Facility (PPRF),
Schulich School of Engineering, University of Calgary, Alberta
Canada
Department of Chemical and Petroleum Engineering,
Schulich School of Engineering, University of Calgary, Alberta,
Canada

Part 1

Challenges and Possibilities – From New Cell Lines to Alternative Uses of Cryopreserved Embryos

1

Embryonic Stem Cells for Therapies – Challenges and Possibilities

Ronne Wee Yeh Yeo and Sai Kiang Lim
*Institute of Medical Biology, Agency for Science, Technology and Research (A*STAR); Yong Loo Lin School of Medicine, National University of Singapore*
Singapore

1. Introduction

The successful establishment of human embryonic stem cells (hESCs) in culture (Thomson et al., 1998) has raised unprecedented public interest and expectation of treating intractable diseases such as diabetes, spinal cord injuries, neurodegenerative and cardiovascular diseases. Much of this enthusiasm was predicated on the unlimited self-renewal capacity of hESCs and their remarkable plasticity in differentiating into every cell type in our body. These features presented the tantalizing possibility of an unlimited cell source in regenerative medicine to generate any tissues to replace injured or diseased tissues. However, translating the potential of hESC into therapies has been challenging. Although translation of hESC has been severely impeded by social and political constraints placed on hESC research through ethical and religious concerns over the destruction of viable blastocysts during hESC isolation, the main challenges have been safety and technical issues.

2. Challenges in ESC therapy

2.1 Overcoming tumor formation

The two defining characteristics of ESCs are: 1) their pluripotency, or the potential to differentiate into all cell types in the adult body; and 2) their unlimited self-renewal capacity, or the ability to remain in an undifferentiated state and divide indefinitely. For mESCs, pluripotency is often demonstrated by the production of mESC-derived animals through germline transmission by chimeras resulting from injection of the cells into blastocysts or through tetraploid complementation. In hESCs, proof of pluripotency has been limited to formation of teratomas or teratocarcinomas, which are tumors composed of randomly distributed tissues from the three primordial germ layers in immunologically incompetent mice (Lensch et al., 2007). Karyotypically normal, low passage hESCs form benign teratomas that do not contain undifferentiated tissues and are less invasive (Blum et al., 2009; Reubinoff et al., 2000; Thomson et al., 1998) while high passage hESCs which have become karyotypically abnormal give rise to highly invasive, malignant teratocarcinomas (Herszfeld et al., 2006; Plaia et al., 2006; Werbowetski-Ogilvie et al., 2009; Yang et al., 2008).
Pluripotency coupled with unlimited self-renewal not only define ESCs, they are also the main appeal of ESC as the cell source for regenerative medicine but at the same time, pose

significant challenges to the transplantation of differentiated ESCs to replace injured or diseased tissues. The propensity of ESC to differentiate into teratomas necessitates the need to eliminate any residual ESCs in the differentiated cell preparation. There have been many strategies to eliminate residual ESCs or enhance the purity of differentiated ESC preparations. The use of heterologous selectable gene markers such as antibiotic resistance gene or fluorescent protein markers (Klug et al., 1996; M. Li et al., 1998; Muller et al., 2000; Soria et al., 2000) is generally not a strategy of choice as this could introduce potentially deleterious gene mutations. Most of the strategies centered around the use of endogenous markers that are unique or highly expressed on ESCs and not on their differentiated progeny. For example, SSEA-4 and TRA-1-60 which are highly expressed on hESCs have shown to be highly efficient in physically removing contaminating ESCs by magnetic or fluorescence-activated cell sorters (MACS or FACS) (Fong et al., 2009b). Another strategy exploit the flotation density of cell on discontinuous density gradients such as Puresperm- or Percoll-based gradients (Fong et al., 2009a). Using a relatively novel strategy, Choo et al. has raised antibodies against undifferentiated hESCs (Choo et al., 2008) and identified an antibody that was cytotoxic against hESCs by oncosis. This antibody was an IgM that recognizes podocalyxin-like protein-1(PODXL). hESCs that were treated with mAB 84 did not form teratoma when transplanted into SCID mice even after 18-24 weeks. Therefore, there are viable technologies to remove or reduce residual hESCs in differentiated hESC preparation and mitigate the risk of teratoma formation in patients receiving hESC-based cell therapy.

2.2 Overcoming immunorejection

Like all tissue transplants, hESC-based cell therapy will have to circumvent host immune rejection to engraft in the recipients. One proposed strategy was to establish ESC repositories with lines expressing the combinations of HLA molecules that are compatible with HLA haplotypes present in the population (Nakajima et al., 2007; Taylor et al., 2005). Alternatively, the host's immune system could be manipulated to induce tolerance to foreign tissues by ablation of donor-reactive T cell in the thymus, generation of tolerogenic dendritic cells and induction of T_{reg} cells [reviewed in (Chidgey et al., 2008)]. However, with the development of induced pluripotent stem cell technology that makes the creation of "patient-specific" pluripotent cells containing the same genetic material as the recipient a highly viable and practical option, the issue of host rejection has become a non issue.

The quest to create "patient-specific" pluripotent cells began with therapeutic cloning or somatic cell nuclear transfer (SCNT) where the diploid nucleus of a somatic cell was injected into a haploid enucleated egg to be reprogrammed by soluble factors in the host cell. Upon stimulation, the re-programmed cell divides to form a blastocyst with an inner cell mass that has identical nuclear genetic composition as the nucleus donor. Although this approach has worked to generate ESCs from different animals such as mice, rabbits, cats, sheep, cattle, pigs, goats [reviewed in (Wilmut et al., 2002)] and even primates (Byrne et al., 2007), no hESC has been generated through this approach as it remains a highly inefficient process and the use of human oocytes is ethically controversial (French et al., 2008; J. Li et al., 2009b). ESCs generated through SCNT are in principle, heterogeneous in their genetic composition as they contain nuclear DNA of the nucleus donor and mitochondrial DNA of the egg donor (Evans et al., 1999). This raises the possibility that SCNT-derived ESCs could be rejected by the innate immune system of the host with which the ESCs share the same nuclear but not mitochondrial genetic material (Ishikawa et al., 2010).

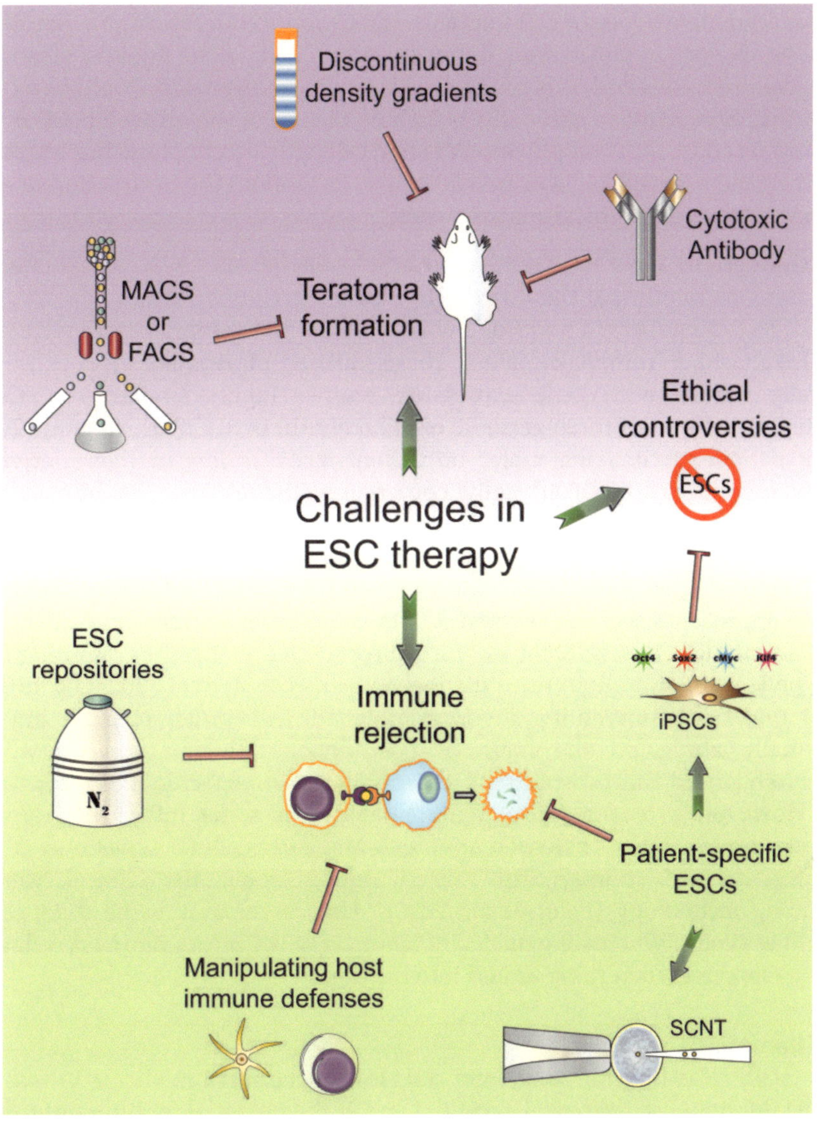

Fig. 1. Mitigating tumor formation and immune rejection. Two of the major challenges to the translation of ESCs into clinical applications are teratoma formation by residual undifferentiated ESCs in the cell preparation and immune rejection of ESC-derived cells or tissues due to incompatible HLA profiles of ESC and recipient. To mitigate the risk of teratoma formation, several methods to remove residual hESCs have been developed using either physical or biological methods. Some of the physical separation methods are based on magnetic- or fluorescence-activated cell sorters (MACS or FACS) that sort against cells with ESC-associated surface markers, SSEA-4 and TRA-1-60 or on cellular density using discontinuous gradients of Percoll or PureSperm. Alternatively, residual ESCs can be destroyed using a cytotoxic antibody (mAb 84) specific for undifferentiated hESCs. To prevent immune rejection, one strategy proposed the establishment of ESC repositories to carry lines expressing HLA combinations compatible with all possible haplotypes in the

population. Alternatively, donor cell tolerance can be induced by manipulating host immune defenses, such as eliminating donor-reactive T cells in the thymus, generating tolerogenic dendritic cells and inducing T_{reg} cells. An ideal approach would be to generate patient-specific ESCs. Some of early efforts include the use of somatic cell nuclear transfer (SCNT). More recently, induced pluripotent stem cell (iPSC) technology has enabled with great ease the generation of self pluripotent stem cells without the destruction of oocytes or embryos, hence bypassing ethical controversies.

The breakthrough in creating "patient-specific" pluripotent cells was achieved when Yamanaka demonstrated that the introduction of transcription factors which regulate ESC self-renewal, including Oct3/4 and Sox2 was sufficient to reprogram somatic cells into ES-like cells (Takahashi & Yamanaka, 2006). These induced pluripotent stem cells (iPSCs) are karyotypically normal with gene expression profiles highly similar to ESCs and can differentiate into cells of all three germ layers (Takahashi et al., 2007; Yu et al., 2007). Apart from being patient-specific, the major attraction of iPSCs lies in their derivation from somatic tissues and not from ethically contentious tissues such as human oocytes or embryos. However, retroviral and lentiviral vectors were required to express the transcription factors for reprogramming of the somatic cells and this carries a risk of insertional mutagenesis. To circumvent the need for viral vectors, non-viral genetic modification approaches were developed (Okita et al., 2008; Soldner et al., 2009; Woltjen et al., 2009). Recently iPSCs were obtained via a direct delivery of reprogramming factors into cells using poly-arginine protein transduction domains (Zhou et al., 2009) or mRNA (Plews et al., 2010), thereby circumventing any form of genetic manipulation. These improvements have essentially abrogated the issue of host/donor cell immune compatibility and considerably enhanced the prospects of generating patient-specific iPSCs for regenerative medicine. However, a recent study demonstrated that some hiPSC derivatives exhibit limited expansion capability, increased apoptosis and early cellular senescence as compared to their hESC-derived counterparts, raising doubts about the clinical value of this reprogramming technology (Feng et al., 2010). Also, it remains to be determined if the progeny of these cells, which are genetically identical to the reprogrammed cell, will trigger any immune response when reintegrated into the donor.

2.3 ESC differentiation

ESC owes its allure as the source of stem cells for regenerative medicine to two important potentials: 1) unlimited self-renewal potential and 2) the potential to differentiate into all the cell types in an adult. Unfortunately, the recent technological advances to circumvent the risks associated with transplantation of ESC-derived cells, namely teratoma formation and host immune rejection, were not matched by similar progress in differentiating hESCs into cells suitable for regenerative medicine. In contrast to adult stem cells where hundreds of clinical trials have been conducted to evaluate their clinical efficacy, the first testing of a hESC-based therapeutic candidate has only just been initiated. In Oct 2010, Geron Corp announced the enrollment of the first patient to test the safety of human embryonic stem cell (hESC)-derived oligodendrocyte progenitor cells, GRNOPC1, in treating spinal cord injury. With the progress made in reducing the risk of teratoma formation by residual ESC in differentiated ESC preparations and the generation of patient-specific iPSC, the major impediment to the development of hESC-based cell therapies remains the general lack of progress in developing protocols for efficient and reproducible differentiation of hESCs into clinically relevant cell types in sufficient quantity and purity suitable for transplantation studies in clinically relevant large animal models.

The pluripotent differentiation potential of hESCs has always been predicated on their ability to form teratomas in immune-compromised animals and embryoid bodies consisting of tissues from the three germ layers. This ability suggest that differentiation of ESC into the various cell types in the adult animal was not contingent on the presence of an embryonic microenvironment. Instead, it relies on a rather minimal environment that did not support the pluripotent and self-renewing state of the ESCs and bore little resemblance to the dynamically evolving microenvironment of a developing embryo. Nevertheless, much effort to direct differentiation of hESCs into potentially therapeutic cell types have focused on the recapitulation of the embryonic microenvironment based on a yet to be tested rationale that the embryonic microenvironment represents the optimal micro-environment for directed *in vitro* differentiation of ESC.

2.3.1 Recapitulating embryonic development to induce lineage commitment

Embryogenesis is a highly dynamic complex process that is still being unraveled despite years of intensive research and much progress in elucidating the molecular and cellular processes involved in formation of an embryo. From a developmental perspective, the ESC represents cells that were frozen in the developmental state of a late-stage embryo just prior to differentiation and lineage commitment. The ability of ESC to re-enter the developmental process and differentiate when returned to the micro-environment of a blastocyst has provided compelling impetus to use the developing embryo to guide and direct *in vitro* differentiation of ESC to a specific cell type. Much effort has therefore been devoted to identifying the molecular cues that were involved in the differentiation of pluripotent cells in the blastocyst into specific terminally differentiated cells. The underlying rationale has always been that a temporal and spatial recapitulation of these cues *in vitro* will direct differentiation of ESC towards a specific cell type.

An early and critical phase of embryogenesis is gastrulation. During this process, the mono-layered blastula undergoes a series of transformation to form the tri-layered gastrula. The formation of these three germ layers (endoderm, mesoderm and ectoderm) marks the first stage of cell fate determination. This is followed by organogenesis when tissues and organs are formed from further differentiation of the germ layers. The endoderm gives rise to the epithelia of the gut and respiratory system, and organs such as liver and pancreas; the mesoderm gives rise to muscles, the circulatory system, bone and connective tissues; and the ectoderm gives rise to the nervous system and the epidermis. Similarly, the initial step towards deriving functional cells and tissues from ESCs may involve germ layer induction *in vitro*.

The first visible sign of gastrulation is the formation of the symmetry-breaking structure called the primitive streak (PS). Epiblast cells, which are derived from the inner cell mass, ingress through the PS to form the mesoderm and definitive endoderm. The remaining epiblast cells that do not ingress form the ectoderm. Many molecular factors have been implicated in this process and they include members of the large transforming growth factor β (TGFβ) and Wnt signaling families (Conlon et al., 1994; Hogan, 1996; Schier, 2003; Yamaguchi, 2001). Painstaking research has revealed some of the temporal and spatial effects of these factors during embryogenesis and many of these factors exerted similar effects on the differentiation of ESC cells. As reviewed by Murry and Keller (Murry &

Keller, 2008)], differentiation of ESCs into each of the three germ layers could be induced by the same factors known to induce them during gastrulation. For example, Wnt, Nodal or BMP4 which have been shown to be important in the formation of epiblast cells in the PS of a developing embryo (Kispert & Herrmann, 1994) could similarly induce the formation of PS-like cells from ESC (Kubo et al., 2004; Lindsley et al., 2006; Ng et al., 2005; Nostro et al., 2008). As in gastrulation, exposure of the PS-like cells to high levels of Nodal further differentiate these cells to a Foxa2^{hi} cells that are comparable to cells in the anterior PS that forms the definitive endoderm (D'Amour et al., 2005; Kubo et al., 2004). In contrast, exposure to Wnt, low level of activin (which activates Nodal) and BMP4 causes the PS-like cells to differentiate into a Flk-1^{+} posterior PS-equivalent population that forms the mesoderm (Nostro et al., 2008). Therefore, the three germ layers can be induced in ESCs by exposing the cells to factors known to be important in the formation of these three germ layers during embryogenesis. Further, by modulating these factors in a concentration and temporal manner that recapitulates early embryonic development, commitment of ESCs to one of the germ layers could be enhanced.

2.3.2 Enhancing lineage commitment

The intensive research efforts to induce a bias in differentiating pluripotent ESCs towards one of the germ layers would, in principle, enhance the subsequent production of specific tissue cell types of this germ layer e.g. muscles from mesoderm. However, enhancing commitment of differentiating ESC to one of the three germ layers may not be the limiting factor in generating clinically useful cell types in sufficient number and purity for therapeutic or screening applications. For example, the most efficient derivation of clinically useful cell types from ESC is neural cell types and not surprisingly, the first ESC-derived cell type to be clinically tested is oligodendrocytes. The relative efficiency of generating neurons, astrocytes and oligodendrocytes from ESC probably lies not in the ease of generating neural progenitor cells but in the relatively high expansion capacity of ESC-derived neural progenitor cells (Dottori & Pera, 2008; Studer, 2009). The high expansion capacity of neural progenitor cells would easily circumvent a limiting supply of rare neural progenitor cells formed during ESC differentiation and obviates the need to first bias differentiation of pluripotent ESCs towards an ectodermal germ lineage. Therefore, the rationale underlying the intensive research efforts to bias differentiating pluripotent ESCs towards one of the germ layers may be redundant at least for the derivation of neural cell types. Unlike ectodermal differentiation which is generally considered the default differentiation pathway for ESC, the derivation of mesodermal or endodermal cell types from ESC could still be enhanced by the recapitulation of early embryonic development processes to enhance mesodermal or endodermal commitment.

2.3.3 Terminal differentiation of ESC

In 2005, D'Amour et al reported the use of a multi-stage protocol that attempts to temporally recapitulate embryonic development for the differentiation of hESC into insulin-producing pancreatic cells for diabetes treatment. During this differentiation regime, they observed the formation of sequential transient cell populations with markers that mapped onto the developmental pathway of pancreatic endoderm. The final cell population representing pancreatic endoderm was transplanted in mice for further differentiation and maturation. When these transplanted animals were treated with streptozotocin, the induction of hyperglycemia was attenuated. It was observed that some of the transplanted mice developed teratomas, suggesting the cell preparation was heterogenous and contaminated with ESCs that could differentiate into all cell types. In contrast, differentiation of mouse ESCs to insulin-producing cells is often a three-step protocol consisting of the formation of embryoid bodies, spontaneous differentiation into ecto-, endo- and mesoderm lineages and finally induction of pancreatic differentiation (Schroeder et al., 2006). However, the cell populations generated using this protocol have low insulin content. Using an approach that combines elements from this protocol and that used in neural differentiation, we first derived highly expansible E-RoSH cell lines with meso-endoderm potential from spontaneously differentiating EBs (Lian et al., 2006; Yin et al., 2004). Like neural stem cells, these E-RoSH cell lines are highly proliferative and provide unlimited supply of cells for differentiation. Serum starvation and nicotinamide supplementation induce differentiation of E-RoSH cells to form a heterogenous, insulin-producing culture. Limiting dilution of such cultures yielded independently derived clonal insulin-producing ERoSHK cell lines. These cells contain equimolar of insulin and C-peptide that was stably maintained over 30 passages at a high concentration of 300-500 pmol/10^6 cells. The insulin-producing ERoSHK cells resemble pancreatic cells and display the defining functional properties of bona fide pancreatic beta cells (G. Li et al., 2009a). They synthesize and store insulin in typical intracellular vesicles. Under stimulation by secretagogues such as glucose, tolbutamide and glibenclamide, these cells close their ATP-sensitive K+ channels, leading to membrane depolarization, opening of Ca2+ channels and the subsequent release of insulin and C-peptide in equimolar ratio, a mechanism resembling that of primary beta cells. Most importantly, these cells can reverse hyperglycemia when grafted into streptozotocin-treated mice. Relative to their progenitor E-RoSH cells, ERoSHK cells also exhibit enhanced activity in biochemical pathways that are also highly characteristic of beta cells such as the pentose phosphate pathway, clathrin-mediated endocytosis and PPAR signaling (T. S. Chen et al., 2010). Importantly, transplantation of ERoSHK cells in hyperglycemic streptozotocin-treated mice reverses the hyperglycemia and removal of the transplanted cells restores the hyperglycemia. The transplanted cells do not form teratomas. Together, these studies illustrated the diversity of approaches that have been taken to differentiate ESCs to insulin-producing cells and the relative potential of each approach in generating the desired end product on a scale to support potential therapeutic application. They also prompted doubts on the need to recapitulate the precise developmental pathway when differentiating ESC.

This question was previously raised by Burns et al (Burns et al., 2004). From their perspective, developmental events directing duodenal endoderm towards an insulin-expressing β-cell phenotype are the result of millions of years of evolutionary selection, driven by environmental pressures rather than by conscious design. Therefore, instead of mapping experimental protocols on to the known developmental pathways of pancreatic endocrine cells, they proposed that conscious design may be a less circuitous route to arrive at the same end-point. However, in lieu of known developmental pathways, there is no obvious source to guide and rationalize such a design. In essence, a conscious design would inevitably have to be an empirical approach of careful observation, trial and error, and high throughput screens.

2.3.4 Empirical differentiation of ESC

Despite a pervasive belief that a high fidelity recapitulation of developmental process represents the best strategy for efficient differentiation of pluripotent stem cells to therapeutically useful cell types , the two human ESC-derived cell types ready for testing in man were derived by empirically formulated protocols. Fortuitously, some elements in these protocols were subsequently found to map onto similar pathways in embryonic development.

In the basic protocol for deriving Geron Corporation's GRNOPC1 which is already in Phase I clinical trial, one of the key elements in inducing neural commitment in ESCs to form neurospheres is retinoic acid (RA) (Nistor et al., 2005). RA was first observed to be an inducer of neural differentiation in embryonal carcinoma cells (ECs) (Jones-Villeneuve et al., 1982) before the first retinoic acid receptor (now known as RARa1) was cloned in 1987 (Giguere et al., 1987; Petkovich et al., 1987). Based on the empirical observation that RA induced neural differentiation in P19 tetracarcinoma cells, RA was used to enhance neural lineage commitment in ESCs (Bain et al., 1995). Today, RA is often used to enhance neural lineage commitment in ESC to generate neurospheres containing neural stem cells and for the subsequent terminal differentiation of neurospheres to produce neurons, oligodendrocytes and astrocytes. Therefore, the use of RA to induce neural differentiation in ESCs was rationalized on empirical observation of their effects on EC cells and this preceded the cloning of RA receptors and our understanding of its role in embryonic development.

To date, there is little evidence that RA plays a significant role in neural differentiation during gastrulation. The first RA signaling in the gastrulating vertebrate embryo occurs in the posterior mesodermal cells when RA is first synthesized by retinaldehyde dehydrogenase 2 (RALDH2) (Niederreither et al., 1997). There is however no RA signaling in the anterior regions of the embryo due to the presence of RA metabolizing enzymes such as CYP26A1 and CYP26C1 (Hernandez et al., 2007; Ribes et al., 2007; Uehara et al., 2007). In fact, RA receptors in the prospective head region of the Xenopus gastrula function as transcriptional repressors to prevent inappropriate activation of genes acting as posterior determinants. Also, the absence of endogenous RA synthesis in mice affect primarily forebrain development but did not compromise the early neural lineage commitment or differentiation (Natalia Molotkova et al., 2007; N. Molotkova et al., 2005; Niederreither et al., 2000; Sirbu et al., 2005). In fact, the pathway for neural differentiation during embryonic development could not have informed on the usefulness of insulin, triiodothyroidin, EGF and FGF in enhancing the *in vitro* proliferation and differentiation of ESC-derived oligodendrocyte precursors and increase oligodendrocyte survival.

The second ESC-derived cell type most likely to be tested in man is Advanced Cell Technology's retinal pigment epithelial (RPE) cells which has been given FDA clearance to initiate a Phase I/II multicenter clinical trial to treat patients with Dry AMD. The derivation of these RPE cells relies primarily on spontaneous differentiation of hESCs (Klimanskaya et al., 2004). RPE cells are formed as colonies of pigmented cells when hESCs undergo spontaneous differentiation by FGF2 withdrawal or embryoid body formation. These colonies of pigmented cells were then picked and expanded using very unremarkable culture medium.

2.3.5 Strategizing differentiation of ESC for therapeutic applications

The progress of ESC-derived oligodendrocytes and RPE cells to clinical testing attests to the robustness and efficiency of the empirically-driven differentiation protocols. In contrast, differentiation of ESCs by meticulous mapping on embryonic development pathway has not yielded cells that are ready for clinical testing. Despite this dichotomy in outcomes, there is

still a prevalent belief that high fidelity recapitulation of embryonic development process is the route to generate the most physiologically relevant cells. Embryonic development is a time tested success with defined milestones. In contrast, an empirically driven differentiation strategy is an inherently inefficient chance event.
The success of a differentiation strategy based on embryonic development is predicated not only on the elucidation but also the recapitulation of the highly dynamic temporal and spatial changes in the embryonic microenvironment that is influenced by both intra- and

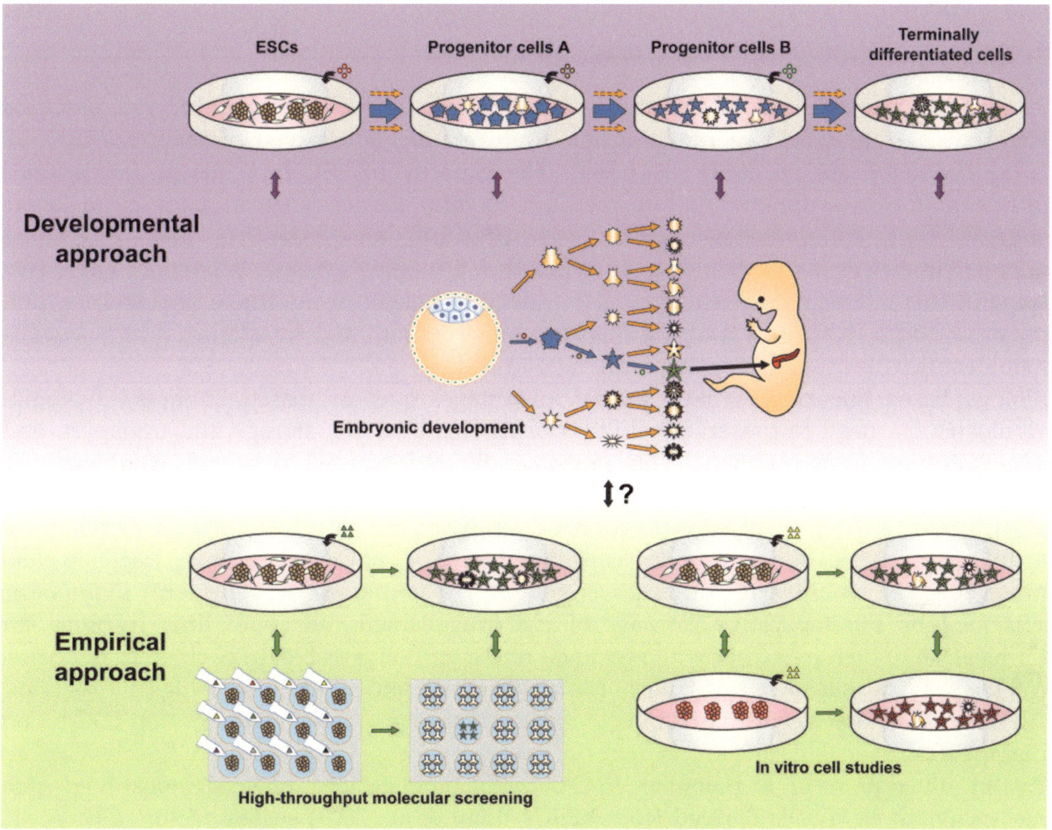

Fig. 2. Strategies for differentiation of ESCs into therapeutically useful cell types. The strategies currently being used could be broadly classified into a developmental or an empirical approach. The developmental approach (upper panel) to produce a desired cell type (green stars) relies on the recapitulation of the developmental pathway (blue arrows) during embryogenesis that produces that desired cell type. The general expectation is that identifying the cues that direct the developmental process during embryogenesis and recapitulating these cues spatially and temporally *in vitro* will be most optimal in yielding physiologically functional cell type (e.g. pancreatic insulin-producing cells). The empirical approach involves the differentiation of ESCs either spontaneously, or using novel factors identified empirically, such as through high-throughput molecular screening or in vitro cell studies (e.g. neural induction of EC cells by RA). These factors may or may not play a role in development.

extra-embryonic factors. Notwithstanding this, translating such a complex differentiation strategy to a scalable commercially viable manufacturing process will be an equally confounding unknown. On the other hand, developing a differentiation strategy using an empirical approach is a chance process of trial and error and fortuitous observation. This inherent inefficiency can be circumvented by high throughput screens to identify inducing molecules or combinations of molecules. There is also a likelihood that such a strategy would provide for a potentially scalable manufacturing process that will support clinical applications.

3. ESC therapeutics: cell versus biologic

A much overlooked form of ESC-derived therapeutics is biological products or biologics from ESC. To date, the predominant or only forms of ESC-derived therapeutics that are being evaluated are primarily cell-based. The capacity of ESC to undergo spontaneous differentiation in a minimal culture medium to form tissues from all three germ layers suggest that differentiating ESCs can produce an inductive and sustaining microenvironment for the various cell types that are being formed. It is conceivable that some of this microenvironment may also induce or sustain some tissue regeneration and repair in adult. However capturing this microenvironment and translating it to a scalable manufacturing process would be a challenge.

Biologic-based therapeutics have several advantages over cell-based therapies. Biologics eliminates the need to preserve viability during manufacture, storage and transport, and administration to the patient. This substantially reduces the cost and complexity of production and delivery. Maintaining cell viability before and after transplantation has always been an important consideration in cell-based therapy. Although preserving the activity of biologics is not a minor consideration, it is, nevertheless more tractable than preserving cell viability. Cell therapy is generally a permanent or long term therapeutic sustained by the replicative capacity of the transplanted cells with little recourse for termination of therapy except when removal of the graft is possible. In contrast to biologics, cell therapy presents increased risks of tumor formation and acute immunological rejections. All things considered, ESC-based biologics is an attractive alternative to develop ESC-based therapeutics.

As an illustration of a potential ESC-derived biologic, we have demonstrated that mesenchymal stem cells derived from hESCs (Lian et al., 2007) secrete factors (Sze et al., 2007) that are cardioprotective in pig and mouse models of myocardial ischemia/reperfusion injury (Timmers et al., 2008). The active component in this secretion was small lipid vesicles of 50-100nm known as exosomes (Lai et al., 2010). Immortalization of these mesenchymal stem cells did not compromise the production or activity of the exosomes (T.S. Chen et al., 2011). These studies provided for the development of a sustainable scalable manufacturing process to produce potentially therapeutic exosomes for testing in the clinic.

4. Conclusion

ESC is a versatile cell that has exerted significant impact on our understanding and investigation of cell biology, differentiation and development. It has provided exciting possibilities for the treatment of highly intractable diseases. As the first ESC-derived cell type makes its way into clinical testing, there is an apprehensive hope that ESC will justify its hype not only as a therapeutic agent but one that will treat a multitude of intractable diseases as wide ranging as its differentiation potential. The establishment of iPSC technology by Shinya Yamanaka (Takahashi & Yamanaka, 2006) represents a paradigm shift in not only our understanding of stem cell biology but also in overcoming the ethical and immune challenges that had stymied the translation of ESC into clinical applications. His approach of re-programming terminally differentiated cells into pluripotent stem cells contradicted fundamental principles in developmental biology. This approach of exploring beyond the obvious and logical using empirical and experimental strategies may be necessary to transcend this current bottleneck in generating the quantity and quality of ESC-derived cells for therapy.

5. References

Bain, G., Kitchens, D., Yao, M., Huettner, J.E., & Gottlieb, D.I. (1995). Embryonic stem cells express neuronal properties in vitro. *Dev Biol*, Vol. 168, No. 2, (Apr), pp. (342-357).

Blum, B., Bar-Nur, O., Golan-Lev, T., & Benvenisty, N. (2009). The anti-apoptotic gene survivin contributes to teratoma formation by human embryonic stem cells. *Nat Biotechnol*, Vol. 27, No. 3, (Mar), pp. (281-287).

Burns, C.J., Persaud, S.J., & Jones, P.M. (2004). Stem cell therapy for diabetes: do we need to make beta cells? *J Endocrinol*, Vol. 183, No. 3, (Dec), pp. (437-443).

Byrne, J.A., Pedersen, D.A., Clepper, L.L., Nelson, M., Sanger, W.G., Gokhale, S., Wolf, D.P., & Mitalipov, S.M. (2007). Producing primate embryonic stem cells by somatic cell nuclear transfer. *Nature*, Vol. 450, No. 7169, (Nov 22), pp. (497-502).

Chen, T.S., Arslan, F., Yin, Y., S.S., T., Lai, R.C., Choo, A., Padmanabhand, J., Lee, C.N., de Kleijn, D.P.V., & Lim, S.-K. (2011). Enabling a robust scalable manufacturing process for therapeutic exosomes through oncogenic immortalization of human ESC-derived MSCs. . *Journal of Translational Medicine* Vol. (in press), No.

Chen, T.S., Tan, S.S., Yeo, R.W., Teh, B.J., Luo, R., Li, G., & Lim, S.K. (2010). Delineating biological pathways unique to embryonic stem cell-derived insulin-producing cell lines from their noninsulin-producing progenitor cell lines. *Endocrinology*, Vol. 151, No. 8, (Aug), pp. (3600-3610).

Chidgey, A.P., Layton, D., Trounson, A., & Boyd, R.L. (2008). Tolerance strategies for stem-cell-based therapies. *Nature*, Vol. 453, No. 7193, (May 15), pp. (330-337).

Choo, A.B., Tan, H.L., Ang, S.N., Fong, W.J., Chin, A., Lo, J., Zheng, L., Hentze, H., Philp, R.J., Oh, S.K., & Yap, M. (2008). Selection against undifferentiated human embryonic stem cells by a cytotoxic antibody recognizing podocalyxin-like protein-1. *Stem Cells*, Vol. 26, No. 6, (Jun), pp. (1454-1463).

Conlon, F.L., Lyons, K.M., Takaesu, N., Barth, K.S., Kispert, A., Herrmann, B., & Robertson, E.J. (1994). A primary requirement for nodal in the formation and maintenance of the primitive streak in the mouse. *Development*, Vol. 120, No. 7, (Jul), pp. (1919-1928).

D'Amour, K.A., Agulnick, A.D., Eliazer, S., Kelly, O.G., Kroon, E., & Baetge, E.E. (2005). Efficient differentiation of human embryonic stem cells to definitive endoderm. *Nat Biotechnol*, Vol. 23, No. 12, (Dec), pp. (1534-1541).

Dottori, M., & Pera, M.F. (2008). Neural Differentiation of Human Embryonic Stem Cells, In: *Neural Stem Cells : Methods and Protocols*, Weiner, L.P., pp. (19-30), Humana Press, 978-1-58829-846-1, Totowa, NJ, USA

Evans, M.J., Gurer, C., Loike, J.D., Wilmut, I., Schnieke, A.E., & Schon, E.A. (1999). Mitochondrial DNA genotypes in nuclear transfer-derived cloned sheep. *Nat Genet*, Vol. 23, No. 1, (Sep), pp. (90-93).

Feng, Q., Lu, S.J., Klimanskaya, I., Gomes, I., Kim, D., Chung, Y., Honig, G.R., Kim, K.S., & Lanza, R. (2010). Hemangioblastic derivatives from human induced pluripotent stem cells exhibit limited expansion and early senescence. *Stem Cells*, Vol. 28, No. 4, (Apr), pp. (704-712).

Fong, C.Y., Peh, G., Subramanian, A., Gauthaman, K., & Bongso, A. (2009a). The use of discontinuous density gradients in stem cell research and application. *Stem Cell Rev*, Vol. 5, No. 4, (Dec), pp. (428-434).

Fong, C.Y., Peh, G.S., Gauthaman, K., & Bongso, A. (2009b). Separation of SSEA-4 and TRA-1-60 labelled undifferentiated human embryonic stem cells from a heterogeneous cell population using magnetic-activated cell sorting (MACS) and fluorescence-activated cell sorting (FACS). *Stem Cell Rev*, Vol. 5, No. 1, (Mar), pp. (72-80).

French, A.J., Adams, C.A., Anderson, L.S., Kitchen, J.R., Hughes, M.R., & Wood, S.H. (2008). Development of human cloned blastocysts following somatic cell nuclear transfer with adult fibroblasts. *Stem Cells*, Vol. 26, No. 2, (Feb), pp. (485-493).

Giguere, V., Ong, E.S., Segui, P., & Evans, R.M. (1987). Identification of a receptor for the morphogen retinoic acid. *Nature*, Vol. 330, No. 6149, (Dec 17-23), pp. (624-629).

Hernandez, R.E., Putzke, A.P., Myers, J.P., Margaretha, L., & Moens, C.B. (2007). Cyp26 enzymes generate the retinoic acid response pattern necessary for hindbrain development. *Development*, Vol. 134, No. 1, (Jan), pp. (177-187).

Herszfeld, D., Wolvetang, E., Langton-Bunker, E., Chung, T.L., Filipczyk, A.A., Houssami, S., Jamshidi, P., Koh, K., Laslett, A.L., Michalska, A., Nguyen, L., Reubinoff, B.E., Tellis, I., Auerbach, J.M., Ording, C.J., Looijenga, L.H., & Pera, M.F. (2006). CD30 is a survival factor and a biomarker for transformed human pluripotent stem cells. *Nat Biotechnol*, Vol. 24, No. 3, (Mar), pp. (351-357).

Hogan, B.L. (1996). Bone morphogenetic proteins in development. *Curr Opin Genet Dev*, Vol. 6, No. 4, (Aug), pp. (432-438).

Ishikawa, K., Toyama-Sorimachi, N., Nakada, K., Morimoto, M., Imanishi, H., Yoshizaki, M., Sasawatari, S., Niikura, M., Takenaga, K., Yonekawa, H., & Hayashi, J. (2010). The innate immune system in host mice targets cells with allogenic mitochondrial DNA. *J Exp Med*, Vol. 207, No. 11, (Oct 25), pp. (2297-2305).

Jones-Villeneuve, E.M., McBurney, M.W., Rogers, K.A., & Kalnins, V.I. (1982). Retinoic acid induces embryonal carcinoma cells to differentiate into neurons and glial cells. *J Cell Biol*, Vol. 94, No. 2, (Aug), pp. (253-262).

Kispert, A., & Herrmann, B.G. (1994). Immunohistochemical analysis of the Brachyury protein in wild-type and mutant mouse embryos. *Dev Biol*, Vol. 161, No. 1, (Jan), pp. (179-193).

Klimanskaya, I., Hipp, J., Rezai, K.A., West, M., Atala, A., & Lanza, R. (2004). Derivation and comparative assessment of retinal pigment epithelium from human embryonic stem cells using transcriptomics. *Cloning Stem Cells*, Vol. 6, No. 3, pp. (217-245).

Klug, M.G., Soonpaa, M.H., Koh, G.Y., & Field, L.J. (1996). Genetically selected cardiomyocytes from differentiating embronic stem cells form stable intracardiac grafts. *J Clin Invest*, Vol. 98, No. 1, (Jul 1), pp. (216-224).

Kubo, A., Shinozaki, K., Shannon, J.M., Kouskoff, V., Kennedy, M., Woo, S., Fehling, H.J., & Keller, G. (2004). Development of definitive endoderm from embryonic stem cells in culture. *Development*, Vol. 131, No. 7, (Apr), pp. (1651-1662).

Lai, R.C., Arslan, F., Lee, M.M., Sze, N.S., Choo, A., Chen, T.S., Salto-Tellez, M., Timmers, L., Lee, C.N., El Oakley, R.M., Pasterkamp, G., de Kleijn, D.P., & Lim, S.K. (2010). Exosome secreted by MSC reduces myocardial ischemia/reperfusion injury. *Stem Cell Res*, Vol. 4, No., (Jan 4), pp. (214-222).

Lensch, M.W., Schlaeger, T.M., Zon, L.I., & Daley, G.Q. (2007). Teratoma formation assays with human embryonic stem cells: a rationale for one type of human-animal chimera. *Cell Stem Cell*, Vol. 1, No. 3, (Sep 13), pp. (253-258).

Li, G., Luo, R., Zhang, J., Yeo, K.S., Lian, Q., Xie, F., Tan, E.K., Caille, D., Kon, O.L., Salto-Tellez, M., Meda, P., & Lim, S.K. (2009a). Generating mESC-derived insulin-producing cell lines through an intermediate lineage-restricted progenitor line. *Stem Cell Res*, Vol. 2, No. 1, (Jan), pp. (41-55).

Li, J., Liu, X., Wang, H., Zhang, S., Liu, F., Wang, X., & Wang, Y. (2009b). Human embryos derived by somatic cell nuclear transfer using an alternative enucleation approach. *Cloning Stem Cells*, Vol. 11, No. 1, (Mar), pp. (39-50).

Li, M., Pevny, L., Lovell-Badge, R., & Smith, A. (1998). Generation of purified neural precursors from embryonic stem cells by lineage selection. *Curr Biol*, Vol. 8, No. 17, (Aug 27), pp. (971-974).

Lian, Q., Lye, E., Suan Yeo, K., Khia Way Tan, E., Salto-Tellez, M., Liu, T.M., Palanisamy, N., El Oakley, R.M., Lee, E.H., Lim, B., & Lim, S.K. (2007). Derivation of clinically compliant MSCs from CD105+, CD24- differentiated human ESCs. *Stem Cells*, Vol. 25, No. 2, (Feb), pp. (425-436).

Lian, Q., Yeo, K.S., Que, J., Tan, E.K., Yu, F., Yin, Y., Salto-Tellez, M., Menshawe El Oakley, R., & Lim, S.K. (2006). Establishing clonal cell lines with endothelial-like potential from CD9(hi), SSEA-1(-) cells in embryonic stem cell-derived embryoid bodies. *PLoS One*, Vol. 1, No., pp. (e6).

Lindsley, R.C., Gill, J.G., Kyba, M., Murphy, T.L., & Murphy, K.M. (2006). Canonical Wnt signaling is required for development of embryonic stem cell-derived mesoderm. *Development*, Vol. 133, No. 19, (Oct), pp. (3787-3796).

Molotkova, N., Molotkov, A., & Duester, G. (2007). Role of retinoic acid during forebrain development begins late when Raldh3 generates retinoic acid in the ventral subventricular zone. *Developmental Biology*, Vol. 303, No. 2, pp. (601-610).

Molotkova, N., Molotkov, A., Sirbu, I.O., & Duester, G. (2005). Requirement of mesodermal retinoic acid generated by Raldh2 for posterior neural transformation. *Mech Dev*, Vol. 122, No. 2, (Feb), pp. (145-155).

Muller, M., Fleischmann, B.K., Selbert, S., Ji, G.J., Endl, E., Middeler, G., Muller, O.J., Schlenke, P., Frese, S., Wobus, A.M., Hescheler, J., Katus, H.A., & Franz, W.M. (2000). Selection of ventricular-like cardiomyocytes from ES cells in vitro. *FASEB J*, Vol. 14, No. 15, (Dec), pp. (2540-2548).

Murry, C.E., & Keller, G. (2008). Differentiation of embryonic stem cells to clinically relevant populations: lessons from embryonic development. *Cell*, Vol. 132, No. 4, (Feb 22), pp. (661-680).

Nakajima, F., Tokunaga, K., & Nakatsuji, N. (2007). Human leukocyte antigen matching estimations in a hypothetical bank of human embryonic stem cell lines in the Japanese population for use in cell transplantation therapy. *Stem Cells*, Vol. 25, No. 4, (Apr), pp. (983-985).

Ng, E.S., Azzola, L., Sourris, K., Robb, L., Stanley, E.G., & Elefanty, A.G. (2005). The primitive streak gene Mixl1 is required for efficient haematopoiesis and BMP4-induced ventral mesoderm patterning in differentiating ES cells. *Development*, Vol. 132, No. 5, (Mar), pp. (873-884).

Niederreither, K., McCaffery, P., Drager, U.C., Chambon, P., & Dolle, P. (1997). Restricted expression and retinoic acid-induced downregulation of the retinaldehyde dehydrogenase type 2 (RALDH-2) gene during mouse development. *Mech Dev*, Vol. 62, No. 1, (Feb), pp. (67-78).

Niederreither, K., Vermot, J., Schuhbaur, B., Chambon, P., & Dolle, P. (2000). Retinoic acid synthesis and hindbrain patterning in the mouse embryo. *Development*, Vol. 127, No. 1, (Jan), pp. (75-85).

Nistor, G.I., Totoiu, M.O., Haque, N., Carpenter, M.K., & Keirstead, H.S. (2005). Human embryonic stem cells differentiate into oligodendrocytes in high purity and myelinate after spinal cord transplantation. *Glia*, Vol. 49, No. 3, (Feb), pp. (385-396).

Nostro, M.C., Cheng, X., Keller, G.M., & Gadue, P. (2008). Wnt, activin, and BMP signaling regulate distinct stages in the developmental pathway from embryonic stem cells to blood. *Cell Stem Cell*, Vol. 2, No. 1, (Jan 10), pp. (60-71).

Okita, K., Nakagawa, M., Hyenjong, H., Ichisaka, T., & Yamanaka, S. (2008). Generation of mouse induced pluripotent stem cells without viral vectors. *Science*, Vol. 322, No. 5903, (Nov 7), pp. (949-953).

Petkovich, M., Brand, N.J., Krust, A., & Chambon, P. (1987). A human retinoic acid receptor which belongs to the family of nuclear receptors. *Nature*, Vol. 330, No. 6147, (Dec 3-9), pp. (444-450).

Plaia, T.W., Josephson, R., Liu, Y., Zeng, X., Ording, C., Toumadje, A., Brimble, S.N., Sherrer, E.S., Uhl, E.W., Freed, W.J., Schulz, T.C., Maitra, A., Rao, M.S., & Auerbach, J.M. (2006). Characterization of a new NIH-registered variant human embryonic stem cell line, BG01V: a tool for human embryonic stem cell research. *Stem Cells*, Vol. 24, No. 3, (Mar), pp. (531-546).

Plews, J.R., Li, J., Jones, M., Moore, H.D., Mason, C., Andrews, P.W., & Na, J. (2010). Activation of pluripotency genes in human fibroblast cells by a novel mRNA based approach. *PLoS One*, Vol. 5, No. 12, pp. (e14397).

Reubinoff, B.E., Pera, M.F., Fong, C.Y., Trounson, A., & Bongso, A. (2000). Embryonic stem cell lines from human blastocysts: somatic differentiation in vitro. *Nat Biotechnol*, Vol. 18, No. 4, (Apr), pp. (399-404).

Ribes, V., Otto, D.M., Dickmann, L., Schmidt, K., Schuhbaur, B., Henderson, C., Blomhoff, R., Wolf, C.R., Tickle, C., & Dolle, P. (2007). Rescue of cytochrome P450 oxidoreductase (Por) mouse mutants reveals functions in vasculogenesis, brain and limb patterning linked to retinoic acid homeostasis. *Dev Biol*, Vol. 303, No. 1, (Mar 1), pp. (66-81).

Schier, A.F. (2003). Nodal signaling in vertebrate development. *Annu Rev Cell Dev Biol*, Vol. 19, No., pp. (589-621).

Schroeder, I.S., Rolletschek, A., Blyszczuk, P., Kania, G., & Wobus, A.M. (2006). Differentiation of mouse embryonic stem cells to insulin-producing cells. *Nat Protoc*, Vol. 1, No. 2, pp. (495-507).

Sirbu, I.O., Gresh, L., Barra, J., & Duester, G. (2005). Shifting boundaries of retinoic acid activity control hindbrain segmental gene expression. *Development*, Vol. 132, No. 11, (Jun), pp. (2611-2622).

Soldner, F., Hockemeyer, D., Beard, C., Gao, Q., Bell, G.W., Cook, E.G., Hargus, G., Blak, A., Cooper, O., Mitalipova, M., Isacson, O., & Jaenisch, R. (2009). Parkinson's disease patient-derived induced pluripotent stem cells free of viral reprogramming factors. *Cell*, Vol. 136, No. 5, (Mar 6), pp. (964-977).

Soria, B., Roche, E., Berna, G., Leon-Quinto, T., Reig, J.A., & Martin, F. (2000). Insulin-secreting cells derived from embryonic stem cells normalize glycemia in streptozotocin-induced diabetic mice. *Diabetes*, Vol. 49, No. 2, (Feb), pp. (157-162).

Studer, L. (2009). The Nervous System, In: *Essentials of Stem Cell Biology*, Lanza, R., Gearhart, J., Hogan, B., Melton, D., Pedersen, R., Thomas, E.D., Thomson, J., and West, M., pp. (169-178), Academic Press, 978-0-12-374729-7, Canada

Sze, S.K., de Kleijn, D.P., Lai, R.C., Khia Way Tan, E., Zhao, H., Yeo, K.S., Low, T.Y., Lian, Q., Lee, C.N., Mitchell, W., El Oakley, R.M., & Lim, S.K. (2007). Elucidating the secretion proteome of human embryonic stem cell-derived mesenchymal stem cells. *Mol Cell Proteomics*, Vol. 6, No. 10, (Oct), pp. (1680-1689).

Takahashi, K., Tanabe, K., Ohnuki, M., Narita, M., Ichisaka, T., Tomoda, K., & Yamanaka, S. (2007). Induction of pluripotent stem cells from adult human fibroblasts by defined factors. *Cell*, Vol. 131, No. 5, (Nov 30), pp. (861-872).

Takahashi, K., & Yamanaka, S. (2006). Induction of pluripotent stem cells from mouse embryonic and adult fibroblast cultures by defined factors. *Cell*, Vol. 126, No. 4, (Aug 25), pp. (663-676).

Taylor, C.J., Bolton, E.M., Pocock, S., Sharples, L.D., Pedersen, R.A., & Bradley, J.A. (2005). Banking on human embryonic stem cells: estimating the number of donor cell lines needed for HLA matching. *Lancet*, Vol. 366, No. 9502, (Dec 10), pp. (2019-2025).

Thomson, J.A., Itskovitz-Eldor, J., Shapiro, S.S., Waknitz, M.A., Swiergiel, J.J., Marshall, V.S., & Jones, J.M. (1998). Embryonic stem cell lines derived from human blastocysts. *Science*, Vol. 282, No. 5391, (Nov 6), pp. (1145-1147).

Timmers, L., Lim, S.-K., Arslan, F., Armstrong, J.S., Hoefler, I.E., Doevendans, P.A., Piek, J.J., El Oakley, R.M., Choo, A., Lee, C.N., Pasterkamp, G., & de Kleijn, D.P.V. (2008). Reduction of myocardial infarct size by human mesenchymal stem cell conditioned medium. *Stem Cell Research*, Vol. 1, No., pp. (129-137).

Uehara, M., Yashiro, K., Mamiya, S., Nishino, J., Chambon, P., Dolle, P., & Sakai, Y. (2007). CYP26A1 and CYP26C1 cooperatively regulate anterior-posterior patterning of the developing brain and the production of migratory cranial neural crest cells in the mouse. *Dev Biol*, Vol. 302, No. 2, (Feb 15), pp. (399-411).

Werbowetski-Ogilvie, T.E., Bosse, M., Stewart, M., Schnerch, A., Ramos-Mejia, V., Rouleau, A., Wynder, T., Smith, M.J., Dingwall, S., Carter, T., Williams, C., Harris, C., Dolling, J., Wynder, C., Boreham, D., & Bhatia, M. (2009). Characterization of human embryonic stem cells with features of neoplastic progression. *Nat Biotechnol*, Vol. 27, No. 1, (Jan), pp. (91-97).

Wilmut, I., Beaujean, N., de Sousa, P.A., Dinnyes, A., King, T.J., Paterson, L.A., Wells, D.N., & Young, L.E. (2002). Somatic cell nuclear transfer. *Nature*, Vol. 419, No. 6907, (Oct 10), pp. (583-586).

Woltjen, K., Michael, I.P., Mohseni, P., Desai, R., Mileikovsky, M., Hamalainen, R., Cowling, R., Wang, W., Liu, P., Gertsenstein, M., Kaji, K., Sung, H.K., & Nagy, A. (2009). piggyBac transposition reprograms fibroblasts to induced pluripotent stem cells. *Nature*, Vol. 458, No. 7239, (Apr 9), pp. (766-770).

Yamaguchi, T.P. (2001). Heads or tails: Wnts and anterior-posterior patterning. *Curr Biol*, Vol. 11, No. 17, (Sep 4), pp. (R713-724).

Yang, S., Lin, G., Tan, Y.Q., Zhou, D., Deng, L.Y., Cheng, D.H., Luo, S.W., Liu, T.C., Zhou, X.Y., Sun, Z., Xiang, Y., Chen, T.J., Wen, J.F., & Lu, G.X. (2008). Tumor progression of culture-adapted human embryonic stem cells during long-term culture. *Genes Chromosomes Cancer*, Vol. 47, No. 8, (Aug), pp. (665-679).

Yin, Y., Que, J., Teh, M., Cao, W.P., El Oakley, R.M., & Lim, S.K. (2004). Embryonic cell lines with endothelial potential: an in vitro system for studying endothelial differentiation. *Arterioscler Thromb Vasc Biol*, Vol. 24, No. 4, (Apr), pp. (691-696).

Yu, J., Vodyanik, M.A., Smuga-Otto, K., Antosiewicz-Bourget, J., Frane, J.L., Tian, S., Nie, J., Jonsdottir, G.A., Ruotti, V., Stewart, R., Slukvin, II, & Thomson, J.A. (2007). Induced pluripotent stem cell lines derived from human somatic cells. *Science*, Vol. 318, No. 5858, (Dec 21), pp. (1917-1920).

Zhou, H., Wu, S., Joo, J.Y., Zhu, S., Han, D.W., Lin, T., Trauger, S., Bien, G., Yao, S., Zhu, Y., Siuzdak, G., Scholer, H.R., Duan, L., & Ding, S. (2009). Generation of induced pluripotent stem cells using recombinant proteins. *Cell Stem Cell*, Vol. 4, No. 5, (May 8), pp. (381-384).

2

Derivation and Characterization of New hESC Lines from Supernumerary Embryos, Experience from Turkey

Zafer Nihat Candan and Semra Kahraman
Memorial Hospital ART & Reproductive Genetics Unit, Istanbul, Turkey

1. Introduction

Human embryonic stem cell (hESC) lines, which are derived from inner cell mass (ICM) of supernumerary blastocysts-stage embryos, have well known unique properties; long-term self-renewing ability with maintenance of an undifferentiated state and pluripotent capacity to differentiate into all derivatives of three embryonic germ layers (Hoffman & Carpenter, 2005; Semb, 2005; Trounson, 2006). Since its first derivation and characterization by Thomson et al in 1998, these intrinsic properties have made hESC very popular worldwide and, thereby, many studies describing isolation and characterization of new hESC lines have been reported (Findikli et al., 2005; Simon et al., 2005; Thomson, 1998). HESCs have been considered very valuable and promising cell source for research involving mainly human embryogenesis, oncology, drug toxicology and developmental biology as well as for cell based regenerative therapies (Edwards, 2004).

Obviously, studies on hESCs mostly focus on their potential use for treatment of degenerative human diseases. However, due to the largely unknown characteristics of established lines and the use of animal based material in their cultures, most of the lines could not be suitable for prospective transplantation studies (Findikli et al., 2006; Rodriguez et al., 2006). Therefore, registration of existing hESC lines with their characteristics in stem cell banks would provide database of cell lines, cooperation and co-regulation for researchers.

In this chapter, it was aimed to report the methods to derive 18 hESC lines which were established and characterized until the declaration of prohibition on hESC research in Turkey by Health Ministry in 2005. Additionally, it was discussed the current legal situation of hESC research and perspectives to that issue in Turkey.

2. Material and methods for derivation and characterization of hESC lines

Derivation and characterization of all hESC lines were undertaken in Memorial Hospital ART &Reproductive Genetics Centre, R&D Laboratory, Istanbul, Turkey between January 2003 and September 2005. All donated supernumerary embryos were used after obtaining written informed consents from couples. All hESC lines were established only for research purpose rather than for any financial interest. This study was approved and controlled by the local ethic committee/Internal Review Board of Istanbul Memorial Hospital.

2.1 Source of supernumerary human embryos

Supernumerary human embryos used for derivation of hESC lines were obtained after in vitro fertilization (IVF) or intracytoplasmic sperm injection (ICSI) and preimplantation genetic diagnosis (PGD) cycles. Most of these embryos had poor quality and thereby, were considered insufficient for replacement or cryopreservation.

Preimplantation genetic diagnosis is applied to three groups of patients with variety of indications in our clinic. The first group of PGD includes patients who have a high risk of transmitting their single gene disorder to their offspring. After diagnosis, abnormal embryos were used to derive hESC lines with specific single gene mutation.

In the second group of PGD cycles, embryos belonging to couples with advanced maternal age, a history of recurrent miscarriages and repeated implantation failures are destined to chromosomal screening (Kahraman et al., 2000, 2006; Lavon et al., 2008; Munne et al., 2005; Verlinsky et al., 2005). Following PGD, chromosomally abnormal embryos were subjected to derivation of hESC lines having chromosomal aneuploidies.

In the third group, PGD is used for identification of embryos for human leukocyte antigen (HLA) matching to an affected older sibling who requires hematopoietic stem cell transplantation. Furthermore, the HLA matching can be combined with mutational analysis for genetic diseases in cases where the sibling is affected with this monogenic disorder and waiting for stem cell transplantation. Therefore, in those cases, embryos having mismatched HLA type or carrying genetic disorder were used for derivation of hESC lines.

2.2 Isolation and preparation of feeder cells

As feeder cells both mouse embryonic fibroblast (MEF) and human foreskin fibroblast (HFF) were used during isolation and long term culture of hESC lines. MEFs were isolated from embryos of the 12- to 14-day pregnant BALb/c mice (Conner, 2000). To isolate single cell suspension of MEF mouse embryos isolated from the sacrificed mice by cervical dislocation were dissociated into small pieces with scissor. Then dissociated tissues were trypsinized in 0.25% trypsin-EDTA (Gibco BRL; Invitrogen, Gaithersburg, MD, USA) for 15min to produce single cell suspension.

Human foreskin fibroblasts were isolated from circumcised tissues of 0-1 year old males. Cell isolation was performed as described previously (Hovatta et al., 2003; Richard et al., 2002). In a brief, following the isolation of dermis from the epidermis by scissor or razor blade, tissue was dissected into small pieces and then trypsinized in 0.05% trypsin-EDTA (Gibco BRL) for approximately 1 h to dissociate into single cells.

Both 25 cm^2 and 75 cm^2 culture flasks were used to culture HFF and MEF lines. These lines were grown in feeder cell culture medium, consisting of 85% high glucose DMEM (Gibco BRL), 10%FBS (Gibco BRL), 1% penicillin streptomycin-amphotericin (Biological Industries) and 2mM L-glutamine (Gibco BRL) at 37°C with 5% CO_2. Supportive medium was changed in every three days. MEF lines could be used in culture of hESC lines up to 6 passages, whereas HFFs had supportive potential up to 15 passages.

Mitotic inactivation of feeder cells was performed after exposing feeder cells to culture medium containing 10µg/ml mitomycin C (Sigma-Aldrich, Poole, Dorset, UK) for 2.5-3h. Inactivated cells were seeded on a 0.1% gelatin-coated at a concentration of 1.5×10^5 cells / ml. After 2 days incubation organ culture dishes were ready to use as feeder plates. Feeder plates could be used for the following 7 d.

2.3 Isolation and long-term culture of hESC lines

Blastocyst stage embryos, that were graded according to the Gardner's scoring criteria, were processed for hESC isolation by either immunosurgery or direct culture (Gardner et al., 2000; Findikli et al., 2005). Prior to the immunosurgery or direct culture, zona pellucida of embryo was removed by the short-term exposure to the 5IU/ml (final conc.) pronase (Sigma) containing embryo culture medium for up to 5 min. Immunosurgery was applied based on the previously published protocol (Solter and Knowles, 1975). After lysis of trophoblastic cells, the resulted intact inner cell mass clumps were placed on feeder cells and cultured until to observe the primary hESC colonies.

In the direct culture method, zona free blastocysts were directly placed on feeder cells and cultured until the appearance of outgrowth, which lasted about 6 to 8 days. Then compacted outgrowths including cells of hESC-like morphology were mechanically split into small clumps. The cell clumps were transferred on new feeder plates. The primary colonies were generally observed after about 5 to 7 days (Figure 1A-B).

HESC lines were cultured at 37 °C in 5% CO2 in the complete stem cell medium (CSCM) with the composition of 85% Ko-DMEM (Gibco BRL), 15% FBS (Hyclone, South, Logan, UT, USA), 1× penicillin/streptomycin/amphotericin B (Biological Industries, Haemek, Israel), 1× non essential amino acid stock solution (Sigma), 0.1% 2mM L Glutamine (Gibco

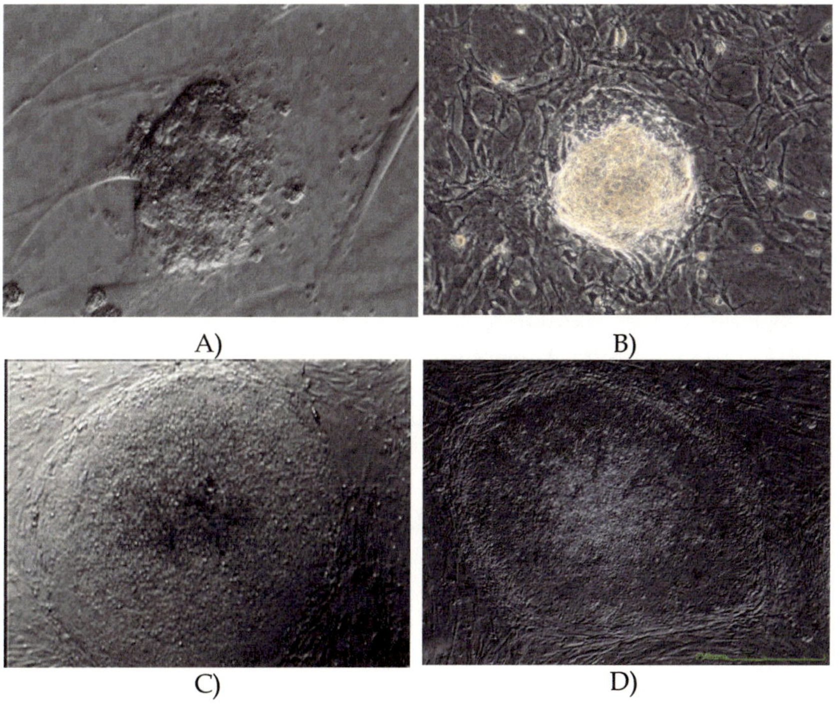

Fig. 1. Establishment of hESC line on HFF and MEF feeder cells. Phase contrast microscopy of NS-10 line at different stages of development. **A)** The formation of outgrowth from the inner cell mass of blastocyst after direct culture of zona-free blastocyst on HFF. **B)** Cell clump, which was formed after mechanically dissociation of outgrowth, included primary hESC like cells. **C)** Circular primary NS-10 colony on MEF. **D)** Polarized colony morphology of NS-10 on HFF. Original magnifications: (A-D) X 200

BRL), 4ng/ml basic fibroblast growth factor (bFGF) (Chemicon, Temecula, CA, USA), β-mercaptoethanol (0.1 mmol/l; Sigma), and 0.1% insulin transferrin selenium complex (Gibco BRL). Culture media was changed daily.
Undifferentiated hESC colonies were split into small colony pieces mechanically with the flame-drawn glass every after 7-8 days of culture. Enzymatic dissociation of hESC was not preferred in this study. Each hESC line was cultured at least up to fifteen passages and was cryopreserved by vitrification technique according to the previously reported protocols (Reubinof et al., 2001; Vanderzwalmen et al., 2003). Briefly, colonies were first mechanically split into small pieces and were sequentially vitrified in two solutions including different concentration of DMSO and ethylene glycol.
Cryopreserved hESC lines were warmed sequentially in solutions including 0.5M and 0.25M sucrose to control efficiency of vitrification and following warming techniques (Reubinoff et al., 2001).

2.4 Karyotyping and immunocytochemistry of hESC lines

G-banding technique was used to karyotype hESC lines. HESC colonies were first incubated with culture medium including 0.1 µg / ml Colcemid (Biological Industries) for 2 h at 37 °C in a %5 CO_2. Then colonies were split into small pieces mechanically and incubated in 0.075M KCI hypotonic solution for 17 minutes at 37°C. Colonies were fixed with methanol-acetic acid solution (3:1) and processed for G-banding analysis. For each line at least 20 metaphases were analyzed for confirmation.
Karyotyping of each line was performed several times to assess whether karyotypes were stable during their long term culture. Confirmation of genetic mutation in hESC line derived from affected embryo was performed by the same procedures applied for single blastomere mutational analysis.
Surface expression markers (SSEA-3, SSEA-4, TRA-1-60 and TRA-1-81), which are unique for undifferentiated hESC lines, were immunochytoemically analyzed according to the manufacturer's instructions (Chemicon) by using flourescein isothiocyanate (FITC)-conjugated goat anti-mouse secondary antibody (anti-IgG) (Santa Cruz Biotechnology, California, USA). Negative controls were performed by addition of phosphate buffer saline instead of the primary antibody. All the other reagents were the same as in the slides run for specific antibodies, except nucleus of cells were stained by 4',6-diamidino-2-phenylindole (DAPI) for visualization.
Alkaline phosphatase activity (Chemicon) of hESC lines was detected by using the Chemicon Alkaline Phophatase Detection kit (Chemicon). Expressions of OCT-4 and housekeeping gene, glyceraldehyde-3 phosphate dehydrogenase (GADPH), genes were detected by reverse transcriptase PCR (RT-PCR). Briefly, extraction of RNA from undifferentiated hESCs and synthesis of cDNA were carried out by using Rneasy Mini Kit (Qiagen Gmbh, Strasse, Germany) and Sensiscript RT Kit (Qiagen), respectively. Then, PCR and following analysis were performed according to the protocol of Amit et al (Amit et al., 2002).

2.5 Differentiation potential of hESC lines

The differentiation ability of hESC lines were analyzed only by *in vitro*. For *in vitro* differentiation, embryoid bodies (EBs) were first generated according to previously published protocol (Itskovits-Eldor et al., 2000; Carpenter et al., 2003). In a brief, small pieces of undifferentiated hESC colonies were transferred into the non adherent bacterial petri dishes and were cultured in CSCM without bFGF for 8-10 days. The resulting EB, which theoretically comprise three embryonic germ layers, were then plated on to 0.1% gelatin coated plastic petri dishes and cultured for long-term for spontaneous differentiation.

Differentiation features of hESC lines was examined under phase contrast microscope and differentiated cells were analyzed by immunocytochemical staining with markers for endoderm (cytokeratin 18, specific marker for epithelial cells, Chemicon), mesoderm (troponin I, specific for cardiac muscle; Chemicon) and ectoderm (Nestin, specific marker for progenitor of neuron, and MAP2AB, specific marker for mature neuron; Chemicon) according to manufacturer's instruction.

Additionally, in these lines (OZ, OZ-1 and OZ-2), rhythmically beating of cardiomyocytes within spontaneously differentiating embryonic stem cells were further analyzed by transmission electron microscopy (TEM). Briefly, cell clumps with spontaneous contractions were gently removed from the culture plate and fixed in 2% glutaraldehyde in 0.1 mol/sodium cacodylate buffer (pH 7.4) for 2 hours. Secondary fixation was performed in 1% OsO4 in the same buffer for 1.5h. The grids were dehydrated in graded ethanol and embedded in Epon 812. The very thin sections about 80 nm were cut and stained with lead citrate for 8 min in order to identify the cellular structures of cardiac muscle cells.

3. Results of hESC study

Experience of hESC from Memorial Hospital comprise three phrases; derivation of first hESC lines in Turkey, which was reported previously (Findikli et al., 2005), using HFF as a feeder cell instead of MEF to derive new hESC lines and derivation of hESC lines from donated embryos from PGD cycles (Candan & Kahraman, 2010).

In the first phase, nine hESC lines, which were named NS-1, NS-2, NS-3, NS-4, NS-5, NS-6, NS-7, NS-8 and MINE, were derived from 26 donated blastocysts stage human embryos with a 34.6% success rate (Table 1). Twenty blastocysts were spare IVF/ICSI embryos and 5 hESC lines (NS-1, NS-2, NS-3, NS-4 and MINE) were derived from these embryos. The remaining 6 embryos had mismatched HLA type and, therefore were not eligible for transferring in PGD cycle. From these embryos, 4 hESC lines (NS-5, NS-6, NS-7 and NS-8)

	No.of blastocysts	No. of stem cell line, n, (%)	Names of derived hESC lines
IVF/ICSI embryos	30	8, (27)	NS-1,2,3,8, MINE, and OZ, OZ-1, OZ-2
PGD for single gene disorder	8	1, (13)	OZ-8
PGD for single gene disorder & HLA typing	6	4, (67)	*NS-4, 5, 6, and 7
PGD for chromosomal screening	42	7, (17)	OZ-3, OZ-4, OZ-5, OZ-6, OZ-7, NS-9, NS-10
Immunosurgery method	15	4, (27)	NS-1,2,3,4
Direct culture method	76	16, (21)	

Table 1. A total number of embryos used for hESC lines and overall outcomes

were derived. Four lines (NS-1, NS-2, NS-3, NS-4) out of 9 hESC lines were obtained by immunosurgey and remaining 5 lines were derived by direct culture (Table 1). Two cell lines (NS-1 and NS-2) were spontaneous differentiated during their first days of *in vitro* culture.

In the second phase of hESC research in Istanbul Memorial Hospital, as an alternative to MEF, HFF was used as a feeder cell for establishment and long-term culture of new hESC lines. Three hESC lines (OZ, OZ-1 and OZ-2) were derived from 10 blastocyst stage spare IVF/ICSI embryos by the direct culture technique with a 30% success rate (Table 1). Unlike to circular colony morphology of hESC colonies on MEF, hESC colonies on HFF had angular shaped morphology, due to the polarity of HFF cells (Figure 1 C-D).

In the final phase, following PGD, embryos diagnosed as having chromosomal abnormalities and single gene mutations were used to establish hESC lines (Table 1). Forty two blastocysts with different chromosomal aneuploidies were directly placed either on MEF or on HFF. From those embryos 7 hESC lines (OZ-3, OZ-4, OZ-5, OZ-6, OZ-7, NS-9 and NS-10) were derived (Table 1 and 2). Of these 7 hESC lines, one line (OZ-3) was derived from biopsied embryo whose diagnosis was suspicious. Although chromosomal content of biopsied blastomere from this embryo was identified as abnormal (trisomy 15) by FISH, because of the fragmentations in nuclear structure of blastomere, we could not interpret the result exactly and thereby assumed it as an abnormal (Candan & Kahraman, 2010).

Three embryos diagnosed as carrying cystic fibrosis and 5 embryos with beta-thalassemia were used to isolate hESC lines with genetic disorder. However, only one hESC (OZ-8), which had a single gene mutation causing beta-thalassemia, was isolated successfully. Only 4 hESC lines were isolated by immunosurgery and the remaining hESC lines were derived after direct culture of blastocysts on feeder cells (Table 2). Following to either direct culture or immunosurgery, the developing three dimensional outgrowths from ICM were split mechanically into small clumps and transferred onto new feeder plate. Duration for successful derivation of first primary hESC colonies was ranged 15 to 20 days, based on the quality of inner cell mass of blastocysts, and application of isolation techniques properly.

Following the first several passages of primary hESC colonies, flat colonies of cells with a distinguishable compacted colony structure, well defined colony border and cellular morphology with higher nucleus to cytoplasm ratio and prominent nucleoli were obtained (Figure 1C-D). Regardless of type of feeder cells, these unique colony features were similar in all hESC. Each of hESC lines were passaged mechanically every after about 7-8d for more than 15 passages (Table 2). While passaging hESC colonies, spontaneously differentiated cells, which were observed frequently in the central or in the periphery part of colonies, were always removed mechanically to maintain undifferentiation state of colonies.

3.1 Unique features of hESC lines

All hESC lines were characterized for cell surface expression markers, which are unique to undifferentiated human embryonic stem cells. As shown in Figure 2A, established hESC lines represented a high level of alkaline phosphates activity. Furthermore, immunocytochemical staining revealed that derived 18 hESC lines were positive for SSEA-3, SSEA-4, TRA-1-60 and TRA-1-81 (Figure 2B). Negative control slides of immunocytochemical staining showed that primary antibodies specifically bound to the certain surface antigens (Figure 2C).

	NS-3	NS-4	NS-5	NS-6	NS-7	NS-8	NS-9	NS-10	MINE
Embryo source	IVF/ICSI	IVF/ICSI	PGD	PGD	PGD	PGD	PGS	PGS	IVF/ICSI
Embryo grades	4BB	3BA	4AA	4AA	4AB	3AB	4BB	4BA	4BB
PGD/PGS results	-	-	HLA mismatched	HLA mismatched	HLA mismatched	HLA mismatched	Tetraploid	Monosomy -13	-
Feeder cell	MEF	MEF	MEF	MEF	MEF	MEF	MEF	MEF	MEF
Alkaline phosphatase	+	+	+	+	+	+	+	+	+
Immuno-staining	+	+	+	+	+	+	+	+	+
SSEA-3	+	+	+	+	+	+	+	+	+
SSEA-4	+	+	+	+	+	+	+	+	+
TRA-1-60	+	+	+	+	+	+	+	+	+
TRA-1-81	+	+	+	+	+	+	+	+	+
OCT-4	+	+	+	+	+	+	+	+	+
In vitro differentiation	+	+	+	+	+	+	+	+	+
Max. passage no. of lines	35	60	39	20	22	22	21	28	49
Karyotype	46 XY	46 XX	46 XX	46 XY	46 XX	46 XX	46 XY	46 XX	46XX

	OZ	OZ-1	OZ-2	OZ-3	OZ-4	OZ-5	OZ-6	OZ-7	OZ-8
Embryo source	IVF/ICSI	IVF/ICSI	IVF/ICSI	PGS	PGS	PGS	PGS	PGS	PGD
Embryo grades	4AA	3AA	4BB	4BB	4AB	3BB	4AC	4BB	4AB
PGD/PGS results	-	-	-	Undiagnosed embryo	Trisomy 21	Monosomy 15+ Trisomy 21	Monosomy 16+ Trisomy 16	Monosomy 16	Beta-Thalasse mia
Type of feeder cell	HFF	HFF	HFF	HFF	HFF	HFF	MEF	MEF	MEF
Alkaline phosphatase	+	+	+	+	+	+	+	+	+
Immuno-staining	+	+	+	+	+	+	+	+	+
SSEA-3	+	+	+	+	+	+	+	+	+
SSEA-4	+	+	+	+	+	+	+	+	+
TRA-1-60	+	+	+	+	+	+	+	+	+
TRA-1-81	+	+	+	+	+	+	+	+	+
OCT-4	+	+	+	+	+	+	+	+	+
In vitro differentiation	+	+	+	+	+	+	+	+	+
Max. passage no. of lines	19	15	15	22	17	20	22	18	18
Karyotype	46 XY	46 XX	46 XX	46 XX	46 XX	46 XY	46 XX	46 XY	46 XX

Table 2. Unique features of 18 hESC lines. (+) represents that HESC line is shown to be positive for those expression markers.

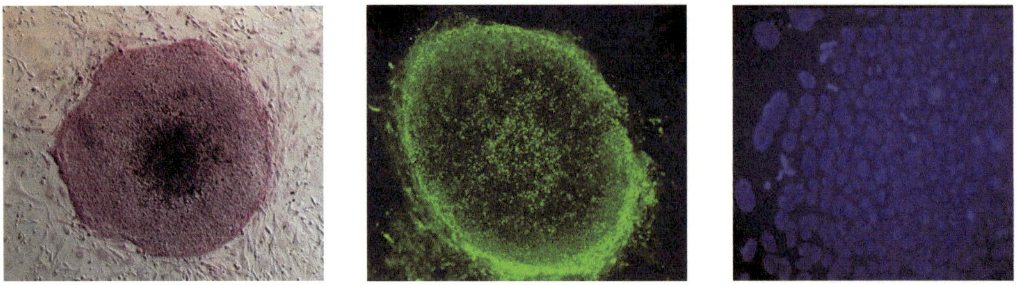

A) B) C)

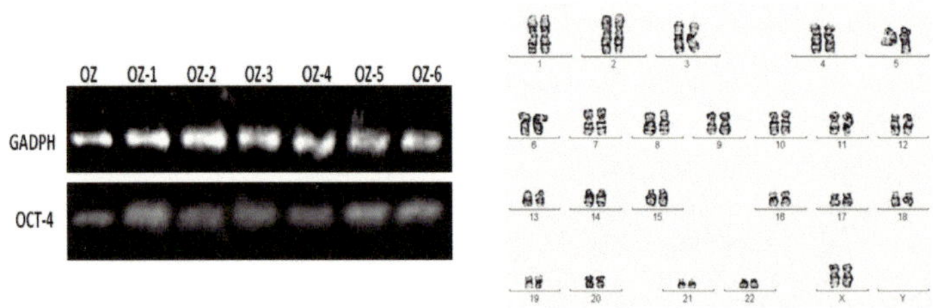

D) E)

Fig. 2. Unique features and karyotyping of OZ-6 hESC line. **A)** alkaline phosphatase staining, **B)** SSEA-3, SSEA-4, TRA-1-60, TRA-1-81and **C)** negative control for immunostaining of hESC lines (cells were stained by DAPI for visualization). Expression of OCT-4 and housekeeping gene GADPH in hESC lines (OZ, OZ-1-7). **E)** Karyotype analyze of OZ-6 line in passage 5, 46XX. Original magnifications: **(A-C)** X 200. Abbreviations: OCT-4, octamer-4 and GADPH, glyceraldehyde-3 phosphate dehydrogenase.

Our hESC lines were not analyzed for SSEA-1 expression, a specific marker for mouse embryonic stem cells. In consistent with the previous reports, expression intensity of SSEA-3 among hESC lines was variable and comparably weaker than SSEA-4 which was consistent and expressed higher in all hESC lines (Oh et al., 2005). Additionally, OCT-4 expressions were higher in all undifferentiated hESC lines when compared to expression level of housekeeping gene GADPH (Figure 2D).

Testing differentiation capacity of each hESC line by embryoid body formation *in vitro* revealed that these lines were capable of differentiating into various cell types derived from the three embryonic germ layers (Figure 3). Spontaneous contracting cell clumps, neural rosette structures, neural-like cells, epithelial like cells were observed under phase microscope and these differentiating cells were discriminated by immunocytochemically (Figure 3B). However, cell lines showed a relatively different capacity or tendency to differentiate into certain type of cell lineage.

Rhythmically contractions in cell clusters were started approximately after ten days from plating EBs on bacterial culture plates and kept continuing up to six weeks. During this period, beating cell clusters, belonging to OZ, OZ-1 and OZ-2 hESC lines, were further analyzed by transmission electron microscopy. Thereby, sarcomere, intercalated disc and myofibril structures were well defined in cardiac muscle cells by TEM analysis (Figure 3E). Following vitrification and thawing procedures, all cell lines could retain their unique properties; long-term extension and pluripotency capacity *in vitro*.

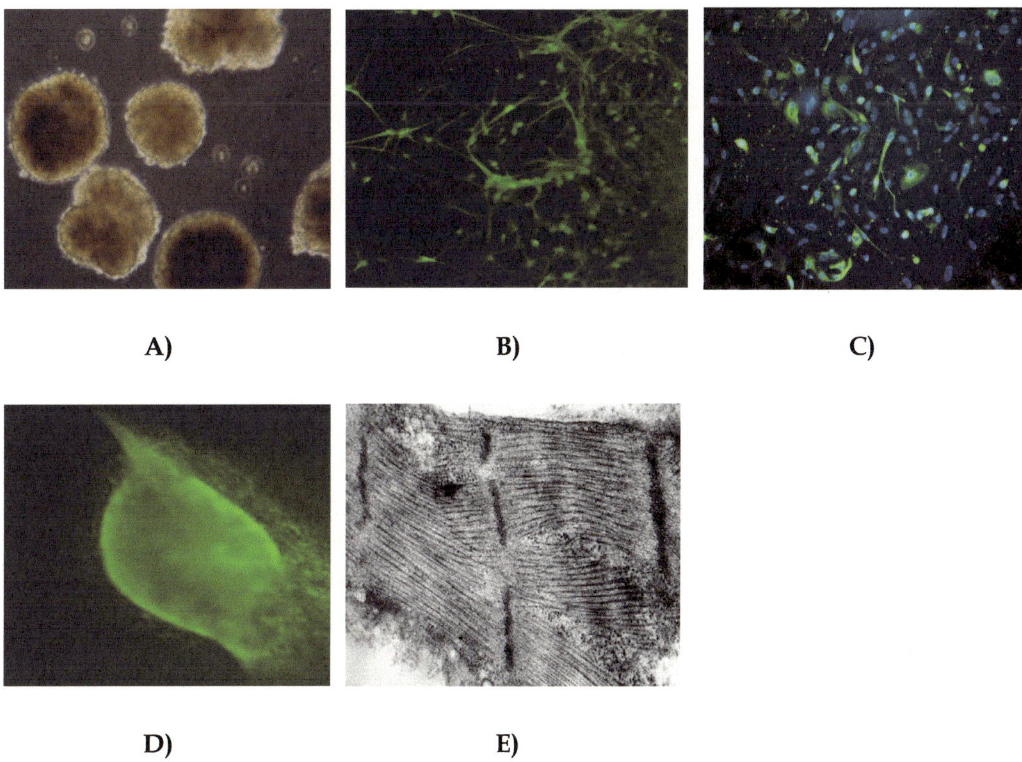

Fig. 3. In vitro differentiation of NS-10 hESC line by embryoid body formation **A)** Phase contrast microscopy of 10-day EBs. Spontaneously differentiating cells **B)** neuron-like cells, positive for neuron specific nestin, **C)** epithelial-like cells, positive for cytokeratin 18 (green) and **D)** cardiac muscle cells, positive for troponin I. **E)** TEM photographs of beating cardiomyocytes. **(A)** X 100, **(B-D)** X200, **(E)** X12000.

3.2 Karyotyping and genetic analysis of hESC lines

Karyotyping analysis of all hESC lines were performed at the 5th passages. HESC lines, which were derived from supernumerary embryos after IVF/ICSI cycles and from PGD embryos, having genetic disorder and/or having mismatched HLA, had normal karyotypes (Figure 2E and Table 2).

HESC lines derived from chromosomally abnormal embryos were first analyzed by FISH at 1st passage whether they had detected chromosomal abnormality. Surprisingly, chromosomal abnormalities were not confirmed in these lines. Contrarily, analyzed chromosomes were euploid in number. Further confirmation was performed by

karyotyping at 5th passages of these lines. In consistent with results after FISH at 1st passage, karyotyping analysis revealed that in contrast to the diagnosis after PGD their karyotypes were normal (Table 2).
Karyotyping of all hESC lines were performed further passages whether cell lines retained normal karyotypes. All hESC lines had stable karyotypes.
Mutation in OZ-8 hESC cell line with beta thalassemia disorder was confirmed by PCR and subsequent sequencing procedures at passage 6. The homozygote single nucleotide transition (guanine to adenine nucleotide transition) in second exon of beta-globin gene was detected.

4. Discussion

Human embryonic stem cell is one of the most contradictory scientific issues since it was first reported by Thomson in 1998. Obviously, this ongoing dispute has been arisen from the use of human embryo for derivation of hESC. In regard to that concern, alternative methods have been proposed by several researches. However, spare embryos generated for reproductive and therapeutic treatments still remain as a main source for hESC derivation. Therefore, registering all existing lines in a database, like stem cell bank, may decrease the necessities to derive new hESC lines worldwide and eventually ethical concern may be alleviated among the public. In that regard, all these hESC lines, which had been established and characterized in Istanbul Memorial Hospital until ruling on ban on hESC research by Turkish Health Ministry, were registered to European hESCreg in 2008.
During the hESC derivation study, 86 donated embryos, which were considered insufficient for transfer and cryopreservation after IVF/ICSI cycles and were diagnosed as having genetic disorder or chromosomal aneuploidies and having mismatched HLA type after PGD cycles, were used.
The derivation efficiency of hESC lines was 20% and success rate was directly related with the quality of blastocysts. Four (NS-1, NS-2, NS-3 and NS-4) out of 20 hESC lines were successfully derived from 15 blastocysts after immunosurgery. However, two of these lines were spontaneously differentiated at the early number of passages. Remaining 16 hESC lines were isolated through direct culture of whole blastocysts on feeder cells. These two methods had comparable success rates (27% vs 21%, $p>0.05$).
All lines described in this chapter had similar colony and cellular morphology. These lines expressed unique cell surface expression markers, including SSEA-3, SSEA-4, TRA-1-60 and TRA-1-81. They also had high level of alkaline phosphatase activity and expressed OCT-4 gene, which keeps pluripotency of hESC lines during long-term culture (Figure 2). Moreover, these 18 hESC lines were proven to have pluripotent capacity *in vitro* by EBs formation (Figure 3). Therefore, these lines have similar unique properties as previously reported existing hESC lines.
Although all lines were capable of differentiating into derivatives of three embryonic germ layers, differentiating characteristic was varied among lines. The prevalence of differentiating cardiomyocytes was higher in EBs generated from OZ-3 line, whereas higher percentage of neuron-like cells were observed in EBs generated from OZ line. The presence of specific cell lines at a various degree in differentiating cell cultures of hESC lines may be attributed to the developmental stage of embryo used for derivation, genomic and epigenetic differences.

In PGD for chromosomal screening cycles, embryos diagnosed as having chromosomal abnormalities were likely to be discarded and not to be considered eligible for transfer. In consistent with the previous results, in which it was aimed to derive chromosomally abnormal hESC lines for investigation of various aspects of early embryonic development, all our hESC lines derived from those embryos had normal karyotypes (Munne et al., 2005; Lavon et al., 2008). As stated in these studies, this unexpected result has been suggested to self correction of embryos during long-term *in vitro* culture (Hazan et al., 2008; Lavon et al., 2008).

As a result of proposed mosaicism and trisomy rescue mechanisms in developing embryos, chromosomal self correction may be occurred. In the mosaic embryos, it was suggested that euploid blastomeres can grow preferentially to abnormal cells during early embryogenesis or alternatively, these normal cells were preferentially allocated to the inner cell mass. Therefore, embryo may be subject to self correction mechanism and may evolve to a fetus with euploid chromosomal number. In trisomic rescue mechanism, it has been speculated that embryos can be corrected in terms of chromosomal number through anaphase-lag, nondisjunction, or chromosomal demolition. However, these mechanisms are still assumptions and exacts mechanisms should be proven by further studies (Hazan et al., 2008).

Furthermore, we speculated that the long-term culture may not only induce the change in chromosomal content but also may result in various genetic or epigenetic modifications in hESC lines which could eventually impair their pluripotent and self renewal capacity. However, due to inured restriction on hESC research, these derived lines could not be analyzed regarding to these aspects (Candan et al 2010).

Embryos, having single gene mutations, are of great importance, while considering as a potential source in genetic based researches on understanding the mechanisms of disease and developing new drugs (Pickering et al., 2003; Verlinsky et al., 2005). We derived one hESC line having mutation, that cause beta thalassemia disorder, from 5 donated embryos after PGD. This cell line can provide as a cell source to study the pathology of beta thalassemia and its effects on different cell types.

5. Conclusions

HESC researches have been increasing continuously since it was first isolated and characterized. Although several problems including derivation clinical grade hESC lines, risk for teratoma formation, HLA incompatibility of cells and establishment of well defined differentiation protocols of hESC lines to certain cell types have been not been solved efficiently, there is still hope for coming day in which hESC lines will be used effectively in cell replacement therapies to treat such devastating human diseases; myocardial infarcts, spinal cord injuries and diabetes. Therefore, hESC lines reported in this chapter could be potential cell source for *in-vitro* studies involving basic molecular and stem cell biology. . Moreover, as in our study, derivation of hESC from PGD embryos with intrinsic genetic content and the disease profile could be an extremely valuable source for research on genetic diseases.

In this chapter, it is aimed to summarized establishment protocols and features of first reported hESC lines in Turkey. Turkey is among the countries in which earlier hESC studies were established first. Today, however, Turkey is one of the countries in which hESC researches were prohibited by Government. Unfortunately, so far new legislation or regulation has not been declared. However, establishment of new guideline is on the agenda of governmental institutions nowadays. Recent lifting of ban on the federal funding for hESC research in US and given permission on first clinical trial of hESC may, in fact, affect the political opinion and hopefully may inure regulations, which resume hESC researches under the control of authority and new guidelines in Turkey.

6. Acknowledgments

We would like to specifically thank to Necati Findikli, Oya Akcin and Ayla Eker Sariboyaci for their previous contributions on the derivation and characterization of hESC lines and all IVF staff for their kindly help in human embryo culture and support. Additionally, we are very thankful to all Genetic laboratory staff for PGD analysis and karyotyping all hESC lines. Study was supported and funded by Istanbul Memorial Hospital.

7. References

Amit M, Margulets V, Segev H. (2002). Human feeder layers for human embryonic stem cells. *Biology of Reproduction,* 68, 2150-2156

Candan NZ & Kahraman S. (2010). Establishment and characterization of human embryonic stem cell lines, Turkey perspectives. *In Vitro Cell.Dev.Biol. – Animal,* 46, 345–355

Carpenter MK, Rosler E, Rao MS. (2003). Characterization and differentiation of human embryonic stem cells. *Cloning and Stem Cells,* 5, 79-88

Conner DA. (2000) Mouse embryo fibroblast (MEF) feeder cell preparation. *Current Protocols in Molecular Biology,* 23.2.1-23.2.7

Edwards RG. (2004). Stem cells today: A. Origin and potential of embryo stem cells. *Reproductive BioMedicine Online,* 8, 275–306

Findikli N, Kahraman S, Akcin O. (2005). Isolation and characterization of new human embryonic stem cell lines. *Reproductive Biomedicine Online,* 10, 617-627

Findikli N, Candan ZN, Kahraman S. (2006). Human embryonic stem cell culture: current limitations and novel strategies. *Reproductive Biomedicine Online,* 13, 581-590

Gardner DK, Lane M, Stevens J. (2000) Blastocyst score affects implantation and pregnancy outcome: towards a single embryo transfer. *Fertility and Sterility,* 73, 1155–1158.

Hazan BS, Frumkin T, Malcov M. (2008). Preimplantation aneuploid embryos undergo self-correction in correlation with their developmental potential. *Fertility and Sterility,* 92, 890- 896

Hoffman LM and Carpenter MK. (2005). Characterization and culture of human embryonic stem cells. *Nature Biotechnology,* 23, 699-708

Hovatta O, Mokkola M, Gertow K. (2003). A culture system using human foreskin fibroblasts as feeder cells allows production of human embryonic stem cells. *Human Reproduction,* 18, 1404-1409

Itskovitz-Eldor J, Schuldiner M, Karsenti D. (2000). Differentiation of human embryonic stem cells into embryoid bodies comprising the three embryonic germ layers, *Molecular Medicine,* 6(2), 88-95.

Kahraman S, Bahce M, Samli H. (2000). Healthy births and ongoing pregnancies obtained by preimplantation genetic diagnosis in patients with advanced maternal age and recurrent implantation failure. *Human Reproduction,* 15, 2003-2007

Kahraman S, Findikli N, Biricik A. (2006) Preliminary FISH studies on spermatozoa and embryos in patients with variable degrees of teratozoospermia and a history of poor prognosis. *Reproductive Biomedicine Online,* 12, 752-761

Lavon N, Narwani K, Golan-Lev T et al. 2008 Derivation of euploid human embryonic stem cells from aneuploid embryos, Stem Cells 26: 1874-1882.

Munne S, Velilla E, Colls P. (2005). Self-correction of chromosomally abnormal embryos in culture and implications for stem cell production. *Fertility and Sterility,* 84, 1328-1334

Oh KS, Kim HS, Ahn JH. (2005). Derivation and Characterization of New Human Embryonic Stem Cell Lines: SNUhES1, SNUhES2, and SNUhES3. *Stem Cells,* 23, 211-219

Pickering SJ, Braude PR, Patel M. (2003). Preimplantation genetic diagnosis as a novel source of embryos for stem cell research. *Reproductive Biomedicine Online,* 7, 353-364.

Reubinoff BE, Pera MF, Vajta G and Trounson AO. (2001). Effective cryopreservation of human embryonic stem cells by the open pulled straw vitrification method. *Human Reproduction,* 16, 2187-2194

Richards M, Fong CY, Chan WK. (2002). Human feeders support prolonged undifferentiated growth of human inner cell masses and embryonic stem cells. *Nature Biotechnology,* 20, 933-936.

Rodriguez CI, Galan A, Valbuena D. (2006). Derivation of clinical grade human embryonic stem cells. *Reproductive Biomedicine Online* 12, 112-118

Semb H. (2005). Human embryonic stem cells: origin, properties and applications. *Apmis,* 113, 743–750

Simon C, Escobedo C, Valbuena D. (2005). First derivation in Spain of human embryonic stem cell lines: use of long-term cryopreserved embryos and animal-free conditions. *Fertility and Sterility,* 83, 246-249

Solter D and Knowles BB. (1975). Immunosurgery of mouse blastocyst. *PNAS,* 72, 5099-5102

Thomson JA, Itskovitz-Eldor J, Shapiro SS. (1998). Embryonic stem cell lines derived from human blastocysts. *Science,* 282, 1145-1147

Trounson A. (2006). The production and directed differentiation of human embryonic stem cells. *Endocr Rev,* 27, 208–219

Van de Stolpe A, van de Brink S, van Rooijen M. (2005). Human embryonic stem cells: towards therapies for cardiac disease. Derivation of a Dutch human embryonic stem cell line. *Reproductive Biomedicine Online,* 11, 476-485

Vanderzwalmen P, Bertin G, Debauche Ch. (2003). Vitrification of human blastocysts with the Hemi-Straw carrier: application of assisted hatching after thawing. *Human Reproduction,* 18, 1504-1511

Verlinsky Y, Strelchenko N, Kukharenko V. (2005). Human embryonic stem cell lines with genetic disorders. *Reproductive Biomedicine Online,* 10, 105-110

3

Cryopreserved Embryos: A Catholic Alternative to Embryonic Stem Cell Research and Adoption

Peter A. Clark
Saint Joseph's University Philadelphia, Pennsylvania
United States of America

1. Introduction

It is estimated that 2.1 million married couples or 5 million people in the United States are affected by infertility.[1] Infertility is defined as failure to get pregnant after one year of unprotected intercourse. About 40% of infertility cases are due to a female factor and 40% due to a male factor. The remaining 20% are the result of a combination of male and female factors, or are of unknown causes.[2] Issues of human infertility are extremely complex physiologically, psychologically, financially, legally and ethically. It is estimated that 85-90% of infertile couples will receive conventional treatment and 10-15% may become candidates for various forms of Assisted Reproductive Technologies (ARTs) to assist them in having their own biological children. In-vitro fertilization (IVF) is one of the most utilized reproductive procedures that has allowed couples to have their own biological children. IVF accounts for 99% of ART. This procedure has been effective but it is still inefficient and expensive. One aspect of the inefficiency is that numerous embryos have been frozen through a process called cryopreservation. It has been estimated that there are 400,000 embryos frozen and stored since the late 1970s.[3] In reality, the actual number of frozen embryos is probably closer to 500,000 with an additional 20,000 embryos added yearly.[4] Freezing these embryos has allowed for a limitation on the number of embryos transferred to a woman's uterus which has decreased the number of multiple gestations. It also allows couples to use the frozen embryos in the future if the initial cycles are unsuccessful. This is not only more effective but also lowers the cost. The issue is now what to do with the 400,000 to 500,000 frozen embryos that remain as "spares." Various alternatives have been suggested. The embryos could be thawed and then destroyed, continued to be cryopreserved indefinitely, used for research, or offered for donation/adoption. All of these options present problems medically, legally and ethically, especially for Roman Catholics.

Medically, the lifespan of a cryopreserved embryo is unknown. The effect of the freezing process is also unknown on the quality of the embryo if brought to term. "Studies have found that babies created through IVF are twice as likely to be born underweight and with major birth defects."[5] With the unknown effects of cryopreservation on embryo development the medical issues become even more complex. Legally, only 2% of frozen embryos are specifically designated for donation/adoption and 5% are specifically designated for destruction or research.[6] The legal issues focus on the applicability of contract law versus family law because frozen embryos are technically considered "property" not "persons." Presently, the applicability of contract law or family law remains unclear. In addition, to date only three states—Florida, Louisiana and New Hampshire—have adopted legislation concerning the disposition or disposal of embryos. Legally and legislatively the issue of embryo donation/adoption is ambiguous at best. Ethically, depending on one's view of when personhood begins, frozen embryos may be considered human persons, which deserve dignity and respect, or they may have less than human status with no particular ethical rights. From an ethical perspective that views personhood beginning at fertilization, one could argue that the "rescue" of these embryos could be considered ethically acceptable. The problem is that in 2008 the Vatican's Congregation for the Doctrine of the Faith issued an Instruction called *Dignitas Personae,* which stated that prenatal adoption, "praiseworthy with regard to intention of respecting and defining human life, presents however various problems" and is considered unethical.[7] If donation/adoption is not acceptable by the Magisterium then the only remaining option would be to stop the process of cryopreservation and allow the thawed embryos to die with dignity and respect under the principle of the extraordinary/ordinary means distinction.

This article will focus on allowing to die under the principle of extraordinary/ordinary means distinction as a viable option to address the 400,000 to 500,000 frozen embryos in the United States. The intended purpose of this article is threefold: first, to examine the medical issues surrounding the cryopreservation of frozen embryos; second, to give an ethical analysis of the arguments for and against allowing to die; and third, to give recommendations on how to avoid the continuation of this problem in the future.

2. Medical aspects

Infertility is a major problem for many couples in the United States. "About one married couple in 12 cannot conceive a child after two years of trying. Infertility stems from many factors, including a woman's age at the first attempt to conceive, damage from pelvic inflammatory disease, previous abortions, uterine abnormalities, and a man's low sperm count or low sperm motility."[8] Individually, male and female factors each account for about 40% of infertility in the United States. Numerous technologies are available to couples from artificial insemination by a husband or a donor, to gamete intrafallopian transfer (GIFT), to zygote intrafallopian transfer (ZIFT), to in-vitro fertilization. Of these reproductive technologies IVF has become the ART of choice for many infertile couples. IVF is an assisted reproductive technology which had its first success in 1978 when Drs. Edwards and Steptoe in Oldham, England created the first "test tube baby" named Louise Brown. Since that first success, IVF technology has been refined and over 3 million babies have been born worldwide.[9]

There are five basic steps to IVF. *1) Harvesting the eggs from the woman's ovaries.* The woman's ovaries are hyperstimulated using fertility drugs that produce numerous eggs. During this period the woman will have regular transvaginal ultrasounds to examine the ovaries and blood tests to check hormone levels. *2) Egg retrieval.* The eggs are removed from the woman's body using follicular aspiration. Using ultrasound images as a guide the physician inserts a thin needle through the vagina and into the ovary and sacs containing the eggs. The needle is connected to a suction device, which pulls the eggs and fluid out of each follicle, one at a time. In rare cases, a pelvic laparoscopy may be used to remove the eggs. *3) Insemination and Fertilization.* The man's sperm is placed with the best quality eggs in a petri dish and stored in an environmentally controlled chamber. The mixing of the sperm and egg is called insemination. The sperm usually enters an egg a few hours after insemination. If there is a low chance for fertilization, one single sperm can be injected into an egg in a procedure called Intracytoplasmic Sperm Injection (ICSI). *4) Embryo culture.* The fertilized eggs remain in the petri dish for 48 to 72 hours to verify that the embryo is not defective and growing properly. If a couple is at high-risk for passing on genetic (hereditary) disorders to a child they may consider using Pre-implantation Genetic Diagnosis (PGD). The procedure is performed 3-4 days after fertilization. A single cell is removed from each embryo to screen it for specific genetic disorders. Those embryos with the genetic disorder are usually destroyed. *5) Embryo transfer.* Anywhere from 1-4 embryos are placed in the woman's womb 3 to 4 days after fertilization. The physician inserts a thin catheter containing the embryos into the woman's vagina, through the cervix, and up into the womb. If the embryo implants in the woman's uterine wall pregnancy will result.[10]

The implantation rate is estimated at 10-25%.[11] The overall birth rate varies from 11% (women over 40) to about 35% (women under 35).[12] This clearly shows that a number of embryos transferred fail to survive, which is why multiple embryos are transferred per cycle and why numerous cycles are required. On average, 2.7 embryos per cycle are transferred in women under 35, with an average of 3 in older women. Depending on the embryo quality, up to 5-6 embryos can be transferred.[13] The average cost of IVF is $12,000-17,000 per cycle. It is estimated that 75% of couples who have tried IVF and who spent from $10,000-100,000 still go home without a baby.[14] Risks include the possibility of ovarian hyperstimulation syndrome (OHSS), risks in the egg retrieval stage which include reactions to anesthesia, bleeding, infection and damage to structures surrounding the ovaries including the bowel and bladder, and finally there are the risks associated with multiple pregnancies. Since 1980 the rate of twins has climbed 70% to 3.2% of births in 2004. Multiple gestations raise the risk of preterm births; low-birth-weight babies, with the possibility of death in very premature infants; long-term health problems; and pregnancy complications, which include pre-eclampsia, gestational diabetes, and Caesarean section. Studies have shown that 56% of IVF twins born in 2004 weighed less than 5.5 pounds, and 65% were born prematurely, before 37 weeks of gestation.[15] Embryos not transferred in a fresh IVF cycle are usually cryopreserved. Freezing these embryos offers individuals the possibility of transferring the frozen embryos for later IVF cycles if the previous cycle does not result in a pregnancy. It is also cost effective and eliminates the need to undergo the steps needed for a fresh IVF cycle. In most cases the best quality embryos are transferred in the fresh cycle and those of a lesser quality are frozen for later transfer. It should be noted that some clinics have individual freezing and thawing to achieve the exact number of embryos desired for transfer. This procedure avoids embryo wastage.

The process of cryopreservation has become an integral part of the IVF procedure. "Cryopreservation is a process of freezing biological tissues for storage, while minimizing cellular damage from freezing and thawing."[16] This technique entails freezing the embryo while simultaneously removing the intracellular water and replacing it with a cryoprotectant solution which help to protect the embryo during the freezing process. The embryos are then placed into cryopreservation straws or vials, which are labeled with the patient's name, the patient's IVF number, and the date of the freeze. Once the process is complete, the embryos are placed in a computer controlled freezing unit. After the freezing run is complete, the straws are stored in a special tank filled with liquid nitrogen at a temperature of minus 196 degrees centigrade.[17] Many storage facilities use a back-up system to minimize the risk of interruption in the freezing process. Liquid nitrogen containers are armed with an automatic alarm system to monitor nitrogen levels and prevent premature thawing.[18] These embryos are looked upon as being in a state of "suspended animation." Cellular activity has ceased, but each embryo is still alive. When the remaining embryos are needed a procedure utilizing rapid thawing and removal of the cryopreservative solution with simultaneous rehydration is used. The embryos are first warmed in a 98.6 F degree solution and the cryoprotectant chemicals are removed.[19]

The embryo thawing process is quite complex. "Embryo survival is based on the number of viable cells in an embryo after thawing. An embryo has 'survived' if >50% of the cells are viable. An embryo is considered to 'partially survive' if <50% of its cells are viable and to be 'atretic' if all the cells are dead at thaw. Approximately, 65-70% of embryos survive thaw, 10% partially survive and 20-25% are atretic. Data suggests that embryos with 100% cell survival are almost as good as embryos never frozen but only about 30-35% survive this fashion. Embryos that are 2, 4 or 8 cells when frozen have about a 5-10% greater survival than embryos with an odd number of cells. Donor egg embryos have a 2-5% greater survival rate than embryos from infertile women when compared by morphology score"[20] The cost of cryopreservation is approximately $600-700 a year. The success rate or pregnancy rate depends on numerous factors: the number of surviving embryos transferred, the number of 100% surviving embryos transferred, and the morphology scores of the transferred embryos. The delivered pregnancy rates range from 5% (a single poor quality embryo) to 36% (4 high quality embryos) when the cycles from 1987 to 2001 were combined. It is estimated that embryo cryopreservation adds about 10-30% more pregnancies per retrieval cycle and the outcomes of the children are normal.[21] The reason for the wide range of costs and success rates is because the Assisted Reproductive Technologies industry in the United States is unregulated. The success rates and costs can vary from clinic to clinic and there is no government oversight examining the widespread differences.

The advantages of embryo freezing are numerous: reducing the risks of multiple gestations potentially increases pregnancy rates, decreasing the number of stimulated treatment cycles needed to achieve pregnancy, decreasing the costs of ARTs, etc. The main disadvantage according to the 2003 RAND/SART Working Group study centers on the approximately 400,000 frozen "spare" embryos stored since the 1970's.[22] More recent numbers have the number of frozen embryos in excess of 500,000. The 500,000 number seems more realistic considering the increase in IVF procedures since 2003. The issue that is confronting parents and fertility clinics is what to do with these "spare" embryos medically, legally and ethically.

The RAND/SART survey in 2003 found that of the 400,000 frozen spare embryos 88.2% were designated for family building and 2.8% (11,000) were designated for research. Those embryos designated for research could produce as many as 275 stem cell lines (cell cultures suitable for further development). However, the number would in reality be much lower. Of the remaining embryos, it is estimated that 2.3% (10,000) are awaiting donation, 2.2% are designated to be discarded, and 4.5% are held in storage for other reasons, including lost contact with a patient, patient death, abandonment, and divorce.[23] There are numerous issues concerning the "spare" frozen embryos. The ART clinics transfer the highest quality embryos (those that grow at a normal rate) to the patient during treatment cycles. The remaining embryos are usually designated as not of the highest quality. In addition, some of the frozen embryos have been in storage for many years, and when these embryos were created the laboratory cultures were not as conducive to preserving embryos as they are today. Some embryos would also die in the freeze-thaw process. Considering all these issues, the question is how many embryos actually are available for research and donation/adoption? The RAND/SART team estimated that 65% of the approximately 11,000 embryos designated for research would survive the freeze-thaw process, resulting in 7,334 embryos. Of those, about 25% (1,834 embryos) would likely be able to survive the initial stages of development to the blastocyst stage (a blastocyst is an embryo that has developed for at least 5 days). Even fewer could be converted into embryonic stem cell lines. Their estimate is about 275 embryonic stem cell lines could be converted from the total number of embryos designated for research. The RAND/SART team also estimates that 2.3% of the 400,000 frozen "spare" embryos designated for donation/adoption, only 23,000-100,000 embryos could be adopted, thawed and successfully born.[24] Having this many children potentially available for adoption would help meet the need of couples seeking adoption in the United States. The problem is that the adoption process for frozen embryos is legally quite ambiguous and very complex. In addition, with the Magisterium of the Catholic Church issuing the Instruction *Dignitas Personae* it appears that donation/adoption is no longer acceptable for Catholics. The only viable option for these spare embryos would be to allow them to be thawed and to die with dignity and respect under the principle of extraordinary/ordinary. The central issue is whether this is an ethically sound viable solution for the Roman Catholic Church.

3. Ethical aspects

Ethically, the concern about spare embryos focuses on the issue of personhood. If embryos are persons then it would be a moral imperative to "rescue" these embryos from their current status of being in "frozen animation." Numerous ethicists, embryologists, legal professionals and specifically, the Roman Catholic Church, argue that personhood begins at conception or what is known as fertilization. Prior to fertilization we have two human gametes—sperm and egg, that are living but are not a living organism. When fertilization occurs, something human and living "in a different sense comes into being."[25] Embryologists argue that "human development begins at fertilization when a male gamete or sperm (spermatozoon) unites with a female gamete or oocyte (ovum) to form a single cell—zygote. This highly specialized, totipotent cell marked the beginning of each of us as a unique individual."[26] The Catholic Church teaches that "human life must be absolutely respected and protected from the moment of conception."[27] "Right from fertilization is begun the adventure of a human life, and each of its great capacities requires time . . . to find its place and to be in a position to act. This teaching remains valid and is further confirmed, if confirmation were needed, by recent findings of human biological science which recognize that in the zygote resulting from fertilization the biological identity of a new human individual is already constituted."[28] The Church argues that at fertilization there is a new genetic individual in its own right, one who is whole, bodily, self-organizing, and genetically distinct from his or her mother and father.[29] Those who argue that personhood begins at fertilization would also argue that there is a moral imperative to give these frozen embryos the opportunity to be born and to develop because they are persons. Ethicist Therese Lysaught believes that embryo donation/adoption is an act that can properly be described as "rescuing a child orphaned before birth."[30] Ethicists arguing for the "rescue" of these children would encourage women to implant these embryos in their wombs in order to bring them to term. Some would permit not only married women to do this but also single women and even lesbian couples. The moral principle of sanctity of human life would overcome any other moral considerations. However, not all, even in the Catholic Church, would agree to this ethical analysis. Opponents of this position argue that this would amount to material cooperation in an objective immoral action. Not only is the process of IVF considered an intrinsic moral evil by the Magisterium of the Catholic Church, but allowing for the donation/adoption of these embryos might condone the objective immoral procedure and may even encourage the creation of additional embryos through the IVF process. The Catholic Church clarified its position on embryo donation/adoption in 2008 in the Instruction from the Congregation of the Faith called *Dignitas Personae.* "The proposal that these embryos could be put at the disposal of infertile couples as a treatment for fertility is not ethically acceptable for the same reasons which make artificial heterologous procreation illicit as well as any form of surrogate motherhood; this practice would also lead to other problems of a medical, psychological and legal nature."[31] This statement by the Magisterium removes donation/adoption as a viable option for Catholics. The only remaining option would be to allow these embryos to die with dignity and respect using the extraordinary/ordinary means distinction. To determine if thawing these embryos and allowing them to die naturally is ethical and to address the ambiguities and unresolved issues surrounding this controversy, the traditional ethical principle of the extraordinary/ordinary means distinction will be examined and applied to this situation.

The history of the Catholic Church's position on the ordinary-extraordinary means distinction dates back to the 16th century Dominican moralists. There are however, some who believe it may go back to Thomas Aquinas (1225-1274) a Dominican Friar and Doctor of the Roman Catholic Church. Thomas' belief in the moral measure of all human activity is whether it leads to God, the final end. Thus, if something was "too difficult" or "too burdensome" what was implied was that it might make loving God too difficult.[32] The general obligation to preserve life and the possible limits to that obligation are also influenced by Thomas' concept of God's dominion over the gift of human life, responsible stewardship and the positive and negative precepts derived from these.[33] Thomas' influence is clearly present, but it is the three Dominican moralists--Francisco De Vitoria, Domingo Soto and Domingo Banez--who articulated the foundation of the ordinary-extraordinary means distinction.

De Vitoria (1486-1546) examined the limits of treatment in regards to nourishment and medicinal drugs. In his seminal work *Relectiones Theologicae* he states:

"If a sick man can take food or nourishment with a certain hope of life, he is required to take food as he would be required to give it to one who is sick. However, if the depression of spirits is so severe and there is present grave consternation in the appetitive power so that only with the greatest effort and as though through torture can the sick man take food, this is to be reckoned as an impossibility and therefore, he is excused, at least from mortal sin."[34]

De Victoria is not condoning suicide here. A healthy person may not starve him-herself because life is problematic. If the means are effective and not burdensome then the person is morally obligated to seek nourishment. However, if the person is so sick or depressed that eating may become a grave burden, then the person is not morally obliged to eat and does not commit a sin. The essential point here is that De Vitoria recognizes both psychological and physiological illness and his notion of grave burden includes both. In regards to medicinal drugs, he argues that they are not *per se* obligatory. The obligation to use them rested on the degree of efficacy. One is not obliged to sacrifice one's whole means of subsistence, nor one's general lifestyle, nor one's homeland in order to acquire a cure or obtain optimum health.[35] It appears that De Vitoria adopted the 16th century's version of the "Reasonable Person" criteria. "To fulfill one's positive obligation to sustain life, it is sufficient to perform 'that by which regularly a man can live.'"[36] The moral components that appear operative here are not natural as opposed to artificial means, but those means that offer a reasonable hope of benefit in regard to cure and return to health. Excessive burdens in terms of financial costs or inconvenience of lifestyle are measured by "the semi-objective standard of the common person regularly considered," or what we refer to as the "reasonable person standard."[37] If the means used to prolong life were ineffective, if the effect was doubtful, or if it involves a grave burden for the person in question, this means need not be morally obligatory.

Prior to the development of modern anesthesia, surgical procedures, especially amputations, were quite painful. Domingo Soto (1494-1560) reasoned that surgery such as amputation of a limb, because of the excessive pain, ought to be considered categorically optional. He argued that such torture was beyond the limits that the "common man" ought to be obliged to suffer for the sake of one's bodily health. Such surgery can make a beneficial surgery "morally impossible" to bear.[38] Besides the question of pain, Soto also recognizes the role that emotions of fear and repugnance could play.[39] Soto incorporates the dimension of optional versus obligatory, adding if a procedure or treatment was too painful or burdensome, it would be morally optional.

In 1595, Domingo Bañez (1528-1604) was the first to articulate the terms "ordinary" and "extraordinary" as they regard obligatory and non-obligatory means of preserving life. He argued that if preserving life was reasonable it was obligatory but insisted that one is "not bound to extraordinary means but to common food and clothing, to common medicines, to certain common and ordinary pain; not, however, to certain extraordinary and horrible pain, nor to expenses which are extraordinary in proportion to the status of this man."[40] One determined if a treatment or medical procedure was ordinary or extraordinary according to whether it was proportionate to one's condition or state in life. "Thus, if something were very costly or burdensome or if it did not offer substantial benefit to the patient, there was no moral obligation to use it. This standard applied to even life-saving measures."[41] The Jesuit moralist Juan Cardinal De Lugo (1583-1660) confirms Bañez's position when he wrote, "...he is not held to the extraordinary and difficult means . . . the 'bonum' of his life is not of such great moment, however, that its conservation must be effected with extraordinary diligence. . ."[42] De Lugo's position, like that of the Dominican moralists, followed the tradition of the Church that states human life is a good but not an absolute good. As a relative good, one's duty to preserve it is a limited duty. While a person has freedom over his or her life, one is never permitted to directly take one's life. The issue becomes to what extent is one obligated to preserve one's life.

The traditional understanding of ordinary-extraordinary means remained basically unchallenged until the mid-1900s with the advent of advances in medicine and technology. How to apply the early distinction of ordinary-extraordinary means to issues like oxygen and feeding tubes, especially with permanently unconscious patients became hotly debated as early as the 1950s. Jesuit moralist Gerald Kelly was one of the first to examine this issue critically. He defined ordinary means of preserving life as "all medicines, treatments, and operations, which offer a reasonable hope of benefit for the patient and which can be obtained and used without excessive expense, pain, or other inconvenience." Extraordinary means would be "all medicines, treatments, and operations, which cannot be obtained or used without excessive expense, pain, or other inconvenience, or which, if used, would not offer a reasonable hope of benefit."[43] The distinctive element of Kelly's interpretation is that it is a patient-centered, quality-of-life approach which is consistent with how the 16th-century-Dominican moralists viewed this distinction. Kelly concludes that no person is morally obligated to use any means, and this would include natural or artificial means, that does not offer a reasonable hope of ameliorating the patient's condition. To clarify this distinction, Kelly was asked if oxygen and intravenous feeding must be used to extend the life of a patient in a terminal coma. He replies: "I see no reason why even the most delicate professional standard should call for their [oxygen and intravenous for a patient in a terminal coma] use. In fact, it seems to me that, apart from very special circumstances, the artificial means not only need not but should not be used, once the coma is reasonably diagnosed as terminal. Their use creates expense and nervous strain without conferring any real benefit."[44]

Many believe that the most authoritative historical study on this topic was done by Daniel Cronin (who later became Archbishop of Hartford) in his 1958 doctoral dissertation at the Gregorian University in Rome entitled, "The Moral Law in Regard to the Ordinary and Extraordinary Means of Preserving Life." After a review of over 50 moral theologians from Aquinas to those writing in the early 1950's Cronin concludes that the Church's teaching is consistent in its view: "Even natural means, such as taking of food and drink, can become optional if taking them requires great effort or if the hope of beneficial results (*spes salutis*) is not present." For a patient whose condition is incurable, he writes, "even ordinary means, according to the general norm, have become extraordinary [morally dispensable] for the patient [so] the wishes of the patient, expressed or reasonably interpreted, must be obeyed."[45] The importance of Cronin's position is that no means—even food and water—can ever be classified as absolutely obligatory regardless of the patient's condition. However, some moralists disputed this fact and claimed that food and water were absolutely ordinary and even tried to say that was what the tradition taught.

On November 24, 1957, in a talk delivered to the International Congress of Anesthesiologists, Pope Pius XII gave papal approbation to the ordinary-extraordinary means tradition that dates back to De Vitoria.

"Natural reason and Christian morals say that man (and whoever is entrusted with taking care of his fellow man) has the right and the duty in case of serious illness to take the necessary treatment for the preservation of life and health . . . But normally one is held to use only ordinary means—according to circumstances of persons, places, times and culture—that is to say, means that do not involve grave burden for oneself or another. A more strict obligation would be too burdensome for most men and would render the attainment of the higher, more important good too difficult. Life, health, and all temporal activities are in fact subordinated to spiritual ends. On the other hand, one is not forbidden to take more than the strictly necessary steps to preserve life and health, as long as he does not fail in some more serious duty."[46]

Pius XII upholds the traditional ordinary-extraordinary means distinction that "involves patient-centered judgments about the quality of life, which must take into account the usefulness of the treatment, one's understanding about death and dying, and the repugnance one may have toward one's life after subjection to a particular medical treatment."[47] It is also important to note that Pius XII emphasized the importance of viewing the person holistically. In an address given to the International Union Against Cancer, in 1956, Pius XII counseled that "before anything else, the doctor should consider the whole man, in the unity of his person, that is to say, not merely his physical condition but his psychological state as well as his spiritual and moral ideals and his place in history."[48] This statement reinforces the traditional understanding of not treating the physiological aspect of the body separate from the person. Benefits of a treatment can only be determined within the context of a person's life.[49] To preserve life at all cost is to risk idolatry and thus would lead a person away from the higher spiritual good which is eternal life.

A contemporary understanding of the ordinary-extraordinary means distinction was given in the 1980 Congregation for the Doctrine of the Faiths' *Declaration on Euthanasia*. The Declaration follows the tradition on the ordinary-extraordinary means distinction since the 16th century, which is based on the effect of the treatment on the patient or those responsible for the care of the patient. The Declaration reminds us of the duty one has to care for one's own life and to seek such care for others. But there are limits to this obligation. One needs to judge the means used by "studying the type of treatment to be used, its degree of complexity or risk, its cost and the possibilities of using it, and comparing these elements with the result that can be expected, taking into account the state of the sick person and his or her physical and moral resources."[50] The Declaration goes on to give four examples: patients are permitted to use experimental, advanced medical techniques, which may be a service to humanity; patients may interrupt treatments if they fall short of expectations; the refusal of a technique that is in use and carries a risk or is burdensome is not equivalent to suicide; finally, when death is imminent in spite of the means used, it is permitted in conscience to make the decision to refuse forms of treatment that would only secure a precarious and burdensome prolongation of life, so long as the normal care due to the sick person in similar cases is not interrupted.[51] Finally, the Congregation for the Doctrine of the Faith reflects the traditional teaching when it writes: "Life is a gift from God, and on the other hand death is unavoidable; it is necessary, therefore, that we, without in any way hastening the hour of death, should be able to accept it with full responsibility and dignity."[52] The only real change is that the document realizes that the terms ordinary and extraordinary are imprecise as terms in regards to the rapid advancement of medicine and technology. More precise terms would be proportionate and disproportionate.[53]

The historical review of the tradition shows a clear distinction between extraordinary and ordinary means that involves patient-centered judgments about the quality of life, which must take into account the usefulness of the treatment, one's understanding about death and dying, and the repugnance one may have toward one's life after subjection to a particular medical treatment.[54] The ethical issue concerning frozen embryos focuses on the foregoing of artificial life support from them, which would allow the embryos to die naturally. Some might argue that this is a form of euthanasia. Pope John Paul II in *Evangelium Vitae* states: "Euthanasia's terms of reference, therefore, are to be found in the intention of the will and in the methods used."[55] The intention here is not to end the life of the embryo but to forego a burdensome treatment and allow the embryo to die naturally with dignity and respect. The Pope himself states clearly that euthanasia must be distinguished from the decision to forego what he refers to as "aggressive medical treatment." "Medical procedures which no longer correspond to the real situation of the patient, either because they are by now disproportionate to any expected results or because they impose an excessive burden on the patient and his family.[56] If the intentionality is to forego a non-beneficial treatment that the surrogate believes is disproportionate and not in the embryo's best interest, then the intentionality is to allow the person to die rather than not to terminate the person directly. These embryos will not be abandoned or discarded. Instead they will be cared for lovingly during the dying process and treated with the utmost dignity and respect.

The benefit of a medical procedure or treatment was traditionally viewed as a prudential judgment of the patient or surrogate on how a particular treatment or procedure would impact on the life of the patient. Benefits and burdens were never judged abstractly. "Not only the means (proposed intervention) but the ends toward which the intervention is aimed are important in moral analysis."[57] The fact that a particular means was able to sustain a human life did not make such a means beneficial to the person. Traditional moralists did not restrict benefits merely to sustaining life, but included broader, more holistic considerations. Improvements in one's condition, relief of pain and suffering, maximization of comfort, restoration of health, among others all were considered beneficial. For DeVitoria and other traditional moralists, the mere preservation of life and vital physiological functions was not sufficient in itself to oblige someone to use a certain means.

The traditional understanding of ordinary-extraordinary means was based on treating the whole person, not one part of the person. Just because a treatment could prolong a life did not mean that a particular treatment was a benefit. Benefits must be considered worthwhile both in quality and duration. In the Catholic moral tradition, a medical treatment was beneficial if it restored a patient to a relative state of health. The frozen embryos will not be implanted into the womb of the biological mother. The options would be to stay frozen in a state of "permanent suspended animation," be used for stem cell research, be used for other forms of experimentation, or be placed for donation/adoption. The Magisterium of the Catholic Church has rejected all of these options as ethically acceptable. To allow these embryos to stay frozen indefinitely violates the basic dignity and respect of the person. Therefore, the only viable option for these frozen embryos would be to stop the process of cryopreservation and allow them to die naturally with dignity and respect. One could equate the process of cryopreservation to maintain the life of the embryo to the use of a mechanical ventilator to maintain the life of a terminal patient. To continue to keep the embryos alive through cryopreservation is a form of extraordinary means that is disproportionate and offers no reasonable hope of benefit for the embryo. Failure to receive a meaningful benefit from a treatment makes said treatment not morally obligatory. Allowing a person to die by foregoing aggressive, non-beneficial treatments is not only morally permissible, it is also treating the person with dignity and respect. Therefore, it is morally and ethically acceptable to allow these embryos to die naturally with dignity and respect under the principle of the extraordinary and ordinary means distinction. However, it is also imperative that safeguards be put in place that would eliminate creating more "spare" embryos in the future.

4. Conclusion & safeguards

Cryopreserved embryos are a complex issue that has medical, legal and ethical dimensions. Allowing these embryos to die naturally is the only viable option that protects and preserves their human dignity. The other viable options: being discarded, destroyed for research, abandoned or kept in "suspended animation" indefinitely, are unacceptable because they have the potential of harming or intentionally killing these embryos that deserve special respect.

To make sure that this situation does not continue in the future, the following recommendations and safeguards are proposed:

1. Only the number of eggs to be placed in the uterus of the mother will be fertilized. Embryos must not be subjected to an intentional interruption of their natural growth and development. There will no longer be "spare" embryos subjected to cryopreservation. Only cryopreservation of gametes would be acceptable.
2. Laws and legislation must be enacted at the federal level that begins to regulate Assisted Reproductive Technologies. Having each state governed by differing sets of legislation could cause potential complications associated with the practice of donation/adoption. How each state defines jurisdiction and how each state interprets at what stage jurisdiction would begin (conception, transfer, or birth) could become highly complex. Specifically, guidelines and safeguards must be put in place that protects donors, parents, providers, and children born of ART.
3. Laws and legislation must be enacted that regulates the creation, destruction and exploitation of human embryos. An example would be the following: legislation established in New Mexico stating that human embryos can only be disposed of through implantation, not intentional destruction or through destructive human embryo research. b) Embryos must not be subjected to non-therapeutic experimentation.

If we believe that human life deserves dignity and respect, then our failure to allow these embryos to die naturally would be medically irresponsible and ethically objectionable, from the Catholic perspective.

5. References

[1] National Center for Health Statistics, "Infertility," October 31, 2007. http://www.cdc.gov/nchs/fastats/fertile.htm.

[2] Center for Disease Control and Prevention, "Assisted Reproductive Technology," June 2008. Http://www.cdc.gov/ART/

[3] Rand Law and Health/Society of Assisted Reproductive Technology (SART) Working Group, "How Many Frozen Human Embryos Are Available For Research?" May 2003. http://www.rand.org/pubs/research_briefs/RB9038/index1.html.

[4] Papanikolaou EG, Camus M, Kolibianakis EM, Van Landuyt L, Van Steirteghem A, Devroey P. "In Vitro Fertilization with Single Blastocyst-Stage versus Single Cleavage-Stage Embryos." *New England Journal of Medicine* 354 (March 16, 2006): 1139-1146.

[5] Lysaught, Therese. "Embryo Adoption?" *Human Life Review* (2003): 1-4. http://findarticles.com/p/articles/mi_qa3798/is_200310/ai_n9297175/print.

[6] Collins, Timothy. "On Abandoned Embryos," *The Linacre Quarterly* 75 (1) (February 2008): 1-15.

[7] Congregation for the Doctrine of the Faith, Instruction *Dignitas Personae* on Certain Bioethical Questions, (September 8, 2008): 1-23, at Section 2, #18.

[8] Pence, Gregory. *Medical Ethics*, 5th edition (Boston, MA.: McGraw Hill, 2008): 100.

[9] Ryan, Caroline. "More Than 3 Million Babies Born From IVF," *BBC News* (June 21, 2006): 1-3. http://news.bbc.co.uk/1/hi/health/5101684.stm.

[10] Medical Encyclopedia, "In-Vitro Fertilization," *Medline Plus,* (January 31, 2007): 1-4. http://0-www.nlm.nih.gov.catalog.llu.edu/medlineplus/ency/article/007279.htm
[11] Scott, J.R. et al. *Danforth's Obstetrics and Gynecology, 9th edition,* (Philadelphia, PA.: Lippincott, Williams & Wilkins, 2003), Ch. 39 "Assisted Reproductive Technology, p.709.
[12] Centers for Disease Control, Department of Health and Human Services, *2003 Assisted Reproductive Technology Report,* (2004). http://www.cdc.gov/ART/ART2003/index.html. See 2003 National Summary Table.
[13] Collins, 3.
[14] Pence, 102.
[15] Tarkan, Laurie. "Lowering Odds of Multiple Births," *The New York Times* (February 19, 2008): 1-4. Http://www.nytimes.com/2008/02/19/healthy/19mult.html.
[16] Collins, 4.
[17] Collins, 4.
[18] Emory Reproductive Center, "Cryopreservation," Emory Health Care, 2008. http://www.emoryhealthcare.org/departments/ivf/services/cryo_preservation.html
[19] Collins, 5. Hoffman, D., Zellman, G., Fair, C. et al. "Cryopreserved Embryos in the United States and their Availability for Research," *Fertility and Sterility* 79 (5) (May 2003): 1063-1069.
[20] Embryo morphology (appearance of the cells/percentage of fragmentation) is one of the most influential factors for embryo survival. Genetics & IVF Institute, "Human Embryo Cryopreservation (Embryo Freezing) and Frozen Embryo Transfer Cycles," (2008): 1-2. http://www.givf.com/fertility/embryofreezing.cfm
[21] Genetics & IVF Institute, 2.
[22] RAND/SART Working Group, 1.
[23] RAND/SART Working Group, 1.
[24] RAND/SART Working Group, 1-2.
[25] Kass, Leon. *Life, Liberty and the Defense of Dignity: The Challenge for Bioethics* (San Francisco: Encounter Books, 2002): 87.
[26] Moore, K, & Persaud, T.V.N. *The Developing Human: Clinically Oriented Embryology* (Philadelphia: Saunders/Elsevier, 2008): 15.
[27] Holy See, *Charter of the Rights of the Family,* no. 4: *L'Osservatore Romano,* November 25, 1983.
[28] Congregation for the Doctrine of the Faith, *Donum Vitae – Instruction on Respect for Human Life in its Origin and on the Dignity of Procreation – Replies to Certain Questions of the Day,* (Rome: Congregation for the Doctrine of the Faith, February 22, 1987): 1-29. http://www.priestsfor life.org/magisterium/donumvitae.htm
[29] Brugger, E. C. "The Principle of Fairness and the Problem of Abandoned Embryos," Panel Discussion at the Conference on Emerging Issues in Embryo Donation and Adoption, (Washington, D.C.: May 30, 2008): 1-6.
[30] Lysaught, 3.
[31] Congregation for the Doctrine of the Faith, Dignitas Personae Section 2, No. 19.

[32] Atkinson, G. Theological history of Catholic teaching on prolonging life, in D. McCarthy and A. Moraczewski, eds., *Moral Responsibility In Prolonging Life Decision* St. Louis: Pope John Center, 1981.
[33] Aquinas, Thomas. (1966). *Summa Theologica* eds. Anthony Ross and PG Walsh, Blackfriars edition New York: McGraw-Hill, II, II, q. 126, a.1 and ad.3; on suicide: II, II, q. 64, a5; On Killing the Innocent: II, II, q. 64, a. 6; On Self-Defense: II, II, q. 64, a. 7 and 8; On Mutilation: II, II, q. 65, a.1
[34] The *Relectio* was a lecture that De Victoria, the preeminent theologian at the University of Salamanca, Spain, would give at the beginning of each school year. These lectures treated some difficult, contemporary dilemma. One can presume that the question of prolonging life must have been a disputed topic at the time. De Vitoria, F. *Relectiones Theologicae* (Lugdini, 1957): Relectio IX, de Temp., n.12. See also Sparks, Richard. *To Treat or Not To Treat.* New York: Paulist Press, 1988): 94-95.
[35] Sparks, 95-96.
[36] Sparks, 95.
[37] Sparks, 96.
[38] Soto, Domingo. *De Justitia et Jure*. (Venice, Italy, 1568): Lib. 5, question 2, article 1; See also Sparks, 96.
[39] Wildes, Kevin. "Ordinary And Extraordinary Means and The Quality Of Life," *Theological Studies* 57 (1996): 500-512 at 504.
[40] Bañez, Domingo. *Scholastica Commentaria in Partem Angelici Doctoris S. Thomae* Tom. IV, "Decisiones de Jure et Justitia," (1614-15): II, IIae, Question 65, Article, 1. Duaci.
[41] Paris, J. and McCormick, R. The Catholic Tradition on the Use of Nutrition and Fluids. *America* 156, (1987): 356-361.
[42] De Lugo, Juan. "Disputationum de justitia et jure" (Lyons, 1642): 10, section 1, n. 9 De Justitia et Jure.
[43] Kelly, G. "The Duty of Using Artificial Means of Preserving Life," *Theological Studies* 11, (1950): 203-220.
[44] Kelly, 210.
[45] Cronin, D. "The Moral Law in Regard to Ordinary and Extraordinary Means of Conserving Life," (Gregorian, Rome, 1958); See also, Paris and McCormick, 360.
[46] Pius XII. "Prolongation of Life," *The Pope Speaks* 4, (1958): 395-396.
[47] Wildes, 512.
[48] Pius XII. "Cancer: A Medical and Social Problem," *The Pope Speaks* 3, (1957): 48.
[49] Wildes, 508.
[50] Congregation for the Doctrine of the Faith. "Declaration on Euthanasia," part 4, *Origins* 10, (1980): 154-157.
[51] Congregation for the Doctrine of the Faith, "Declaration on Euthanasia,"156.
[52] Congregation for the Doctrine of the Faith, "Declaration on Euthanasia,"156.
[53] Congregation for the Doctrine of the Faith, "Declaration on Euthanasia," 157.
[54] Wildes, 512.
[55] John Paul II. "Evangelium Vitae," no. 65. *Origins* 24, (1995): 687-727 at 62-63.
[56] John Paul II. "Evangelium Vitae," no. 65, at 62-63.

[57] Shannon, T. and Walter, J. "Artificial Nutrition & Hydration: Assessing Papal Statement," National *Catholic Reporter* (2004): 1-4. http://www.natcath.com/NCR_online/archives2/2004b/041604/041604i.php

Part 2

Methods, Tools and Technologies for Embryonic Stem Cell Culture, Manipulation and Clinical Application

4

Bioprocess Development for the Expansion of Embryonic Stem Cells

Megan M. Hunt[1,3], Roz Alfred[1,3], Derrick E. Rancourt[2],
Ian D. Gates[3] and Michael S. Kallos[1,3]
[1]*Pharmaceutical Production Research Facility (PPRF), Schulich School of Engineering, University of Calgary, Alberta*
[2]*Department of Biochemistry and Molecular Biology, Faculty of Medicine, University of Calgary, Alberta*
[3]*Department of Chemical and Petroleum Engineering, Schulich School of Engineering, University of Calgary, Alberta, Canada*

1. Introduction

As the North American population demographic shifts towards an older profile, the demand for regenerative therapies will rise as more people face age-related degenerative diseases ranging from osteoarthritis to neurological disorders such as Parkinson's disease. Current treatment options focus on palliative measures to alleviate symptoms rather than addressing the underlying cause. Tissue engineering has emerged as an interdisciplinary field of work that aims to restore function to the tissues identified as the underlying cause of illness. The most important building blocks used in tissue engineering are living cells. These cells are generally obtained from one of two sources – adult tissues or embryos. Adult tissues have very sparse and difficult-to-isolate stem cell populations. Further, their use for large scale treatment is not realistic due to their limited proliferation capabilities (Thomson & Odorico, 2000). Embryonic stem cells (ESCs), in contrast, are rapidly emerging as a promising cell source due to their unique characteristics – namely their ability to readily proliferate in culture as well as their potential to differentiate into all cell types of the adult body.

A major roadblock that has been anticipated on the path to clinical implementation of ESC derived cellular therapies is the labour intensive and highly variable nature of small scale cultures. It has been estimated that as many as 26 billion human ESCs may be required as a starting point for treating one patient taking into account losses during culture, inefficient differentiation protocols, downstream processing and purification of mature cell types (Ouyang & Yang 2008). Based on current small scale culture techniques, thousands of static tissue culture flasks and several weeks to months of culture time would be required to generate this number of cells (Ouyang & Yang 2008). There are notable disadvantages in producing this quantity of cells in static tissue culture including heterogeneity between flasks, lack of environmental controls, large quantities of materials, large amounts of incubator space and the many hours of labour required to maintain the cultures. Several automated bioprocessing systems exist which aim to reduce the amount of labour required to maintain static cultures, however these systems are still limited by surface area which is a key control on mass transfer of nutrients (Terstegge et al., 2007; Terstegge et al., 2009; Thomas et al., 2008). More effective methods are required to generate the large numbers of pluripotent ESCs needed for subsequent differentiation into functional tissue types. Suspension bioreactors offer a means to scale-up production of ESCs in a controlled culture environment not bound by the same surface area limitations as static culture techniques.

The development of protocols for bioreactor expansion of stem cells has progressed rapidly in the past decade. Groups have successfully cultured hematopoietic stem cells, neural stem cells, human pancreatic progenitor cells and more recently both mouse and human embryonic stem cells in suspension (Zandstra et al., 1994; Kallos et al., 1999; Chawla et al., 2006; Cormier et al., 2006; Krawetz et al., 2009; Kehoe et al., 2010). Despite the advances in bioreactor design, there is still a need for optimization and standardization of suspension culture protocols to ensure reproducible and predictable cell populations for use in clinical applications. In this chapter our aim is to review recent progress in the large scale culture of both murine and human ESCs as well as discuss fundamental bioprocess issues to be considered when using stirred suspension bioreactor culture systems.

2. Mouse embryonic stem cells as a model

The use of animal models in medical research goes back at least a century (Mouse Genome Sequencing Consortium, 2002). New therapies and drugs must be assessed for safety and efficacy; however, testing them on humans is not an ethical or realistic option. While single cells have proven to be invaluable resources for many studies, intact living animals have full organ systems which can undergo complex disease progression dynamics. Different animal models serve different purposes – larger animals models such as pigs and dogs allow for studies complex systems (i.e. joint loading) whereas smaller ones such as mice are useful for studies of genetics and progression of degenerative diseases. Mice have been shown to have similar genetics and physiological structures to humans. They naturally develop several diseases (cancer and diabetes for example) typically attributed to humans and can also be induced to present symptoms of neurological disorders such as Alzheimer's (see Bedell et al., 1997 for a review of mice models for many human diseases). Animal models such as mice provide an excellent means for evaluating cell therapies: it is cost effective and provides fast results (the murine gestational period is a matter of weeks which allows for quick assessment of germline transmission which is paramount in murine ESC derivation evaluation).

Knowledge gained from isolation of mouse ESCs has laid a solid foundation for isolation of ESCs from other animals including non-human primates and ultimately humans (Bongso et al., 1994; Thomson et al., 1995; Thomson et al., 1998; Reubinoff et al., 2000). Within our own laboratory, work with mouse cell lines has established foundation protocols for human studies although the methods have not been directly transferable. Specifically, successful expansion of murine neural stem cells as aggregates in suspension bioreactors (Kallos & Behie 1999; Gilbertson et al., 2006) enabled us to develop protocols for human neural stem cells (Baghbaderani et al., 2010). Similarly, expanding murine ESCs as aggregates (Cormier et al., 2006; zur Nieden et al., 2007) allowed us to successfully expand pluripotent human ESCs as aggregates in suspension bioreactors (Krawetz et al., 2010). The fact that protocols are not directly transferrable is expected since cells from different species have been observed to present different growth morphologies, growth kinetics (doubling times, for example), as well as different markers for cell pluripotency and differences in associated signaling pathways (illustrated in Table 1). However, from a bioengineering point of view, there is still tremendous value in conducting murine experiments first.

	Murine ESCs	Human ESCs
Pluripotency markers		
SSEA-1	+	-
SSEA-3	-	+
SSEA-4	-	+
TRA-1-60	-	+
TRA-1-81	-	+
Oct-4	+	+
Factors that affect self renewal	LIF + Serum—activates JAK/STAT3 pathway [1] LIF+BMP (serum free) – activates SMAD and/or MAPK pathways [1]	bFGF+MEF+SR or high bFGF+SR (no MEFs)- suppresses BMP signalling and/or upregulates expression of TGFβ ligands, activation of ERK &PI3K[1] TGFβ/Activin/Nodal[1]
Teratoma formation in vivo	+	+
Colony morphology	Tight rounded multilayered	Lose rounded monolayers
Passaging	Single cells	Clumps

Table 1. Summary of culture similarities and differences between mouse and human embryonic stem cells. Adapted from National Institutes of Health, 2009; [1] Yu & Thomson, 2008. bFGF= basic fibroblast growth factor, BMP= bone morphogenetic protein, LIF= leukemia inhibitory factor, MEF= mouse embryonic fibroblasts, SR= serum replacement, TGFβ= transforming growth factor beta.

3. Bioprocess development for suspension culture of ESCs

Many types of cell culture systems can produce cells and tissues in culture but they can generally be grouped into two categories: 1. adherent culture (tissue culture flasks) and 2. non-adherent culture. Both cultures can further be classified as stationary (or static), where the culture media is not moving, or suspension, where the medium is agitated or perfused (Sen et al., 2010). Conventional culture methods for propagation of ESCs use static tissue culture flasks (stationary adherent culture). As this is the current standard method for ESC culture, it is necessary to understand the effect of static culture environment on ESCs before successful development of suspension culture protocols. Static culture also provides a baseline for comparison of cell production from suspension culture.

3.1 Adherent culture

In 1981, Evans and Kaufman reported that isolation and culture of pluripotent cells were dependent on: 1) the stage at which pluripotent cells exist in the embryo (in other words, day at which cells are harvested from the blastocyst), 2) explantation of sufficiently large numbers of these cells from the embryo and 3) tissue culture conditions conducive to propagation (multiplication) of the cells rather than differentiation. These considerations have been the basis for isolation of mammalian ESC lines. Currently, however, the focus of many groups is on the third point: trying to determine the optimum combination of factors to support long term expansion of these cells while maintaining functionality (pluripotency and self-renewal). Based on their knowledge of embryonic carcinoma cells, which are pluripotent cells isolated from germ cell tumors, Evans and Kaufman isolated the inner cell mass of a mouse blastocyst and cultured the cells on a feeder layer of mitotically inactivated mouse embryonic fibroblasts (MEFs) in 20% serum containing medium. The resulting cells exhibited a normal karyotype in contrast to embryonic carcinoma cells, which did not, and thus the first line of mouse ESCs was derived. This reliance on a feeder layer initially restricted ESC research to static adherent culture conditions. Some years later it was found that conditioned medium was able to support the growth of murine ESCs in the absence of a feeder layer (Yu & Thomson, 2008). Subsequent fractionation of conditioned medium led to identification of leukemia inhibitory factor (LIF) as one of the key ingredients that supported viability (Williams et al., 1988; Smith et al., 1988 as cited in Yu & Thomson 2008). Standard lab scale cultures today use gelatin as a substrate for mouse ESC adherence together with DMEM-based medium supplemented with LIF. This combination has repeatedly demonstrated long term propagation of stable pluripotent cells. Table 2 summarizes static culture conditions described in recent publications.

Based on previous methods of ESC derivation, Bongso et al. (1994) isolated the inner cell mass of a human blastocyst on a feeder layer of human oviductal epithelial cells in a medium containing human LIF and 10% human serum. While initial colonies had promising morphologies, the cells either differentiated or died after only two passages. This is an example of difficulties encountered in transferring protocols directly between species. Shortly after, another group successfully derived a non-human primate ESC line (Thomson et al., 1995). The cells isolated from a rhesus monkey embryo were cultured on irradiated mouse embryonic fibroblasts in DMEM based medium supplemented with human LIF and 20% fetal bovine serum (FBS). Although human LIF was used in the initial derivation of these cells, it was subsequently removed and the cells were successfully cultured for over a year while maintaining a stable karyotype and the ability to differentiate into all three germ layers. Conversely, when human LIF was used and MEFs removed, the cells differentiated (Thomson et al., 1995). Based on this success, Thomson then used a similar protocol to derive the first human ESC lines using irradiated MEFs and DMEM medium with 10% FBS (Thomson et al., 1998). Subsequent work has led to the refinement of medium and identification of optimum substrates with many groups now reporting the use of DMEM with serum replacement medium or the commercially available defined medium, mTeSR™ with matrigel™ (an ECM extract of mouse sarcoma) as the substrate. These culture protocols have proven to give rise to stable pluripotent cell populations (Xu et al., 2005; Ludwig et al., 2006; Levenstein et al., 2006). Table 3 provides a brief summary of static culture conditions reported in recent publications of human ESCs. It is apparent that these small scale adherent culture conditions are well suited to most laboratory research. However, if we look to the future of stem cell based therapies in clinical settings, these small scale culture techniques are simply not feasible on a large scale.

Ouyang & Yang (2008) provided a useful summary of cell numbers estimated to be required for different clinical applications. For treatment of an adult with leukemia, they calculated

Reference	Cell Lines	Substrate	Medium	Passaging
Kehoe 2008	E14Tg2a	0.1% Gelatin	DMEM+10%FBS	TrypLE
Veraitch 2008	E14Tg2a	0.1% Gelatin	GMEM+10%FBS	Trypsin
Abranches 2009	E14Tg2a 46C S25	0.1% Gelatin	GMEM+10%FBS	Not specified for maintenance conditions
Fernandes, TG 2009	46C	0.1% Gelatin	DMEM+10%FBS ESGRO complete KO-DMEM+15%KSR	Trypsin, Accutase
Marinho 2009	USP-1	Gelatin MEFs	DMEM/F12+15%KSR+CHO-CM (for LIF)- changed every 2 days	TrypLE
Alfred 2010	D3	Gelatin	DMEM+15%FBS	Not specified for static
Jing 2010	E14Tg2a	0.1% Gelatin	DMEM+10%FBS	TrypLE
Taiani 2010	D3	MEFs	DMEM+15%FBS	Not specified for static
Ito 2010	E14Tg2a J1	Gelatin MEFs	GMEM+15%FBS	Not specified

Table 2. Summary of recent publications with static maintenance of murine ESCs. DMEM= Dulbecco's Modified Eagle's Medium, GMEM= Glasgow's Minimal Essential Medium, KO-DMEM= KnockOut™ DMEM (Inivitrogen), KSR= KnockOut™ Serum Replacement (Invitrogen), ESGRO Complete= defined serum-free medium (Chemicon/Millipore)-contains LIF, CHO-CM= Chinese Hamster Ovary Conditioned Medium. Unless otherwise stated, culture medium also included LIF, 2-mercaptoethanol and non-essential amino acids.

Reference	Cell Lines	Substrate	Medium	Passaging
Peerani 2008	H9, CA1, I6	MEFs Matrigel	KO-DMEM+20%KSR+bFGF XVIVO10+bFGF+TGFb1	Collagenase
Bauwens 2008	H9, H2B	MEFs	KO-DMEM+20%KSR+bFGF	Collagenase
Niebruegge 2008	H9, HES2	Matrigel MEFs	XVIVO10+bFGF+TGFb1 KO-DMEM+20%KSR+bFGF	Collagenase
Phillips 2008a	ESI-017	HFF+ Fibronectin Fibronectin	KO-DMEM+20%KSR+bFGF HFF-CM	Collagenase NB6
Bendal 2009	H1, H9	Matrigel	MEF-CM with and without bFGF	Collagenase
Fernandes 2009	H9	MEFs	DMEM/F12+20%KSR+bFGF	TrypLE
Gibson 2009	H9	MEFs	DMEM/F12+20%KSR+bFGF	Collagenase
Hentze 2009	HES2, HES3, HES4, ESI-014, 017, 035, 049, 051, 053	HFFs	KO-DMEM+20%KSR+bFGF	Collagenase, TrypLE
Lee 2009	H9, I6, HES2	MEFs	KO-DMEM+20%KSR+bFGF	Not specified
Lock 2009	H1, H9	MEFs Matrigel	DMEM/F12+20%KSR+bFGF MEF-CM+bFGF	Collagenase
Montes 2009	HSI81, SHEF1	Matrigel	MSC-CM + bFGF HFF-CM +bFGF	Collagenase
Nie 2009	H1, H9	MEFs Matrigel	DMEM/F12+20%KSR+bFGF MEF-CM + bFGF	Collagenase
Oh 2009	HES2, HES3	Matrigel	MEF-CM + bFGF	Collagenase
Amit 2010	I3, I4, I6, H9.2	MEF	DMEM/F12+15%KSR+bFGF	Collagenase
Chen 2010	HES2, HES3	Matrigel	mTeSR, MEF-CM, StemPRO	TrypLE
Krawetz 2010	H9	HFF Matrigel	mTeSR	Collagenase+ TrypLE

Reference	Cell Lines	Substrate	Medium	Passaging
Singh 2010	HES2, HES3, ES1049	HFF	KO-DMEM+20%KSR+bFGF	TrypLE
Larijani 2011	Royan H5, H6, hiPSC1, hiPSC4	Matrigel	DMEM/F12+20%KSR+bFGF	Not specified for adherent cultures
Leung 2011	HES2, HES3	Matrigel	MEF-CM+bFGF	Enzymatic

Table 3. Summary of recent publications with static maintenance of human ESCs. bFGF = basic fibroblast growth factor, TGFb1= transforming growth factor beta 1, XVIVO10= serum free medium (Lonza), HFF-CM= human foreskin fibroblast conditioned medium, MEF-CM= mouse embryonic fibroblast conditioned medium, mTeSR= serum-free defined medium (STEMCELL Technologies), StemPRO= serum-free defined medium (Invitrogen).

that $2.1x10^{10}$ cells would be required per treatment. If one T-75 culture flask supports the production of $2x10^6$ cells in 2-3 days, production of $2.1x10^{10}$ cells would require over 10,000 tissue culture flasks and up to 5 weeks of culture time (Ouyang & Yang, 2008). They also determined flask count and time estimates for Parkinson`s disease and diabetes. In both cases similarly large numbers of tissue culture flasks were calculated. The handling of this number of tissue culture flasks would be extremely labor-intensive. Additionally, this type of culture lacks continuous monitoring and environmental controls that may result in spontaneous stem cell differentiation. Alternatively, suspension bioreactors provide a controlled environment to produce the same number of cells. Studies from our lab have produced densities of mouse ESCs in suspension of approximately $1\text{-}2x10^6$ cells/mL (Alfred et al., 2010; Cormier et al., 2006). As such, production of clinical numbers would require suspension volumes of the order of liters.

3.2 Suspension bioreactor culture

There have been arguments that not enough is known about ESCs to take them out of adherent cultures and culture them in suspension conditions as reduced adhesion in anchorage dependent cells has been associated with disorganized growth patterns and changes in cell-to-cell contact (Freshney, 2000). However, the final test of success for any new culture format is the functionality of the cells. Refinement of suspension culture protocols to ensure stable karyotypes, continued expression of pluripotency markers as well as demonstration of pluripotency through embryoid body and/or teratoma formation is the goal of embryonic stem cell bioprocess development.

It is clear upon review of recent publications that culture conditions vary a great deal between mouse and human ESCs as illustrated in Tables 4 and 5, respectively. Mouse ESCs were able to readily form aggregates in suspension whereas human ESCs required the addition of ROCK inhibitor to survive non-adherent conditions and form aggregates. As for differences within each cell type, it is clear that there is greater variation between different human ESC protocols than between different mouse ESC protocols. As murine ESCs were established years before human ESCs, culture medium and passaging techniques are far more standardized and suspension protocols appear to be fairly similar between publications.

While these studies illustrate that it is possible to culture human ESCs in suspension (i.e. proof of concept), work must be done with respect to optimization and standardization of protocols. An understanding of key culture variables is necessary to achieve this.

Reference	Cell Lines	Inoculation Density (cells/mL)	Medium	Passaging	Spinner Type (Volume)	Agitation Rate (rpm)
Fok 2005	R1, CCE	$5x10^4$	DMEM+15%FBS	Trypsin	Bellco (50 mL)	60,100
Cormier 2006	R1	$3.75\text{-}10x10^4$	DMEM+15%FBS	Trypsin	Corning (100 mL)	60-120
Zur Neiden 2007	R1	$3.75x10^4$	DMEM+15%FBS	Trypsin	NDS (100 mL)	100
Hwang 2008	E14Tg2a	$3.0x10^5$	DMEM+10%FBS	N/A	Synthecon (50 mL)	25
Kehoe 2008	E14Tg2a	$1\text{-}7.5x10^4$	DMEM+10%FBS DMEM+10-20% KSR	TrypLE	Corning (100 mL)	60-120
Tsuji 2008	D3, E14, EB5	$(1\text{-}500)x10^4$	DMEM+15%FCS	Trypsin	96 well plates	N/A
Alfred 2010	D3	$3.75x10^4$	DMEM+15%KSR	Trypsin	NDS (100 mL)	100

Table 4. Summary of suspension culture conditions for undifferentiated murine ESCs as aggregates. Unless otherwise stated, culture medium also included LIF, 2-mercaptoethanol and non-essential amino acids.

Reference	Cell Lines	Inoculation Density (cells/mL)	Medium	Passaging	Bioreactor Type	Agitation Rate (rpm)
Kehoe 2010	H1	$6.0x10^4$	MEF-CM+ Matrigel+ROCK	Accutase+ ROCK	Not specified	60
Krawetz 2010	H9	$1.8x10^4$	mTeSR+ROCK+ rapamycin	Accutase+ ROCK	NDS (100mL)	100

Reference	Cell Lines	Inoculation Density (cells/mL)	Medium	Passaging	Bioreactor Type	Agitation Rate (rpm)
Amit 2010	I3, I4, I6, H9.2	1-5x10^{6}	DMEM/F12+KSR+ bFGF DMEM/F12+KSR+ bFGF+IL6RIL6 DMEM/F12+KSR+ bFGF+LIF DMEM/F12+KSR+ bFGF+IL6	Collagenase Trypsin+ ROCK	Erlenmeyer (25mL)	90
Olmer 2010	HES3	0.3x10^{5}	mTeSR+/-ROCK KO-DMEM+ KSR+bFGF+/-ROCK KO-DMEM+FBS +/- ROCK	Collagenase	6well plate	NA
Singh 2010	HES2, HES3, ES1049	2.5x10^{5} 1x10^{6} 1x10^{6}	KO-DMEM+20%KSR +bFGF mTeSR	Collagenase +TrypLE+ ROCK	Low att. plate Stirred dish (10mL) CELLSPN (100mL)	NA 35 40
Steiner 2010	HES1, HES2, H7	0.37-1.2x10^{6}	KO-DMEM+14%KSR +bFGF +Activin A+ fibronectin+ laminin+ gelatin+BDNF+NT3 +NT4+Nutridoma-CS	Collagenase Trituration+ ROCK	12well plate	NA
Larijani 2011	Royan H5, H6, hiPSC1, hiPSC4	15x10^{4}	DMEM/F12+20%KSR + bFGF MEF-CM+bFGF MEF-CM+bFGF+ N2+B27 (all with ROCK)	Trypsin+ ROCK	6well plate	NA

Table 5. Summary of suspension culture conditions for human ESCs as aggregates. ROCK = p160-Rho-associated coiled kinase inhibitor (Y-27632) (Watanabe et al., 2007).

3.2.1 Suspension bioreactor design considerations

Traditional reactor types that may be used in biological processes include batch reactors, continuous stirred tank reactors, and plug flow reactors. Bioreactors support and control living biological entities and therefore require process control and stringent steps to eliminate contamination (Williams, 2002). Mammalian cells are far more sensitive to culture conditions and far less stable with respect to maintenance of cell functionality than many other cell types such as yeast, bacteria and fungi. Additionally, shear is normally not a concern for these cell types whereas ESCs are far more sensitive to their environment. Human ESCs for example will spontaneously transform, differentiate or undergo apoptosis as a reaction to small changes in culture conditions (Chu & Robinson, 2001). This becomes a challenge as mixing is an integral consideration in reactor design for nutrient and oxygen mass transfer and the requirement of achieving target shear places an additional constraint on agitation speed (Schmidt, 2005). The aim of bioreactor design is to minimize the cost of production while retaining the desired quality of the product all within biological, fluid mechanical, and mass transfer constraints (van't Riet & Temper, 1991). With stem cells, the objective is to robustly produce large quantities of viable undifferentiated cells for further differentiation and purification steps. Downstream processing will be the cost determining step and therefore the goal is maximization of product concentration obtained from the bioreactors (van't Riet & Temper, 1991). To accomplish this, some key design areas must be considered including: materials (bioreactor walls and agitators), medium (chemical composition, pH, temperature), rheological conditions (fluid dynamics and mass transfer systems), and residence time of cells within the bioreactor. Issues such as temperature, pH, and medium formulation are fairly well defined for human ESC culture (King & Millar, 2007). Areas that are less understood with respect to suspension culture are those of the physical and geometric properties of the bioreactor itself. Parameters falling within this area are broad ranging from rheology and hydrodynamics, to mixing and agitator design, to heat and mass transfer, to issues of scale up and process control. Throughout the work within our lab group we have observed that culture parameters with the largest impact on cell populations are agitation rate, inoculation density and oxygen transfer.

3.2.2 Agitation speed

The rate of agitation within a stirred suspension bioreactor is an important consideration as it affects not only the mixing of oxygen and nutrients within the medium, but it also maintains cell aggregate sizes and keeps cells in suspension. However, development of effective bioprocesses for suspension culture requires an understanding of the nature of the biomass within the reactor. Embryonic stem cells along with many other mammalian cell types have been observed as being extremely sensitive to shear (Garcia-Cardena & Adamo, 2011; Toh & Voldman, 2011). While the medium must be agitated at a speed sufficient to ensure nutrient mixing, this speed must also be kept within limits so as not to exert undue shear on the sensitive biomass. We therefore need to understand the forces caused by agitation, how these forces act on cells, how we can quantify these forces, and the relationship between the forces and cell viability. There have been a very few reports on the effect of shear in stirred bioreactors on murine ESC growth. Cormier et al., (2006) found that aggregate diameter correlated with agitation rate but did not quantify the correlation. To date there have not been any comprehensive investigations into shear effects on growth kinetics in human ESC suspension cultures.

3.2.3 Inoculation density

Inoculation density is an important parameter that tends to be taken for granted in many studies. In our lab, we have seen that when inoculating suspension spinners, too few cells will fail to initiate agglomeration and therefore aggregates do not form and cells die due to lack of cell-to-cell interaction. Too many cells may form massive aggregates that create issues with oxygen transfer to cells at the aggregate centre (necrosis). Additionally, rapid depletion of nutrients in the medium and subsequent rapid build-up of waste materials can adversely affect growth and expansion of the cell population (Sen et al., 2010).

In general, inoculating at the lowest possible cell density allows for maximum cell number amplification, however many have observed that this also causes a longer lag phase which results in an extended culture period required to reach maximum cell density. Extremely low and high initial densities have been observed to affect the exponential growth phase with a reduction in specific growth rates (Cormier et al., 2006; Fernandes, TG et al., 2009).

A review of literature fails to turn up a great deal of further information on inoculation densities with respect to ESC culture. With human ESCs the passaging techniques used (passaged as clumps until recently) prevented quantification of cells.

3.2.4 Nutrient and oxygen uptake

Oxygen is essential for cell proliferation and viability. In smaller vessels, surface aeration as the main mode of oxygen supply is typically sufficient and commonly used due to its simplicity. However, as culture volumes increase and cell densities increase, the surface area to volume ratios decrease and surface aeration may no longer be sufficient to ensure oxygen transport to the cells (Gilbertson et al., 2006; Baghbaderani et al., 2008). Options to increase oxygen transport include increasing agitation rate of the impeller, sparging, or medium additives such as Perfluorocarbons (PFCs) which increase oxygen solubility within the liquid medium. To determine which method is best, the oxygen requirements of the system must be determined.

Starting with the basic mass conservation equation for a specific component within a control volume (in this case the oxygen within the liquid cell culture medium), we have:

$$\text{Input} + \text{Generation} = \text{Output} + \text{Accumulation} \tag{1}$$

The input of oxygen into the medium, also known as the oxygen transfer rate (OTR), by using surface aeration only (no sparging) is controlled by the concentration difference between the headspace of the vessel and bulk medium oxygen concentration. Since no oxygen is released from the system, the output term is zero. The accumulation term refers to the rate of change in oxygen concentration within the medium while the generation term actually refers to the oxygen uptake rate (OUR) of the cells which is negative to indicate the cells are consuming oxygen rather than producing it. With this in mind, the material balance becomes:

$$\text{OTR} - \text{OUR} = 0 + dC_{O2}/dt \tag{2}$$

where OTR and OUR are as follows:

$$\text{OTR} = k_L a \, \Delta C_{O2} \tag{3}$$

$$\text{OUR} = q_{O2} X \tag{4}$$

where k_La is the volumetric mass transfer coefficient, ΔC_{O2} is the difference in oxygen concentration between oxygen saturation concentration in the bulk medium (C_{O2}^* which may be determined via Henry's law for an assumed partial pressure of 20% O_2 in normal incubator air with 5% CO_2) and the measured bulk medium concentration (C_{O2}). The term q_{O2} is the specific oxygen uptake rate of the cells and X is the cell density within the medium (Garcia-Ochoa & Gomez., 2009).
Rearranging Equation 2 yields an expression for the rate of change in oxygen concentration within the bulk medium:

$$dC_{O2}/dt = k_La\, \Delta C_{O2} - q_{O2}X \tag{5}$$

As can be seen from the preceding equations, the k_La value is the unknown when determining the rate of oxygen transfer. There have been numerous methods described for the determination of the mass transfer coefficient depending on the nature of the system: factors such as aeration, vessel design, medium composition and the effect of the presence of a microorganism must all be taken into account.
Throughout the literature there have been many attempts to develop empirical relationships for the determination of k_La. These relationships are sometimes based on both dimensional and dimensionless values. Garcia-Ochoa & Gomez (2009) have presented a very comprehensive summary of a number of published correlations for the determination of k_La in stirred suspension bioreactors. Upon review, it is apparent there is considerable variation between the correlations presented. For example, in 1979, Van't Riet & Temper proposed a correlation based on the power input per unit volume (P/V) but stated there was no influence of the impeller geometry and placement within the bioreactor. Subsequent studies contradicted this statement by showing that changing the impeller geometry alone caused an increase in mass transfer rate and a change in impeller placement also had an effect (See Garcia-Ochoa & Gomez, 2009 for a full review). To date there still does not appear to be a firm consensus as to which correlation is best suited for the determination of k_La in a stirred vessel.
The wide range of empirical relationships developed may in part be due to the similarly wide range of protocols for the experimental determination of k_La. For culturing stem cells, the presence of the cells themselves is assumed to have a large impact on mass transfer in the system. As such, the biomass present within the system is the main consideration when determining k_La values experimentally. There have been thorough reviews elsewhere which have described several approaches to experimentally determine k_La (Garcia-Ochoa & Gomez., 2009). One example that we have used, described by Baghbaderani et al. (2008), is as follows. Briefly, the assumption is made that at some point in time the reactor reaches steady state during which any oxygen entering the medium would be immediately consumed by the cells (that is, C_{O2} equals zero in the medium). At this point the maximum cell density is attained and the limiting rate of oxygen transfer is reached. In this instance oxygen transfer into the medium equals the oxygen uptake rate of the cells and there would be no change in the oxygen concentration in the bulk liquid. In other words the rate of change in oxygen concentration is zero (dC/dt=0). Equation 5 then simplifies to:

$$k_La = q_{O2}\, X \;/\; C_{O2}^* \tag{6}$$

The specific oxygen uptake (q_{O2}) of the cells in the system can be determined by measuring the oxygen concentration in medium without cells and then placing a known number of cells in the medium and sealing off the vessel with no headspace to replenish the oxygen. Measurements of the oxygen concentration in the medium are taken over time and the resulting rate of oxygen decrease is directly attributed to the uptake by the cells in the medium. That is:

$$dC_{O2}/dt = q_{O2}X \quad (7)$$

Once the specific uptake rate for a cell type is known, the limiting k_La value for a desired cell concentration can be determined. This method, however, obviously does not take into account any dimensional effects of the system. As such, Baghbaderani et al. (2008) have recommended a correlation put forth by Aunins et al. (1989) for a 500mL Corning spinner which takes the following form:

$$k_La = 1.08\, Re^{0.78}\, [D_{O2}\, a \,/\, D_T] \quad (8)$$

Where Re and D_{O2} are the Reynolds' number (dimensionless) and diffusion coefficient of oxygen ($m^2\, s^{-1}$) in the bioreactor medium, a, is the specific mass transfer interfacial area (taken as surface area/volume, m^{-1}) and D_T is the tank diameter (m). Determining this value based on system parameters and then comparing to the k_La determined experimentally for the cell type allows for an indication of whether or not oxygen transfer limitations exist within the system.

3.3 Suspension bioreactor culture of ESCs on microcarriers

Microcarriers are small, usually spherical or nearly spherical beads on which adherent cell types may grow. These beads are available in a multitude of materials including gels, polymers, and collagen. Based on their surface topography, they are generally described as macroporous (allowing cells to expand within the microcarrier), microporous (cell attachment occurs on the surface of the microcarrier, however, cells are exposed to medium on the attached surface as well), or non-porous microcarriers (cells are exposed to medium only on surfaces not attached to the microcarrier). When added to a suspension bioreactor, microcarriers provide high surface area to volume ratio which enables higher cell densities compared to that obtained in static culture. This area can be adjusted by varying the number of microcarriers in the culture (see Table 6 for a summary of microcarrier types and specifications). Microcarriers in cell culture offer several advantages including: better control of culture macro-environment within bioreactors compared to static tissue cultures and roller bottles, a reduction in labor costs, ease of downstream clinical applications as cells can be transplanted while on microcarriers, significant reduction in the space required for a given-sized operation and hence higher cell densities per unit volume. In addition, by allowing cells to grow on a surface, microcarrier cultures harness all the advantages of a static tissue culture system, as well as the controlled environment of a bioreactor system. This provides the cells direct exposure to the medium and reduces mass transfer limitations of oxygen and nutrients. Comprehensive reviews on microcarrier materials and specifications as well as their role in tissue engineering can be found elsewhere (GE Healthcare, 2005; Martin et al., 2011).

Though it has been over four decades since microcarriers were developed (Van Wezel, 1967), only recently have microcarriers been investigated as suitable scaffold materials for

Type	Company	Core Material	Surface Coating	Density (g/cm^3)	Bead Diameter (μm)	Surface Area (cm^2/g)
CultiSpher S	Percell Biolytica	Crosslinked pharmaceutical grade gelatin	Porcine gelatin	1.04	130 – 380	**7500**
Collagen	SoloHill Engineering	Crosslinked Polystyrene, modified with gelatin	Type1 porcine gelatin	1.02	90-150	**480**
Fact III	SoloHill Engineering	Crosslinked Polystyrene, modified with cationic gelatin	Cationic, type1 porcine gelatin	1.02	90-150	**480**
Glass	SoloHill Engineering	Crosslinked Polystyrene, modified with high silica glass	High silica glass	1.02	125-212	**360**
Pronectin F	SoloHill Engineering	Crosslinked Polystyrene, modified with recombinant fibronectin	Recombinant fibronectin	1.02	125-212	**360**
Hillex II	SoloHill Engineering	Modified Polystyrene, modified with cationic trimethyl-ammonium	Cationic, trimethyl ammonium	1.11	160 - 180	**515**
Cytodex 3	Amersham Biosciences	Crosslinked dextran, Denatured Collagen on the surface	Porous Porcine gelatin	1.04	141 - 211	**2700**
Cytodex 1	Amersham Biosciences	Crosslinked Dextran with N,N-diethylaminoethyl groups	Porous Porcine gelatin Cationic	1.03	147-248	**4400**

Table 6. Summary of some of the commercially available microcarriers for cell culture (adapted from Alfred et al., 2011).

cultivation of ESCs (as shown in Tables 7 and 8). However, several issues need to be resolved before ESCs produced on microcarriers and their progeny can be used in drug discovery and regenerative medicine applications, respectively. These issues include excessive agglomeration, which has been shown to be detrimental to cell expansion as a result of necrotic centers (Borys and Papoutsakis 1992), mediocre to low cell yields (especially for human ESC cultures), as well as the elimination of serum and MEFs used in microcarrier cell cultures. Finally, lack of attachment of cells to the microcarriers and the formation of cystic structures in suspension (Abranches et al., 2007; Nie et al., 2009; Storm et al., 2010) must also be addressed to eliminate heterogeneities in cultures.

4. Future directions for large scale production of human ESCs

The publications summarised in Tables 6 and 7 coupled with our own initial successes culturing human ESCs in suspension are a promising step towards development of strategies for implementation of stem cell therapies on a large scale. However, it is also quite apparent that major discrepancies exist in protocols between lab groups. Significant cell-manufacturing and regulatory challenges must be overcome before clinical application of stem cell therapies will be viable. It must be noted that standards and methodologies are only *just* being developed for efficacy evaluation, product characterization and process validation and control. As such, human ESC culture must be approached as a multidimensional optimization problem with the goal to increase target cell output while decreasing cost and occurrences of adverse events (Kirouac & Zandstra, 2008). In 2008, Kirouac & Zandstra suggested that process design and optimization should incorporate: i) assessment of relevant cell properties, ii) measurement and control of key parameters, iii) robust predictive strategies for evaluating the parameters that may impact culture output, iv) approaches to test these many different parameters in a high throughput and scale relevant manner.

Development of predictive strategies includes computer modeling methods to assist in prediction of culture outcomes based on various input parameters with their inherent uncertainty. To date, many groups have used mathematical modeling techniques to describe cell proliferation in a variety of systems (Mantzaris et al., 2001; Galban & Locke., 1999;Lemon et al., 2007) as well as differentiation (Lemon et al., 2007; Yener et al., 2008; Prudhomme et al., 2004). This modeling approach is limited as only a small number of input variables can be considered to maintain the practicality of derived relationships and ease of computation. Empirical or correlation models on the other hand, do not rely on mathematically describable relationships between input and output variables. Some researchers have used factorial design for process optimization by using response surface maps that approximate relationships between variables and outcomes (Chung et al., 2006; Audet, 2010).

Both of these modeling approaches are reasonable and practical when the input variables are limited and generated data sets are of a manageable size. When these data sets become increasingly large, for example, using tools such as microarrays, other methods of analysis are beginning to surface. Methods such as neural networks and statistical learning methods may prove to be very useful. These algorithms learn by example, or are trained by a data set, to assign labels to objects and recognize patterns within very large amounts of data. The use of these methods has yet to be fully defined within biological systems but it is apparent that the process is underway (Nobel, 2006).

Reference	Cell Lines	Medium	Passaging/ Harvesting	Inoculation Density (cells/mL)	Microcarrier Type	Agitation Rate (rpm)
Fok 2005	CCE, R1, M8, 9J	DMEM+15%FBS	Trypsin	$5x10^4$	Cytodex 3, Glass coated styrene	60
Abranches 2007	S25	GMEM+10%FBS	Trypsin	$1x10^4$, $5x10^4$ or $1x10^5$	Cytodex 3	60
Fernandes 2007	46C	DMEM+10%FBS ESGRO Complete	Trypsin	$5x10^4$	Cytodex 3, Cultispher S	40
Phillips 2008b	R1	MEF-CM	Trypsin	$1x10^5$	Hillex II	40
Marinho 2009	USP-1	MEF-CM + DMEM/F12 + 15%KSR + CHO-CM	TrypLE	$2x10^5$	Cytodex 3	70
Storm 2010	E13tg2a	KO-DMEM+ 20%KSR GMEM+10%Serum	Trypsin	$6x10^4$	Collagen, FACT, Cultispher S	45

Table 7. Summary of suspension culture conditions for mouse ESCs on microcarriers. Medium components listed are only the base. See references for full medium components.

Reference	Cell Lines	Medium	Passaging/ Harvesting	Inoculation Density (cells/mL)	Microcarrier Type	Agitation Rate (rpm)
Phillips 2008b	ESI-017	HFF-CM	TrypLE	$6.25x10^4$	Hillex II	Not specified
Fernandes 2009	H9	MEF-CM	TrypLE	$2\text{-}2.5x10^5$	Cytodex 3	60
Lock 2009	H1, H9	DMEM/F12+ 20%KSR+bFGF	TrypLE	$5\text{-}20x10^4$	Collagen+ Matrigel coated	45-80
Nie 2009	H1, H9	MEF-CM + bFGF	Trypsin	$3\text{-}7x10^4$ cells/cm^2	Cytodex 3+ MEFs or matrigel	6well plate on rocker

Reference	Cell Lines	Medium	Passaging/ Harvesting	Inoculation Density (cells/mL)	Microcarrier Type	Agitation Rate (rpm)
Oh 2009	HES2, HES3	MEF-CM, mTeSR1, StemPRO	TrypLE Collagenase Mechanical	$2x10^5$(6well) $6x10^5$(spinner)	Matrigel coated cellulose	100 (6well) 25(spinner)
Chen 2010	HES2, HES3	MEF-CM mTeSR StemPRO	Mechanical	$2x10^5$	DE-53+ matrigel	120(6well)
Storm 2010	SHEF-3	MEF-CM+bFGF KO-DMEM+ 20%KSR	Trypsin+ ROCK	$6x10^4$	Cultispher S	45
Serra 2010	SCED-461	MEF-CM+ROCK	TrypLE	1.5,3,4.5x10^5 4.5x10^5	Cytodex 3	100mL Wheaton 24rpm 300mL BIOSTAT 50-65rpm
Leung 2011	HES2, HES3	MEF-CM	Mechanical	0.8-6x10^5 (6well) 3.2-6x10^5 (spinner)	DE-53	100-120(6well) 25(spinner)

Table 8. Summary of suspension culture conditions for human ESCs on microcarriers. Medium components listed are only the base. See references for full medium components.

As computational methods become more refined along with our understanding of biological processes, it is clear that regenerative medicine will benefit. The use of modeling and scale up techniques has the potential to bring stem cell based therapies into mainstream application and greatly benefit those suffering from debilitating degenerative diseases.

5. Conclusions

Overall, though considerable progress has been made in the development of bioprocesses for the production of ESCs and their progeny, it is obvious that more basic research is needed prior to downstream application of these cells. While protocols developed so far have focused on scaling up production of ESCs and their derivatives, other fundamental issues including elimination of animal derived products in ESC cultures, immune rejection and tumor formation upon transplantation of ESC derivatives must also be addressed. Utilization of modeling techniques to address possible outcomes of interaction effects of various variables involved will be a cost effective method to address some of these issues. Finally, it will be necessary to develop optimized robust, controllable systems in production facilities designed to meet the manufacturing requirements established by various governing bodies.

6. References

Abranches, E., Bekman, E., Henrique, D., & Cabral, JM. (2007). Expansion of mouse embryonic stem cells on microcarriers. *Biotechnology & Bioengineering*. Vol. 96, No. 6, pp.1211-1221. eISSN 1097-0290.

Abranches, E., Silva, M., Pradier, L., Schulz, H., Hummel, O., Henrique, D., & Bekman, E. (2009). Neural differentiation of embryonic stem cells in vitro: a road map to neurogenesis in the embryo. *PLoS One*. Vol. 4, No. 7, e6286. eISSN 1932-6203.

Ahsan, T., & Nerem, RM. (2010). Fluid shear stress promotes an endothelial-like phenotype during the early differentiation of embryonic stem cells. *Tissue Engineering Part A*. Vol. 16, No. 11, pp. 3547-53. eISSN 1937-335X.

Alfred, R., Radford, J., Fan, J., Boon, K., Krawetz, R., Rancourt, D., & Kallos, MS. (2011). Efficient suspension bioreactor expansion of murine embryonic stem cells on microcarriers in serum-free medium. *Biotechnology Progress*. [ePub ahead of print, 27 April 2011] . eISSN 1520-6033.

Alfred, R., Gareau, T., Krawetz, R., Rancourt, D., & Kallos, MS. (2010). Serum-free scaled up expansion and differentiation of murine embryonic stem cells to osteoblasts in suspension bioreactors. *Biotechnology & Bioengineering*. Vol. 106, No. 5, pp. 829-840. eISSN 1097-0290.

Amit, M., Chebath, J., Margulets, V., Laevsky, I., Miropolsky, Y., Shariki, K., Peri, M., Blais, I., Slutsky, G., Revel, M., & Itskovitz-Eldor, J. (2010). Suspension culture of undifferentiated human embryonic and induced pluripotent stem cells. *Stem Cell Reviews and Reports*. Vol. 6, No. 2, pp. 248-59. eISSN 1558-6804.

Audet, J. (2010). Adventures in time and space: Nonlinearity and complexity of cytokine effects on stem cell fate decisions. *Biotechnology & Bioengineering*. Vol. 106, No. 2, pp. 173-82. eISSN 1097-0290.

Aunins, JG., Woodson, BA Jr., Hale, TK., & Wang, DI. (1989). Effects of paddle impeller geometry on power input and mass transfer in small-scale animal cell culture vessels. *Biotechnology & Bioengineering*. Vol. 34, No. 9, pp. 1127-32. eISSN 1097-0290.

Bauwens, CL., Peerani, R., Niebruegge, S., Woodhouse, KA., Kumacheva, E., Husain, M., & Zandstra, PW. (2008). Control of human embryonic stem cell colony and aggregate size heterogeneity influences differentiation trajectories. *Stem Cells*. Vol. 26, No. 9, pp. 2300-10. Epub 2008 Jun 26. eISSN 1549-4918.

Baghbaderani, BA., Behie, LA., Sen, A., Mukhida, K., Hong, M., & Mendez, I. (2008). Expansion of human neural precursor cells in large-scale bioreactors for the treatment of neurodegenerative disorders. *Biotechnology Progress*. Vol. 24, No. 4, pp. 859-70. eISSN 1520-6033.

Baghbaderani, BA., Mukhida, K., Sen, A., Kallos, MS., Hong, M., Mendez, I., & Behie, LA. (2010). Bioreactor expansion of human neural precursor cells in serum-free media retains neurogenic potential. *Biotechnology & Bioengineering*. Vol. 105, No. 4, pp.823-833. eISSN 1097-0290.

Bedell, MA., Jenkins, NA., & Copeland NG. (1997). Mouse models of human disease. Part I: techniques and resources for genetic analysis in mice. *Genes & Development*. Vol. 11, No. 1, pp. 1-10. eISSN 1549-5477.

Bedell, MA., Largaespada, DA., Jenkins, NA., Copeland, & NG. (1997). Mouse models of human disease. Part II: recent progress and future directions. *Genes & Development*. Vol. 11, No. 1, pp. 11-43. eISSN 1549-5477.

Bendall, SC., Hughes, C., Campbell, JL., Stewart, MH., Pittock, P., Liu, S., Bonneil, E., Thibault, P., Bhatia, M., & Lajoie, GA. (2009). An enhanced mass spectrometry approach reveals human embryonic stem cell growth factors in culture. *Molecular & Cellular Proteomics*. Vol. 8, No. 3, pp. 421-32. Epub 2008 Oct 20. eISSN 1535-9484.

Bongso, A., Fong, CY., Ng, SC., & Ratnam, S. (1994). Isolation and culture of inner cell mass cells from human blastocysts. *Human Reproduction*. Vol. 9, No. 11, pp. 2110-7. eISSN 1460-2350.

Borys, MC., & Papoutsakis, ET. (1992). Formation of bridges and large cellular clumps in CHO-cell microcarrier cultures: effects of agitation, dimethyl sulfoxide and calf serum. *Cytotechnology*. Vol. 8, No. 3, pp. 237-48. eISSN 1573-0778.

Chawla, M., Bodnar, CA., Sen, A., Kallos, MS., & Behie, LA. (2006). Production of islet-like structures from neonatal porcine pancreatic tissue in suspension bioreactors. *Biotechnology Progress*. Vol. 22, No. 2, pp. 561-567. eISSN 1520-6033.

Chen, X., Chen, A., Woo, TL., Choo, AB., Reuveny, S., & Oh, SK. (2010). Investigations into the metabolism of two-dimensional colony and suspended microcarrier cultures of human embryonic stem cells in serum-free media. *Stem Cells and Development*. Vol. 19, No. 11, pp. 1781-92. Epub 2010 Sep 10. eISSN 1557-8534.

Cherry, RS., & Kwon, KY. (1990). Transient shear stresses on a suspension cell in turbulence. *Biotechnology & Bioengineering*. Vol. 36, No. 6, pp. 563-71. eISSN 1097-0290.

Chin, AC., Fong, WJ., Goh, LT., Philp, R., Oh, SK., & Choo, AB. (2007). Identification of proteins from feeder conditioned medium that support human embryonic stem cells. *Journal of Biotechnology*. Vol. 130, No. 3, pp. 320-328. eISSN 1573-0972.

Chu, L., & Robinson, DK. (2001). Industrial choices for protein production by large-scale cell culture. *Current Opinion in Biotechnology*. Vol. 12, No. 2, pp. 180-187. ISSN 0958-1669.

Chung, CA., Yang, CW., & Chen, CW. (2006). Analysis of cell growth and diffusion in a scaffold for cartilage tissue engineering. Biotechnology & Bioengineering. Vol. 94, No. 6, pp. 1138-1146. eISSN 1097-0290.

Cormier, JT., zur Nieden, NI., Rancourt, DE., & Kallos, MS. (2006). Expansion of undifferentiated murine embryonic stem cells as aggregates in suspension culture bioreactors. *Tissue Engineering*. Vol. 12, No. 11, pp. 3233–3245. ISSN 2152-4955.

Evans, MJ., & Kaufman, MH. (1981). Establishment in culture of pluripotential cells from mouse embryos. *Nature*. Vol. 292, No. 5819, pp. 154–156. ISSN 0028-0836.

Fernandes, AM., Fernandes, TG., Diogo, MM., da Silva, CL., Henrique, D., & Cabral, JM. (2007). Mouse embryonic stem cell expansion in a microcarrier-based stirred culture system. *Journal of Biotechnology*. Vol. 132, No. 2, pp. 227-36. Epub 2007 Jun 7. eISSN 1573-0972.

Fernandes, AM., Marinho, PA., Sartore, RC., Paulsen, BS., Mariante, RM., Castilho, LR., & Rehen, SK. (2009). Successful scale-up of human embryonic stem cell production in a stirred microcarrier culture system. *Brazilian Journal of Medical and Biological Research*. Vol. 42, No. 6, pp. 515-22. eISSN 1678-4510.

Fernandes, TG., Fernandes-Platzgummer, AM., da Silva, CL., Diogo, MM., & Cabral, JM. (2009). Kinetic and metabolic analysis of mouse embryonic stem cell expansion under serum-free conditions. *Biotechnology Letters*. Vol. 32, No. 1, pp. 171-9. Epub 2009 Aug 25. eISSN 1573-6776.

Fok, EY., & Zandstra, PW. (2005). Shear-controlled single-step mouse embryonic stem cell expansion and embryoid body-based differentiation. *Stem Cells*. Vol. 23, No. 9, pp. 1333–1342. eISSN 1549-4918.

Freshney, RI. (2000). Transformation, In: *Culture of animal cells: a manual of basic technique (4th Edition)*, pp. 269-84, John Wiley & Sons, ISBN 9780-471-34889-9, New York.

Galban, CJ., & Locke, BR. (1999) Analysis of cell growth kinetics and substrate diffusion in a polymer scaffold. Biotechnology & Bioengineering. Vol. 65, No. 2, pp. 121-132. eISSN 1097-0290.

Garcia-Cardena, G., & Adamo, L. (2011). Directed Stem Cell Differentiation by Fluid Mechanical Forces. *Antioxidants & Redox Signaling*. 2011 Feb 5. [Epub ahead of print]. eISSN 1557-7716.

Garcia-Ochoa, F., & Gomez, E. (2009). Bioreactor scale-up and oxygen transfer rate in microbial processes: an overview. *Biotechnology Advances*. Vol. 27, No. 2, pp. 153-76. Epub 2008 Nov 12. ISSN 0734-9750.

GE Healthcare. (2005). *Microcarrier Cell Culture: Principles and Methods*. Ed. Piscataway. Retrieved from: <http://www.gelifesciences.com/aptrix/upp00919.nsf/Content/180CFF1511F005C9C1257628001CBE22/$file/18114062AB.pdf>

Gibson, JD., Jakuba, CM., Boucher, N., Holbrook, KA., Carter, MG., & Nelson, CE. (2009). Single-cell transcript analysis of human embryonic stem cells. *Integrative Biology (Camb)*. Vol. 1, No. 8-9, pp. 540-51. Epub 2009 Jul 30. eISSN 1757-9708.

Gilbertson, JA., Sen, A., Behie, LA., & Kallos, MS. (2006). Scaled-up production of mammalian neural precursor cell aggregates in computer-controlled suspension bioreactors. *Biotechnology & Bioengineering*. Vol. 94, No. 4, pp. 783-792. eISSN 1097-0290.

Hentze, H., Soong, PL., Wang, ST., Phillips, BW., Putti, TC., & Dunn, NR. (2009). Teratoma formation by human embryonic stem cells: evaluation of essential parameters for future safety studies. *Stem Cell Research*. Vol. 2, No. 3, pp. 198-210. Epub 2009 Feb 12. ISSN 1873-5061.

Ito, S., D'Alessio, AC., Taranova, OV., Hong, K., Sowers, LC., & Zhang, Y. (2010). Role of Tet proteins in 5mC to 5hmC conversion, ES-cell self-renewal and inner cell mass specification. *Nature*. Vol. 466, No. 7310, pp. 1129-33. ISSN 0028-0836.

Jing, D., Parikh, & A., Tzanakakis, ES. (2010). Cardiac cell generation from encapsulated embryonic stem cells in static and scalable culture systems. *Cell Transplantation*. Vol. 19, No. 11, pp. 1397-412. Epub 2010 Jun 29. ISSN 0963-6897.

Kallos, MS., Sen, A., & Behie, LA. (2003). Large-scale expansion of mammalian neural stem cells: a review. *Medical & Biological Engineering and Computing*. Vol. 41, No. 3, pp.271-282. eISSN 1741-0444.

Kallos, MS., & Behie, LA. (1999). Inoculation and growth conditions for high-cell-density expansion of mammalian neural stem cells in suspension bioreactors. *Biotechnology & Bioengineering*. Vol. 63, No. 4, pp. 473-483. eISSN 1097-0290.

Kehoe, DE., Lock, LT., Parikh, A., & Tzanakakis, ES. (2008). Propagation of embryonic stem cells in stirred suspension without serum. *Biotechnology Progress*. Vol. 24, No. 6, pp. 1342–1352. eISSN 1520-6033.

Kehoe, DE., Jing, D., Lock, LT., & Tzanakakis, ES. (2010). Scalable stirred-suspension bioreactor culture of human pluripotent stem cells. *Tissue Engineering Part A*. Vol. 16, No. 2, pp. 405-21. eISSN 1937-335X.

King, JA., & Miller, WM. (2007) Bioreactor development for stem cell expansion and controlled differentiation. *Current Opinion in Chemical Biology*. Vol. 11, pp. 394-398. ISSN 1367-5931.

Kirouac, DC., & Zandstra, PW. (2008). The systematic production of cells for cell therapies. *Cell Stem Cell*. Vol. 3, No. 4, pp. 369-81. ISSN 1934-5909

Krawetz, R., Taiani, JT., Liu, S., Meng, G., Li, X., Kallos, MS., & Rancourt, D. (2009). Large-scale expansion of pluripotent human embryonic stem cells in stirred suspension bioreactors. *Tissue Engineering Part C: Methods*. [Epub ahead of print]. DOI: 10.1089/ten.tec.2009.0228. eISSN 1937-3392.

Larijani, MR., Seifinejad, A., Pournasr, B., Hajihoseini, V., Hassani, SN., Totonchi, M., Yousefi, M., Shamsi, F., Salekdeh, GH., & Baharvand, H. (2011). Long-Term Maintenance of Undifferentiated Human Embryonic and Induced Pluripotent Stem Cells in Suspension. *Stem Cells and Development*. 2011 Feb 24. [Epub ahead of print]. eISSN 1557-8534.

Lee, LH., Peerani, R., Ungrin, M., Joshi, C., Kumacheva, E., & Zandstra, P. (2009). Micropatterning of human embryonic stem cells dissects the mesoderm and endoderm lineages. *Stem Cell Research*. Vol. 2, No. 2, pp. 155-62. Epub 2008 Dec 3. ISSN 1873-5061.

Lemon, G., Waters, L., Rose, FRAJ., & King, JR. (2007) Mathematical modeling of human mesenchymal stem cell proliferation and differentiation inside artificial porous scaffolds. Journal of Theoretical Biology. Vol. 249, No. 3, pp. 543-553. ISSN 0022-5193.

Leung, HW., Chen, A., Choo, AB., Reuveny, S., & Oh, SK. (2011). Agitation can induce differentiation of human pluripotent stem cells in microcarrier cultures. *Tissue Engineering Part C: Methods*. Vol. 17, No. 2, pp. 165-72. Epub 2010 Nov 4. eISSN 1937-3392.

Levenstein, ME., Ludwig, TE., Xu, RH., Llanas, RA., VanDenHeuvel-Kramer, K., Manning, D., & Thomson, JA. (2006). Basic fibroblast growth factor support of human embryonic stem cell self-renewal. *Stem Cells*. Vol. 24, No. 3, pp. 568-74. Epub 2005 Nov 10. eISSN 1549-4918.

Lock, LT., & Tzanakakis, ES. (2009). Expansion and differentiation of human embryonic stem cells to endoderm progeny in a microcarrier stirred-suspension culture. *Tissue Engineering Part A*. Vol. 15, No. 8, pp. 2051-63. eISSN 1937-335X.

Ludwig, TE., Levenstein, ME., Jones, JM., Berggren, WT., Mitchen, ER., Frane, JL., Crandall, LJ., Daigh, CA., Conard, KR., Piekarczyk, MS., Llanas, RA., & Thomson, JA. (2006). Derivation of human embryonic stem cells in defined conditions. *Nature Biotechnology*. Vol. 24, No. 2, pp. 185-7. Epub 2006 Jan 1. eISSN 1546-4696.

Mantzaris, NV., Daoutidis, P., & Srienc, F. (2001) Numerical solution of multi-variable cell population balance models: I. Finite difference methods. Computers & Chemical Engineering. Vol. 25, No. 11-12, pp. 1411-1440. ISSN 0098-1354.

Marinho, PA., Fernandes, AM., Cruz, JC., Rehen, SK., & Castilho, LR. (2010). Maintenance of pluripotency in mouse embryonic stem cells cultivated in stirred microcarrier cultures. *Biotechnology Progress*. Vol. 26, No. 2, pp. 548-55. . eISSN 1520-6033.

Martin, GR. (1981). Isolation of a pluripotent cell line from early mouse embryos cultured in medium conditioned by teratocarcinoma stem cells. *Proceedings of the National Academy of Sciences of the United States of America*. Vol. 78, No. 12, pp. 7634–7638. ISSN 0027-8424.

Martin, Y., Eldardiri, M., Lawrence-Watt, DJ., & Sharpe, JR. (2011). Microcarriers and their potential in tissue regeneration. *Tissue Engineering Part B: Reviews*. Vol. 17, No. 1, pp. 71-80. Epub 2010 Dec 29. eISSN 1937-3376.

Montes, R., Ligero, G., Sanchez, L., Catalina, P., de la Cueva, T., Nieto, A., Melen, GJ., Rubio, R., García-Castro, J., Bueno, C., & Menendez, P. (2009). Feeder-free maintenance of

hESCs in mesenchymal stem cell-conditioned media: distinct requirements for TGF-beta and IGF-II. *Cell Research*. Vol. 19, No. 6, pp. 698-709. eISSN 1748-7838.
Mouse Genome Sequencing Consortium, et al. (2002). Initial sequencing and comparative analysis of the mouse genome. *Nature*. Vol. 420, No. 6915, pp. 520-62. ISSN 0028-0836.
National Institutes of Health, U.S. Department of Health and Human Services. (2009). Appendix B: Mouse Embryonic Stem Cell Cultures . In Stem Cell Information, Accessed Wednesday, April 20, 2011. Available from: <http://stemcells.nih.gov/info/scireport/appendixb>
Nelson, TJ., Behfar, A., Yamada, S., Martinez-Fernandez, A., & Terzic, A. (2009). Stem cell platforms for regenerative medicine. *Clinical and Translational Science*. Vol. 2, No. 3, pp. 222–227. eISSN 1752-8062.
Nie, Y., Bergendahl, V., Hei, DJ., Jones, JM., & Palecek, SP. (2009). Scalable culture and cryopreservation of human embryonic stem cells on microcarriers. *Biotechnology Progress*. Vol. 25, No. 1, pp. 20-31. eISSN 1520-6033.
Niebruegge, S., Bauwens, CL., Peerani, R., Thavandiran, N., Masse, S., Sevaptisidis, E., Nanthakumar, K., Woodhouse, K., Husain, M., Kumacheva, E., & Zandstra, PW. (2009). Generation of human embryonic stem cell-derived mesoderm and cardiac cells using size-specified aggregates in an oxygen-controlled bioreactor. *Biotechnology & Bioengineering*. Vol. 102, No. 2, pp. 493-507. eISSN 1097-0290.
Noble, WS. (2006). What is a support vector machine? *Nature Biotechnology*. Vol. 24, No. 12, pp. 1565-7. eISSN 1546-4696.
Oh, SK., Fong, WJ., Teo, Y., Tan, HL., Padmanabhan, J., Chin, AC., & Choo, AB. (2005). High density cultures of embryonic stem cells. *Biotechnology Bioengineering*. Vol. 91, No. 5, pp. 523–533. eISSN 1097-0290.
Oh, SK., Chen, AK., Mok, Y., Chen, X., Lim, UM., Chin, A., Choo, AB., & Reuveny, S. (2009). Long-term microcarrier suspension cultures of human embryonic stem cells. *Stem Cell Research*. 2(3):219-30. Epub 2009 Mar 4. ISSN 1873-5061.
Olmer, R., Haase, A., Merkert, S., Cui, W., Palecek, J., Ran, C., Kirschning, A., Scheper, T., Glage, S., Miller, K., Curnow, EC., Hayes, ES., & Martin, U. (2010). Long term expansion of undifferentiated human iPS and ES cells in suspension culture using a defined medium. *Stem Cell Research*. Vol. 5, No. 1, pp. 51-64. Epub 2010 Mar 30. ISSN 1873-5061.
Ouyang, A., & Yang, ST. (2008). A two-stage perfusion fibrous bed bioreactor system for mass production of embryonic stem cells. *Expert Opinion on Biological Therapy*. Vol. 8, No. 7, pp. 895-909. eISSN 1744-7682.
Papoutsakis, ET. (1991). Fluid-mechanical damage of animal cells in bioreactors. *Trends in Biotechnology*. Vol. 9, No. 12, pp. 427-37. ISSN 0167-7799.
Peerani, R., Rao, BM., Bauwens, C., Yin, T., Wood, GA., Nagy, A., Kumacheva, E., & Zandstra, PW. (2007). Niche-mediated control of human embryonic stem cell self-renewal and differentiation. *European Molecular Biology Organisation Journal*. Vol. 26, No. 22, pp. 4744-55. Epub 2007 Oct 18. ISSN 0261-4189.
Phillips, BW., Lim, RY., Tan, TT., Rust, WL., & Crook, JM. (2008a). Efficient expansion of clinical-grade human fibroblasts on microcarriers: cells suitable for ex vivo expansion of clinical-grade hESCs. *Journal of Biotechnology*. Vol. 134, No. 1-2, pp. 79-87. Epub 2008 Jan 8. eISSN 1573-0972.

Phillips, BW., Horne, R., Lay, TS., Rust, WL., Teck, TT., & Crook, JM. (2008b). Attachment and growth of human embryonic stem cells on microcarriers. *Journal of Biotechnology*. Vol. 138, No. 1-2, pp. 24-32. Epub 2008 Aug 14. eISSN 1573-0972.

Prudhomme, WA., Duggar, KH., & Lauffenburger, DA. (2004) Cell population dynamics model for deconvolution of murine embryonic stem cell self renewal and differentiation responses to cytokines and extracellular matrix. Biotechnology & Bioengineering. Vol. 88, No. 3, pp. 264-72. eISSN 1097-0290.

Reubinoff, BE., Pera, MF., Fong, CY., Trounson, A., & Bongso, A. (2000). Embryonic stem cell lines from human blastocysts: somatic differentiation in vitro. *Nature Biotechnology*. Vol. 18, No. 4, pp. 399-404. eISSN 1546-4696.

Schmidt, FR. (2005) Optimization and scale up of industrial fermentation processes. *Applied Microbiology and Biotechnology*. Vol. 68, No. 4, pp. 425-435. eISSN 1432-0614.

Sen, A., Kallos, MS., & Behie, LA. 2002. Expansion of mammalian neural stem cells in bioreactors: effect of power input and medium viscosity. *Brain Research: Developmental Brain Research*. Vol. 134, No. 1-2, pp. 103-113. ISSN 0165-3806.

Sen, A., Kallos, MS., & Behie, LA. (2010). Bioprocess Engineering of Neural Stem Cells. In: *Wiley Encyclopedia of Industrial Biotechnology*, Ed. MC Flickinger, Publ. John Wiley & Sons, ISBN 978-0-471-79930-6.

Serra, M., Brito, C., Sousa, MF., Jensen, J., Tostões, R., Clemente, J., Strehl, R., Hyllner, J., Carrondo, MJ., & Alves, PM. (2010). Improving expansion of pluripotent human embryonic stem cells in perfused bioreactors through oxygen control. *Journal of Biotechnology*. Vol. 148, No. 4, pp. 208-15. Epub 2010 Jun 17. eISSN 1573-0972.

Singh, H., Mok, P., Balakrishnan, T., Rahmat, SN., & Zweigerdt, R. (2010). Up-scaling single cell-inoculated suspension culture of human embryonic stem cells. *Stem Cell Research*. Vol. 4, No. 3, pp. 165-79. Epub 2010 Mar 12. ISSN 1873-5061.

Steiner, D., Khaner, H., Cohen, M., Even-Ram, S., Gil, Y., Itsykson, P., Turetsky, T., Idelson, M., Aizenman, E., Ram, R., Berman-Zaken, Y., & Reubinoff, B. (2010). Derivation, propagation and controlled differentiation of human embryonic stem cells in suspension. *Nature Biotechnology*. Vol. 28, No. 4, pp. 361-4. Epub 2010 Mar 28. eISSN 1546-4696.

Storm, MP., Orchard, CB., Bone, HK., Chaudhuri, JB., & Welham, MJ. (2010). Three-dimensional culture systems for the expansion of pluripotent embryonic stem cells. *Biotechnology & Bioengineering*. Vol. 107, No. 4, pp. 683-95. eISSN 1097-0290.

Taiani, J., Krawetz, RJ., Nieden, NZ., Wu, YE., Kallos, MS., Matyas, JR., & Rancourt, DE. (2009). Reduced differentiation efficiency of murine embryonic stem cells in stirred suspension bioreactors. *Stem Cells and Development*. Vol. 19, No. 7, pp. 989-98. eISSN 1557-8534.

Terstegge, S., Laufenberg, I., Pochert, J., Schenk, S., Itskovitz-Eldor, J., Endl, E., & Brüstle, O. (2007). Automated maintenance of embryonic stem cell cultures. *Biotechnology & Bioengineering*. Vol. 96, No. 1, pp. 195-201. eISSN 1097-0290.

Terstegge, S., Rath, BH., Laufenberg, I., Limbach, N., Buchstaller, A., Schütze, K., & Brüstle, O. (2009). Laser-assisted selection and passaging of human pluripotent stem cell colonies. *Journal of Biotechnology*. Vol. 143, No. 3, pp. 224-30. Epub 2009 Jul 15. eISSN 1573-0972.

Thomas, RJ., Anderson, D., Chandra, A., Smith, NM., Young, LE., Williams, D., & Denning, C. (2009). Automated, scalable culture of human embryonic stem cells in feeder-free conditions. *Biotechnology & Bioengineering*. Vol. 102, No. 6, pp. 1636-44. eISSN 1097-0290.

Thomson, JA., Kalishman, J., Golos, TG., Durning, M., Harris, CP., Becker, RA., & Hearn, JP. (1995). Isolation of a primate embryonic stem cell line. *Proceedings of the National Academy of Sciences of the United States of America*. Vol. 92, No. 17, pp. 7844-8. ISSN 0027-8424.

Thomson, JA., Itskovitz-Eldor, J., Shapiro, SS., Waknitz, MA., Swiergiel, JJ., Marshall, VS., & Jones, JM. (1998). Embryonic stem cell lines derived from human blastocysts. *Science*. Vol. 282, No. 5391, pp. 1145-7. eISSN 1095-9203.

Toh, YC., & Voldman, J. (2011). Fluid shear stress primes mouse embryonic stem cells for differentiation in a self-renewing environment via heparan sulfate proteoglycans transduction. *Journal* of the Federation of American Societies for Experimental Biology. Vol. 25, No. 4, pp. 1208-17. Epub 2010 Dec 23. eISSN 1530-6860.

van't Reit, & K., Temper, J. (1991) *Basic bioreactor design*. (7th Edition). Dekker, ISBN 978-0-8247-8446-1, New York.

Veraitch, FS., Scott, R., Wong, JW., Lye, GJ., & Mason, C. (2008). The impact of manual processing on the expansion and directed differentiation of embryonic stem cells. *Biotechnology & Bioengineering*. Vol. 99, No. 5, pp. 1216-29. eISSN 1097-0290.

Watanabe, K., Ueno, M., Kamiya, D., Nishiyama, A., Matsumura, M., Wataya, T., Takahashi, JB., Nishikawa, S., Nishikawa, S., Muguruma, K., & Sasai, Y. (2007). A ROCK inhibitor permits survival of dissociated human embryonic stem cells. *Nature Biotechnology*. Vol. 25, No. 6, pp. 681-6. Epub 2007 May 27. eISSN 1546-4696.

Williams, JA. (2002). Keys to bioreactor selections. *Chemical Engineering Progress Magazine*, March:24-41.

Xu, RH., Peck, RM., Li, DS., Feng, X., Ludwig, T., & Thomson, JA. (2005). Basic FGF and suppression of BMP signaling sustain undifferentiated proliferation of human ES cells. *Nature Methods*. Vol. 2, No. 3, pp. 185-90. Epub 2005 Feb 17. eISSN 1548-7105.

Yener, B., Acar, E., Aguis, P., Bennett, K., Vandenberg, SL., & Plopper, GE. (2008). Multiway modeling and analysis in stem cell systems biology. *BMC Systems Biology*. Vol. 2, No. 63. ISSN 1752-0509

Yu, J., & Thomson, JA. (2008). Pluripotent stem cell lines. *Genes & Development*. Vol. 22, No. 15, pp. 1987-97. eISSN 1549-5477.

Zandstra, PW., Eaves, CJ., & Piret, JM. (1994). Expansion of hematopoietic progenitor cell populations in stirred suspension bioreactors of normal human bone marrow cells. *Biotechnology (N Y)*. Vol. 12, No. 9, pp. 909-14. ISSN 0733-222X.

zur Nieden, NI., Cormier, JT., Rancourt, DE., & Kallos, MS. (2007). Embryonic stem cells remain highly pluripotent following long term expansion as aggregates in suspension bioreactors. *Journal of Biotechnology*. Vol. 129, No. 3, pp. 421–432. eISSN 1573-0972.

5

Small-Scale Bioreactors for the Culture of Embryonic Stem Cells

Allison Van Winkle[1,2], Ian D. Gates[2] and Michael S. Kallos[1,2]
[1]*Pharmaceutical Production Research Facility (PPRF), Schulich School of Engineering, University of Calgary, Alberta,*
[2]*Deparment of Chemical and Petroleum Engineering, Schulich School of Engineering, University of Calgary, Alberta, Canada*

1. Introduction

Stem cells have great potential for use as a regenerative therapy for degenerative diseases, such as diabetes or Parkinson's. However, to create a large-scale and reproducible protocol, large homogeneous cell populations are required. For example, it is estimated that to treat one patient with stem cell-derived functional beta-cell equivalents for diabetes would require approximately 1 billion homogeneous beta-cell equivalents (Docherty et al. 2007). To reproducibly expand ESCs to this order remains a major research hurdle. Suspension bioreactors offer major advantages over traditional, static culture methods, including the ability to monitor and control important bioprocess parameters such as dissolved oxygen, pH, and temperature. In addition, for clinical implementation of a stem cell therapy, automation associated with bioreactors will aid in compliance with regulatory protocols (Martin et al. 2009). Other advantages of suspension bioreactors over traditional (static) cell culture methods, include scalability, enhanced oxygen and nutrient transfer, homogeneity, and increased reproducibility. However, the use of larger-scale bioreactors (for example, greater than 100 mL working volume) incurs significant expenses as considerable amounts of media, cells, and other supplies are required. In addition, extensive time and handling is necessary to generate enough cells for inoculation. Small-scale bioreactors (less than 100 mL working volume) require fewer cells, are more economical, and require less labour than larger bioreactors. The use of small-scale bioreactors potentially permits high-throughput experimentation to test operating and growth conditions (media components, agitation rate, cell density) and the resulting interactions.

In a suspension bioreactor, cells are suspended in liquid medium which consists of a mixture of water, glucose, amino acids and dissolved oxygen, among other factors. As the suspension bioreactor is agitated, the environment within the bioreactor is more homogeneous than traditional culture environments, such as a T-flask, where gradients occur in the static media. The hydrodynamic environment created by agitation of the suspension bioreactor is known to influence cell survival. Excessive amounts of shear stress can lead to damage to cell membranes (Betts and Baganz 2006) whereas insufficient amounts of shear stress can cause excessive agglomeration (Li et al. 2009). In addition, manipulating the amount of shear stress on cells has been found to influence the resulting diameter of mammalian neural stem cell aggregates (Sen et al. 2001) which in turn also influences the distances through which nutrients must diffuse.

2. Stem cell culture in suspension bioreactors

It is well established in literature that stem cells can be cultured in suspension bioreactors (Kallos and Behie 1999; Krawetz et al. 2009; Kwon et al. 2003; Niebruegge et al. 2009; Schroeder et al. 2005; Youn et al. 2005). As an example, murine embryonic stem cells have been successfully expanded in 100 mL suspension bioreactors within specific ranges of agitation rates and corresponding shear stresses. The mESCs were expanded over a period of 6 days. Three agitation rates were tested: 80, 100 and 120 rpm. mESC expansion was found to be greatest at 100 rpm reaching a peak viable cell density of 10^6 cells/mL. It was also observed that aggregate diameters in the 100 rpm bioreactors were smaller than that of the 80 rpm bioreactors whereas the 120 rpm bioreactors resulted in excessive cell debris and no proliferation indicating substantial damage to the cells (Cormier et al. 2006).

Cormier et al. also examined the effect of cell culture in suspension bioreactors on pluripotency. Immunocytochemistry was examined for several mESC pluripotency markers: Oct-4, Nanog and SSEA-1, which are expressed in undifferentiated mESC cells, and downregulated once cells differentiate (Murphy and Polak 2002; Tavares et al. 2007). Samples were taken from static culture (prior to suspension culture) and on Day 4 of the suspension culture (within the exponential growth phase) and positive expression of all pluripotency markers were observed in both samples (Cormier et al. 2006).

mESCs have also been serially passaged in suspension bioreactors (zur Nieden et al. 2007). Aggregates were harvested on Day 4 (within the exponential growth phase), dissociated into single cells and re-inoculated at a cell density equal to 3.75×10^4 cells/mL. This procedure was repeated for seven subsequent passages; static cultures were maintained over the same time period. The suspension culture had a cumulative 2.5 billion-fold expansion over the 28 day period, whereas a control static culture had a 76.4 trillion fold expansion. However, as the static cultures were passaged every two days, there were twice as many static passages as suspension passages. Per passage, the suspension culture was found to have a 31-fold expansion whereas the static culture had a smaller 10-fold expansion. In addition, 37 static culture flasks would be required to obtain the total cell number as obtained from one suspension bioreactor (zur Nieden et al. 2007). While the amount of static culture flasks could be reduced by using larger static flasks, the homogeneity, reproducibility and reduced amount of labour motivate the use of suspension bioreactors for clinical applications.

Suspension bioreactors have been used to produce mesoderm and cardiac cells from hESCs however these cells were first cultured under static conditions on a layer of feeder cells and transferred into suspension culture after differentiation had begun (Niebruegge et al. 2009). The use of feeder cells decreases the amount of reproducibility and increases the risk of contamination and immune response; this protocol could not be used for clinical application.

The expansion of hESCs in a suspension bioreactor has been successful with the addition of Rho-associated kinase (ROCK) inhibitor which has been found to diminish dissociation-induced apoptosis for single cells (Li et al. 2009; Watanabe et al. 2007). Krawetz et al. exposed H9 hESCs to ROCK inhibitor for 24 hours and obtained a cell fold expansion of 25 over 6 days (Krawetz et al. 2009). Kehoe et al. exposed H1 hESCs to ROCK inhibitor for only

30 minutes and obtained a cell fold expansion of 5.6 over 7 days (Kehoe et al. 2009). These researchers also used different media for the cell culture. In a 50 mL suspension bioreactor, Singh et al. (2010) cultured various hESC lines: hES2, hES3, ESI04 and found that culturing the cells in the presence of ROCK and a heat shock treatment resulted in the greatest amount of cell survival (60% cell survival with ROCK and heat shock treatment, compared to 2% in the control case). It was also found that all three cell lines formed comparably sized aggregates and resulted in approximate 2-fold cell expansion per passage. However, Singh et al. noted cell line differences in the maintenance of pluripotency within the suspension bioreactors; pluripotency markers remained high for the hES2 and hES3 lines, but Oct4, Tra-160 and SSEA4 down regulated for the ESI049 hESC line (Singh et al. 2010).

Other research groups have reported successful expansion of hESCs in suspension culture where the suspension culture in this case is not a suspension bioreactor but rather a shake flask (Amit et al. 2010), low-attachment well plate (Olmer et al. 2010) or in some cases, not described (Rezaei Larijani et al. 2011; Steiner et al. 2010). The data obtained from these researchers provides valuable information about the hESCs ability to differentiate in suspension where hESCs were differentiated to neural cells in suspension (Steiner et al. 2010) and to neural, cardiomyocytes and hepatocytes in suspension (Rezaei Larijani et al. 2011). Additionally, the ability of hESCs to maintain stable karyotypes and maintain pluripotency in suspension was established (Amit et al. 2010; Olmer et al. 2010; Rezaei Larijani et al. 2011; Steiner et al. 2010). However, these cell culture strategies are not amenable to substantial scale-up for clinical purposes due to heterogeneity and lack of control over bioprocess parameters.

Another option for suspension culture of hESCs is using microcarriers which provide an adherent surface for the cells to grow on. hESCs have been successfully expanded in a long-term (6 months) suspension culture grown on microcarriers; the cells retained their pluripotency (as indicated by > 95% expression levels of SSEA4 and TRA-1-60 markers) reaching a cell density twice that of a static culture: 1.6 million cells/mL as opposed to 0.8 million cells/mL (Oh et al. 2009). However, microcarrier cultures are disadvantageous for large scale production since the cells must be dissociated from the surface of the microcarrier and/or the microcarrier dissolved before clinical use.

While the results of hESCs cultured in suspension and in suspension bioreactors are promising, there is still substantial research to be performed regarding agitation rates, shear stress, and media components before hESCs can be successfully expanded in controlled bioprocesses to large clinically meaningful populations.

3. Small-scale bioreactors

Small-scale systems have traditionally been used for preliminary research such as screening a large number of experimental conditions. However, for process scale up, where the biological product obtained for a large-scale operation (5 to 100 L for mammalian cells) is equivalent to the small scale bioreactor product; many experiments are required to determine desired media components and operating conditions. However, the use of standard-scale bioreactors (100 mL working volume) incurs significant cost since considerable amounts of media, cells, and other supplies are required. In addition, extensive time and handling is necessary to generate enough cells for inoculation. A scalable, small-scale suspension bioreactor system would require fewer cells, and less labour than was previously possible with larger suspension bioreactors. High-throughput cell culture experiments at a more economical scale could then be performed for culture optimisation (Kostov et al. 2001) and allow for factorial experimental design.

While substantial research has been completed in the area of scaled-up production (Baghbaderani et al. 2008; Fernandes-Platzgummer et al. 2011; Garcia-Ochoa and Gomez 2009; Gilbertson et al. 2006; Park et al. 2010; Youn et al. 2005; Yu et al. 2009) to increase cell cultures to clinically meaningful numbers such as the estimated 1 billion cells per patient required for stem cell therapy (Docherty et al. 2007), there has been little development in scaling down mammalian cell bioreactor systems. Many scaled-down bioreactors in the literature have been developed to culture *Escherichia coli* (bacteria) or *Saccharomyces cerevisiae* (yeast). Some consist of shaken well plate systems (Duetz 2007; Elmahdi et al. 2003; Micheletti et al. 2006). In these systems, bacteria and yeast cultures are more mechanically srong and thus they can be agitated at substantially higher rates than mammalian cells (Betts and Baganz 2006). In addition, surface tension effects are more pronounced in shaken systems than in stirred configurations (Betts and Baganz 2006). Recently, many of the small-scale systems in literature have been changed from initial shaken models to a stirred design (Gill et al. 2008; Islam et al. 2007; Kusterer et al. 2008; Micheletti et al. 2006; Weuster-Botz et al. 2005).

There are many inconsistencies in the literature on the definition of a microbioreactor. This report will use the definitions of small-scale bioreactors as outlined in Table 1.

Bioreactor	**Working Volume**
Nanobioreactor	$< 1\ \mu L$
Microbioreactor	$1 \leq v < 1000\ \mu L$
Minibioreactor	$1 \leq v < 100$ mL
Standard bioreactor	$100 \leq v < 500$ mL
Large bioreactor	≥ 500 mL

Table 1. Working Volumes of Bioreactors

Industrial sized bioreactors can reach sizes much greater than 1L, but currently, large bioreactors of this size and beyond have not been needed for embryonic stem cell culture.

3.1 Bacteria and yeast culture in small-scale bioreactors

There are a number of different small-scale stirred bioreactor systems currently available in the literature all of which contain oxygen and pH sensors. The first was developed by the Lye group at University College, London. This bioreactor system was designed to be geometrically similar to large scale bioreactors, and driven by a miniature turbine impeller. They found that to ensure the impeller was submerged in liquid, and to achieve an agitation rate of up to 2000 rpm, the bioreactor was restricted to be 100 mL in size. The impeller was magnetically driven, and up to 16 reactors could be operated in parallel with continuous on-line monitoring and control of pH, dissolved oxygen and temperature. The calculated maximum specific growth rate (of *E. coli*) and final biomass concentrations achieved were similar between the small-scale bioreactor and a conventional large-scale 2 L bioreactor. The group concluded that results obtained from small-scale parallel experiments in the miniature bioreactors can thus be converted to large-scale bioreactors (Gill et al. 2008). While this conclusion is valid, the work was performed on a bioreactor which is considered standard-sized for lab-scale mammalian cell experimentation, and the bioreactor system presented by the Lye group is not suitable for high-throughput analysis.

A second small-scale bioreactor system was developed by the Weuster-Botz group from Technische Universitat, Germany. This minibioreactor system uses a 48-reactor block of magnetically stirred bioreactors each one of the order of 10 mL scale. Dissolved oxygen and pH are measured by chemical sensors. Sampling and acid or base additions were carried out by a robotic liquid handling system with up to 20 μL sampled every 1 to 4 hours. This sample was read by a microtitre plate fluorescence reader, in which the optical density was correlated to cell density. The growth characteristics of *E. coli* and *S. cerevisiae* were similar at the 10 mL and 20 L scale (Kusterer et al. 2008). The above system has recently been modified with a new impeller for the culture of mycelium forming microorganisms. These cultivations are more prominently affected by shear forces than bacteria or yeast cultures as shear stress effects the morphology and productivity of these cells (Hortsch et al. 2010). Additionally, the culture broth is a non-Newtonian, shear-thinning fluid (Pollard et al. 2002) with lower viscosity at higher shear rates which influences mixing, heat and mass transfer processes within the bioreactor. The new impeller developed by this group aimed to ensure sufficient gas-liquid mass transfer for the cultured *Streptomyces tendae*. The impeller is a vertical, one-sided paddle which distributes power throughout the reactor, resulting in consistent viscosity throughout the reactor.

The 10 mL bioreactors were agitated at 1200 rpm. Reference cultivations were cultured in a 3L stirred tank bioreactor (impeller speed was equal to 800 rpm) where the power input was held constant between the two bioreactor sizes. The production of nikkomycin Z was compared between the two scales and it was found that approximately 300 mg L^{-1} were produced at each size scale (Hortsch et al. 2010). However, while these agitation rates are lower than that used for bacterial or yeast cultures, they are still much higher than that which mammalian cells could sustain.

A third small-scale bioreactor system was developed by the Rao group of the University of Maryland. A 2 mL minibioreactor was developed, and the volumetric mass transfer coefficient (k_La) was kept constant from the small-scale bioreactor up to 1 L, based on the rational that higher k_La may yield increased cell growth due to oxygen availability, and k_La should thus be kept constant for scalability between the two systems. It was observed that oxygen depletion patterns between the two systems were similar during exponential growth. The total cost of the bioreactor system was less than US$400 (in 2001); the Rao group proposes that it seems possible to develop a bioprocess system where many bioprocesses can run in parallel (96 or more) for lower cost than a bench scale bioreactor. However, the system was magnetically stirred by a stir-bar located on the bottom of the vessel (Kostov et al. 2001). This arrangement, with the stir bar free-floating in the system, has been found to be detrimental to mESC expansion; it is hypothesized that the stir bar, sitting on the bottom of the vessel, effectively crushes the cells as they are pulled underneath (Millar 2009).

The Rao group has also developed an additional minibioreactor with a working volume of 10 to 35 mL agitated with an impeller, for the cultivation of bacteria and yeast (Harms et al. 2006). The design of the bioreactor agitation system has not been published other than to specify that the small-scale bioreactors are stirred and each well is controlled by an individual stepper motor. The use of an individual motor for each well would provide opportunity for agitation rate testing in multiples on a small scale, but not yet to the small micro-scale desired in this study. In addition, increasing the number of motors used vastly increases the cost and operational complexity of the system.

A fourth small-scale reactor, developed by the Jensen group at the Massachusetts Institute of Technology was also designed for bacteria and yeast culture, and is the smallest stirred small-scale reactor to be found in the literature at 0.1 to 1.0 mL in working volume. The most recent form of this microbioreactor has the impeller extend upward from the base of the reactor; 100 μL volumes were found to replicate the growth kinetics and gene expression profiles of *E. coli* as in a 0.5 L bioreactor (Zhang et al. 2007). While these results are promising, an array for parallel experimentation has not yet been developed (a single reactor only) and the location of the impeller at the base of the reactor would make operation (cleaning and sterilizing) difficult.

3.2 Mammalian cell culture in small-scale bioreactors

To date, there has not been any scaled-down stirred, suspension bioreactors published for mammalian cell culture. However, there have been developments relating to small-scale mammalian cell culture devices to investigate the cell microenvironment (Cimetta et al. 2009; Figallo et al. 2007; Wu et al. 2006). These devices are not amenable to scaled-up production, or high-throughput screening for bioreactor cultivations, since bioreactor hydrodynamics are not designed to be similar to a standard bioreactor but they are designed rather to study the cell microenvironment. In addition, these systems are perfusion based allowing for constant exchange of nutrients for waste; while this is beneficial for cell culture, it is in contrast to the batch cultivations of a stirred-tank bioreactor which this chapter has focused on.

Small-scale hollow-fiber devices for use as a bioartificial liver are also being developed (Gramer and Britton 2002; Rodriguez et al. 2008). However, these reactors are designed for cell product secretion, and not for cell expansion, and again are not amenable to scaled-up expansion of stem cells.

There have been developments for mammalian cell culture in small-scale shaken bioreactors. Chinese Hamster Ovary (CHO) cells were cultivated in a small-scale shaken bioreactor, equipped with a ventilation cap. The small-scale bioreactor has a footprint of a 24 well-plate, and each vial has a working volume of 6 mL (Isett et al. 2007). The cell growth and metabolite profiles in th 6 mL bioreactor were found to be similar when compared to a large-scale 2 L bioreactor culture where 85% of the viable cell density of CHO cells cultured in the small-scale bioreactor was within 20% of the viable cell density of CHO cells cultured in the 2L bioreactor (Chen et al. 2008). While this is a promising result, the size scale remains a barrier to high throughput screening, and no engineering calculations were done to show a comparison of the bioreactor hydrodynamics between the two size scales.

Girard et al. (2001) also cultured CHO cells in a small-scale shaken bioreactor; the cells were cultured in the wells of a twelve-well microtiter plate and shaken on a rotational shaker plate. As the well plates are stackable, several hundred small-scale bioreactor experiments were able to be reproducibly completed. Cell density was measured with the use of fluorescence measurements, which allows for non-invasive sampling. In this study, the formation of cellular aggregates was compared between the shaken well plate small-scale bioreactor (2 mL working volume) and a 3 L stirred bioreactor and the resulting cell aggregate size was found to be comparable. The two systems were also compared with respect to production of protein where the small-scale bioreactor resulted in double the

production of protein from the standard bioreactor. However, the results also indicated that evaporation was a problem within the small-scale bioreactor which led to inconsistencies in the results (Girard et al. 2001).
CHO cells have also been cultivated in a 700 μL microbioreactor; cultures were inoculated in small chambers of a plastic card with diffusible walls for oxygen and carbon dioxide mass transfer. The cards were then rotated. The cell culture behaved similarly to a 3 L large-scale bioreactor, where the correlation factor, R^2, between the two scales for the resulting viable cell density was equal to 0.84. In addition, the small-scale bioreactor was used to analyse the process in a high throughput manner; finding that supplementing the media substantially ($p < 0.05$) affected the cells resulting in a higher cell mass. However, due to the small volumes used, Legmann et al. noticed significant differences in the cell counts from the microbioreactors (Legmann et al. 2009).
By using agitation induced by thermal convection, a 250 μL microbioreactor has been developed for expansion of human hematopoietic stem cells. The temperature gradient within the microbioreactor was small ranging from 36.0 to 36.015°C, where the heat source was provided at the base of the reactor. This resulted in temperatures suitable for mammalian cell culture. Experimental tests comparing the microbioreactor culture to a static control resulted in no statistically significant effect on the cell proliferation rate (Luni et al. 2010). Thus, while this microbioreactor yields successful cell expansion, it does not result in the large degree of enhanced cell expansion seen in suspension bioreactors as compared to static controls such as the 31-fold expansion observed in a standard 100 mL bioreactor as compared to 10-fold expansion in a static control (Cormier et al. 2006). Thermal convection mixing may provide a cell culture environment not significantly different from the static control. In addition, this system would be difficult to scale up because increasing large temperature gradients would be required which could be higher than viability threshold temperatures for mammalian cells.

3.3 Small-scale bioreactors operating in parallel

While the creation of a small-scale bioreactor for cell culture is indeed beneficial for research, small-scale bioreactors must be capable of operating in parallel with many other small-scale bioreactors for high-throughput experimentation and factorial analysis. The degree of parallel application found in small-scale bioreactors throughout the literature is presented in Table 2. Most small-scale bioreactors found in the literature are not yet amenable to high-throughput analysis while maintaining consistent hydrodynamic environments as a stirred standard scale bioreactor.

Research	Bioreactor Scale	Number of Bioreactors Operating in Parallel
Gill et al. 2008	Standard (100 mL)	4-16
Kusterer et al. 2008	Minibioreactor (12 mL)	48
Isett et al. 2007	Minibioreactor (6 mL)	24
Girard et al. 2001	Minibioreactor (2 mL)	12-144
Legmann et al. 2009	Microbioreactor (700 μL)	180

Table 2. Parallel Application of Small-Scale Bioreactors. The number of bioreactors at each size scale is presented.

Applying parallel operation to bioreactors from a single prototype results in a number of complications. For example, Gill et al. found that the magnetically agitated bioreactor impeller design must be altered when creating a parallel application to avoid interacting magnetic fields. Due to the impeller design changes, the researchers found that the bioreactors must be agitated at speeds greater than 1000 rpm to maintain a homogeneous gas bubble distribution within the culture media (Gill et al. 2008). Maintaining sterility has also been noted as an obstacle to parallelization, in particular during culture sampling. Kusterer et al. found that washing the components encountering sampling needles with a mixture of 70% ethanol (v/v) and 20% acetone (v/v) to be sufficient (Kusterer et al. 2008). However, cross-contamination between samples must also be avoided where each vessel must be treated as an individual sample with individual sampling ports. Girard et al. accomplished this through measuring cell growth via green fluorescent protein expressing cells and a fluorescence plate reader (Girard et al. 2001). Success with this method is thus dependent on a green fluorescent protein expressing cell line. Isett et al. also used fluorescence to monitor dissolved oxygen level (Isett et al. 2007). While Legmann et al. were able to successfully culture the mammalian cells in parallel microbioreactors, the hydrodynamics of the microbioreactor developed would not relate to a standard bioreactor.

3.4 Hydrodynamic environment of small-scale bioreactors

The hydrodynamic environment in a bioreactor is known to be a function of the bioreactor and impeller geometries, agitation rate, medium viscosity and volume (Flickinger and Drew 1999; King and Miller 2007; Marks 2003; Sen et al. 2002). When scaling reactors up, from lab scale to an industrial scale, it is common to hold several parameters constant to maintain a consistent hydrodynamic environment. This method may also be used for scaling down the size of reactors. It is common to maintain dimensional similarity (Flickinger and Drew 1999) such as a liquid height to reactor diameter ratio. However, these calculations do not consider the effect of shear stress on mammalian cells. These cells are very sensitive to shear stress due to the absence of a cell wall; high agitation rates can adversely affect mammalian cell growth (Betts and Baganz 2006).

While research results from the literature suggest small-scale bioreactors as potential technology for use in high throughput screening (Barrett et al. 2010; Betts and Baganz 2006; Kumar et al. 2004) and cells cultured in small-scale reactors have reached cell densities similar to that of large-scale bioreactors (Gill et al. 2008; Kusterer et al. 2008; Zhang et al. 2007), there remains a knowledge gap of the fluid behaviour between the two scales. For example, correlations and dimensionless numbers used to characterize systems at the large-scale may not be applicable at the small scale. The Reynolds number, often used to characterize fluid systems, has been found to not be applicable at the micro-scale (Betts et al. 2006; Micheletti and Lye 2006; Vallejos et al. 2006). In addition, shear stress has been found to be a critical factor to maintaining cell cultures in suspension in bioreactors (Sen et al. 2001). The maximum shear stress on an aggregate in suspension is calculated as a function of fluid density and power dissipated per unit mass (Cherry and Kwon 1990; Sen et al. 2002) which is in turn a function of the Power Number. A correlation between the Power Number, Reynold's Number, and the reactor geometry has been developed (Nagata 1975). The Reynold's Number and Power Number are dimensionless numbers which act as a representation of physical forces. The Reynold's Number (Re) is a ratio of inertial to viscous forces, and is calculated as:

$$Re = \frac{\rho U D}{\mu} \tag{1}$$

Where ρ is the fluid density (kg/m³), U is the fluid velocity (m/s), D is a characteristic length of the system (m) and μ is the fluid viscosity (Pa*s). The Power Number relates the resistance force to the intertial force, and is calculated for a reactor system as follows, using the shear stress within the system.

The maximum shear stress for a suspended sphere (aggregate) in turbulence can be found from the following equation (Cherry and Kwon 1990; Sen et al. 2002):

$$\tau_{max} = 5.33\rho\sqrt{\varepsilon\vartheta} \tag{2}$$

where τ_{max} is the maximum shear stress on an aggregate (Pa), and ε is the power dissipated per unit mass (m^2/s^3), which is calculated as:

$$\varepsilon = \frac{P}{V_L \rho} \tag{3}$$

where P is the power consumed ($kg\ m^2/s^3$), and V_L is the liquid volume (m^3). The power consumed can be calculated as:

$$P = P_N N^3 D^5 \rho \tag{4}$$

where P_N is the Power Number (dimensionless). The power number has been correlated to be a function of the Reynolds number and the physical properties of the bioreactor. The empirical correlation is as follows for an unbaffled tank with paddle impellers (Nagata 1975):

$$P_N = \frac{K_1}{Re} + K_2\left[\frac{10^3+1.2Re^{0.66}}{10^3+3.2Re^{0.66}}\right]^{K_4} \tag{5}$$

where

$$K_1 = 14 + \frac{W}{D_t}\left[670\left(\frac{D_i}{D_t} - 0.6\right)^2 + 185\right] \tag{6}$$

$$K_2 = 10^{K_3} \tag{7}$$

$$K_3 = 1.3 - 4\left[\frac{W}{D_t} - 0.5\right]^2 - 1.14\frac{D_i}{D_t} \tag{8}$$

$$K_4 = 1.1 + 4\frac{W}{D_t} - 2.5\left[\frac{D_i}{D_t} - 0.5\right]^2 - 7\left[\frac{W}{D_t}\right]^4 \tag{9}$$

where W is the impeller width (m), D_t is the tank diameter (m) and D_i is the impeller diameter (m).

When calculations are performed with these correlations to try to determine the shear forces on cells within the small-scale bioreactors, the resulting agitation rate needed to maintain the same shear as present in the larger bioreactor is much too high to sustain mammlian cell growth. Specifically, these correlations strongly depend on the Reynolds number. The use of a specific Reynolds number as a guide for the transition from laminar to turbulent flow has been found inapplicable to small-scale systems such as a microbioreactor (Betts et al. 2006; Micheletti and Lye 2006; Vallejos et al. 2006). At larger scales, the Reynold's number clearly represents the transition from laminar to turbulent flow. In addition, as these correlations were developed for a large-scale reactor, they may also not be applicable on a very small-scale.

3.5 Oxygen environment in small-scale bioreactors

Typically, stem cell bioreactors use surface aeration to supply oxygen to the suspended cells. The surface area to volume ratio in smaller bioreactors is much larger than the standard 100 mL bioreactors, as shown in Table 3. For embryonic stem cells in suspension culture, the mass transfer of oxygen has been found not to be a limiting factor (Cormier et al. 2006).

Dimension	Standard Bioreactor (100 mL)	Microbioreactor (250 μL)
Volume (m^3)	1.00E-04	2.50E-07
Surface Area (m^2)	2.68E-03	3.12E-05
Surface Area/Volume (m^{-1})	2.68E+01	1.25E+02

Table 3. Surface area to volume ratio of the standard 100 mL bioreactor and an example microbioreactor.

The following calculations were performed to characterize the mass transfer of oxygen to mESCs in a 250 μL microbioreactor. The oxygen consumption rate of mESCs in suspension has been previously measured (zur Nieden et al. 2007). Assuming a peak cell density of 10^6 cells/mL and a 250 μL working volume, the peak consumption rate of oxygen of the cells in a microbioreactor vial can be determined. A total vessel volume of 1 mL is assumed. By using the ideal gas law, the amount of oxygen in the headspace (0.75 mL) is equal to $2.95\text{x}10^{-5}$ mol (0.94 mg). Assuming no oxygen transfer from the surrounding environment, where the oxygen in the headspace is the only oxygen available to the cells, and knowing the total oxygen consumption of the microbioreactor (Table 4), the amount of time required for the oxygen in the headspace to completely deplete was found to be 34 days, given a constant cell density of 10^6 cells/mL. While it is unlikely that the cells would remain at a constant cell density, this density is the maximum found in standard 100 mL bioreactors, and thus represents a worst case scenario. It is more likely that the cell density would initially begin at a much smaller value, and then as the cells expanded the cell density would increase to 10^6 cells/mL. In addition, as the microbioreactor vial would be opened much more frequently than this for sampling or media changes, there is thus an abundant supply of oxygen available to the cells.

Description	Value
Specific Oxygen Consumption Rate	4 x 10^{-17} $\frac{mol\ O_2}{cell \cdot s}$
Cell Density	10^6 cells/mL
Media Volume	0.25 mL
Total Cells	2.5 x 10^5 cells
Total Oxygen Consumption	1 x 10^{-11} $\frac{mol\ O_2}{s}$

Table 4. Oxygen consumption by murine embryonic stem cells in a microbioreactor. The cell density is assumed to be the peak cell density found by Cormier et al. 2006 and the oxygen consumption rate is as determined by zur Nieden et al. 2007.

The above calculations assume no mass transfer resistance from the bulk gas, through the liquid media to the cells, and that the consumption of oxygen by the cells is the limiting step. To instead consider the mass transfer of oxygen from the headspace to the liquid media as the limiting step, the following equation may be used to calculate the oxygen transfer rate:

$$OTR = k_L a(C_{sat} - C_B) \tag{10}$$

Where $k_L a$ is the volumetric mass transfer coefficient (h^{-1}), C_{sat} is the saturation concentration of oxygen in the liquid (mol/L) and C_B is the steady state bulk concentration of oxygen in the liquid (mol/L). The value for the mass transfer coefficient was the same as used by Millar (2009) ($k_L a = 2h^{-1}$). The solubility of oxygen (from standard air) in water used was $C_{sat} = 2.5 \times 10^{-4}$ mol/L or 8 mg/L (Lenntech 2009). The steady state bulk concentration was calculated using a mass balance, and found to be equal to 8.85 x 10^{-6} mol/L (0.28 mg/L). With these values, an oxygen transfer rate of 4.82 x 10^{-4} mol/Lh was found. Using this oxygen transfer rate, in combination with the amount of oxygen in the headspace calculated earlier in this section (2.95x10^{-5} mol), it was calculated that with the mass transfer of oxygen as the limiting step, there is enough oxygen in the microbioreactor to supply the media with oxygen for 244 hours (10 days). This value, while smaller than that found when the cell consumption was considered the limiting step, is still long enough for cell culture experiments, where cell counts are normally taken once per day, where the oxygen supply to the headspace would then be replenished.

Thus for the majority of cell culture experiments, a closed microbioreactor system provides a sufficient supply of oxygen. For long term experiments, or if cell samples were not taken more frequently than once per week, an oxygen supply may need to be considered as an alternative design.

4. Conclusions

Embryonic stem cells are considered to be a source of cells with tremendous potential for regenerative medicine, as they are capable of self-renewal and can differentiate into any cell type. The bioprocessing of these cells (i.e. creating reliable protocols for cell expansion and differentiation) is a critical component of developing therapies ready for clinical implementation. Suspension bioreactors are understood to be a scalable, reproducible and reliable method for cell expansion in the biotechnology industry. It is well established that a small-scale bioreactor system would be beneficial for stem cell research. Microbioreactor systems available in the literature, primarily developed for bacterial and yeast cultures, were reviewed and the applications for mammalian, or stem cell cultures were considered. However, the fluid environment at this scale remains largely uncharacterised, and bioreactor scale-down for cell culture has been shown to not be a linear process. Hydrodynamic studies and modeling of the shear and mass transfer environment at this small-scale would benefit this field in the development of a small-scale bioreactor applicable for embryonic stem cell expansion and differentiation.

5. References

Amit M, Chebath J, Margulets V, Laevsky I, Miropolsky Y, Shariki K, Peri M, Blais I, Slutsky G, Revel M and others. 2010. Suspension culture of undifferentiated human embryonic and induced pluripotent stem cells. Stem Cell Rev 6(2):248-59.

Baghbaderani B, A. , Mukhida K, Sen A, Hong M, Mendez I, Behie L, A. 2008. Expansion of Human Neural Precursor Cells in Large-Scale Bioreactors for the Treatment of Neurodegenerative Disorders. Biotechnology Progress 24(4):859-870.

Barrett TA, Wu A, Zhang H, Levy MS, Lye GJ. 2010. Microwell engineering characterization for mammalian cell culture process development. Biotechnol Bioeng 105(2):260-75.

Betts JI, Baganz F. 2006. Miniature bioreactors: current practices and future opportunities. Microb Cell Fact 5:21.

Betts JI, Doig SD, Baganz F. 2006. Characterization and application of a miniature 10 mL stirred-tank bioreactor, showing scale-down equivalence with a conventional 7 L reactor. Biotechnol Prog 22(3):681-8.

Chen A, Chitta A, Chang D, Amanullah A. 2008. Twenty-four well plate miniature bioreactor system as a scale-down model for cell culture process development. Biotechnology and bioengineering 102(1):148-160.

Cherry R, S., Kwon K-Y. 1990. Transient shear stresses on a suspension cell in turbulence. Biotechnology and bioengineering 36(6):563-571.

Cimetta E, Figallo E, Cannizzaro C, Elvassore N, Vunjak-Novakovic G. 2009. Micro-bioreactor arrays for controlling cellular environments: Design principles for human embryonic stem cell applications. Methods 47(2):81-89.

Cormier JT, zur Nieden NI, Rancourt DE, Kallos MS. 2006. Expansion of undifferentiated murine embryonic stem cells as aggregates in suspension culture bioreactors. Tissue Eng 12(11):3233-45.

Docherty K, Bernardo AS, Vallier L. 2007. Embryonic stem cell therapy for diabetes mellitus. Semin Cell Dev Biol. England. p 827-38.

Duetz WA. 2007. Microtiter plates as mini-bioreactors: miniaturization of fermentation methods. Trends Microbiol 15(10):469-75.

Elmahdi I, Baganz F, Dixon K, Harrop T, Sugden D, Lye GJ. 2003. pH control in microwell fermentations of S. erythraea CA340: Influence on biomass growth kinetics and erythromycin biosynthesis. Biochemical Engineering Journal 16(3):299-310.

Fernandes-Platzgummer A, Diogo MM, da Silva CL, Cabral JM. 2011. Large-scale expansion of mouse embryonic stem cells on microcarriers. Methods Mol Biol 690:121-34.

Figallo E, Cannizzaro C, Gerecht S, Burdick JA, Langer R, Elvassore N, Vunjak-Novakovic G. 2007. Micro-bioreactor array for controlling cellular microenvironments. Lab Chip 7(6):710-9.

Garcia-Ochoa F, Gomez E. 2009. Bioreactor scale-up and oxygen transfer rate in microbial processes: An overview. Biotechnology Advances 27(2):153-176.

Gilbertson J, A. , Sen A, Behie L, A., Kallos M, S. 2006. Scaled-up production of mammalian neural precursor cell aggregates in computer-controlled suspension bioreactors. Biotechnology and bioengineering 94(4):783-792.

Gill NK, Appleton M, Baganz F, Lye GJ. 2008. Design and characterisation of a miniature stirred bioreactor system for parallel microbial fermentations. Biochemical engineering journal 39(1):164-176.

Girard P, Jordan M, Tsao M, Wurm FM. 2001. Small-scale bioreactor system for process development and optimization. Biochem Eng J 7(2):117-119.

Gramer M, J., Britton T, L. 2002. Antibody production by a hybridoma cell line at high cell density is limited by two independent mechanisms. Biotechnology and bioengineering 79(3):277-283.

Harms P, Kostov Y, French JA, Soliman M, Anjanappa M, Ram A, Rao G. 2006. Design and performance of a 24-station high throughput microbioreactor. Biotechnol Bioeng 93(1):6-13.

Hortsch R, Stratmann A, Weuster-Botz D. 2010. New milliliter-scale stirred tank bioreactors for the cultivation of mycelium forming microorganisms. Biotechnol Bioeng 106(3):443-51.

Isett K, George H, Herber W, Amanullah A. 2007. Twenty-four-well plate miniature bioreactor high-throughput system: assessment for microbial cultivations. Biotechnol Bioeng 98(5):1017-28.

Islam RS, Tisi D, Levy MS, Lye GJ. 2007. Framework for the rapid optimization of soluble protein expression in Escherichia coli combining microscale experiments and statistical experimental design. Biotechnology Progress 23(4):785-793.

Kallos MS, Behie LA. 1999. Inoculation and growth conditions for high-cell-density expansion of mammalian neural stem cells in suspension bioreactors. Journal of Engineering and Applied Science 63(4):473-483.

Kehoe DE, Jing D, Lock LT, Tzanakakis ES. 2009. Scalable Stirred-Suspension Bioreactor Culture of Human Pluripotent Stem Cells. Tissue Engineering Part A 0(0).

Kostov Y, Harms P, Randers-Eichhorn L, Rao G. 2001. Low-cost microbioreactor for high-throughput bioprocessing. Biotechnol Bioeng 72(3):346-52.

Krawetz R, Taiani JT, Liu S, Meng G, Li X, Kallos MS, Rancourt DE. 2009. Large-Scale Expansion of Pluripotent Human Embryonic Stem Cells in Stirred-Suspension Bioreactors. Tissue Engineering Part C: Methods 16(4):573-582.

Kumar S, Wittmann C, Heinzle E. 2004. Minibioreactors. Biotechnol Lett 26(1):1-10.

Kusterer A, Krause C, Kaufmann K, Arnold M, Weuster-Botz D. 2008. Fully automated single-use stirred-tank bioreactors for parallel microbial cultivations. Bioprocess Biosyst Eng 31(3):207-15.

Kwon J, Kim BS, Kim MJ, Park HW. 2003. Suspension culture of hematopoietic stem cells in stirred bioreactors. Biotechnol Lett 25(2):179-82.

Legmann R, Schreyer HB, Combs R, G., McCormick E, L., Russo AP, Rodgers S, T. 2009. A predictive high-throughput scale-down model of monoclonal antibody production in CHO cells. Biotechnology and bioengineering 104(6):1107-1120.

Li X, Krawetz R, Liu S, Meng G, Rancourt DE. 2009. ROCK inhibitor improves survival of cryopreserved serum/feeder-free single human embryonic stem cells. Hum. Reprod. 24(3):580-589.

Luni C, Feldman HC, Pozzobon M, De Coppi P, Meinhart CD, Elvassore N. 2010. Microliter-bioreactor array with buoyancy-driven stirring for human hematopoietic stem cell culture. Biomicrofluidics 4(3).

Martin I, Smith T, Wendt D. 2009. Bioreactor-based roadmap for the translation of tissue engineering strategies into clinical products. Trends Biotechnol. England. p 495-502.

Micheletti M, Barrett T, Doig SD, Baganz F, Levy MS, Woodley JM, Lye GJ. 2006. Fluid mixing in shaken bioreactors: Implications for scale-up predictions from microlitre-scale microbial and mammalian cell cultures. Chemical Engineering Science 61(9):2939-2949.

Micheletti M, Lye GJ. 2006. Microscale bioprocess optimisation. Curr Opin Biotechnol. England. p 611-8.

Millar VF. 2009. Development of Microbioreactors for Embryonic Stem Cell Bioprocess Design [MSc. Thesis]. Calgary, Alberta: University of Calgary.

Murphy CL, Polak JM. 2002. Differentiating Embryonic Stem Cells: GAPDH, But Neither HPRT Nor Î²-Tubulin Is Suitable as an Internal Standard for Measuring RNA Levels. Tissue engineering 8(4):551-559.

Nagata S. 1975. Mixing: Principles and Applications: Wiley: New York

Niebruegge S, Bauwens C, Peerani R, Thavandiran N, Masse S, Sevaptisidis E, Nanthakumar K, Woodhouse K, Husain M, Kumacheva E and others. 2009. Generation of human embryonic stem cell-derived mesoderm and cardiac cells using size-specified aggregates in an oxygen-controlled bioreactor. Biotechnology and bioengineering 102(2):493-507.

Oh SKW, Chen AK, Mok Y, Chen X, Lim UM, Chin A, Choo ABH, Reuveny S. 2009. Long-term microcarrier suspension cultures of human embryonic stem cells. Stem Cell Research 2(3):219-230.

Olmer R, Haase A, Merkert S, Cui W, Palecek J, Ran C, Kirschning A, Scheper T, Glage S, Miller K and others. 2010. Long term expansion of undifferentiated human iPS and ES cells in suspension culture using a defined medium. Stem Cell Res. England: 2010 Elsevier B.V. p 51-64.

Park Y, Subramanian K, Verfaillie CM, Hu WS. 2010. Expansion and hepatic differentiation of rat multipotent adult progenitor cells in microcarrier suspension culture. Journal of Biotechnology 150(1):131-139.

Pollard DJ, Hunt G, Kirschner TK, Salmon PM. 2002. Rheological characterization of a fungal fermentation for the production of pneumocandins. Bioprocess and Biosystems Engineering 24(6):373-383.

Rezaei Larijani M, Seifinejad A, Pournasr B, Hajihoseini V, Hasani N, Totonchi M, Yousefi M, Shamsi F, Salekdeh GH, Baharvand H. 2011. Long-term Maintenance of

Undifferentiated Human Embryonic and Induced Pluripotent Stem Cells in Suspension. Stem Cells Dev.

Rodriguez J, V. , Pizarro MD, Scandizzi A, L., Guibert E, E., Almada L, L., Mamprin M, E. . 2008. Construction and Performance of a Minibioreactor Suitable as Experimental Bioartificial Liver. Artificial Organs 32(4):323-328.

Schroeder M, Niebruegge S, Werner A, Willbold E, Burg M, Ruediger M, Field LJ, Lehmann J, Zweigerdt R. 2005. Differentiation and lineage selection of mouse embryonic stem cells in a stirred bench scale bioreactor with automated process control. Biotechnology and Bioengineering 92(7):920-933.

Sen A, Kallos MS, Behie LA. 2001. Effects of Hydrodynamics on Cultures of Mammalian Neural Stem Cell Aggregates in Suspension Bioreactors. Industrial & Engineering Chemistry Research 40(23):5350-5357.

Sen A, Kallos MS, Behie LA. 2002. Expansion of mammalian neural stem cells in bioreactors: effect of power input and medium viscosity. Developmental Brain Research 134(1-2):103-113.

Singh H, Mok P, Balakrishnan T, Rahmat SN, Zweigerdt R. 2010. Up-scaling single cell-inoculated suspension culture of human embryonic stem cells. Stem Cell Res. England: 2010 Elsevier B.V. p 165-79.

Steiner D, Khaner H, Cohen M, Even-Ram S, Gil Y, Itsykson P, Turetsky T, Idelson M, Aizenman E, Ram R and others. 2010. Derivation, propagation and controlled differentiation of human embryonic stem cells in suspension. Nat Biotechnol. United States. p 361-4.

Tavares R, Xu K, Zhang C, de Freitas V. 2007. A specific and quick gene expression study in mouse ES cells. Journal of Assisted Reproduction and Genetics 24(8):366-372.

Vallejos JR, Kostov Y, Ram A, French JA, Marten MR, Rao G. 2006. Optical Analysis of Liquid Mixing in a Minibioreactor. Biotechnol Bioeng 93:906-911.

Watanabe K, Ueno M, Kamiya D, Nishiyama A, Matsumura M, Wataya T, Takahashi JB, Nishikawa S, Nishikawa S-i, Muguruma K and others. 2007. A ROCK inhibitor permits survival of dissociated human embryonic stem cells. Nat Biotech 25(6):681-686.

Weuster-Botz D, Puskeiler R, Kusterer A, Kaufmann K, John GT, Arnold M. 2005. Methods and milliliter scale devices for high-throughput process design. Bioproc Biosys Eng 28:109-119.

Wu MH, Urban JP, Cui Z, Cui ZF. 2006. Development of PDMS microbioreactor with well-defined and homogenous culture environment for chondrocyte 3-D culture. Biomed Microdevices 8(4):331-40.

Youn B, S., Sen A, Kallos M, S., Behie L, A., Girgis-Gabardo A, Kurpios N, Barcelon M, Hassell J, A. 2005. Large-Scale Expansion of Mammary Epithelial Stem Cell Aggregates in Suspension Bioreactors. Biotechnology Progress 21(3):984-993.

Yu Y, Li K, Bao C, Liu T, Jin Y, Ren H, Yun W. 2009. Ex vitro expansion of human placenta-derived mesenchymal stem cells in stirred bioreactor. Appl Biochem Biotechnol 159(1):110-8.

Zhang Z, Perozziello G, Boccazzi P, Sinskey AJ, Geschke O, Jensen KF. 2007. Microbioreactors for Bioprocess Development. JALA - Journal of the Association for Laboratory Automation 12(3):143-151.

zur Nieden NI, Cormier JT, Rancourt DE, Kallos MS. 2007. Embryonic stem cells remain highly pluripotent following long term expansion as aggregates in suspension bioreactors. Journal of Biotechnology 129(3):421-432.

6

Synthetic Surfaces for Human Embryonic Stem Cell Culture

Andrei G. Fadeev and Zara Melkoumian
Corning Incorporated
USA

1. Introduction

Human embryonic stem cells (hESCs) have two properties that distinguish them from other cell types: self-renewal, the ability to propagate indefinitely in culture, and pluripotency, the ability to differentiate into any type of specialized cells found in the human body. These properties provide the foundation for the development of hESC-derived cell-based therapeutics, where specific cell types derived by differentiation of hESCs become a therapeutic agent that cures the disease or restores the function of damaged organs or tissue. To make this a reality, several technologies must be developed to provide an unlimited and consistent supply of hESC-derived cells for clinical use. These include robust and scalable methods for production of undifferentiated hESCs, differentiation of the hESCs into desirable cell types, recovery, purification, storage and transportation of the derived cells to the location of use, and methods and techniques for delivery of the therapeutic cells to a human body to provide health benefits.

Since the derivation of the first hESC lines by Thomson, J. et al. (Thomson, 1998) and Reubinoff, B. et al. (Reubinoff et al., 2000), hundreds of new lines have been established and propagated under various cell culture conditions. Historically, hESCs were maintained in complex culture systems under poorly defined conditions comprising mouse or human feeder cell layers and medium containing fetal bovine serum (FBS) or serum replacement to provide an extracellular matrix (ECM)-rich environment for cell adhesion, as well as soluble growth factors for self-renewal. It is highly desirable that the cell culture systems utilized for therapeutic cells, including cell culture surfaces and the media, are well defined (all components are known and characterized and their abundance is controlled) and of non-animal origin or xeno-free (do not contain biological materials of a non-human nature).

Establishment of the first human embryonic stem cell line (Thomson, 1998) was accomplished by extending to hESCs a cell culture system developed for culturing mouse embryonic stem cells that is based on inactivated mouse embryonic fibroblasts (MEF) as a feeder layer. Soon after the first reports on isolation of human pluripotent cells came realization that feeder-free cell culture is essential for production of cells for transplantation (Donovan & Gearhart, 2001; Pera et al., 2000); (Pedersen, 2002). Back in 2000 this looked like a challenge that would require a very long time to overcome, as 19 prior years of using MEFs to support stem cell culture of non-human cells did not result in significant understanding of what exactly MEFs provide for stem cells. To make matters worse, there was experimental evidence showing that neither MEF conditioned medium nor ECM secreted by MEFs alone was sufficient to produce the same outcomes (Thomson, 1998). Ten years later, a fully synthetic cell culture surface (Corning® Synthemax™ Surface), which allows xeno- and serum-free culture and differentiation of therapeutic quality stem cells under chemically defined conditions, became commercially available to address this problem.

In this chapter, we will review the evolution of hESC culture conditions from complex un-defined feeder layer surfaces and FBS-containing medium used initially to derive and maintain hESC in culture, to fully defined, xeno-free culture systems described by multiple independent studies in the past few years. The focus of our review will be a development of an element of the cell culture system, a fully defined, xeno-free cell culture surface, and the benefits of a fully synthetic cell culture surface for propagation and differentiation of human embryonic stem cells.

Authors of this chapter are Corning Incorporated employees and have financial conflicts of interest.

2. Importance and significance of a fully defined cell culture system

The envisioned applications of hESCs, such as cell therapy and drug discovery, require growth and differentiation of hESCs on a large scale. Despite numerous ongoing clinical trials involving adult and human embryonic stem cells as a source of cells for a variety of therapies, there are fundamental challenges in large scale manufacture of hESCs that need to be addressed. These challenges include measurement metrics defining purity and the quality of the produced cells for a particular therapy application, complexity and variability of cell culture environment for different stem cells and their derivatives, and a need for automation of hESCs culture to bring the required consistency and repeatability into the cell therapy process (Thomas & Williams, 2009).

A recent review of cell culture procedures for hESCs (Fernandes et al., 2010) reports that in the vast majority (>80%) of cell culture studies published in 2009, hESCs were grown either on MEFs and cell lines derived from MEFs, or human feeder cells. In less than 18% of the studies the cells were cultured on other surfaces, ~16.5% of which used Matrigel™, an extracellular matrix isolated from Engelbreth-Holm-Swarm (EHS) mouse sarcoma (Becton Dickinson, Bedford, MA).

In the US, regulation of production and marketing of stem cell based therapies falls under FDA (Food and Drug Administration) jurisdiction. FDA repeatedly expressed concerns in regards to the potential use of stem-cell products derived from hESC that were isolated and cultured on MEFs (Halme & Kessler, 2006) and indicated the necessity of testing these cells for adventitious agents to meet FDA xenotransplantation guidelines. Another source of animal-derived biological material is non-human serum, which is frequently used in cell culture. FDA views fetal bovine serum (FBS) as a possible source of the prion that causes bovine spongiform encephalopathy and requires that FBS is produced in a country certified to be free of this disease (Kirschstein & Skirboll, 2001).

Cell culture surfaces, as a part of the cell culture system, should meet FDA guidelines and address hESCs process scalability issues. The desirable attributes for hESC cell culture surfaces for an animal product-free cell culture are an absence of animal derived materials and ideally even human derived materials (as those too may be subject to batch-to-batch variability and contamination with human pathogens), and compatibility with a defined, serum-free media. Other typical cell culture surface attributes include absence of extractables, particulates, and other contaminants, compatibility with conventional sterilization techniques (typically gamma irradiation), stability at room temperature, similarity to common tissue culture treated (TCT) polystyrene and other cell culture surfaces, and, ideally, the surface needs to be ready for use, rather than require further preparation and manipulation.

3. Progress in hESC culture conditions

The first significant improvement in hESC culture conditions was reported by Geron Corporation in 2001 (Xu et al., 2001). When Matrigel and gelatine were tested as surfaces for hESC lines H1, H7, H9 and H14 using MEF conditioned medium, all cells attached to Matrigel, formed tight colonies separated by differentiated cells, and propagated. In contrast, the cells on gelatine showed poor survival and appeared differentiated after one passage. Since the major constituents of Matrigel are laminin, collagen IV, heparan sulfate proteoglycans, and entactin/nidogen (Kleinman et al., 1982, Kleinman, 1986 #244), the authors then explored individual ECM components of Matrigel and observed great similarity between Matrigel and laminin cultured cells.

Further investigation of hESC culture using media conditioned with cells other than MEFs (STO, an immortal mouse embryonic fibroblast; NHG190, a transfected mouse embryonic cell line; BJ5ta, an immortalized human foreskin fibroblast cell line; an immortalized human retinal epithelial cell line) revealed that only Matrigel (or laminin) and MEF conditioned medium maintain long-term propagation of undifferentiated hESCs, as cells under other conditions differentiated within 1 to 7 passages. There was no difference found in integrin expression profiles between cells cultured on MEFs, Matrigel, or laminin, and cells under all three of these cell culture conditions expressed $\alpha6$ and $\beta1$ integrins consistent with integrin mediated hESC attachment to laminin. The cultured cells were positive for Tra-1-60 and Tra-1-81, negative for SSEA-1, and formed teratomas with various differentiated cells in immunodefficient mice. These data demonstrated the capability of a MEF-free surface to support proliferation of hESCs in an undifferentiated state.

While Matrigel was a significant improvement over MEFs, its mouse origin and undefined and variable nature makes it highly undesirable as a substrate for therapeutic hESC scale up. It was also found that current hESC lines derived on MEF and cultured on Matrigel in the presence of animal serum have nonhuman sialic acid Neu5Gc, a potential immunogen, incorporated into their cell membranes (Martin et al., 2005). Exposure of the cells to human sera with antibodies specific for Neu5Gc resulted in binding of immunoglobulin and complement deposition, which would lead to cell killing in vivo, thus significantly reducing the efficacy of cell therapy. After reviewing several approaches, Martin et al. concluded that it is easier and safer to start over and derive hESC lines without exposing cells to animal products containing Neu5Gc.

The elimination of animal derived products from hESC culture requires a xeno-free cell culture surface and medium. The interdependence of the cell culture surface and medium in providing optimal conditions for cell culture performance made progress in this field iterative. Short term solutions involved human feeder cells (Genbacev et al., 2005), (Choo et al., 2004), and serum replacement or serum-free medium (Amit et al., 2000).

The next significant milestone in development of hECS culture was achieved in 2005 (Xu et al., 2005), when Geron demonstrated proliferation of H7 and H9 hESCs in serum-free medium on Matrigel. Geron evaluated 15 growth factors and their combinations at various doses in a medium containing 80% Knockout–Dulbecco's modified Eagle's medium, 20% Knockout™ serum replacement (Invitrogen, Carlsbad, CA), 1 mM L-glutamine, 0.1 mM β-mercaptoethanol, 1% nonessential amino acids. Although the exact composition of Knockout™ formulation is a commercial secret, the product is a more-defined growth supplement (Price et al., 1998) that reduces the spontaneous differentiation of ESCs. Basic fibroblast growth factor (bFGF) was found to support growth of hESCs without conditioned medium.

This was followed up by a study conducted at Prof. J. A. Thomson's lab that disclosed a complete composition of serum- and xeno-free medium, TeSR1, for hESC culture (Ludwig et al., 2006). The medium is based on DMEM/F12 medium supplemented with bFGF, TGF-β1, LiCl, γ-aminiobutyric acid, and pipecolic acid. It contains only six polypeptides (Table 1) and none of them is of animal origin. Ludwig et al. also screened a variety of human ECM proteins individually and in combination to establish that a substrate coated with a mixture of collagen IV, fibronectin, laminin, and vitronectin can support growth of H1 and H9 hESC lines for at least 7 months. The stem cells cultured under these conditions did not test positive for the presence of nonhuman sialic acid. Despite these advances, the importance of further optimization of the cell culture surface was recognized, as the purified human matrix components are expensive and could potentially be contaminated with human pathogens.

When compared to different human ECM proteins, as well as human and animal sera matrices (Hakala et al., 2009), Matrigel was found to be a superior hESC culture surface. The comparison was performed using conventional and modified hESC culture media that included human foreskin fibroblast-conditioned culture medium, chemically defined mTeSR®1 medium and its xeno-free counterpart, TeSR1 medium. Judging by the maximum passage number attainable, hESC morphology, and expression of stem cell markers, Hakala concluded that a xeno-free, fully defined, and reproducible feeder cell-free hESC culture method still remained to be developed.

Polypeptide	Conc. mM
bFGF	$5.77\ 10^{-6}$
TGF β1	$2.35\ 10^{-8}$
Human Insulin	$3.92\ 10^{-3}$
Human Holo-Transferrin	$1.37\ 10^{-4}$
Human Serum Albumin	$1.95\ 10^{-1}$
Glutathione	$6.38\ 10^{-3}$

Table 1. Polypeptides in TeSR1, (Ludwig et al., 2006)

4. Synthetic surfaces for hESCs

Typically cells are cultured on a polystyrene surface that is rendered hydrophilic (e.g. TCT and Corning® CellBIND® culture ware) by various plasma treatments. It is recognized that cell attachment occurs through interactions between integrins, cell adhesion receptors, and extracellular matrix proteins. To enable cell attachment to a surface, ECM proteins need to be immobilized on that surface first.

The proteins adsorb onto a surface from the culture medium, in which case serum supplement is a source of ECM proteins. However, when hESCs are cultured on a TCT

surface in serum supplemented medium (80% Dulbecco modified Eagle's medium supplemented with 20% fetal bovine serum), they undergo differentiation within a passage or two (Thomson, 1998). Alternatively, the ECM could be secreted by the cultured cells. Unlike other cell types, hESCs do not secrete a sufficient amount of ECM to sustain themselves in an undifferentiated state (Braam et al., 2008). Therefore, development of a synthetic surface for hESC culture under serum free conditions is not a trivial task (Couture, 2010), (Elefanty & Stanley, 2010).
The approaches to develop a solution to the problem described above can be grouped into three categories. First is to design a surface that facilitates ECM protein adsorption. The second is to identify a specific extracellular matrix protein that could be manufactured at a significant scale and coated onto conventional cell culture surfaces. The third approach entails development of a biomimetic surface that mimics the function of extracellular matrix proteins.

4.1 Polymer surfaces facilitating ECM protein adsorption.

Attempts to develop a surface for hESC culture using conventional polymers involved investigation of improved processing methods for polystyrene (Mahlstedt et al.). When TCPS (tissue culture polystyrene) was etched using radio frequency oxygen plasma for 5 min at pressures below 20 mT, HUES7 and NOTT1 hESC lines attached and proliferated on the surface for at least 10-14 passages in MEF conditioned medium showing doubling times similar to those for cells cultured on Matrigel and expressing hESC pluripotency markers. BD Primaria™ and Corning® CellBIND® surfaces were tested in a similar experiment as controls and also demonstrated the ability to support hESC culture to some extent. Quartz crystal microbalance analysis showed an increase in the amount of adsorbed proteins from conditioned medium onto the plasma etched surface, which was related to the improved cell culture performance. The proposed underlying mechanism is an increased hydrophilicity of the surface that enables replacement of small albumin, which is the first protein to adsorb onto the surface due to its abundance in serum, over time with less abundant, higher molecular weight, cell adhesive proteins. Attempts to culture hESCs on this surface using mTeSR1 or StemPro® hESC defined media were unsuccessful, as these media do not contain cell adhesion ECM proteins.
Similar material properties were achieved through organic synthesis. Villa-Diaz (Villa-Diaz et al., 2010) identified a polymer, poly[2-(methacryloyloxy)ethyl dimethyl-(3-sulfopropyl)ammonium hydroxide, which, when coated on polystyrene, supported the undifferentiated proliferation of BG01 and H9 hESC lines for over 25 passages in MEF-conditioned medium. The surface did not support hESC culture in mTeSR1 medium, while mixed result were observed in StemPro medium, the composition of which is not disclosed.
A more recent study (Brafman et al., 2010) described screening of 90 polymers for their ability to support hESC culture. Brafman identified only one polymer, poly(methyl vinyl ether-alt-maleic anhydride) (PMVE-alt-MA), that HUES1 and HUES9 cells attached to and proliferated for five passages in StemPro medium. The expression levels of endogenous ECM proteins and integrins in hESCs grown on PMVE-alt-MA were found to be significantly higher compared to hESCs grown on Matrigel, suggesting that the cells were generating ECM proteins required for their attachment to the surface.

4.2 Extracellular matrix protein coatings

Multiple publications demonstrated that Matrigel could be replaced, under certain conditions, with a mixture of collagen, fibronectin, laminin, and vitronectin (Ludwig et al.,

2006), or even individual ECM proteins including laminin, fibronectin, or vitronectin (Table 2). As evidenced by the data, conditioned medium can support hESC proliferation in an undifferentiated state on most common ECM proteins. However, transition to defined media such as mTeSR1, TeSR1, X-VIVO™ 10 seems to restrict the choice to laminin or vitronectin.

Laminin is a major constituent of Matrigel, and it was the first pure ECM protein identified to support hESC attachment and proliferation of the hESCs (Xu et al., 2001). Fairly recently, a detailed study by Braam et al. demonstrated the ability of vitronectin to support hESC culture in mTeSR1 medium and investigated the underlying mechanism (Braam et al., 2008). E-Cadherin, a cell-cell adhesion glycoprotein, also was shown to enable attachment and propagation of H9 cells in mTeSR1 medium (Nagaoka et al., 2010).

Protein	Cell line	Medium	Passages	Reference
Laminin Collagen Fibronectin	H1,H9 H1 H1	MEF conditioned	6-7 6 6	(Xu et al., 2001),
Laminin	H1	X-VIVO 10 + GF	40	(Li et al., 2005)
Laminin	KhES-1, KhES-2, hES-3	MEF conditioned	10	(Miyazaki et al., 2008)
Laminin	HS420, HS207, HS401, H1, H9	Variants of TeSR1 and mTeSR1	5-20	(Rodin et al., 2010)
Fibronectin	H9, BG01	HESCO	8	(Lu et al., 2006)
Fibronectin	I-3, I-6, H9	Knockout SR	>50	(Amit et al., 2004)
Collagen I	H1, H9	SDEC conditioned	5	(Jones et al. 2010)
Collagen I	HUES-1 Shef 1	hESF9	21 15	(Furue et al., 2008)
Vitronectin	MEL1, MEL2, hES1	StemPro	>10	(Prowse et al., 2010)
Vitronectin	HUES1, HES2, HESC-NL3	mTeSR1	12	(Braam et al., 2008)
Vitronectin	H9	mTeSR1	10	(Rowland et al., 2010)
Vitronectin	HES-3, H1	mTeSR1	>30	(Yap et al., 2011)
Vitronectin	CHA6, H9	mTeSR1	>30	(Yoon et al., 2010)
E-Cadherin	H9	mTeSR1	37	(Nagaoka et al., 2010)

Table 2. ECM surfaces for hESC culture.

Miyazaki et al., (Miyazaki et al., 2008) investigated attachment of three hESC lines (KhES-1, KhES-2, and KhES-3) to various recombinant laminins (511, 411, 332, 211, and 111) in MEF conditioned medium and observed a significant discrepancy in cell attachment to these laminins with laminin 332 showing the best performance across these cell lines. Others reported recombinant laminin-511 to be a good substrate for HS420, HS207, and HS401 cell lines (Rodin et al., 2010).

Although extracellular matrix protein coatings are frequently used for hESC culture in research labs, application of these coatings to hESC scale up is not straightforward. Recombinant proteins are fairly expensive to produce and purify. They are prone to batch-to-batch variabity, need to be coated onto a surface under aseptic conditions, and they can degrade or denature upon dehydration. Published protocols indicate that coatings of purified proteins require optimization (Yap et al., 2011); for example, a defined thickness threshold needs to be achieved to enable hESCs attachment and proliferation rates similar to those observed on Matrigel coatings. Formation of such a coating requires a significant period of time (hours) under aseptic conditions. Utilization of the protein for the coating is very inefficient, as >50% of it remains in the solution. All this makes purified protein cell culture surfaces expensive and limits their scalability.

4.3 Biomimetic surfaces for stem cells

The interest in synthetic biomimetic surfaces was initiated by a discovery of the RGD (arginine-glycine-aspartic acid) peptide sequence present in most ECM proteins. When coupled to a surface, the RGD sequence promotes cell attachment (Pierschbacher & Ruoslahti, 1984). A large variety of polymer materials incorporating RGD peptides have been designed and studied (Hersel et al., 2003).

Cells attach to the ECM through integrins, cell adhesion receptors (Humphries et al., 2006), (Hynes, 2002). Integrins support a broad spectrum of cellular functions including proliferation and differentiation, and can bind to such ECMs as collagen, fibronectin, laminin, vitronectin, and N-linked glycoproteins, (e.g. osteopontin and bone sialoprotein). They can also interact with other cells through vascular- or intracellular- cell adhesion molecules (VCAM and ICAM). The integrins are heterodimeric molecules consisting of α and β subunits. There are twenty four known $\alpha\beta$ combinations, which can be divided into several subfamilies based on evolutionary relationships. The orthologs of human integrins recognizing the RGD amino acid sequence and laminins can be traced back to primitive organisms, while other integrins must have evolved with increasing organism complexity.

From such a perspective, it is not surprising that hESCs primarily express integrins recognising the RGD sequence and laminin (Braam et al., 2008), (Meng et al., 2010), (Rowland et al., 2010), (Prowse et al., 2011). Braam et al., evaluated integrin expression in three hESC lines (HES2, HUES1, and HESCNL3) using fluorescence-activated cell sorting analysis. While integrin chains $\alpha1$, $\alpha4$, $\alpha10$, and $\beta3$ and integrins $\alpha9\beta1$ and $\alpha V\beta6$ were either not detected or detected at very low levels, cells expressed integrin chains $\alpha2$, $\alpha3$, $\alpha5$, $\alpha6$, $\alpha11$, $\beta1$ and integrin $\alpha V\beta5$. This leaves the following combinations of integrin chains to enable cell attachment: $\alpha2\beta1$, $\alpha3\beta1$, $\alpha5\beta1$, $\alpha6\beta1$, $\alpha11\beta1$, $\alpha V\beta5$, and $\alpha V\beta1$. These combinations, designated in Figure 1 as small orange circles, comprise cell surface receptors for all major ECM components.

It was also found, as discussed in the previous section, that attachment through either laminin or RGD binding integrins (Li et al., 2005), (Rodin et al., 2010), (Wong et al., 2010), (Braam et al., 2008), (Rowland et al., 2010), (Yoon et al., 2010), (Yap et al., 2011) is sufficient to support undifferentiated proliferation of hECS in serum-free, chemically defined medium.

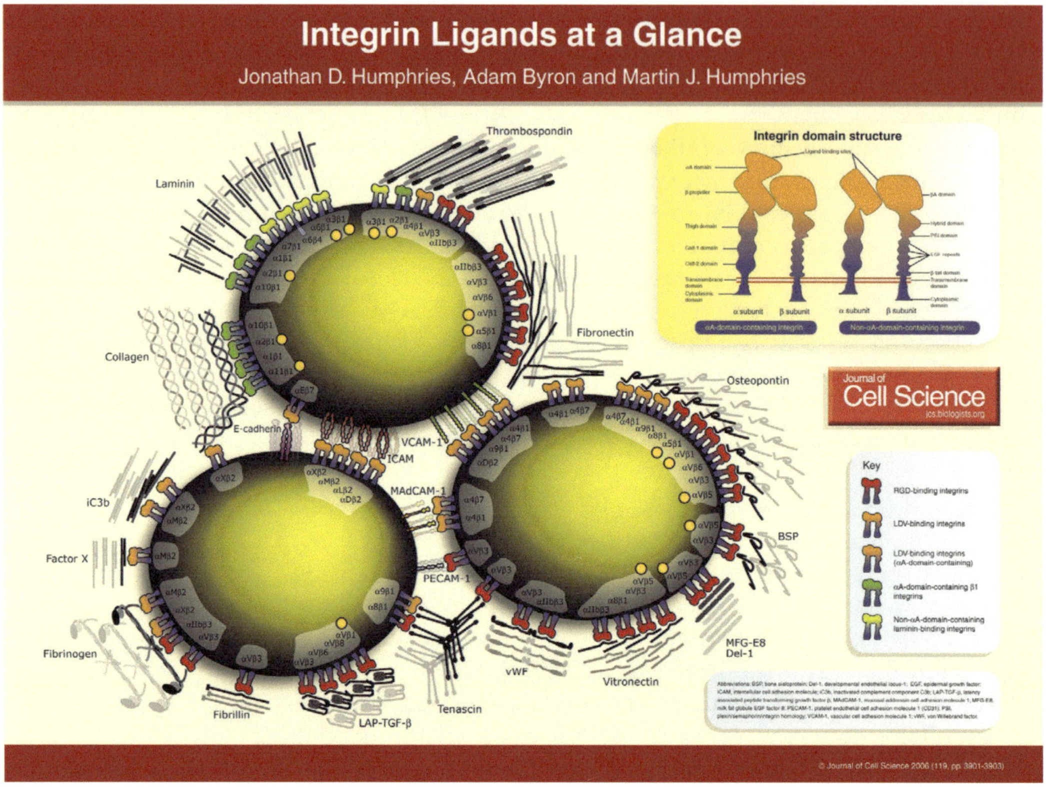

Fig. 1. Mammalian intergrins and their ligands (Humphries et al., 2006)

For these reasons, surfaces with immobilized laminin and RGD peptides were studied as synthetic surfaces for hESC proliferation (Li et al., 2006), (Derda et al., 2007), (Melkoumian et al., 2010) (Derda et al., 2010), (Klim et al., 2010). Li et al. was first to demonstrate that grafting the RGD containing peptide, Ac-KGGNGEPRGDTYRAY, from bone sialoprotein to acrylamide-acrylic acid copolymer can support short term attachment and proliferation of HSF-6 hESCs in MEF conditioned KSR-supplemented medium. The colonies of hESCs expressed Oct-4 and SSEA-4 markers of undifferentiated hESCs after five days in culture and maintained the same morphology as those cultured under the same conditions on Matrigel.

Derda et al. extensively studied laminin-derived peptides (Derda et al., 2007) in an attempt to identify specific sequences supporting hESC attachment. Using a self-assembled monolayers technique, the authors discovered several peptides enabling attachment and short term proliferation (6 days) of hESCs, again though, in MEF conditioned medium. The most important finding, which facilitated further advancement in the field, was an observation that hESCs, in comparison to other cells, require relatively high surface peptide densities due to low expression levels of corresponding integrins.

This suggests a hypothesis where a critical number of integrins on the cell surface are needed to be simultaneously engaged to enable optimal hESC adhesion. At low surface peptide density the probability of such engagement is reduced. We (Melkoumian et al., 2010) focused efforts on peptide-acrylate surfaces (PASs), as did others (Li et al., 2006), and pursued high peptide densities through optimization of acrylate polymer and peptide conjugation. The developed poly(hydroxyethyl methacrylate-co-carboxyethyl acrylate) coating provides a

significant amount of functional groups for peptide conjugation, while its swellable nature enhances presentation and accessibility of the conjugated peptide for integrin binding.
Co-conjugating a small amount of fluorescently labelled peptide together with a cell binding peptide and comparing the fluorescence intensities to a calibration curve generated by drying down the same mixture of peptides onto the acrylate surface, we estimated that a BSP peptide (Table 3) density of 6-9 pmol/mm^2 is sufficient to enable cell culture performance in X-VIVO 10 medium similar to Matrigel (Melkoumian et al., 2010), (Figure 2).
An alternative approach in enabling hESC adhesion and proliferation was demonstrated by Kohlar, et al., (Kolhar et al., 2010), who employed cyclic RGD peptides to improve the strength of peptide - integrin binding to achieve a similar outcome. The reported peptide densities, 0.1–0.3 fmol/mm^2, are significantly lower. Nevertheless, stronger integrin binding to cyclic peptide allows development of a sufficient number of peptide-integrin interactions to enable cell attachment and proliferation.

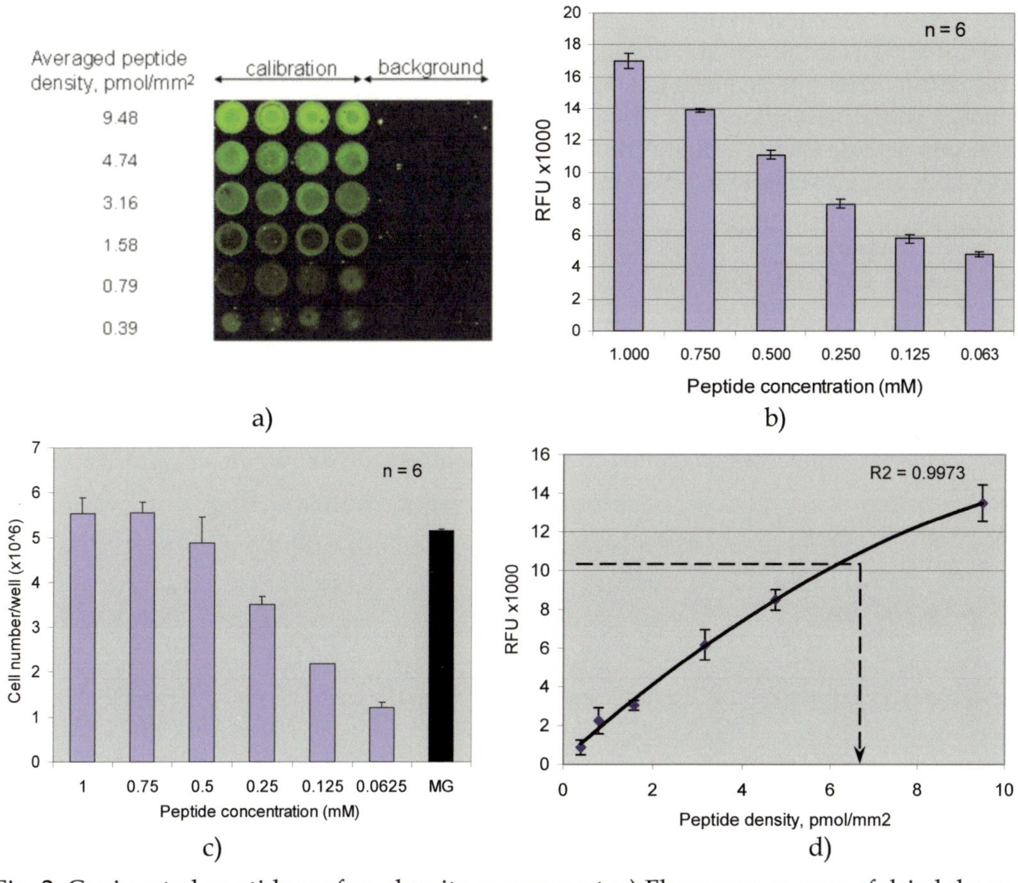

Fig. 2. Conjugated peptide surface density assessment; a) Fluorescence scan of dried down peptides representing various peptide densities; b) Fluorescence of conjugated peptide mixture; c) BSP peptide concentration (mM)-dependent H7 cell number after 5 days in culture in X-VIVO™ 10 + 80 ng/ml bFGF + 0.5 ng/ml TGFβ1 medium on PAS in 6-well plate, conjugated with serial dilution of BSP peptide. Cell seeding density was 1 x 10^6 cells per well; d) Peptide density calibration curve; (Melkoumian et al., 2010), supplementary Figure 1A and Figure 1B.

When a series of RGD peptides (Table 3) from various ECM proteins (bone sialoprotein, vitronectin, and fibronectin) was conjugated to the acrylate coating and tested for their ability to support hESC culture, only BSP and VN peptides supported attachment and proliferation of H1 and H7 hESCs similar to Matrigel (Melkoumian et al., 2010). None of the fibronectin peptides provided adequate cell attachment. According to Humphries et al., BSP protein engages the same integrins, αVβ5 and αVβ3, as vitronectin suggesting the same integrin activation by both BSP and VN peptides, (see Figure 1). Braam et al. (Braam et al., 2008) reported binding of hESCs to vitronectin through αVβ5 integrin and showed it being sufficient for cell self-renewal under chemically defined conditions. Since αVβ3 integrin was not detected in that study, it is highly likely that BSP and VN peptides enable hESC attachment through αVβ5 integrin, as it is the only remaining choice for BSP peptide, (Figure 1). This interaction is fairly specific and that specificity is governed by RGD flanking sequences.

Peptide based cell culture surfaces are not limited to integrin interactions, but can also be designed to take advantage of multiple attachment mechanisms. Recently Klim at el., (Klim et al., 2010) demonstrated self-assembled monolayer peptide biomimetic surfaces that included peptides binding to glycosaminoglycans. The cell culture surface presenting vitronectin peptide GKKQRFRHRNRKG was found to support attachment and proliferation of H9, H13, and H14 cells in mTeSR1 medium in the presence of ROCK inhibitor.

Sequence	Abbreviation	Reference
Ac-KGGNGEPRGDTYRAY	BSP	(Oldberg et al., 1988)
Ac-KGGPQVTRGDVFTMP	VN	(Suzuki et al., 1985)
GRGDSPK	FN-1	(Pierschbacher & Ruoslahti, 1984)
Ac-KGGAVTGRGDSPASS	FN-2	(Pierschbacher & Ruoslahti, 1984)
BSP, bone sialoprotein; VN, vitronectin; FN, fibronectin.		

Table 3. Peptide sequences conjugated to acrylate coatings; (Melkoumian et al., 2010)

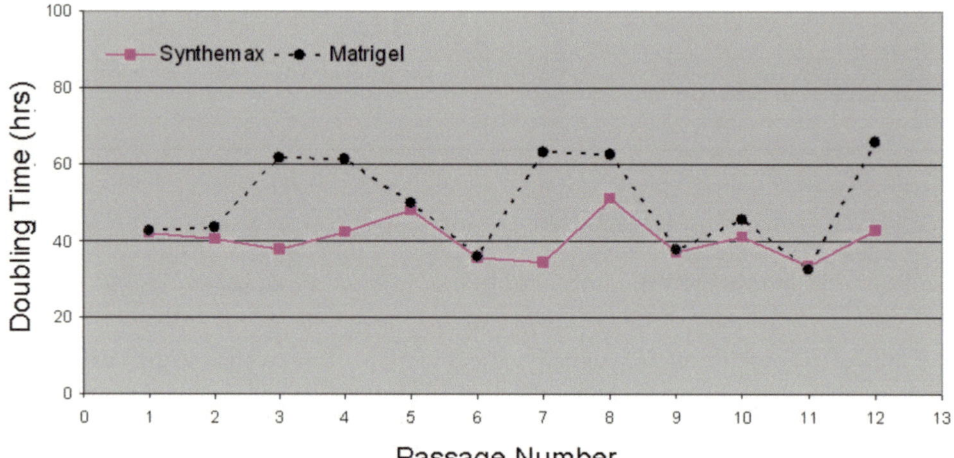

Fig. 3. Consistency of doubling time for H7 cells grown on Corning® Synthemax™ and Matrigel™.

The fact that hESCs can be cultured under defined conditions on a variety of substrates, from complex ECM protein mixtures to single peptide surfaces, demonstrates the native ability of undifferentiated hESCs to propagate without the need for a very precise and complex signalling through adhesion receptors (Klim et al., 2010), (Ying et al., 2008).
One of the desirable and anticipated advantages of synthetic surfaces is improved consistency and reproducibility of cell culture performance. Non-biological, small molecular weight compounds required for manufacture of synthetic surface can be consistently obtained in high quantity and purity and can be manipulated in a conventional industrial setting. When compared to proteins, short peptides can be easily manufactured via chemical synthesis. Purification and quality assurance are dramatically less complex for peptides. Short peptides do not denature and exhibit higher stability towards chemical and physical agents, thus, overall providing a scalable manufacturing platform for hESC cell culture surfaces, enabling cell therapy applications. This will translate into industrial processes resulting in consistent cell culture surface, as it did for tissue culture treated polystyrene surfaces. We observed a clear improvement in cell culture consistency at the early stages of surface development (Figure 3).

Cell Line	Culture Medium	Surface	Doubling Time (h)	%Oct-4 + Cells	%Tra1-60 + cells	%SSEA-4 + cells	Passages
H7	X-VIVO 10	Synthemax	41 ± 5	92 ± 4	85	100	21
		Matrigel	50 ± 12	92 ± 3	96	100	
	mTeSR1	Synthemax	55 ± 21	90 ± 4	66	-	11
		Matrigel	69 ± 26	83 ± 8	56	-	
	StemPro	Synthemax	31 ± 6	96 ± 0	-	-	5
		Geltrex	31 ± 5	97 ± 0	-	-	
	NutriStem™	Synthemax	32 ± 9	92 ± 3	-	-	5
		Matrigel	35 ± 4	92 ± 6	-	-	
H1	X-VIVO 10	Synthemax	46 ± 8	82 ± 5	78 ± 4	91	14
		Matrigel	53 ± 16	80 ± 8	83 ± 6	84	
H9	mTeSR1	Synthemax	43 ± 3	93 ± 5	73 ± 4	-	5
		Matrigel	44 ± 6	95 ± 4	76 ± 2	-	
BG01v	mTeSR1	Synthemax	40 ± 6	86 ± 6	-	-	11
		Matrigel	44 ± 6	89 ± 5	-	-	

Table 4. hESC long-term expansion on Synthemax™ in defined media: X-VIVO™ 10 (Lonza, + 80ng/ml bFGF + 0.5ng/ml TGFb1); mTeSR®1 (Stem Cell Technologies); NutriStem™ (StemGent); StemPro® hESC SFM (Invitrogen).

Since the publication in Nature Biotechnology (Melkoumian et al., 2010) gamma sterilization methods were developed for the peptide acrylate surface and the surface was commercialized under the Corning® Synthemax™ Surface trade name. It is being offered in two grades (therapeutic and research) in a number of formats including 6-well plates, T-75 and T-225 flasks. The utility of the surface has been demonstrated for additional cell lines including H9 and BGO1v. In addition to X-VIVO 10 and mTeSR1, the surface was found compatible with StemPro and NutriStem media. Multipassage hESC data on Synthemax surface is summarized in Table 4.

For hESC-based therapeutics it is critical to have a defined, scalable culture system for both the expansion and differentiation stages of cell production. The picture would not have been complete without testing Synthemax surface for hESC differentiation. After passaging cells 14 times on PAS surface, cells were treated with Activin A and BMP4 as previously described (Laflamme et al., 2007) to investigate differentiation into cardiomyocytes. The experiment resulted in spontaneously beating cell aggregates expressing the cardiomyocyte-specific markers, Nkx2.5 and α-actinin, and shows that the BSP peptide acrylate surface supports both expansion and differentiation of hESC to therapeutic progenitor cells, (see Figure 4).

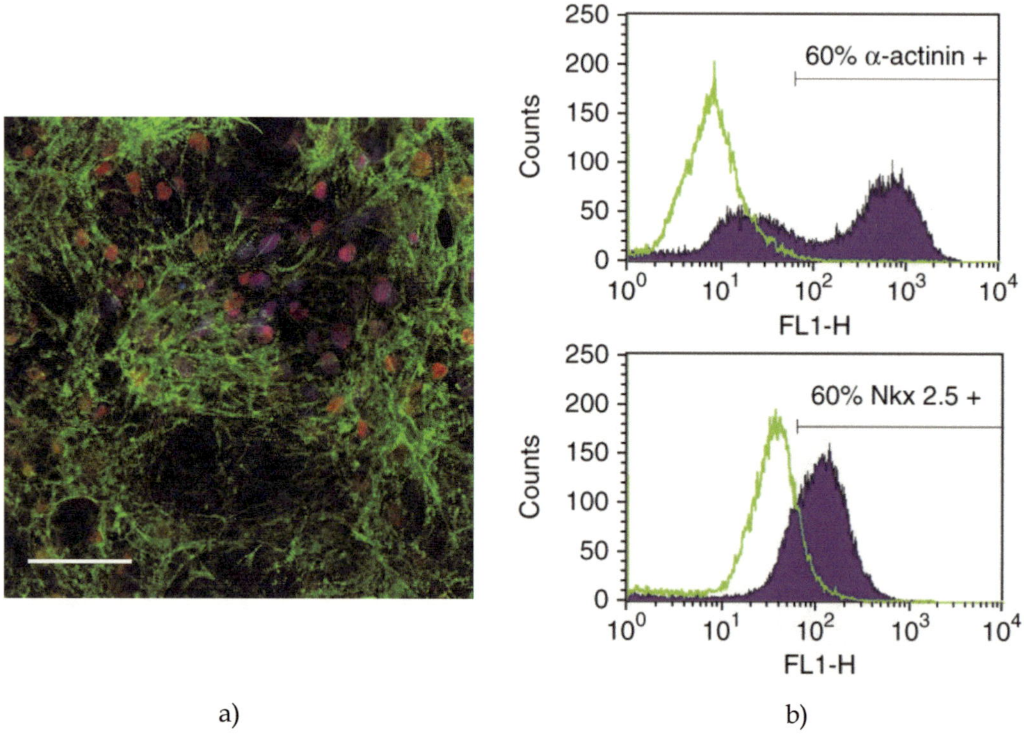

Fig. 4. Cardiomyocyte differentiation of H7 hESC on BSP-PAS surface: Immunofluorescent staining (a) and flow cytometry analyasis (b) of differentiated cells for cardiomyocyte-specific markers, Nkx 2.5 shown in red and α-actinin shown in green.

5. Conclusions

Although a variety of recombinant protein coatings in combination with a defined serum-free cell culture medium have been shown to provide a good cell culture system for hESCs, the limited scalability of most of these systems limits their use for cell therapy applications. Peptide acrylate surfaces based on a pure synthetic chemistry approach offer consistency, reproducibility, scalability, safety, and eventually lower cost for hESC-derived cell therapy applications.

6. Acknowledgements

The authors acknowledge the contribution of the following people in various aspects of the Corning® Synthemax™ surface development and commercialization: Y. Zhou, D. M. Weber, J. L. Weber, P. J. Dolley-Sonneville, J. Yang, L. Qiu, C. A. Priest, C. B. Shogbon, A. W. Martin, J. Nelson, P. West, J. P. Beltzer, S. Pal, R. Brandenberger, S. Caracci, C. Philip, J. Mooney, T. Goodrich, P. Knowles, J. Botelho, M. Rothenberg, J. Eickmann, A. Ferrie, M. McFarland, J. Lebkowski, A. H. Davis, D. J. Earp, A. Frutos, M. Lewis, J. Morley, T. Garvey, D. Davenport, P. Gagnon, P. Szlosek, V. Ravichandran, C. Wolcott, A. Fusco, D. T. Henk, K. Delavan-Boorsma, S. Edell,

7. References

Amit, M., M. K. Carpenter, M. S. Inokuma, C. P. Chiu, C. P. Harris, M. A. Waknitz, J. Itskovitz-Eldor & J. A. Thomson (2000). Clonally derived human embryonic stem cell lines maintain pluripotency and proliferative potential for prolonged periods of culture. *Developmental Biology* 227(2): 271-278.

Amit, M., C. Shariki, V. Margulets & J. Itskovitz-Eldor (2004). Feeder Layer- and Serum-Free Culture of Human Embryonic Stem Cells. *Biology of Reproduction* 70(3): 837-845.

Braam, S. R., L. Zeinstra, S. Litjens, D. Ward-van Oostwaard, S. van den Brink, L. van Laake, F. Lebrin, P. Kats, R. Hochstenbach, R. Passier, A. Sonnenberg & C. L. Mummery (2008). Recombinant Vitronectin Is a Functionally Defined Substrate That Supports Human Embryonic Stem Cell Self-Renewal via αVβ5 Integrin. *Stem Cells* 26(9): 2257-2265.

Brafman, D. A., C. W. Chang, A. Fernandez, K. Willert, S. Varghese & S. Chien (2010). Long-term human pluripotent stem cell self-renewal on synthetic polymer surfaces. *Biomaterials* 31(34): 9135-9144.

Choo, A. B. H., J. Padmanabhan, A. C. P. Chin & S. K. W. Oh (2004). Expansion of pluripotent human embryonic stem cells on human feeders. *Biotechnology and Bioengineering* 88(3): 321-331.

Couture, L. A. (2010). Scalable pluripotent stem cell culture. *Nature Biotechnology* 28(6): 562-563.

Derda, R., L. Li, B. P. Orner, R. L. Lewis, J. A. Thomson & L. Kiessling (2007). Defined substrates for human embryonic stem cell growth identified from surface arrays. *ACS Chemical Biology* 2(5): 347-355.

Derda, R., S. Musah, B. P. Orner, J. R. Klim, N. Li & L. L. Kiessling (2010). High-throughput discovery of synthetic surfaces that support proliferation of pluripotent cells. *Journal of the American Chemical Society* 132(4): 1289-1295.

Donovan, P. J. & J. Gearhart (2001). The end of the beginning for pluripotent stem cells. *Nature* 414(6859): 92-96.

Elefanty, A. G. & E. G. Stanley (2010). Defined substrates for pluripotent stem cells: Are we there yet? *Nature Methods* 7(12): 967-968.

Fernandes, A. M., T. Meletti, R. Guimarães, M. P. Stelling, P. A. N. Marinho, A. S. Valladão & S. K. Rehen (2010). Worldwide survey of published procedures to culture human embryonic stem cells. *Cell Transplantation* 19(5): 509-523.

Furue, M. K., J. Na, J. P. Jackson, T. Okamoto, M. Jones, D. Baker, R. I. Hata, H. D. Moore, J. D. Sato & P. W. Andrews (2008). Heparin promotes the growth of human embryonic stem cells in a defined serum-free medium. *Proceedings of the National Academy of Sciences of the United States of America* 105(36): 13409-13414.

Genbacev, O., A. Krtolica, T. Zdravkovic, E. Brunette, S. Powell, A. Nath, E. Caceres, M. McMaster, S. McDonagh, Y. Li, R. Mandalam, J. Lebkowski & S. J. Fisher (2005). Serum-free derivation of human embryonic stem cell lines on human placental fibroblast feeders. *Fertility and Sterility* 83(5): 1517-1529.

Hakala, H., K. Rajala, M. Ojala, S. Panula, S. Areva, M. Kellomäki, R. Suuronen & H. Skottman (2009). Comparison of biomaterials and extracellular matrices as a culture platform for multiple, independently derived human embryonic stem cell lines. *Tissue Engineering - Part A* 15(7): 1775-1785.

Halme, D. G. & D. A. Kessler (2006). FDA regulation of stem-cell-based therapies *New England Journal of Medicine* 355(16): 1730-1735.

Hersel, U., C. Dahmen & H. Kessler (2003). RGD modified polymers: Biomaterials for stimulated cell adhesion and beyond. *Biomaterials* 24(24): 4385-4415.

Humphries, J. D., A. Byron & M. J. Humphries (2006). Integrin ligands at a glance. *Journal of Cell Science* 119(19): 3901-3903.

Hynes, R. O. (2002). Integrins: Bidirectional, allosteric signaling machines. *Cell* 110(6): 673-687.

Jones, M. B., C. H. Chu, J. C. Pendleton, M. J. Betenbaugh, J. Shiloach, B. Baljinnyam, J. S. Rubin & M. J. Shamblott (2010). Proliferation and pluripotency of human embryonic stem cells maintained on type I collagen. *Stem Cells and Development* 19(12): 1923-1935.

Kirschstein, R. & L. R. Skirboll (2001). Stem Cells: Scientific Progress and Future Research Directions, National Institutes of Health

Kleinman, H. K., M. L. McGarvey, L. A. Liotta, P. G. Robey, K. Tryggvason & G. R. Martin (1982). Isolation and characterization of type IV procollagen, laminin, and heparan sulfate proteoglycan from the EHS sarcoma. *Biochemistry* 21(24): 6188-6193.

Klim, J. R., L. Li, P. J. Wrighton, M. S. Piekarczyk & L. L. Kiessling (2010). A defined glycosaminoglycan-binding substratum for human pluripotent stem cells. *Nature Methods* 7(12): 989-994.

Kolhar, P., V. R. Kotamraju, S. T. Hikita, D. O. Clegg & E. Ruoslahti (2010). Synthetic surfaces for human embryonic stem cell culture. *Journal of biotechnology* 146(3): 143-146.

Laflamme, M. A., K. Y. Chen, A. V. Naumova, V. Muskheli, J. A. Fugate, S. K. Dupras, H. Reinecke, C. Xu, M. Hassanipour, S. Police, C. O'Sullivan, L. Collins, Y. Chen, E. Minami, E. A. Gill, S. Ueno, C. Yuan, J. Gold & C. E. Murry (2007). Cardiomyocytes derived from human embryonic stem cells in pro-survival factors enhance function of infarcted rat hearts. *Nature Biotechnology* 25(9): 1015-1024.

Li, Y., S. Powell, E. Brunette, J. Lebkowski & R. Mandalam (2005). Expansion of human embryonic stem cells in defined serum-free medium devoid of animal-derived products. *Biotechnology and Bioengineering* 91(6): 688-698.

Li, Y. J., E. H. Chung, R. T. Rodriguez, M. T. Firpo & K. E. Healy (2006). Hydrogels as artificial matrices for human embryonic stem cell self-renewal. *Journal of Biomedical Materials Research - Part A* 79(1): 1-5.

Lu, J., R. Hou, C. J. Booth, S. H. Yang & M. Snyder (2006). Defined culture conditions of human embryonic stem cells. *Proceedings of the National Academy of Sciences of the United States of America* 103(15): 5688-5693.

Ludwig, T. E., M. E. Levenstein, J. M. Jones, W. T. Berggren, E. R. Mitchen, J. L. Frane, L. J. Crandall, C. A. Daigh, K. R. Conard, M. S. Piekarczyk, R. A. Llanas & J. A. Thomson (2006). Derivation of human embryonic stem cells in defined conditions. *Nature Biotechnology* 24(2): 185-187.

Mahlstedt, M. M., D. Anderson, J. S. Sharp, R. McGilvray, M. D. Barbadillo Muñoz, L. D. Buttery, M. R. Alexander, F. R. A. J. Rose & C. Denning Maintenance of pluripotency in human embryonic stem cells cultured on a synthetic substrate in conditioned medium (2009). *Biotechnology and Bioengineering* 105(1): 130-140.

Martin, M. J., A. Muotri, F. Gage & A. Varki (2005). Human embryonic stem cells express an immunogenic nonhuman sialic acid. *Nature Medicine* 11(2): 228-232.

Melkoumian, Z., J. L. Weber, D. M. Weber, A. G. Fadeev, Y. Zhou, P. Dolley-Sonneville, J. Yang, L. Qiu, C. A. Priest, C. Shogbon, A. W. Martin, J. Nelson, P. West, J. P. Beltzer, S. Pal & R. Brandenberger (2010). Synthetic peptide-acrylate surfaces for long-term self-renewal and cardiomyocyte differentiation of human embryonic stem cells. *Nature Biotechnology* 28(6): 606-610.

Meng, Y., S. Eshghi, Y. J. Li, R. Schmidt, D. V. Schaffer & K. E. Healy (2010). Characterization of integrin engagement during defined human embryonic stem cell culture. *FASEB Journal* 24(4): 1056-1065.

Miyazaki, T., S. Futaki, K. Hasegawa, M. Kawasaki, N. Sanzen, M. Hayashi, E. Kawase, K. Sekiguchi, N. Nakatsuji & H. Suemori (2008). Recombinant human laminin isoforms can support the undifferentiated growth of human embryonic stem cells. *Biochemical and Biophysical Research Communications* 375(1): 27-32.

Nagaoka, M., K. Si-Tayeb, T. Akaike & S. A. Duncan (2010). Culture of human pluripotent stem cells using completely defined conditions on a recombinant E-cadherin substratum. *BMC Developmental Biology* 10.

Oldberg, A., A. Franzen & D. Heinegard (1988). The primary structure of a cell-binding bone sialoprotein. *Journal of Biological Chemistry* 263(36): 19430-19432.

Pedersen, R. A. (2002). Feeding hungry stem cells. *Nature Biotechnology* 20(9): 882-883.

Pera, M. F., B. Reubinoff & A. Trounson (2000). Human embryonic stem cells. *Journal of Cell Science* 113(1): 5-10.

Pierschbacher, M. D. & E. Ruoslahti (1984). Cell attachment activity of fibronectin can be duplicated by small synthetic fragments of the molecule. *Nature* 309(5963): 30-33.

Price, P. J., M. D. Goldsborough & M. L. Tilkins (1998). Embryonic Stem Cell Serum Replacement. *Patent Application WO9830679 (A1)*

Prowse, A. B. J., F. Chong, P. P. Gray & T. P. Munro (2011). Stem cell integrins: Implications for ex-vivo culture and cellular therapies. *Stem Cell Research* 6(1): 1-12.

Prowse, A. B. J., M. R. Doran, J. J. Cooper-White, F. Chong, T. P. Munro, J. Fitzpatrick, T. L. Chung, D. N. Haylock, P. P. Gray & E. J. Wolvetang (2010). Long term culture of human embryonic stem cells on recombinant vitronectin in ascorbate free media. *Biomaterials* 31(32): 8281-8288.

Reubinoff, B. E., M. F. Pera, C. Y. Fong, A. Trounson & A. Bongso (2000). Embryonic stem cell lines from human blastocysts: Somatic differentiation in vitro. *Nature Biotechnology* 18(4): 399-404.

Rodin, S., A. Domogatskaya, S. Ström, E. M. Hansson, K. R. Chien, J. Inzunza, O. Hovatta & K. Tryggvason (2010). Long-term self-renewal of human pluripotent stem cells on human recombinant laminin-511. *Nature Biotechnology* 28(6): 611-615.

Rowland, T. J., L. M. Miller, A. J. Blaschke, E. L. Doss, A. J. Bonham, S. T. Hikita, L. V. Johnson & D. O. Clegg (2010). Roles of integrins in human induced pluripotent stem cell growth on matrigel and vitronectin. *Stem Cells and Development* 19(8): 1231-1240.

Suzuki, S., A. Oldberg, E. G. Hayman, M. D. Pierschbacher & E. Ruoslahti (1985). Complete amino acid sequence of human vitronectin deduced from cDNA. Similarity of cell attachment sites in vitronectin and fibronectin. *EMBO Journal* 4(10): 2519-2524.

Thomas, R. J. & D. J. Williams (2009). Large-scale manufacturing of therapeutic human stem cells. *Pharmaceutical Technology* 33(7): 74-79.

Thomson, J. A. (1998). Embryonic stem cell lines derived from human blastocysts. *Science* 282(5391): 1145-1147.

Villa-Diaz, L. G., H. Nandivada, J. Ding, N. C. Nogueira-De-Souza, P. H. Krebsbach, K. S. O'Shea, J. Lahann & G. D. Smith (2010). Synthetic polymer coatings for long-term growth of human embryonic stem cells. *Nature Biotechnology* 28(6): 581-583.

Wong, J. C. Y., S. Y. Gao, J. G. Lees, M. B. Best, R. Wang & B. E. Tuch (2010). Definitive endoderm derived from human embryonic stem cells highly express the integrin receptors αV and β5. *Cell Adhesion and Migration* 4(1): 39-45.

Xu, C., M. S. Inokuma, J. Denham, K. Golds, P. Kundu, J. D. Gold & M. K. Carpenter (2001). Feeder-free growth of undifferentiated human embryonic stem cells. *Nature Biotechnology* 19(10): 971-974.

Xu, C., E. Rosler, J. Jiang, J. S. Lebkowski, J. D. Gold, C. O'Sullivan, K. Delavan-Boorsma, M. Mok, A. Bronstein & M. K. Carpenter (2005). Basic fibroblast growth factor supports undifferentiated human embryonic stem cell growth without conditioned medium. *Stem Cells* 23(3): 315-323.

Yap, L. Y. W., J. Li, I. Y. Phang, L. T. Ong, J. Z. E. Ow, J. C. H. Goh, V. Nurcombe, J. Hobley, A. B. H. Choo, S. K. W. Oh, S. M. Cool & W. R. Birch (2011). Defining a threshold surface density of vitronectin for the stable expansion of human embryonic stem cells. *Tissue Engineering - Part C: Methods* 17(2): 193-207.

Ying, Q. L., J. Wray, J. Nichols, L. Batlle-Morera, B. Doble, J. Woodgett, P. Cohen & A. Smith (2008). The ground state of embryonic stem cell self-renewal. *Nature* 453(7194): 519-523.

Yoon, T.-M., B. Chang, H.-T. Kim, J.-H. Jee, D.-W. Kim & D.-Y. Hwang (2010). Human Embryonic Stem Cells (hESCs) Cultured Under Distinctive Feeder-Free Culture Conditions Display Global Gene Expression Patterns Similar to hESCs from Feeder-Dependent Culture Conditions. *Stem Cell Reviews and Reports* 6(3): 425-437.

7

Efficient Integration of Transgenes and Their Reliable Expression in Human Embryonic Stem Cells

Kenji Sakurai[1,2,a], Miho Shimoji[1], Kazuhiro Aiba[1,3] and Norio Nakatsuji[2,3]

[1]Stem Cell and Drug Discovery Institute

[2]Institute for Frontier Medical Sciences, Kyoto University

[3]Institute for Integrated Cell-Material Sciences (iCeMS), Kyoto University

[a]Present affiliation: Kyowa Hakko Kirin Co., Ltd.

Japan

1. Introduction

Human embryonic stem cells (hESCs) are promising materials for both basic and applied research fields such as human developmental biology and drug discovery because of their capacity for long-term self-renewal and differentiation into virtually all types of cells and tissues (Thomson et al., 1998; Hoffman & Carpenter, 2005). For example, with regard to their applications for drug development research, these cells can be used test whether drugs under development are efficacious against specific diseases by using hESCs-derived disease models which show disease phenotypes (Barbaric et al., 2010; Laustriat et al., 2010).

Genetic manipulation is a useful strategy for generating cellular disease models. Random integration of exogenous genes into the genome of hESCs is the most straightforward and easiest method. Random integration, however, often results in inactivation or silencing of integrated genes in hESCs (Ellis, 2005; Liew et al., 2007; Xia et al., 2007) and also might alter the cellular phenotypes due to insertional mutagenesis, which disrupts gene functions in undefined gene regions (Hacein-Bey-Abina et al., 2003; Nienhuis et al., 2006). Furthermore, the reliability of data may be compromised by differences in the transgene integration sites when comparing multiple transgenic cell lines. Introducing transgenes by gene targeting is one way to avoid undesired gene silencing and insertional mutagenesis, but the targeting efficiency in hESCs is notoriously low (Zwaka and Thomson, 2003; Urbach et al, 2004; Irion et al, 2007; Di Domenico et al, 2008; Ruby et al, 2009). Therefore, a lot of time and effort may be required to obtain gene-targeted hESC clones. To address these issues, several strategies have been explored. Researchers have identified usable native sites for gene integration in the human genome. Bacteriophage phiC31 integrase, or adeno-associated virus type 2 (AAV2), can mediate plasmid integration into pseudo-attP sites or the adeno-associated virus integration site 1 locus in the genome, respectively (Thyagarajan et al., 2008; Smith et al., 2008). However, the human genome possesses 23 different pseudo-attP sites, so the integration site cannot be specifically controlled, and the AAV2-mediated targeting efficiency is still low (4.16%). In contrast, foreign sequences such as bacterial loxP and yeast FRT are useful for gene integrations into the genome. Both loxP and lox2272 sequences can be introduced at the silence-resistance sites of hESCs, but the selection efficiency of recombination-mediated cassette exchange is relatively low (two out of 92 clones) (Du et al., 2009). As a result, it is thus considered that there is room for improvement in the targeting and the selection of loxP technology.

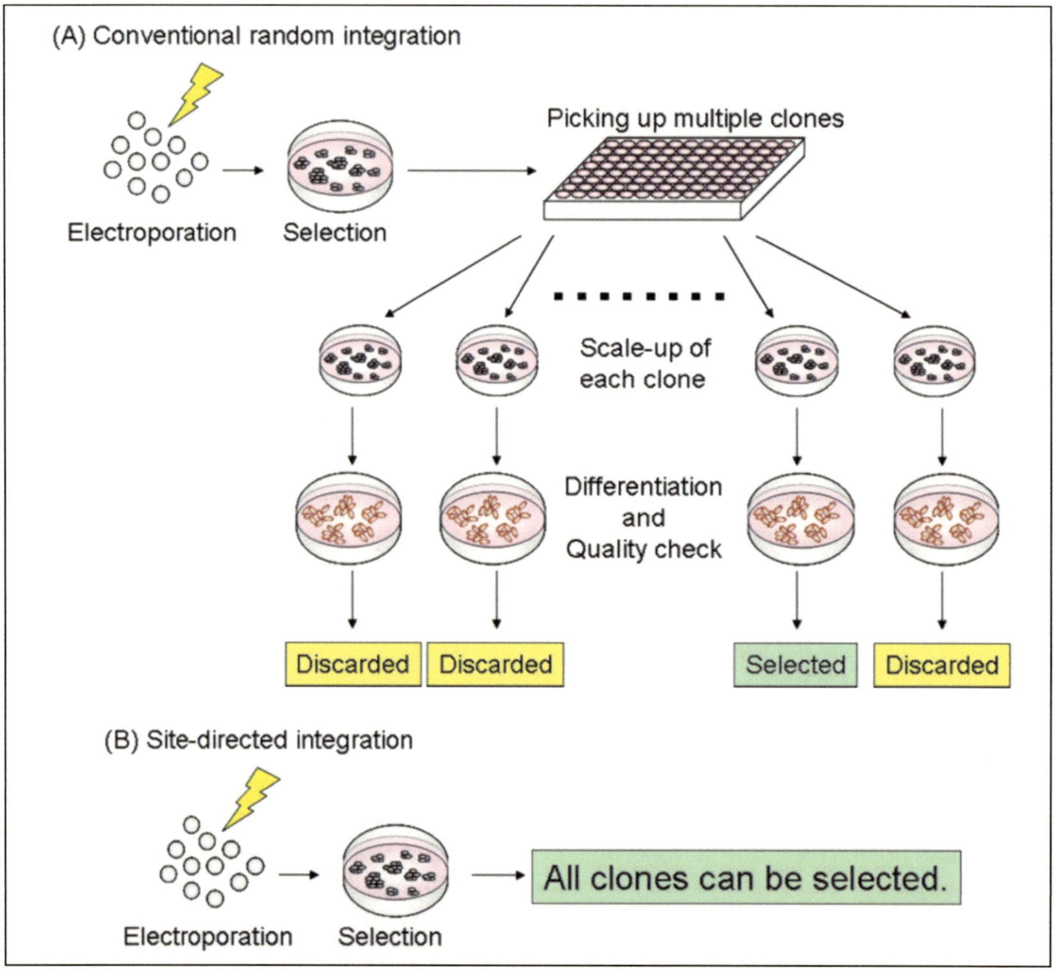

Fig. 1. Schematic diagrams of the conventional random integration and site-directed integration. (A) Conventional random integration requires selection of the proper clone(s) from among a large number of clones. This puts a burden on researchers to scale-up, differentiate and check the quality of each individual clone. (B) The site-directed integration produces clones by manipulating the only docking site. Hence, all clones can be used.

In this chapter, a site-directed integration system based on homologous recombination (HR) and Cre recombinase-mediated site-specific integration is described. First, gene targeting was performed to introduce a docking site containing two loxP sites and a hygromycin resistance gene without the first methionine codon into the hypoxanthine phosphoribosyltransferase 1 (HPRT) locus. Next, a gene of interest was inserted into the docking site by Cre recombinase. Correct insertion into the docking site confers hygromycin resistance. Using this strategy, when an EGFP expression cassette (CAG promoter-driven EGFP) was inserted into the HPRT-docking site of the hESC line, the EGFP fluorescence was detected in all of the hygromycin-resistant clones. Furthermore, when a single vector carrying both an rtTA expression cassette and a Tetracycline (Tet) Response Element (TRE)-driven EGFP was introduced into the docking site, doxycyclin (Dox) could induce EGFP expression in the hygromycin resistant clones in a dose-dependent manner, and the clones did not leak EGFP expression in the absence of Dox.

Every time a transgene is inserted into the HPRT locus, carrying out conventional gene targeting is impractical because of the low HR efficiency, whereas the new site-directed integration system has greatly improved efficiency. Once the hESC lines with the docking site in the HPRT locus are generated by HR, any gene of interest can be integrated into the HPRT locus with nearly 100% efficiency. Furthermore, all clones created by this system have the same genetic background. This enables the effects among different genes integrated at the HPRT locus to be evaluated without consideration of data from multiple cell lines, unlike the usual need for comparison among randomly integrated hESC lines. Therefore, the new site-directed integration system makes it possible to produce transgenic hESC lines quickly and thus obtain reliable research results (Fig. 1).

2. Site-directed integration of transgenes for hESCs and its future applications

2.1 Homologous recombination for hESCs

To integrate the docking site for site-directed gene integration, we first carried out HR for hESCs. We chose the HPRT locus as the transgene integration site because, along with ROSA26 and others, this locus is considered to always allow transgene expression in the mouse genome (Bronson et al., 1996; Zambrowicz et al., 1997; McCreath et al., 2000). In addition, previous reports have shown the success of HR to the HPRT locus of hESCs (Zwaka & Thomson, 2003).

To generate the HPRT-targeting vector, the 5′- and 3′- homologous arms (7.0 kb and 2.0 kb respectively) were amplified from KhES-1 genomic DNA by polymerase chain reaction (PCR) using KOD FX (TOYOBO, Japan, also known as KOD Xtreme, supplied from Merck). This targeting vector contained a loxP site flanked by a neomycin resistance gene expression cassette, and another loxP site was located 5′ to a promoter-less hygromycin resistance gene lacking the start ATG codon. These elements thus became the docking site for gene integration. To exclude the effects of external transcription factors, the docking site was inserted between two DNase I-hypersensitive site 4 (HS4) insulators (Chung et al., 1993) (Fig. 2). After the targeting vector was linearized by cutting the NotI site at the distal end of the 5′arm, it was then delivered to the female hESC KhES-1 cells (Suemori et al., 2006) by electroporation.

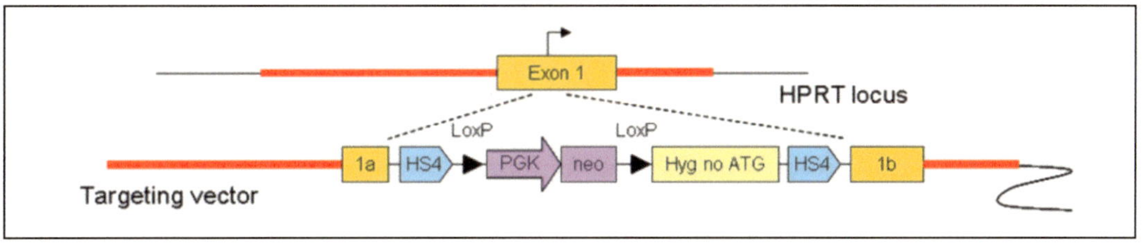

Fig. 2. The structure of the targeting vector for the human HPRT locus.

For electroporation, two confluent 100 mm dishes of KhES-1 (2-4 x 10^6) cells were completely dissociated using 0.05% trypsin and 0.2 mM EDTA. Cells were mixed with 10 μg of linearized targeting vector, electroporated in GenePulser Xcell (BioRad) using the following settings: square mode, 250 V, 4 ms x 2, 5 s interval, and then dispensed into a 60 mm dish plated with neomycin-resistant feeders (mouse embryonic fibroblasts (MEFs)). Two days after electroporation, the cells were treated with 50 μg/ml G418 (Sigma). The next day, the G418 concentration was increased to 100 μg/ml. Selection was performed for 10-14 days post-electroporation.

Out of the 424 G418-resistant clones screened, six (1.42%) had undergone the desired homologous recombination event (Table 1) (Sakurai et al., 2010). HR was confirmed by PCR and a Southern blotting analysis (Sakurai et al., 2010). One targeted clone, named K1-HS, was randomly selected and used for subsequent gene replacement. We also carried out gene targeting for KhES-1 sub-line 1 (Hasegawa et al., 2006) and this targeting efficiency was 0.89% (three clones underwent HR out of 336 G418-resistant clones) (Table 1).

hESC line	G418 resistance	Homologous recombination	Efficiency
KhES-1	424	6	1.42%
Sub-line 1	336	3	0.89%

Table 1. HPRT gene targeting experiments in hESCs.

2.2 Site-directed gene integration

To perform site-directed integration for the HPRT locus, we constructed the pInsert vector as a basic vector, which contained the EF1α promoter, a Kozak sequence (Kozak, 1987), an ATG codon and a loxP site in this order (Fig. 3A). To examine whether this site-directed integration system was functional, we tried to integrate an EGFP expression vector into the HPRT locus and constructed the pInsert-Tif-CAG-EGFP vector, which carries an EGFP transgene under the control of the CAG promoter and a Cor insulator, with the tandem core element of the chicken HS4 beta globin insulator (Otsuki et al., 2005) (Fig. 3B).

Electroporation was used to introduce the plasmids into the cells. Five micrograms of a Cre expression vector (pEF1α-Cre) and 20 μg of the pInsert-Tif-CAG-EGFP vector were used for each electroporation procedure. The number of cells (K1-HS), the methods used for cell dissociation, and the electroporation settings were the same as those used for the gene targeting experiments. The ESCs were then plated onto hygromycin-resistant MEFs (Dainippon Sumitomo Pharma, Japan). Hygromycin selection (40 μg/ml, Invitrogen) was started two days after electroporation, and was performed for 10-14 days post-electroporation.

In the presence of hygromycin, survival is highly skewed towards clones that have undergone the correct recombination event. This is because our method requires the reconstitution of the hygromycin resistance gene by Cre-mediated recombination between the EF1α promoter-ATG sequence in a plasmid vector carrying the transgene of interest and a promoter-less hygromycin resistance gene lacking a start codon in the targeted HPRT

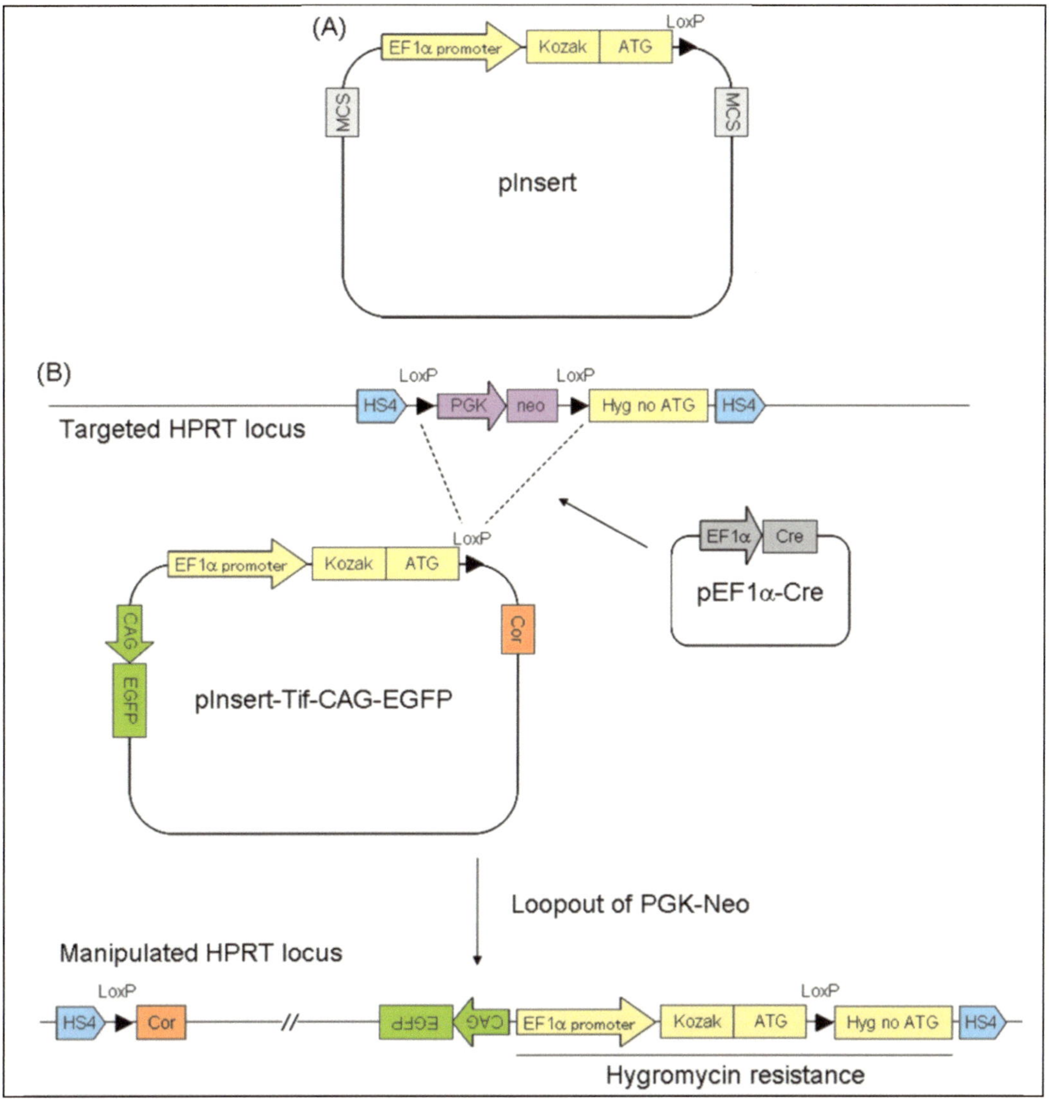

Fig. 3. (A) A vector developed for site-directed integration, pInsert. (B) The strategy used for site-directed integration. Cre recombinase-mediated integration of the pInsert vectors carrying a gene of interest confers hygromycin resistance.

locus (Fukushige & Sauer, 1992; Beard et al., 2006). In this case, only clones in which the CAG-EGFP cassette is integrated into the docking site on the HPRT locus can survive under hygromycin selection. All hygromycin resistant clones therefore theoretically express EGFP, because the HPRT locus supports strong, ubiquitous expression of inserted sequences, and it is not subject to any gene-silencing effects.

In five transfection experiments, a total of 186 hygromycin-resistant clones were obtained, and all of them were EGFP-positive (Table 2) (Sakurai et al., 2010). Eight clones were randomly selected from the hygromycin-resistant clones, and all of them expressed similar levels of EGFP. The correct and single-copy integration were confirmed by PCR and a

	Exp 1	Exp 2	Exp 3	Exp 4	Exp 5	Total
Hygromycin resistant	45	22	13	26	80	186
EGFP$^+$	45	22	13	26	80	186
Efficiency	100%	100%	100%	100%	100%	100%

Table 2. Efficiency of EGFP expression in the site-directed integration using a pInsert-Tif-CAG-EGFP vector.

Southern analysis (Sakurai et al., 2010). The correctly integrated clones were expected to have lost their neomycin resistance due to cassette exchange. As expected, all clones (80 clones that were randomly chosen from the initial 186) were neomycin-sensitive (Sakurai et al., 2010). The pInsert-Tif-CAG-EGFP integrated clones maintained their expression of EGFP for at least 6 months in an undifferentiated state in the presence of hygromycin (data not shown). When these cells were differentiated into neurons, it was possible to detect EGFP expression in the resulting neurons (Fig. 4).

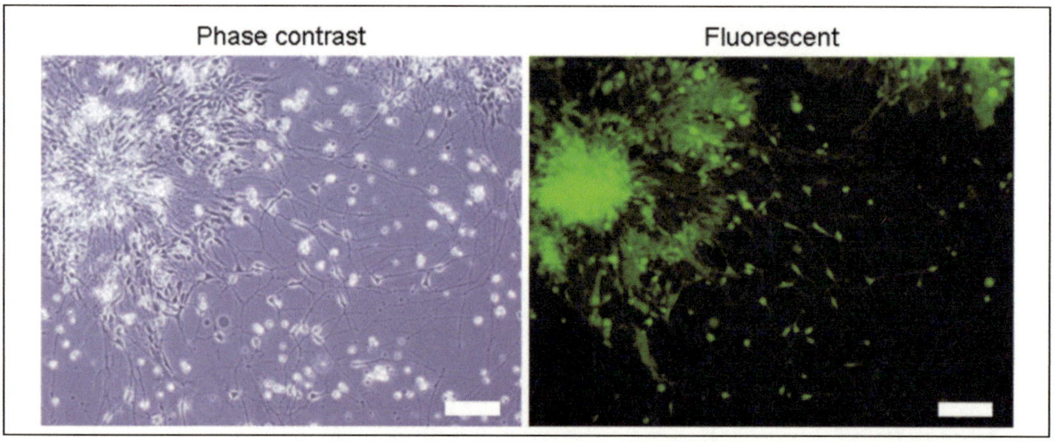

Fig. 4. Neural differentiation of a pInsert-Tif-CAG-EGFP integrated clone. Bar = 100 μm.

2.3 Inducible gene expression based on the site-directed integration

We attempted to introduce all components of the Tet-inducible gene expression system into the docking site at the HPRT locus. Site-directed integration was performed by co-transfection of a pInsert vector carrying CAG-rtTA, the Cor insulator and TREtight-EGFP (pInsert-CTOR-EGFP) (Fig. 5), with a pEF1α-Cre. Fifty micrograms of pInsert-CTOR-EGFP was used for this experiment, and the other methods were the same as were used for the site-directed integration of pInsert-Tif-CAG-EGFP (Sakurai et al., 2010).

Treatment with Dox for 2 days after the emergence of hygromycin-resistant colonies resulted in a dose-dependent induction of EGFP expression (Sakurai et al., 2010). Two clones

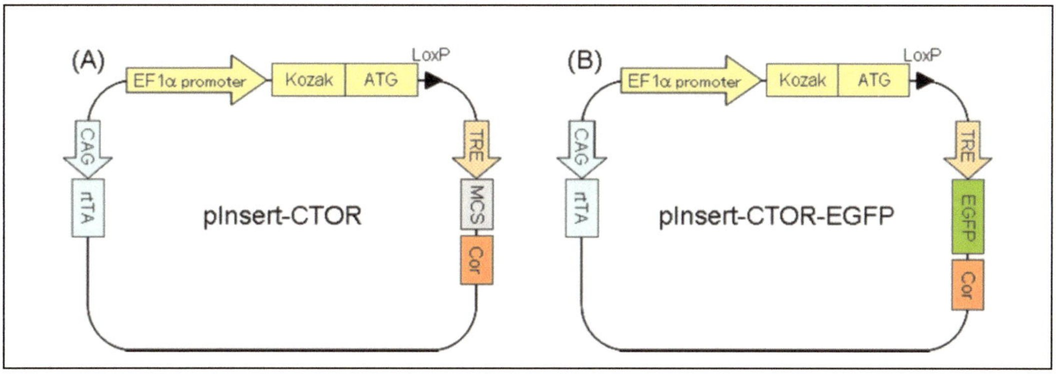

Fig. 5. (A) pInsert-CTOR and (B) pInsert-CTOR-EGFP vectors for the first-generation Tet-inducible gene expression system based on the site-directed integration.

that were randomly selected exhibited similar responsiveness to Dox (Sakurai et al., 2010). Their correct integration of pInsert-CTOR-EGFP was confirmed by PCR and a Southern blotting analysis (Sakurai et al., 2010). One of the two clones was further analyzed in additional experiments, and it could form embryoid bodies (EBs) and retained the ability to differentiate into representative cells of all three germ layers (Sakurai et al., 2010). Taken together, these results indicate that these ESCs maintain their pluripotency after a second round of genetic modification and clonal selection.

Although we put two insulators on both sides of the TRE-EGFP cassette in the pInsert-CTOR-EGFP integrants to avoid the leaky expression of EGFP, the results indicated that the faint EGFP signals could still be detected in the absence of Dox (Sakurai et al., 2010). Furthermore, although a FACS analysis showed that most of the Dox-treated live ESCs expressed EGFP, the fluorescent micrographs displayed their EGFP expression in a patchy fashion (Sakurai et al., 2010). Therefore, we tried to improve the pInsert-CTOR vector for practical use.

2.4. pInsert-Tet5 vector: the reliable Tet-on system

We investigated methods to diminish the leaky expression of transgenes downstream of the TRE. We examined the orientation of the TRE-EGFP cassette and the types of insulators. We did not modify the CAG-rtTA cassette because its structure was the same as the CAG-EGFP cassette of the pInsert-Tif-CAG-EGFP vector that demonstrated a robust expression of EGFP. As a result, we developed a pInsert-Tet5 vector with undetectable expression of EGFP in the absence of Dox.

The pInsert-Tet5 vector was generated by exchanging the Cor insulator of pInsert-CTOR for the HS4 insulator with orientation in opposition to the HS4 insulators of the K1-HS cells (Fig. 6A). The pInsret-Tet5-EGFP was introduced into K1-HS cells with pEF1α-Cre, and transfected cells were selected by hygromycin treatment (Fig. 6B). The experimental method was the same as that used for the pInsert-CTOR-EGFP. After the emergence of hygromycin-resistant cells, Dox was added for induction of EGFP expression (Fig. 7A). Unlike the pInsert-CTOR-EGFP integrated clones with their patchy expression, the fluorescent micrographs indicated that EGFP was expressed in most of the cells in the colonies in response to Dox treatment (Fig. 7A).

We picked up three pInsert-Tet5-EGFP integrated clones randomly and analyzed their responsiveness to Dox. All three clones expressed a similar level of EGFP when they were treated with Dox for 3 days (Fig. 7B). In the absence of Dox, the fluorescence intensities of all clones were the same as their parental K1-HS, indicating that the pInsert-Tet5-EGFP integrants do not leak any significant level of EGFP expression in the absence of Dox (Fig. 7B). Moreover, we further analyzed clone 3, and could induce EGFP expression in a Dox dose-dependent manner (Fig. 7C). Single-copy integration of a pInsert-Tet5-EGFP vector was confirmed by a Southern analysis (Fig. 7D).

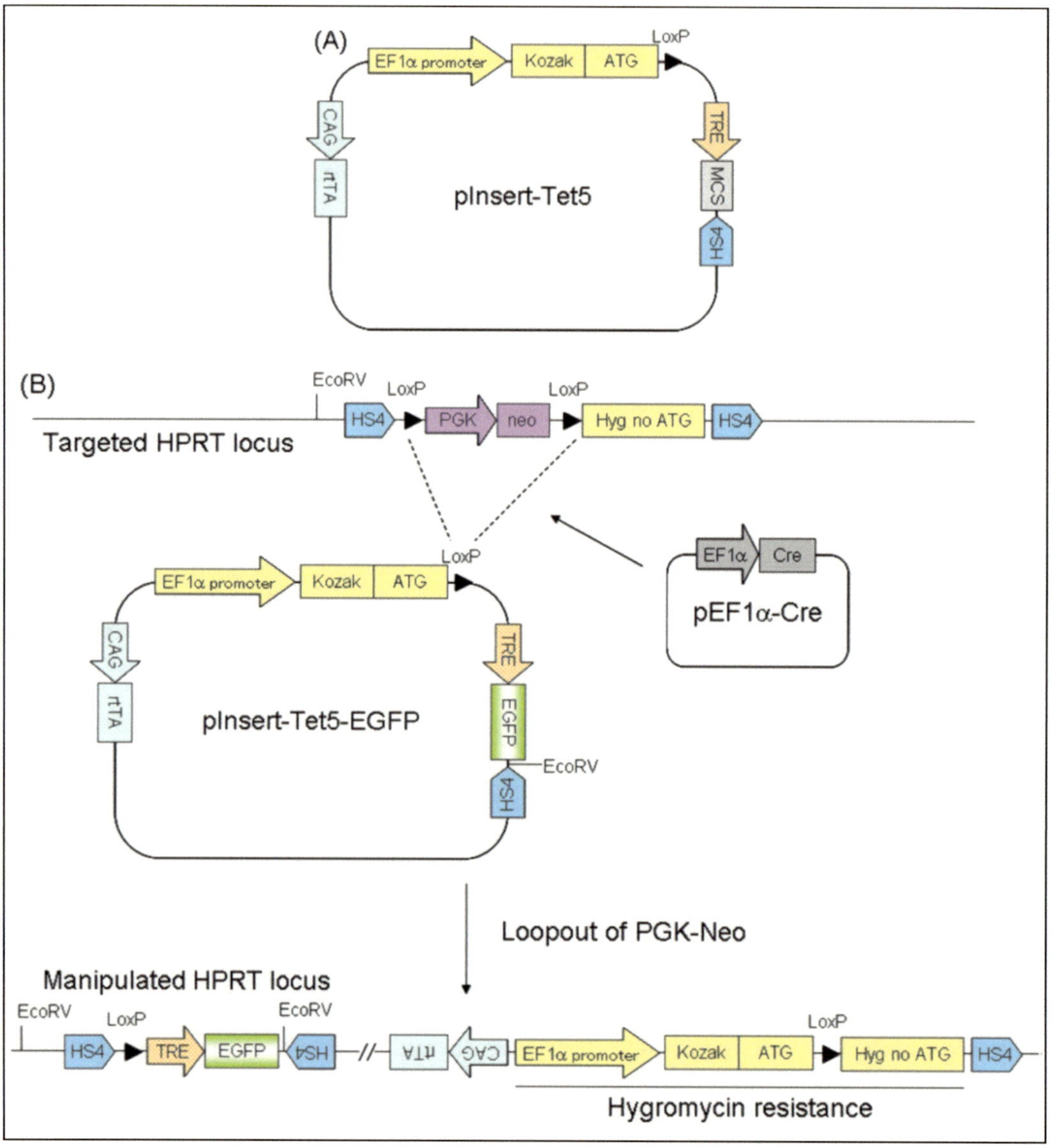

Fig. 6. (A) The pInsret-Tet5 vector used to generate a reliable Tet-On system. The insulator of this vector (HS4) was changed from that of the pInsert-CTOR vector (Cor). (B) The strategy for site-directed integration of pInsert-Tet5-EGFP. The TRE-EGFP cassettes are wedged between two HS4 insulators to avoid the leaky expression of EGFP.

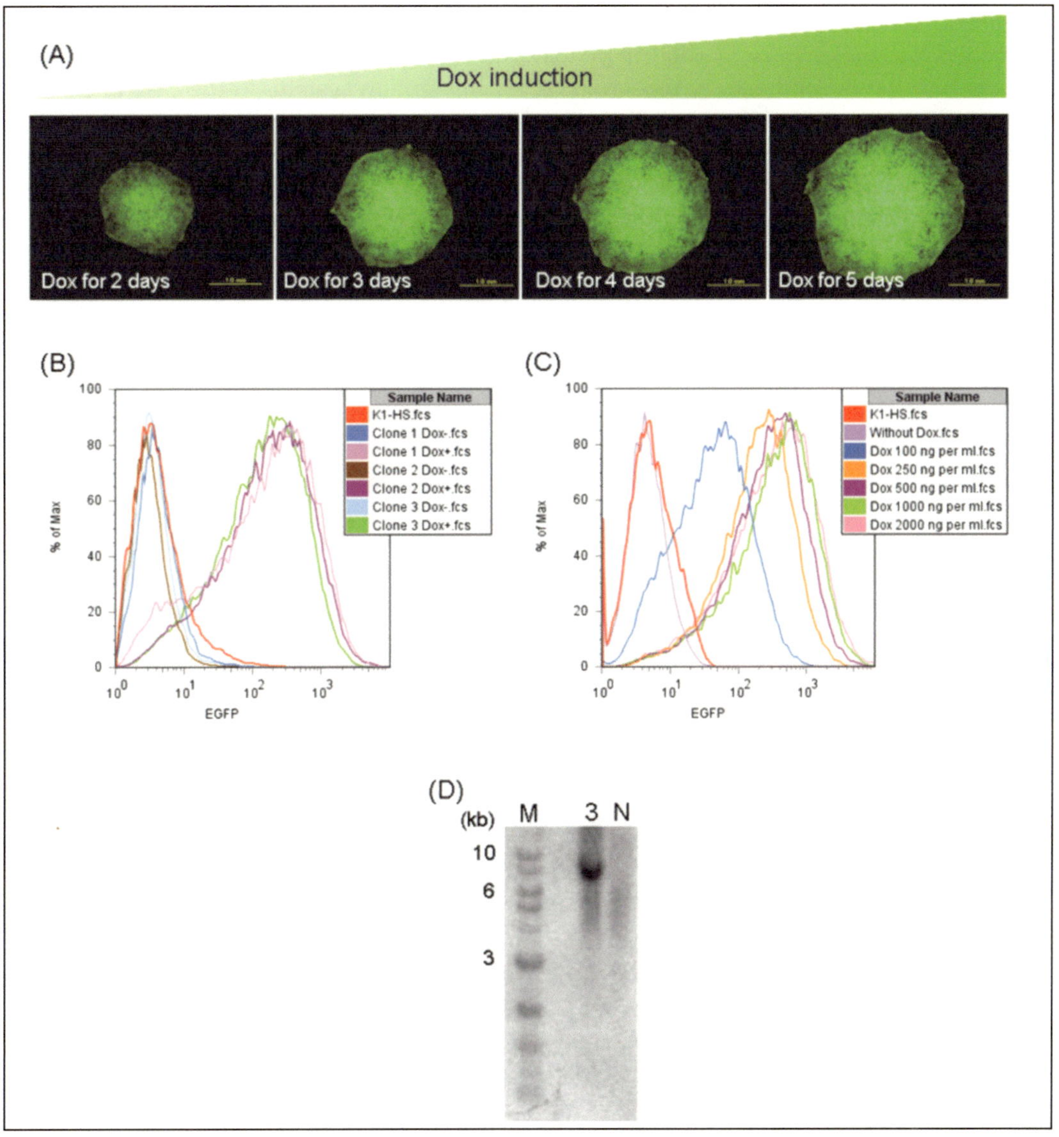

Fig. 7. (A) Dox-inducible EGFP induction of a pInsert-Tet5-EGFP integrated hygromycin-resistant clone. (B) The FACS analysis of three pInsert-Tet5-EGFP integrated clones. All of them expressed a similar level of EGFP when they were treated with 2 μg/ml of Dox for 3 days. Furthermore, the pInsert-Tet5-EGFP integrants did not leak any detectable EGFP in the absence of Dox. (C) The dose-dependent induction of EGFP expression by Dox. Cells were treated with 100, 250, 500, 1000 or 2000 ng/ml Dox for 3 days. (D) The Southern blotting analysis. Genomic DNA was digested by EcoRV and hybridized with an EGFP probe. M: Size marker, 3: Clone 3, N: Non-integrated genome. The expected band size was 7.5 kb.

We believe that the combination of K1-HS and pInsert-Tet5 is a feasible Dox-inducible gene expression system. To establish even greater feasibility for using this system, we plan to examine to accuracy of the transgene integration, Dox responsiveness in differentiated cells, epigenetic modification near the docking site, and genomic structure of pInsert-Tet5 integrated locus. There is a possibility that several copies of the pInsert vector may insert in tandem into the docking site in the site-directed integration system. In the case of pInsert-Tet5, tandem integration of pInsert-Tet5 vectors may abolish the reliable regulation by Dox, because the gene of interest would be downstream of the EF1α promoter, without intervening insulators. Therefore, it will be necessary to pick up a few hygromycin-resistant clones and to confirm their correct integration of the pInsert-Tet5 vector.

2.5 Future applications of the site-directed integration system in human pluripotent stem cells

The drug discovery and development process is time-consuming and costly. Nevertheless, the average success rate of approval is approximately 11% after clinical trials (Kola and Landis, 2004). The major causes of failure are a lack of efficacy and toxicity in humans (Kola and Landis, 2004). Animal disease models can provide a wealth of information about the complexity of various disease processes, but efficacious lead compounds for these models are often found to be inefficacious during human clinical trials (Kola and Landis, 2004). Moreover, animal models do not always indicate the toxicity of compounds in humans because of the differences in the biological processes in humans and animals (Martignoni et al., 2006). These facts demonstrate the need for more predictive drug screening systems (Kola and Landis, 2004; Ebert and Svendsen, 2010). In this regard, human pluripotent stem cells are considered to be a promising cell source for drug screening because of the virtually limitless supply of normal human cells that can be differentiated into any specific cell type.

2.5.1 Disease modeling and efficacy screening

There are several approaches that have been designed to create *in vitro* disease models using pluripotent stem cells. One is that ESCs carrying serious genetic disorders can be differentiated into cells with disease-related phenotypes. Preimplantation genetic diagnosis (PGD) can identify embryos with genetic defects. The generation of disease-specific hESC lines via PGD has already been reported. For example, there are hESC lines with cystic fibrosis, myotonic dystrophy type I, Huntington's disease, adrenoleukodystrophy, Dushenne muscular dystrophy, Becker muscular dystrophy, fragile-X syndrome and thalassaemia (Pickering et al., 2005; Verlinsky et al., 2005; Mateizel et al., 2006; Stephenson et al., 2009). Differentiation of these disease-specific hESC lines into each of the specific cell types that manifest disease phenotypes may make it possible to screen for factors that restore the normal phenotype and to identify factors associated with disease initiation or progression. Furthermore, many induced pluripotent stem cells (iPSCs) have been generated from adult patients with incurable diseases and it is also hoped that they will provide new sources for drug screening (Dimos et al., 2008; Park et al., 2008; Ebert et al., 2009; Lee et al., 2009; Soldner et al., 2009; Ye et al., 2009; Maehr et al., 2009; Ye et al., 2009).

Using pluripotent stem cells carrying disease-linked genes and a high-throughput screening system, small-molecule compounds or humoral factors can be identified as potential therapeutic drugs (Barbaric et al., 2010). However, in order to more effectively screen for intracellular proteins or artificial peptides that are involved in or affect the disease process, the site-directed integration system may be more useful. In this section, we describe an example of an application of the site-directed integration system for drug screening and disease mechanism research.

First, as shown in Fig. 8, the docking site was targeted to the HPRT locus of diseased ES/iPS cells. Although the targeting efficiency of the docking site by electroporation is relatively low (Table 1), more efficient gene targeting method for human ESC/iPS cells using helper-dependent adenoviral vectors or zinc-finger nuclease has recently been reported (Suzuki et al., 2008; Lombardo et al., 2007; Zou et al., 2009) and applying these targeting methods may resolve these technical problems. Next, libraries of pInsert vectors which contain expression cassettes of cDNAs coding for various proteins or artificial peptides are constructed. Insertion of each pInsert vector into the targeted docking site confers stable expression of exogenous proteins in the diseased ES/iPS cells. These ES/iPS cells should show the phenotypes of the disease after differentiation. However, if differentiated cells show normal phenotypes, this means that the disease phenotype has been resolved, suggesting that expressing the specific exogenous protein influences the disease process. Such exogenous proteins can be identified by analyzing the sequence of the integrated pInsert vector.

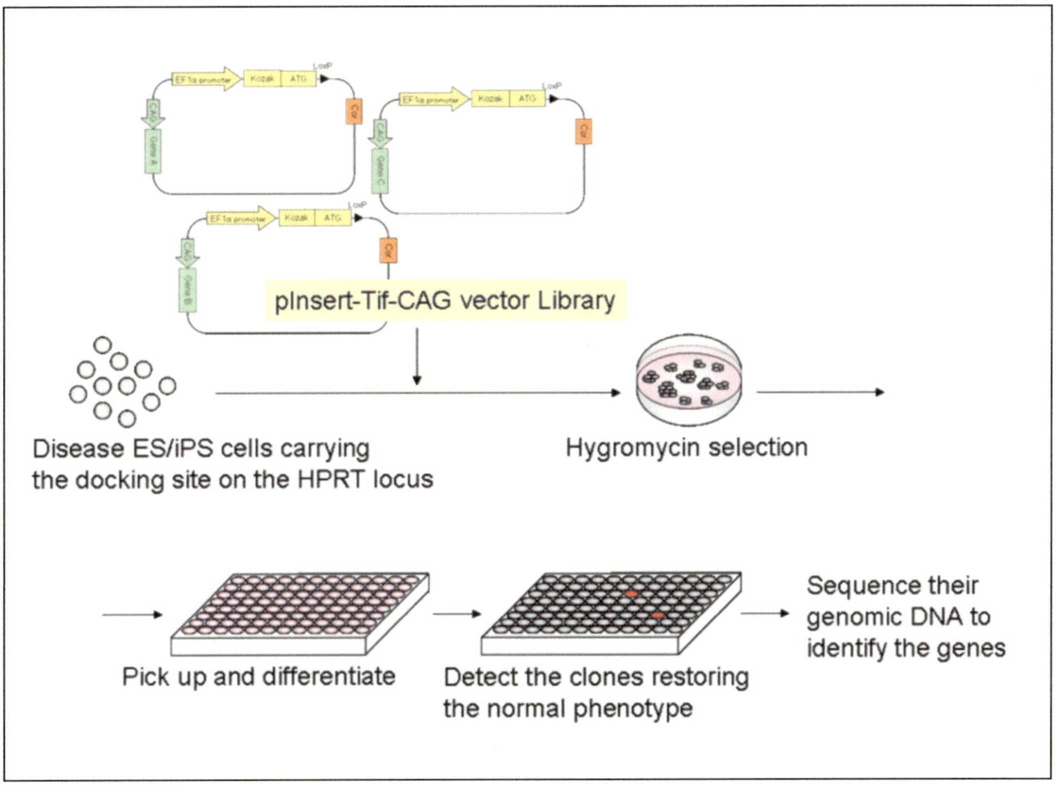

Fig. 8. A schematic diagram for discovering potential disease-curing proteins or peptides using pInsert vector libraries and disease-specific ES/iPS cells.

However, although a number of disease-specific ES/iPS cell lines have been established, only a few studies have demonstrated that the differentiated cells showed disease phenotypes (Ebert et al., 2009; Lee et al., 2009), while many disease-specific ES/iPS cells did not show any disease phenotypes, perhaps because of the late onset or multi-factorial nature of such diseases. Therefore, long-term culture might be needed before the disease phenotypes will emerge or the disease will be initiated. However, overexpression of disease-linked genes in K1-HS cells using the site-directed integration system might enable an accelerated onset of disease phenotypes, and if overexpression of the disease-related gene leads to cell death, the pInsert-Tet5 vector system represents a suitable method because it can control gene expression by Dox, allowing for short-term gene expression.

2.5.2 Toxicity screening

Toxicity screening may be the most advanced application of pluripotent stem cells for drug screening. Public-private collaborations such as Stem Cells for Safer Medicine or Stem Cells and Drug Discovery Institute have been launched to enable consistent differentiation of stem cells into particular cell types with physiologically relevant phenotypes suitable for toxicology testing. Cardiomyocytes derived from human ES/iPS cells have already begun to be used for drug-induced cardiotoxicity screening (Asai et al., 2010).

In the drug development process, it is necessary to evaluate the metabolism of the novel compound in hepatocytes, because the liver is the main detoxification organ in the body. The models used at present are primary hepatocytes or immortalized cell lines such as HepG2, which present major limitations in terms of supply and their relevance to normal metabolic reactions, respectively (Wilkening et al., 2003; Laustriat et al., 2010). Recently, Inamura et al. reported the efficient generation of hepatoblasts from human pluripotent stem cells (Inamura et al., 2011). However, it is still difficult to obtain homogeneous populations of mature hepatocytes, and contamination with other unwanted cell types decreases the reliability of the toxicity data. Using a pInsert vector equipped with a lineage-specific selection marker and an HS4 insulator downstream of loxP may enable the reproducible production of concentrated hepatocytes.

2.5.3 Applications for cellular medicine

Human pluripotent stem cells are expected to provide sources for cellular therapy as well as drug screening. Geron Corporation already started a phase I clinical trial of transplantation of oligodendrocyte progenitor cells derived from hESCs to patients with spinal cord injury (http://www.geron.com/GRNOPC1Trial/). However, when transplanting differentiated cells from human ES/iPS cells, it is necessary to consider the potential for uncontrollable overproliferation and tumor formation due to implantation of small numbers of undifferentiated cells. Using the pInsert-Tet5-like vector equipped with an undifferentiated state-specific promoter instead of the CAG promoter and a proper suicide gene downstream of TRE may resolve this problem (Fig. 9). Because the undifferentiated cells will die in the presence of Dox, administration of Dox to transplant recipients can remove the undifferentiated cells.

When differentiated cells are provided for cellular therapy, it is important to determine how to differentiate the hES/iPS cells efficiently into a single type of cells. If differentiation promoting factors are identified, they could be useful in the production of hES/iPS cell-derived differentiated cells. However, there have been only a few studies using high-throughput screening methods to discover small molecules that increase the differentiation of hES/iPS cells into specific cell lineages (Borowiak et al., 2009; Zhu et al., 2009). Our pInsert-Tet5 vector can also identify proteins with activities promoting differentiation. First, pInsert-Tet5 libraries which are equipped with a variety of cDNAs downstream of TRE can be introduced into K1-HS cells. Hygromycin-resistant clones can then be induced to desired differentiation lineages, and then Dox can be added at various stages of differentiation. Differentiation-promoting factors at the different stages can then be identified by sequencing the cDNAs from clones whose differentiation efficiency has been increased by the administration of Dox.

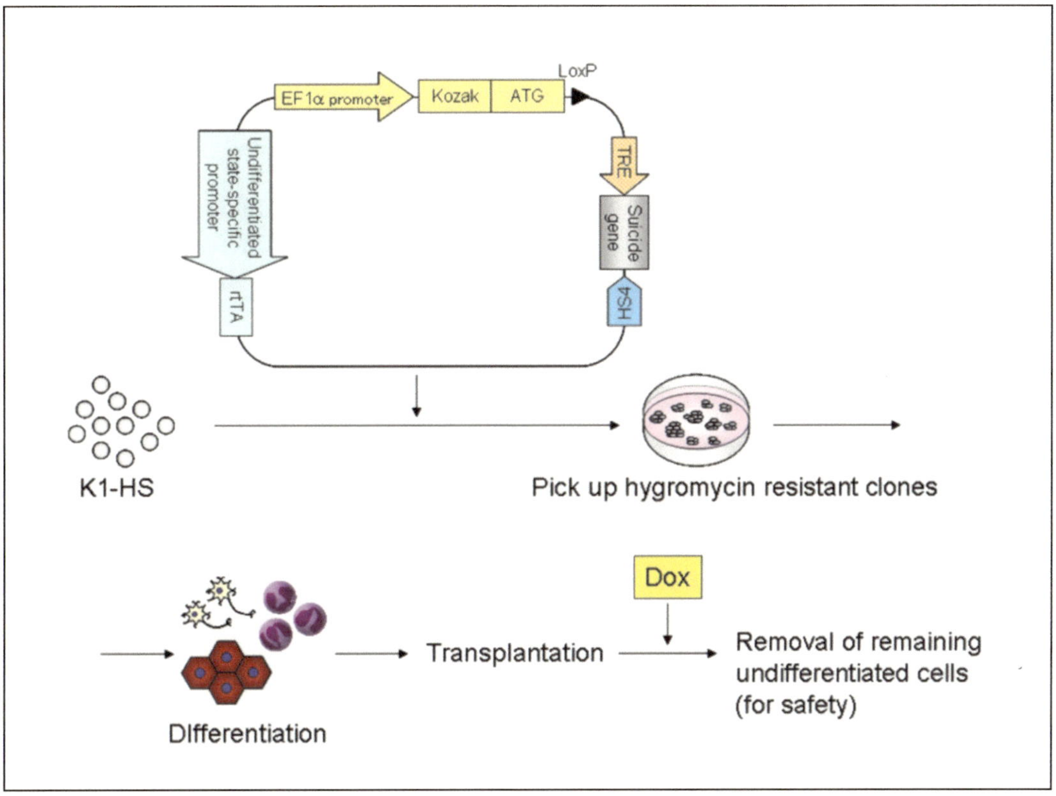

Fig. 9. A theoretic method to remove any remaining undifferentiated cells after transplantation using the inducible gene expression system based on the pInsert vector.

3. Conclusion

We have performed gene targeting in hESCs to introduce a transgene docking site into the HPRT locus, and have demonstrated high efficiency Cre-mediated integration of a given construct into this docking site. Moreover, we have developed a Tet-On system with undetectable background expression based on this site-directed integration. This review also discussed the potential applications of our site-directed integration system for drug discovery, toxicity screening and cell-based therapy.

4. Acknowledgment

We thank all members of the Stem Cell and Drug Discovery Institute for their excellent technical assistance, fruitful discussions and valuable suggestions.

5. References

Asai, Y.; Tada, M.; Otsuji, T.G. & Nakatsuji, N. (2010). Combination of functional cardiomyocytes derived from human stem cells and a highly-efficient microelectrode array system: an ideal hybrid model assay for drug development. *Curr. Stem Cell Res. Ther.*, Vol. 5, No. 3, pp. 227-232, ISSN 1574-888X

Barbaric, I.; Gokhale, P.J. & Andrews, P.W. (2010). High-content screening of small compounds on human embryonic stem cells. *Biochem. Soc. Trans.*, Vol. 38, No. 4, pp. 1046-50, ISSN 0300-5127

Beard, C.; Hochedlinger, K.; Plath, K.; Wutz, A. & Jaenisch, R. (2006). Efficient method to generate single-copy transgenic mice by site-specific integration in embryonic stem cells. *Genesis*, Vol. 44, No. 1, pp. 23-28, ISSN 1526-954X

Borowiak, M.; Maehr, R.; Chen, S.; Chen, A.E.; Tang, W.; Fox, J.L.; Schreiber, S.L. & Melton, D.A. (2009). Small molecules efficiently direct endodermal differentiation of mouse and human embryonic stem cells. *Cell Stem Cell*, Vol. 4, No. 4, pp. 348-358, ISSN 1934-5909

Bronson, S.K.; Plaehn, E.G.; Kluckman, K.D.; Hagaman, J.R.; Maeda, N. & Smithies, O. (1996). Single-copy transgenic mice with chosen-site integration. *Proc. Natl. Acad. Sci. U S A*, Vol. 93, No. 17, pp. 8804-8808, ISSN 0027-8424

Chung, J.H.; Whiteley, M. & Felsenfeld, G. (1993). A 5' element of the chicken beta-globin domain serves as an insulator in human erythroid cells and protects against position effect in Drosophila. *Cell*, Vol. 74, No. 3, pp. 505-514, ISSN 0092-8674

Di Domenico, A.I.; Christodoulou, I.; Pells, S.C.; McWhir, J. & Thomson, A.J. (2008). Sequential genetic modification of the hprt locus in human ESCs combining gene targeting and recombinase-mediated cassette exchange. *Cloning and stem cells*, Vol. 10, No. 2, pp. 217-230, ISSN 1536-2302

Dimos, J.T.; Rodolfa, K.T.; Niakan, K.K.; Weisenthal, L.M.; Mitsumoto, H.; Chung, W.; Croft, G.F.; Saphier, G.; Leibel, R.; Goland, R.; Wichterle, H.; Henderson, C.E. & Eggan, K. (2008). Induced pluripotent stem cells generated from patients with ALS can be differentiated into motor neurons. *Science*, Vol. 321, No. 5893, pp. 1218-1221, ISSN 0036-8075

Du, Z.W.; Hu, B.Y.; Ayala, M.; Sauer, B. & Zhang, S.C. (2009). Cre Recombination-Mediated Cassette Exchange for Building Versatile Transgenic Human Embryonic Stem Cells Lines. *Stem Cells*, Vol. 27, No. 5, pp. 1032-1041, ISSN 1066-5099

Ebert, A.D.; Yu, J.; Rose, F.F. Jr.; Mattis, V.B.; Lorson, C.L.; Thomson, J.A. & Svendsen, C.N. (2009). Induced pluripotent stem cells from a spinal muscular atrophy patient. *Nature*, Vol. 457, No. 7227, pp. 277-280, ISSN 0028-0836

Ebert, A.D. & Svendsen, C.N. (2010). Human stem cells and drug screening: opportunities and challenges. *Nat. Rev. Drug Discov.*, Vol. 9, No. 5, pp. 367-372, ISSN 1474-1776

Ellis, J. (2005). Silencing and variegation of gammaretrovirus and lentivirus vectors. *Hum. Gene. Ther.*, Vol. 16, No. 11, pp. 1241-1246, ISSN 1043-0342

Fukushige, S. & Sauer, B. (1992). Genomic targeting with a positive-selection lox integration vector allows highly reproducible gene expression in mammalian cells. *Proc. Natl. Acad. Sci. U S A*, Vol. 89, No. 17, pp. 7905-7909, ISSN 0027-8424

Hacein-Bey-Abina, S.; Von Kalle, C.; Schmidt, M.; McCormack, M.P.; Wulffraat, N.; Leboulch, P.; Lim, A.; Osborne, C.S.; Pawliuk, R.; Morillon, E.; Sorensen, R.; Forster, A.; Fraser, P.; Cohen, J.I.; de Saint Basile, G.; Alexander, I.; Wintergerst, U.; Frebourg, T.; Aurias, A.; Stoppa-Lyonnet, D.; Romana, S.; Radford-Weiss, I.; Gross, F.; Valensi, F.; Delabesse, E.; Macintyre, E.; Sigaux, F.; Soulier, J.; Leiva, L.E.; Wissler, M.; Prinz, C.; Rabbitts, T.H.; Le Deist, F.; Fischer, A. & Cavazzana-Calvo, M. (2003). LMO2-associated clonal T cell proliferation in two patients after gene therapy for SCID-X1. *Science*, Vol. 302, No. 5644, pp. 415-419, ISSN 0036-8075

Hasegawa, K.; Fujioka, T.; Nakamura, Y.; Nakatsuji, N. & Suemori, H. (2006). A method for the selection of human embryonic stem cell sublines with high replating efficiency after single-cell dissociation. *Stem Cells*, Vol. 24, No. 12, pp. 2649-2660, ISSN 1066-5099

Hoffman, L.M. & Carpenter, M.K. (2005). Characterization and culture of human embryonic stem cells. *Nat. Biotechnol.*, Vol. 23, No. 6, pp. 699-708, ISSN 1087-0156

Inamura, M.; Kawabata, K.; Takayama, K.; Tashiro, K.; Sakurai, F.; Katayama, K.; Toyoda, M.; Akutsu, H.; Miyagawa, Y.; Okita, H.; Kiyokawa, N.; Umezawa, A.; Hayakawa, T.; Furue, M.K. & Mizuguchi, H. (2011). Efficient Generation of Hepatoblasts From Human ES Cells and iPS Cells by Transient Overexpression of Homeobox Gene HEX. *Mol. Ther.*, Vol. 19, No. 2, pp. 400-407, ISSN 1525-0016

Irion, S.; Luche, H.; Gadue, P.; Fehling, H.J.; Kennedy, M. & Keller, G. (2007). Identification and targeting of the ROSA26 locus in human embryonic stem cells. *Nat. Biotechnol.*, Vol. 25, No. 12, pp. 1477-1482, ISSN 1087-0156

Kola, I. & Landis, J. (2004). Can the pharmaceutical industry reduce attrition rates? *Nat. Rev .Drug Discov.*, Vol. 3, No. 8, pp. 711-715, ISSN 1474-1776

Kozak, M. (1987). An analysis of 5'-noncoding sequences from 699 vertebrate messenger RNAs. *Nucleic Acids Res.*, Vol. 15, No. 20, pp. 8125-8148, ISSN 0305-1048

Laustriat, D.; Gide, J. & Peschanski, M. Human pluripotent stem cells in drug discovery and predictive toxicology. *Biochem. Soc. Trans.*, Vol. 38, No. 4, pp. 1051-57, ISSN 0300-5127

Lee, G.; Papapetrou, E.P.; Kim, H.; Chambers, S.M.; Tomishima, M.J.; Fasano, C.A.; Ganat, Y.M.; Menon, J.; Shimizu, F.; Viale, A.; Tabar, V.; Sadelain, M. & Studer, L. (2009). Modelling pathogenesis and treatment of familial dysautonomia using patient-specific iPSCs. *Nature*, Vol. 461, No. 7262, pp. 402-406, ISSN 0028-0836

Liew, C.G.; Draper, J.S.; Walsh, J.; Moore, H. & Andrews, P.W. (2007). Transient and stable transgene expression in human embryonic stem cells. *Stem Cells*, Vol. 25, No. 6, pp. 1521-1528, ISSN 1066-5099

Lombardo, A.; Genovese, P.; Beausejour, C.M.; Colleoni, S.; Lee, Y.L.; Kim, K.A.; Ando, D.; Urnov, F.D.; Galli, C.; Gregory, P.D.; Holmes, M.C. & Naldini, L. (2007). Gene editing in human stem cells using zinc finger nucleases and integrase-defective lentiviral vector delivery. *Nat. Biotechnol.*, Vol. 25, No. 11, pp. 1298-1306, ISSN 1087-0156

Maehr, R.; Chen, S.; Snitow, M.; Ludwig, T.; Yagasaki, L.; Goland, R.; Leibel, R.L. & Melton, D.A. (2009). Generation of pluripotent stem cells from patients with type 1 diabetes. *Proc. Natl. Acad. Sci. U S A*, Vol. 106, No. 37, pp. 15768-15773, ISSN 0027-8424

Martignoni, M.; Groothuis, G.M. & de Kanter, R. (2006). Species differences between mouse, rat, dog, monkey and human CYP-mediated drug metabolism, inhibition and induction. *Expert Opin. Drug Metab. Toxicol.*, Vol. 2, No. 6, pp. 875-894, ISSN 1742-5255

Mateizel, I.; De Temmerman, N.; Ullmann, U.; Cauffman, G.; Sermon, K.; Van de Velde, H.; De Rycke, M.; Degreef, E.; Devroey, P.; Liebaers, I. & Van Steirteghem, A. (2006). Derivation of human embryonic stem cell lines from embryos obtained after IVF and after PGD for monogenic disorders. *Hum. Reprod.*, Vol. 21, No. 2, pp. 503-511, ISSN 0268-1161

McCreath, K.J.; Howcroft, J.; Campbell, K.H.; Colman, A.; Schnieke, A.E. & Kind, A.J. (2000). Production of gene-targeted sheep by nuclear transfer from cultured somatic cells. *Nature*, Vol. 405, No. 6790, pp. 1066-1069, ISSN 0028-0836

Nienhuis, A.W.; Dunbar, C.E. & Sorrentino, B.P. (2006). Genotoxicity of retroviral integration in hematopoietic cells. *Mol. Ther.*, Vol. 13, No. 6, pp. 1031-1049, ISSN 1525-0016

Otsuki, A.; Tahimic, C.G.; Tomimatsu, N.; Katoh, M.; Chen, D.J.; Kurimasa, A. & Oshimura, M. (2005). Construction of a novel expression system on a human artificial chromosome. *Biochem. Biophys. Res. Commun.*, Vol. 329, No. 3, pp. 1018-1025, ISSN 0006-291X

Park, I.H.; Arora, N.; Huo, H.; Maherali, N.; Ahfeldt, T.; Shimamura, A.; Lensch, M.W.; Cowan, C.; Hochedlinger, K. & Daley, G.Q. (2008). Disease-specific induced pluripotent stem cells. *Cell*, Vol. 134, No. 5, pp. 877-886, ISSN 0092-8674

Pickering, S.J.; Minger, S.L.; Patel, M.; Taylor, H.; Black, C.; Burns, C.J.; Ekonomou, A. & Braude, P.R. (2005). Generation of a human embryonic stem cell line encoding the cystic fibrosis mutation deltaF508, using preimplantation genetic diagnosis. *Reprod. Biomed. Online*, Vol. 10, No. 3, pp. 390-397, ISSN 1472-6483

Ruby, K.M. & Zheng, B. (2009). Gene targeting in a HUES line of human embryonic stem cells via electroporation. *Stem cells*, Vol. 27, No. 7, pp. 1496-1506 ISSN 1066-5099

Sakurai, K.; Shimoji, M.; Tahimic, C.G.; Aiba, K.; Kawase, E.; Hasegawa, K.; Amagai, Y.; Suemori, H. & Nakatsuji, N. (2010). Efficient integration of transgenes into a defined locus in human embryonic stem cells. *Nucleic Acids Res.*, Vol. 38, No. 7 e96, ISSN 0305-1048

Smith, J.R.; Maguire, S.; Davis, L.A.; Alexander, M.; Yang, F.; Chandran, S.; ffrench-Constant, C. & Pedersen, R.A. (2008). Robust, persistent transgene expression in human embryonic stem cells is achieved with AAVS1-targeted integration. *Stem Cells*, Vol. 26, No. 2, pp. 496-504, ISSN 1066-5099

Soldner, F.; Hockemeyer, D.; Beard, C.; Gao, Q.; Bell, G.W.; Cook, E.G., Hargus, G.; Blak, A.; Cooper, O.; Mitalipova, M.; Isacson, O. & Jaenisch, R. (2009). Parkinson's disease patient-derived induced pluripotent stem cells free of viral reprogramming factors. *Cell*, Vol. 136, No. 5, pp. 964-977, ISSN 0092-8674

Stephenson, E.L.; Mason, C. & Braude, P.R. (2009). Preimplantation genetic diagnosis as a source of human embryonic stem cells for disease research and drug discovery. *BJOG*, No. 116, Vol. 2, pp. 158-165, ISSN 1470-0328

Suemori, H.; Yasuchika, K.; Hasegawa, K.; Fujioka, T.; Tsuneyoshi, N. & Nakatsuji, N. (2006). Efficient establishment of human embryonic stem cell lines and long-term maintenance with stable karyotype by enzymatic bulk passage. *Biochem. Biophys. Res. Commun.*, Vol. 345, No. 3, pp.926-932 ISSN 0006-291X

Suzuki, K.; Mitsui, K.; Aizawa, E.; Hasegawa, K.; Kawase, E.; Yamagishi, T.; Shimizu, Y.; Suemori, H.; Nakatsuji, N. & Mitani K. (2008). Highly efficient transient gene expression and gene targeting in primate embryonic stem cells with helper-dependent adenoviral vectors. *Proc. Natl. Acad. Sci. U S A*, Vol. 105, No. 37, pp. 13781-13786, ISSN 0027-8424

Thomson, J.A.; Itskovitz-Eldor, J.; Shapiro, S.S.; Waknitz, M.A.; Swiergiel, J.J.; Marshall, V.S. & Jones, J.M. (1998). Embryonic stem cell lines derived from human blastocysts. *Science*, Vol. 282, No. 5391, pp. 1145-1147, ISSN 0036-8075

Thyagarajan, B.; Liu, Y.; Shin, S.; Lakshmipathy, U.; Scheyhing, K.; Xue, H.; Ellerström, C.; Strehl, R.; Hyllner, J.; Rao, M.S. & Chesnut, J.D. (2008). Creation of engineered human embryonic stem cell lines using phiC31 integrase. *Stem Cells*, Vol. 26, No. 1, pp. 119-126, ISSN 1066-5099

Urbach, A.; Schuldiner, M. & Benvenisty, N. (2004). Modeling for Lesch-Nyhan disease by gene targeting in human embryonic stem cells. *Stem Cells*, Vol. 22, No. 4, pp. 635-641, ISSN 1066-5099

Verlinsky, Y.; Strelchenko, N.; Kukharenko, V.; Rechitsky, S.; Verlinsky, O.; Galat, V. & Kuliev, A. (2005). Human embryonic stem cell lines with genetic disorders. *Reprod. Biomed. Online*, Vol. 10, No. 1, pp. 105-110, ISSN 1472-6483

Wilkening, S.; Stahl, F. & Bader A. (2003). Comparison of primary human hepatocytes and hepatoma cell line Hepg2 with regard to their biotransformation properties. *Drug Metab. Dispos.*, Vol. 31, No. 8, pp. 1035-1042, ISSN 0090-9556

Xia, X.; Zhang, Y.; Zieth, C.R. & Zhang, S.C. (2007). Transgenes delivered by lentiviral vector are suppressed in human embryonic stem cells in a promoter-dependent manner. *Stem Cells Dev.*, Vol. 16, No. 1, pp. 167-176, ISSN 1547-3287

Ye, L.; Chang, J.C.; Lin, C.; Sun, X.; Yu, J.; & Kan, Y.W. (2009). Induced pluripotent stem cells offer new approach to therapy in thalassemia and sickle cell anemia and option in prenatal diagnosis in genetic diseases. *Proc. Natl. Acad. Sci. U S A*, Vol. 106, No. 24, pp. 9826-9830, ISSN 0027-8424

Ye, Z.; Zhan, H.; Mali, P.; Dowey, S.; Williams, D.M.; Jang, Y.Y.; Dang, C.V.; Spivak, J.L.; Moliterno, A.R. & Cheng, L. (2009). Human-induced pluripotent stem cells from blood cells of healthy donors and patients with acquired blood disorders. *Blood*, Vol. 114, No. 27, pp. 5473-5480, ISSN 0006-4971

Zambrowicz, B.P.; Imamoto, A.; Fiering, S.; Herzenberg, L.A.; Kerr, W.G. & Soriano, P. (1997). Disruption of overlapping transcripts in the ROSA beta geo 26 gene trap strain leads to widespread expression of beta-galactosidase in mouse embryos and hematopoietic cells. *Proc. Natl. Acad. Sci. U S A*, Vol. 121, No. 5, pp. 3789-3794. ISSN 0027-8424

Zhu, S.; Wurdak, H.; Wang, J.; Lyssiotis, C.A.; Peters, E.C.; Cho, C.Y.; Wu, X. & Schultz, P.G. (2009). A small molecule primes embryonic stem cells for differentiation. *Cell Stem Cell*, Vol. 4, No. 5, pp. 416-426, ISSN 1934-5909

Zou, J.; Maeder, M.L.; Mali, P.; Pruett-Miller, S.M.; Thibodeau-Beganny, S.; Chou, B.K.; Chen, G.; Ye, Z.; Park, I.H.; Daley, G.Q.; Porteus, M.H.; Joung, J.K. & Cheng, L. (2009). Gene targeting of a disease-related gene in human induced pluripotent stem and embryonic stem cells. *Cell Stem Cell*, Vol. 5, No. 1, pp. 97-110, ISSN 1934-5909

Zwaka, T.P. & Thomson, J.A. (2003). Homologous recombination in human embryonic stem cells. *Nat. Biotechnol.*, Vol. 21, No. 3, pp. 319-21, ISSN 1087-0156

8

Embryonic Stem Cells: Introducing Exogenous Regulators into Embryonic Stem Cells

Yong-Pil Cheon
Sungshin Women's University, Dongseondong 3ga, Seoul, Korea

1. Introduction

The application of embryonic stem (ES) cells to research and therapy has been a landmark development in science. Cell therapy using ES cells depends on the progress of the stable culture conditions and differentiation induction methods. ES cells were first obtained directly from inner cell masses (ICMs) of blastocysts by Evans and Kaufman (1981), and Martin (1981). These cells can self renew to produce stem cell itself and repopulate into many different tissues, including the somatic- and germ-cell lineages in chimeras (Bradley et al., 1984). After developing the methods for establishing the ES (mES) cell lines of mice, ES cell lines of other species including primates and human were also established (Thomson et al., 1995, 1998). Although ES cells can be obtained from 8-cell stage embryos or morula, they are more easily obtained from blastocyst stage embryos (Delhaise et al., 1996; Strelchenko et al., 2004). Now the sources of ES cells have been extended to include epiblast cells (Nichols and Smith, 2011) (Fig. 1).

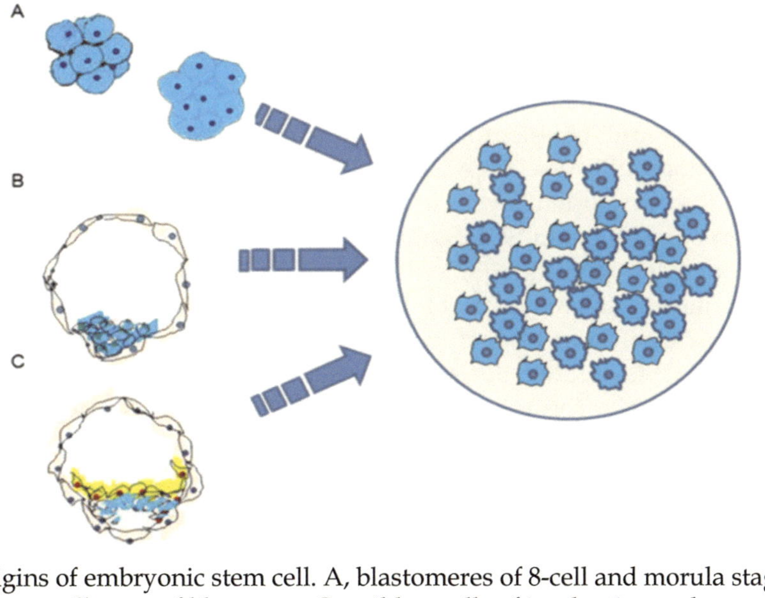

Fig. 1. The origins of embryonic stem cell. A, blastomeres of 8-cell and morula stage embryos; B, inner cell mass of blastocyst; C, epiblast cells of implanting embryo.

2. Human and mouse embryonic stem cells

2.1 Mouse embryonic stem cell

mES cells have been used mainly in the generation of mutant mice and in investigating cellular differentiation and the physiological role of genes during embryogenesis and development. These things can be accomplished through gene targeting and transmission of the acquired characters. Transmission of foreign genes to the next generation can be accomplished with incorporation of foreign genes into the germ cell-lineage (Bradley et al., 1984; Wang et al., 1996). Specific gene deficient mice can be developed by manipulation of the genes in ES cells and introduce of ES cells into ICM (Hooper et al., 1987; Kuehn et al., 1987; Wang et al., 1996). Mutations targeting specific genes by homologous recombination in ES cells have been designed and initially achieved using hypoxanthine-guanine phosphoribosyl transferase (HPRT) (Doetschmann et al., 1987). Germ line transmission of a targeted mutation in a gene other than HPRT was first obtained with c-abl (Schwartzberg et al., 1989) and β2-microglobulin (Zijlstra et al., 1989). To date, several hundred genes have been disrupted by homologous recombination in ES cells. The application of these techniques provided a way to evaluate the physiological roles of specific genes in ontogeny. In addition, conditional knockout mice are developed using ES cells and are used to analyze gene functions in a specific tissue or developmental stage. These approaches have provided much novel information and several important tools to study human development including ES cell therapies.

2.2 Early data from embryonic carcinoma cells

Human ES (hES) cells have become a key word in basic sciences and applied sciences. Before the optimal conditions were established for ES cell cultures, most of the basic knowledge about ES cells came from studies on embryonic carcinoma (EC) cells (Jakob, 1984; Rossant and Papaioannou, 1984; Smolich and Papkoff, 1994). The first human EC cell line was established in vitro in 1985 (Sekiya et al., 1985). EC cells can spontaneously differentiate into somatic cells, germ cells and extraembryonic cells (Zákány et al. 1984), and have self-renewal ability (Andrews, 1987; Rosenstraus et al., 1984). During differentiation of EC cells into other cell types, proliferation is prerequisite, and apoptosis of EC cell derivatives accompany this process (Azizi et al., 2010; Mummery et al., 1984). After starting differentiation, the expression levels of some genes are determined at the levels of transcription and translation. In 1984, Schindler and Sherman reported the profile of protein and mRNA expression in murine EC cells, and the expression patterns of specific genes were explored during differentiation of EC cells (Scott et al., 1984; Silvan et al., 2009). Additionally, studies on proteins and carbohydrates have been performed to characterize the EC cells and their induced descendants (Amano et al., 2010; Andrews et al., 1984; Jemmerson et al., 1985). The role of the extracellular matrix is also examined during differentiation induction of EC cells (Grabel, 1984; Wartiiovaara et al., 1984) as the karyoplasm in EC cell gene expression (Gautsch, 1982). Differentiation induction of EC cells has been successful with various substances, such as retinoic acid, dibutyryl cyclic AMP (Jones-Villeneuve et al., 1983; Muramatsu and Muramatsu, 1983), dimethyl sulfoxide (Edwards et al., 1983) and coculture with specific cell types (Allin, 1984). Bell et al (1984) used growth factors to induce EC cell differentiation in target tissues.

2.3 Stemness and the differentiation of embryonic stem cells

Although ES cells share some characteristics with EC cells, they have important differences (Chamber and Smith, 2004). The in vitro developmental potency and differentiation of ES cells have been studied since the mid-1980s. The developmental potency of ES cells was studied in vitro by Doetschman et al (1985). Additionally, the pluripotency and the characteristics of ES cells in vitro were analyzed in depth. ES cells lack differentiation-inhibitory activity (DIA) unlike cells of cell lines, and they can differentiate into a wide variety of cell types. ES cells have an apparently normal diploid karyotype during long-term culture, can extensively colonize embryos without causing tumors or developmental anomalies, and can form normal gametes when differentiated into the germ line (Kaufman et al., 1983; Martin, 1981; Suda et al., 1987). The purified and cloned myeloid leukemia inhibitory factor (LIF) can exhibit DIA in the maintenance of competent mES cell lines (Smith et al., 1988; Williams et al., 1988). However, the requirement for maintaining stemness varies with the source of the ES cells. Primate ES cells stay in an undifferentiated state when grown on embryonic fibroblast feeder layers but differentiate or die in the absence of feeder cells, despite the presence of recombinant leukemia inhibitory factor (Thomson et al., 1995).

Studies on the differentiation of stem cells during organogenesis have a long history (Ruch, 1967). ES cell differentiation in vitro is controlled by exogenous factors, particularly paracrine factors (Heath and Smith, 1986; Spence et al., 2010). Various cell types can be induced from embryoid bodies using paracrine factors (Risau et al., 1988). During differentiation induction, various genes, such as genes of transcription factors, are up-regulated or down-regulated (Dyson et al., 1989). The up-regulated transcription factors trigger a series of gene expression changes for the formation of a specific cell type (Kopp et al., 2008).

3. Potency of embryonic stem cell in clinic

3.1 Sources of human embryonic stem cells

Through the accumulating knowledge on pluripotency and on the control of cellular differentiation, a fundamental understanding of developmental biology at the cellular and molecular levels has been expanded. It represents a gateway to major future clinical applications of these principles. The levels of transcription factors including nuclear receptors in ES cells are critical in the maintenance of stemness or differentiation of ES cells (Jeong and Mangelsdorf, 2009; Redshaw and Strain, 2010). Certain nuclear receptors are involved in the maintenance of ES cells in cooperation with Oct3/4, Nanog and Klf4 upon exogenous signals. Liver receptor homolog 1 (LRH1) and estrogen-related receptor beta (ERRβ) have functional roles in the maintenance of stemness of ES cells. On the other hand, germ cell nuclear factor (GCNF) and retinoic acid receptors (RARs) promote ES cell differentiation (Jeong and Mangelsdorf, 2009). The *Sall* gene family that encodes a group of developmental transcription factors controls the embryonic development and is also involved in the determination of ES cell fate. *Sall1* knockout mice die perinatally with kidney dysgenesis or agenesis. *Sall3* knockout mice have plate deficiency and abnormalities in cranial nerves. Sall4 mutation causes Acro-renal-ocular syndrome and Duane-radial ray syndrome. *Sall4* also plays a role in the maintenance of stemness of ES cells through controlling its own expression and the expression of OCT4 (Yang et al., 2010). Identifying such switching molecules that regulate ES cell self-renewal versus differentiation can provide insights into the nature of the ES cell (Navarro et al., 2010). Chromatin methylation is one example of how an epigenetic modification can modulate ES cell fate (Christophersen and Helin, 2010).

In the late of 1990s the possibility of using ES cells for medicine was suggested (Gearhart, 1998). The use of stem cells in regenerative medicine already has a long history, for example, in bone marrow transplantation and skin grafting. With the establishment of human embryonic stem (hES) cell lines, future clinical applications based on their developmental and regenerative abilities will become possible. hES cells were first established in 1998 by Thomson and colleagues. As in primate ES cells, the feeder layer seems essential for maintaining the hES cell lines. It is suggested that TGF-beta and myofibroblasts can support the propagation of hES cells in vitro (Kumar et al., 2010).

Previously, the source of hES cells was restricted to morula and blastocysts but this is no longer the case. In 2006, induced pluripotent stem (iPS) cells were generated from mouse embryonic or adult fibroblasts by the retrovirus-mediated introduction of four transcription factors, Oct3/4, sox2, c-Myc, and Klf4 (Takahashi and Yamanaka, 2006). In 2007, iPS cells were established from human fibroblasts via introduction by the same four factors (Takahashi et al., 2007), and by using a slightly different combination of genes (Yu et al., 2007). Human iPS cells are similar to human ES cells in many aspects, e.g., morphology and proliferation ability. Furthermore, mouse iPS cells are competent for adult chimeric mice and germline transmission. Growth potential, gene expression patterns, and the epigenetic status of iPS cells are similar to those observed in hES cells. iPS cells can differentiate into all three germ layers through embryoid bodies and to teratomas. In addition, they have been differentiated directly into neurons, beating cardiomyocytes in vitro (Carvajal-Vergara et al., 2010) and others (Liu et al., 2011).

3.2 Applied developmental biology

A fertilized mammalian oocyte is totipotent and develops into the fetus and placenta concurrently losing its totipotency. During cleavage, the transition from maternal to embryonic genome activation encompassed in early stage embryos and cleaving embryos can gain the competence to become blastocysts (Hamatani et al., 2004). The competency to accomplish implantation and further development is gained during the periimplantation stages. For that process, the epithelium like structure, trophectoderm is formed and interacts with the epithelium of the uterine endometrium. On the other hand, the cells isolated within the blastoceol become inner cell mass (ICM) (Johnson and Mcconnell, 2004). ICM is a mass of cells having pluripotency and its cells can differentiate into all cell types of adult and extraembryonic membranes (Fernanez-Tresquerres et al., 2010). In specific in vitro conditions, ICM cells can keep their potency and maintain their proliferation ability (Yamanakaet al., 2006).

Grouped cell movement and sequential spatio-temporal expressions of gene sets was observed after implantation. As a consequence of gastrulation, the embryos have three germ layers, endoderm, mesoderm and ectoderm (Hassoun et al., 2010; Luxardi et al., 2010). This process is accomplished with the forming of the axis (Rocha et al., 2010). For example, Med12, a large protein complex of 30 subunits interacting with transcription factors, is essential in correct Wnt/β-catenin and Wnt/planar cell polarity signaling (Rocha et al., 2010; Taatjes et al., 2004).

Most cells of embryos are eventually committed through triggering to a particular developmental pathway from which they rarely depart (Belting et al., 2004). Cell determination is a stepwise process, and that full determination is preceded by a state of bias. This bias is either conformed or modified through cell interactions later in development. It is accomplished through community effect, lateral inhibition among equivalent cells, and embryonic induction which is the interaction between nonequivalent cells (Artavanis-Tsakonas et al., 1999). The determined state is almost stably passed on during mitosis. These are matched with the default progression and/or induced progression.

A specific type cell is derived from a preexisting progenitor or stem cell for the specific area (Belting et al., 2004; Frieberg, 1996). Designed modification of a group of ES cells to a specific cell type is mostly achieved by step by step induction through morphogens or intracellular signaling molecules as shown during organogenesis (Osakada and Takahashi, 2011). The accumulation of knowledge regarding tissue genesis and organogenesis during gastrulation will give rise to the accomplishment of goals across other fields.

The revolutionary advances in developmental biology have extended the utility of ES cells to both therapy and industry. As applied to the study of developmental biology (Lehtonen et al., 1989), the stemness and developmental potency of ES cells can be applied to various fields of medicine (Gearhart, 1998; Saxena et al., 2010). Thus far, stem cell biology has focused on identifying novel pathways such as those that maintain pluripotency and induce specific cell types. However, to date, it is not yet feasible to use ES cells in medicine or to strictly control their differentiation to a specific cell type in humans. To readily apply ES cells in therapy or other industries, the improvement of efficient protocols to direct stem cell differentiation into well-defined lineages is critical (Heng et al., 2004).

4. Approaches to modulate embryonic stem cell properties

To obtain a specific cell type, it is critical to understand the interactions of multiple genes and the associated factors that are involved in the differentiation and de-differentiation of ES cells. When we use inducers to generate a specific cell type from the embryoid body (EB) or directly from ES cells, it is difficult to generate single cell types at one time (Lu et al., 2009; Park and Lee, 2007). As shown in gastrulation and organogenesis, most morphogens and inducers are secreted into the intercellular fluid and into induced concentration-dependent specific gene expressions in neighboring cells (Wolpert, 1978). Therefore, in most cases of induction by inducer, unexpected results can occur during the induction of a cell type from an ES cell or embryoid bodies (EBs).

Selective activation of the gene sets is essential to get a unique function of a specific cell. One of the best ways to accomplish this is through the delivery of a construct which contains target genes and expression-control sequences (Ishizaka et al., 2002). Among them intracellular signaling mediators and transcription factors are key molecules to get a specific cell type and they are main target for differentiation induction. To meet these requirements, many experimental conditions and gene delivery systems have been used. However, ES cells cannot be translocated to high affinity variable regulators such as a molecule of DNA, RNA or proteins. To overcome this problem, various approaches have been developed. When we classified according to their characteristics, mediators are either chemical, physical, or viral.

Induced pluripotent stem cell (iPS) is a good example of a cell type in which transgene delivery and induction by transcription factors has been successful (Meissner et al., 2007; Park et al., 2008). However, those strategies introduce changes in DNA sequences or disorder in genomic equivalence. Moreover, viral systems are suspected to have life-threatening effects of immunogenicity and carcinogenicity. In addition, the efficiency of gene transfer in hES cells is still poor compared with other cell lines (Cao et al., 2010; Wasungu and Hoekstra, 2006).

For these reasons, many scientists hesitate to use these strategies in medicine and have searched for new ways to control ES cell differentiation. Functional protein is easily turned over in the cytoplasm and regulates the function of a cell without disturbing the genetic background. Therefore, if we develop an easy way of introducing the transcription factors, intracellular signaling molecules or drug molecules into ES cells, ES cells will become a safe medicine.

4.1 Delivery of nucleotides in to ES cells

Introduction of regulating DNA or RNA into ES cells can be mediated by chemicals. Leaped progress in chemical mediated method has led to successful expression-construct delivery. Chemical mediators include diethylaminoethyl-dextran (DEAE-D), calcium phosphate, cationic lipid (liposome), cationic polymer, polylysine, histone, chitosan and peptides.

The delivery of DNA or RNA with calcium phosphate was developed by Graham and van der Eb (1973) and has become a common method (Marucci et al., 2011). The mixture of DNA construct, calcium ion and phosphate are presented to cells in culture and the cells import the mixture through endocytosis. DEAE-D was developed by Pagano and Vaheri in 1965 for enhancing the infectivity of poliovirus RNA for cell culture. DEAE-D is a polycation, and the mixture of DNA and DEAE-D is positively charged. It is known that the mixtures are transferred into cytoplasm through endocytosis. Cationic lipids for DNA-transfection procedures were developed in 1987 (Felgner et al., 1987) and have become one of the most common methods (Templeton et al., 1997). Cationic lipids form liposome and the surface of liposome is positively charged. Cationic polymers are a group of highly water-soluble molecules such as polyethyleneimine (PEI) and dendrimers (Boussif et al., 1995; Dunlap et al., 1997). There are several types of cationic polymers; linear (polylysine, spermine, and histone), branched (polyethyleneimine, dendrimers) and spherical (Chitosan) (Zhao et al., 2006). Cationic polymer self-assemble with DNA and generate tortoidal or spherical particles (Tang and Szoka, 1997). Polyplexes are engulfed by cells. Cationic peptide carriers are a new development and are expected to have an important role in gene delivery in vivo and in vitro. The peptides bind with DNA through ionic interaction to the phosphate backbone and additional noncovalent bonds. The transport mechanisms are endosomolytic (Chen et al., 2001) or membrane-penetrating (Xia et al., 2001). This method is not commonly used to carry the DNA or RNA constructs but the modular design of proteins is a good method for gene delivery (Xavier et al., 2009).

Among the chemical mediated methods, mostly cationic lipids are employed in ES cells (Liou et al., 2010; Ma et al., 2004). One of the merits of the cationic lipid methods is that it is simple to apply. Methods which are optimized to ES cells or EB are a developing field (Liou et al., 2010; Ma et al., 2004; Villa-Diaz et al., 2010). In the case of the other chemical mediators, the efficiency is quite low or inadaptable to ES cells (Hong et al., 2004). On the other hand, combined methods are developed in ES cells such as peptide-liposome (Torchilin et al., 2003)

On the other hand, physical forces such as, pressure, sound, shock, wave, or electrical pulse are used to deliver constructs. The history of using physical methods for gene delivery is relatively short. The good points of these methods are saving cost, reducing risk through the large amount of DNA, and standardizing the quantity and procedure of gene delivery. The restriction of physical delivery includes restriction of delivery into the nucleus and internal organs or tissues that are difficult to reach. The microinjection method is one widely used procedure to delivery DNA or RNA constructs directly into the cells. Using injection pipette the constructs are introduced into the cytosol or microorganells of the target cell and it makes it possible to study the complicated cellular processes, structure, and function. Usually direct DNA inoculation by conventional needle injection and hydrodynamics deliver the naked plasmid DNA into cells, tissues or organs in vivo. These are very restricted, so far (Davis et al., 1993; Zhang et al., 1999). The genetic shotgun was introduced for gene delivery in 1990 (Armaleo et al., 1990). In this protocol, microscopic tungsten spheres are used to deliver DNA or RNA constructs. This method has been used to transfect plants cells, muscle cells, and various cultured cells including epithelial cells, endothelial cells, and monocytes (Chou et al., 2004). Gene delivery using electroporation is the most versatile method and can be applied to a wide variety of cell types. A short-pulsed electric field can result in the cellular uptake of DNA. Ultrasound is another physical energy source to open the membrane. Using this energy, new gene delivery methods have been developed. It facilitates the transfer of DNA into cells and across tissues. Its efficiency is very high and intraturmoral injection of DNA followed by focused ultrasound increased the expression of the target gene (Watanabe et al., 2010). The polymer-encapsulated microsphere delivery system developed the DNA construct in a manner of controlled drug-delivery. It can deliver the constructs in a manner that has site-specificity, is nuclease-safe, and provides a sustained release of DNA without repeated administration (Little et al., 2004). Polymer-encapsulated microsphere delivery may be useful in therapy in vivo (Little et al., 2004).

Among the physical force mediated transduction methods, electroporation has been used as a standard transfection method for mES cells because its efficiency is quite high (Ma et al., 2004). However, its use has also been limited because the viability of the transfected cells is quite low (Matsuoka et al., 2007; Svingen et al., 2009). The other limit is that ES cells have to located in a spedific chamber to get treatment. However, with the progression of the electroporation system, it is possible to get a large fraction of transiently transfected cells with minimal loss of cell viability and pluripotency (Moore et al., 2010). The microinjection method, conventional needle injection, hydrodynamics deliver, genetic shotgun, ultrasound and polymer-encapsulated microsphere delivery systems are not yet commonplace and, although further research is needed, it is expected that physical mediators will be good tools for working with ES cells.

Virus mediate foreign gene delivery is observed in nature. Viruses have host specificity and the efficiency of viral infection can be increased by higher titers while avoiding immunosurveillance by an infected host. Using those characteristics, viral genomes have been developed as gene-delivery vehicles, since the early 1980s (Berkner, 1988; Shimotohno and Temin, 1981). Viral vectors include the adenoviruses, adeno-associated viruses, herpes simplex viruses, baculoviruses, lentiviruses, retroviruses and alphaviruses (Robbins and Ghivizzani, 1998; Couto and High 2010). The characteristics of vectors are dependent on their origins. Varial mediated gene delivery is a powerful method used with ES cells. iPS cells are a good example of viral mediate gene delivery of functional DNA constructs (Meissner et al., 2007; Park et al., 2008).

4.2 Delivery of functional proteins

DNA constructs are powerful mediators for induction of ES cell differentiation. However, genetic modification by exogenous DNA construct can cause unidentified side effects. The development of a safe and efficient differentiation controller is, therefore, an urgent requirement for the effective implementation of stem cells in therapy and industry. As an efficient differentiation controller, functional proteins such as signal transduction proteins and transcription factors are suggested, because they degrade rapidly in physiological condition and can maintain or modulate the cell types.

The molecular size of the protein is huge compared with the amino acid and cannot freely pass the cell membrane or nuclear membrane. To deliver proteins, cationic lipids and cationic polymers also have been applied (Murthy et al., 2003; Verdurmen and Brock, 2011). However, their translocation efficiency is variable, depending on the cell type and the size and quality of the mixtures. Besides, the functional proteins cannot be translocated into the karyoplasms. Peptides are also a useful candidate for the translocation of peptides into cytoplasm but it also has limitations in medical or biological application. However, they have aroused great interest and have continuously progressed.

Cell-penetrating peptides (CPPs) are synthesized peptides derived from various proteins. CPPs can transport small molecules, peptides and proteins in the form of recombinants and mixtures (Sawant and Torchilin, 2010). Some transcription factors have a domain with strongly basic heteropeptides, containing at least four arginines and lysines, which functions as a nuclear localization signal (Boulikas, 1994; Henkel et al., 1992). *Drosophila* transcription factor Antennapedia (Antp) also has the ability to translocate across the cell membranes (Balayssac et al., 2006). A mildly hydrophobic segment located approximately 30 residues downstream of the signal peptide is critical in membrane permeability in an *E. coli* protein (Nilsson et al., 1993). Those sequences interacting with lipid bilayers are called CPPs. They can carry functional cargo in the form of transcription factors or drugs (Lin et al., 1995). CPPs are also referred to as cell-permeable peptides, protein transducing domains (PTDs), or Trojan horse peptides. Thus far, the designed CPPs are more than two hundred including a galparan analog (Pooga et al., 1998), poly-arginine (Wadia and Dowdy, 2003), TAT-PTD, penetratin, buforin 2, transportan, transportan 10, MAP, pVec, R8, R11, KALA, K-FGF, PEP-1, and Synb1 (Chugh et al., 2010; Dietz and Bähr, 2005; Pooga and Langel, 2005) (Table 1). CPPs are able to ferry much larger proteins, chemicals, DNA or RNA, directly into living cells (Dietz, 2010; Jung et al., 2007; Kwon et al., 2009).

Applying CPPs to basic sciences, medicine and industry has been attempted. CPPs can translocate a specific protein into the cytoplasm in mixed form or recombinant protein form. Using a recombinant cell-permeable Cre protein (His-TAT-NLS-Cre), recombination can be efficiently induced in mES cell-derived cells. Fortunately, the permeable Cre protein has no overt side effects on proliferation and neural differentiation (Haupt et al., 2007). Yang and colleagues (2009) worked with a library of poly-(beta-amino ester) end-modified derivatives. These derivatives were developed and optimized for high transfer efficiency in hES cell-derived cells, human adipose-derived stem cells (hADSCs) and human mesenchymal stem cells (hMSCs). In the presence of 10% serum, the transfer efficiency was 27 ± 2% in hMSCs, 24 ± 3% in hADSCs and 56 ± 11% in hSECds (Yang et al., 2009). Recently human iPS cells were successfully established with CPP-conjugated reprogramming proteins (Kim et al., 2009). However, the importation efficiency was poor.

CPP	Cys-CPP Sequence	Numer of		Class of CPP
		Lys	Arg	
Tat-PTD	GRKKRRQRRRPPQ	2	6	Cationic
Penetratin (ATF)	RQIKIWFQNRRMKWKK	4	3	Cationic
M918	MVTVLFRRLRIRRASGPPRVRV	0	7	Cationic
KALA	WEAKLAKALAKALAKHLAKAL AKALKACEA	7	0	Cationic
R7	RRRRRRR	0	7	Cationic
R8	RRRRRRRR	0	8	Cationic
R11	RRRRRRRRRRR	0	11	Cationic
MAP	KLALKLALKALKAALKLA	5	0	Amphipathic
pEP-1	KETWWETWWTEWSQPKKKRKV	5	1	Amphipathic
Buforin 2	TRSSRAGLQFPVGRVHRLLRK	1	5	Amphipathic
Transportan 10 (P10)	AGYLLGKINLKALAALAKKIL	4	0	Chimeric
Transportan	GWTLNSAGYLLGK^{b}INLKALAAL AKKIL	4	0	Chimeric
pVec	LLIILRRRIRKQAHAHSK	2	4	Chimeric
K-FGF	AAVALLPAVLLALLAP	0	0	Hydrophobi c sequences
β3-S	AAVALLPAVLLALLAP	0	0	Hydrophobi c sequences
SynB1	RGGRLSYSRRRFSTSTGR	0	6	Antimicrobia l sequence
NLS	ALWKTLLKKVLKAPKKKRKVC	8	1	Antimicrobia l sequence

b: attachment of Cys(Npys) on the

Table 1. Sequences and features of some cell-penetrating peptides

To promote the transduction of functional proteins into cytoplasm and karyoplasms of ES cells, the transduction abilities of various types of CPPs have been analyzed in mES cells (Jung et al., 2007). In these studies, the effects of the site of the CPP and the His tag in functional recombinant protein were first analyzed. We chose two expression vectors, one with an N-terminal His tag (6xHis-tag) and the other with a C-terminal tag (Fig. 2). Buforin 2 cannot mediate the translocation of conjugated EGFP into the cytoplasm of R1 mES cells whether the His tag is located N-terminally (Buforin 2-EGFP-N) or C-terminally (Buforin 2-EGFP-C). In contrast, pEP-1-EGFP-N translocates into the cytoplasm of R1 ES cells but pEP-1-EGFP-C does not. pEP-1-EGFP-C mainly localized at the plasma membrane. The translocational efficiency of a CPP depends on the amino acid sequence and the length of the CPP. Buforin 2 is a 21-residue peptide containing a +7 net charge (TRSSRAGLQFPVGRVHRLLRK) and is a membrane-permeabilizing antimicrobial peptide (Takeshima et al., 2003). Its structure is amphipathic, consisting of an N-terminal random coil region, an extended helical region, a hinge, and a C-terminal regular α-helical region. The membrane permeability is dependent on the α-helical content of the peptides (Park et al., 2000). Although it efficiently translocates into HeLa cells (Takeshima et al., 2003), it cannot penetrate ES cells whether the 6xHis-tag is located N-terminally or C-terminally (Fig. 3).

In HeLa cells, buforin 2-mediated translocation of proteins is concentration-dependent but temperature-independent (Takeshima et al., 2003). pEP-1 is a 21-residue peptide with a +6 net charge (KETWWETWWTEWSQPLKKRKV), and is a membrane-permeabilizing antimicrobial peptide and is amphipathic. pEP-1 can be translocated into cytoplasm and karyoplasms of R1 mES cells when the 6x His is located N-terminally (pEP-1-EGFP-N), as in COS-7 cells (Petrescu et al., 2009), but pEP-1-EGFP-C cannot penetrate the R1 cell membrane (Fig. 3) (Jung et al., 2007). The translocation of Buforin 2-EGFP-N, Buforin 2-EGFP-C, pEP-1-EGF-N and pEP-1-EGF-C does not depend on their concentrations (Fig. 4) (Jung et al., 2007). Translocation into the cell by pEP-1 dose not depend on energy (Henriques et al., 2007). It is suggested that electrostatic interactions are most important in the pEP-1 membrane interactions (Henriques et al., 2007). Although both Buforin 2 and pEP-1 are amphipathic, their translocation effects in ES cells differ. Additionally, the penetration affinity of pEP-1 is changed by the 6xHis-tag. Based on these observations, it is suggested that the N-terminal location of His on pEP-1 is responsible for its cell-penetrating efficiency. Therefore, the penetration effects of CPPs in ES cells are different from other cell types, such as epithelial-derived cell lines or stroma-derived cell lines (Jung et al., 2007; Petrescu et al., 2009; Takeshima et al., 2003).

Although 6xHis-tag is important in the efficiency of pEP-1 penetration into ES cells, the best strategy to express functional proteins in cells is to use a vector with a C-terminal tag. Based on the previous analysis, R7, MAP, pVec, K-FGF and yPFY were chosen as CPP and analyzed for their ability to transport EGFP through the plasma membrane of ES cells. R7 is an arginine-rich peptide with a +7 net charge. pVec is an 18-amino-acid peptide with a +8 net charge and is derived from the murine vascular endothelial-cadherin protein (Elmquist et al. 2010; Hällbrink et al., 2001). N-terminal hydrophobic part of pVEC is crucial for efficient cellular translocation (Elmquist et al., 2006). MAP (KLAL, model amphipathic

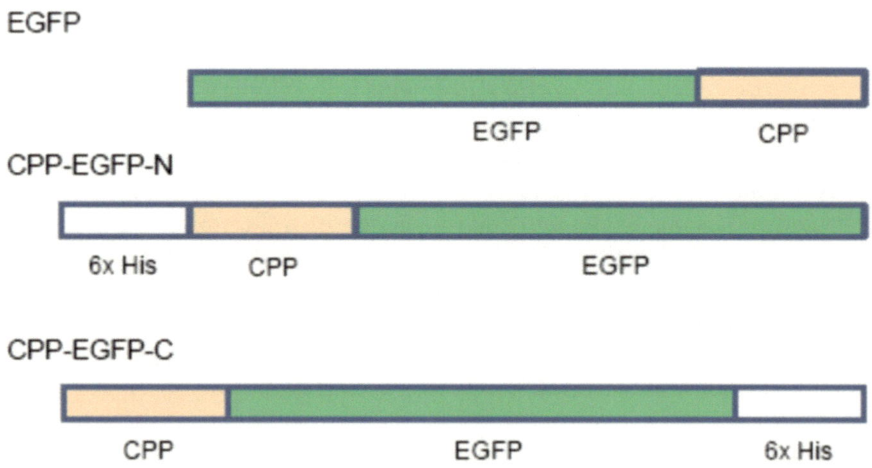

Fig. 2. Constructs of recombinant EGFP, CPP-EGFP-N and Pep-1-EGFP-C. In this construct, we wanted to use 6xHis-tag for purify and identification. Because it is known that 6xHis-tag usually do not disturb the formation of 3-dimensional structure and function of gene products. It is possible that His has + charge and it can effect on the role of CPP. Therefore we tested the possible effects of 6x His on the ability of CPPs.

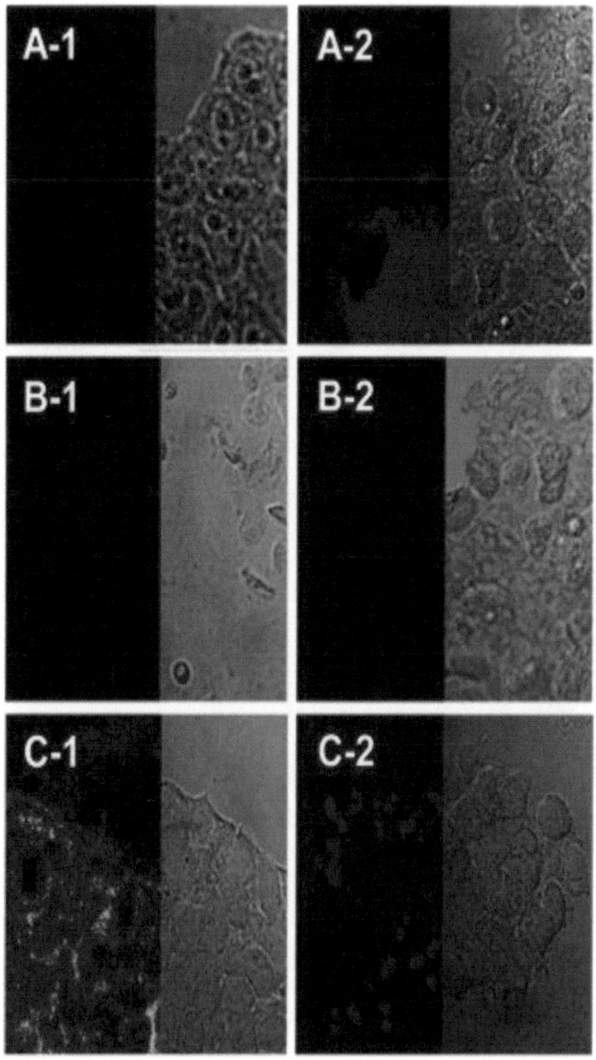

Fig. 3. Translocation of EGFP-N and -C (A_{-1} and A_{-2}, respectively), Buforin 2-EGFP-N and -C (B_{-1} and B_{-2}, respectively), and pEP-1-EGFP-N and -C (C_{-1} and C_{-2}, respectively) into R1 mES cells 24 hr after treatment. In natural culture condition, EGFP could not pass the cell membrane and the R1 ES cell did not take it through endocytosis. Site of 6xHis-tad did not give effect on the Buforin-2 transduction ability in mES cells but it gave effect on the pEP-1 transduction ability in mES cells.

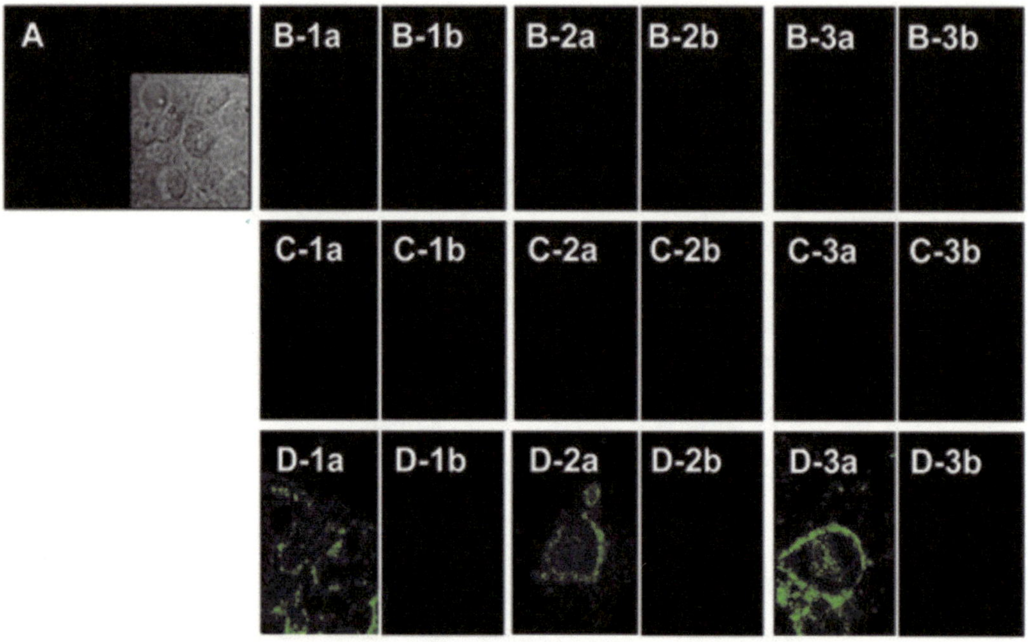

Fig. 4. Efficiency of transfection depends on the kinds of but independent on the concentration of Buforin 2 and pEP-1-EGFPs.
CPP-EGFPs were administered at various concentrations (1; 5μg/ml, 2; 50 μg/ml; 3; 500 μg/ml) and observed after 12 hr. Vehicle (A), EGFP (B) and EGFP (B-b), Buforin 2-EGFP (C), pEP-1-EGFP (D) was administered to R1 mES cells and detected using confocal microscopy. -a means the N terminal 6xHis-tag and –b means the C terminal 6xHis-Tag.

peptide) is an amphipathic helical peptide with a +5 net charge (Hällbrink et al., 2001). A positive charge as well as helicity and amphipathicity are all required for efficient translocation (Wolf et al., 2006). K-FGF and yPFY have no net positive charge. In our laboratory, pET-20b(+)deXhoI modified of pET-20b was used to construct CPP-EGFPs. To get the recombinant CPP-EGFPs, *BL21 E. coli* were used and purified with Ni^{2+}-affinity column (Qiagen). R1 mES cells were maintained in ES medium containing 1000 IU LIF. R7-EGFP-C, MAP-EGFP-C, pVec-EGFP-C, K-FGF-EGFP-C and yPFY-EGFP-C (each 10 ng/ml) were incubated with R1 mES cells for 6 hr in ES medium containing 10% fetal calf serum and then analyzed for translocation with confocal microscope. Propidium iodide was used to stain the nucleus. Interestingly, all of them were localized both to the cytoplasm and karyoplasms (Fig. 5) (Cheon et al., manuscript in preparation). Based on these results, it is suggested that all these carriers can be applied in ES cells.

Recently, it was discovered that R7 mediates the translocation of functional proteins into the cytoplasm of stem cells (Jo et al., 2010). Short-form human ESRRB was cloned in R7-pET-20b(+)deXhoI, and recombinant RT-ESRRB was purified in BL21(DE3)pLysS. R7-ESRRB-6xHis successfully translocated into the cytoplasm and karyoplasm (Fig. 6), and increased the expression of OCT4, NANOG and SOX2 (Fig. 7).

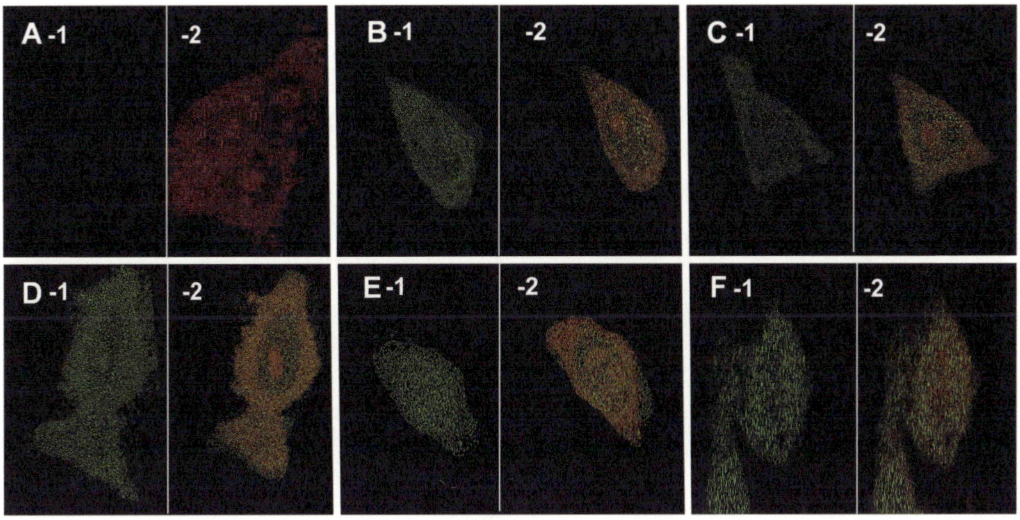

Fig. 5. Translocation of CPP-eGFP into R1 ES cells.
(A-F) mES cells were treated with 10ng/ml EGFP (A), R7-EGFP (B), yPFY-EGFP (C), MAP-EGFP (D), K-FGF-EGFP (E), or pVec-EGFP (F) for 6 hr at 37°C. After treatment, all cells were stained with propidium iodide. 1 is eGFP images and 2 is the merged images.

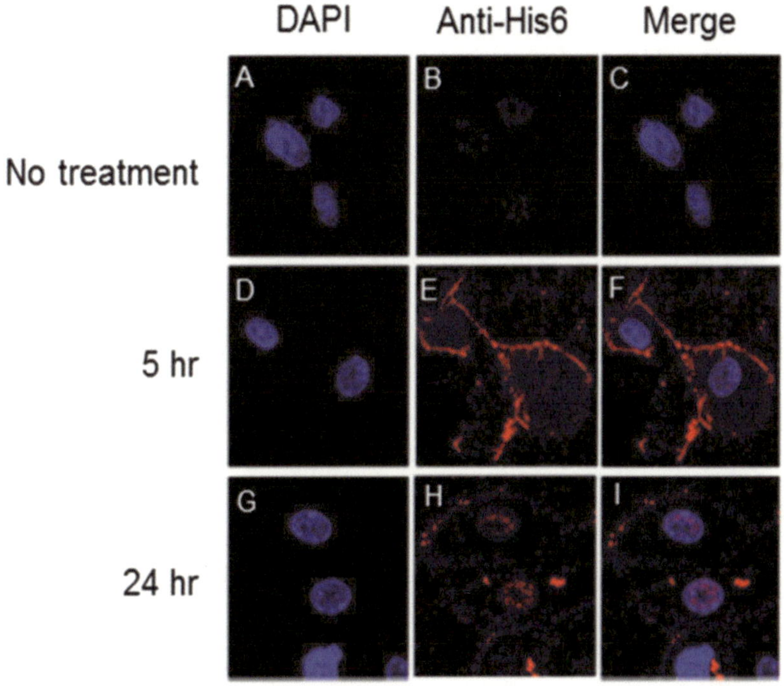

Fig. 6. R7-hESRRB-His6 protein can penetrate the plasma membrane within 5 hr after treatment and are completely localized in the nucleus within 24 hr after treatment.
(A-C) the cells were not treated with R7-hESRRB-His6 protein but were immunostained

with anti-His-Taq(6X) antibody; (D-F) the cells were immunostained with anti-His-Taq(6X) antibody 5 hours after treatment with R7-hESRRB-His6 protein; (G-I) the cells were immunostained with anti-His-Taq(6X) antibody 24 hours after treatment with R7-hESRRB-His6 protein. Images were captured using a confocal microscope (Carl Zeiss LSM 510 META, Zeiss, Jena, Germany) at ×1,000 magnification.

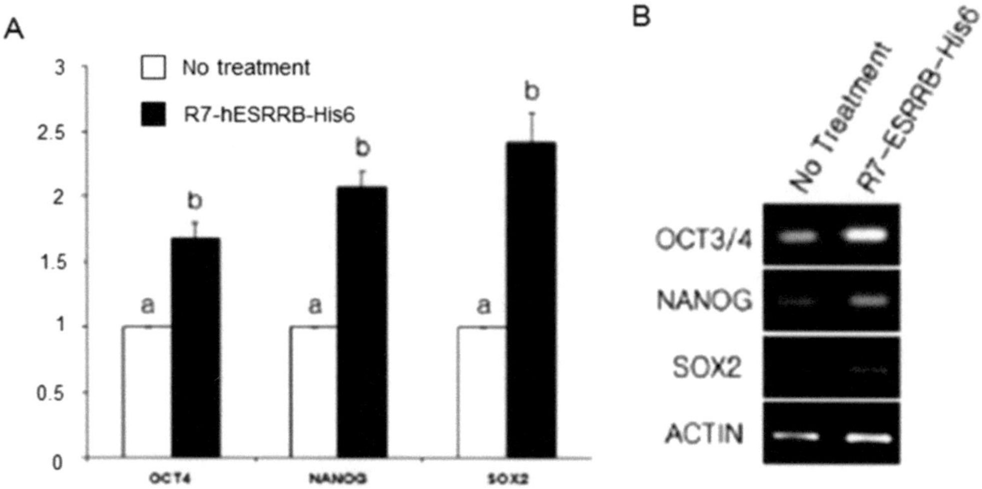

Fig. 7. Treatment with R7-hESRRB-His6 protein for 8 days (192 hr) increases the expression of OCT4, NANOG and SOX2.
A, real-time PCR for the expression of OCT4, NANOG and SOX2; B, agarose gel electrophoresis and EtBr staining after real-time PCR. Statistically significant differences between each group are denoted by different letters (P<0.05)

4.3 Usage of nanobeads as carriers of functional proteins

The amount of translocated functional protein is also an important factor in directing the induction of ES cells into specific cell types or maintaining ES cell stemness. In addition, the stability of the CPP-proteins is limited under physiological conditions. The translocation of CPP-proteins into cytoplasm can be inhibited by serum (Furuhata et al., 2006). Therefore, to control ES cell differentiation, it is necessary to regulate the amount of translocated functional protein. To prolong the lifespan of CPPs or CPP-conjugates, they have been modified but this approach has limits (Grunwald et al., 2009).

Recently, nanoparticle techniques have been applied to medicine. Due to the physical and chemical characteristics of nanoparticles, they are more effective at targeting difficult-to-reach sites and have a better side effect profile. Hence, smaller doses of nanomedicines are needed to achieve the same therapeutic effect (Hock et al., 2011; Kuai et al., 2010; Neundorf, 2008). These advantages of nanoparticles as carriers are also evident in the study of ES cells. Poly(β-amino esters) can fulfill the role of translocation mediator of a VEGF plasmid into hES cells (Yang et al. 2009). A chimeric protein, GFP-FRATtide, attached to a hydrophobically modified 15-nm silica nanoparticle is efficiently delivered to the cytosol of human embryonic kidney cells and rat neural stem cells (Shah et al., 2011). Tran and colleagues studied the effects of nanoparticles on morphological differentiation in mES cells (Tran et al., 2007). However, so far, the efficiency of this technique is not high in ES cell studies.

Another strategy to use nanoparticles is to apply the metabolic characteristics of ES cells and the characteristics of membranes to the nanoparticles. ES cells can use glucose and require glucose for further development. During the development of preimplantation embryos, the mammalian embryo needs a constant supply of energy substrates to remain viable. Usually, mammalian oocytes receive substrates, especially energy substrates, from the oviduct and uterus because they do not store much substrate in the cytoplasm during oogenesis. Glucose, lactate and pyruvate are essential components in preimplantation embryo culture media, and there are stage-specific preferences for them. Glucose may permit the expression of metabolic enzymes and transporters in compacting morula that are capable of generating the energy required for blastocyst formation. In addition, metabolites of glucose may be involved in cellular activity during the development of preimplantation embryos. Periimplantation embryos need glucose (Biggers and Summers, 2008). Therefore, ES cells can take up glucose as well as glycated proteins. Glucose-coated polymeric nanobeads have been prepared by dispersion polymerization (Jung et al., 2009). As a model protein, enhanced green fluorescent protein was ligated to the nanobeads and successfully translocated into mES cells (Fig. 8, Fig. 9). It is suggested that glucose-coated nanobeads could be a general cargo for the intracellular delivery of various macromolecules in ES cells.

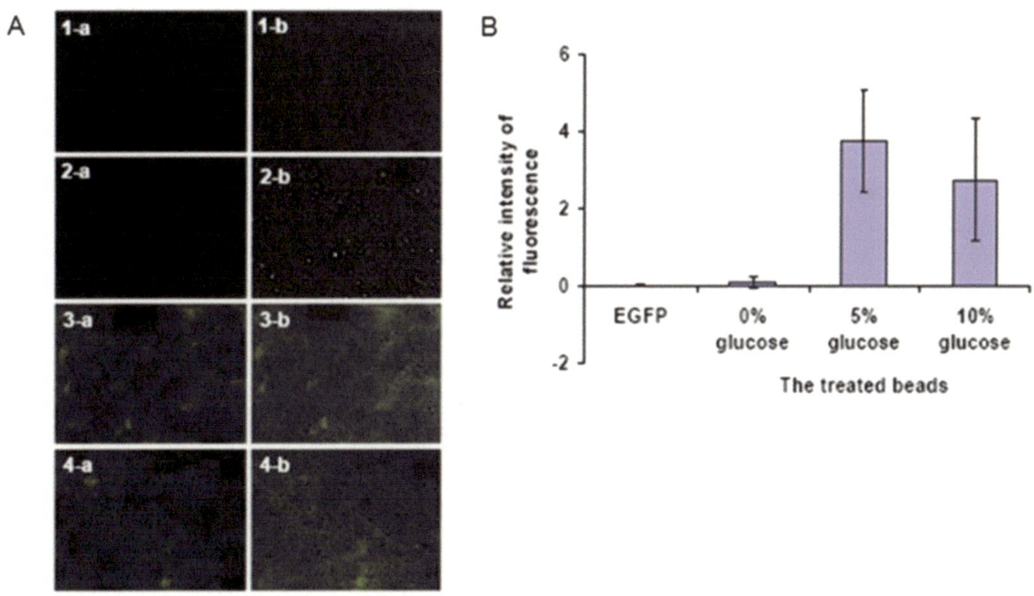

Fig. 8. Translocation experiments of the EGFP-ligated beads into R1 mouse embryonic stem cells.
A, the confocal microscopic images. 1-a and 1-b, cells treated with EGFP alone; 2-a and 2-b, cells treated with non-coated beads; 3-a and 3-b, cells treated with 5%-glucose-coated beads; 4-d and 4-m, cells treated with 10%-glucose-coated beads; 1-a, 2-a, 3-a, and 4-a are dark field images; 1-b, 2-b, 3-b, and 4-b are merged images (dark-bright fields). The merged images show that the cells are localized at the fluorescent regions. B, the relative intensities of the translocated EGFP. The intensities were measured using confocal microscopy (average ± SD; the measurements were repeated seven times and averaged).

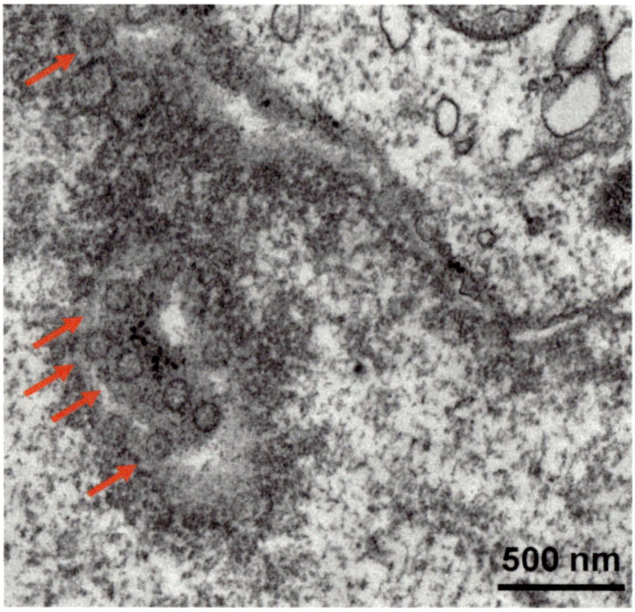

Fig. 9. TEM photomicrograph of EGFP-ligated glucose beads in R1 mouse embryonic stem cells. Arrows indicate the translocated beads.

In addition, these nanobeads could be more stable in the blood stream and tissue fluids compared than CPPs. On the other hand, if we add more functions in nanobeads, it can be carry a designed amount of functional protein. For example, shuttle signaling and self degradation can be helpful. To apply glucose nanobeads to in vitro and in vivo systems, further studies are needed, such as optimizing their homing and controlling the transporting amounts in karyoplasms and cytoplasm.

5. Summary

ES cells have been used to study the differentiation of various cell types and tissues in vitro, such as neural cells, hematopoietic lineages, cardiomyocytes, hepatocytes, β-cells, and epithelium. In addition, ES-derived cells have been successfully transplanted into fetal and adult mice, where they have demonstrated morphological and functional integration (Kennedy et al., 1997). However, their use is still in the beginning stages in humans. It is difficult to translate laboratory advancements to human therapy because there are many obstacles to using ES cells in medicine (Piscaglia et al., 2010; Rossi et al., 2010). DNA construct mediated induction of ES cell-differentiation can cause unexpected side effects in the body. Therefore the best way is to use the functional proteins which regulate cellular characteristics such as transcription factors. Those factors can express their functional role, when they are located in proper sites, cytoplasm or karyoplasms. CPPs can be transported into cytoplasm and karyoplasms.

This means that a membrane penetrating carrier is needed for the protein-application to get a specific cell type from ES cells. Based on this, CPP may be a clue to get such a goal. A nanobead which can easily translocate the cell membrane is also a clue to achieve such a goal. Although to apply the CPPs or specific nanoparticle-cargo in cell therapy using protein, further studies are needed. These studies would open up new methods of therapy which are safe and economic. However, research on all aspects of ES cell biology will soon overcome these obstacles and make their use in therapy more safe and effective.

6. References

Allin EP. 1984. Induction of differentiation in mouse embryonal carcinoma cell lines by coculture with rat glioma cells. Cell Mol Biol 30:377-384.

Amano M, Yamaguchi M, Takegawa Y, Yamashita T, Terashima M, Furukawa J, Miura Y, Shinohara Y, Iwasaki N, Minami A, Nishimura S. 2010. Threshold in state-specific embryonic glycotypes uncovered y a full portrait of dynamic N-glycan expression during cell differentiation. Mol Cell Proteomics 9:523-537.

Andrews PW. 1987. Human teratocarcinoma stem cells: glycolipid antigen expression and modulation during differentiation. J Cell Biochem 35:321-332.

Andrews PW, Banting G, Damjanov I, Arnaud D, Avner P. 1984. Three nonoclonal antibodies defining distinct differentiation antigens associated with different high molecular weight polypeptides on the surface of human embryonal carcinoma cells. Hybridoma 3:347-361.

Armaleo D, Ye GN, Kein TM, Shark KB, Sanford JC, Johnston SA. 1990. Biolistic nuclear transformation of Saccharomyces cerevisiae and other fungi. Curr Genet 17:97-103.

Artavanis-Tsakonas S, Rand Md, Lake RJ. 1999. Notch signaling: Cell fate control and signal intergration in development .Science 284:770-776.

Azizi H, Mehrjardi NZ, Shahbazi E, Hemmesi K, Bahmani MK, Baharvand H. 2010. Dehydroepiandrosterone stimulates neurogenesis in mouse embryonal carcinoma cell- and human embryonic stem cell-derived neural progenitors and induces dopaminergic neurons. Stem Cells Dev 19:809-818.

Balayssac S, Burlina F, Convert O, Bolbach G, Chassaing G, Lequin O. 2006. Comparison of penetratin and other homeodomain-derived cell-penetration peptides: interaction in a membrane-mimicking environment and cellular uptake efficiency. Biochemistry 45:1408-1420.

Bell SM, Aspinall R, Stern PL. 1984. Enrichment of rat NK cytotoxicity for H2-negative murine embryonal carcinoma cells by panning and short-term culture in TCGF. 53:23-32.

Belting M, Dorrell MI, Sandgren S, Aguilar E, Ahamed J, dorfleutner A, Carmeliet P, Mueller BM, Friedlander M, Ruf W. 2004. Regulation of angiogenesis by tissue factor cytoplasmic domain signaling. Nat med 10:502-509.

Berkner KL. 1988. Development of adenovirus vectors for the expression of heterologous genes. Biotechniques 6:616-629.

Biggers ID, Summers MC. 2008. Choosing a culture medium: making informed choices. Fertil Steril 90:473-483.

Boulikas T. 1994. Putative nuclear localization signals (NLS) in protein transcription factors. J Cell Biochem 55:32-58.

Boussif O, Lezoualch F, Zanta MA, Mergny MD, Scherman D, Demeneix B Behr JP. 1995. A versatile vector for gene and oligonucleotide transfer into cells in culture and in vivo: polyethylenimine. Proc Natl Acad Sci U S A 92:7297-7301.

Bradley A, Evans M, Kaufman MH, Rovertson E. 1984. Formation of germ-line chimaeras from embryo-derived teratocarcinoma cell lines. Nature 309:255-256.

Cao F, Xie X, Gollan T, Zhao L, Narsinh K, Lee RJ, Wu JC. 2010. Comparison of gene-transfer efficiency in human embryonic stem cells. Mol Imaging Bio 12:15-24.

Carvajal-Vergara X, Sevilla A, D'Souza SL, Ang YS, Schaniel C, Lee DF, Yang L, Kaplan AD, Adler ED, Rozov R, Ge Y, Cohen N, Edelmann LJ, Chang B, Waghray A, Su J, Pardo S, Lichtenbelt KD, Tartaglia M, Gelb BE, Lemischka IR. 2010. Patient-specific induced pluripotent stem-cell-derived models of LEOPARD syndrome. Nature 465:808-812.

Chambers I, Smith A. 2004. Self-renewal of teratocarcinoma and embryonic stem cells. Oncogene 23:7150-7160.

Chen QR, Zhang l, Stass A, Mixson AJ. 2001. Branched co-polymers of histone and lysine are efficient carriers of plasmids. Nucleic Acids Res 29:1334-1340.

Chou TH, Biswas S, Lu S. 2004. Gene delivry using physical methods; an overview. Methods Mol Biol 245:147-166.

Christophersen NS, Helin K. 2010. Epigenetic control of embryonic stem cell fate. J Exp Med 207:2287-2295.

Chugh A, Eudes F, Shim YS. 2010. Cell-penetrating peptides: nanocarrier for macromolecule delivery in living cells. IUBMB Life 62:183-193.

Couto LB, High KA. 2010. Viral vector-mediated RNA interference. Curr Opin Pharmacol . 10:534-542.

Davis HL, Whalen RG, Demeneix BA. 1993. Direct gene transfer into skeletal muscle in vivo; factors affecting efficiency of transfer and stability of experession. Hum Gene Ther 4:151-159.

Delhaise F, Bralion V, Schuurbiers N, Dessy F. 1996. Establishment of an embryonic stem cell line from 8-cell stage mouse embryos. Eur J Morphol 34:237-243.

Dietz GP. 2010. Cell-penetrating peptide technology to deliver chaperones and associated factors in diseases and basic research. Curr Pharm Biotechnol 11:167-174.

Dietz GPH, Bähr M. 2005. Peptideenhanced cellular internalization of proteins in neuroscience. Brain Res Bull 68:103-114.

Doetschman TC, Eistetter H, Katz M, Schmidt W, Kemler R. 1985. The in vitro development of blastocyst-derived embryonic stem cell lines: formation of visceral yolk sac, blood islands and myocardium. J Embryol Exp Morpho 87:27-45.

Doetschman T, Gregg RG, Maeda N, Hooper ML, Melton DW, Thompson S, Smithies O. 1987. Targeted correction of a mutant HPRT gene in mouse embryonic stem cells. Nature 330:576-578.

Dunlap DD, Maggi A, Soria MR, Monaco L. 1997. Nanoscopic structure of DNA condensed for gene delivery, Nucleic Acids Res 25:3095-3101.

Dyson PJ, Poirier F, Watson RJ. 1989. Expression of c-myb in embryonal carcinoma cells and embryonal stem cells. Differentiation 42:24-27.

Edwards MK, Harris JF, McBurney MW. 1983. Induced muscle differentiation in an embryonal carcinoma cell line. Mol Cell Biol 3:2280-2286.

Elmquist A, Hansen M, Langel U. 2006. Structure-activity relationship study of the cell-penetrating peptide pVEC. Biochem Biophys Acta 1758:721-729.

Elmquist A, Lindgren M, Bartfai T, Langel U. 2010. VE-cadherin derived cell-penetrating peptide, pVEC, with carrier functions. Exp Cell Res 269:237-244.

Evans MJ, Kaufman MH. 1981. Establishment in culture of pluripotential cells from mouse embryos. Nature 292:154-156.

Felgner PL, Gadek TR, Holm M, roman R, Chan HW, Wenz M, Northrop JP, Ringold GM, Danilsen M. 1987. Lipofection: a highly efficient, lipid-meidated DNA-transfection procedure. Proc Natl Acad Sci U S A 84:7413-7417.

Fernanez-Tresquerres B, Canon S, Rayon T, Parnaute B, Crespo M, Torroja C, Manzanares M. 2010. Evolution of the mammalian embryonic pluripotency gene regulatory network. Proc Natl Acad Sci USA 107:19955-19960.

Freiberg RA, Spencer DM, Choate KA, Peng PD, Schreiber SL, Crabtree GR, Khavari PA. 1996. Specific triggering of the Fas signal transduction pathway in normal human keratinocytes. J Biol Chem 271:31666-31669.

Furuhata M, Kawakami H, Toma K, Hattori Y, Maitani Y. 2006. Intracellular delivery of proteins in complexes with oligoarginine-modified liposomes and the effect of oligoarginine length. Bioconjug Chem 17:935-942.

Gautsch JW. 1982. Lack of retrovirus gene expression in teratocarcinoma stem cells is limited to nucleus. Somatic Cell Genet 8:143-149.

Gearhart J. 1998. New potential for human embryonic stem cells. Science 282:1061-1062.

Grabel LB. 1984. Isolation of a putative cell adhesion mediating lectin from teratocarcinoma stem cells and its possible role in differentiation. Cell Differ 15:121-124.

Graham FL, van der Eb AJ. 1973. Transformation of rat cells by DNA of human adenovirus 5. Virology 54:536-539.

Grunwald J, Rejtar T, Sawant R, Wang Z, Torchilin VP. 2009. TAT peptide and its conjugates: proteolytic stability. Bioconjugate Chem 20:1531-1537.

Hällbrink M, Floren A, Elmquist A, Pooga M, Bartfai T, Langel U. 2001. Cargo delivery kinetics of cell-penetrating peptides. Biochim Biophys Acta 1515:101-109.

Hamatani T, Carter MG, Sharov AA, Ko MS. 2004. Dynamics of global gene expression changes during mouse preimplantation development. Dev Cell 6:117-131.

Hassoun R, Puschel B, Viebahn C. 2010. Sox17 expression patterns during gatrulation and early neurulation in the rabbit suggest two sources of endoderm formation. Cells Tissues Organs 191:68-83.

Haupt S, Edenhofer F, Peitz M, Leinhaas A, Brustle O. 2007. Stage specific conditional mutagenesis in mouse embryonic stem cell-derived neural cells and post-mitotic neurons by direct delivery of biologically active Cre recombinase. Stem Cells 25:181-188.

Heath JK, Smith AG. 1988. Regulatory fators of embryonic stem cells. J Cell Sci Suppl 10:257-266.

Heng BC, Haider HKh, Sim EK, Cao T, Ng SC. 2004. Strategies for directing the differentiation of stem cells into the cardiomyogenic lineage in vitro. Cardiovasc Res 62:34-42.

Henkel T, Zael U, van Zee K, Müller JM, Fanning E, Baeuerle PA. 1992. Intramolecular masking of the nuclear location signal and dimerization domain in the precursor for the p50 NF-kappa B subunit. Cell 68:1121-1133.

Henriques ST, Quimtas A, Bagatolli LA, Homblé F, Castanho MA. 2007. Energy-independent translocation of cell-penetrating peptides occurs without formation of pores. A biophysical study with pep-1. Mol Membr Biol 24:282-293.

Hock SC, Ying YM, Wah CL. 2011. A review of the current scientific and regulatory status of nanomedicines and the challenges ahead. DPA J Pharm Sci Technol 65:177-195.

Hong Y, Chen S, Gui J, Schartl M. 2004. Retention of the developmental pluripotency in medaka embryonic stem cells after gene transfer and long-term drug selection for gene tarteting in fish. Transgenic Res 13:41-50.

Hooper MI, Hardy K, Handyside A, Hunter S, Monk M. 1987. HPRT-deficient (Lesch-Nyhan) mouse embryos derived from germ-line colonization by cultured cells. Nature 326:292-294.

Ishizaka S, Shiroi A, KandaS, Yoshikawa M, Tsujinoue H, Kuriyama S, hasuma T, Nakatani K, Takahashi K. 2002. Development of hepatocytes from ES cells after transfection with the HNF-3beta gene. FASEB J 16:1444-1446.

Jakob H. 1984. Stem cells and embryo-derived cell lines: tools for study of gene expression. Cell Differ 15:77-80.

Jemmerson R, Shah N, Takeya M, Fishman WH. 1985. Characterization of the placental alkaline phosphatase-like (Nagao) isozyme on the surface of A431 human epidermoid carcinoma cells. Cancer Res 45:282.287.

Jeong Y, Mangelsdorf DJ. 2009. Nuclear receptor regulation of stemness and stem cell differentiation. Exp Mol Med 41:525-537.

Jo J, Lee Y, Oh MH, Ko JJ, Cheon YP, Lee DR. 2010. Up-regulation of pluripotency-related genes in human amniotic fluid-derived stem cells by ESRB conjugated with cell-penetrating peptide. Dev Reprod 14:243-251.

Johnson MH, McConnell JM. 2004. Lineage allocation and cell polarity during mouse embryogenesis. Semin Cell Dev Biol 15:583-597.

Jones-Villeneuve EM, Rudnicki MA, Harris JF, McBurney MW. 1983. Retinoic acid-induced neural differentiation of embryonal carcinoma cells. Mol Cell Biol 3:2271-2279.

Jung S, Huh S, Cheon YP, Park S. 2009. Intracellular protein delivery by glucose-coated polymeric beads. Chem Commun 33:5003-5005.

Jung S, Park S, Lim H, Cheon Y. 2007. Mouse embryonic stem cell uptakes of Buforin 2 and pEP-1 conjugated with EGFP. Dev Reprod 11:111-119.

Kaufman MH, Robertson EJ, Handyside AH, Evans MJ. 1983. Establishment of pluripotential cell lines from haploid mouse embryos. J Embryol Exp Morphol 73:249-261.

Kennedy M, Firpo M, Choi K, Wall C, Robertson S, Kabrun N, Keller G. 1997. A common precursor for primitive erythropoiesis and definitive haematopoiesis. Nature 386:488-493.

Kim D, Kim CH, Moon JI, Chung YG, Chang MY, Han BS, Ko S, Yang E, Cha KY, Lanza R, Kim KS. 2009. Generation of human induced pluripotent stem cells by direct delivery of reprogramming proteins. Cell Stem Cell 4:472-476.

Kopp JL, Ormsbe BD, Desler M, Rizzino A. 2008. Small increases in the level of sox2 trigger the differentiation of mouse embryonic stem cells. Stem Cells 26:903-911.

Kuai R, Yuan W, Qin Y, Chen H, Tang J, Yuan M, Zhang Z, He Q. 2010. Efficient delivery of payload into tumor cells in a controlled manner by TAT and thiolytic cleavage PEG co-modified liposomes. Mol Pharm [Epub ahead of print].

Kuehn M, Bradley A, Rovertson EJ, Evans MJ. 1987. A potential animal model for Lesch-Nyhan syndrome through the introduction of HPTR mutations in mice. Nature 326:295-298.

Kumar N, Pethe P, Bhartiya D. 2010. Role of TGFbeta and myofibroblasts in supporting the propagation of human embryonic stem cells in vitro. Hnt J Dev Biol 54:1329-1336.

Kwon SJ, Han K, Jung S, Lee JE Park S, Cheon YP, Lim HJ. 2009. Transduction of the MPG-tagged fusion protein into mammalian cells and oocytes depends on amiloride-sensitive endocytic pathway. BMC Biotechnol 9:73.

Lehtonen E, Laasonen A, Tienari J. 1989. Teratocarcinoma stem cells as a model for differentiation in the mouse embryo. Int J Dev Biol 33:105-115.

Lin YZ, Yao SY, Veach RA, Torgerson TR, Hawiger J. 1995. Inhibition of nuclear translocation of transcription factor NF-kB by a synthetic peptide containing a cell membrane-permeable motif and nuclear localization sequence. J Biol Chem 270:14255-14258.

Liou JY, Ko BS, Chang TC. 2010. An efficient transfection method for mouse embryonic stem cells. Methods Mol Biol 650:145-153.

Little SR, Lynn DM, Ge Q, Anderson DG, Puram SW, Chen J, Eisen HN, Langer R. 2004. Poly-beta amino ester-containing microparticles enhance the activity of nonviral genetic vaccines. Proc natl Acad Sci U S A 101:9534-9539.

Liu H, Kim Y, Sharkis S, Marchionni L, Jang YY. 2011. In vivo liver regeneration potential of human induced pluripotent stem cells from diverse origins. Sci Transl Med 3:82ra39.

Lu M, Kardel MD, O'Connor MD, Eaves CJ. 2009. Enhanced generation of hematopoietic cells from human hepatocarcinoma cell-stimulated human embryonic and induced pluripotent stem cells. Exp Hematol 37:924-936.

Luxardi G, Marchal L, Thome V, Kodjabachian L. 2010. Distinct Xenopus Nodal ligands sequentially induce mesendoderm and contro gastrulation movements in parallel to the Wnt/PCP pathway. Development 137:417-426.

Ma H, Liu Q, Diamond SL, Pierce EA. 2004. Mouse embryonic stem cells efficiently lipofacted with nuclear localization peptide result in a high yield of chimeric mice and retain germline transmission potency. Methods 33:113-120.

Martin GR. 1981. Isolation of a pluripotent cell line from early mouse embryos culturedin medium conditioned by teratocarcinoma stem cells. Proc Natl Acad Sci U S A 78:7634-7638.

Marucci G, Lammi C, Buccioni M, Dal Ben D, lambertucci C, Amantini C, Santoni G, kandhavelu M, Abbracchio MP, Lecca D, volpini R, Cristalli G. 2011. Comparison and optimization of transient transfection methods at human astrocytoma cell line 1321N1. Anal Biochem 414:300-302.

Matsuoka H, Shimoda S, Ozaki M, Mizukami H, Shibusawa M, Yamada Y, Saito M. 2007. Semi-quantitative expression and knockdown of a target gene in single-cell mouse embryonic stem cells by high performance microinjection. Biotechnol Lett 29:341-350.

Meissner A, Wernig M, Jaenisch R. 2007. Direct reprogramming of genetically unmodified fibroblasts into pluripotent stem cells. Nat Biotechnol 25:1177-1181.

Moore JC, Atze K, yeung PL, Toro-Ramos Aj, Camarillo C, Thompson K, Ricupero CL, Brenneman MA, Cohen RI, Hart RP. 2010. Efficient, high-throughput transfection of human embryonic stem cells. Stem Cell Res Ther 1:23.

Mummery CL, van den Brink CE, van der Saag PT, de Laat SW. 1984. The cell cycle, cell death, and cell morphology during retinoic acid-induced differentiation of embryonal carcinoma cells. Dev Biol 104:297-307.

Muramatsu H, Muramatsu T. 1983. A fucosyltransferase in teratocarcinoma stem cells. Decreased activity accompanying differentiation to parietal endoderm cells. FEBS Lett 163:181-184.

Murthy N, Xu M, Schuck S, Kunisawa J, Shastri N, Frehet Mj. 2003. A macromolecular delivery vehicle for protein-based vaccines; acid-degradable protein-loaded microgels. Proc Natl Acad Sci U S A 100:4995-5000.

Navarro P, Oldfield A, Legoupi J, Festuccia N, Dubois A, Attia M, Schoorlemmer J, Rougeulle C, Chambers I, Avner P. 2010. Molecular coupling of Tsix regulation and pluripotency. Nature 468:457-460.

Neundorf I, Rennert R, Franke J, Közle I, Bergmann R. 2008. Detailed analysis concerning the biodistribution and metabolism of human calcitonin-derived cell-penetrating peptides. Bioconjug Chem 19:1596-1603.

Nichols J, Smith A. 2011. The origin and identity of embryonic stem cells. Development 138:3-8.

Nilsson IM, Gafvelin G, von Heijne G. 1993. Different sec-requirements for signal peptide cleavage and protein translocation in a model E. coli protein. FEBS Lett 318:7-10.

Osakada F, Takahashi M. 2011. Neural induction and patterning in mammalian pluripotent stem cells. CNS neurol Disord Drug Targets 10:419-432.

Pagano JS, Vaheri A. 1965. Enhancement of infectivity of poliovirus RNA with diethylaminoethyl-dextran (DEAE-D). Arch Gesamte Virusforsch 17:456-464.

Park CH, Lee SH. 2007. Efficient generation of dopamine neurons from human embryonic stem cells. Methods Mol Biol 407:311-322.

Park CB, Yi KS, Matsuzaki K, Kim MS, Kim SC. 2000. Structure-activity analysis of buforin 2, a histone H2A-derived antimicrobial peptide: the proline hinge is responsible for the cell-penetrating ability of buforin 2. Proc Natl Acad Sci U S A 97:8245-8250.

Park IH, Zhao R, West JA, Yabuuchi A, Huo H, Ince TA, Lerou PH, Lensch MW, Daley GQ. 2008. Reprogramming of human somatic cells to pluripotency with defined factors. Nature 451:141-146.

Petrescu AD, Vespa A, Huang H, Mclntosh AL, Schroeder F, Kier AB. 2009. Fluorescent sterols monitor cell penetrating peptide Pep-1 mediated uptake and intracellular targeting of cargo protein in living cells. Biochem Biophys Acta 1788:425-441.

Piscaglia AC, Campanale M, Gasbarrini A, Gasbarrini G. 2010. Stem cell-based therapies for liver diseases: state of the art and new perspectives. Stem Cells 2010:259461.

Pooga M, Hällbrink M, Zorko M, Laggel U. 1998. Cell penetration by transportan. FASEB J 12:67-77.

Pooga M, Langel U. 2005. Synthesis of cell-penetrating peptides for cargo delivery. 298:77-89.

Redshaw Z, Strain AJ. 2010. Human haematopoietic stem cells express Oct4 pseudogenes and lack the ability to initiate Oct4 promoter-driven gene expression. J Negat Results Biomed 9:2.

Risau W, Sariola H, Zerwes HG, Sasse J, Ekblom P, Kemler R, Doetschman T. 1988. Vasculogenesis and angiogenesis in embryonic-stem-cell-derived embryoid bodies. Development 102:471-478.

Robbins PD, Ghivizzani SC. 1998. Viral vectors for gene therapy. Pharmacol Ther 80:35-47.

Rocha PP, Scholze M, Bleiss W, Schrewe H. 2010. Med12 is essential for early mouse development and for canonical Wnt and Wnt/PCP signaling. Development 137:2723-2731.

Rosenstraus MJ, Sterman B, Carr A, Brand L. 1984. Fibroblast feeder layers inhibit differentiation of retinoic acid-treated embryonal carcinoma cells by increasing the probability of stem cell renewal. Exp Cell Res 152:378-389.

Rossant J, Papaioannou VE. 1984. The relationship between embryonic, embryonal carcinoma and embryo-derived stem cells. Cell Differ 15:155-161.

Rossi SL, Nistor G, Wyatt T, Yin H, Poole AJ, Weiss JH, Gardener MJ, Dijkstra S, Fischer DF, Keirstead HS. 2010. Histological and functional benefit following transplantation of motor neuron progenitors to the injured rat spinal cord. PLoS One 5:e11852.

Ruch JV. 1967. Effect of dexamethasone phosphate on the differentiation of embryonic stem bronchi in vivo and in vitro. CR Seances Soc Bio Fil 161:1339-13342.

Sawant R, Torchilin V. 2010. Intracellular transduction using cell-penetrating peptides. Mol biosyst 6:628-640.

Saxena AK, Singh D, Gupta J. 2010. Role of stem cell research in therapeutic purpose-a hope for new horizon in medical biotechnology. J Exp Ther Oncol 8:223-233.

Schindler J, Sherman MI. 1984. Changes in protein synthetic profiles during retinoic-acid induction of differentiation of murine embryonal carcinoma cells. 28:78-85.

Schwartzberg PL, Goff SP, Robertson EJ. 1989. Germ line transmission of a c-abl mutation produced by targeted gene disruption in ES cells. Science 246:799-803.

Scott RW, Vogt TF, Croke ME, Tilghman SM. 1984. Tissue-specific activation of a cloned alpha-fetoprotein gene during differentiation of a transfected embryonal carcinoma cell line. Nature 310:562-567.

Sekiya S, Kawata M, Iwasawa H, Inaba N, Sugita M, Suzuki N, Motoyama T, Yamamoto T, Takamizawa H. 1985. Characterization of human embryonal carcinoma cell lines derived from testicular germ-cell tumors. Differentiation 29:259-267.

Shah DA, Kwon SJ, Bale SS, Banerjee A, Dordick JS, Kane RS. 2011. Regulation of stem cell signaling by nanoparticle-meidated intracellular protein delivery. Biomaterials 32:3210-3219.

Shimotohno K, Temin HM. 1981. Formation of infectious progeny virus after insertion of hepes simplex thymidine kinase gene into DNA of an avian retro virus. Cell 26:67-77.

Silvan U, Diez-Torre A, Arluzea J, Andrade R, Silio M, Arechaga J. 2009. Hypoxia and pluripotency in embryonic and embryonal carcinoma stem cell biology. Differentiation 2009 78:159-168.

Smith AG, Heath JK, Donaldson DD, Wong GG, Moreau J, Stahl M, Rogers D. 1988. Inhibition of pluripotential embryonic stem cell differentiation by purified polypeptides. Nature 336:688-690.

Smolich BD, Papkoff J. 1994. Regulated expression of Wnt family members during neuroectodermal differentiation of P19 embryonal carcinoma cells: overexpression of Wnt-1 perturbs normal differentiation-specific properties. Dev Biol 166:300-310.

Spence JR, Mayhew CN, Rankin SA, Kuhar MF, Vallance JE Tolle K, Hoskins EE, Kalinichenko W, Wells SI, Zorn AM, Shroyer NF, Wells JM. 2011. Directed differentiation of human pluripotent stem cells into intestinal tissue in vitro. Nature 470:105-109.

Strelchenko N, Verlinsky O, Kukharenko V, Vrlinsky Y. 2004. Morual-derived human embryonic stem cells. Reprod Biomed Online 9:623-629.

Suda Y, Suzuki M, Ikawa Y, Aizawa S. 1987. Mouse embryonic stem cells exhibit indefinite proliferative potential. J Cell Physiol 133:197-201.

Svingen T, Wilhelm D, Combes AN, Hosking B, Harley VR, Sinclair AH, Koopman P. 2009. Ex vivo magnetofection: A novel strategy for the study of gene function in mouse organogenesis. Dev Dyn 238:956-964.

Taatjes DJ, Marr MT, Jjian R. 2004. Regulatory diversity among metazoan co-activator complexs. Nat Rev Mol Cell Biol 5:403-410.

Takahashi K, Yamanaka S. 2006. Induction of pluripotent stem cells from mouse embryonic and adult fibroblast cultures by defined factors. Cell 126:663-676.

Takahashi K, Ohnuki M, Narita M, Ichisaka T, Tomoda K, Yamanaka S. 2007. Strategies and new developments in the generation of patient-specific pluripotent stem cells from adult human fibroblasts by defined factors. Cell 131:861-872.

Takeshima K, Chikushi A, Lee KK, Yonehara S, Matsuzaki K. 2003. Translocation of analogues of the antimicrobial peptides magainin and buforin across human cell membranes. J Biol Chem 278:1310-1315.

Tang MX, Szoka FC. 1997. The influence of polymer structure on the interations of cationic polymers with DNA and morphology of the resulting complexes. Gene Ther 4:823-832.

Templeton NS, Lasic DD, Frederik PM, Strey HH, Roberts DD, Pavlakis GN. 1997. Improved DNA: liposome complexes for increased systemic delivery and gene expression. Nat Biotechnol 15:647-652.

Thomson JA, Itskovitz-Eldor J, Shapiro SS, Waknitz MA, Swiergiel JJ, Marshall VS, Jones JM. 1998. Embryonic stem cell lines derived from human blastocysts. Science 282:1145-1147.

Thomson JA, Kalishman J, Golos TG, Durning M, Harris CP, Becker RA, Hearn JP. 1995. Isolation of a primate embryonic stem cell line. Proc Natle Acad Sci USA 92:7844-7848.

Torchilin VP, Levchenko TS, Rammohan R, Volodina N, Papahadjopoulos-Sternberg B, D'Souza GGM. 2003. Cell transfection in vitro and in vivo with nontoxic TAT peptide-liposome-DNA complexes. Proc Natl Acad Sci U S A 100:1972-1977.

Tran DN, Ota LC, Jacobson JD, Paaton WC, Chang PJ. 2007. Influence of nanoparticles on morphological differentiation of mouse embryonic stem cells. Fertil Steril 87:965-970.

Verdurmen WP,. Brock R. 2011. Biological responses towards cationic peptides and drug carriers. Trends Pharmacol Sci 32:116-124.

Villa-Diaz LG, Garcia-Perez JL, Krebsbach PH. 2010. Enhanced transfection efficiency of human embryonic stem cells by the incorporation of DNA liposomes in extracellular matrix. Stem Cells Dev 19:1949-1957.

Wadia JS, Dowdy SF. 2003. Modulation of cellular function by TAT mediated transduction of full length proteins. Curr Protein Pept Sci 4:97-104.

Wang Y, Krushel LA, Edelman GM. 1996. Targeted DNA recombination in vivo using an adenovirus carrying the cre recombinase gene. Proc Natl Acad Sci USA 93:3932-3936.

Wartiovaara J, Liesi P, Rechardt L. 1984. Expression of laminin and fibronectin in endodermal and neural differentiation of F9 embryonal carcinoma cells. Prog Clin Biol Res 151:233-247.

Wasungu L, Hoekstra D. 2006. Cationic lipids, lipoplexes and intracellular delivery of genes. J Control Release 116:255-264.

Watanabe Y, Horie S, Funaki Y, Kikuchi Y, Yamazaki H, Ishii k, Mori S, Vassaux G, Kodama T. 2010. Delivery of Na/I symporter gene into skeletal muscle using nanobubbles and ultrasound: visualization of gene expression by PET. J Nucl Med 51:951-958..

Williams RL, Hilton DJ, Pease S, Willson TA Stewart CL, Gearing DP, Wagner EF, Metcalf D, Nicola NA, Gough NM. 1988. Myeloid leukaemia inhibitory factor maintains the developmental potential of embryonic stem cells. Nature 336:684-687.

Wolf Y, Pritz S, Abes S, Bienert M, Lebleu B, Oehlke J. 2006. Structural requirements for cellular uptake and antisense activity of peptide nucleic acids conjugated with various peptides. Biochemistry 45:14944-14954.

Wolpert L. 1978. Pattern formation in biological development. Sci Am 239:154-164.

Xavier J, Singh S, Dean DA, Rao NM, Gopal . 2009. Designed multi-domain protein as a carrier of nucleic acids into cells. J Control Release 133:154-160.

Xia H, Mao Q, Davidson BL. 2001. The HIV Tat protein transduction domain improves the biodistribution of beta-glucuronidase expressed from ecombinant viral vetors. Nat Biotechnol 19:640-644.

Yamanaka Y, Ralston A, Stephenson RO, Rossant J. 2006. Cell and molecular regulation of the mouse blastocyst. Dev Dyn 235:2301-2314.

Yang F, Green JJ, Dino T, Keung L, Cho SW, Park H, Langer R, Anderson DG. 2009. Gene delivery to human adult and embryonic cell-derived stem cells using biodegradable nanoparticulate polymeric vectors. Gene Ther 16:533-546.

Yang J, Gao C, Chai L, Ma Y. 2010. A novel SALL4/OCT4 transcriptional feedback network for pluripotency of embryonic stem cells. PLoS One 5:e10766.

Yu J, Vodyanik MA, Smuga-Otto K, Antosiewicz-Bourget J, Frane JL, Tian S, Nie J, Jonsdottir GA, Ruotti V, Stewart R, Slukvin II, Thomson JA. 2007. Induced pluripotent stem cell lines derived from human somatic cells. Science 318:1917-1920.

Zákány J, Burg K, Raskó I. 1984. Spontaneous differentiation in the colonies of a nullipotent embryonal carcinoma cell line (F9). Differentiation 27:146-151.

Zhang G, Budker V, Wolff JA. 1999. High levels of foreign gene expression in hepatocytes after tail vein injections of naked plasmid DNA. Hum Gene Ther 10:1735-1737.

Zhao X, Yu SB, Wu FL, Mao ZB, Yu CL. 2006. Transfection of primary chondrocytes using chitosan-pEGFP nanoparticles. J Control Release 112:223-228.

Zijlstra M, Li E, Sajjai F, Subramani S, Jaenish R. 1989. Germ-line transmission of a disrupted β2-microglobulin gene produced by homologous recombination in embryonic stem cells. Nature 342:435-438.

9

Functional Control of Target Single Cells in ES Cell Clusters and Their Differentiated Cells by Femtoinjection

Hideaki Matsuoka, Mikako Saito and Hisakage Funabashi
Department of Biotechnology and Life Science,
Tokyo University of Agriculture and Technology
Japan

1. Introduction

Microinjection is well recognized as an ideal but technically difficult method for the direct introduction of any molecules into single cells (King, 2004; Matsuoka et al., 2006). The difficulty depends on the cell size, intracellular structure, and the physical properties of the cell surface. Egg cells are an easy target because of their large size of 100–200 μm. Blastocysts are also a large target for the injection of embryonic stem (ES) cells (Fig. 1 (A)). For blastocysts, the tip diameter of an injector is 5 μm or greater. On the other hand, fibroblasts are as small as 20–30 μm but comparatively easy targets because their cell surface appears to be firm. Moreover they appear to be resilient against deep insertion of an injection capillary. By contrast, plant cells are larger than fibroblasts but much more difficult targets. Plant cells usually have large vacuoles in the intracellular space and therefore microinjection should be performed into a thin space between the cell membrane and the vacuole. Typical examples are tobacco cultured cells, BY-2, with a size of 40×80 μm (Fig. 1 (B)) and rice protoplasts with a diameter of 30–50 μm. By comparison, ES cells are extraordinarily difficult targets because of their small size (15–20 μm) (Fig. 1 (C)) and sticky cell surface. In fact, microinjection of a plasmid vector into ES cells was not successfully performed by a microinjection expert until our experiment performed in 2005 (Matsuoka et al., 2005).

Before then, microinjection speed was no higher than 10 cells per h and the success rate was 7–8%. In the case of ES cells, the success rate was only 0.2%. Therefore, it was essential to increase the throughput of the method for the purpose of single-cell studies. Thus, we developed a useful robot that could support the microinjection operator; a single-cell manipulation supporting robot (SMSR) (Matsuoka et al., 2005). Using the SMSR, the injection speed increased to 100 cells per h and its success rate reached as high as 10%, even when operated by non-specialist personnel. The quantity of DNA ejected from the injection capillary was estimated to be no greater than 50 fg, and consequently the quantity of DNA actually introduced into a cell was in the fg range. Moreover, when an enhanced green fluorescent protein (EGFP) expression vector was injected into ES cells, the EGFP expression intensity changed in line with the varied concentration of the vector in the injection capillary. This method has been termed femtoinjection because it has enabled semi-quantitative injection at the femtogram level (Matsuoka et al., 2007).

In this chapter we describe the potential applications of femtoinjection to single-cell analysis in ES cell studies. The single cell indicates an isolated cell as well as a particular target cell within a cluster of cells.

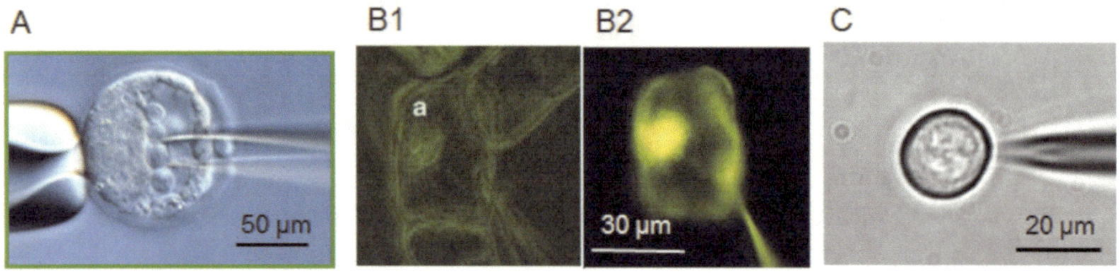

Fig. 1. Microinjection into various cells. (A) Injection of mouse ES cells into a blastocyst; (B) injection of a fluorescent dye (lucifer yellow) into the cytosol of a tobacco cultured cell, BY-2, showing phase-contrast image (B1) and fluorescence image (B2); (C) injection into a mouse ES cell.

2. Why is femtoinjection important?

One of the important objectives of bio-research is to understand the functional role of biomolecules that impact on cellular life as a whole. Many biological functions such as enzymatic activities, signal transduction, and transcriptional and translational activities have been individually proven by employing conventional molecular biological methods. However, in many cases, these methods require destruction of the cells and thus it is difficult to observe the final impact on a living cell. Moreover, in many cases, investigational assays are carried out in vitro which provides a completely different environment to that inside a cell.

For instance, the concentrations of macromolecules such as proteins and nucleic acids inside living cells are extremely high (Zimmerman & Trach, 1991), and under such specific conditions biomolecules behave differently, when compared with in vitro experiments. This is known as the macromolecular crowding effect, and it has been shown that a wide variety of intracellular processes such as protein folding, the association of ribosomes, and the binding of polymerase to DNA are affected by this effect (Hancock, 2004; Zimmerman & Minton, 1993; Zimmerman & Trach, 1988). Furthermore, the localization of biomolecules, which creates the gradation of concentrations and separated reactions in a cell, must be considered as an essential factor for investigating the original activities, despite being very difficult to reconstruct in vitro. Moreover, molecules do not work alone but interact with other molecules, some of which may be unknown factors, leading to sequential reactions.

Thus, conventional methods are indeed limited for the investigation of the functional roles in a living cell and so the development of a new method has been required. One of the most straightforward methods is the microinjection of molecules of interests into a living cell. In fact, microinjection was used to obtain specific data about their functional roles in living cells. However, the performance of ordinary microinjection was insufficient because it was difficult to produce quantitative data. Thus, we have long been engaged in the development of a high-performance microinjection method as described below.

3. SMSR and "suguwaculture"

The SMSR was designed according to the experience of microinjection experts. The most important functions are the operation modes of an injection capillary. The speed and timing of a push-pull motion is dependent on the cell size, the physical properties of the cell surface, and the experience of the operator. Thus, the operational parameters can be adjusted to optimum values for different conditions. The most recent version of the SMSR is shown in Fig. 2.

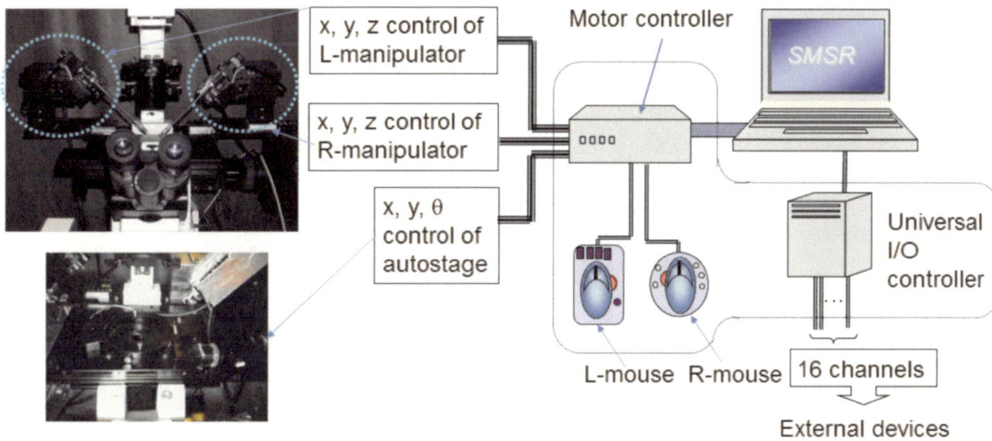

Fig. 2. Schematic of the SMSR.

Microinjection includes not only the insertion of a micropipette into a target cell but also various associated operations such as X-Y stage manipulation, microscope focus adjustment, selection and transformation of a target single cell, exchange of a micropipette, and recording of microscopic images. These operations may be classified into two categories: operations performed by observation with a microscope (on-microscope operations) and operations performed without a microscope (off-microscope operations). Frequently changing between on-microscope and off-microscope operations is annoying and time-consuming. The concept of SMSR is to enable the operator to concentrate their attention only on the microinjection by facilitating associated operations.

To realize this concept, an X-Y coordinate registration system for each target cell was essential. We developed a disposable coordinate standard (CS) chip by ejecting melted polystyrene into a metal mold (Yamada et al., 2008). The chip size was 16.0 mm × 6.0 mm, and the chip surface was divided into four parts of different height and width (Fig. 3 (A), (B)). An adhesive tape was pasted onto the chip (Fig. 3 (C)) so that it could be attached to a culture dish simply by removing a cover sheet and placing it beneath the dish. The edges of these parts could be recognized as straight lines 2 μm in width under the microscope (Fig. 2 (D), (E)). The culture dish depicted in Fig. 2 (F) is a popular one and one which is used in our laboratory. The thickness of the CS chip is no thicker than 0.3 mm because the height of the bottom rim of commercially available culture dishes is in the range 0.4–1.4 mm.

Many cell scientists are concerned about the dish material and its surface treatment, because theses factors can influence the growth and physiological properties of the test samples (cells and tissues). Once they are able to obtain successful results using a particular type of dish, they will continue to use dishes of the same type. Therefore, it is extremely important that the CS chip can be applied to any type of dish. Moreover the CS chip must not contact the test samples directly. Our new CS chip meets these requirements and cannot be replaced by other devices such as multiple microwells and multiple pore plates, some of which are already commercially available.

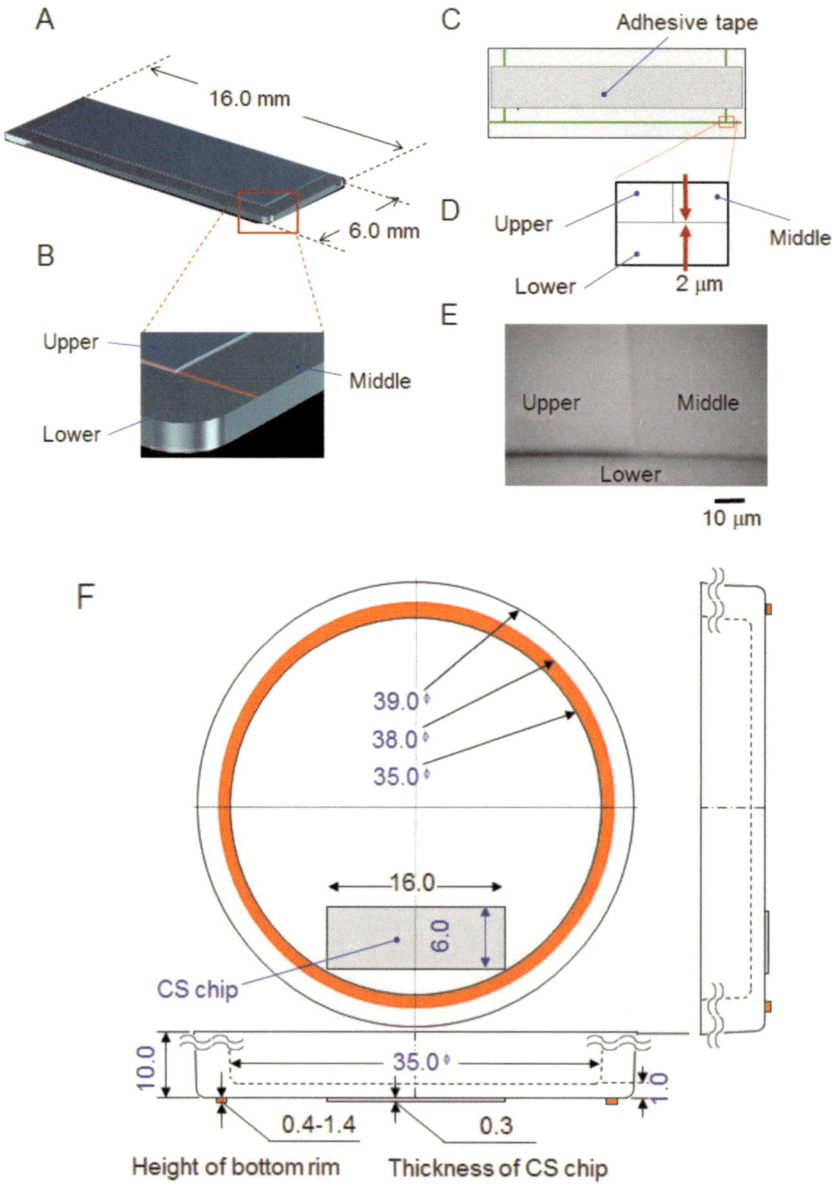

Fig. 3. Coordinate standard chip. Reproduced from *Micros. Microanal.* (2008) 14, 236-42 with permission from Cambridge University Press.

A standard protocol for single-cell experiments such as microinjection or image capture is as follows. Initially, the X–Y coordinates of 50–100 target single cells per culture dish are consecutively registered, after which an operator can signify $Cell_1$, $Cell_2$, ..., and $Cell_n$ in the relevant order to the microscopic view center simply by clicking a foot switch. Alternatively, an operator can select a particular cell number, $Cell_i$, directly by selecting the i-th "Cell number" with a mouse. If the dish is removed from the automatic stage and replaced some time later, the rapid cell search function can be used because of the registered X–Y axes. Such applications include, for instance, the simultaneous use of multiple culture dishes, cell culture in a separate incubator, and image capture with a different microscope (Fig. 4). Therefore, operators can easily search any single cell in any culture dish on any microscope at any time. Such a system is remarkably useful for various modes of single-cell experiment. We have named this system "Suguwaculture", which means instantly (*sugu* in Japanese) recognizable (*wakaru* in Japanese) culture.

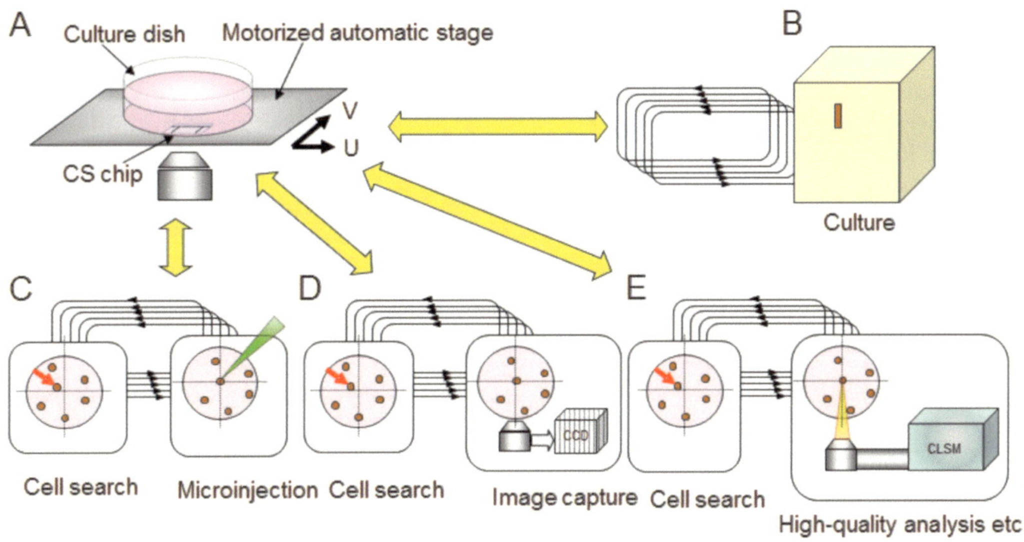

Fig. 4 Time-lapse single-cell experiments with multiple apparatus using the Suguwaculture system. (A) single-cell experiments such as cell registration and cell manipulation on the automatic stage of an inverted microscope, (B) culture in an incubator, (C) microinjection, (D), and (E) time-lapse image captures with other optical apparatus. Reproduced from *Micros. Microanal.* (2008) 14, 236-42 with permission from Cambridge University Press.

4. What can be done by femtoinjection?

4.1 Semi-quantitative introduction of macromolecules of any size

DNA solution in a glass capillary was ejected into a micro water drop and the quantity of DNA in the water drop was determined by quantitative real-time polymerase chain reaction. The ejected quantity was influenced by the diameter of the capillary tip and the intensity and period of pressure. The optimum conditions of the tip diameter, the pressure strength, and the pressure application period were typically 0.5–0.8 μm, 0.70 kgf/cm^2, and 30 ms, respectively. For DNA concentrations of 10, 50, and 100 ng/μl under optimum conditions, the ejected quantity changed accordingly within the range 5–50 fg. Therefore, the quantity of DNA actually introduced into an ES cell was estimated to be no greater than the fg level (Matsuoka et al., 2007).

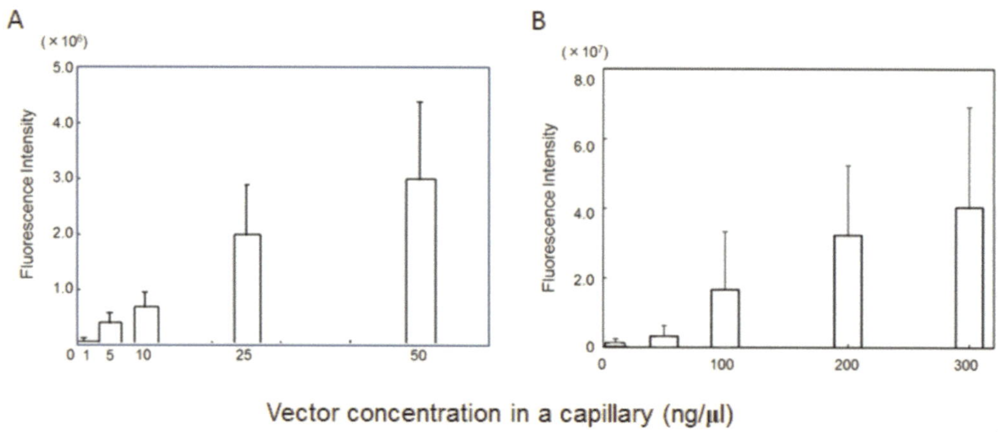

Fig. 5. Semi-quantitative relationship between the *EGFP* expression intensity and the vector concentration in a capillary in the range 1–50 ng/μl (A), and a higher range from 10–300 ng/μl (B). Error bar indicates mean ± SD (n=10).

An EGFP expression vector was injected into ES cells to visualize the gene expression. The fluorescence of EGFP appeared after 24 h of culture not only in the injected cell but also in the neighboring cells that were daughter cells or their further descendent cells. The fluorescence intensity was averaged over the whole area of the target single cell or the single cluster containing the initial target cell. Thus, the calculated intensity of EGFP fluorescence was thought to be proportional to the quantity of the gene successfully introduced into the cell. In fact, the EGFP fluorescence intensity increased as the vector concentration increased from 1 to 300 ng/μl (Fig. 5) (Saito & Matsuoka, 2010)

If the average of multiple cells is used, methods such as lipofection and electroporation might show a semi-quantitative introduction performance. By comparison, we investigated the performance of lipofection using an EGFP expression vector (pCAG-EGFP). Mouse ES cells were suspended in a solution containing pCAG-EGFP (5 μg DNA/50 μl medium) and a cell fusion-inducing agent, lipofectamine 2000 (15 μl/35 μl medium). After incubation at 37°C for 24 h, fluorescent cell colonies were selected arbitrarily for microscopic image analysis. The fluorescence intensity was integrated and averaged over each colony. Thus, the obtained values were then averaged to quantify the gene expression intensity. Then the concentration of pCAG-EGFP or lipofectamine was varied from 1/100 to 100 times the initial concentrations. As depicted in Fig. 6, the optimum condition could be determined but it was difficult to find a proper range in which a quantitative or semi-quantitative relationship could be satisfied.

4.2 Semi-quantitative introduction of multiple factors

Control of the relative expression intensity of multiple genes is of great importance. To demonstrate such semi-quantitative control, we investigated a particular case of two genes: *EGFP* and *DsRed*. We prepared pCAG-EGFP, a DsRed expression vector (pCAG-DsRed), and a composite vector, pCAG-EGFP-IRES-DsRed, in which both genes were constructed in tandem (Fig. 7 (A)). When a solution containing pCAG-EGFP and pCAG-DsRed at the same concentration (5 ng/μl) was injected into an ES cell, the fluorescence intensity of EGFP appeared to be much higher than that of DsRed (Fig.7(B)). In the case of pCAG-EGFP-IRES-DsRed, the fluorescence image was similar (Fig. 7 (C)).

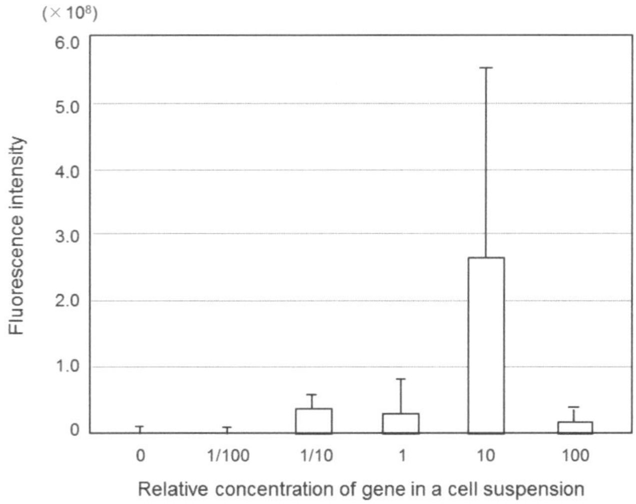

Fig. 6. Relationship between the *EGFP* expression intensity and the vector concentration in a cell suspension for lipofection. Error bar indicates mean ± SD (n=20).

The molecular size of EGFP protein is 25.4 kDa and the EGFP can pass through the nuclear membrane, because the cut-off size of the nuclear membrane is thought to be 50 kDa. Therefore, EGFP distributes both in the cytosol and the nucleus. By contrast, DsRed needs to form a tetramer to become a fluorescent protein. The molecular size of the DsRed tetramer (DsRedT4) is 103.6 kDa. The tetramer is likely to be formed immediately in the cytosol, because fluorescent DsRed was distributed predominantly in the cytosol. Assuming that the fluorescence intensities of the respective fluorescent units, EGFP and DsRedT4, are the same, the relative fluorescence intensity of EGFP to DsRed should be 4:1 if the gene copy number is the same.

In fact, however, the intensity of EGFP expression was roughly 10 times higher than that of DsRed in the case of either separate vectors (Fig. 7 (D)) or the composite vector (Fig. 7 (E)). Next we mixed pCAG-EGFP and pCAG-DsRed at a ratio of 1:10 and injected the mixture into ES cells. Consequently the gene expression intensity of both genes was the same (Fig. 7 (F)). This result demonstrates that it is feasible to control the relative expression intensity of multiple genes easily as desired by regulating a ratio of the mixture.

4.3 Temporal and spatial control of injection in a cell colony

In order to analyze the functional role of a single cell in a cluster or colony of multiple cells, it is necessary to introduce a gene or other marker into a particular target single cell and then continue microscopic observation. In the case of the ES cell, an isolated single cell grows to be a colony with a diameter of approximately 50 μm after culture for 72 h. Therefore, we prepared such colonies of ES cells and demonstrated temporal and spatial control of femtoinjection into a target single cell.

The first case was the injection of a fluorescent dye, Dextran-Alexa 488 (1 μg/μl, MW 70 kDa). The success of physical introduction into only a single cell was confirmed by the diffusion pattern of the dye (Fig. 8 (A)). The culture was continued for a further 42 h. Then injection of the same dye into two single cells was performed. The physical introduction of the dye was successful at this time too (Fig. 8 (B)). Such an injection into any single cell located at the surface of a colony was possible.

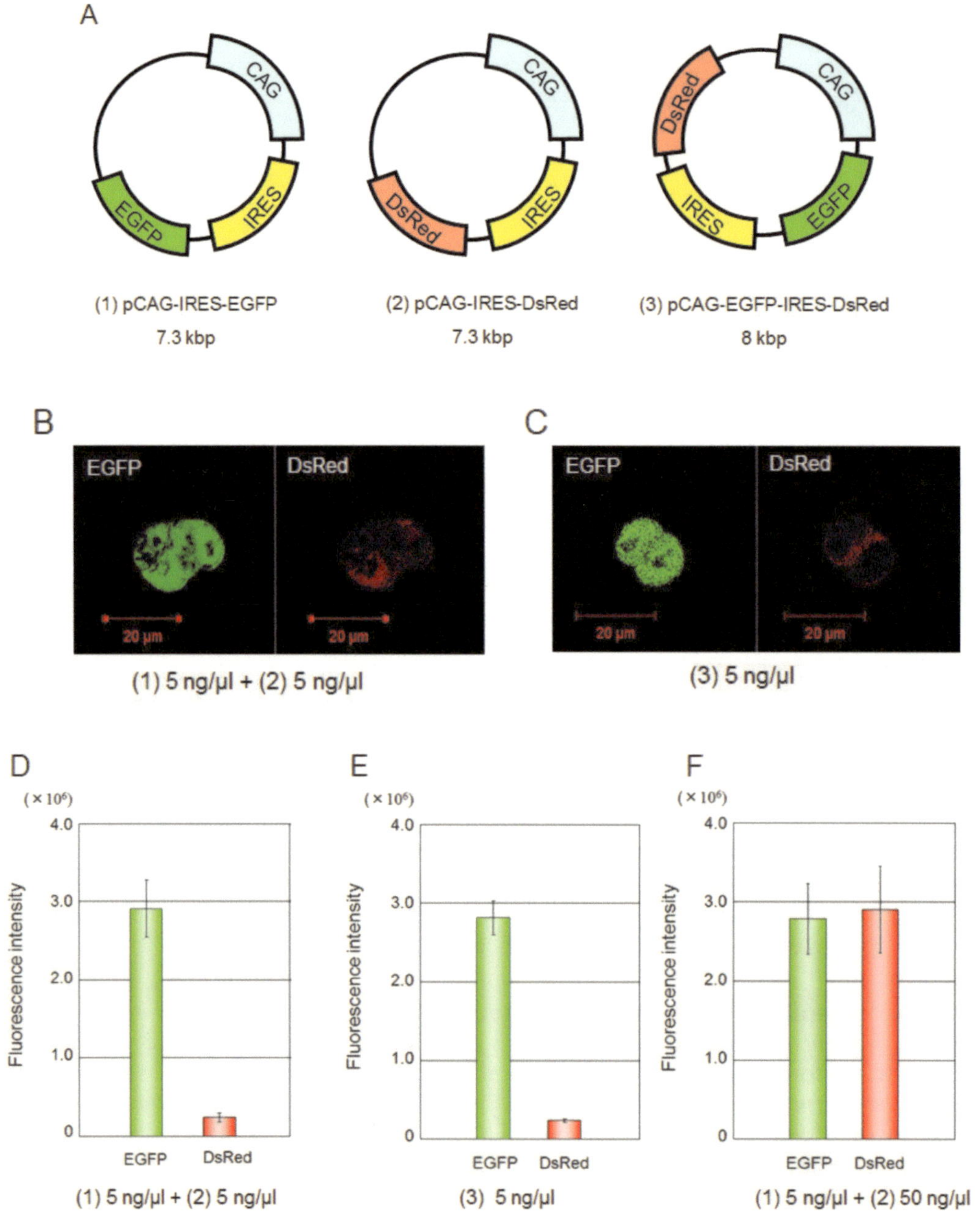

Fig. 7. Quantitative relationship in simultaneous introduction of EGFP and DsRed expression vectors. (A) Constructs of vectors; (B), (C) Examples of fluorescent images of EGFP and DsRed, respectively; (D), (E), (F) Relative intensities of EGFP and DsRed fluorescence. Error bar indicates mean ± SD (n=10).

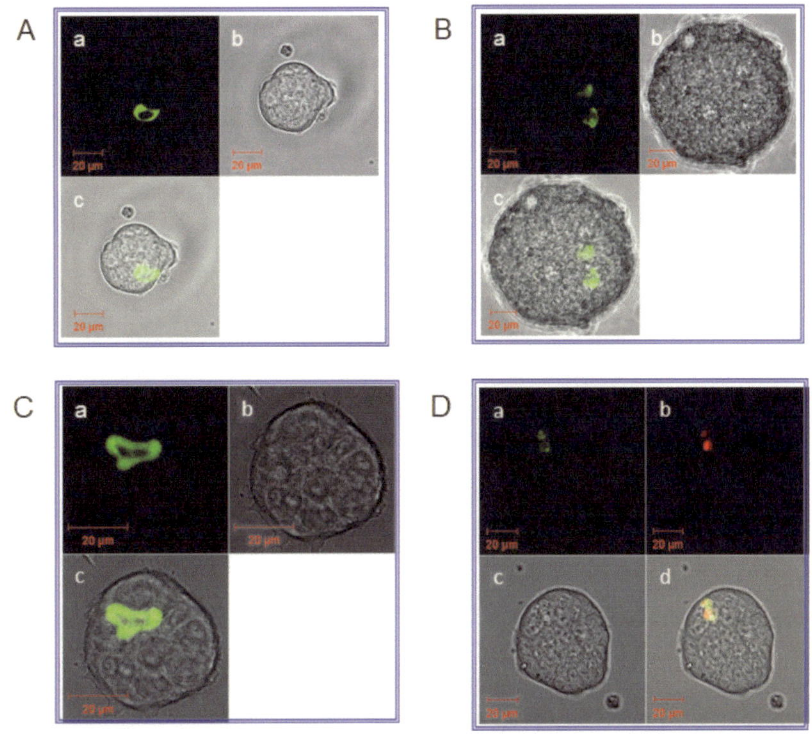

Fig. 8. Injection of Dextran-Alexa 488 (A, B) or Dextran-Alexa 488 and pCAG-DsRed (C, D) into a target single cell of an ES cell colony. Injection was performed after culture for 72 h. (A, C) Immediately after injection; (B, D) 42 h after injection. (A–C) (a) Dextran-Alexa 488, (b) phase-contrast image, (c) superposition of a and b; (D) (a) Dextran-Alexa 488, (b) DsRed, (c) phase-contrast image, (d) superposition of a, b, and c.

The next case was the simultaneous injection of a gene expression vector together with a fluorescent dye. A solution containing pCAG-DsRed (50 ng/µl) and Dextran-Alexa 488 was injected into a single cell in a colony (Fig. 8 (C)). The diffusion pattern of Dextran-Alexa 488 became a marker of the success of physical introduction. After incubation for 24 h, the expression of DsRed was observed in the same cell in which the diffusion of Dextran-Alexa 488 was observed (Fig. 8 (D)). This indicates the physical and physiologic success of injection.

4.4 Spatial control of injection site in a cell: cytosol or nucleus

When a plasmid vector is introduced into a cell, the vector has to be transported to the nucleus. Therefore, its expression efficiency will be enhanced if the vector is introduced directly into the nucleus rather than into the cytosol. The nucleus is located in the center of a cell. The direct injection into the nucleus requires a deeper insertion of an injection capillary. Such an insertion, however, seems to decrease the likelihood of cell viability. Next we investigated the possibility of non-lethal insertion into the nucleus of ES cells.

Dextran-Alexa 488 was injected into single cells of a DsRed-expressing ES cell line. This cell line shows red fluorescence only in the cytosol, because DsRedT4 cannot pass through the nuclear membrane (Fig. 9 (A)). For the same reason, Dextran-Alexa 488 cannot either.

Consequently if Dextran-Alexa 488 is introduced into the cytosol, the cytosol will become yellow. On the other hand, if the dye is introduced into the nucleus, the cytosol will remain red and the nucleus will become green. Indeed, two patterns were observed as depicted in Fig. 9 (B) and (C). These results indicate direct injection into the cytosol and nucleus, respectively.

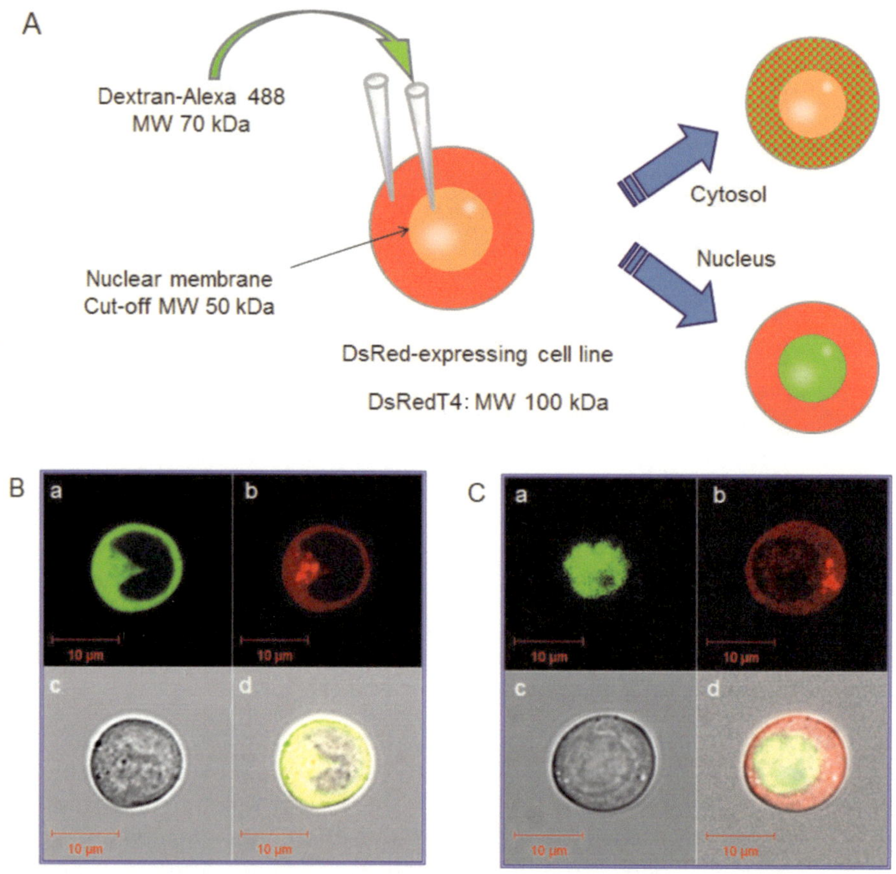

Fig. 9. Injection of Dextran-Alexa 488 into a single cell of a DsRed-expressing ES cell line. (A) Experimental protocol; (B) example of introduction into the cytosol; (C) example of introduction into the nucleus. (a) Dextran-Alexa 488, (b) DsRed, (c) phase-contrast image, (d) superposition of a, b, and c.

Next, we investigated the effect of direct injection on the expression efficiency of an injected vector. The yellowish green fluorescent protein Venus was selected as a demonstrative gene production because Venus fluorescence was expected more stable and intense than EGFP. The Venus-expression vector and Dextran-Fluorescein (MW 70 kDa) were injected into DsRed-expressing ES cells. The results are summarized in Fig. 10.

Injection was performed into 306 single cells, and 180 cells maintained viability. Dextran-Fluorescein was observed in 158 of 180 cells. This indicates that the success rate of physical introduction of the fluorescent dye was 52% (158/306). In these 158 cells, 117 injections were into the cytosol, whereas 41 were injected into the nucleus. The expression rate of the *Venus* gene was no higher than 27% (31/117) in the case of cytosolic injection. By contrast, the rate reached 83% (34/41) for nucleic injection. These results strongly suggest that the direct introduction of a vector into the nucleus is effective for the enhancement of gene expression efficiency in spite of a high risk of loss of cell viability due to the deep insertion.

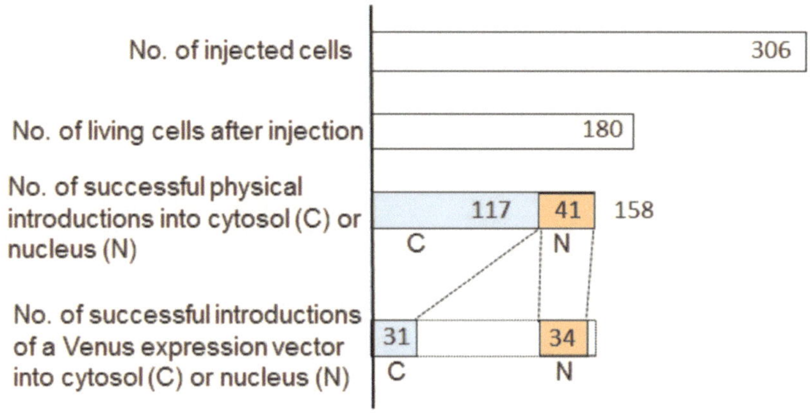

Fig. 10 Comparison of gene expression efficiencies attained by injection into the cytosol and nucleus.

5. Analysis of a DNA nuclear targeting sequence as a potential enhancer of the expression of a femtoinjected exogenous gene in ES cells

In order to enhance the expression efficiency of an injected gene, it is effective to inject it directly into the nucleus, as described above. However, deep insertion into the nucleus is likely to cause serious damage to the cell. Therefore, an alternative method is required to promote transportation of the gene toward the nucleus. A simple and promising approach is to add a specific sequence into a vector, using a so-called DNA nuclear targeting sequence (DTS) (Dean, 1997; Dean et al., 1999a, 1999b) (Fig. 11).

A DTS is a specific sequence recognized by transcriptional factors. Therefore, the injected vector is expected to form a complex with the specific or general transcriptional factor. Since the transcriptional factors are imported into the nucleus by an intrinsic active transport system, the vector–DTS complex may also be imported into the nucleus via the same system. Consequently the transport of the vector into the nucleus is expected to be promoted to result in the enhancement of its expression.

For instance, the 72 base pairs of the simian virus 40 (SV40) promoter/enhancer sequence (SV40-DTS) was reported to be a DTS (Dean, 1997), which enhanced the nuclear import so that the efficiency of the gene expression increased compared to without the DTS. Since this sequence has the binding site of ubiquitously expressed transcriptional factors, it has been suggested that the introduced vector DNA forming a complex with these transcriptional factors passes through a nuclear pore with the help of the importin family which recognizes nuclear localization signals of transcriptional factors (Miller & Dean, 2009; Miller et al., 2009). Although it has been revealed that SV40-DTS functions as a DTS in various types of cells, the function in ES cells remains unclear, probably due to the difficulty of the injection of a vector into the cytoplasm.

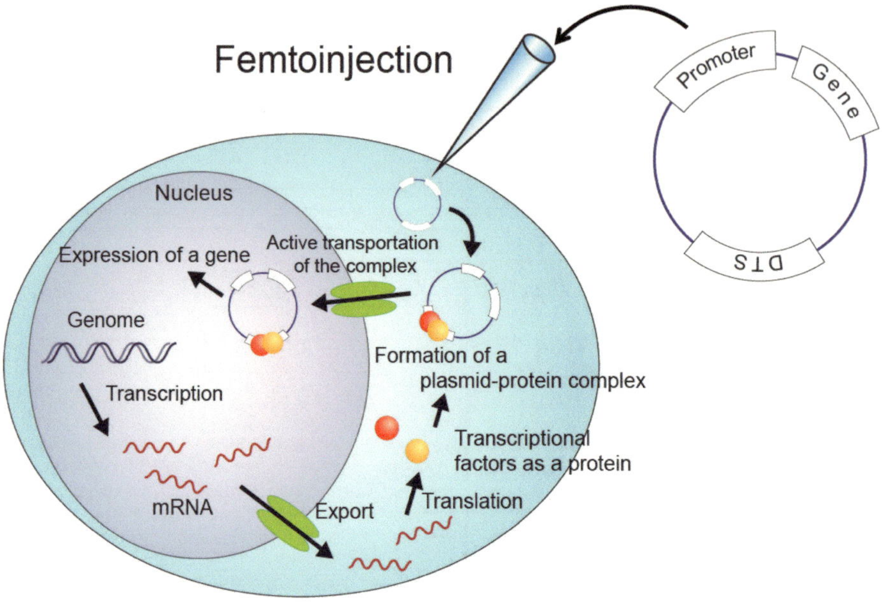

Fig. 11. The enhancement of gene expression efficiency of femtoinjected plasmid by utilizing protein life dynamics. The femtoinjected plasmid is drugged into a nucleus via the intrinsic transportation system recognizing a nuclear localization signal of transcriptional factors after forming a plasmid-protein complex. Thus the plasmids are actively delivered in nucleus, expressing the gene efficiently. Reproduced from *Biochem. Biophys. Res. Commun.* (2010) 400), 554-8 with permission from Elsevier.

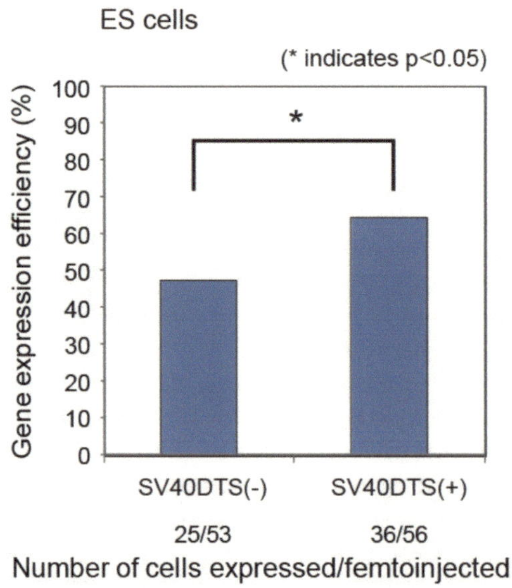

Fig. 12. Effect of SV40-DTS on the gene expression efficiency in ES cells. The plasmids containing SV40-DTS express a probe gene (EGFP) efficiently in ES cells. Reproduced from *Biochem. Biophys. Res. Commun.* (2010) 400), 554-8 with permission from Elsevier.

However, with femtoinjection, it is now feasible to investigate the effect of a DTS on the efficiency of gene expression in ES cells (Funabashi et al., 2010). An EGFP expression vector (SV40-DTS positive vector or control vector encoding no DTS) and Dextran-Texas Red (MW 40 kDa) were injected into ES cells. According to the protocol described in section 4.4, the injected site was determined (i.e., cytosol or nucleus). The MW of Dextran-Texas Red is similar to the cut-off size of the nuclear membrane (MW ca. 50 kDa), Dextran-Texas Red can hardly diffuse into the nucleus. The efficiency of gene expression was defined as "the number of cells in which EGFP expression was observed per number of cells in which the *EGFP* gene was femtoinjected into the cytoplasm determined by the observation of Dextran-Texas Red". As shown in Fig. 12, comparison of gene expression efficiency between the SV40-DTS positive vector-injected ES cells and ES cells injected with the control vector clearly revealed the effect of SV40-DTS. This was the first report to identify the role of DTS in ES cells (Funabashi et al., 2010).

Inspired by the success of the investigation, we hypothesized that a sequence recognized by ES cell-specific transcriptional factors should function as an ES cell-specific DTS. Oct3/4 binding Sox2 regulatory region 2 (SRR2) has been reported to be a distal enhancer for Sox2 expression where the Oct3/4 and Sox2 complex is bound (Tomioka et al., 2002). As both Sox2 and Oct3/4 are well-known ES cell-specific transcriptional factors for maintaining pluripotency (Masui et al., 2007; Niwa et al., 2000; Niwa, 2007), it is expected that the 81 bp of SRR2 (Fig. 13) functions as a DTS (termed SRR2-DTS) and that it promotes the transportation of the plasmid giving rise to the efficiency of gene expression.

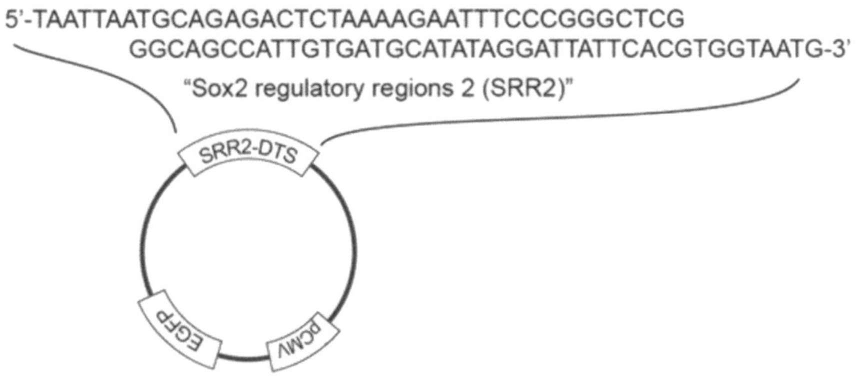

Fig. 13. Sequence of Sox2 regulatory regions 2 (SRR2). SRR2 is an 81 bp sequence containing Oct3/4 and Sox2 binding sequences. Reproduced from *Biochem. Biophys. Res. Commun.* (2010) 400), 554-8 with permission from Elsevier.

In the case of MIN6 cells (mouse insulinoma cells (Miyazaki et al., 1990)) which express neither Sox2 nor Oct3/4, no difference is observed in terms of the gene expression efficiency regardless of the types of introduced plasmids as shown in Fig. 14 (A), whereas ES cells femtoinjected with the SRR2-DTS-positive plasmid exhibit higher gene expression efficiency than ES cells injected with the SRR2-DTS-negative plasmid (Fig. 14 (B)). This result supports our hypothesis that the enhancement of gene expression observed is due to the expression of these ES cell-specific transcriptional factors.

The observed phenomenon can possibly be explained as follows. The mRNAs of ES cell-specific transcriptional factors Oct3/4 and Sox2 were exported to the cytoplasm, and then translated into proteins. These proteins met the femtoinjected plasmids which had SRR2-

DTS forming the complex on the sequence. This complex was drugged into the nucleus via the nucleic pore transportation system in which the nuclear localization signals on these transcriptional factors were recognized by importin family members. This active transportation of the plasmid resulted in the higher gene expression efficiency. Indeed, it has been reported that Oct3/4 and Sox2 are actively transported into the nucleus by importin α1/β1 complex and possibly by importin β1, respectively (Yasuhara et al., 2007, 2009), in ES cells. It has also been suggested that these importins will function work on the SV40-DTS depending transportation (Dean, 1997; Dean et al., 1999; Lam & Dean, 2010; Miller et al., 2009). Thus, both sequences function as DTSs in ES cells. In particular, it has been suggested that SRR2-DTS functions as an ES cell-specific DTS.

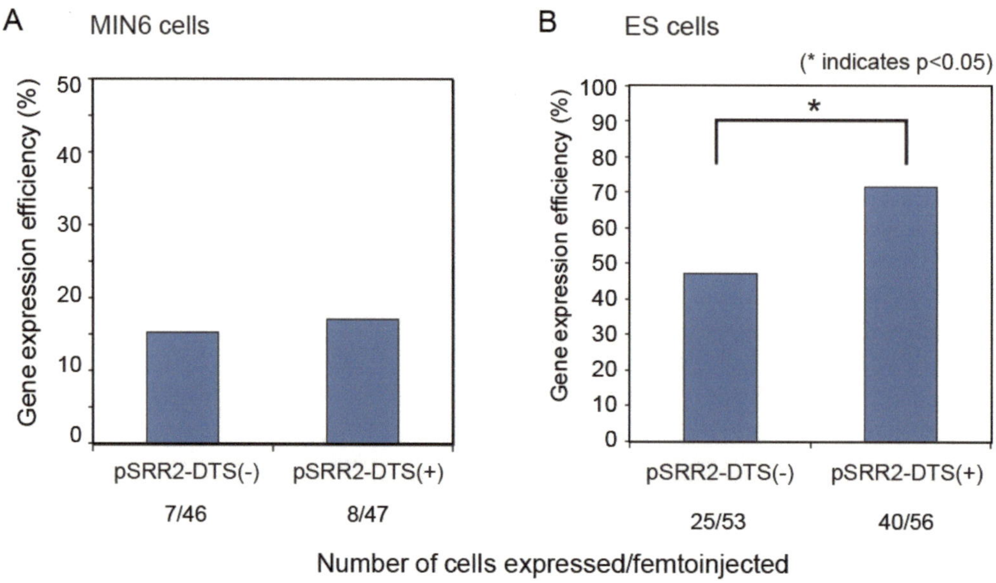

Fig. 14. Effect of SRR2 region on the gene expression efficiency in MIN6 cells (A) and ES cells (B). There was no effect of SRR2 sequence on the gene expression efficiency in the case of MIN6 cells while clear enhancement was observed in case of ES cells. Reproduced from *Biochem. Biophys. Res. Commun.* (2010) 400), 554-8 with permission from Elsevier.

Femtoinjection has enabled injection into living single ES cells of all substances including ions, fluorescent dyes, and nucleic acids, which range from low molecular weight chemicals to macromolecules, as long as they are soluble. On the other hand, it is not yet feasible to aim for high resolution injection sites such as small organelles in mammalian cells. The intrinsic transport system inside living cells such as the protein life dynamics utilized here is expected to help the distribution of introduced substances to the place where one desire to send for.

6. Femtoinjection assay for differentiation-inducing factors based on a morphological indicator

With regard to quantitative gene expression control, how to maintain the undifferentiated state of ES cells is of particular interest. Several genes such as *Oct3/4*, *Sox2*, and *Nanog* are well recognized as undifferentiated cell markers (Masui et al., 2007; Niwa, 2007). However, it is also recognized that none of them alone is an exclusive marker. Rather the undifferentiated state seems to be maintained by an appropriate balance of their expression intensities. To prove this supposition, the simultaneous control of expression intensities of these genes is thought to be an effective method. Femtoinjection is particularly useful for this purpose.

First, we varied the expression intensity of Oct3/4 alone by femtoinjection of Oct3/4 expression vector into ES cells (Matsuoka & Saito, 2010). As a non-invasive marker of differentiation, we introduced the perimeter-to-radius ratio (PR ratio) as a novel morphologic indicator (Fig. 15). Generally ES cells are circular but change into a morphologically complex shape once differentiated. The PR ratio can sensitively reflect such a change. If the cell shape is a true circle, the PR ratio is 6.28; however, it becomes greater as the cell shape becomes more complex.

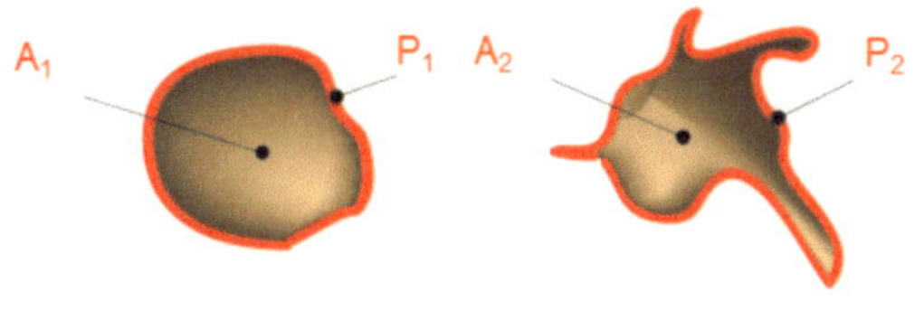

Fig. 15. Definition of PR ratio for quantitative expression of morphologic complexity. Reproduced from *ECS Transactions* (2009) 16, 9-14 with permission from The Electrochemical Society.

As shown in Fig. 16, an Oct3/4 expression vector containing an EGFP reporter gene (pCAG-Oct3/4-IRES-EGFP) was injected into single cells of mouse ES cells. As a control, an EGFP expression vector was injected into single cells in the same manner. At the same time, no injection control cells were prepared and observed in parallel. After injection, cell morphology was recorded every 24 h and the PR ratio was determined. In some cases, the PR ratio increased at 24 h but returned to the initial level at 48 h or later. In other cases, it increased continuously throughout the observation period for 72 h. Both with and without injection control, the PR ratio was 7.3–8.0 until 72 h. Based on these results, we defined the differentiated state as follows: PR ratio became greater than 8.0 and remained at this level at 72 h. According to this definition, the differentiation rate was determined as (number of differentiated cells/number of EGFP positive cells at 24 h) × 100%.

Figure 17 shows an example of injection of pCAG-Oct3/4-IRES-EGFP at a concentration of 600 ng/µl in the injection capillary. At a concentration of 300, 600, or 1200 ng/µl, the differentiation rate was 2/6 (33%), 5/7 (71%), and 6/7 (86%), respectively. Though the number of test samples is small, this result suggests that the increase in Oct3/4 expression intensity is effective for the promotion of differentiation of ES cells.

For the up-regulation of a target gene expression or a specific protein effect, the introduction of the over-expression vector is thought to be a useful method, as demonstrated above. The introduction of mRNAs or target proteins is thought to be a more direct way to cause the up-regulation effects. On the other hand, the down-regulation of a target gene expression may be produced by the introduction of siRNA or shRNA. The introduction of inhibitors against the molecules of interests including antibodies and dominant negatives may also be an effective way to produce the down-regulation effects. In either case, the quantitative introduction of respective factors is essential for the regulation.

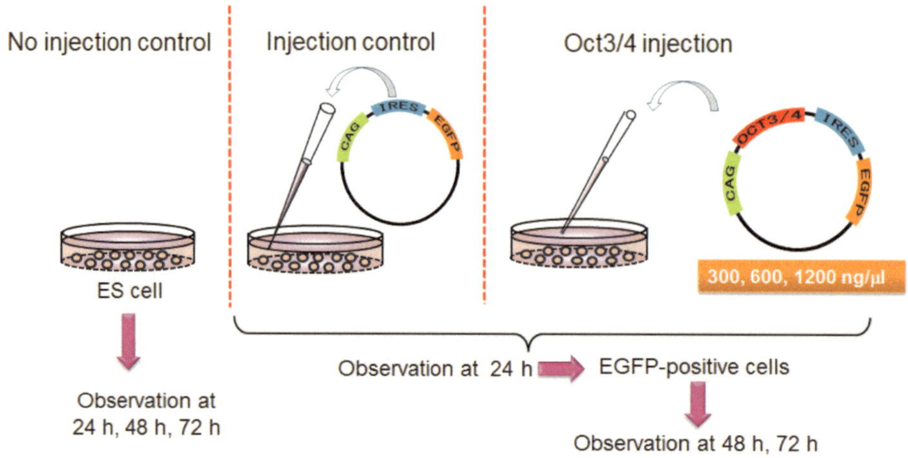

Fig. 16. Protocol for femtoinjection of Oct3/4 expression vector.

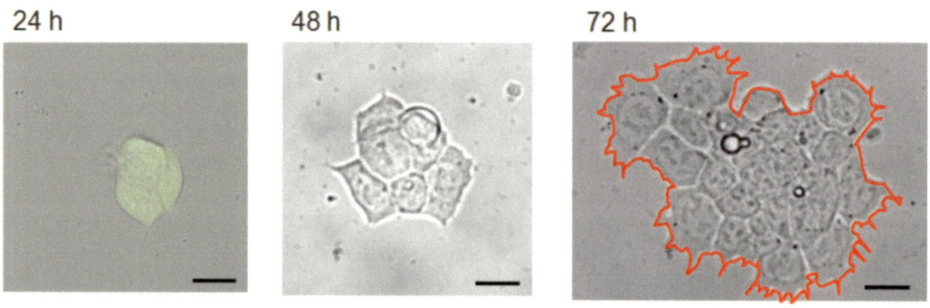

Fig. 17. Morphologic changes in the ES cell colony induced by injection of Oct3/4 expression vector. The concentration of pCAG-Oct3/4-IRES-EGFP in the injection capillary was 600 ng/μl. Scale bar indicates 20 μm. Reproduced from *ECS Transactions* (2009) 16, 9-14 with permission from The Electrochemical Society.

7. Marking vibrating muscle fiber cells with non-harmful fluorescent dyes

We have previously cultured mouse ES cells and tried to prepare muscle fibers. In the course of the study, it was interesting to find that the polyamines spermine could induce remarkable growth of multi-layer muscle fibers (Fig. 18) (Sasaki et al., 2008). The growth process was so unique that we have been deeply engaged in gene expression and functional analyses. With regard to the pulsation function, it should be of fundamental significance to analyze how to acquire synchronously pulsation potency. Therefore, we considered it necessary to analyze the cell-to-cell communication in a cell sheet during synchronous pulsation and during rest. Thus, we intended to challenge the injection into single cell in muscle fiber during its pulsation as well as when rest.

Femtoinjection into a particular target cell in an ES cell colony was demonstrated in section 4.3. Such injection into a target single cell is also possible in a cell cluster composed of differentiated cells of different sizes and shapes, as long as the outline of a target single cell can be observed microscopically. In the present case of pulsating cells, however, it seemed to be almost impossible. Next we re-examined various parameters relevant to injection

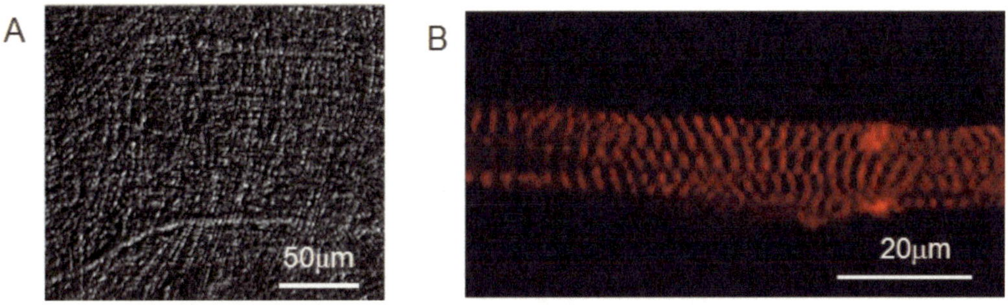

Fig. 18. Muscle fiber prepared from mouse ES cells with spermine treatment. (A) Multilayer sheet of contractile muscle fiber; (B) fluorescence image of striated muscle fibers stained using an antibody to α-actinin that was detected with an Alexa Fluor-labeled secondary antibody.

performance and finally succeeded in injecting a pulsating single cell in the muscle fiber, as described below.

We aimed to suppress the pulsation intensity by cooling the muscle fiber for a while and to then perform injection, before warming it to room temperature. To confirm the cooling effect, it was necessary to monitor the pulsation intensity. The optical intensity at a particular position in the muscle fiber was measured continuously and its relative change was quantified to express the pulsation intensity at the measured position. Figure 19 shows

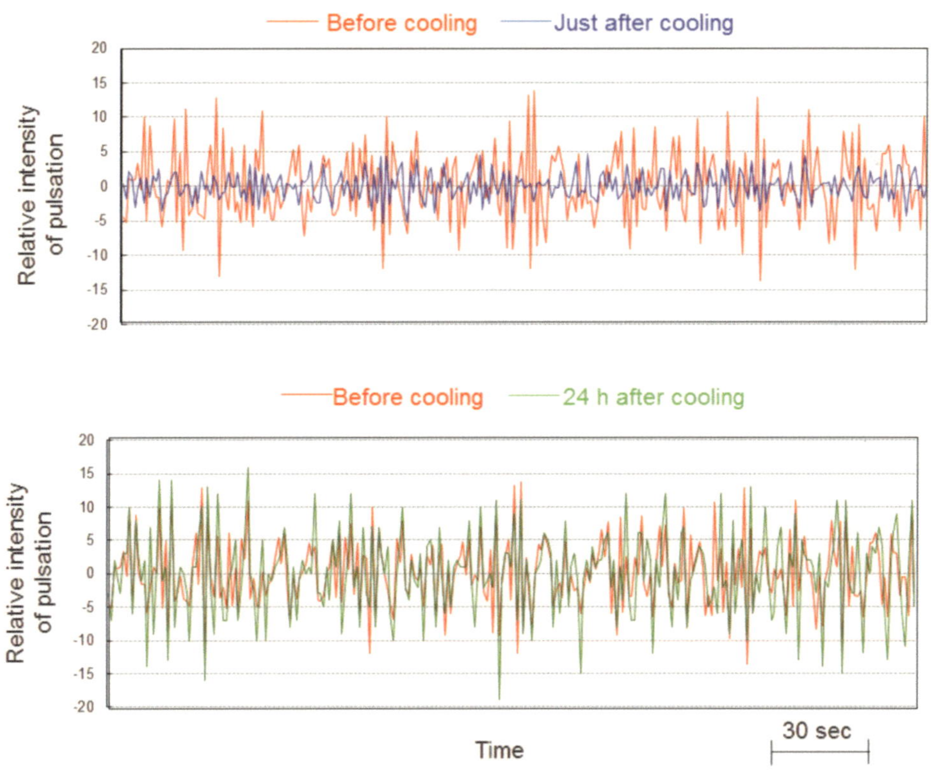

Fig. 19. Influence of cooling on the pulsation of a muscle fiber.

a record obtained during the temperature variations. The cells cultured at 37°C were rapidly cooled to 4°C. The pulsation intensity decreased to between a half and one-third. Then the cells were warmed gradually to room temperature. After 24 h from the start of warming, the cells recovered their initial pulsation intensity to the original level. Therefore, cooling to 4°C for a short period of time had no obvious effect on the intrinsic pulsation function.

Another point to be considered with regard to muscle fibers is the physical properties of the cell membrane. Then the tip parameters were modified to adapt to muscle fiber. Finally the tip diameter of an injection capillary was made smaller and the tip length was made shorter than for capillaries used for ES cells. Using this capillary, a solution of Alexa 488 (1 µg/µl, MW 643) and/or Dextran-Texas Red (10 µg/µl, MW 3,000) was injected into the target single cells simultaneously. Dextran-Texas Red stayed within a small area around the injection point that was supposedly within the target cell (Fig. 20). Therefore, only the injection into the target cell was considered successful. On the other hand, Alexa 488 distributed to a wider area (approximately 200 µm×100 µm). However, the diffusion area was not a simple circular pattern but an irregular one. Such an irregular diffusion pattern was unexpected. The present results are still preliminary but suggest a heterogeneous distribution of gap junctions with different opening diameters.

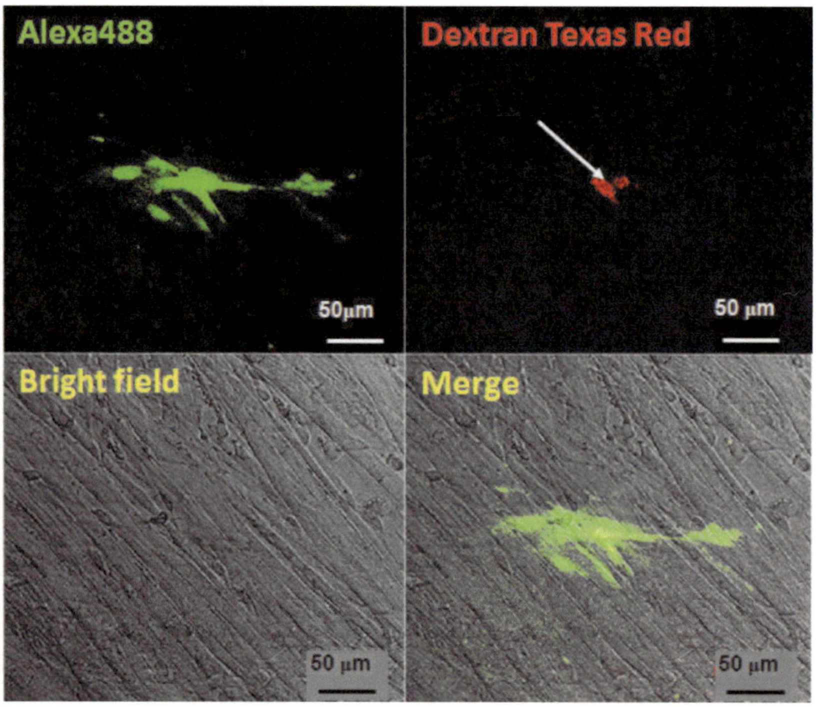

Fig. 20. Femtoinjection of Alexa 488 and Dextran-Texas Red into a single pulsating cell of a muscle fiber.

8. Automatic injection system

SMSR is a useful machine but its further improvement is required to produce a fully automatic system. To accomplish this, it is essential to detect a stop signal for the injector when the injector tip has just penetrated the cell membrane. Next we devised an electric circuit to measure the intracellular potential sensitively without a high degree of electrical noise. The measured potential change was thought to be a useful control signal for the injector drive. In order to measure the electrical potential and at the same time introduce a marker dye into a target cell, a double barrel capillary was used (Matsuoka et al., 2006).

A glass capillary with a θ-shaped cross section (outer diameter 1 mm) was heated and pulled to make the tip diameter smaller than 1 μm using a laser puller. One channel of the capillary was filled with a solution of 0.5 M KCl. A Ag/AgCl wire was then inserted into the capillary to produce a potential measuring electrode (MeaE) (Fig. 21). The reference electrode for MeaE was a commercially available Ag/AgCl electrode (RE). The other channel was filled with a solution containing 0.5 M KCl and 1 mM lucifer yellow. A Pt wire was inserted into the capillary to produce a dye-introducing electrode (IntE). The counter electrode (CE) for the IntE was prepared by filling a glass tube with 3% agar gel containing 0.5 M KCl.

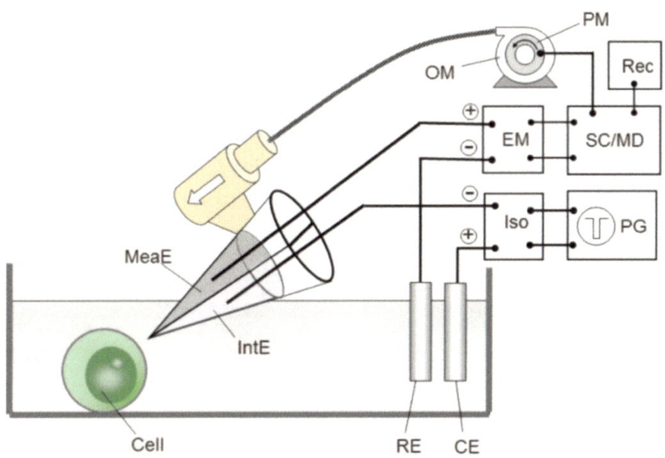

Fig. 21. Schematic diagram of an automatic microinjector positioning system. MeaE: potential measuring electrode, IntE: dye-introducing electrode, RE: reference electrode for MeaE, CE: counter electrode for IntE, Iso: isolator, PG: pulse generator, EM: electrometer, SC/MD: signal converter and pulse motor driver, PM: pulse motor, OM: oil pressure manipulator, Rec: recorder. Reproduced from *Bioelectrochemistry* (2006) 69, 187-92 with permission from Elsevier.

A microinjector was propelled manually and carefully penetrated into a mouse ES cell. The output potential of the MeaE changed sharply to -16 mV (Fig. 22 (A)). It returned to the initial level when removed from the cell. The value of -16 mV is defined as the intracellular potential and is equivalent to the cell membrane potential if the intracellular space is assumed to be homogeneous. The mean and standard deviation (SD) of the intracellular potential for 35 sample ES cells were -16.3 mV and 5.6 mV, respectively (Fig. 22 (B). A steady intracellular potential is available but the level is variable from cell to cell. Therefore, an absolute value cannot be used as the sign of cell membrane penetration.

Instead, we intended to use the differential of potential change as a stop signal for the pulse motor (Fig. 23). Thus, the microinjector was correctly positioned in the cell without losing cell viability. Its success rate was 73% for ES cells. After positioning the microinjector in the cell, lucifer yellow was introduced by electrophoresis via the IntE. Under this condition, the rate of viable cells became 16%. Such a loss of viability was thought to be due to an electrical potential that was inevitably loaded onto the cell membrane during the electrophoretic introduction of lucifer yellow.

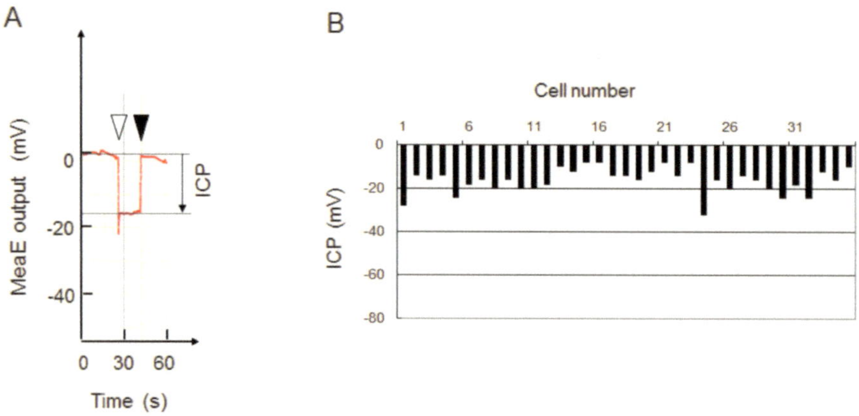

Fig. 22. Intracellular potential (ICP) of mouse ES cells. (A) ICP change profile obtained during the penetration of an injector into the cell (white triangle) and the removal of the injector from the cell (black triangle); (B) steady-state value of ICP from 35 cells. Reproduced from *Bioelectrochemistry* (2006) 69, 187-92 with permission from Elsevier.

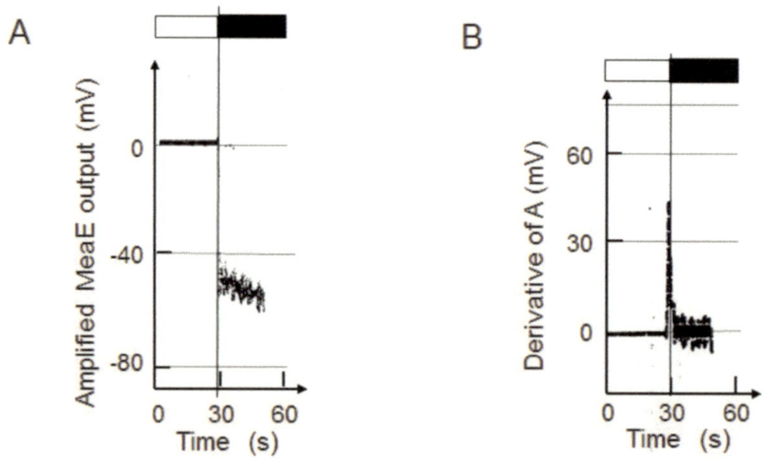

Fig. 23. Potential change (A) and its differential (B) during the penetration of an injector through the cell membrane. White box: an injector is moving during this period; black box: the injector stops and stayed there during this period. The ordinate of (B) indicates the differentiated and properly amplified value of (A). Reproduced from *Bioelectrochemistry* (2006) 69, 187-92 with permission from Elsevier.

9. Conclusion

As a result of the development of SMSR, femtoinjection is now a practical methodology. Femtoinjection has enabled the semi-quantitative introduction of any size of macromolecules, simultaneous introduction of multiple factors, temporal and spatial control of injection in a cell colony, and spatial control of the injection site whether into the cytosol or nucleus. The results described in this chapter provide some understanding of the significance of single-cell studies. Technological progress will further improve the spatio-temporal precision. Consequently this method can provide us with novel insight into cell functions, especially in ES cell studies, with regard to how to maintain the undifferentiated state and how to produce specific functional cells.

10. Acknowledgments

The studies described in this chapter were funded from CREST of Japan Science and Technology Agency and Grants-in-Aid for Scientific Research, from the Ministry of Education, Culture, Sports, Science, and Technology. We thank Soichiro Shimoda, Meiri Shibusawa, Masakazu Ozaki, Hajime Mizukami, Marina Kawazoe, Makoto Takatsu, and Yohei Yamada for their contribution to these studies. We appreciate kind donation of mouse ES cell lines by H. Niwa (RIKEN, CDB), MIN6 by J. Miyazaki (Osaka University), and Venus expression vector by A. Miyawaki (RIKEN, BSI).

11. References

Dean, D.A. (1997). Import of plasmid DNA into the nucleus is sequence specific, *Exp. Cell Res.* Vol. 230(No. 2): 293-302.

Dean, D.A., Byrd Jr., J.N. & Dean, B.S. (1999a).Nuclear targeting of plasmid DNA in human corneal cells, Curr. Eye Res. Vol. 19(No. 1): 66-75.

Dean, D.A., Dean, B.S., Muller, S. & Smith, L.C. (1999b). Sequence requirements for plasmid nuclear import, *Exp. Cell Res.* Vol. 253(No. 2): 713-722.

Funabashi, H., Takatsu, M., Saito, M. & Matsuoka, H. (2010). Sox2 regulatory region 2 sequence works as a DNA nuclear targeting sequence enhancing the efficiency of an exogenous gene expression in ES cells, *Biochem. Biophys. Res. Commun.*, Vol. 400(No. 4): 554-558.

Hancock, R. (2004). A role for macromolecular crowding effects in the assembly and function of compartments in the nucleus, *J. Struct. Biol.* Vol. 146(No. 3): 281-290.

King, R. (2004). Gene delivery to mammalian cells by microinjection, *Methods Mol. Biol.* Vol. 245: 167–174.

Lam, A.P. & Dean, D.A. (2010). Progress and prospects: nuclear import of nonviral vectors, *Gene Ther.* Vol. 17(No. 4): 439-447.

Masui, S., Nakatake, Y., Toyooka, Y., Shimosato, D., Yagi, R., Takahashi, K., Okochi, H., Okuda, A., Matoba, R., Sharov, A.A., Ko, M.S. &Niwa, H. (2007). Pluripotency governed by Sox2 via regulation of Oct3/4 expression in mouse embryonic stem cells, *Nat. Cell Biol.* Vol. 9(No. 6): 625-635.

Matsuoka, H., Komazaki, T., Mukai, Y., Shibusawa, M., Akane, H., Chaki, A., Uetake, N. & Saito, M. (2005). High throughput easy microinjection with a single-cell manipulation supporting robot, *J. Biotechnol.*, Vol. 116(No. 2): 185-194.

Matsuoka, H., Yamada, Y., Matsuoka, K.,&Saito, M. (2006) High Throughput Microinjection Technology for the Single-cell Analysis of BY-2 in Vivo. *inTobacco BY-2 Cells: From Cellular Dynamics to Omics,* Eds.Nagata, T., Matsuoka, K. &and Inzé, D., pp339-346,Springer-Verlag Berlin-Heidelberg.

Matsuoka, H.,Shimoda, S., Miwa, Y. & Saito, M. (2006).Automatic positioning of a microinjector in mouse ES cells and rice protoplasts, *Bioelectrochemistry*Vol. 69(No. 2): 187-192.
Matsuoka, H., Shimoda, S., Ozaki, M., Mizukami, H., Shibusawa, M., Yamada, Y. & Saito, M. (2007). Semi-quantitative expression and knockdown of a target gene in single-cell mouse embryonic stem cells by high performance microinjection, *Biotechnol. Lett.*, Vol. 29(No. 3): 341-350.
Matsuoka, H. & Saito, M. (2009). Single-cell Injectoassay for ES Cell Engineering, *ECS Transactions* Vol. 16(No. 17): 9-14
Miller, A.M., & Dean, D.A. (2009a).Tissue-specific and transcription factor-mediated nuclear entry of DNA, *Adv. Drug Deliv. Rev.* Vol. 61(No. 7-8): 603-613.
Miller, A.M., Munkonge, F.M., Alton, E.D. & Dean, D.A. (2009b). Identification of protein cofactors necessary for sequence-specific plasmid DNA nuclear import, *Mol. Ther.* Vol. 17(No. 11) 1897-1903.
Miyazaki, J., Araki, K., Yamato, E., Ikegami, H., Asano, T., Shibasaki, Y., Oka, Y. & Yamamura, K. (1990). Establishment of a pancreatic beta cell line that retains glucose-inducible insulin secretion: special reference to expression of glucose transporter isoforms, *Endocrinology* Vol. 127(No. 1): 126-132.
Niwa, H., Miyazaki, J. & Smith, A.G. (2000). Quantitative expression of Oct-3/4 defines differentiation, dedifferentiation or self-renewal of ES cells, *Nat. Genet.* Vol. 24(No. 4): 372-376.
Niwa, H. (2007). How is pluripotency determined and maintained?,*Development* Vol. 134(No. 4): 635-646.
Saito, M. & Matsuoka, H. (2010). Semi-quantitative analysis of transient single-cell gene expression in embryonic stem cells by femtoinjection, in *Methods Mol. Biol.* Vol. 650 (Part 2), *RNAi and microRNA-Mediated Gene Regulation in Stem Cells*, Eds. Zhang, B. &Stellwag, E.J., pp. 155-170, Humana Press.
Sasaki, T., Matsuoka,H. &Saito, M. (2008).Generation of a multilayer muscle fiber sheet from mouse ES cells by the spermine action at specific timing and concentration, *Differentiation*Vol. 76(No. 10): 1023-1030.
Tomioka, M., Nishimoto, M., Miyagi, S., Katayanagi, T., Fukui, N., Niwa, H., Muramatsu, M. & Okuda, A. (2002). Identification of Sox-2 regulatory region which is under the control of Oct-3/4-Sox-2 complex, *Nucleic Acids Res.* Vol. 30 (No. 14): 3202-3213.
Yamada, Y., Yamaguchi, N., Ozaki, M., Shinozaki, Y., Saito, M. & Matsuoka, H. (2008).An instant cell recognition system using a microfabricated coordinate standard chip useful for combinable cell observation with multiple microscopic apparatuses, *Microsc.Microanal.*, Vol. 14(No. 3): 236-242.
Yasuhara, N., Shibazaki, N., Tanaka, S., Nagai, M., Kamikawa, Y., Oe, S., Asally, M., Kamachi, Y., Kondoh, H. &Yoneda, Y. (2007). Triggering neural differentiation of ES cells by subtype switching of importin-alpha, *Nat. Cell Biol.* Vol. 9 (No. 1): 72-79.
Yasuhara, N., Oka, M. &Yoneda, Y. (2009).The role of the nuclear transport system in cell differentiation, *Semin. Cell Dev. Biol.* Vol. 20 (No. 5): 590-599.
Zimmerman, S.B. &Trach, S.O. (1988). Effects of macromolecular crowding on the association of E. coli ribosomal particles, *Nucleic Acids Res.* Vol. 16(No. 14A): 6309-6326.
Zimmerman S.B. &Trach, S.O. (1991). Estimation of macromolecule concentrations and excluded volume effects for the cytoplasm of Escherichia coli, *J. Mol. Biol.* Vol. 222(No. 3): 599-620.
Zimmerman S.B. & Minton, A.P. (1993). Macromolecular crowding: biochemical, biophysical, and physiological consequences, *Annu. Rev. Biophys. Biomol.Struct.* Vol. 22: 27-65.

10

From Pluripotency to Early Differentiation of Human Embryonic Stem Cell Cultures Evaluated by Electron Microscopy and Immunohistochemistry

Janus Valentin Jacobsen, Claus Yding Andersen,
Poul Hyttel and Kjeld Møllgård
University of Copenhagen
Denmark

1. Introduction

Under appropriate culture conditions human embryonic stem cells (hESCs) can retain an undifferentiated state during numerous passages (Thomson et al., 1998). In the undifferentiated state, hESCs express characteristic markers like NANOG, OCT4, TDGF1, DNMT3B, GABRB3, and GDF3, and are maintained by plating undifferentiated cells or colonies of cells into new culture dishes with fresh medium every 7 to 10th day (Adewumi et al., 2007). Following periods exceeding 7 to 10 days in culture without passage, the cell population tends to become heterogeneous with differentiation starting to occur within a given colony or in various parts of a culture dish. The tendency for undergoing differentiation is independent of whether feeder cells, protein matrixes, or special plastic surfaces are used and what specific hESC medium is employed. Although an increasing density of cells during culture has been suggested to be one reason for spontaneous differentiation of cells to occur, it is also well known that morphologically perfect undifferentiated hESCs often appear in very high density in the same culture dish, even when differentiation has started to occur (Laursen et al., 2007). The transition from the undifferentiated state to more differentiated cell types appears to take place as a gradual process in colonies of hESCs, and it is currently not known how the ultrastructural organization of cells changes along with the differentiation process as defined from immunohistochemical differentiation markers. In the present study we have performed a spatiotemporal investigation on the differentiation of hESC colonies by electron microscopical and immunohistochemical approaches.

2. Materials and methods

2.1 Culture of human embryonic stem cell lines LRB008, LRB010 and LRB017

The hESC lines LRB008, LRB010, and LRB017 were established at the Laboratory of Reproductive Biology, Copenhagen University Hospital from surplus embryos donated by couples undergoing IVF treatment after informed consent. The study was approved by the regional ethical committee of Copenhagen and Frederiksberg Municipalities (permission no.

KF 01-188/03). Donated embryos that developed to the blastocyst stage, were used to derive hESC lines following isolation of the inner cell mass isolated by manual dissection using hypodermic needles. The zona pellucida was digested by pronase (1mg/mL) (Sigma-Aldrich, P8811), and the inner cell masses (ICMs) were isolated by immunosurgery. The immunosurgery was performed by incubation in rabbit anti-whole serum antibody (Sigma-Aldrich, H8765) diluted 1:3 in KnockOut Dulbeccos's modified Eagles medium (KO-DMEM) (Invitrogen, 10829-018) for 30 min, followed by three successive washes in KO-DMEM and incubation in guinea pig serum (State Serum Institute, Copenhagen) diluted in 1:5 KO-DMEM. The ICMs and the derived cell lines were cultured on mitotically inactivated mouse embryonic fibroblast (MEF) cells in hESC medium based on KO-DMEM supplemented with knockout serum replacement (15% of final concentration), L-glutamine (2 mmol/L), β-mercaptoethanol (0.1 mmol/L), MEM non-essential aminoacids (0.1 mmol/L), Penicillin/Streptomycin (50U/g/mL), and basic fibroblast growth factor (bFGF, 4 ng/mL). The cells were cultured in a humidified atmosphere consisting of 6% CO_2, 7% O_2 and 87% N_2 at 37°C. Cell lines were normally passaged onto fresh MEF feeders once a week, when confluence reached around 70-80%. Passage of cells was carried out by mild trypsin treatment (0.05% trypsin in EDTA). An exception to MEFs as supportive feeders was the use of mitotically inactivated human foreskin fibroblast (hFF) in experiments where cell lines LRB008 and LRB017 were harvested for scanning electron microscopy (SEM).

2.2 Parallel sampling of colonies from the same culture dishes

For optimal concurrency in the collection of material for electron microscopy and immunohistochemistry, the experimental design was based on parallel sampling from the same culture dishes as illustrated in Figure 1. Colonies from the hESC line LRB010 were cultured for a period of either 6, 8, 11, 14 or 17 days. Samples were pooled in groups of *early* stage (6 and 8 days), *intermediate* stage (11 and 14 days), and *late* stage (17 days) colonies, according to duration of cultures at harvest. In addition, colonies of two other hESC lines, LRB008 and LRB017, were cultured and harvested after 7 and 14 days, for immunohistochemistry and investigation of colony surface ultrastructure by scanning electron microscopy (SEM).

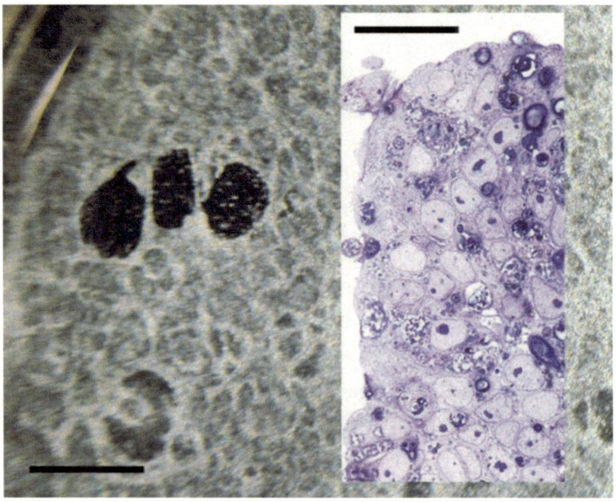

Fig. 1. The principle of parallel sampling.

The large greyscale phase contrast micrograph shows the bottom of a culture dish with growth of the hESC line LRB010 on a HFF feeder cell layer. Three black holes can be observed next to each other, where hESC colonies have been removed manually by hypodermic needles for fixation in glutaraldehyde and subsequent ultrastructural studies. The light micrograph shown in the inset depicts a semithin survey section (2 µm) with part of a fixed colony embedded in epoxy resin and stained with toluidine blue. After removal of tissue samples for electron microscopy, the remaining layer within the culture dish was fixed with Bouin's fixative for immunohistochemistry. Phase contrast micrograph: Scale bar = 5 mm. Light micrograph insert: Scale bar = 25 µm.

2.3 Processing of tissue for transmission electron microscopy (TEM)

Cultured hESC lines were detached using hypodermic needles and fixed in 3% glutaraldehyde in 0.1 M Na-phosphate buffer (pH 7.3) for 1-2 h at 4°C. Following fixation, specimens were transferred into 0.1 M Na-phosphate buffer (pH 7.3) and stored at 4°C for later processing. Specimens were embedded in 4% agar at 45°C (Bacto-agar, Difco Laboratories, Detroit, USA) under stereo microscopy and post-fixed in 1% OsO_4 in 0.1 M Na-phosphate buffer (pH 7.3) for 1 h at room temperature followed by wash in 0.1 M Na-phosphate buffer (pH 7.3) for 5 min. Tissue samples were stained en-bloc with 1% uranyl acetate in MilliQ water (MilliRO Plus and MilliQ PF Plus Water Purifications Systems, Millipore A/S, Hedehusene, Denmark), dehydrated in a series of ascending concentrations of ethanol (50% for 10 min, 70% for 10 min, 96% for 10 min, 99% for 3x20 min). Following the dehydration steps, samples were embedded in Epon (TAAB 812 Embedding resin, Medium), using propylene oxide as an intermedium, and polymerized for 48 h at 60°C. Semithin sections (2 µm) were cut on an ultramicrotome (Reichert Ultracut UCT, Leica) using glass knifes prepared on a knifemaker (LKB Bromma 7800). Sections were then stained with 1% basic toluidine blue for evaluation by bright-field light microscopy until a satisfactory part of the colony was exposed. Sections of interest were re-embedded (Hyttel and Madsen, 1987) for further ultrathin sectioning (70 nm) on an ultramicrotome (Reichert Ultracut UCT, Leica) using a diamond knife (Jumdi, 2 mm). The ultrathin sections were contrast stained using 2% uranyl acetate in MilliQ water and lead citrate (Reynolds, 1963), collected on 150 mesh copper grids covered with a parlodion/amylacetate film, and examined using a transmission electron microscope (CM100, Philips, Darmstadt, Netherlands).

2.4 Processing of tissue for scanning electron microscopy (SEM)

Cultures of hESCs grown on glass coverslips were fixed in 3% glutaraldehyde diluted in 0.1 M Na-phosphate buffer (pH 7.3) for approximately 1 h at 4°C, then transferred into 0.1 M Na-phosphate buffer (pH 7.3) and stored at 4°C. On the day of further processing the specimens were washed 3x5 min in 0.1 M Na-phosphate buffer (pH 7.3), post-fixed in 1% OsO_4 in 0.1 M Na-phosphate buffer (pH 7.2) followed by additional 3x5 min washings in 0.1 M Na-phosphate buffer (pH 7.3). Glass coverslips with hESC colonies were then transferred from plastic growth plates into glass dishes and dehydrated in a series of ascending concentrations of acetone (25% for 10 min, 40% for 10 min, 60% for 10 min, 75% for 10 min, 90% for 10 min, 100 % for 3x20 min). Glass coverslips with colonies were too large to fit into chambers of the critical point dryer and therefore had to be fractioned into smaller pieces under stereo microscope. The intermediate fluid acetone, was eliminated from the cells using a critical point dryer (EMS850, Electron Microscopy Sciences, Hatfield, Pennsylvania,

USA) by several flushings with liquid CO_2. Subsequently, critical point drying was performed. Pieces of coverslips with dried colonies were then mounted on specimen holders and coated with 5 nm gold/palladium in a sputter coater (SC7640 Suto/Manual High Resolution Sputter Coater, Quorum Technologies, Newhaven, UK). Specimens were evaluated using a scanning electron microscope (FEI Quanta 200, FEI Company, Eindhoven, The Netherlands).

2.5 Processing of tissue for immunohistochemistry

Colonies of the three hESC lines (LRB008, LRB010 and LRB017) were cultured under conditions as described above for 6, 7, 8, 11, 14 and 17 days and fixed in Bouin's fixative *in situ* in the culture dish following removal of material intended for electron microscopy. After 1 to 2 hours of fixation the fixative was replaced with 70 % ethanol and 24 - 48 hours later the 70 % ethanol was replaced with 90 % ethanol. Following overnight dehydration in 90 % ethanol, the samples were exposed to 99 % ethanol for 12 hours. Then the colonies were gently dissected free from the bottom using a Cell Scraper (Nunc) starting from the periphery. Colonies were then lifted carefully from the bottom of the culture dish to a small metal embedding mould into which xylene was pipetted. After 1 hour's exposure to xylene, paraffin was gently added to the embedding mould. Following paraffin embedding specimens were cut in 3-5 μm thick serial horizontal sections, strictly in parallel to the bottom of the cultures (Brøchner et al., 2010). Every 10th section was stained by toluidine blue.

Prior to immunohistochemistry, non-specific binding was inhibited by incubation for 30 minutes with blocking buffer (ChemMate antibody diluent S2022, DakoCytomation, Glostrup, Denmark) at room temperature. The sections were incubated overnight at 4°C with rabbit polyclonal or mouse monoclonal antibodies against OCT4, occludin, ZO-1, claudin-5, β-catenin, E-cadherin, N-cadherin, vimentin and nestin (details in Table 1). The sections were then washed with TBS and incubated for 30 minutes with the REAL EnVision™ Detection System, Peroxidase/DAB+, Rabbit/Mouse, (K5007) from DakoCytomation. The sections were counterstained with Mayer's hematoxylin and dehydrated in graded alcohols followed by xylene and coverslipped with DPX mounting media. Control sections were incubated with mouse IgG1, IgG2a or irrelevant rabbit antibodies, as well as subjected to omission of primary or secondary antibodies. Regarding OCT4, preincubation was performed 1 hour before incubation with the corresponding peptide in the proportion 1 to 5.

Antigen	Manufacturer	Code nr.	Antibody-species		
β-catenin	TriChem	17C2	Mouse IgG2a	÷	1:1000
Claudin-5	Abcam	Ab 5106	Rabbit	M6	1:100
E-cadherin	Abcam	Ab1416	Rabbit	M6	1:50
N-cadherin	Abcam	Ab12221	Rabbit	M6	1:500
Nestin	Chemicon	MAB 5326	Mouse, IgG1	÷	1:500
Occludin	Abcam	Ab64482	Rabbit	M6	1:75
OCT4	Abcam	Ab 19857	Rabbit	TEG	1:400
Vimentin	Dako	M0725	Mouse	M6	1:100
ZO-1	Zymed	Z-R1	Rabbit	÷	1:400

Table 1. Primary antibodies, manufactures, code numbers and dilutions.

3. Results

The cells and colonies grown on MEF feeders or hFF feeders in a culture medium with bFGF added developed well, with a typical morphology of undifferentiated stem cells showing areas of large single nucleated cells with scanty cytoplasm interspersed with cells of more differentiated morphology depending on culture age.

3.1 Transmission electron microscopy

Early stage colonies: In 2 µm survey sections from Epon-embedded material, the colonies appeared multi-layered with the highest numbers of cell layers in the center, flattening towards the periphery. Three different cell compartments within the colony were chosen for closer ultrastructural examinations, as illustrated by rectangles in Fig. 2, and consisted of: 1) an apical compartment with cells facing the culture medium; 2) a basal compartment containing cells anchoring the colony to the feeder cell layer, and; 3) a compartment in the center of the colony demarcated by apical and basal cells. In the apical compartment (Fig. 2C-E), cells were polarized with a large apical cytoplasm, rich in organelles, and nuclei located basally. Cells were connected to each other by apical junctional complexes - zonulae occludentes (tight junctions) and zonulae adherentes (adherens junctions), with abundant association of intermediary filaments of the cytoskeleton. Microvilli were observed projecting from the cell surfaces. Below the apical epithelial covering of the colony, a population of cells with a large nucleus to cytoplasm ratio and poor in organelles, representing a typical undifferentiated ESC-like morphology, resided in the center of the colony (Fig. 2B). Cell membranes of the basal cell compartment did not show any sign of junctional specializations, and higher magnification confirmed a continuous even spacing between the plasma membranes of the basal cells (Fig. 2F and G).
Intermediate stage colonies: The 2 µm survey sections showed a colony thickness of one or few cell layers. The colony harvested the 11th day of culture was still attached to a monolayer of MEF feeder cells with a distinct basement membrane separating them. Towards the apical surfaces, well developed tight and adherens junctions at the lateral plasma membrane connected cells of the apical compartment. Below the junctional complex, well-developed gap junctions and maculae adherentes (desmosomes) could be observed. All junctional complexes were closely associated with extensive bundles of intermediate filaments (Fig. 3C-D and 4D). In basal regions, adjacent cells were interdigitating and showed many gap junctions, whereas tight junctions, adherens junctions, and desmosomes were lacking (Figs. 3A, B, E & F and 4A, B, C & E).
Late stage colonies: In 2 µm survey sections, basal and center compartments were easily identified. Remnants of the basement membrane, but no feeder cells, were observed beneath the basal cells of the colony (Fig. 5A). The apical compartment contained a population of cells with obvious epithelial specializations such as junctional complexes with associated desmosomes, polarized organelle rich cytoplasm with basally located nuclei, and microvilli covering the apical surfaces. As was the case for early and intermediate stage colonies, a network of intermediary filaments radiated from sites of junctional complexes, where gap junctions were also present (Fig. 5C-E). At this late stage, a population of cells with ESC-like undifferentiated morphology similar to the ones previously described was still present (Fig. 5B).
Specializations linking adjacent plasma membranes of the basal compartment apart from obvious gap junctions were not observed (Fig. 5F-G).

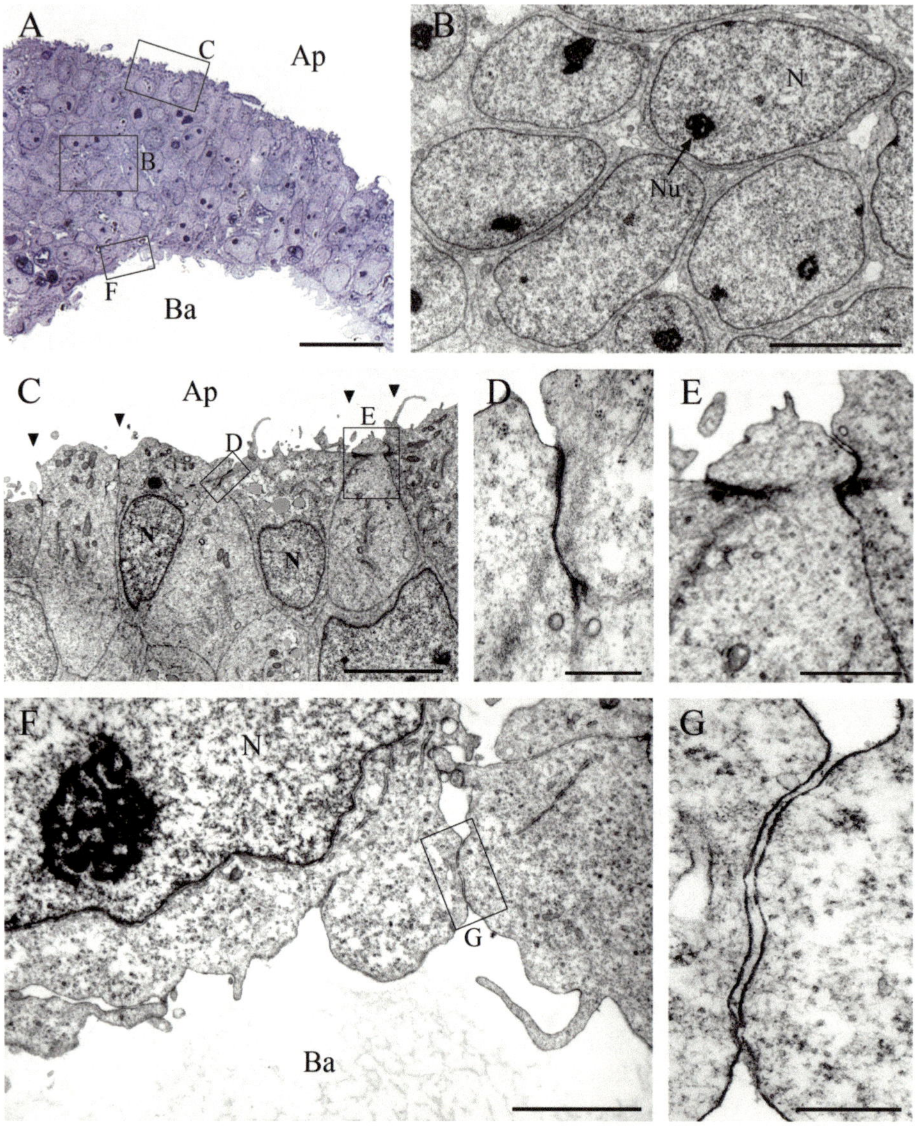

Fig. 2. Transmission electron micrographs of hESC line LRB010 cultured for 6 days (early). **(A)** Semithin cross section illustrating 3 different spatial compartments within the colony: Upper rectangle represents the localization of an apical compartment facing the culture medium *in vitro*, lower rectangle illustrates a population of basal cells anchoring the colony to the feeders, and the rectangle in the centre represents cells in between the apical and the basal compartment. Stained with toluidine blue. Ap, apical; Ba, basal. Scale bar = 25 μm. **(B)** Representative area from the center compartment with cells of typical undifferentiated morphology. N, nucleus; Nu, nucleolus. Scale bar = 5 μm. **(C)** Apical region of the colony with tight junction specializations shown in inserts. Scale bar = 5 μm. **(D)** Tight junction complex from insert in C (left). Scale bar = 500 nm. **(E)** Tight junctions from inset in C (right). Scale bar = 1 μm. **(F)** Basal region of the colony with a representative junction framed. Scale bar = 2 μm. **(G)** High magnification of inset in (F) showing no signs of junction specializations. Scale bar = 250 nm.

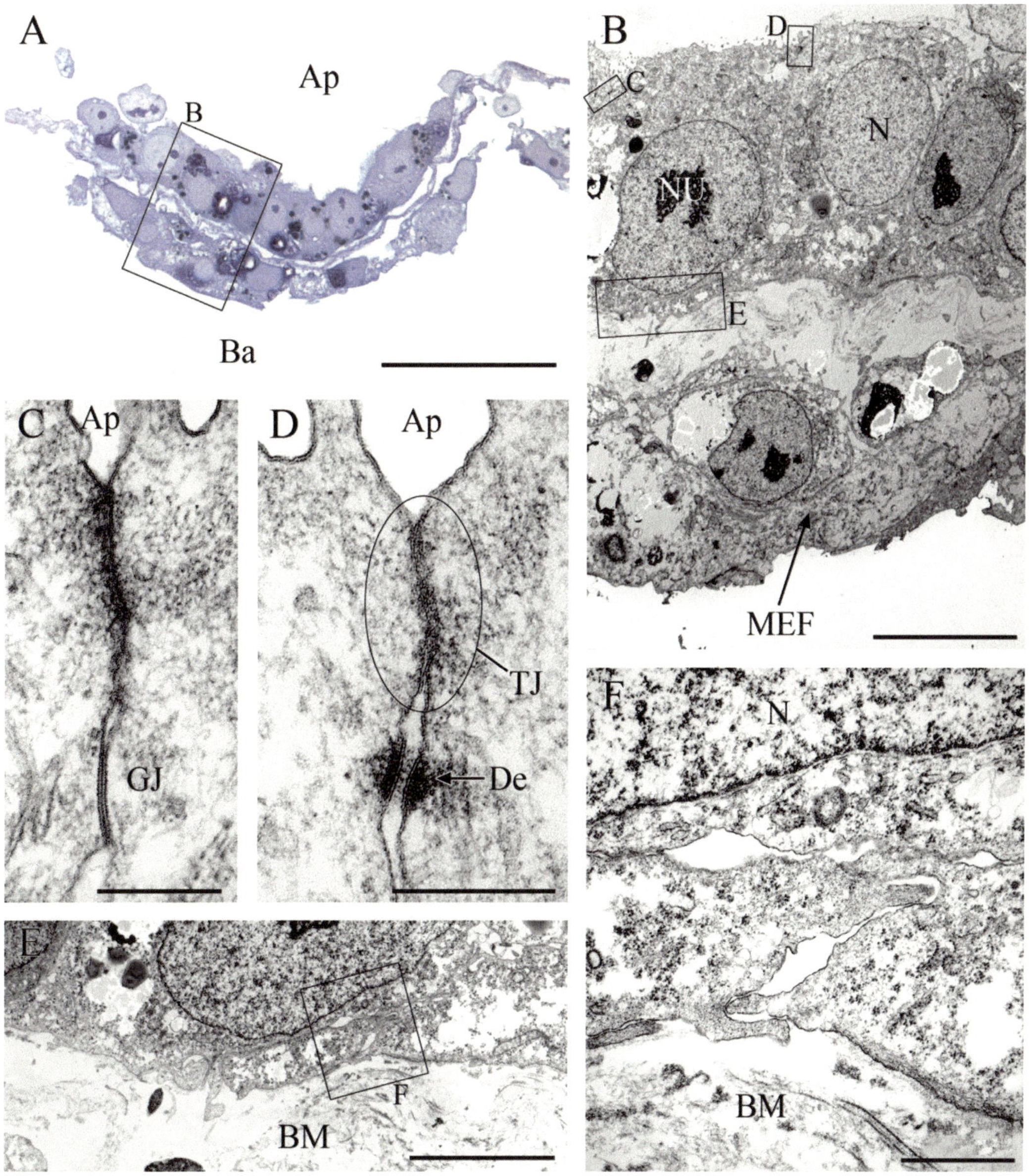

Fig. 3. Transmission electron micrographs of hESC line LRB010 cultured for 11 days (intermediate).

(A) Light microscopy of semithin cross section of colony with framed area spanning the apical and basal compartments magnified in B-F. Ap, apical; Ba, basal. Scale bar = 50 µm. **(B)** Apical and basal regions further magnified are framed. MEF, mouse embryonic fibroblast feeder layer; N, nucleus; Nu, nucleolus. Scale bar = 10 µm. **(C)** Junctional complex with distinct gap junction (GJ). Scale bar = 250 nm. **(D)** Tight- and adherens junction (TJ) with associated desmosome (De). Scale bar = 250 nm. **(E)** Basal region showing interdigitating processess without junctional specializations. BM, basement membrane. Scale bar =10 µm. **(F)** Higher magnification of insert in E. Scale bar = 1 µm.

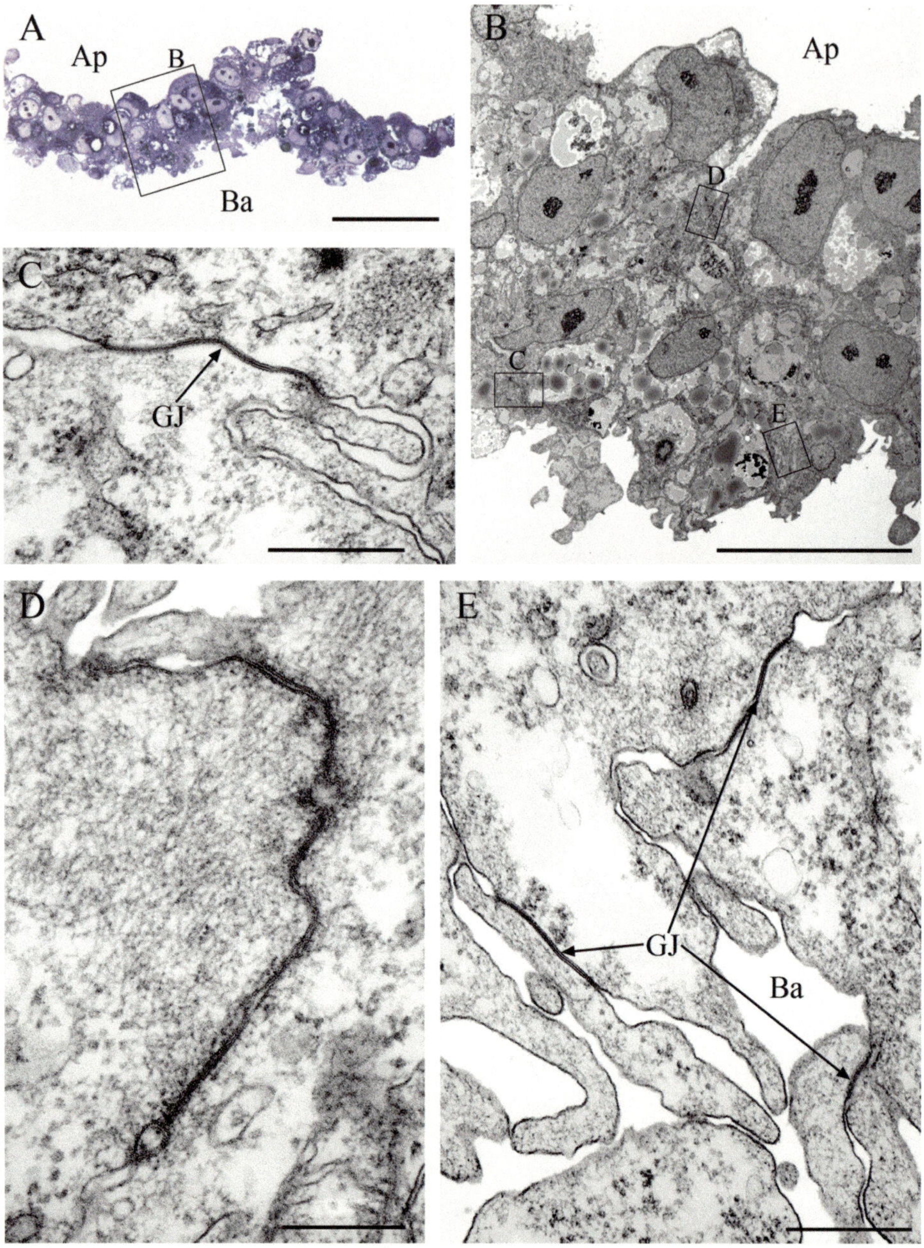

Fig. 4. Transmission electron micrographs of hESC line LRB010 cultured for 14 days (intermediate).

(A) Light microscopy of semithin cross section of colony marking representative area (frame) chosen for further analysis. Ap, apical; Ba, basal. Scale bar = 50 µm. **(B)** Marked locations (frame) of junctional areas seen in (C-E). BM, basement membrane. Scale bar = 20 µm. **(C)** Gap junction (GJ) in connection to inderdigitations in the basal region of the colony. Scale bar = 500 nm. **(D)** Tight- and adherens junction with prominent intermediary filament cytoskeleton. Scale bar = 250 nm. **(E)** Abundant gap junctions (GJ) at the basal region. Scale bar = 500 nm.

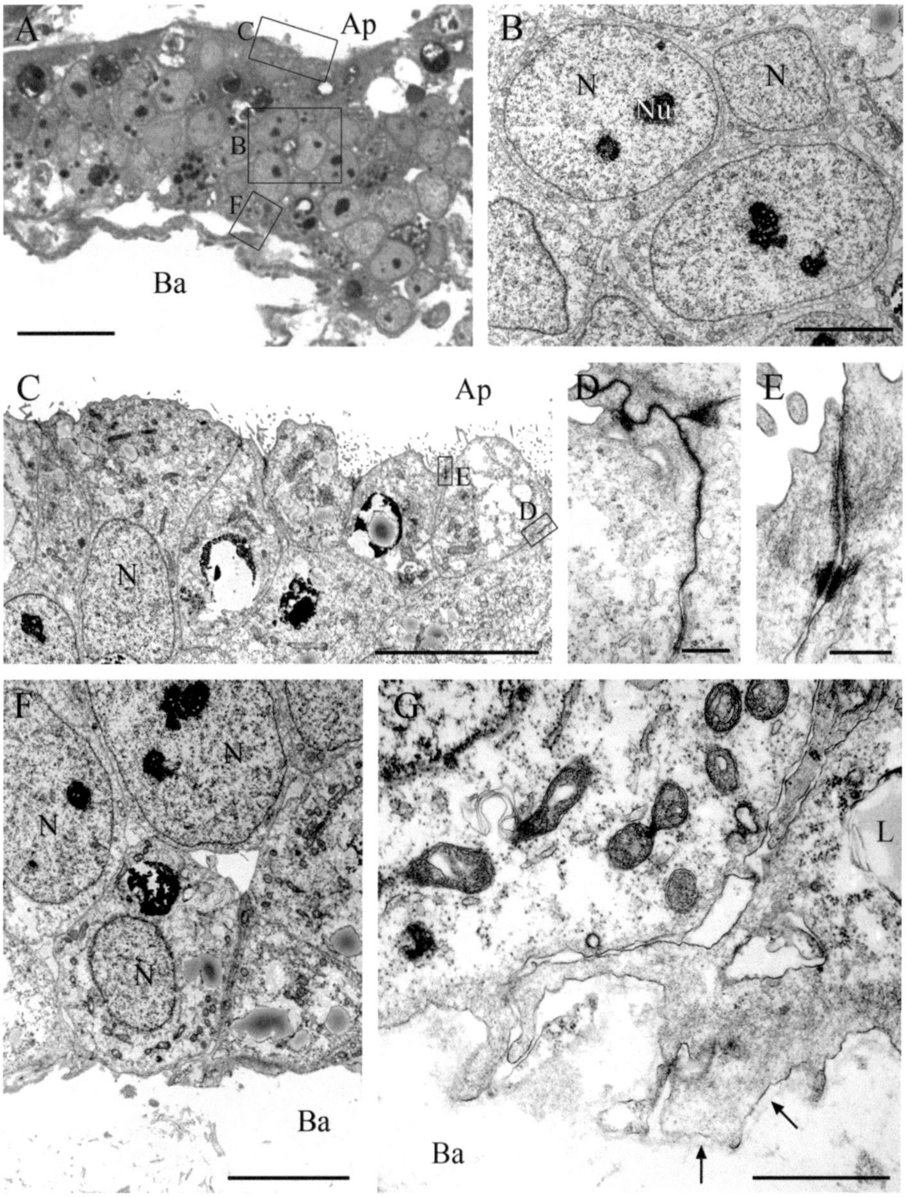

Fig. 5. Transmission electron micrographs of hESC line LRB010 cultured for 17 days (late). **(A)** Light microscopy of semithin overview section showing areas representative of apical, central and basal compartments of the colony. Ap, apical; Ba, basal. Scale bar = 20 μm. **(B)** Central compartment containing cells of typical undifferentiated morphology. N, nucleus. Scale bar = 5 μm. **(C)** Apical compartment where top layer of cells are bound together by apical junctional complexes (arrows). Two of the junctions can be seen in D-E. Scale bar = 10 μm. **(D)** Tight junction (C, left). Scale bar = 500 nm. **(E)** Tight- and adherens junction (C, right) with distinct desmosome. Scale bar = 250 nm. **(F)** Basal compartment. Scale bar = 5 μm. **(G)** Basal region without specializations. Basement membrane is marked by arrows. M, mitochondria; L, lipid droplet. Scale bar = 1 μm.

3.2 Scanning electron microscopy

Although 7 day-old colonies (LRB008 and LRB017) and feeders presented cracks resulting from the dehydration and critical drying processes, plenty of undamaged colony surface area was available for further examination. At low magnification, the colonies appeared as a flattened sheet of cells with a distinct boundary to the feeder cell layer they were grown on (Fig. 6A-B). The majority of the colony surface consisted of a fairly homogenous, flattened cell layer, but marked differences could be noted between individual cells in the flattened cell layer, with some cells having smooth surfaces, while others appeared with more rough surfaces covered by microvilli (Fig. 6C). Other surface areas displayed a more uniform

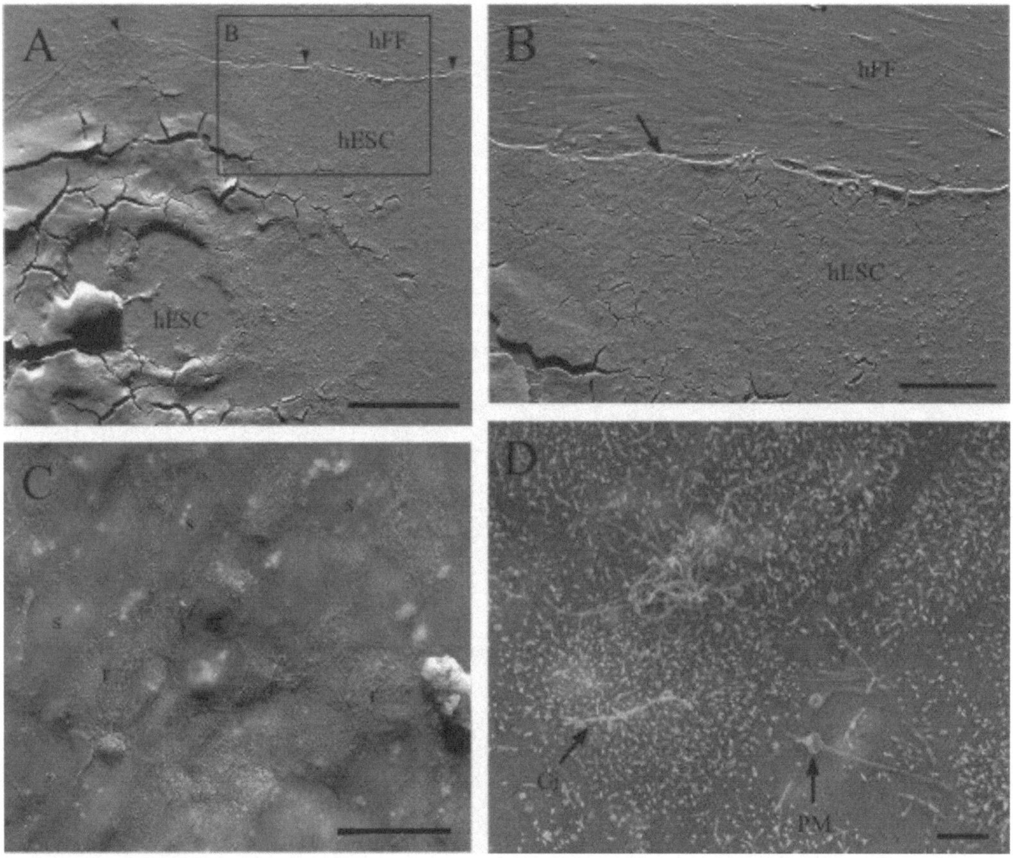

Fig. 6. Scanning electron micrographs of hESC lines LRB008 and LRB017 grown on human foreskin fibroblasts for 7 days in culture.

(A) Colony with arrowheads indicating border between colony and feeder layer. hFF, human foreskin fibroblasts; hESC, human embryonic stem cells. Scale bar = 500 µm. **(B)** Inset in (A) showing higher magnification of the periphery with arrow indicating border between colony and feeder layer. Scale bar = 200 µm. **(C)** Colony showing marked differences in cell surfaces. Some cells appear smooth surfaced (s), while others have a rougher surface (r) coved by microvillous extensions. Scale bar = 20 µm. **(D)** Microvillous surface in the periphery of the colony showing a cilium (Ci) extending from it, and a polar midbody (PM) still connecting dividing daughter cells. Scale bar = 2 µm.

distribution of microvilli, and the presence of individual cilia could be documented (Fig. 6D). Commonly, both cell lines exhibited superficial cytoplasmic bridges still connecting dividing daughter cells, with distinct polar midbodies (Fig. 6D).

Low magnification images of the 14 day-old colonies revealed great differences in the overall morphology of the two different cell lines investigated. One colony (LRB008) formed big bulging areas of cells growing in layers into the culture medium, while the other (LRB017) retained the flattened sheet-like appearance even after 14 days in culture (Fig. 7A-D). In the bulging areas of LRB008, separate cells clearly grew on top of each other, some forming structures protruding from the surface (Fig. 7A-B). Superficial cytoplasmic bridging between dividing cells was seen in LRB008, but not in LRB017. The rather uniform cell layer of LRB017 occasionally showed rounded protrusions of the plasma membrane, often in clusters (Fig. 7D). Furthermore, a varied microvillous covering could be observed through out the colonies of both cell lines on bulging as well as homogenous flat areas.

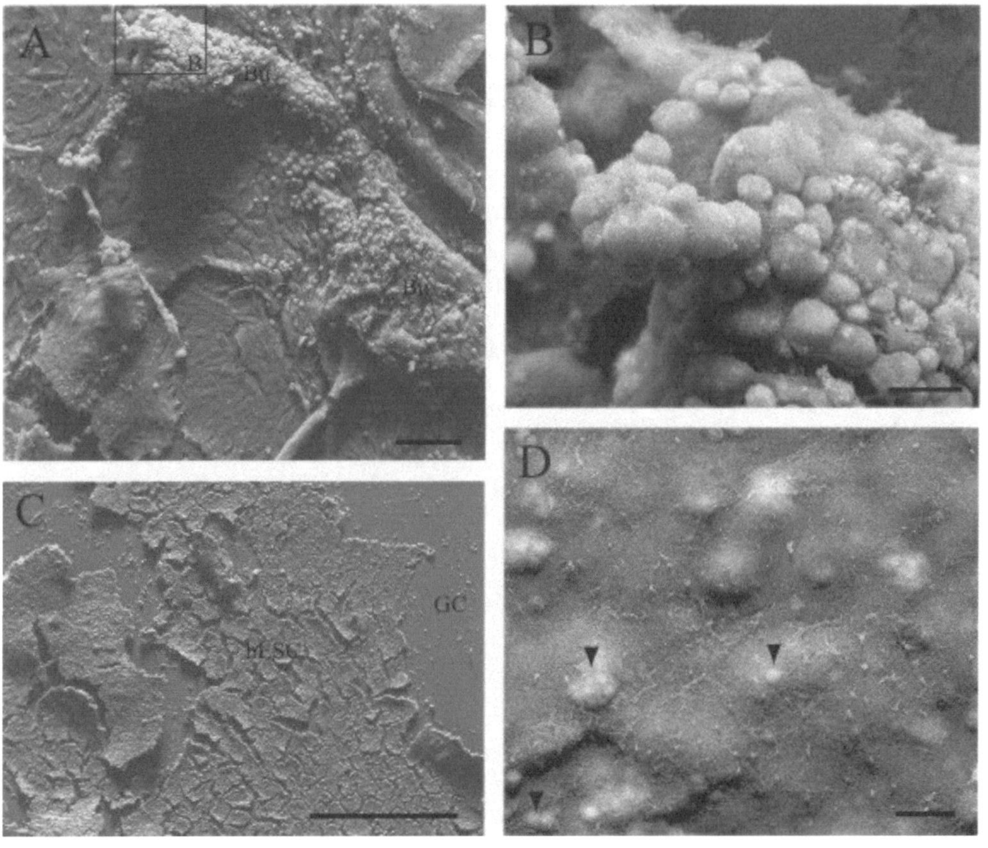

Fig. 7. Scanning electron micrographs of hESC lines LRB008 and LRB017 grown on human foreskin fibroblasts for 14 days in culture.

(A) Bulging (Bu) areas of differentiation on LRB008 after 14 days of culture. Scale bar = 100 µm. **(B)** Higher magnification of inset in A showing differentiating cells protruding from the surface. Scale bar = 20 µm. **(C)** Colony of LRB017 partly detached from the glass coverslip (GC) they were grown on. Scale bar = 400 µm. **(D)** Higher magnification of hESCs from (C) with areas of membrane protrusions marked by arrowheads. Scale bar = 5 µm.

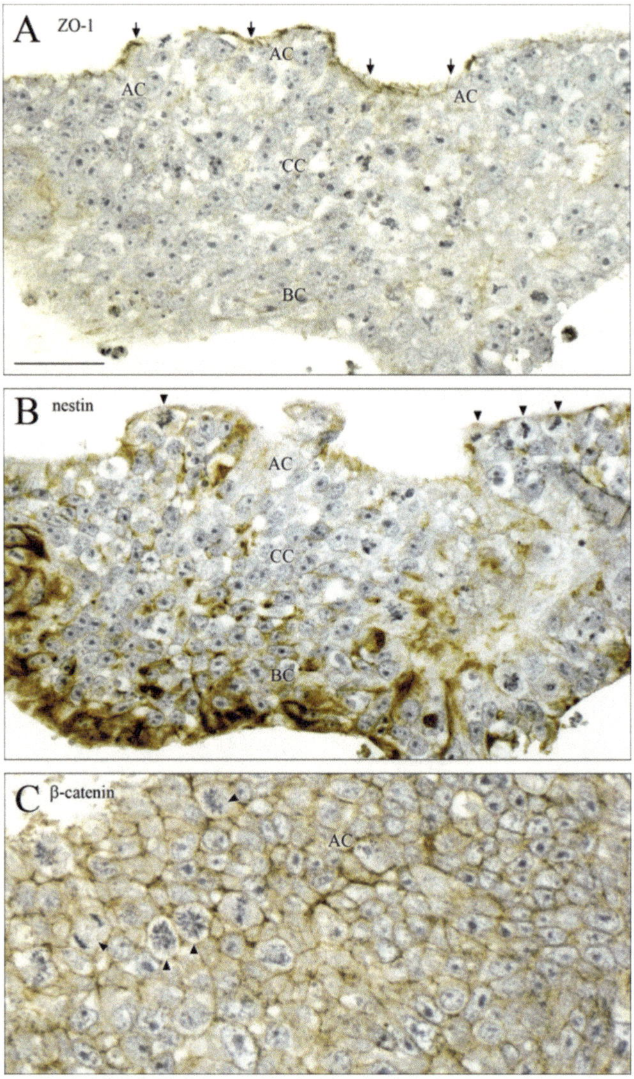

Fig. 8. Expression of ZO-1, nestin and β-catenin shown by immunohistochemistry in sections from hESC line LRB010 cultured for 6 days.
In **(A)** stained for ZO-1 the apical compartment (AC) is separated from the culture medium by a tight junctional complex (arrows). The central compartment (CC) and the basal compartment (BC) show no distinct reactivity. **(B)** The basal compartment (BC) is strongly positive for nestin, whereas the central compartment (CC) is characterized by cells of typical undifferentiated morphology. Note the pattern of differentiation from the center towards the basal compartment indicating a gradual change from hESCs to nestin-positive neuroectodermal cells. Some of the many mitotic figures in the apical compartment (AC) situated at the apical surface facing the culture medium are indicated by arrowheads. **(C)** A tangential section through the apical compartment (AC) illustrates the strong β-catenin-reactivity associated with the adherens junctions. Note the many mitotic figures (arrowheads), which are also seen in the same region in cross-sectioned colonies (arrowheads in **B**). A, B & C: same magnification. Scale bar = 25 μm.

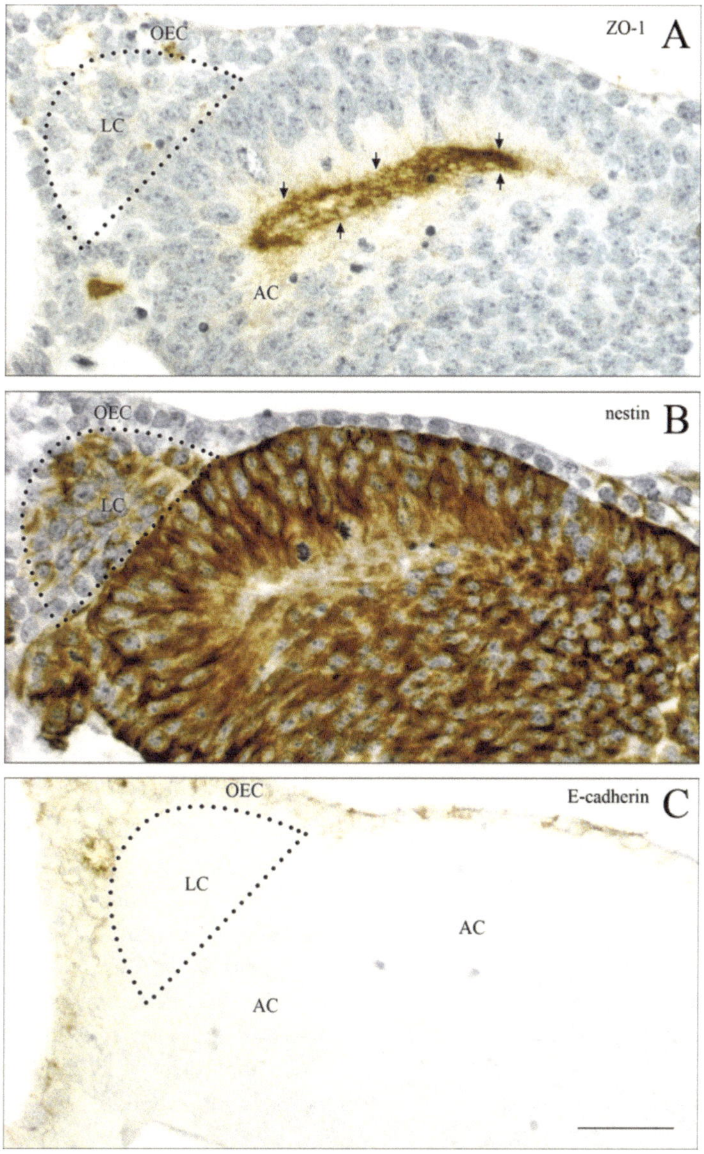

Fig. 9. Expression of ZO-1, nestin and E-cadherin shown by immunohistochemistry in tangential sections from hESC line LRB010 cultured for 17 days.
(A): The apical compartment (AC) in the center of the figure, stained for ZO-1 shows a well-developed tight junctional complex (arrows). A lateral compartment (LC) marked by a punctuated line and an outer epithelial compartment (OEC) are not stained. **(B)** The entire pseudo-stratified epithelium of the apical compartment is strongly positive for nestin. A new lateral compartment (LC) marked by a punctuated line is slightly positive for nestin, but strongly positive for vimentin and N-cadherin (not shown). **(C)** A single-layered outer epithelial compartment (OEC) and its extension towards the upper left of the figure show positive reactivity for E-cadherin, whereas both apical (AC) and lateral compartments (LC) are unstained. A,B & C: same magnification. Scale bar = 25 μm.

3.3 Immunohistochemistry

It was possible to distinguish three different compartments in most colonies already from day 6: (1) An apical epithelial-like polarized layer facing the culture medium; (2) a basal cell layer facing the feeder cell layer; and (3) a central compartment enclosed by the apical and basal cell layers consisting of 'classical' hESCs with 'undifferentiated morphology' (Fig. 8 A-B). The youngest colony showed the presence of markers for tight junctions (occludin, claudin-5 and ZO-1 - Fig. 8A), and adherens junctions (E-cadherin associated with β-catenin - Fig. 8C) in the apical polarized epithelial-like cell layer. This was in marked contrast to the epithelioid 'classical' hESCs of the central compartment characterized by absence of markers for tight junctions. All cells exhibited positive OCT4-staining of their nuclei pointing to heterogeneity of the early hESCs, i.e. an epithelial-like cell type with junctional complexes facing the outside compartment (the culture medium) and an epithelioid cell type without classical junctional complexes, but with strong OCT4 reactivity in their nuclei. The basal cell layer, on the other hand, exhibited a distinct reactivity for the neuroectodermal marker nestin (Fig. 8B), but markers for tight junctions were absent. A gradual change in the pattern of differentiation from a nestin-negative, but OCT4-positive central compartment towards a strongly nestin-positive basal compartment indicating a gradual change from an OCT4-positive hESC population to a neuroectodermal cell population existed as illustrated in Fig. 8B.

In older colonies (17 days) immunohistochemical analysis revealed that the majority of the differentiating cells had lost their OCT4-positivity, indicating that the central compartment with 'classical' hESCs showing 'undifferentiated morphology' had disappeared in most colonies. An apical epithelial-like pseudostratified layer facing the culture medium was still present. The apical regions of the lateral cell membranes were joined by tight junctions as shown by a positive reactivity for occludin, claudin-5, and ZO-1 (Fig. 9A). In some cases the entire apical layer was densely stained for nestin (Fig. 9B), whereas the apical regions of other colonies showed a very strong staining for vimentin with some more basally located nestin-positive cells (not shown). A single layered outer compartment was characterized by a lack of staining for vimentin and nestin but by a positive staining reaction for E-cadherin (Fig. 9C). In a newly developed lateral compartment adjacent to the apical and above the central compartment cells showed a strongly positive N-cadherin and vimentin-staining but no staining for E-cadherin (Fig. 9C) suggesting an epithelial-mesenchymal transition (EMT) resulting in migratory vimentin-positive mesenchymal-like cells.

4. Discussion

Although living hESCs often appear homogenously undifferentiated in their colony formation, as observed in phase contrast microscopy, the present study demonstrates that various differentiation events occur already within a few days after passage both at the ultrastructural and light microscopic level. Usually it is sufficient to maintain hESCs in an undifferentiated state by a passage of cells every 5-7 days. However, here we demonstrate that already within the early stage colonies, around the time of normal passage, a pronounced compartmentalization has taken place within colonies, suggesting spatial and temporal dimensions of differentiation. In all colonies examined, irrespective of the culture period and spatial configuration of the colony, the apical layer of cells, which is in direct contact with the culture medium, formed an apico-basolateral polarized sheet of tightly interconnected cells resembling a columnar epithelium and thus providing solid documentation for a morphological specialization linking adjacent cells by junctional complexes consisting of tight and adherens junctions, shown by EM and immunohistochemistry. Such an epithelial coating, forming a barrier to the external environment, might provide deeper lying cells with a selective transcellular uptake of nutrients and other media components, thus creating a microenvironment supporting the undifferentiated state. A central compartment, containing cells with typical undifferentiated ESC-morphology, was indeed observed with EM and immunohistochemistry in both *early* and *medium* stage colonies, indicating long-term retention of an environment permissive of undifferentiated cell growth, regardless of surrounding peripheral differentiation. Indeed these observations might also explain why morphological undifferentiated hESC in culture dishes that has been grown for a prolonged period of time of perhaps two weeks or more is often found in the central and basal compartments which are in direct contact with bottom of the culture dish without tight junctions underneath layers of more differentiated cells on top of the colony.

Based upon studies of short-term feeder free cultures, Ullmann *et al.* (2007) proposed a colony structure almost identical to the *early* stage colony in this study. The immunostainings for E-cadherin, a marker for epithelial adherens junctions, were localized only to the uppermost cell layer of a multilayer, whereas it disappeared peripherally, where vimentin positive cells began to appear. This provides evidence for an epithelial-mesenchymal transition (EMT) in the peripheral regions where the colonies flattens into a monolayer. However, according to our TEM investigations, these peripheral monolayers of the *intermediate* stage colonies grown on MEF exhibited junctional complexes, hence suggesting an epithelial nature.

Colonies organized with confined undifferentiated cells and peripheral differentiation was confirmed by a shift in immunoreactivity from the pluripotency marker SSEA-4 to a marker of differentiation SSEA-1 towards the periphery (Johkura *et al.*, 2004). This down-regulation in pluripotency markers was to some extent also indicated in the short-term feeder free cultures referred to above, where the intensity in immunostainings of the key pluripotency markers OCT4 and NANOG decreased in peripheral regions (Ullmann *et al.*, 2007). Despite an apparent decrease in the amount of the key transcription factors, expression levels were still substantial, raising the question whether a deviation from the undifferentiated morphology to a more epithelial-like morphology is accompanied by decrease in potency. Additionally, increased amounts of laminin in the periphery of the colonies confirmed accumulation of extracellular matrices under standard culture conditions (Johkura *et al.*, 2004), most likely as a constituent of the basement membrane secreted by differentiated epithelial cells. One of the *intermediate* (11 days in culture) and the *late* stage colonies had distinct basement membranes. The basement membrane of the *intermediate* stage colony harvested at the 14th day of culture was supposedly teared of at harvest. This was indicated by the damaged appearance basally. Neither did our observations indicate presence of a basement membrane beneath the most basal cells of the multilayered *early* stage colony, possibly explaining colony detachment from feeders without damaging structures of the basal cell layer or lifting up accompanying feeders. The location of the basement membrane in the *late* stage colony (Fig. 5), with its multilayered appearance, suggests that it could be a single-cell layer cut in a transverse section deviating somewhat from the perpendicular plane, thus increasing the interface of the section. On the other hand, a model of a multilayered colony consisting of several layered compartments cannot be excluded.

Gap junctions were identified both as part of the apical junctional complex (Fig. 4D), spanning all stages, but also in basal regions of the presumed monolayer in one of the *intermediate* stage colonies (Fig. 4E). A definite cell-to-cell coupling is demonstrated between epithelial cells of the apical compartment of murine cells of the ICM which are connected to each other (Lo & Gilula, 1979). The present study suggests that this also may apply to human ICMs. In accordance, Sathananthan *et al.* (2002) points out that a high degree of ultrastructural similarity exists between cells of the ICM and hESCs with undifferentiated morphology. Hence a gap junctional coupling could be expected between cells within the compartment of cells with undifferentiated morphology. Indeed, gap junctions were found by TEM within the undifferentiated compartment (micrographs not included), also demonstrating cell-to-cell coupling to exist between these cells. Whether the apical epithelial compartment is gap junctionally coupled to the underlying cells with undifferentiated morphology remains to be determined. Differentiation processes are normally associated with the progressive loss of communication between different tissues, for example as observed in between the ICM and the trophectoderm (Lo & Gilula, 1979). This suggests the possibility of two separate compartments intra-connected by gap junctions without coupling to each other. Huettner *et al.* (2006) demonstrated extensive gap junction communication in hESC colonies by dye injections and physiological recordings together with documentation of different connexins by immunocytochemistry and RT-PCR, assuming colonies to consist of uniform monolayers of undifferentiated cells corresponding to apical epithelial cells.

In order to document possible differences among different hESC lines, two other cell lines, LRB008 and LRB017, were included to improve our knowledge about the general colony outline and surface structures from a SEM point of view. At least after 7 days, this analysis confirmed that the periphery is composed of a primarily homogenous flat monolayer comparable to single-cell layers visualised by TEM. Cell line LRB017 showed a distinct accumulation of cells in the center of the colony. This was more difficult to identify in cell line LRB008, because the area in question had suffered considerable damage from processing of the sample, which in itself could indicate a thicker cell layer more vulnerable to shrinkage during dehydration. These observations agreed very well with the previously proposed colony structure described by Ullmann *et al.* (2007), but contradicted the saucer-shaped colony structure described by Sathananthan *et al.* (2002), in spite of identical culture conditions and time of harvest for the latter. The impact of duration of cultures is well illustrated by the 14-day old cultures. These showed a more chaotic growth of differentiating cells protruding from the surface in a highly irregular pattern in LRB008 (Fig. 7A-B), while LRB017 maintained a flattened growth. Single or clustered rounded protrusions in the cell membrane, in particular the 14-day-old cultures (Fig. 7D), showed similarity to the coarse particles and halos previously identified by phase contrast microscopy consisting of apoptotic cells and bodies (Johkura *et al.*, 2004). This suggests a pronounced cell death at this late stage. The fact that the entire colonies, in general, were covered with microvilli with varying heterogeneity, and often with microvilli-free cells interspersed, suggests that the extent of epithelial specialization of the medium-facing cell layer. Heterogenous cell surfaces between groups of cells, as well as between single adjacent cells, very well correspond to results from a study demonstrating regional differences in expression of specific markers for hESCs (Laursen *et al.,* 2007). Recently, one group published findings of primary cilia on surfaces of 33 and 90 percent of the hESCs of two different cell lines, H1 and LRB003, respectively (Kiprilov *et al.,* 2008). Additionally, cilia were shown to co-localize with OCT4 staining, thus indicating their pluripotency with the assumption of the associated hedgehog signalling machinery to be important in the maintenance of the undifferentiated/self-renewable state. Following 7 days in culture, with the same culture conditions and using two other hESC lines derived also derived in our laboratory, only one single cilium was found. This raises the question of importance of cilia in regulation of hESCs as proposed by Kiprilov *et al.* (2008), and definitely demonstrates differences among hESC lines.

To EMT, or not to EMT - in vivo and in culture. During *in vivo* embryogenesis and organ development the process initially termed epithelial-mesenchymal transformation (Hay, 2005), now referred to as epithelial-mesenchymal transition (EMT), (Kalluri & Weinberg, 2009; Thiery et al., 2009) is a key event. EMT is a biological process that allows an epithelial cell to undergo multiple biochemical changes that enable it to assume a mesenchymal cell phenotype, which includes enhanced migratory capacity, invasiveness, elevated resistance to apoptosis, and increased production of extracellular matrix components (Kalluri and Weinberg, 2009). Early in the third week of human embryonic development, epiblast cells lateral to the primitive streak begin to move into the streak, where they undergo an EMT, i.e. the epithelial epiblast cells elongate and detach from their neighbours and then migrate away below the primitive streak as individual loosely organized mesendodermal progenitors. This process of gastrulation represents the first EMT after implantation (Thiery et al., 2009). After gastrulation the remaining epithelial epiblast cells, which did not enter the primitive streak and therefore did not undergo the initial EMT, form epidermal and neural territories progressively along the rostrocaudal axis. The differentiating neural plate is flanked by the prospective neural crest and further laterally by the prospective surface ectoderm, which constitutes the remaining areas of the ectoderm. Neural crest cells from the neural crest territory in the dorsal neural tube then delaminate in a process of primary EMT (Thiery et al., 2009). However, what has not really been emphasized so far, is the lack of EMT in the initial differentiation of epiblast cells directly into the remaining ecto- and neuroectodermal germ layer - an *in vivo* process of **E**piblast-**e**ctodermal **t**ransformation (EET) corresponding to an *in vitro* process of ESC-ectodermal transformation.

In a previous *in vitro* study of regional differences in the expression of markers during differentiation of hESC lines (LRB001-004), we noticed a distinct and direct transformation from undifferentiated OCT4 positive hESCs to SSEA1-positive fully differentiated epithelial cells (Laursen et al., 2007), and in a very recent investigation of six different LRB-cell lines we observed a similar abrupt transformation between OCT4-positive hESC regions and areas with either ectodermal p63-positive cells or neuroectodermal nestin-positive cells, while a direct transition into either meso- or endodermal marker positive cells, was not seen (Brøchner et al., 2011). In the present *in vitro* study we found an OCT4-postive epithelial-like cell type with a well-defined junctional complex facing the outside compartment which strongly resemble an *in vivo* embryonic epiblast cell type, whereas the pluripotent cells without junctional complexes in the central compartment of early cultures resemble an ICM-like cell type. The gradual pattern of differentiation from the center towards both apical and basal compartments indicates a direct shift from hESCs to columnar epithelial epiblast-like cells and to nestin-positive neuroectodermal cells in the basal compartment.

5. Conclusion

In conclusion, the epithelial coating of the colony might serve two different purposes: (1) The tight junction component of the junctional complex forms a barrier for transepithelial transport between the culture medium and the internal environment of the colony; thus the epithelial cells could provide cells in central and basal compartments with nutrients and other media components, probably via a specific receptor mediated transcellular epithelial transport, in order to create a microenvironment supporting the undifferentiated stage of underlying cells within the central compartment. (2) There seems to be a direct transformation of pluripotent hESCs into ecto- and neuroectodermal germ layer cells, while the adherens component of the junctional complex is instrumental in directing the further differentiation of hESCs into mesodermal and endodermal lineages via an epithelial-mesenchymal transition (EMT).

6. Acknowledgements

The expert technical assistance of Sussi Forchhammer, Hanne Hadberg, Pernille S. Froh, Ha Nguyen (Department of Cellular and Molecular Medicine, The Panum Institute, University of Copenhagen, Copenhagen), Marjo Westerdahl (Laboratory of Reproductive Biology, Rigshospitalet, University Hospital of Copenhagen, Copenhagen), and Hanne Marie Mølbak Holm (Department of Basic Animal and Veterinary Sciences, Faculty of Life Sciences, University of Copenhagen) through all stages of the project is gratefully acknowledged. Keld B. Ottosen (Department of Cellular and Molecular Medicine, The Panum Institute, University of Copenhagen, Copenhagen) is thanked for the final layout of several figures.

7. References

Adewumi, O.; Aflatoonian, B.: Ahrlund-Richter, L.; Amit, M.; Andrews, P.W.; Beighton, G.; Bello, P.A.; Benvenisty, N.; Berry, L.S.; Bevan, S.; Blum, B.; Brooking, J.; Chen, K.G.; Choo, A.B.; Churchill, G.A.; Corbel, M.; Damjanov, I.; Draper, J.S.; Dvorak, P.; Emanuelsson, K.; Fleck, R.A.; Ford, A.; Gertow, K.; Gertsenstein, M.; Gokhale, P.J., Hamilton, R.S.; Hampl, A.; Healy, L.E.; Hovatta, O.; Hyllner,J.; Imreh, M.P.; Itskovitz-Eldor, J.; Jackson, J.; Johnson, J.L.; Jones, M.; Kee, K.; King, B.L.; Knowles, B.B.; Lako, M.; Lebrin, F.; Mallon, B.S.; Manning, D.; Mayshar, Y.; McKay, R.D.; Michalska, A.E.; Mikkola, M.; Mileikovsky, M.; Minger, S.L.; Moore, H.D.; Mummery, C.L.; Nagy, A.; Nakatsuji, N.; O'Brien, C.M.; Oh, S.K.; Olsson, C.; Otonkoski, T.; Park, K.Y.; Passier, R.; Patel, H.; Patel, M.; Pedersen, R.; Pera, M.F.;

Piekarczyk, M.S.; Pera, R.A.; Reubinoff, B.E.; Robins, A.J.; Rossant, J.; Rugg-Gunn, P.; Schulz, T.C.; Semb, H.; Sherrer, E.S.; Siemen, H.; Stacey, G.N.; Stojkovic, M.; Suemori, H.; Szatkiewicz, J.; Turetsky, T.; Tuuri, T.; van den, B.S.; Vintersten, K.; Vuoristo, S.; Ward, D.; Weaver, T.A.; Young, L.A. & Zhang, W. (2007). Characterization of human embryonic stem cell lines by the International Stem Cell Initiative. Nat. Biotechnol. *25*, 803-816.

Brøchner, C.B.; Johansen, J.S.; Larsen, L.A.; Bak, M.; Mikkelsen, H.B.; Byskov, A.G.; Andersen, C.Y. & Møllgård, K. (2011). YKL-40 is expressed in undifferentiated human embryonic stem cells and in cell progeny of all three germ layers with varying intensity (submitted)

Brøchner, C.B.; Vestentoft, P.S.; Lynnerup, N.; Andersen, C.Y. & Møllgård, K. (2010). A two- and three-dimensional approach for visualizing human embryonic stem cell differentiation. Methods Mol Biol. 584, 179-93.

Hay, E.D. (2005). The mesenchymal cell, its role in the embryo, and the remarkable signaling mechanisms that create it. Dev. Dyn. 233:706-720.

Huettner, J.E.; Lu, A.; Qu, Y.; Wu, Y.; Kim, M. & McDonald, J.W. (2006). Gap junctions and connexon hemichannels in human embryonic stem cells. Stem Cells *24*, 1654-1667.

Hyttel, P. & Madsen, I. (1987). Rapid method to prepare mammalian oocytes and embryos for transmission electron microscopy. Acta Anat. (Basel) *129*, 12-14.

Johkura, K.; Cui, L.; Asanuma, K.; Okouchi, Y.; Ogiwara, N. & Sasaki, K. (2004). Cytochemical and ultrastructural characterization of growing colonies of human embryonic stem cells. J. Anat. *205*, 247-255.

Kalluri R. & Weinberg R.A. (2009). The basics of epithelial-mesenchymal transition. J. Clin. Invest 119:1420-1428.

Kiprilov, E.N.; Awan, A.; Desprat, R.; Velho, M.; Clement, C.A.; Byskov, A.G.; Andersen, C.Y.; Satir, P.; Bouhassira, E.E.; Christensen, S.T. & Hirsch, R.E. (2008). Human embryonic stem cells in culture possess primary cilia with hedgehog signaling machinery. J. Cell Biol. *180*, 897-904.

Laursen, S.B.; Møllgård, K.; Olesen, C.; Oliveri, R.S.; Brøchner, C.B.; Byskov, A.G.; Andersen, A.N.; Hoyer, P.E.; Tommerup, N. & Andersen, C.Y. (2007). Regional differences in expression of specific markers for human embryonic stem cells. Reprod. Biomed. Online. *15*, 89-98.

Lo, C.W. & Gilula, N.B. (1979). Gap junctional communication in the post-implantation mouse embryo. Cell *18*, 411-422.

Reynolds, E.S. (1963). Liver parenchymal cell injury. I. Initial alterations of the cell following poisoning with carbon tetrachloride. J. Cell Biol. 19, 139-157.

Sathananthan, H.; Pera, M. & Trounson, A. (2002). The fine structure of human embryonic stem cells. Reprod. Biomed. Online. *4*, 56-61.

Thiery, J.P.; Acloque, H.; Huang, R.Y.J. & Nieto, M.A. (2009) Epithelial-Mesenchymal Transitions in Development and Disease. Cell 139, 871-890.

Thomson, J.A.; Itskovitz-Eldor, J.; Shapiro, S.S.; Waknitz, M.A.; Swiergiel, J.J.; Marshall, V.S. & Jones, J.M. (1998). Embryonic stem cell lines derived from human blastocysts. Science 282, 1145-1147.

Ullmann, U.; In't, V.P.; Gilles, C.; Sermon, K.; De Rycke, M.; Van de Velde. H.; Van Steirteghem A. & Liebaers, I. (2007). Epithelial mesenchymal transition process in human embryonic stem cells cultured in feeder-free conditions. Mol. Hum. Reprod. *13*, 21-32.

Part 3

Applications of Embryonic Stem Cells in Research and Development

11

Methods to Generate Chimeric Mice from Embryonic Stem Cells

Kun-Hsiung Lee
Division of Biotechnology
Animal Technology Institute Taiwan
Chunan, Miaoli,
Taiwan

1. Introduction

A chimera is an animal that has two or more populations of genetically distinct cells that originated in different embryos, fetuses, or individuals of the same or different species. During recent decades, embryos, the inner cell mass (ICM), teratocarcinoma stem cells, embryonal carcinoma stem cells, embryonic stem (ES) cells (ESCs), primordial germ cells, spermatogonial stem cells, extraembryonic endoderm (XEN) cells (Kunath et al., 2005), induced pluripotent stem cells (iPSCs) (Boland et al., 2009; Kang et al., 2009; Takahashi & Yamanaka, 2006; Zhao et al., 2009), epiblast stem cells (EpiSCs) (Brons et al., 2007; Tesar et al., 2007), and other cells have been used to generate chimeric embryos that can develop into chimeras once transferred to a foster mother.

To date, three methods can be applied to produce ESC-derived chimeric embryos: (1) aggregation, (2) microinjection, and (3) coculture. This review uses mouse ES cells as an example to describe and compare existing methods for generating chimeric embryos.

2. Methods for generating chimeric embryos

Long before successful generation of chimeric embryos using mouse ES cells (Evans & Kaufman, 1981; Martin, 1981), large chimeric morulae were first generated using zona pellucida (ZP)-free (denuded) pre-implantation embryos aggregated mechanically in a small drop of medium (Tarkowski, 1961). However, Tarkowski's mechanical method (via a pipette) is technically difficult and tedious for broken and removed the ZP one by one. Later, a study indicated that mouse ZP can be digested and removed easily using pronase (~1 min) and pipettes (Mintz, 1962). A subsequent study demonstrated that using acidic Tyrode's solution (pH 2.5) to dissolve mouse ZP is a relatively simpler and cheaper method (Nicolson et al., 1975). A batch of intact whole embryos (~25) submerged in the acidic Tyrode's solution for approximately 10 seconds is sufficient to partially dissolve the ZP, and the embryos can then be transferred to a buffered medium to wash away the denuded embryos via pipettes. Aggregation has since become one of the major methods for generating chimeric embryos.

Unfortunately, once blastocysts form, generating chimeric blastocysts via the aggregation method is generally impossible. However, one may introduce cells into the cavities of blastocysts to obtain chimeric embryos. It was the first report that chimeric embryos produced using the five-instrument microsurgical method to introduce ICMs or cells into the blastocyst cavity (Gardner, 1968). Apparently, the five-instrument method is too complex for routine operation. Thereafter, the two-micropipette microinjection method was developed (Moustafa & Brinster, 1972).

Since aggregation and microinjection methods are commonly used to produce chimeric embryos; techniques, equipment, and protocols have been modified and improved. For technical details of current methods, including cells, embryos, instruments, and equipment for making micropipettes and generating chimeric embryos, see previous articles (Bradley, 1987; Nagy et al., 2003; Nagy et al., 2010; Papaioannou & Johnson, 2000; Papaioannou & Dieterlen-Lievre, 1984; Pluck and Klasen, 2009). The website (http://www.mshri.on.ca/nagy/default.htm#) of Professor Nagy at the Samuel Lunenfeld Research Institute, Mount Sinai Hospital, Toronto, Ontario, Canada, also provides technical protocols for generating chimeric mouse and ESC-derived mouse via aggregation between ES cells and diploid (2n) or tetraploid (4n) 2.5-day post-coitum (dpc) embryos.

Although the aggregation and microinjection methods are effective for producing chimeric embryos, due to instrumental and technical limitations, they are unsuited to mass production. Thereafter, the coculture method was developed (Wood et al., 1993a; Wood et al., 1993b).

2.1 Comparison of conventional methods for generating chimeric mice

Currently, the most common technique for generating chimeric embryos is direct microinjection of ES cells into the cavity of 3.5-dpc blastocysts. Microinjection is a highly stable and reproducible method that can generate good germline transmitted chimeras. However, this method has various limitations. First, an expensive micromanipulation system is required. Second, intensive training is needed to master micromanipulation skills. Third, the microinjection step is time-consuming, averaging only 20–40 blastocysts/h, limiting production to 50–100 blastocysts daily (Bradley, 1987; Hogan et al., 1994; Nagy et al., 2003). Therefore, generating chimeras usually requires pay-based services. Although using Piezo-driving (Kawase et al., 2001) to introduce ES cells into the cavity of blastocysts may have relatively better efficiency in producing chimeras, it is rarely applied as it requires a high skill level and an extra expensive device.

Well sandwich aggregation is the second most popular method for generating chimeric embryos. This method is also a highly stable, reproducible, and easy method for generating chimeras and has a germline transmission efficiency nearly equivalent to that of blastocyst microinjection (Bradley, 1987; Hogan et al., 1994; Nagy et al., 2003; Papaioannou & Johnson, 2000). Well sandwich aggregation does not require expensive and sophisticated instruments, and is easily learned and implemented. Those familiar with using a mouth pipette can use this method routinely in a laboratory. However, two or more embryos (either XX or XY) are required for aggregation to create a single reconstructed embryo, which is disadvantageous for inbred mice, as only 6–10 embryos can be recovered from each mouse through superovulation. Although single embryo aggregation is a viable option, its efficiency in generating chimeras varies and is inferior to methods using two or more embryos. Therefore, very few studies have used single embryo aggregation to generate chimeric embryos.

Another alternative for generating chimeric embryos is coculturing 2.5-dpc denuded single 4-cell embryos to morulae with ES cells on dish surfaces (Shimada et al., 1999; Wood et al., 1993a) or in droplets (Ueda et al., 1995). However, the efficiency of generating chimeras via this method is far inferior to that of microinjection and well sandwich aggregation. Only a few studies have used this method to generate chimeric embryos.

Table 1 summarizes and compares conventional methods for generating chimeric embryos.

	Microinjection	Well sandwich aggregation[1]	Single embryo aggregation[1]	Coculture[2]
Equipment	Very expensive	Inexpensive	Inexpensive	Very cheap
Skill level	Very high	Low	Low	Very low
Time needed to learn the technique[3]	2–3 months	2–3 weeks	2–3 weeks	1–2 weeks
Time needed to produce chimeric embryos[4]	20–40 blastocysts/h and not more than 50–100 blastocysts/d	~30 pairs/h	~40 embryos/h	>100 embryos/h

[1]Cultured overnight then reconstructed chimeric embryos are recovered for transfer or other treatments.
[2]Cocultured for 3–4 h, the embryo-ESC aggregates are recovered and cultured overnight to produce chimeric embryos.
[3]Persons must be familiar with cell and embryo culturing as well as mouth pipetting.
[4]Enriched ES cells and recovered intact embryos are ready for use.

Table 1. Comparison of conventional methods for generating chimeric embryos

2.2 Vial coculture method for generating chimeric mice

Although the microinjection method produced good and reliable results, it is hard to practice by a laboratory. Therefore, outsourcing to a core facility or commercial company is common. Unfortunately, service charges are high at approximately US$1,000–3,000/case. Conversely, the aggregation method is easily applied and inexpensive. However, this method must be applied in a one-by-one manner and is tedious. The conventional coculture method is also easily applied and is inexpensive. Furthermore, this method facilitates routine mass production of chimeric embryos. Unfortunately, outcomes are not as reliable and good as those of the microinjection and well sandwich aggregation methods. In conventional coculture protocols, denuded embryos on a dish surface have only two-dimensional ES-cell contact surfaces, resulting in only 55–64% of denuded embryos adhering to ES cells (Ueda et al., 1995; Wood et al., 1993a). Obviously, an improved coculturing method is needed that can achieve results as good as or better than those by the microinjection and well sandwich aggregation methods. Moreover, an improved coculturing method should be easily applied, cheap, and suited to mass production.

Recently, my laboratory developed an alternative simple, very cheap, and reproducible method for mass production of chimeric embryos by coculturing 2.5-dpc denuded 8-cell embryos and compacting morulae with ES cells in 1.7-mL Eppendorf vials (micro test tube); this method has fewer technological and instrument-based limitations than conventional methods. Although depressed microwells made by a darning needle had three-dimensional possibility for denuded embryos and ES cells to contact each other, however, in the vial coculturing system, the large number of enriched ES cells surrounding the denuded embryos from every direction may improve the overall adherence. Furthermore, gravity may also contribute to enhanced ES cell adherence via this method. The resulting chimeras show significantly high levels of chimerism and high germline transmission rates (Lee et al., 2007). Table 2 lists an example schedule and protocol for the vial coculture method. Figures 1 and 2 show vial coculturing results.

Date	Time	Target	Treatment
Friday	~16:30	Donor females superovulated	PMSG 5–10 units/ip
Sunday	~16:30		hCG 5–10 units/ip; donor females are mated with studs
Monday	am	Donor females	Plug checked
		ES cells	Thawed or pass in high density to 0.1% gelatin-coated dishes
Tuesday	~16:30	Recipients (ICR, CD-1, or F1 hybrid)	Estrus females mated with vasectomized males
Wednesday	am	Recipients	Plug checked
	~08:30	ES cells	Enriched via the double plating method
	~08:40	35-mm cell culture dish	HK (20.85 mM hepes-buffered KSOM, 285 ± 10 mOsm/kg H_2O) droplets under light-weight mineral oil prepared; room temperature (RT)
	~09:00	Donor females	Recovery of the 2.5-dpc embryos, which are kept in HK at RT until used
	~10:00	ES cells	The first round of enriched cells harvested and re-suspended to a cell medium for the second standing
	~10:25	Enriched ES cells	Cells are harvested and stored at 4°C until the concentration is adjusted for coculturing
	~10:40	6-cell embryos to compacting morulae	ZP are removed using the acidic Tyrode's solution
	~11:00	1.7-mL vials	Approximately 0.8 mL enriched ES cells are transferred to vials; after ~5 min, denuded embryos are added for coculturing; 5% CO_2, 37°C incubator
	~11:15	60-mm bacteriological dish	10-μL droplets of KSOM-AA or KSOM-AA containing 1% FBS are prepared and put in an incubator under 5% CO_2 at 37°C
	~14:00	1.7-mL vials	Vial coculturing ends, embryo-ESC aggregates are recovered
	~14:30	60-mm bacteriological dish	embryo-ESC aggregates are washed to droplets of KSOM-AA or KSOM-AA containing 1% FBS for culturing overnight; 5% CO_2, 37°C incubator
Thursday	am or pm	Recipients	Chimeric embryos are transferred to uterus horns of 2.5-dpc pseudopregnant recipients
Sunday		Recipients	Pups are born after ETed for 17 days

[1]One person can finish the vial coculturing easily.

Table 2. Schedule and protocol for the vial coculture method for generating chimeric mice[1]

2.2.1 Technical considerations for vial coculture method

To ensure that cells adhered to the denuded embryos are the ES cells mainly, any enrichment method for ES cells can be used. When using the double-plating selection method, approximately 96% of harvested cells expressed bright-green fluorescence, and approximately 92% of these cells were <12 μm in diameter (Lee et al., 2007). The method is easily implemented, and is selective, effective, and reproducible in removing debris, dead cells, and feeder cells from the prepared ES single-cell suspension.

Different volumes of Eppendorf vials are available for coculturing. I recommend the 1.7-mL vial due to the size is good for handling. Approximately 0.8-mL aliquots of enriched ES cells, either fresh or thawed; ~5.0×10^5 cells/mL in KSOM-AA alone or KSOM-AA containing 1% fetal bovine serum (FBS) are added to sterile polypropylene 1.7-mL vials with snap caps. After left to stand for 5 min, ≦200 denuded 6-cell embryos to morulae are gently and circularly blown from beneath the medium surface into the vial via a mouth pipette, and then coculture in an incubator under 5% CO_2 at 37°C for 3 ± 1 h. Denuded 2- to 4-cell embryos are unsuitable for coculturing, as blastomeres sometimes can easily separate during coculturing. Additionally, the relatively smaller diameters of blastomeres make recovery difficult. However, adherent cells are consistently observed in separated blastomeres. Conversely, segregation of blastomeres of electrofused tetraploid (4n) 3- to 4-cell embryos does not occur.

After the coculturing is ended, precipitate in vials is aspirated gently and loose cells on embryonic surfaces are removed by washing using a mouth pipette. Denuded embryos adhered to ES cells are recovered. Following coculturing for 3 h in 1.7-mL Eppendorf vials, >90% of denuded 6-cell embryos to morulae adherent ES cells can be recovered. Moreover, approximately 90% of recovered embryos adhered tightly to ES cells and approximately 5–10% (the percentage increases as the number of embryos in the same vial increases) of recovered aggregates had 2 or 3 embryos clustered around and adhered to ES cells. Aggregates have >3 embryos, if necessary, can be separated by gentle pipetting using a mouth pipette. During coculturing, approximately 10% of denuded 8-cell embryos developed into compacting morulae. Experimental data show that cell adherence could reach 100% when ES cell concentrations are increased, or when the coculturing period is extended. The embryo-ESC aggregates are washed directly in either KSOM-AA or KSOM-AA containing 1% FBS droplets (10 μL) under light-weight paraffin oil on bacteriological dishes. Two to four embryo-ESC aggregates are carefully allocated to different corners of the same droplet to prevent possible adhesion and are cultured overnight in an incubator at 37°C under 5% CO_2 until transfer to uterine horns of 2.5-dpc pseudopregnant recipients.

Mouse preimplantation embryos and ES cells require different in vitro culture requirements. Essentially, ES cells require at least 5% FBS (Wakayama et al., 1999) to maintain survival and possible pluripotency unless cultured in knockout serum replacement (KSR) (Goldsborough et al., 1998). It has been showed that poor ES cell viability after overnight coculturing in M16 or KSOM media (Huang et al., 2008; Kondoh et al., 1999). Although, M16 supplemented with FBS, could enhance chimera generation (Kondoh et al., 1999). Unfortunately, the viable fetuses derived from the FBS (5-15%) groups were significantly fewer than that derived from the FBS-free control group (Arny et al., 1987; Caro and Trounson, 1984; Khosla et al., 2001). Khosla et al. (2001) indicated that FBS exerts a direct adverse effect on genes responsible for postimplantation development.

To summarize, coculturing 2.5-dpc denuded 6-cell embryos to morulae with ES cells in 1.7-mL Eppendorf vials for approximately 3 h is a simple and effective alternative method for mass production of chimeric embryos. Table 3 compares conventional methods and the vial coculture method for generating chimeric embryos and germline transmitted chimeric mice.

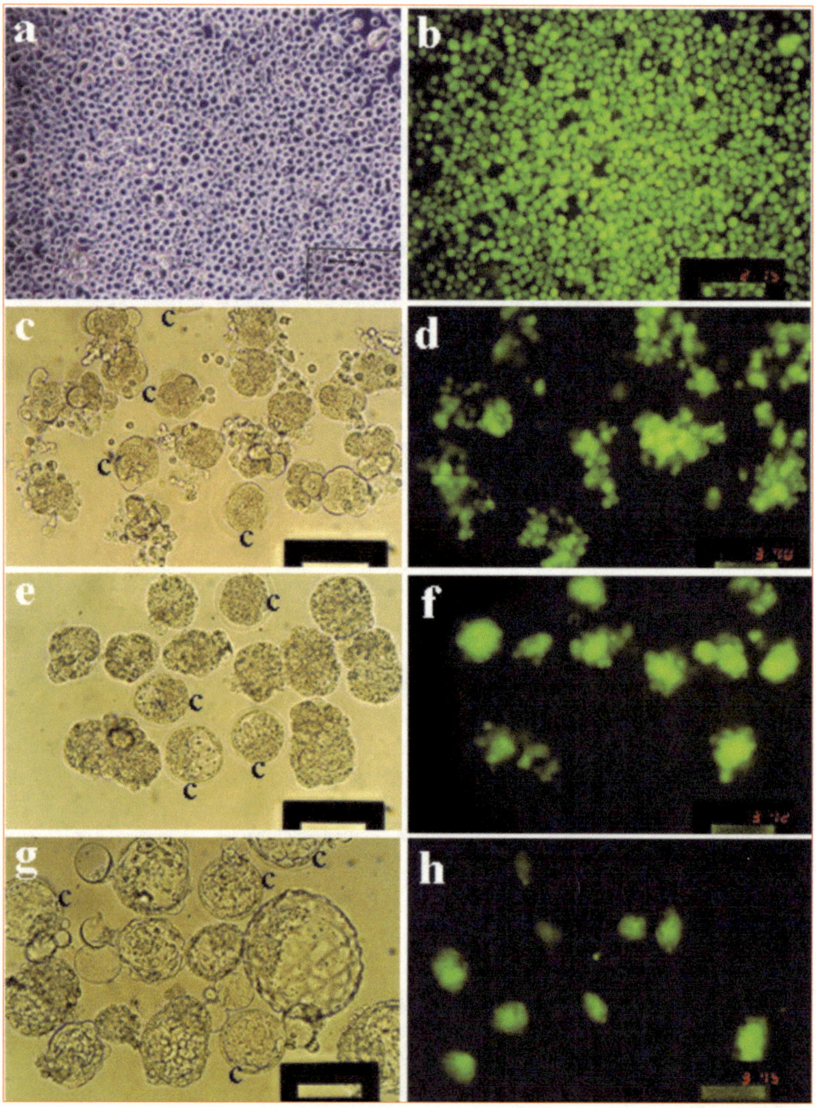

Fig. 1. Enrichment of ES cells and development of denuded embryo-ESC aggregates. a: Attaching and/or attached cells were recovered after the original single-cell suspension of the mouse ES cell, ESC 26GJ9012-8-2, cultured on a 100-mm dish in a 5% CO_2 incubator at 37°C for 80 min. b: Over 94% of cells expressed the green fluorescence protein (GFP). c, d: After denuded 8-cell embryos, morulae, and ESC 26GJ9012-8-2 cells were cocultured in an 1.7-mL Eppendorf vial for 2 h, recovered embryos had adherent green fluorescing ES cells on their surfaces. The left bright and right green fluorescent images show the same view of 14 embryos, including 2 8-cell embryos, 2 compacting morulae with zona pellucida (as the control; c), 5 single-embryo-cell aggregates, 3 2-embryo-cell aggregates, and 2 3-embryo-cell aggregates (some kind sandwich aggregation). e, f: After culturing aggregates from panels c and d overnight in droplets of KSOM-AA containing 1.0% FBS, the aggregates had cells with surface green fluorescence mingling in the developing compacting and compacted

chimeric morulae. g, h: After further overnight culturing, chimeric morulae from panels e and f developed into chimeric blastocysts displaying green fluorescence cells primarily in the ICM. Scale in panel a: bar = 50 µm. Scale in panels b–h: bar = 100 µm. (Reproduced with permission from Lee et al., 2007. Theriogenology 67:228–237.)

Fig. 2. Chimeras with high percentages of coat color distribution and germline transmission generated using the Eppendorf vial coculture method. In this experiment, chimeric morulae and/or blastocysts developed from aggregates of ICR × B6CBAF1 embryos and ESC 26GJ9012-8-2 cells were transferred to pseudopregnant ICR 2.5-dpc uterine horns. The pups born alive had high percentages of coat color and green fluorescence expression. a, b: Pups derived from fresh ESC 26GJ9012-8-2 cells. c, d: Pups derived from thawed ESC 26GJ9012-8-2 cells. e: Green fluorescence was widely expressed on internal organs of mature chimeras. The control mouse had no green fluorescence expression. f: Pups expressing green

fluorescence, following phenotypically normal male chimeras, were naturally mated with ICR to achieve high germline transmission rates. (Reproduced with permission from Lee et al., 2007. Theriogenology 67:228–237.)

	Microinjection	Well sandwich aggregation	Single embryo aggregation and coculture	Vial coculture
Pups born alive after chimeric embryos ETed[2] (A)	~60%	~30%	~20%	~25%
Chimeras (B)	~50% (B/A)	~55% (B/A)	~45% (B/A)	~60% (B/A)
Male chimeras (C)	~50% (C/B)	~50% (C/B)	~40% (C/B)	~60% (C/B)
Male chimeras with germline transmission (D)	~30% (D/C)	~40% (D/C)	~30% (D/C)	~50% (D/C)
Total efficiency of germline transmission[3] (A × B × C × D)	~4.5%	~3.3%	~1.1%	~4.5%

[1]Data are compiled from previous studies.
[2]Pups born alive/chimeric embryos transferred.
[3]Germline transmitted male chimeras/chimeric embryos transferred. The total efficiency of germline transmission is highly variables depends on ES cells used, donor embryos used, persons did, and mouse facilities managed.

Table 3. Comparative efficiency of different methods for generating germline transmitted chimeric mice[1]

2.3 Generation of ESC-derived mice

Authentic ES cells are defined by three cardinal properties: unlimited symmetrical self-renewal in vitro; comprehensive contribution to primary chimeras; and generation of functional gametes for genome transmission (Buehr et al., 2008). However, using 4n complementation for generating ESC-derived mice is regarded as the most solid criterion for ES cell pluripotency. The criterion is also accepted for generation of iPSC-derived mice (Boland et al., 2009; Kang et al., 2009; Zhao et al., 2009).

The mouse pluripotent ES cells can be established from zygotes, blastomeres, and ICMs (Lee et al., 2011). However, the originate of ES cells limit its developmental potential to embryo proper mainly (Beddington & Robertson, 1989). Tetraploid embryos typically stop their development during the early stage of gestation and do not develop beyond the mid-gestation period due to a lack of the embryo proper, but persist in extraembryonal membranes (Eakin & Behringer, 2003; Kaufman & Webb, 1990; Snow, 1975; Tarkowski et al., 1977). Combining 4n embryos and 2n ES cells is logical as 4n embryos may complement the deficient extraembryonal differentiation of ES cells while allowing full expression of their potential for fetal development (Nagy et al., 1990). Thereafter, a viable and fertile ESC-derived mouse was generated (Nagy et al., 1993).

Since ESC-derived mice were generated successfully using 4n 4-cell embryos aggregated with 2n ES cells (Nagy et al., 1990; Nagy et al., 1993), well sandwich aggregation is the main choice to do that (Eggan et al., 2002; Li et al., 2005; Ohta et al., 2008; Schoonjans et al., 2003; Ueda et al., 1995). The second popular method is microinjecting ES cells into 4n blastocysts (Eggan et al., 2001; Kirchain et al., 2008; Li et al., 2005; Lin et al., 2010; Schwenk et al., 2003; Wang et al., 1997; Wang & Jaenisch, 2004).

Most tetraploid-ESC neonates derived from hybrid ES cells developed into fertile adults (Eggan et al., 2001). Conversely, most studies revealed that ESC-derived pups from inbred ES cells died shortly after delivery. However, one study demonstrated that inbred ES cells can generate ESC-derived mice (Schoonjans et al., 2003). Furthermore, the Caesarean section is required to overcome the failure of respiratory problems of ESC-newborns (Nagy et al., 1993). Apparently, this is a tedious work and is not a practical protocol for routine operation. Previous studies suggested that the recipient mothers can be subject to natural delivery instead of Caesarean section (Lee et al., 2003; Li et al., 2005). However, this is not suitable for any ES cell.

Previous studies also revealed that an ESC-derived mouse have host embryo contamination (Eggan & Jaenisch, 2003; Li et al., 2005), poor viability, and other minor abnormalities such as altered growth rate and body weight. However, adults had normal morphological, physiological, and neurological characteristics (Schwenk et al., 2003).

Using 4n complementation method, 4n embryos require produced of 2n 2-cell embryos usually by electrofusion and are less viable than normal 2n embryos. Thus, the generation efficiency of ESC-derived mice is relatively low at approximately 1-5%. One reason for this low efficiency may be the low cell number of 4n blastocysts. Using 3–5 aggregated 4n embryos, the efficiency in generating ESC-derived mice can be increased 2–3 times (Ohta et al., 2008). Additionally, the modified method would be applicable to any ES cell, including general ES cells used for gene targeting (Ohta et al., 2008).

Although ES cells can produce viable and fertile ESC-derived mice, this is an inefficient process; many tetraploid-ESC aggregates die before reaching term, even when early passage ES cells are used (Nagy et al., 1993). Therefore, this approach cannot be considered as a feasible approach for routinely achieving germline transmission from ES cells (Nagy et al., 1993). Reasonably, other alternatives might be developed.

2.4 Generation of ESC-derived F0 mice

After the blastocyst microinjected with ES cells that can adhere to and mingle in ICMs, then co-develop to an embryo proper including germ cells. The same phenomenon was also observed when 2.5-dpc embryos aggregated with ES cells. The mechanisms underlying this phenomenon have been investigated. Unfortunately, this mechanism remains unclear. However, data from studies of chimeric embryos suggest a combined contribution of multiple factors, including geometrical effects of cell size and polar or apolar positioning (Hillman et al., 1972; Johnson & Ziomek, 1981; Saburi et al., 1997; Tarkowski & Wroblewska, 1967). Notably, this phenomenon may also reflect the possibility that ES cells are naturally more committed to an ICM fate (Wood et al., 1993a).

Previous studies have demonstrated that ES cells adhering to the surfaces of 8-cell embryos or compacting morulae are generally localized in the ICM of blastocysts following culturing (Lee et al., 2007; Shimada et al., 1999; Wood et al., 1993a; Wood et al., 1993b). Repentigny and Kothary (2010) recently microinjected ES cells into the perivitelline space (PVS) of zygotes. They showed that at the 2- to 4-cell stage embryos, injected ES cells remain in the PVS and are not incorporated into embryos. The ES cells begin partial blastomere incorporation into an embryo at the 8-cell embryo. Finally, at the compacted morula, ES cells are almost completely incorporated into an embryo. At the blastocyst, ES cells form an ICM (Repentigny & Kothary, 2010). Whether ES cells can replace ICMs completely and develop thereafter as an ESC-derived mouse is worthy of investigation.

Compared with 3.5-dpc blastocyst microinjection, an alternative method that microinjects ES cells into the tight space between ZP and blastomeres (the subzonal cavity) of 2.5-dpc 8-cell embryos has been reported (Tokunaga & Tsunoda, 1992). Experimental results showed that the proportion of male chimeric mice capable of germline transmission increased significantly. Furthermore, 100% coat color chimeric mice with germline transmission were produced. Unfortunately, the meaning of 100% coat color chimerism was not investigated (Tokunaga & Tsunoda, 1992). Papaioannou and Johnson (1993, 2000) have been mentioned that the result of microinjecting ES cells into 2.5-dpc 8-cell embryos was comparable to but not better then microinjection of 3.5-dpc blastocysts. Notably, microinjecting 2.5-dpc embryos are more difficult than microinjecting 3.5-dpc blastocysts due to the tight and small subzonal cavity and possible damage to blastomeres, explaining why only a few follow-up studies exist.

Laser-assisted microinjection of 7–9 ES cells into the subzonal cavity of 2n 8-cell embryos may enhance microinjection and efficiently yield F0 generation mice (100% coat color chimerism) that are fully ESC-derived and healthy, exhibiting 100% germline transmission and containing no more than 0.1% host embryo contamination (Poueymirou et al., 2007). They suggested that the F0 mouse is equivalent to the ESC-derived mouse. A subsequent study indicated that 8–10 ES cells Piezo (toxic mercury in a microinjection pipette used) microinjected into the subzonal cavity of 2n 4- or 8-cell embryos also generated F0 ES cell offspring (Huang et al., 2008).

The efficiency of generating ESC F0 mice is much better than using ESC-derived mice via ES cell assemble with 4n embryos (Eakin & Hadjantonakis, 2006; Eggan et al., 2001; Li et al., 2005; Nagy et al., 1993; Ueda et al., 1995; Wang & Jaenisch, 2004). However, these methods need an expensive laser or Piezo-driving equipment and additional training is required to acquire the necessary skills. The disadvantages of both methods limit their applications. Recent studies, which used conventional microinjection to introduce ES cells into 2- to 8-cell embryos, produced 100% coat color chimeras (Kraus et al., 2010; Ramirez et al., 2009). However, the technical problems still exist.

The Eppendorf vial coculture method can generate massive amounts of chimeric embryos. The resulting chimeric mice show approximately 40% of pups born alive with almost 100% ES cells coat color distribution (Lee et al., 2007). The major disadvantage of the coculture method is variable adhesion of ES cells onto the surfaces of denuded embryos. In practice, 4-cell embryos to morulae are recovered from superovulated 2.5-dpc donor mice. The denuded 8-cell embryos and morulae are good for vial coculturing. However, the 4-cell embryo is not suitable for vial coculturing, because blastomeres usually separate during coculturing.

The 2n ES cells microinjected into, aggregated with, or cocultured with 2.5-dpc denuded 2n 8-cell embryos and/or morulae can generate germline transmitted F0 mice. New methods with the higher efficiency may be worth developing to overcome the limitations and disadvantages of existing approaches.

2.5 The hypertonic microinjection method for generating chimeric mice

For conventional microinjecting ES cells to 2.5-dpc 3-cell embryos to morulae, the tight subzonal cavity is a major technical hurdle. Theoretically, increasing the space of the subzonal cavity can solve this problem.

Zona pellucida is a rigid glycoprotein that resists both hypertonic and hypotonic solutions. In contrast, the volume of an embryo proper changes in proportion to osmolarity of solutions (Leibo, 1980). In other words, the space of the subzonal cavity increases when embryos are in hypertonic solutions. Notably, a high sucrose concentration is virtually non-toxic to embryos and oocytes (Kasai et al., 1983; Kasai et al., 1992; Kuleshova et al., 1999). Therefore, pre-blastocyst embryos in a microinjection medium can increase the space of the subzonal cavity proportionally to the added sucrose concentration and that may pose no threat to embryos for hours. Accordingly, my laboratory is developing a method in which ES cells are hypertonically microinjected into 2.5-dpc embryos. Table 4 gives an example schedule and protocol for this method. The hypertonic microinjecting ES cells into 2.5-dpc embryos and fertile chimeras are shown at Fig. 3.

Date	Time	Target	Treatment
Friday	~16:30	Donor females superovulated	PMSG 5–10 units/ip
Sunday	~16:30		hCG 5–10 units/ip; Donor females are mated with studs
Monday	am	Donor females	Plug checked
		ES cells	Thawed or pass in a high density to 0.1% gelatin-coated dishes
Tuesday	~16:30	Recipients (ICR, CD-1, or F1 hybrid)	Estrus females mated with vasectomized males
Wednesday	am	Recipients	Plug checked
	~08:00	ES cells	Enriched via the double plating method
	~08:20	60-mm cell culture dish	HK, EK[2], and KSOM-AA droplets under light-weight mineral oil prepared; RT
	~08:30	Donor females	Recovered 2.5-dpc embryos and stay in HK at RT until used
	~09:30	Making microinjection and holding pipettes	Microinjection pipette with a beveled tip and spike: outer diameter, 18–20 µm; inner diameter, 13–15 µm. Holding pipette: outer diameter, 50–70 µm; inner diameter, 22–25 µm
	~09:50	ES cells	The first round of enriched cells harvested and re-suspended in a cell medium for the second standing
	~10:20	Enriched ES cells	Harvested cells are kept at 4°C until re-suspended in high density in a hepes-buffered microinjection medium[3] containing 0.2 M sucrose[4]
	~10:40	Microinjection chambers	A ~80-µL droplet of 0.2 M sucrose microinjection medium under light-weight mineral oil is prepared.

Date	Time	Target	Treatment
			Thousands of enriched ES cells, then 3-cell embryos to compacting morulae added to the droplet of microinjection medium in a row.
	~10:55	60-mm cell culture dish	HK, EK, and KSOM-AA droplets under light-weight mineral oil to 5% CO_2, 37°C incubator
	~11:00	Microinjection chamber and micropipettes	Setup for microinjection
	~11:20	Microinjection chamber	Sucking hundreds of ES cells into a microinjection pipette within 5~10 min[5]
	~11:30	Microinjection begins	Approximately 5–30 ES cells are microinjected at RT into the large subzonal cavity of 3-cell embryos to compacting morulae
	~12:30	Microinjection finishes	After 60–150 2.5-dpc embryos are microinjected within 1 h[6], injected embryos are washed to and culture in EK droplets for ~1 h; 5% CO_2, 37°C incubator
	~13:50	microinjected embryos	Wash to fresh EK or KSOM-AA droplets for overnight culturing; 5% CO_2, 37°C incubator
Thursday	am or pm	Recipients[7]	Chimeric embryos are transferred to uterus horns of 2.5-dpc pseudopregnant recipients
Sunday		Recipients[7]	Pups born after ETed for 17 days

[1]Base on one person finish the hypertonic microinjection. However, two persons are more efficiently and practically.
[2]EK (6.5% KSR without FBS): 37.5% KSR ESC medium (20% KSR) and 62.5% KSOM-AA (285 ± 10 mOsm/kg H2O)
[3]Any hepes-buffered microinjection medium can be used. My laboratory usually uses EHK (37.5% KSR ESC medium and 62.5% HK) as the microinjection medium no matter how ES cells are culturing either in FBS- or KSR-ESC media.
[4]Most enriched ES cells in 0.2 M sucrose EHK (500 ± 10 mOsm/kg H_2O) are approximately 9–11 µm in diameter. Therefore, the diameter of microinjection pipette is smaller than a conventional pipette.
[5]Cleaned and enriched ES cells are very important when sucking cells into a microinjection pipette. Otherwise, a single cell suspension, which has many cells with large diameters or sticky debris, will slow loading, generating a microinjection bottleneck.
[6]The randomly sucked embryos by a holding micropipette do not need to adjust the position for microinjecting ES cells into the subzonal cavity, which can be completed in less then 30 seconds.
[7]Embryos with ZP can be transferred into the oviducts of 0.5-dpc recipients (Ramirez et al., 2009). Pups will be born after being subjected to ET for 19 days.

Table 4. Schedule and protocol for the hypertonic microinjection method for generating chimeric mice[1]

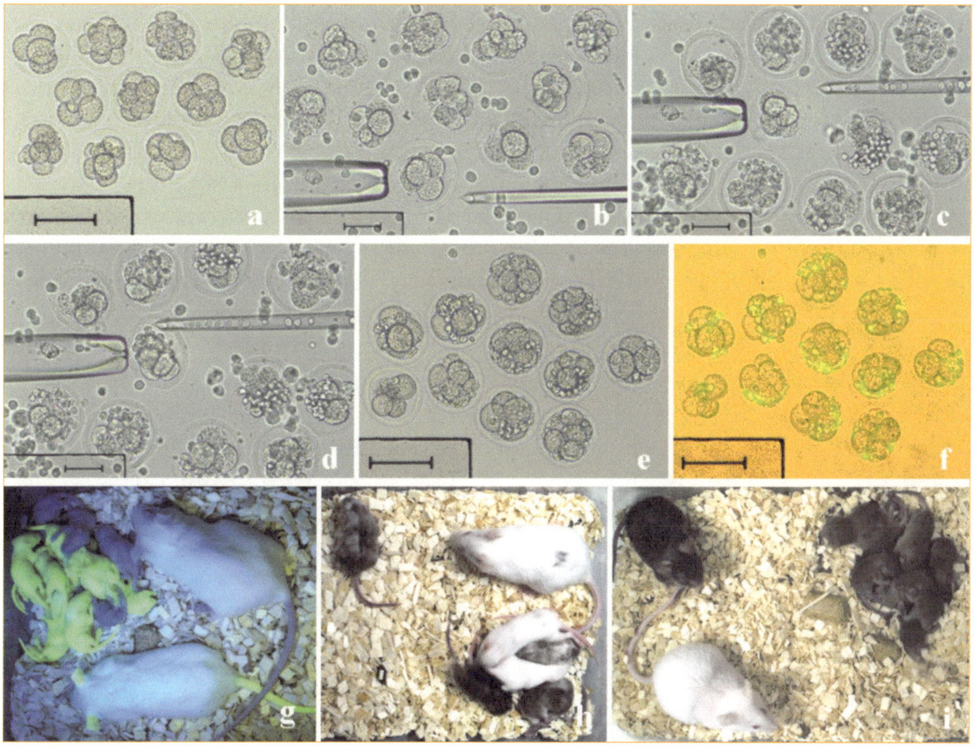

Fig. 3. The generation of germline transmitted chimeric mice via the hypertonic microinjecting ES cells into a subzonal cavity of ICR × ICR 2.5-dpc embryos. The chimeric embryos were transferred into the pseudopregnant ICR 2.5-dpc uterine horns. a: The 2.5-dpc 4- to 8-cell stage embryos in isotonic KSOM-AA (285 ± 10 mOsm/kg H_2O) have tight subzonal cavities. b: Embryos and enriched ES cells (ESC 26GJ9012-8-2, P14) in 0.2 M sucrose EHK (37.5% KSR ESC medium and 62.5% hepes-KSOM) hypertonic injection medium (500 ± 10 mOsm/kg H_2O) show large subzonal cavities. c, d: Injecting and injected embryos in hypertonic injection medium. e, f: After injected approximately 25 ES cells into embryos, which were washed to isotonic EK (37.5% KSR ESC medium and 62.5% KSOM-AA). The bright (e) and bright plus green fluorescent (f) images show the same view of 11 injected embryos. g: The green fluorescence expressing germline transmitted chimeric mouse was generated by injecting ESC 26GJ9012-8-2 cells to 4-cell embryos. h: Chimeric pups born alive after approximately 15 ESC98B33 cells (P5) were injected into 4-cell embryos and cultured overnight in EK. The ESC98B33 cell was derived from C57BL/6J 0.5-dpc denuded zygotes cultured on human foreskin fibroblast (Hs68) feeders and KSR ESC medium containing 2i (0.5 μM PD0325901 and 3 μM CHIR99021) and 10 μM ACTH fragments 1–24. i: A chimeric mouse with 100% coat color distribution was generated after approximately 20 ESC98B27 cells (P8) were injected into compacting morulae and cultured overnight in KSOM-AA. The ESC98B27 cell was derived from an isolated single blastomere of a C57BL/6J 1.5-dpc 2-cell embryo cultured on Hs68 feeders and KSR ESC medium containing 2i. Scale in panels a, e, f: bar = 100 μm. Scale in panels b–d: bar = 50 μm.

Preliminary data demonstrate that ES cells microinjected into the subzonal cavity of 2.5-dpc embryos in a microinjection medium containing 0.2 M sucrose (~500 mOsm/kg H_2O) can generate chimeric embryos with high percentages of chimerism (including 100% coat color and/or GFP expression) and viable, healthy, germline transmitted mice (Fig. 3). These preliminary results also indicate that hypertonic microinjection is at least comparable to conventional, laser, and Piezo microinjection methods for generating germline-transmitted chimeras.

Technically, the differences between conventional microinjection and hypertonic microinjection are that later method uses the 0.2 M sucrose microinjection medium, a microinjection pipette with a smaller diameter, and very fast microinjection. The developing hypertonic microinjection method may be an useful alternative for generating chimeric or F0 mice.

2.6 Optimal method of generating chimeras depends on embryo stage

Depending on embryo stage, germline transmitted chimeric mouse or F0 mouse are generated by microinjection, well sandwich aggregation, or coculture methods. Typically, embryos recovered from 0.5-dpc (zygotes) and 1.5-dpc (2-cell embryos) mice for use in generating chimeras are usually no better than 2.5-dpc and 3.5-dpc (Repentigny & Kothary, 2010). In practice, therefore, 3.5-dpc and 2.5-dpc embryos are more commonly used. For blastocysts, conventional microinjection is the only choice. For pre-blastocyst embryos, almost all methods are possible. However, the optimal method may differ. The comparative efficiency of methods for generating germline transmitted chimeric mouse or F0 mouse is summarized at the Table 5, which also might be adopted for generation of ES cell-derived mouse via 4n embryos.

Embryo stage	2.5-dpc			3.5-dpc	
	3-cell to 5-cell embryo	6-cell to 8-cell embryo	Compacting morula[3]	Compacted morula[3]	Blastocyst
Conventional microinjection	+	+	+	+	+++
Hypertonic microinjection	++	+++	+++	+	+−
Laser-assisted microinjection	+	+	+	+	+−
Piezo-driving microinjection	+	+	+	+	+−
Well sandwich aggregation	++	+++	+++	−	−
Single embryo aggregation	+−	+	+	−	−
Conventional coculture	+−	+	+	−	−
Vial coculture	+	+++	+++	−	−

[1]Methods are compared in terms of efficiency in generating germline transmitted chimeras, required equipment, required skills, and operational time.

[2]+++: best; ++: 2nd choice; +: 3rd choice; +−: not suggested −: not good at all.

[3]Compacting and compacted morulae can be cultured in Ca++-, Mg++-free PBS for 30–60 min to reveal the blastomeres. In some cases, approximately 10 ES cells can be directly microinjected into an embryo proper.

Table 5. Comparative efficiency of methods for generating germline-transmitted chimeric mouse or F0 mouse[1,2]

3. Conclusion

Currently, the most common techniques for generating chimeric mice or ESC-derived mice are microinjection or well sandwich aggregation. Single denuded embryo aggregation or coculturing with ES cells are less common alternatives because the efficiency in generating chimeras is inferior to that of microinjection and well sandwich aggregation. Thus, both methods are rarely employed. However, due to systemic limitations and the disadvantages of conventional microinjection, aggregation, and coculturing, new methods are needed.
Recently, my laboratory developed an alternative simple, inexpensive, and reproducible method for mass production of chimeric embryos by coculturing 2.5-dpc denuded 8-cell embryos and/or compacting morulae with ES cells in 1.7-mL Eppendorf vials (micro test tube). This vial coculture method has significantly fewer technological and instrumental problems than existing methods. The resulting chimeras have significant levels of chimerism (including 100% coat color chimerism) and high germline transmission rates.
Previous studies showed that microinjecting ES cells into 2.5-dpc 8-cell embryos could produce 100% coat color chimerism. However, due to the tight space between ZP and blastomeres, one must be very careful to avoid damaging blastomeres while microinjecting. Thus, the method is rarely adopted. Using a laser pulse or Piezo-driving equipment to assist introducing ES cells into the subzonal cavity of 8-cell embryos could have superior efficiency in generating ESC-derived F0 chimeras (100% coat color chimerism), which are equivalent to ESC-derived mice. However, only few studies have adopted either method due to the skill and/or extra expensive instruments needed.
Recently, my laboratory revealed that ES cells microinjected into the subzonal cavity of 2.5-dpc embryos in a microinjection medium supplemented with 0.2 M sucrose could efficiently generate chimeric embryos with high percentages of chimerism and viable, healthy, germline transmitted F0 ES-cell mice.
Both vial coculture and hypertonic microinjection methods are useful alternatives for producing germline chimeric or F0 mice effectively, efficiently, and reliably.

4. Acknowledgments

Parts of the contents have been published at the journal of Theriogenology (Lee et al., 2007; 67:228-237). Special thanks to co-authors. Also, I would like to thank the National Science Council of the Republic of China, Taiwan, for financially supporting my research under Contract Nos. NSC99-2324-B-059-001, NSC97-2317-B-059-004, and NSC95-2317-B-059-001. Drs. C.F. Tu, C.k. Juang, and S.F. Guo as well as Ms. H.R. Chang and T.L. Hsu at Animal Technology Institute Taiwan (ATIT) are commended for their critical comments, technical assistance, and routine maintenance in the laboratory and mouse facility.

5. References

Arny, M., L. Nachtigall, & J. Quagliarello. (1987). The effect of preimplantation culture conditions on murine embryo implantation and fetal development. *Fertility and Sterility* 48:861-865.

Beddington, R. S. & E. J. Robertson. (1989). An assessment of the developmental potential of embryonic stem cells in the midgestation mouse embryo. *Development* 105:733-737.

Boland, M. J., J. L. Hazen, K. L. Nazor, A. R. Rodriguez, W. Gifford, G. Martin, S. Kupriyanov, & K. K. Baldwin. (2009). Adult mice generated from induced pluripotent stem cells. *Nature* 461:91-94.

Bradley, A. (1987). Production and analysis of chimeric mice. In: *Teratocarcinomas and Embryonic Stem Cells, a Practical Approach.* E. J. Robertson, (Ed.), pp. 113-151. IRL Press, ISBN 1-85221-004-4, Oxford, UK.

Brons, I. G., L. E. Smithers, M. W. B. Trotter, P. Rugg-Gunn, B. Sun, S. Chuva de Sousa Lopes, S. K. Howlett, A. Clarkson, L. hrlund-Richter, R. A. Pedersen, & L. Vallier. (2007). Derivation of pluripotent epiblast stem cells from mammalian embryos. *Nature* 448:191-195.

Buehr, M., S. Meek, K. Blair, J. Yang, J. Ure, J. Silva, R. McLay, J. Hall, Q. L. Ying, & A. Smith. (2008). Capture of authentic embryonic stem cells from rat blastocysts. *Cell* 135:1287-1298.

Caro, C. M. & A. Trounson. (1984). The effect of protein on preimplantation mouse embryo development in vitro. *Journal of In Vitro Fertilization and Embryo Transfer* 1:183-187.

Eakin, G. S. & R. R. Behringer. (2003). Tetraploid development in the mouse. *Developmental Dynamics* 228:751-766.

Eakin, G. S. & A. K. Hadjantonakis. (2006). Production of chimeras by aggregation of embryonic stem cells with diploid or tetraploid mouse embryos. *Nature Protocols* 1:1145-1153.

Eggan, K. & R. Jaenisch. (2003). Differentiation of F1 embryonic stem cells into viable male and female mice by tetraploid embryo complementation. *Methods in Enzymology* 365:25-39.

Eggan, K., H. Akutsu, J. Loring, L. Jackson-Grusby, M. Klemm, W. M. Rideout, R. Yanagimachi, & R. Jaenisch. (2001). Hybrid vigor, fetal overgrowth, and viability of mice derived by nuclear cloning and tetraploid embryo complementation. *Proceedings of the National Academy of Sciences USA* 98:6209-6214.

Eggan, K., A. Rode, I. Jentsch, C. Samuel, T. Hennek, H. Tintrup, B. Zevnik, J. Erwin, J. Loring, L. Jackson-Grusby, M. R. Speicher, R. Kuehn, & R. Jaenisch. (2002). Male and female mice derived from the same embryonic stem cell clone by tetraploid embryo complementation. *Nature Biotechnology* 20:455-459.

Evans, M. J. & M. H. Kaufman. (1981). Establishment in culture of pluripotential cells from mouse embryos. *Nature* 292:154-156.

Gardner, R. L. (1968). Mouse chimeras obtained by the injection of cells into the blastocyst. *Nature* 220:596-597.

Goldsborough, M. D., M. L. Tilkins, P. J. Price, J. Lobo-Alfonso, J. R. Morrison, M. E. Stevens, J. Meneses, R. Pederson, R. Koller, B. Koller, & A. Latour. (1998). Serum-free culture of murine embryonic stem (ES) cells. *Focus (Gibco)* 20:8-12.

Hillman, N., M. I. Sherman, & C. Graham. (1972). The effect of spatial arrangement on cell determination during mouse development. *Journal of Embryology and Experimental Morphology* 28:263-278.

Hogan, B., F. Costantini, R. Beddington, & E. Lacy. (1994). In vitro manipulation of preimplantation embryos. In: *Manipulating the Mouse Embryo, a Laboratory Manual.* B. Hogan, F. Costantini, R. Beddington, & E. Lacy, (Eds.), pp. 189-216. Cold Spring Harbor Lab. Press, ISBN 0-87969-384-3, Plainview, USA.

Huang, J., K. Deng, H. Wu, Z. Liu, Z. Chen, S. Cao, L. Zhou, X. Ye, D. L. Keefe, & L. Liu. (2008). Efficient production of mice from embryonic stem cells injected into four- or eight-cell embryos by piezo micromanipulation. *Stem Cells* 26:1883-1890.

Johnson, M. H. & C. A. Ziomek. (1981). Induction of polarity in mouse 8-cell blastomeres: Specificity, geometry, and stability. *Journal of Cell Biology* 91:303-308.

Kang, L., J. Wang, Y. Zhang, Z. Kou, & S. Gao. (2009). iPS cells can support full-term development of tetraploid blastocyst-complemented embryos. *Cell Stem Cell* 5:135-138.

Kasai, M., M. Nishimori, S. E. Zhu, T. Sakurai, & T. Machida. (1992). Survival of mouse morulae vitrified in an ethylene glycol-based solution after exposure to the solution at various temperatures. *Biology of Reproduction* 47:1134-1139.

Kasai, M., K. Niwa, & A. Iritani. (1983). Protective effect of sucrose on the survival of mouse and rat embryos stored at 0°C. *Journal of Reproduction and Fertility* 68:377-380.

Kaufman, M. H. & S. Webb. (1990). Postimplantation development of tetraploid mouse embryos produced by electrofusion. *Development* 110:1121-1132.

Kawase, Y., T. Iwata, M. Watanabe, N. Kamada, O. Ueda, & H. Suzuki. (2001). Application of the piezo-micromanipulator for injection of embryonic stem cells into mouse blastocysts. *Contemporary Topics in Laboratory Animal Science* 40:31-34.

Khosla, S., W. Dean, D. Brown, W. Reik, & R. Feil. (2001). Culture of preimplantation mouse embryos affects fetal development and the expression of imprinted genes. *Biology of Reproduction* 64:918-926.

Kirchain, S. M., A. M. Hayward, J. M. Mkandawire, P. Qi, & A. A. Burds. (2008). Comparison of tetraploid blastocyst microinjection of outbred Crl:CD1(ICR), hybrid B6D2F1/Tac, and inbred C57BL/6NTac embryos for generation of mice derived from embryonic stem cells. *Comparative Medicine* 58:145-150.

Kondoh, G., Y. Yamamoto, K. Yoshida, Y. Suzuki, S. Osuka, Y. Nakano, T. Morita, & J. Takeda. (1999). Easy assessment of ES cell clone potency for chimeric development and germ-line competency by an optimized aggregation method. *Journal of Biochemical and Biophysical Methods* 39:137-142.

Kraus, P., G. Leong, V. Tan, X. Xing, J. W. Goh, S. P. Yap, & T. Lufkin. (2010). A more cost effective and rapid high percentage germ-line transmitting chimeric mouse generation procedure via microinjection of 2-cell, 4-cell, and 8-cell embryos with ES and iPS cells. *Genesis* 48:394-399.

Kuleshova, L. L., D. R. MacFarlane, A. O. Trounson, & J. M. Shaw. (1999). Sugars exert a major influence on the vitrification properties of ethylene glycol-based solutions and have low toxicity to embryos and oocytes. *Cryobiology* 38:119-130.

Kunath, T., D. Arnaud, G. D. Uy, I. Okamoto, C. Chureau, Y. Yamanaka, E. Heard, R. L. Gardner, P. Avner, & J. Rossant. (2005). Imprinted X-inactivation in extra-embryonic endoderm cell lines from mouse blastocysts. *Development* 132:1649-1661.

Lee, K. H., H. R. Chang, H. W. Wang, C. J. Lin, S. C. Wu, & C. F. Tu. (2003). Natural farrowed fertile mice derived from green fluorescent embryonic stem cells. *Journal of the Chinese Society of Animal Science* 32(4):91 (Abstr.)

Lee, K. H., C. k. Chuang, H. W. Wang, L. Stone, C. H. Chen, & C. F. Tu. (2007). An alternative simple method for mass production of chimeric embryos by coculturing denuded embryos and embryonic stem cells in Eppendorf vials. *Theriogenology* 67:228-237.

Lee, K. H., C. k. Chuang, S. F. Guo, & C. F. Tu. (2011). Simple and efficient derivation of mouse embryonic stem cell lines using differentiation inhibitors or proliferation stimulators. *Stem Cells and Development* (In Press; DOI: 10.1089/scd.2011.0021)

Leibo, S. P. (1980). Water permeability and its activation energy of fertilized and unfertilized mouse ova. *Journal of Membrane Biology* 53:179-188.

Li, X., Y. Yu, W. Wei, J. Yong, J. Yang, J. You, X. Xiong, T. Qing, & H. Deng. (2005). Simple and efficient production of mice derived from embryonic stem cells aggregated with tetraploid embryos. *Molecular Reproduction and Development* 71:154-158.

Lin, C. J., T. Amano, J. Zhang, Y. E. Chen, & X. C. Tian. (2010). Acceptance of embryonic stem cells by a wide developmental range of mouse tetraploid embryos. *Biology of Reproduction* 83:177-184.

Martin, G. R. (1981). Isolation of a pluripotent cell line from early mouse embryos cultured in medium conditioned by teratocarcinoma stem cells. *Proceedings of the National Academy of Sciences USA* 78:7634-7638.

Mintz, B. (1962). Formation of genotypically mosaic mouse embryos. *American Zoologist* 2:432 (Abstr.)

Moustafa, L. A. & R. L. Brinster. (1972). Induced chimaerism by transplanting embryonic cells into mouse blastocysts. *Journal of Experimental Zoology* 181:193-201.

Nagy, A., M. Gertsenstein, K. Vintersten, & R. Behringer. (2003). Production of Chimeras. In: *Manipulating the Mouse Embryo, a Laboratory Manual.* A. Nagy, M. Gertsenstein, K. Vintersten, & R. Behringer, (Eds.), pp. 453-506. Cold Spring Harbor Lab. Press, ISBN 0-87969-574-9, New York, USA.

Nagy, A., E. Gocza, E. M. Diaz, V. R. Prideaux, E. Ivanyi, M. Markkula, & J. Rossant. (1990). Embryonic stem cells alone are able to support fetal development in the mouse. *Development* 110:815-821.

Nagy, A., J. Rossant, R. Nagy, W. bramow-Newerly, & J. C. Roder. (1993). Derivation of completely cell culture-derived mice from early-passage embryonic stem cells. *Proceedings of the National Academy of Sciences USA* 90:8424-8428.

Nagy, A., K. Nagy, & M. Gertsenstein. (2010). Production of mouse chimeras by aggregating pluripotent stem cells with embryos. In: *Methods in Enzymology, Vol. 476: Guide to Techniques in Mouse Development, Part A: Mice, Embryos, and Cells,* 2nd Ed. P. Wassarman (Ed.). pp. 123-149. Academic Press, ISBN 9780123747754, USA.

Nicolson, G. L., R. Yanagimachi, & H. Yanagimachi. (1975). Ultrastructural localization of lectin-binding sites on the zonae pellucidae and plasma membranes of mammalian eggs. *Journal of Cell Biology* 66:263-274.

Ohta, H., Y. Sakaide, K. Yamagata, & T. Wakayama. (2008). Increasing the cell number of host tetraploid embryos can improve the production of mice derived from embryonic stem cells. *Biology of Reproduction* 79:486-492.

Papaioannou, V. & R. Johnson. (1993). Production of chimeras and genetically defined offspring from targeted ES cells. In: *Gene Targeting, a Practical Approach.* A. L. Joyner, (ed.), pp. 107-146. IRL Press, Oxford Univ. Press, ISBN 0-19-963406-8, Oxford, UK.

Papaioannou, V. & R. Johnson. (2000). Production of chimeras by blastocyst and morula injection of targeted ES cells. In: *Gene Targeting, a Practical Approach.* 2nd Ed. A. L. Joyner, (ed.), pp. 133-175. Oxford Univ. Press, ISBN 0-19-963792-X, Oxford, UK.

Papaioannou, V. E. & F. Dieterlen-Lievre. (1984). Making Chimeras. In: *Chimeras in Developmental Biology*. N. Douarin & A. McLaren, (eds.), pp. 3-37. Academic Press, ISBN 0-12-440580-0, USA.

Pluck, A. & C. Klasen. (2009). Generation of chimeras by morula aggregation. *Methods in Molecular Biology* 561:219-229.

Poueymirou, W. T., W. Auerbach, D. Frendewey, J. F. Hickey, J. M. Escaravage, L. Esau, A. T. Dore, S. Stevens, N. C. Adams, M. G. Dominguez, N. W. Gale, G. D. Yancopoulos, T. M. DeChiara, & D. M. Valenzuela. (2007). F0 generation mice fully derived from gene-targeted embryonic stem cells allowing immediate phenotypic analyses. *Nature Biotechnology* 25:91-99.

Ramirez, M. A., R. Fernandez-Gonzalez, M. Perez-Crespo, E. Pericuesta, & A. Gutierrez-Adan. (2009). Effect of stem cell activation, culture media of manipulated embryos, and site of embryo transfer in the production of F0 embryonic stem cell mice. *Biology of Reproduction* 80:1216-1222.

Repentigny, Y. & R. Kothary. (2010). Production of mouse chimeras by injection of embryonic stem cells into the perivitelline space of one-cell stage embryos. *Transgenic Research* 19:1137-1144.

Saburi, S., S. Azuma, E. Sato, Y. Toyoda, & C. Tachi. (1997). Developmental fate of single embryonic stem cells microinjected into 8-cell-stage mouse embryos. *Differentiation* 62:1-11.

Schoonjans, L., V. Kreemers, S. Danloy, R. W. Moreadith, Y. Laroche, & D. Collen. (2003). Improved generation of germline-competent embryonic stem cell lines from inbred mouse strains. *Stem Cells* 21:90-97.

Schwenk, F., B. Zevnik, J. Bruning, M. Rohl, A. Willuweit, A. Rode, T. Hennek, G. Kauselmann, R. Jaenisch, & R. Kuhn. (2003). Hybrid embryonic stem cell-derived tetraploid mice show apparently normal morphological, physiological, and neurological characteristics. *Molecular Cell Biology* 23:3982-3989.

Shimada, H., T. Kaname, M. Suzuki, Y. Hitoshi, K. Araki, T. Imaizumi, & K. Yamamura. (1999). Comparison of ES cell fate in sandwiched aggregates and co-cultured aggregates during blastocyst formation by monitored GFP expression. *Molecular Reproduction and Development* 52:376-382.

Snow, M. H. (1975). Embryonic development of tetraploid mice during the second half of gestation. *Journal of Embryology and Experimental Morphology* 34:707-721.

Takahashi, K. & S. Yamanaka. (2006). Induction of pluripotent stem cells from mouse embryonic and adult fibroblast cultures by defined factors. *Cell* 126:663-676.

Tarkowski, A. K. (1961). Mouse chimaeras developed from fused eggs. *Nature* 190:857-860.

Tarkowski, A. K., A. Witkowska, & J. Opas. (1977). Development of cytochalasin B-induced tetraploid and diploid/tetraploid mosaic mouse embryos. *Journal of Embryology and Experimental Morphology* 41:47-64.

Tarkowski, A. K. & J. Wroblewska. (1967). Development of blastomeres of mouse eggs isolated at the 4- and 8-cell stage. *Journal of Embryology and Experimental Morphology* 18:155-180.

Tesar, P. J., J. G. Chenoweth, F. A. Brook, T. J. Davies, E. P. Evans, D. L. Mack, R. L. Gardner, & R. D. G. McKay. (2007). New cell lines from mouse epiblast share defining features with human embryonic stem cells. *Nature* 448:196-199.

Tokunaga, T. & Y. Tsunoda. (1992). Efficacious production of viable germ-line chimeras between embryonic stem (ES) cells and 8-cell stage embryos. *Development, Growth and Differentiation* 34:561-566.

Ueda, O., N. Kamada, K. Jishage, S. Uchida, & H. H.Suzuki. (1995). Factors affecting the efficiency of chimera production by coculture of zona-free embryos with frozen-thawed embryonic stem cells in mice. *Journal of Reproduction and Development* 41:181-186.

Wakayama, T., I. Rodriguez, A. C. F. Perry, R. Yanagimachi, & P. Mombaerts. (1999). Mice cloned from embryonic stem cells. *Proceedings of the National Academy of Sciences USA* 96:14984-14989.

Wang, Z. Q., F. Kiefer, P. Urbanek, & E. F. Wagner. (1997). Generation of completely embryonic stem cell-derived mutant mice using tetraploid blastocyst injection. *Mechanisms of Development* 62:137-145.

Wang, Z. & R. Jaenisch. (2004). At most three ES cells contribute to the somatic lineages of chimeric mice and of mice produced by ES-tetraploid complementation. *Developmental Biology* 275:192-201.

Wood, S. A., W. S. Pascoe, C. Schmidt, R. Kemler, M. J. Evans, & N. D. Allen. (1993a). Simple and efficient production of embryonic stem cell-embryo chimeras by coculture. *Proceedings of the National Academy of Sciences USA* 90:4582-4585.

Wood, S. A., N. D. Allen, J. Rossant, A. Auerbach, & A. Nagy. (1993b). Non-injection methods for the production of embryonic stem cell-embryo chimaeras. *Nature* 365:87-89.

Zhao, X. Y., W. Li, Z. Lv, L. Liu, M. Tong, T. Hai, J. Hao, C. l. Guo, Q. W. Ma, L. Wang, F. Zeng, & Q. Zhou. (2009). iPS cells produce viable mice through tetraploid complementation. *Nature* 461:86-90.

12

Embryonic Stem Cells in Toxicological Studies

Carmen Estevan, Andrea C. Romero, David Pamies,
Eugenio Vilanova and Miguel A. Sogorb
Universidad Miguel Hernández de Elche
Spain

1. Introduction

1.1 Toxicity assays

For most of their lives, humans are exposed to a variety of chemicals from different sources. The characterization of chemicals' toxicity implies the necessity to have available reliable testing methods to assess their capability to produce adverse effects on living organisms. Thus, it is necessary to acquire a battery of relevant internationally agreed methods to be employed by governments, industry and independent laboratories to assess the safety of chemicals. Presently, a set of validated tests is widely used to analyze the toxicity of chemicals and pharmaceutical drugs and to ensure their safety, and is recognized for this purpose (OECD, 2008; Stummann et al., 2009).

OECD Guidelines require *in vivo* studies to comply with the human health risk assessment process before registering and authorizing the use of chemicals. Toxicity guidelines characterize the adverse effects against the target organism that are attributable to exposure to the tested chemical with two main purposes in mind: the hazard identification and risk assessment deriving from exposure to the assessed chemical (Estevan et al., 2011).

Currently, the OECD has at least one *in vivo* validated guideline for testing the following endpoints: skin sensitization, skin and eye irritation and corrosion, mutagenicity, acute toxicity, target organ systemic toxicity (repeated daily dose for 28 days, 90 days, and for more than 12 months), carcinogenicity and toxicity to reproduction (screening, teratogenicity and two generations). All these OECD Guidelines have to be usually applied for a safe and consistent risk assessment, even when the chemical's hazard is not related to the tested target system. In this way, some of the most relevant tests are reproduction/developmental studies because the endpoints assessed in these protocols and the non-observed-adverse-effect levels (NOAELs) obtained are typically taken into account not only for reproductive toxicity testing, but also for the risk assessment of general systemic effects. For example, the NOAELs of systemic maternal toxicity in the teratogenicity study are frequently considered for risk characterization in short-term exposure scenarios.

Guidelines for testing toxicity to reproduction are also highly relevant because the effects of chemicals or drugs on germ cells or early embryos may lead to infertility or impaired development of pre-implantation embryos, and might result in embryotoxic or teratogenic effects in the progeny (Spielmann, 2005; Krtolica et al., 2009). This is the reason why toxicity to reproduction is required in all government regulations for the authorization and regulation of most chemicals. For example, extensive screening and multi-generation studies (including teratogenicity, fertility and development) are required for registering agrochemicals.

1.2 Necessity of cellular assays in toxicology

It is obvious that the application of an *in vivo* test is very expensive, time-consuming and involves lots of ethical concerns relating to the use of animals for experimentation purposes. Nevertheless, we now go on to illustrate these concerns with some figures. Höfer and coworkers (2004) estimated the number of animals needed to apply each *in vivo* test. It can be easily concluded that the number of animals needed to test a single chemical with all the above-mentioned available tests is around 4,710. Fleischer (2007) also estimated that the cost of this set of tests would be around € 1,800,000.

Apart from this, the relevance of many *in vivo* toxicological assays is questionable as regards human health because there are times when the animal model is not representative because of inter-species differences in pharmacokinetics and toxicokinetics (Wobus and Löser, 2011). Such inter-species differences have been detected, for example, in the cases of 13-cis retinoic acid (Hendrickx, 1998) and thalidomide (Nau, 1993; Tzimas et al., 1994). Specifically, the dramatic consequences of the teratogenic effects of thalidomide on humans urgently lead to the need for new, optimized human-specific test systems.

Therefore due to the aforementioned economical, logistical and ethical concerns, it is obvious that the availability of fast, safe, reliable and high throughput alternatives to animal experimentation methods would be extremely welcomed by industry, researchers and regulatory bodies. Cellular methods fit all these requirements and, due to the availability of cellular cultures for all tissues, these methods are good candidates to be employed for testing toxicity in all the possible target organ systems (liver, skin, heart, reproduction, etc.).

1.3 Roles of stem cells in toxicological studies

Stem cells may play several relevant roles in cellular assays for testing toxicity, and these roles can be played in either embryotoxicity or in organogenesis for testing toxicity in adult tissues (Figure 1). Indeed, embryonic stem cells (ESC) or reprogrammed-induced pluripotent stem cells (iPSC) can be employed through forced directed differentiation protocols as a source of adult human cells cultures (hepatocytes, neurons, and others) that would be further used for the *in vitro* testing of respective target organ toxicity (hepatotoxicity, neurotoxicity, etc.). This application is useful because it avoids the sacrifice of animals for experimentation purposes, allows the use of human cells which usually have a limited availability and avoids the need for the above-mentioned interspecies extrapolation. This application of stem cells related with organogenesis will be commented and explained in more detail in another section of this chapter. Nonetheless, it is relevant to remark that the recent iPSC applications in toxicology and drug research provide new alternatives to the standard routine tests performed by industry and offer new chemical safety assessment strategies (Laustriat et al., 2010; Trosko and Chang, 2010).

The other role of stem cells in toxicological studies relates to embryotoxicity. Indeed, the testing methods relating to these applications take advantage of ESCs to differentiate in adult tissues and help the studies into how these differentiation processes are altered by exposure to the tested chemical (Figure 1). The most important and relevant factor for embryotoxicity testing is that ESCs mimic early *in vivo* embryonic development processes and show tissue-specific expression profiles (Wobus and Löser, 2011). In combination with three-dimensional cultures supported by extracellular matrix proteins, stem cell-based systems can mimic the microenvironment of the *in vivo* niche, which is especially relevant in humans for testing embryotoxicity (Trosko and Chang, 2010) and for organogenesis (Miki et al., 2011; Lock and Tzanakakis., 2009) and further assay of toxicity on the generated cell cultures.

Developmental toxicity assays with stem cells occupy a prominent position within the available battery of cellular assays since some studies have demonstrated that stem cells can be used to understand the processes underlying organogenesis, as shown in the case of the heart (Miller et al., 2008). Therefore, the study of the interferences in stem cell differentiation caused by chemicals can be used to detect potential developmental toxicants. An additional advantage of these models is that genetic molecular approaches enable not only for an exhaustive analysis of which genes are affected by each substance, but also for the subsequent research into the mechanisms of action underlying the teratogenic or embryotoxic effects caused by the assessed chemical (Pamies et al., 2011).

Details of ESCs-based embryotoxicity testing methods will be provided in the following sections of this chapter, but can be divided into four categories (Table 1): the validated Embryonic Stem cell Test (EST); variations of the EST using molecular endpoints as indicators of altered differentiation; the ACDC method, and human cell-based embryotoxicity methods.

Methods	References
EST	ESAC, 2002; Glenschow et al., 2000, 2002 and 2004
Variations of validated EST with molecular end-points	Bremer et al., 2001; Buesen et al., 2009; Osman et al., 2010; Romero-Lucena., 2010; Seiler et al., 2004; Stummann et al., 2007; van Dartel et al., 2009; 2010a, 2010b, 2011a, 2011b
ACDC	Barrier et al., 2010; Jeffay et al., 2010
Human cell-based embryotoxicity	Addler et al., 2008a, 2008b

Table 1. Embryotoxicity testing methods employing embryonic stem cells

The cellular methods for testing embryotoxicity deserve special mention because, despite the validated *in vivo* OECD guidelines for testing toxicity to reproduction covering fertility, teratogenicity and development, there are no specific guidelines available for exclusively testing embryotoxicity (Pamies et al., 2011). Indeed, this is a major gap because a guideline for this purpose would allow the detection of developmental toxins in early development stages without awaiting teratogenicity.

There are other well-established alternative methods for testing embryotoxicity with endpoints relating to differentiation, such as the frog embryo teratogenesis assay on xenopus (Bantle et al., 1990), the chicken embryo toxicity screening test on chicken embryos (Jelinek et al., 1985), the micromass assay using mouse embryonic mesenchymal cells (Flint, 1993; Spielmann, 2005), and the mammalian whole embryo culture assay using mouse (Sadler et al., 1982) or rat (Schmidt, 1985; Cockroft and Steele, 1987) embryos. Nevertheless, all these methods are not based on stem cells and are not of interest in this book.

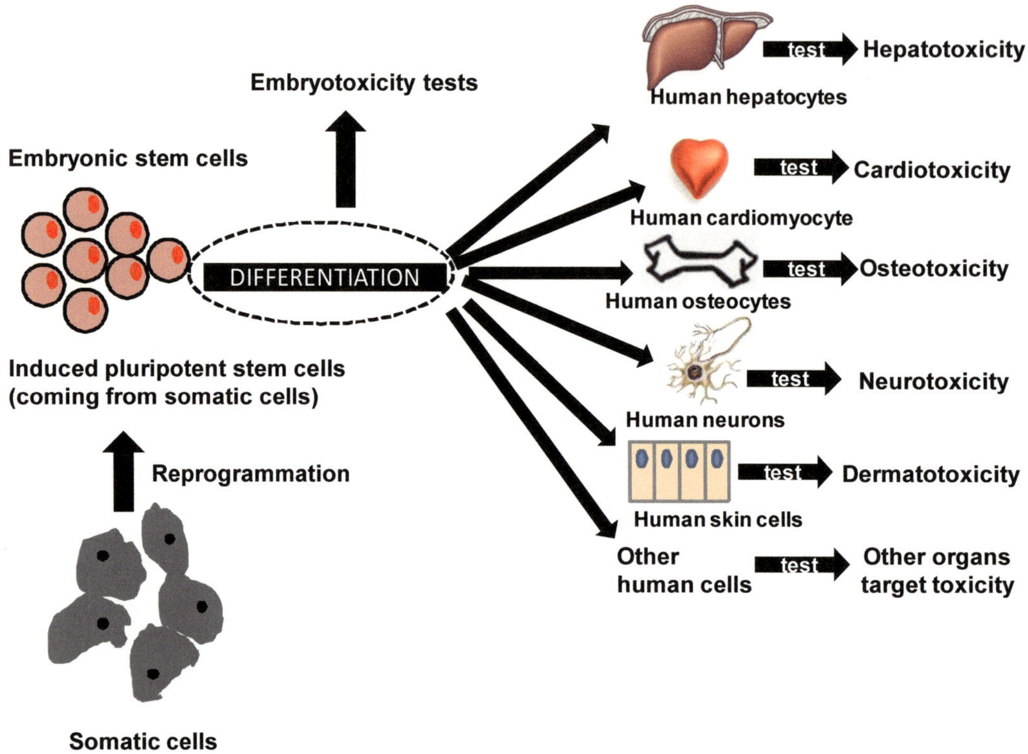

Fig. 1. Role of stem cells in toxicity testing.

1.4 Earlier toxicity assays with stem cells

Some of the earliest attempts for testing cytotoxicity, embryotoxicity or teratogenicity screening using mammalian cells date back to between two and three decades (Kimmel et al., 1982; Schwetz et al., 1991). The most important drawback of all these assays is that they rely on somatic cells, which do not entirely reflect the reaction of embryonic cells to toxic compounds (Wobus and Löser, 2011).

Embryotoxicity tests include the analysis of the toxic effects of chemicals on embryonic cells in differentiation, whereas cytotoxicity assays measure the chemical-induced alterations of metabolic pathways or the structural integrity of cells, which may or may not be directly related to cell differentiation (Wobus and Löser, 2011). Both cytotoxicity and embryotoxicity assays have been applied with pluripotent stem cells. The determination of cytotoxic effects of chemicals based on the colorimetric MTT test for cellular growth and survival (Mosmann, 1983) was performed with murine ESCs (Laschinski et al., 1991), showing greater sensitivity to selected embryotoxic agents than mouse adult cells (fibroblasts).

In order to overcome the limitations of cytotoxicity assays to determine embryotoxic effects, the ESC-derived embryoid body (EB) model was introduced (Wobus et al., 1988; Wobus et al., 1991). The ESCs differentiated in EBs for 5-7 days develop *in vitro* into the progenitor cells of all three primary lineages of ectoderm, mesoderm and endoderm (Wobus and Boheler, 2005). The fundamentals of the EB model were that early developmental processes simulated by EB differentiation would be affected by exposure to toxic chemicals and, consequently, specific differentiation processes would be inhibited or accelerated.

One of the first experiences to illustrate how embryonic cells could be used for testing embryotoxicity was the case of retinoic acid. There is evidence that retinoic acid *in vivo* acts as morphogen by forming a concentration-dependent gradient along embryonic axes, thus affecting the antero-posterior patterning of the body axis and limbs (Kessel and Gruss, 1991). High concentrations of retinoic acid applied to mouse ESCs during early EB differentiation (days 0-2) revealed increased neural-specific transcript levels and significantly induced neuronal differentiation (Fraichard et al., 1995; Strubing et al., 1995), whereas retinoic acid application at lower concentrations applied between days 2 and 5 induced mesodermal, specifically skeletal muscle and partially cardiac differentiation (Wobus et al., 1994). This is in line with the developmental processes in the EB where the maximum expression levels of neural plate morphogenesis genes like *brachyury* and the expression of genes associated with dorsal mesoderm specification were observed at around day 3 of EB formation (Wobus et al., 1994; Dani et al., 1997). Cardiac and vascular smooth muscle cells both originated (at least partially) from the lateral plate mesoderm, and were induced by retinoic acid when applied on days 5-7 and 7-11, respectively (Drab et al., 1997). Obviously, the specific temporal response of the different cell types within the EB to retinoic acid correlated with the retinoic acid receptor gene expression (Rohwedel et al., 1999). These early studies provided the basis to subsequently establish the so-called EST method.

2. The embryonic stem cell method test (EST)

In the past, several groups have used murine ESC to establish an *in vitro* embryotoxicity assay. Laschinsky et al. (1991) compared cytotoxicity in both ESC and mouse fibroblasts to assess the embryotoxic potential of teratogenic agents. The data showed that ESC were more sensitive to toxic agents than adult cells. Newall and Beedles (1994) measured both the cytotoxicity and the colony-forming potential of ESC after 7 days of culture in the presence of teratogenic agents. In these assays, only a few embryotoxic agents could be correctly classified. This was probably due to the fact that only two endpoints were selected for the biostatistical assessment in the ESC assays, which seems insufficient. To overcome the limitations of the previously mentioned ESC tests, a third endpoint value, alterations in cell differentiation, was introduced and linked to the two former ones (cytotoxicity in embryonic and adult cells). These three endpoints allow the test compounds to be classified into three *in vivo* embryotoxicity categories.

2.1 Basis of the method

The fundamentals for the test are that the *in vitro* tests of basal cytotoxicity are sufficiently predictive for the rodent *in vivo* LD_{50} assay (Ekwall, 1999; Spielmann et al., 1999) and that ESCs show alterations in their *in vitro* differentiation pattern when exposed to embryotoxic chemicals during EB differentiation (Wobus et al., 1994). The EST method is based on the potential of D3 mouse ESC to differentiate into beating cardiomyocytes and in determining how this differentiation is altered by exposure to chemicals, together with the cytotoxic insults in these cells and in a mouse adult cells model.

Two permanent mouse cell lines are used: D3 embryonic stem cells to represent embryonic tissue and 3T3 fibroblasts to represent adult tissue. The test was developed when D3 mouse ESC were discovered to be able to form EBs in the absence of the cytokine leukemia inhibiting factor which, after 10 days of culture, spontaneously differentiated into cardiomyocytes.

The basis of the EST protocol is summarized in Figure 2. The test comprised the morphological analysis of beating clusters in differentiating EBs (seeded in single wells of multiwell plates) and the determination of those concentrations of the test substance at which cardiac differentiation was inhibited by 50% after 10 days of exposure (ID_{50}). The cytotoxic effects on D3 and 3T3 cells were estimated by determining those concentrations of the test substance at which proliferation was inhibited by 50% after 10 days of exposure, yielding IC_{50} D3 and IC_{50} 3T3, respectively. On the basis of these 3 endpoints, a biostatistical prediction model formed by three different functions helped assign the tested chemical to one embryotoxicity category (strong, weak or non embryotoxic) (Genschow et al., 2000; Genschow et al., 2002; Spielmann et al., 2001) (Figure 1). This method is currently validated by the European Centre for Validation of Alternative Methods (ECVAM) as a screening assay for potentially embryotoxic chemicals (ESAC, 2002).

2.2 EST performance

The testing of the 20 test chemicals employed in the EST validation study provided 78% accuracy (correct classifications related to the *in vivo* data) (Genschow et al., 2004). The highest precisions were detected for weak and strong embryotoxic chemicals, which were correctly detected in 83% and 81% of cases, respectively (Genschow et al., 2000). The poorest precision was recorded for the detection of non embryotoxic compounds (70%). Finally, the predictability for strongly embryotoxic chemicals was 100%, while that for non- and weakly embryotoxic compounds was similar (70 and 72%, respectively), which is considered sufficiently high.

2.3 Limitations of EST

EST has several limitations. Since D3 mouse ESCs are differentiated as EBs in "hanging drops", this test is laborious and time-consuming. EST does not consider molecular endpoints to detect cardiac differentiation (formation of beating cardiomyocytes) (Schmidt et al., 2001; Piersma, 2004), and it is an assessment based exclusively on histological considerations which introduces a high degree of variability. The inclusion of differentiation markers of the ectoderm, mesoderm and endoderm lineages and the analysis of the tissue-specific gene expression have been recommended as additional endpoints for the purpose of improving EST's predictability on the basis of better differentiation assessment.

As with most *in vitro* methods, the EST is limited by lack of metabolic competence. This lack has been recognized as another potential factor to affect the EST's experimental outcome and the correct classification of tested substances (Verwei et al., 2006). Experiments which attempted to add the S9 mix, the mix usually applied as a source of exogenous metabolic activation enzymes, failed because it is toxic for D3 mouse ESC. Therefore, the consortium of the ReProTect project (Integrated Project of the European Union funded within the 6th Framework Programme, including 35 partners focused on alternatives to animal tests for reproductive toxicity with the scope of pre-validating and validating the brightest ones) recommended other test systems to analyze the potential toxicity of those compounds that require metabolic activation (Marx-Stoelting et al., 2009). Other relevant limitation of the EST is that this is also incapable of detecting teratogenic chemicals with mechanisms acting beyond the initial embryo differentiation steps.

This method needs an intermediate level of technical difficulty because it requires two endpoints and the cultures of two different cell lines. In addition, it does not yield information about the morphological alterations caused by the teratogen as other methods do (e.g., whole embryo culture) (Pamies et al., 2011).

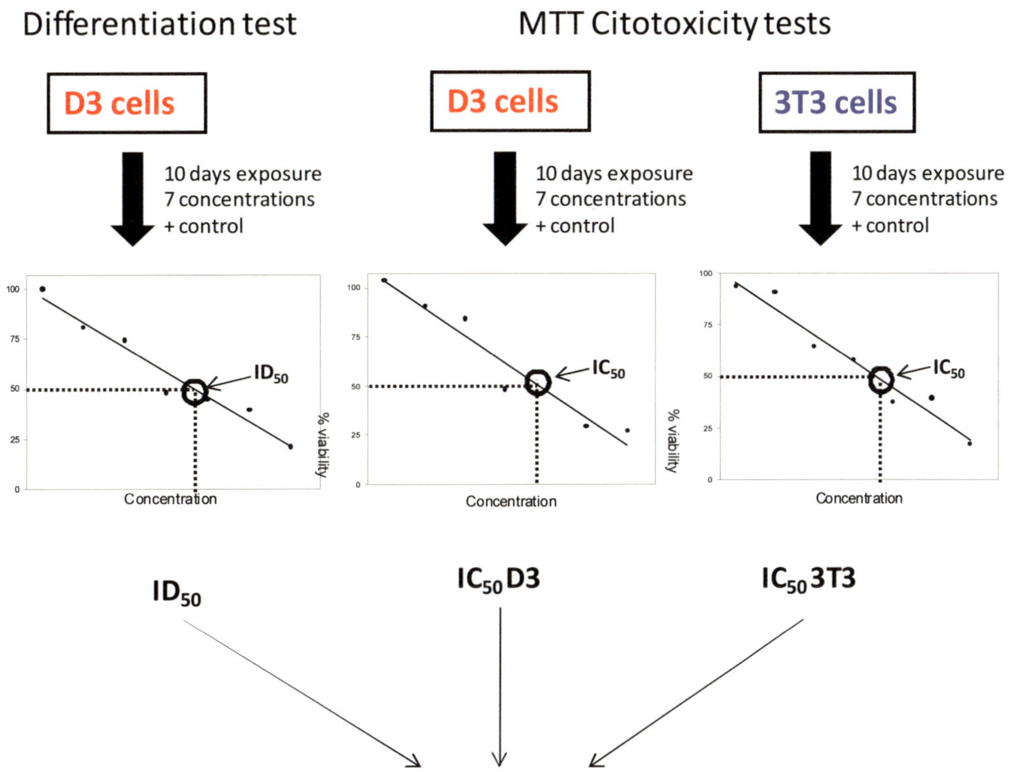

Linear discriminant functions I, II and III for classification

I: $5.916 \log(IC_{50}\,3T3) + 3.500 \log(IC_{50}\,D3) - 5.307\,[(IC_{50}\,3T3 - ID_{50}) / IC_{50}\,3T3] - 15.27$

II: $3.651 \log(IC_{50}3T3) + 2.394 \log(IC_{50}\,D3) - 2.033\,[(IC_{50}\,3T3 - ID_{50}) / IC_{50}\,3T3] - 6.85$

III: $-0.125 \log(IC_{50}3T3) - 1.917 \log(IC_{50}D3) + 1.500\,[(IC_{50}3T3 - ID_{50}) / IC_{50}\,3T3] - 2.67$

Classification criteria

If I>II and I>III	**Class 1**	**Not embryotoxic**
If II>I and II>III	**Class 2**	**Weak embryotoxic**
If III>I and III>II	**Class 3**	**Strong embryotoxic**

Fig. 2. Embryonic Stem cell test (EST) method.

3. Proposed variations of the EST based on molecular endpoints

As stated in the previous section, the EST has several limitations, one of which is the most relevant: lack of molecular endpoints for a reliable assessment of the degree of differentiation achieved by EBs. Indeed, ECVAM has recommended the inclusion of molecular endpoints in the EST with a view to improving its performance (Spielman et al., 2006). A great deal of work has been done to follow this recommendation in an attempt to introduce gene expression as biomarkers of differentiation. A second line of work has also been introduced, that of reducing the technical complexity of the EST using a single cell line.

3.1 Use of biomarkers of differentiations as endpoints

Previous studies about lineage-dependent effects of retinoic acid on ESC differentiation (Wobus et al., 1994) and the application of reporter gene constructs controlled by cardiac-specific promoters (Kolossov et al., 1998) have demonstrated that an ESC differentiation model is suitable to be analyzed at a cellular level. Genes coding for tissue-specific proteins are expressed in the course of ESC differentiation in a pattern which closely resembles the time pattern observed during mouse embryogenesis (Rohwedel et al., 2001). For example, cardiac-specific transcription factor Nkx2.5, expressed in cardiac precursor cells during embryogenesis *in vivo,* is the first cardiac-specific gene expressed during EB development in parallel to the α1 subunit of the L-type Ca^{2+} channel (Fässler et al., 1996), followed by the expression of α- and β-myosin heavy chain isoforms.

The recent progress made in improving the EST has underlined previous demands to implement new embryotoxicity testing strategies using ESCs by analyzing tissue-specific genes via the reporter gene expression, automated high-throughput screening for changes in gene and protein expression patterns using microchip arrays for transcriptome and proteome analyses, and the application of human ESCs (Rohwedel et al., 2001).

Reporter gene assays for developmental toxicity, specifically for cardiac toxicity (Bremer et al., 2001), have been included in the EST. Cardiac markers α-myosin heavy chain and α-actin (both quantified by fluorescence-activated cell sorting (FACS)) have been successfully employed instead of the microscopic observation of beating EBs to determine cardiac differentiation after exposing D3 EBs to model embryotoxic test substances (Seiler et al., 2004; Seiler et al., 2006). The FACS-based EST has been successfully adopted for assessing developmental toxicity (Buesen et al., 2009). While the molecular FACS-EST shows the same sensitivity as the validated EST for the classification of chemicals, test duration is reduced and almost identical ID_{50} values were obtained for ten representative compounds of the three classes, indicating that FACS analyses values can serve as a new EST toxicological endpoint.

Other potential biomarker genes of embryotoxicity that may be used in early development stages are *Pnpla6, a-fetoprotein, nestin and Vgfa*. All these genes displayed statistically significant reductions in their respective expressions after 5 days of exposure to a non cytotoxic dose of 50 ng/ml of embryotoxic 5-fluorouracil (Table 2) (Romero, 2010). These marker genes also offer an advantage over the conventional EST method as they are recordable after 5 days of exposure instead of the 10 days of differentiation needed for obtaining a beating cardiomyocytes. Alpha-fetoprotein has also been independently reported as a very sensitive biomarker of exposure to 5-fluorouracil (Pamies et al., 2010).

Gene	%
Pnpla6	80±4
AFP	26±7
Nes	54±11
Vgfa	46±17

Table 2. Expression of several marker genes after 5 days exposure of D3 mouse embryonic stem cells in differentiation to 50 ng/ml embryotoxic 5-fluorouracil. The gene expression was indicated as a percentage as regards a differentiated control under the same conditions, but in the absence of 5-fluorouracil. In all cases, differences in expression as regards the control were statistically significant for at least $p<0.05$. (Data taken from Romero, 2010).

Inclusion of additional differentiation endpoints, specifically of the neural lineage, has been requested by researchers and authorities (Spielmann et al., 2006). The purpose of using such systems is to correctly classify the substances that were not identified as embryotoxic in the conventional EST. Inclusion of parameters for neuronal differentiation into the EST would allow, for example, classification in accordance with the *in vivo* data of methylmercury as a highly embryotoxic, whereas the conventional EST failed to correctly classify this compound (Stummann et al., 2007). In order to increase the number of molecular markers as endpoints of embryotoxicity testing, real-time Taqman RT-PCR analyses have been adopted for the EST in pilot studies (zur Nieden et al., 2004).

Peters and co-workers determined the relative embryotoxic potential by using a modified EST screening system (Peters et al., 2008). In this study, 12 compounds were investigated in a modified EST performed in 96-well plates. Test substances were applied on day 3 of culture for 10 days, and the assay did not involve the preparation of EBs. This newly revised high throughput EST allowed the analysis of a larger number of substances, less manual work, yet it yielded data comparable to those obtained with the conventional EST.

Present efforts to improve the EST now address the application of additional lineage-specific markers to define additional toxicological endpoints. Other than cardiac markers, markers of neuronal, bone and cartilage development are included. However, this will prolong the test duration from 7-10 to 30, or even 32 days (Marx-Stoelting et al., 2009). In the future, novel molecular endpoints and reporter-based systems will have to be included in the EST (Rohwedel et al., 2001; Marx-Stoelting et al., 2009; Spielmann, 2009). Such improvements will also be required to establish EST-like tests with human ESCs.

3.2 Use of transcriptomics and proteomics as endpoints

The identification of embryotoxic compounds on the basis of analyzing the alterations they cause in the expression of genes and proteins relating to the differentiation of embryonic stem cells has been proved possible (van Dartel et al., 2009; 2010a; 2010b; 2011a; 2011b; and Oshman et al., 2010). By following this methodology in a procedure that includes a 4-day assay (three days for embryonic body formation and one additional day to expose cells to the assessed chemical), the set of genes called "van_Dartel_heartdiff_24h" was able to correctly predict the embryotoxicity of 83% of the assessed chemicals (ten correct predictions of a total of 12 assessed chemicals, where two non embryotoxic chemicals were wrongly classified as embryotoxic) (van Dartel et al., 2011a). In the same study, the group of genes called "EST biomarker genes" correctly predicted the embryotoxicity of eight of the 12 assessed chemicals (67%). Here the mistakes corresponded to one non embryotoxic and three embryotoxic chemicals, respectively (van Dartel et al., 2011a).

Protein markers can also be used to detect exposure to embryotoxic chemicals. Osman and coworkers found that the embryonic bodies formed from day 3 of differentiation of D3 cells, which were further exposed to embryotoxic monobutyl phthalate for 24 hours, expressed 33 proteins in a differential way, including cardiomyocytes biomarkers (whose expression was repressed if compared with controls) and chromatin modulator enzymes (Osman et al., 2010).

3.3 The ACDC method

The Adherent Cell Differentiation and Cytotoxicity (ACDC) assay is a test to establish a model system to assess the chemical effect using a single cell culture (instead of two as the EST does) in order to improve feasibilities for throughput assays (Barrier et al., 2010). The ACDC assay uses quantitative markers for both the differentiation degree and cell proliferation. In this assay, pluripotent J1 mouse ESCs are plated in a 96-multiwell plate and further cultured in differentiation medium for 9 days. Afterward, each well is assessed for the cell number and differentiation to cardiomyocytes (using quantitative in-cell Western analysis for myosin heavy chain protein normalized with the cell number). This method has already proved suitable for testing the effects of haloacetic acids and their major metabolites (Jeffay et al., 2010), but has still not been validated.

4. Human stem cells for screening cytotoxicity and embryotoxicity

Development of human ESC-based *in vitro* systems for testing the embryotoxicity of chemicals implies significant progress. The use of human ESCs in *in vitro* methods would enhance their predictability and avoid problems associated with the interpretation of the results obtained with animal-based assays in a human context (Wobus and Löser, 2011). For example, species-specific differences between mouse and human pre-implantation development, as in DNA methylation, DNA repair and the expression of those genes involved in drug metabolism, may interfere with the correct interpretation of animal studies regarding their significance to humans (Krtolica et al., 2009). The use of human ESC-based test systems could avoid the incorrect classification of chemicals due to inter-species variations and would, consequently, improve consumers' safety.

Adler and co-workers (2008a) provided proofs that human ESCs are a relevant *in vitro* model for developmental toxicity testing. They studied the cytotoxic effects of well-known embryotoxicants on human ESC, human ESC-derived progenitors and human foreskin fibroblasts. As observed in the respective murine cells, all-trans retinoic acid and 13-cis retinoic acid had a stronger cytotoxic effect on pluripotent stem cells than on fibroblasts, while the mesenchymal progenitors deriving from human ESC displayed the strongest sensitivity to both compounds. While all-trans retinoic acid and 13-cis retinoic acid substances revealed comparable cytotoxic effects on human ESC in this study, only all-trans retinoic acid was seen to be cytotoxic in mouse ESC in previous analyses. This suggests the necessity of developing human ESC-based assays to assess human-specific developmental toxicity.

In a second study, Adler and co-workers arranged the test system according to the EST, but used human instead of mouse ESC (Adler et al., 2008b). The cytotoxic effects of the two well-known developmental toxicants, 5-fluorouracil and all-trans retinoic acid, on human ESCs and human fibroblasts were similar to those previously observed in mice systems. However, testing for the potential cytotoxic effects on pluripotent stem cells does not suffice to detect the developmental toxicity of chemicals that affect developmental processes (Rohwedel et al., 2001). These authors proposed markers of undifferentiated cells such as *Oct4, hTert and Dusp6*, as well as markers of neural plate morphogenesis and early cardiogenesis, such as *Brachyury* and *GATA-4* for monitoring cardiac differentiation, thus demonstrating that these markers may have the potential to serve as endpoints for developmental toxicity studies using a humanized EST (Adler et al., 2008b).
Despite the progress made, numerous problems have to be solved before human ESC-based assays can be implemented into routine procedures for the developmental toxicity testing of drugs and chemicals. These problems include the establishment of reliable and reproducible differentiation procedures that can be performed in a high-throughput format. The predictability, the sensitivity and specificity of the respective test systems for a wider panel of chemicals are still to be shown. In addition, very few substances have been tested in the different approaches. Furthermore, human ESC-based systems may have the capacity to predict human-specific embryotoxic effects which cannot be measured with mouse cell systems due to species-specific differences. However, the superiority of human ESC-based systems over existing animal *in vitro* tests for developmental toxicity has still to be shown and, in this context, the use of different human ESC lines would prove advantageous.

5. Induced pluripotent stem cells in toxicological studies

iPSC play a prominent role in their use as a tool in toxicity tests in the embryotoxicity field or as a source of cellular tissues for testing toxicity in adult tissues (Figure 1). These two applications are particularly interesting when used with human cells because they will eliminate uncertainties deriving from the necessary inter-species interpolation when animal cells are used.

5.1 Applications in embryotoxicity studies

The successful reprogramming of adult somatic cells into a ESC-like state through genetic manipulation (Yu et al., 2007; Takahashi et al., 2007) offers a new opportunity for toxicology assay development as reprogrammed somatic cells possess very similar characteristics to those of human ESC. Nevertheless, some questions arise as to the use of iPSC for toxicological assays. One is whether the different epigenetics of reprogrammed iPSC (Surani et al., 2008; Han and Sidhu, 2008) might influence responses to a toxic insult. The other challenge is that the derivation of iPSC involves permanent genetic modifications to somatic cells due to the use of the viral transduction of recombinant DNA (Yu et al., 2007; Takahashi et al., 2007). Integration of recombinant DNA into the genome of iPSC lines might lead these cells or their differentiated progenies to behave differently from normal human cells, particularly when exposed to toxic challenge.
The application of human iPSC in embryotoxicity testing might be a promising tool, as well as ESC (Heng et al., 2009). However on the basis of the recently reported inconveniences

caused by reprogramming, future work has to show whether human iPSC will be applicable and offer advantages over human ESC in the developmental toxicology field (Gore et al., 2011).

In the future, it is possible that iPSC lines for toxicology screening may also be derived from reprogramming primary explanted hepatocytes and cardiomyocytes because such somatic lineages have wide-ranging applications in both toxicology and pharmacology screenings (Heng et al., 2009).

5.2 Applications in organogenesis and target organ toxicity testing

The ability to generate iPSC from somatic cells by the forced expression of reprogramming factors Oct3/4 and Sox2, along with either Klf4 or Nanog and Lin28, raises the possibility of generating patient-specific cell types of all lineages. Differentiated cell types produced from patient's iPS cells have many potential therapeutic applications, including their use in tissue replacement and gene therapy (Si-Tayeb et al., 2010). They also provide a platform for drug toxicity screenings. Generation of hepatocytes from iPSC is a particularly appealing goal because this parenchymal cell of the liver is associated with several congenital diseases (Burlina, 2004) and is the main site of xenobiotic control.

Mouse iPSC were generated from fibroblasts (Si-Tayeb et al., 2010), and embryos were then produced from this iPSC tetraploid complementation using transgenic mice which ubiquitously express enhanced green fluorescent protein (EGFP). When embryos were generated from mouse iPSC, from which EGFP is absent, all the embryos, including their livers, lacked the EGFP expression except in extra embryonic tissues. A thorough examination of iPSC-derived embryos and their livers revealed that they appeared to be identical to controls. These livers were examined, and the expression of those proteins that are characteristic of specific cell types was identified to reveal that, like the control fetal livers, the iPSC-derived livers contained hepatocytes (HNF4a positive), endothelial cells (GATA4 positive), sinusoidal cells (LYVE1 positive) and Kuppfer cells/macrophage (F4/80 positive).

In the cardiogenesis field, the latest studies have revealed that mouse iPSC can differentiate into cardiomyocytes by EBs formation or through the use of collagen IV-coated dishes and feeder cells (Narazaki et al., 2008; Mauritz et al., 2008). However, it is unknown whether iPSC can differentiate into cardiomyocytes without EBs formation and without using collagen IV-coated dishes or feeder cells. In addition, myocardial cell differentiation efficiency has still not been determined as being uniform or diverse in different iPS cell lines (Kaichi et al., 2010).

The transcription factors, such as GATA4 and Nkx2.5, expressed in the lateral mesoderm play important roles in the following heart developmental processes (Molkentin et al., 1997; Kasahara et al., 1998). The activities of cardiac transcription factors are regulated, in part, by histone acetyltransferases and histone deacetylases (HDACs). In the treatment of ES cells with trichostatin A (TSA), a specific HDAC inhibitor has been reported which not only induces the acetylation of both GATA4 and histones, but facilitates their differentiation into cardiomyocytes (Kawamura et al., 2005).

In the very near future, it is expected that other cellular types other than hepatocytes and cardiocytes can be successfully obtained, as shown in Figure 1. These cell cultures would also be suitable for use in toxicity studies as platforms for testing *in vitro* toxicity on adult cells.

6. Conclusions and final remarks

Stem cells are an excellent platform for testing embryotoxicity *in vitro* on the basis of these cells' capacity to mimic the embryo formation process. Nevertheless, much effort and hard work is still needed to find good suitable endpoints for monitoring different differentiation stages. There is also the urgent need to go ahead with the characterization of those models employing human cells, either ESC or those from reprogrammed iPSC.
Stem cells also represent a good opportunity to generate adult tissues to test toxicity in such tissues, which again is especially relevant in humans. Nevertheless, the development of good differentiation protocols to allow the generation of a variety of cell types apart from hepatocytes and cardiocytes is still needed.

7. References

Adler, S., Lindqvist, J., Uddenberg, K., Hyllner, J. & Strehl, R. (2008a). Testing potential developmental toxicants with a cytotoxicity assay based on human embryonic stem cells. *Altern Lab Anim* 36:129–140.

Adler, S., Pellizzer, C., Hareng, L., Hartung, T. & Bremer, S. (2008b). First steps in establishing a developmental toxicity test method based on human embryonic stem cells. *Toxicol In Vitro* 22:200–211.

Bantle, J.A., Fort, D.J., Rayburn, J.R., DeYoung, D.J. & Bush, S.J. (1990). Further validation of FETAX: evaluation of the developmental toxicity of five known mammalian teratogens and non-teratogens. *Drug Chem Toxicol* 13:267–282.

Barrier, M., Jeffay, S., Nichols, H. & Hunter, S. (2010). Evaluation of a mouse embryonic stem cell adherent cell differentiation and cytotoxicity (ACDC) assay. *The Toxicologist* 114: 358 - 9.

Bremer, S., Worth, A.P., Paparella, M., Bigot, K., Kolossov, E., Fleischmann, B.K., Hescheler, J. & Balls, M. (2001). Establishment of an in vitro reporter gene assay for developmental cardiac toxicity. *Toxicol In Vitro* 15:215–223.

Buesen, R., Genschow, E., Slawik, B., Visan, A., Spielmann, H., Luch, A. & Seiler, A. (2009). Embryonic stem cell test remastered: comparison between the validated EST and the new molecular FACS-EST for assessing developmental toxicity in vitro. *Toxicol Sci* 108:389–400.

Burlina, A.B. (2004). Hepatocyte transplantation for inborn errors of metabolism. *J Inherit Metab Dis* 27:373–383.

Cockroft, D.L. & Steele, C.E. (1987). Postimplantation Embryo Culture and its Application to Problems in Teratology, In: *In vitro methods in toxicology*, C.K. Atterwill, C.E. Steele (Ed.), Cambridge University Press, ISBN: 978-0-52-132684-1, Cambridge.

Dani, C., Smith, A.G., Dessolin, S., Leroy, P., Staccini, L., Villageois, P., Darimont, C. & Ailhaud, G. (1997). Differentiation of embryonic stem cells into adipocytes in vitro. *J Cell Sci* 110(Pt 11):1279–1285.

Drab, M., Haller, H., Bychkov, R., Erdamnn B., Lindschau, C., Haase, H., Morano, I., Lufth F.C. & Wobus, A.M. (1997). From totipotent embryonic stem cells to spontaneously contracting smooth muscle cells: a retinoic acid and db-cAMP in vitro differentiation model. *FASEB J* 11:905–915.

Ekwall, B. (1999). Overview of the final MEIC results: II. The in vitro- in vivo evaluation, including the selection of a practical battery of cell tests for prediction of acute lethal blood concentrations in humans. *Toxicol In Vitro* 13:665–673.

ESAC (European Centre for Validation of Alternative Methods (ECVAM) Scientific Advisory Committee) (2002). The Use of Scientifically-Validated *in Vitro* Tests for Embryotoxicity. Available at: http://ecvam.jrc.ec.europa.eu/

Estevan, C., Pamies, D., Sogorb, M.A. & Vilanova, E. (2011). OECD guidelines and validated methods for in vivo testing of reproductive toxicity. In: *Reproductive and developmental toxicology*, R.C. Gupta (Ed.), Academic Press, ISBN 978-0-12-382032-7, Hopkinsville.

Fässler, R., Rohwedel, J., Maltsev, V., Bloch, W., Lentini, S., Guan, K., Gullberg, D., Hescheler, J., Addicks, K. & Wobus, A.M. (1996). Differentiation and integrity of cardiac muscle cells are impaired in the absence of β1 integrin. *Journal of Cell Science* 109: 2989–2999.

Fleischer, M. (2007). Testing costs and testing capacity according to the REACH requirements - results of a survey of independent and corporate GLP laboratories in the EU and Switzerland. *J Bus Chem* 4: 96-114.

Flint, O.P. (1993). In vitro tests for teratogens: desirable endpoints, test batteries and current status of the micromass teratogen test. *Reprod Toxicol* 7 (Suppl 1):103–111.

Fraichard, A., Chassande, O., Bilbaut, G., Dehay, C., Savatier, P. & Samarut, J. (1995). In vitro differentiation of embryonic stem cells into glial cells and functional neurons. *J Cell Sci* 108(Pt 10):3181–3188.

Genschow, E., Scholz, G., Brown, N., Piersma, A., Brady, M., Clemann, N., Huuskonen, H., Paillard. F., Bremer, S., Becker, K. & Spielmann, H. (2000). Development of prediction models for three in vitro embryotoxicity tests in an ECVAM validation study. *In Vitro Mol Toxicol* 13: 51 - 66.

Genschow, E., Spielmann, H., Scholz, G., Seiler, A., Brown, N., Piersma, A., Brady, M., Clemann, N., Huuskonen, H., Paillard, F., Bremer, S. & Becker, K. (2002). The ECVAM international validation study on in vitro embryotoxicity tests: results of the definitive phase and evaluation of prediction models. European Centre for the Validation of Alternative Methods. *Altern Lab Anim* 30: 151 - 76.

Genschow, E., Spielman, H., Scholz, G., Pohl, I., Seiler, A., Clemann, N., Bremer, S. & Becker, K. (2004). Validation of the embryonic stem cell test in the international ECVAM validation study on three in vitro embryotoxicity tests. *Altern Lab Anim* 32: 209-44.

Gore, A., Li, Z., Fung, H.L., Young, J.E., Agarwal, S., Antosiewicz-Bourget, J., Canto, I., Giorgetti, A., Israel, M.A., Kiskinis, E., Lee, J.H., Loh, Y.H., Manos, P.D., Montserrat, N., Panopoulos, A.D., Ruiz, S., Wilbert, M.L., Yu, J., Kirkness, E.F., Izpisua Belmonte, J.C., Rossi, D.J., Thomson, J.A., Eggan, K., Daley, G.Q., Goldstein, L.S. & Zhang, K (2011). Somatic coding mutations in human induced pluripotent stem cells. *Nature* 471(7336):63-67.

Han, J. & Sidhu, K.S. (2008). Current concepts in reprogramming somatic cells to pluripotent state. *Curr Stem Cell Res Ther* 3(1):66–74.

Hendrickx, A.G. (1998). Teratogenicity of Retinoids in Rodents, Primates and Humans. *J Toxicol Sci* 23:272.

Heng, B.C., Richards, M., Shu, Y. & Gribbon, P. (2009). Induced pluripotent stem cells: a new tool for toxicology screening? *Arch Toxicol* 83(7):641-4.

Höfer, T., Gerner, I., Gundert-Remy, U., Liebsch, M., Schulte, A., Spielmann, H., Vogel, R. & Wettig, K. (2004). Animal testing and alternative approaches for the human health risk assessment under the proposed new European chemicals regulation. *Arch Toxicol* 78: 549-564.

Jeffay, S., Nichols, H., Barrier, M. & Hunter, S, (2010), Effects of haloacetic acids and their major metabolites in a mouse embryonic stem cell adherent cell differentiation and cytotoxicity (ACDC) assay. *The Toxicologist* 114: 359.

Jelinek, R., Peterka, M. & Rychter, Z. (1985). Chick embryotoxicity screening test—130 substances tested. *Indian J Exp Biol* 23:588–595.

Kaichi, S., Hasegawa, K., Takaya, T., Yokoo, N., Mima, T., Kawamura, T., Morimoto, T., Ono, K., Baba, S., Yamanaka, S., Nakahata, T. & Heike, T. (2010). Cell line-dependent differentiation of induced pluripotent stem cells into cardiomyocytes in mice. *Cardiovasc Res* 88(2):314-23.

Kasahara, H., Bartunkova, S., Schinke, M., Tanaka, M. & Izumo, S. (1998). Cardiac and extracardiac expression of Csx/Nkx2.5 homeodomain protein. *Circ Res* 82:936-946.

Kawamura, T., Ono, K., Morimoto, T., Wada, H., Hirai, M., Hidaka, K., Morisaki, T., Heike, T., Nakahata, T., Kita, T. & Hasegawa K. (2005). Acetylation of GATA-4 is involved in the differentiation of embryonic stem cells into cardiac myocytes. *J Biol Chem* 20:19682-19688.

Kessel, M. & Gruss, P. (1991). Homeotic transformations of murine vertebrae and concomitant alteration of Hox codes induced by retinoic acid. *Cell* 67:89–104.

Kimmel, G.L., Smith, K., Kochhar, D.M. & Pratt, R.M. (1982). Overview of in vitro teratogenicity testing: aspects of validation and application to screening. *Teratog Carcinog Mutagen* 2:221–229.

Kolossov, E., Fleischmann, B.K., Liu, Q., Bloch, W., Viatchenko-Karpinski, S., Manzke, O., Ji, G.J., Bohlen, H., Addicks, K. & Hescheler, J. (1998). Functional characteristics of ESC-derived cardiac precursor cells identified by tissue-specific expression of the green fluorescent protein. *J Cell Biol* 143:2045–2056.

Krtolica, A., Ilic, D., Genbacev, O. & Miller, R.K. (2009). Human embryonic stem cells as a model for embryotoxicity screening. *Regen Med* 4:449–459.

Laschinski, G., Vogel, R. & Spielmann, H. (1991). Cytotoxity test using blastocyst-derived euploid embryonal stem cells: a new approach to *in vitro* teratogenesis screening. *Reproductive Toxicol* 5: 57-64.

Laustriat, D., Gide, J. & Peschanski, M. (2010). Human pluripotent stem cells in drug discovery and predictive toxicology. *Biochem Soc Trans* 38:1051–1057.

Lock, L.T., & Tzanakakis, ES. (2009). Expansion and differentiation of human embryonic stem cells to endoderm progeny in a microcarrier stirred-suspension culture. *Tissue Eng Part A.* 15(8):2051-2063.

Marx-Stoeling, P., Adriaens, El., Ahr, H.J. Bremer, S., Garthoff, B., Gelbke, H.P., Piersma, A., Pellizzer, C., Reuter, U., Rogiers, V., Schenk, B., Schwengberg, S., Seiler, A., Spielmann, H., Steemans, M., Stedman, D.B., Vanparys, P., Vericat, J.A., Verwei, M., van der Water, F., Weiman, M. & Schwarz, M., (2009). A review of the implementation of the embryonic stem cell test (EST). The report and recommendations of an ECVAM/ReProTect Workshop. *Altern Lab Anim* 37:313–328.

Mauritz, C., Schwanke, K., Reppel, M., Neef, S., Katsirntaki, K., Maier, L.S., Nguemo, F., Menke, S., Haustein, M., Hescheler, J., Hasenfuss, G. & Martin, U. (2008). Generation of functional murine cardiac myocytes from induced pluripotent stem cells. *Circulation* 118(5):507-517.

Miki, T., Ring, A., & Gerlach, J. (2011). Hepatic Differentiation of Human Embryonic Stem Cells Is Promoted by Three-Dimensional Dynamic Perfusion Culture Conditions. *Tissue Eng Part C Methods.* 17(5): 557-568.

Miller, R.A., Christoforou, N., Pevsner, J., McCallion, A.S. & Gearhart, J.D. (2008). Efficient array-based identification of novel cardiac genes through differentiation of mouse ESCs. PLoS One 3: e2176.

Molkentin, J.D., Lin, Q., Duncan, S.A. & Olson, E.N. (1997). Requirement of the transcription factor GATA4 for heart tube formation and ventral morphogenesis. *Genes Dev* 11:1061-1072.

Mosmann, T. (1983). Rapid colorimetric assay for cellular growth and survival: application to proliferation and cytotoxicity assays. *J Immunol Methods* 65:55–63.

Narazaki, G., Uosaki, H., Teranishi, M., Okita, K., Kim, B., Matsuoka, S., Yamanaka, S. & Yamashita, J.K. (2008). Directed and systematic differentiation of cardiovascular cells from mouse induced pluripotent stem cells. *Circulation* 118(5):498-506.

Nau, H. (1993). Embryotoxicity and teratogenicity of topical retinoic acid. *Skin Pharmacol* 6(Suppl 1):35–44.

Newall, D.R. & Beedles, K.E. (1994). The stem-cell test - a novel *in vitro* assay for teratogenic potential. *Toxicol In Vitro* 8: 697-701.

OECD (2008). Guidance document on mammalian reproductive toxicity testing and assessment. Series on testing and assessment no. 43.

Osman, A.M., van Dartel, D.A., Zwart, E., Blokland, M., Pennings, J.L. & Piersma, A.H. (2010). Proteome profiling of mouse embryonic stem cells to define markers for cell differentiation and embryotoxicity. *Reprod Toxicol* 30(2): 322-332.

Pamies, D., Vicente-Salar, N., Sogorb, M.A., Roche, E. & Reig, J.A. (2010). 5-fluorouracil on α-fetoprotein gene expression during the in vitro mouse embryonic stem cell differentiation. *Int J Toxicol* 29(3): 297-304.

Pamies, D., Estevan, C., Sogorb, M.A. & Vilanova, E. (2011). Mechanism-based models in reproductive and developmental toxicology. In: *Reproductive and developmental toxicology*, R.C. Gupta (Ed.), Academic Press, ISBN 978-0-12-382032-7, Hopkinsville.

Peters, A.K., Steemans, M., Hansen, E., Mesens, N., Verheyen, G.R. & Vanparys, P. (2008). Evaluation of the embryotoxic potency of compounds in a newly revised high throughput embryonic stem cell test. *Toxicol Sci* 105:342–350.

Piersma, A.H. (2004). Validation of alternative methods for developmental toxicity testing. *Toxicol Lett* 149:147–153.

Rohwedel, J., Guan, K. & Wobus, A.M. (1999). Induction of cellular differentiation by retinoic acid in vitro. *Cells Tissues Organs* 165:190–202.

Rohwedel, J., Guan, K., Hegert, C. & Wobus, A.M. (2001). Embryonic stem cells as an in vitro model for mutagenicity, cytotoxicity and embryotoxicity studies: present state and future prospects. *Toxicol In Vitro* 15:741–753.

Romero-Lucena, A.C. (2010). "Gen codificante de la esterasa diana de neuropatía (NTE) como indicador de embriotoxicidad en ensayos *in vitro*". Master in Bioengineering 2010. Universidad Miguel Hernández de Elche.

Sadler, T.W., Horton, W.E. & Warner, C.W. (1982). Whole embryo culture: a screening technique for teratogens? *Teratog Carcinog Mutagen* 2:243–253.

Schmidt, B.P. (1985). Teratogenicity testing of new drugs with the postimplantation embryo culture system. Homburger F (ed) Concepts in toxicology. Karger Publ, Basel, pp 46–57.

Schmidt, M.M., Guan, K. & Wobus, A.M. (2001). Lithium influences differentiation and tissue-specific gene expression of mouse embryonic stem (ES) cells in vitro. *Int J Dev Biol* 45:421–429.

Schwetz, B.A., Morrissey, R.E., Welsch, F. & Kavlock, R.A. (1991). In vitro teratology. Environ *Health Perspect* 94:265–268.

Seiler, A., Visan, A., Buesen, R., Genschow, E. & Spielmann, H. (2004). Improvement of an *in vitro* stem cell assay for developmental toxicity: the use of molecular endpoints in the embryonic stem cell test. *Reprod Toxicol* 18:231–240.

Seiler, A., Buesen, B., Hayess, K., Schlechter, K., Visan, A., Genschow, E., Slawik, B. & Spielmann, H. (2006). Current status of the embryonic stem cell test: the use of recent advances in the field of stem cell technology and gene expression analysis. ALTEX: Alternativen zu Tierexperimenten 23:393–399.

Si-Tayeb, K., Noto, F.K., Nagaoka, M., Li, J., Battle, M.A., Duris, C., North, P.E., Dalton, S, & Duncan, S.A. (2010). Highly efficient generation of human hepatocyte-like cells from induced pluripotent stem cells. *Hepatology* 51(1):297-305.

Spielmann, H., Genschow, E., Liebsch, M. & Halle, W. (1999). Determination of the starting dose for acute oral toxicity (LD50) testing in the up and down procedure (UDP) from cytotoxicity data. *Alt Lab Anim* 27:957–966.

Spielmann, H. & Liebsch, M. (2001). Lessons learned from validation of in vitro toxicity test: from failure to acceptance into regulatory practice. *Toxicol In Vitro* 15(4-5):585-90.

Spielmann, H. (2005). Predicting the risk of developmental toxicity from in vitro assays. *Toxicol Appl Pharmacol* 207:375–380.

Spielmann, H., Seiler, A., Bremer, S., Hareng, L., Hartung, T., Ahr, H., Faustman, E., Haas, U., Moffat, G.J., Nau, H., Vanparys, P., Piersma A., Sintes, J.R. & Stuart, J. (2006). The practical application of three validated in vitro embryotoxicity tests. The report and recommendations of an ECVAM/ZEBET workshop (ECVAM workshop 57). *Altern Lab Anim* 34:527–538.

Spielmann, H. (2009). The way forward in reproductive/developmental toxicity testing. *Altern Lab Anim* 37:641–656.

Strubing, C., Ahnert-Hilger, G., Shan, J., Wiedenmann, B., Hescheler, J. & Wobus, A.M. (1995). Differentiation of pluripotent embryonic stem cells into the neuronal lineage in vitro gives rise to mature inhibitory and excitatory neurons. *Mech Dev* 53:275–287.

Stummann, T.C., Hareng, L. & Bremer, S. (2007). Embryotoxicity hazard assessment of methylmercury and chromium using embryonic stem cells. *Toxicology* 242:130–143.

Stummann, T.C., Beilmann, M., Duker, G., Dumotier, B., Fredriksson, J.M. Jones, R.L., Hasiwa, M., Kang, Y.J., Mandenius, C.F., Meyer, T., Minotti, G., Valentin, Y.J., Zünkler, B.J. & Bremer, S. (2009). Report and recommendations of the workshop of the European centre for the validation of alternative methods for drug-induced cardiotoxicity. *Cardiovasc Toxicol* 9:107–125.

Surani, M.A., Durcova-Hills, G., Hajkova, P., Hayashi, K. & Tee, W.W. (2008). Germ line, stem cells, and epigenetic reprogramming. *Cold Spring Harb Symp Quant* 73:9-15.

Takahashi, K., Tanabe, K., Ohnuki, M., Narita, M., Ichisaka, T., Tomoda, K. & Yamanaka, S. (2007). Induction of pluripotent stem cells from adult human fibroblasts by defined factors. *Cell* 131(5):861–872

Trosko, J.E. & Chang, C.C. (2010). Factors to consider in the use of stem cells for pharmaceutic drug development and for chemical safety assessment. *Toxicology* 270:18–34.

Tzimas, G., Sass, J.O., Wittfoht, W., Elmazar, M.M., Ehlers, K. & Nau, H. (1994). Identification of 9, 13-dicis-retinoic acid as a major plasma metabolite of 9-cis-retinoic acid and limited transfer of 9-cis-retinoic acid and 9, 13-dicis-retinoic acid to the mouse and rat embryos. *Drug Metab Dispos* 22:928–936.

van Dartel, D.A., Pennings, J.L., Hendriksen, P.J., van Schooten, F.J. & Piersma, A.H. (2009). Early gene expression changes during embryonic stem cell differentiation intocardiomyocytes and their modulation by monobutyl phthalate. *Reprod Toxicol* 27(2):93-102.

van Dartel, D.A., Pennings, J.L., de la Fonteyne, L.J., van Herwijnen, M.H., van Delft, J.H., van Schooten, F.J. & Piersma, A.H. (2010a). Monitoring developmental toxicity in the embryonic stem cell test using differential gene expression of differentiation-related genes. *Toxicol Sci* 116(1): 130-139.

van Dartel, D.A., Pennings, J.L., van Schooten, F.J. & Piersma, A.H. (2010b). Transcriptomics-based identification of developmental toxicants through their interference with cardiomyocyte differentiation of embryonic stem cells. *Toxicol Appl Pharmacol* 243(3):420-428.

van Dartel, D.A., Pennings, J.L., Robinson, J.F., Kleinjans, J.C. & Piersma, A.H. (2011a). Discriminating classes of developmental toxicants using gene expression profiling in the embryonic stem cell test. *Toxicol Lett* 201(2): 143-151.

van Dartel, D.A., Pennings, J.L., de la Fonteyne, L.J., Brauers, K.J., Claessen, S., van Delft, J.H., Kleinjans, J.C. & Piersma, A.H. (2011b). Evaluation of developmental toxicant identification using gene expression profiling in embryonic stem cell differentiation cultures. *Toxicol Sci* 119(1):126-134.

Verwei, M., van Burgsteden, J.A., Krul, C.A., van de Sandt, J.J. & Freidig, A.P. (2006). Prediction of in vivo embryotoxic effect levels with a combination of in vitro studies and PBPK modelling. *Toxicol Lett* 165:79–87.

Wobus, A.M., Grosse, R. & Schoneich, J. (1988). Specific effects of nerve growth factor on the differentiation pattern of mouse embryonic stem cells in vitro. *Biomed Biochim Acta* 47:965–973.

Wobus, A.M., Wallukat, G. & Hescheler, J. (1991). Pluripotent mouse embryonic stem cells are able to differentiate into cardiomyocytes expressing chronotropic responses to adrenergic and cholinergic agents and Ca2 channel blockers. *Differentiation* 48:173–182.

Wobus, A.M., Rohwedel. J., Maltsev. V. & Hescheler, J. (1994). In vitro differentiation of embryonic stem cells into cardiomyocytes or skeletal muscle cells is specifically modulated by retinoic acid. *Roux's Arch Dev Biol* 204:36–45.

Wobus, A.M. & Boheler, K.R. (2005). Embryonic stem cells: prospects for developmental biology and cell therapy. *Physiol Rev* 85:635–678.

Wobus, A.M. & Löser, P. (2011). Present state and future perspectives of using pluripotent stem cells in toxicology research. *Arch Toxicol* 85:79–117.

Yu, J., Vodyanik, M.A., Smuga-Otto, K., Antosiewicz-Bourget, J., Frane, J.L., Tian, S., Nie, J., Jonsdottir. G.A., Ruotti, V., Stewart, R., Slukvin, I.I. & Thomson, J.A. (2007). Induced pluripotent stem cell lines derived from human somatic cells. *Science* 318(5858):1917–1920

zur Nieden, N.I., Ruf, L.J., Kempka, G., Hildebrand, H. & Ahr, H.J. (2001). Molecular markers in embryonic stem cells. *Toxicol In Vitro* 15(4-5):455-461.

zur Nieden, N.I., Kempka, G. & Ahr, H.J. (2004). Molecular multiple endpoint embryonic stem cell test–a possible approach to test for the teratogenic potential of compounds. *Toxicol Appl Pharmacol* 194:257–269.

13

Teratomas Derived from Embryonic Stem Cells as Models for Embryonic Development, Disease, and Tumorigenesis

John A. Ozolek[1] and Carlos A. Castro[2]
[1]Division of Pediatric Pathology, Department of Pathology, Children's Hospital of Pittsburgh, University of Pittsburgh School of Medicine, Pittsburgh, Pennsylvania,
[2]Division of Developmental and Regenerative Medicine, Departments of Obstetrics, Gynecology, and Reproductive Sciences; Cell Biology-Physiology; and Bioengineering; Pittsburgh Development Center; Magee-Womens Research Institute, Pittsburgh, Pennsylvania,
United States of America

1. Introduction

The word teratoma is derived from the Greek teratos which means "deformity" or "monster" and -oma which means "tumor". Thus, the teratoma is a monster tumor. A more recent definition which is more elaborate and perhaps more politically correct is an encapsulated tumor with tissue or organ components that can be traced to derivatives of the three primordial germ layers; ectoderm, mesoderm, and endoderm (Rosai and Ackerman 2004). It is easy to understand why this tumor was coined "monster" since macroscopically teratomas can appear as a conglomeration of tissue with different colors and textures some of which may be recognizable as gross anatomical structures such as hair, teeth, and limbs. The most extreme examples actually resemble a distorted fetus (fetiform). The initial descriptions of sacrococcygeal teratomas date back to ancient tablets written by Egyptian fetoscopists around 2500 B.C. These oddities were often attributed great significance when interpreted in concert with the concurrent cultural and religious beliefs. For example, ancient Egyptians interpreted a third foot in the middle as signifying great prosperity for the land (Oosterhuis et al. 2007). These cases likely represented rare sacrococcygeal teratomas containing a fully formed foot which has been reported in the modern medical literature (Legbo, Opara, and Legbo 2008). Seventeenth century illustrations demonstrate hair sprouting from a teratoma and at that time teratomas were thought to arise from nightmares or witchcraft while in the 19th century they were thought to arise from perverse sexual practices (Oosterhuis et al. 2007; Gatcombe et al. 2004). More recent (late nineteenth and early twentieth century) theories invoke cell biology principles purporting the derivation of teratomas from primitive germ cells. Indeed some current definitions will expand the above general definition to include their histogenesis from pluripotent stem cells (Oosterhuis et al. 2007).

While at first glance, most teratomas microscopically appear as disorganized masses with recognizable tissue types contributed from the three basic germ layers, very little is really known about their composition and how they form and this information may hold important clues to normal and abnormal development. Studying the cellular and tissue milieu within teratomas both in vivo and ex vivo would help in beginning to answer several important questions that we will address in the following sections. With this in mind, the objectives for this chapter are:

Derivation of teratomas from stem cells and factors that influence their growth and development: This section will address the tumorigenicity of embryonic stem cells and their relationship to cancer cells, some reasons why experimental teratomas have not been a hot focus of investigation, and factors that influence experimental teratoma formation (stem cell phenotype [e.g. embryonic, induced pluripotent stem cells (ES cells, iPS cells], genetic differences in stem cell lines, cellular and metabolic factors (e.g. mitochondrial metabolism), nature of the host where the teratoma grows, and site of injection (microenvironment)).

Current technology for tracking stem cell fate, monitoring teratoma formation, and delineation of specific tissue types in vivo: Ensuring embryonic cell transplantation safety by non-invasive in vivo detection and monitoring is a pressing need if ES cells will be used in the clinic. This section will review current methodologies for imaging of teratomas both in vivo and ex vivo. We will present some of our data using high-resolution MRI to quantify and delineate specific tissue types within teratomas.

Histopathology of teratomas: This section will discuss the myriad of tissues present in teratomas and their significance. An overview of the most common tissue types, tissues rarely found, and primitive organs seen microscopically in experimentally derived teratomas will be presented.

Teratomas in the study of embryonic development: We have examined teratomas derived from mouse, non-human primate, and human embryonic stem cells and have observed similarities and differences in quantities of tissue types derived from the three germ layers that in all likelihood have developmental implications across species. We will discuss our methods for semiquantifying specific tissue types from serial histological sections. We will present some data resulting from collaborative efforts with biomedical engineers at Carnegie Mellon University on the automated identification and quantification of tissue types from digital images of histological sections of teratomas. This section will also discuss the use of teratomas as potential models for providing insight into the molecular and genetic mechanisms of development and tissue lineage commitment.

Teratomas as models of disease: This section will outline how teratomas may be used as sources of tissue for studying specific diseases. Disease specific genetic alterations of ES cells allow for production of tissue that can recapitulate in vivo diseased tissue/organs. We will discuss the use of teratomas as platforms for studying the toxic effects of compounds/molecules/intrauterine environment on embryonic development. We present some of our recent data using ES cells as a model for Alzheimer's disease.

Teratomas in the study of tumorigenesis: This section will address the recent novel unifying theory of teratoma, germ cell tumor, and other tumor formation. We also introduce teratomas as substitute platforms for studying tumor growth and behavior.

2. Derivation of teratomas from stem cells and factors that influence their growth and development

2.1 Roots of the monster: Embryonic stem cells and cancer cells: One in the same?

In the experimental setting, teratomas can be derived from several cell types including embryonic stem cells (ES cells), embryonic carcinoma cells (EC cells), embryonic and primordial germ cells (PGCs), and induced pluripotent cells (iPS cells) (Aleckovic and Simon 2008; Blum and Benvenisty 2008, 2009; Kooreman and Wu). Embryonic stem cells share phenotypic and genetic characteristics of cancer cells including the ability for self renewal, prolonged proliferation in vitro, lack of contact inhibition, telomerase activity and the ability to invoke angiogenesis. However, human ES cells (hES cells) are karyotypically normal (i.e. diploid), often a major difference between cancer cells. In order to maintain their ability to self renew and perpetuate indefinitely, they require special culture conditions (specialized feeder cells or conditioned media on feeder free substrates such as matrigel). HES cells have certain molecular and genetic features used to characterize them including genetic and immunohistochemical expression of pluripotency transcription factors Nanog, Oct4, and Sox2, and expression of surface antigens SSEA3, SSEA4, TRA-1-60, TRA-1-81. The triumvirate of Nanog, Oct4, and Sox2 act to promote expression of self-renewal genes while repressing differentiation genes. Loss of Oct4 results in failure of the inner cell mass to develop; while loss of Sox2 and Nanog results in defective/lack of epiblast formation and subsequent differentiation towards trophectoderm and primitive endoderm. Several genes used to generate iPS cells are also linked to cancer. These include Oct4 and Sox2 which in combination with either Nanog and Lin28 or Myc and Klf4 can transform the somatic cell back to pluripotency. (Blum and Benvenisty 2008; Chambers and Tomlinson 2009; Knoepfler 2009; Kooreman and Wu ; Lanza 2006; Loh, Ng, and Ng 2008).

ES cells have unique epigenetic features. They maintain unique demethylated CpG islands allowing for increased expression of genes compared to differentiated cells. A similar phenomenon is present in cancer cells where a small reduction in methylation is sufficient to induce cancer and conversely many cancers show hypermethylation leading to inhibition of gene expression. Many tumor suppressor genes are also hypermethylated and silenced in hES cells. Certain onco-fetal genes such as survivin (BIRC5) are enhanced in undifferentiated hES cells and teratomas derived from them. Deletion of survivin promotes apoptosis in cultured hES cells and in the teratomas derived from them. Survivin is a protein which imparts anti-apoptotic activity and may be involved in protecting cells in harsh environments and is expressed in many cancers. The anti-apoptotic activity of survivin is generated by inhibition of caspase activity. Caspases have a key role in destructing Nanog implying that survivin by inhibiting caspase activity may have a role in hES cell self-renewal. These findings suggest that untransformed hES cells are possibly tumorigenic since they already have genetic hallmarks of tumors despite having no mutational transformation. (Blum and Benvenisty 2008, 2009; Chambers and Tomlinson 2009; Knoepfler 2009; Kooreman and Wu ; Loh, Ng, and Ng 2008; Yu and Thomson 2008).

2.2 Teratomas: The reviled monster

Since first derived from the blastocyst of the developing mouse by Martin Evans and Mathew Kaufman at Cambridge and Gail Martin at UCSF independently in 1981, ES cell lines have been derived from non-human primates (1995) and then from humans by James Thomson in 1998 at the University of Wisconsin (Yu and Thomson 2008). In the mouse the same cells that produce teratomas in the experimental setting when placed back into the developing blastocyst (the blastocyst complementation assay) integrate into the embryo without tumor production and contribute to the formation of tissues derived from the three primordial germ layers (by definition pluripotent) and also the germ line. In fact, this is the test of pluripotency for potential mouse ES cells (mES cells)-the ability to form a chimera (Prelle, Zink, and Wolf 2002). Primarily due to ethical restraints, the chimera test of pluripotency is not the method used to assess whether given cells are pluripotent for non-human primate and human cells. The gold standard has been the ability of a given collection of cells to form teratomas when injected into immunodeficient mice. This test and the production of this tumor has produced great consternation in the evolving world of regenerative medicine for several reasons that predominantly include time and cost (Dolgin). The main goal of regenerative medicine thus far has been to guide stem cells (whatever their origin) to differentiate into specific tissue types (e.g. cartilage, islets, heart muscle) to be used as new tissue for repair of injured or aging tissue. The teratoma is the antithesis of this philosophy; once again living up in a more figurative sense to its word origin roots--a monster!

Teratoma formation by stem cells remains the last frontier to conquer before effective clinical trials using stem cells can be widely accepted. Already several examples exist of the clinical use of stem cells in animals resulting in teratoma formation (Knoepfler 2009). Cells for clinical use can be grown and issues concerning immune rejection of cells have been alleviated due to the emergence of iPS cells, cells derived from nuclear transfer, and human Wharton jelly stem cells (Menendez, Bueno, and Wang 2006). Several papers have addressed specific measures developed at the genetic and cell sorting level to combat this undesirable "side effect" if you will (Koch, Jordan, and Platt 2006; Kooreman and Wu ; Cao et al. 2007; Bulic-Jakus et al. 2006). Recently, some sentiment has arisen amongst stem cell investigators to abandon the teratoma assay as the measure of pluripotency in favor of molecular/bioinformatics approaches. In Dolgin's recent article, Owen Witte, Director of the Broad Stem Cell Research Center at UCLA calls the teratoma assay; "the most ridiculous assay on the planet".

Most literature centered on teratomas derived experimentally from embryonic stem cells portrays teratomas as a means to an end or as an unwanted side effect in the clinic. This may explain the relative paucity of literature that specifically examines the factors that influence teratoma development from stem cells in the experimental setting. However, the specific articles that have focused on the teratoma shed valuable attention on how specific variables seem to affect the growth and characteristics of teratomas.

2.3 Stem cell phenotype

A single paper comparing the rates of teratoma formation using several embryonic stem cell types known to produce teratomas (e.g. mES, non-human primate (nhpES), hES cells) while controlling for variables such as site of injection, number of cells injected, phenotype of the host has not been published. The paper by Hentze et al comes close as it includes several hES cell lines, injection sites, cell numbers, and included semiquantitative histopathological analysis of the teratomas (Hentze et al. 2009). In a brief report, Gutierrez-Aranda et al (Gutierrez-Aranda et al.) demonstrated that human iPS cells produced teratomas sooner and at a higher rate per injection than human embryonic stem cells. In their study they used seven different hES cell lines and 4 iPS cell lines. They performed subcutaneous or intratesticular injections of 1×10^6 cells suspended in 30% matrigel. All mice (100%) injected with iPS cells developed teratomas regardless of injection site compared to 81% and 94% efficiency of teratoma production for subcutaneous and intratesticular injection of hES cells respectively. In addition, iPS cells developed tumors much more quickly at both sites of injection compared to hES cells. These authors claimed no differences in histological composition of teratomas based on site of injection. However, specific notation of any histological differences in teratomas when compared between cell types (hES and iPS cells) was not made but presumably was not different. Histological analysis was defined by microscopic morphology and immunohistochemical markers of specific tissue lineages, but was not quantified. They did not report any teratocarcinomas or islands of undifferentiated cells in any of their teratomas. In our empiric evaluation of several teratomas (Castro and Ozolek unpublished observations) derived from iPS cells we have noted significant areas of undifferentiated, malignant appearing cells that would qualify the lesions as teratocarcinomas. The karyotype of the iPS cells used to derive the teratomas that we have examined was not available. Differences in mitochondrial metabolism within undifferentiated embryonic stem cells may also influence their ultimate fate. Schieke et al demonstrated that mES cells with the highest resting mitochondrial membrane potential had high metabolic rates, high resting oxygen consumption, lower rates of mesodermal differentiation in the absence of pluripotency culture conditions, and produced larger teratomas by weight compared to genetically identical mES cells with the lowest resting mitochondrial membrane potentials. Rapamycin treatment to inhibit mTOR activity produced the same results as the population of mES cells with the lowest resting mitochondrial membrane potential indicating a direct effect of rapamycin on mitochondrial function. The authors speculate that since mitochondrial metabolism is directly related to the production of reactive oxygen species and genomic instability, stem cell populations within organisms might show preferential selection of stem cells with low metabolism as a means to prevent deleterious genomic events over the lifespan of the organism (Schieke et al. 2008). This concept is important and may have implications for tumor development in humans as we will discuss in section 7.

2.4 Genetic aberrations of stem cell lines and teratoma formation

In the above article by Gutierrez-Aranda, they speculate the difference in teratoma production efficiency and latency between iPS and hES cells was due to subtle genetic differences not able to be detected by conventional G-banding karyotyping and that more sensitive techniques would need to be employed to answer this question (Gutierrez-Aranda et al.) . Two of their lines (one iPS and one hES cell line) were aneuploid, but apparently did not show any differences in the histological composition of teratomas derived from them compared to euploid cell lines. The aneuploidy was not further specified. HES cells can undergo chromosomal derangements when cultured for extended periods. The most common aberrations noted include gains of chromosomes 12, 17, and X but others have been noted (Blum and Benvenisty 2008). However few studies have shown or addressed the issue of teratoma production from stem cells known to harbor genetic aberrations. An interesting paper by Karin Gertow et al at the Karolinska Institute from 2007 demonstrated hES cells with trisomy 12 derived from the HS181 line in culture maintained under pluripotent conditions (Gertow et al. 2007). HES cells with trisomy 12 were found intermingled with diploid cells or could be found as the predominant cell type in some colonies. Several interesting results from this paper: The percentage of trisomy 12 cells in culture increased with passage number. The overall amount of tissues present in teratomas was roughly the same between teratomas derived from predominantly diploid versus predominantly trisomy 12 hES cells with the exception of increased renal tissue in trisomy 12 teratomas. The authors postulated that since approximation of metanephric mesenchyme with neural tubes induces development of renal tubules, reciprocal interactions between the metanephric mesenchyme and abundant neural elements might be responsible for renal development. Interestingly, Nanog, whose expression is in large part responsible for maintaining pluripotency in ES cells is located on chromosome 12 and overexpression has been shown to induce primitive ectoderm (of which neuroectoderm is a constituent) (Darr, Mayshar, and Benvenisty 2006).

In a study by Herszfeld et al at the Monash Institute of Medical Research and the Australian Stem Cell Centre from 2006, they examined the expression of CD30 in cell lines from germ cell tumors and in hES cells grown in culture (Herszfeld et al. 2006). CD30 is a member of the TNF receptor superfamily, is present on Reed-Sternberg cells, activated lymphocytes, decidual cells, and is a biomarker of embryonal carcinoma in diagnostic pathology. As expected, RT-PCR products and expression of CD30 was found in the embryonal carcinoma line cells while none was present in the hES cell lines. However, under certain culture conditions (serum free) hES cell lines demonstrated karyotypic abnormalities and overgrew the diploid population. These karyotypic abnormalities included duplications of 1q, trisomy 12, balanced translocation between chromosomes 1 and 6 and a lesion on the long arm of chromosome 1 involving duplications of q2 and an inverted duplication of q3-4 (Interestingly, many of these chromosomal aberrations are present in human germ cell tumors). Teratomas formed by these abnormal cell lines still differentiated into all three tissue types, but had a higher proportion of primitive undifferentiated cells compared to diploid ES cells; a feature seen in teratocarcinomas. In one hES cell line mosaic for trisomy 12, combined indirect immunofluorescence for CD30 and FISH for the alpha-centromeric probe for chromosome 12 showed clear concordance between CD30 expression and the presence of an extra copy of chromosome 12. In addition, CD30 positive cells had lower levels of both spontaneous and induced apoptosis. The authors conclude that the emergence of CD30 expression in aneuploid cell lines may be an adaptive response of the cells to stress permitting survival under suboptimal conditions at the expense of DNA damage.

Prokhorova et al showed defective pluripotency in a hES cell line that had acquired trisomy 20 after 30 passages (Prokhorova et al. 2009). Teratomas (and embryoid bodies in culture) from this line were significantly smaller in size and had a decreased ratio of differentiated to undifferentiated tissues. Tissues present were more primitive and showed ill-defined glandular structures compared to teratomas derived from euploid hES cells. Similarly, Bloch et al demonstrated that 129/Sv mice injected with mES cells with a homozygous deficiency of β-1 integrin developed very small teratomas or did not develop teratomas compared to heterozygous deficient β-1 integrin or wild-type ES cells (Bloch et al. 1997). Other findings in β1-null teratomas included abnormal distribution of extracellular matrix proteins, partially detached basement membranes, and lack of ES cell derived endothelial cells in teratoma blood vessels. Interestingly, a consistent finding mentioned in a few studies is that blood vessels within teratomas are a chimera between host-derived and ES cell derived endothelial cells. This is in contrast to other tissues present in teratomas that are invariably derived nearly entirely from the exogenous ES cells (Gerecht-Nir et al. 2004; Gerecht-Nir et al. 2003; Gertow et al. 2004; Goldberg Cohen et al. 2006).

2.5 Site of injection

Several anatomical sites of the immunodeficient mouse are used for injection of stem cells for the teratoma assay. These have been chosen for their accessibility to injection, immune privilege status, and ample space to contain a large tumor that still allows the host animal to live and function. The most common sites for injection include intratesticular, subcutaneous, intramuscular, and kidney capsule. Again, only a few studies have specifically addressed comparison of teratomas from different graft sites. Prokhorova et al injected several hES cells lines at four different sites (subcutaneous, intramuscular, kidney capsule, intratesticular) (Prokhorova et al. 2009). Subcutaneous injections were done with cells suspended in 30% matrigel and without matrigel. The highest efficiency of teratoma formation occurred for kidney capsule injections (10/10, 100%) followed by subcutaneous injections using cells suspended in 30% matrigel (64/68, 94%), followed by intratesticular (60%), subcutaneous without matrigel (33%), and intramuscular (12.5%). It should be noted, however, that all injections were done with approximately 1 x 10^6 cells although for the kidney capsule injections the number of cells injected was not explicitly defined. The number of injections and efficiencies were difficult to compare due to the wide variation in the number of injections for each site (range 5 [intratesticular] to 68 [subcutaneous with matrigel]). Several cell lines were used and it is unclear if all cell lines were used for all injection sites. This study also did not find any differences in teratoma composition by histological examination, but again tissue types were not quantified. Cooke et al demonstrated that intrahepatic injections of hES cells and EC cells produced rapid tumors 5-8 times larger than those obtained by subcutaneous injections (Cooke, Stojkovic, and Przyborski 2006). Intrahepatic tumors displayed more cystic areas with immature and mature tissue types, higher expression of SSEA3 (embryonic stem cell marker), and lower expression of nestin (intermediate filament present in early tissues with neuronal specification and other tissues) compared to subcutaneous tumors. They postulated that the microanatomy of the liver with open vascular spaces (sinusoidal system) and growth factors (e.g. stem cell factor) produced in the liver contribute to the rapid growth of teratomas. In our experience, we have achieved between 67% and 86% efficiency of teratoma formation when injecting between 5 x 10^5 and 1 x 10^6 nhpES cells using the intratesticular location. Some evidence suggests that a post-ischemic environment at least in the central nervous system promotes teratoma formation. Seminatore et al transplanted neuroprogenitors (NPGs) derived from hES cells at various stages of differentiation ranging from early NPGs to differentiated NPGs into rats with either no ischemic lesion or rats with small or large ischemic lesions (Seminatore et al.). The early NPGs had embryoid body-like structures that exhibited characteristics of pluripotent stem cells and expressed markers of pluripotent stem cells. Teratomas formed in the formed in the central nervous system of rats of rats injected with NPGs with embryoid-body like structures regardless of the presence or absence of an ischemic lesion. In addition, large ischemic lesions significantly promoted the survival of transplanted early NPGs compared to sham. The authors propose that the post-ischemic environment produces a variety of factors that promote the survival, growth, and differentiation into teratomas from an undifferentiated population of ES cells. In the study by Hentze et al from Singapore, they injected several hES cells lines into kidney capsule, intramuscular, subcutaneous, intraperitoneal, testis, liver, and epididymal fat pad. These authors confirmed the necessity of matrigel for development of teratomas in subcutaneous locations and the teratomas derived from these locations usually lacked large cystic cavities compared to other sites. In their experiments with a limited number of teratomas, intramuscular injections also produced tumors rapidly without significant cyst formation (Hentze et al. 2009).

2.6 Number of cells injected

A few studies have been specifically designed in attempts to answer the question of how many undifferentiated ES cells have to be present in an injection of differentiated cells destined for therapeutic purposes to produce a teratoma. Cao et al in 2007 demonstrated that a minimum of 500-1000 mouse ES cells containing a double fusion reporter construct of Firefly luciferase enzyme (Fluc) and green fluorescent protein (eGFP) were necessary for teratoma formation after subcutaneous injection (Cao et al. 2007). This same number of cells was also able to generate a faint signal allowing for in vivo bioluminescence live imaging. Fewer cells resulted in no detectable bioluminescence on in vivo imaging and no teratoma formation. For the subcutaneous injections, the number of ES cells injected ranged from 1 to 10,000 increasing by a factor of 10. The ES cells were combined with mouse embryonic fibroblasts and matrigel to equal 100,000 cells for injection into adult nude mice. This same group two years later performed similar experiments using undifferentiated H9 hES cells. In this study at least 100,000 (1×10^5) cells was necessary for teratoma formation after intramyocardial injection into adult into adult severe combined immunodeficient (SCID) mice and at least 10,000 cells were necessary for teratoma formation after intramuscular injection (5/7 tumors).

While not specifically addressed in most studies as a variable, the manner of dissociation of ES cells in preparation for injection or transplantation likely plays a key role in the final number of cells necessary to produce a teratoma. Hentze et al also demonstrated that an increasing number of injected hES cell colonies and single cells produced teratomas more quickly in a linear fashion. In their experiments, hES cell colonies were injected in conjunction with approximately 10^6 human foreskin fibroblasts. Teratoma formation was dependent on the cell line used and its adaptation to trypsin digestion. They found that a collagenase passaged hES cell line did not form teratomas after single cell trypsin digestion (Hentze et al. 2009). It is known that ES cells show increased apoptosis and cell death when dissociated into single cells compared to leaving cells in clusters or clumps. Single cell suspensions may require higher cell numbers to compensate for cell loss.

2.7 Host immunological status

Much less literature has specifically addressed the role of the host animal in teratoma formation. Most investigators use some form of immunodeficient mouse. For our teratoma injections, we have used a severe combined immunodeficient mouse strain that is superimposed on a non-obese diabetic background (NOD-SCID) from Jackson Laboratory. These mice lack T and B cell function within a normal hematopoietic background. Of the studies referenced in this section that specifically address teratoma formation from ES cells, little consistency is present in the strain of immunodeficient mouse used for cell transplantation or injection. This is illustrated in the following table:

Study	**Host**
(Lee et al. 2009)	SCID mouse
(Cao et al. 2007)	Nude athymic rats
(Kishi et al. 2008)	NOD-SCID and NOD-SCID/γ^{null} (NOG)
(Gutierrez-Aranda et al.)	NOD-SCID IL2R$\gamma^{-/-}$
(Seminatore et al.)	Sprague-Dawley rats (exogenous immunosuppression with cyclosporine, azathioprine and methylprednisolone
(Prokhorova et al. 2009)	NOD/MrkBomTac-Prkdc^{scid} and NOD/LtSz-Prkdc^{scid}
(Cooke, Stojkovic, and Przyborski 2006)	Adult male nude (nu/nu) mice

Table 1. Immunodeficient hosts used for experimental teratoma formation

The SCID mouse has functional deficits in T and B cells, the NOD/SCID IL2Rγ (NOG) mouse lacks functional T, B, and NK cells as well as lacking response to cytokine signals, and the nude mouse lacks T cells. The effect of these immune capabilities on undifferentiated ES cells that are transplanted or injected into them is unknown. As a related aside, work by Koch et al demonstrated that only 20% of undifferentiated ES cells treated with human serum formed teratomas in immunocompromised mice compared to ES cells not treated or treated with heat inactivated serum. They show that the alternate complement pathway is responsible at least in part for dampening the tumorigenic potential of undifferentiated ES cells (Koch, Jordan, and Platt 2006).

In summary, it is clear that both endogenous and exogenous factors contribute to efficiency of teratoma production and composition. It is also clear that more work needs to be done to unravel how specific conditions affect the formation and composition of teratomas. This will include carefully controlled studies and an efficient means to carefully compare the histological composition of the resulting teratomas.

3. Current technology for tracking stem cell fate, monitoring teratoma formation, and delineation of specific tissue types in vivo

3.1 Tracking stem cells and monitoring teratoma formation in vivo

Many investigations producing teratomas have been motivated by finding ways to detect their formation in vivo and molecular modifications to ES cells to prevent their occurrence. This makes obvious sense since their occurrence in the therapeutic setting is undesirable and would not be detectable for weeks to perhaps months after placement or injection of stem cells. The field of molecular imaging has emerged as a means to study organs and track lesions (such as teratomas) both structurally and functionally. In the field of tissue regeneration/engineering it has been used to track the fate of exogenous stem cells or their differentiated progeny. In vivo imaging technologies include magnetic resonance imaging (MRI), computed tomography (CT), positron emission tomography/CT (PET/CT), single photon emission computed tomography (SPECT), charged coupled device (CCD) camera, and ultrasound. Molecular imaging has evolved to allow better understanding of disease using imaging probes tied to a molecular target that can subsequently be followed and imaged in vivo. Two main methods exist for tracking cells or molecular/genetic functions of cells: One, labeling cells directly and two, via use of reporter gene constructs (Kooreman and Wu).

The first labeling modality involves the use of nanoparticles or quantum dots which emit specified wavelengths of light and are photostable. However, delivery to cells can be difficult, nanoparticles tend to aggregate in the cytosol, and can bind to other molecules. Another modality is iron particle imaging where superparamagnetic iron oxide particles (SPIOs) or ultrasmall superparamagnetic iron oxide particles (USPIOs) can be incubated with cells in culture, taken up by cells, and can be followed after injection using MRI. Cells labeled in such a way have been detected in vivo for up to two months. The major drawback is that these particles can be taken up by macrophages and therefore false positive signals can be detected by MRI for extended periods. A third labeling modality is radionuclide imaging. Here cells are given a radioactive compound and imaged using PET or other radioactive detection methods. The main drawback is that cells can be tracked for at most several days and the radionuclide can leak into other cells.

Reporter gene imaging has shown great promise since the stable integration of reporter genes allows progeny of cells to be tracked. These can take the form of an enzyme that interacts with an exogenous probe to give rise to a signal, a cell surface receptor, a transmembrane protein that mediates intracellular uptake of radioisotopes, or storage protein that concentrates endogenous contrast elements. The expression of reporter genes could be linked to genes of therapeutic or developmental pathway interest and cell progeny tracked by in vivo imaging. Depending on the reporter gene construct, detection can be through ultrasensitive CCD cameras (bioluminescence imaging; BLI), PET, and MRI. In the realm of small animal imaging (e.g. immunodeficient mice), several reporter gene constructs have been developed. The double fusion construct of The double fusion construct of Fluc and eGFP is commonly (eGFP) is commonly used. The animal is given a reporter probe, D-luciferin, which in combination with ATP is oxidized to oxyluciferin, AMP, and light at 560 nm. This reaction is catalyzed by the luciferase enzyme (Fluc) (Marques and Esteves da Silva 2009). The emitted low energy photons are detected by ultrasensitive CCD cameras. The green fluorescence protein allows for detection in postmortem histology. Triple fusion reporter constructs have also been designed that use Fluc, a monomeric red fluorescent protein (mRFP) (for cell sorting purposes), and a herpes simplex virus truncated thymidine kinase (HSVttk) for deep tissue PET imaging. The HSVttk phosphorylates its substrate; a fluoridated (^{18}F) hydroxymethyl butyl guanine generating photons detected by PET. Reporter-suicide gene constructs could also theoretically allow detection and elimination of transplanted cells targeting those destined for teratoma formation. In the case of the reporter construct containing HSVttk, administration of ganciclovir to mice for several weeks can result in lack of teratoma formation. Ganciclovir makes use of its specificity for the viral thymidine kinase and is converted to a toxic drug by phosphorylation by the viral thymidine kinase. Thus, cells infected with the virus produce highly-toxic triphosphates that lead to cell death (Kooreman and Wu).

Transgenes specifically designed for MRI have some advantages in that MRI produces three-dimensional images compared to BLI and ultrasensitive CCD camera. A number of MRI reporter genes have been developed including tyrosinase, transferrin receptor, β-galactosidase, and ferritin (Gilad et al. 2007). Ferritin reporter genes have shown no effect on ES cell tumorigenicity or pluripotency and in comparison to SPIOs can be tracked much longer in vivo although the signal is orders of magnitude less than SPIOs and thus might be better suited for applications requiring long term monitoring (i.e. gene therapy, tracking cell differentiation or other pathways) (Liu et al. 2009; Cohen et al. 2009). A theoretical drawback of reporter gene constructs is interference of normal cellular function by integration of genes into the nuclear genome.

PET imaging has the advantages of spatial/temporal resolution and high sensitivity. Cao et al were able to visualize teratomas in vivo using a radiolabeled RGD (arginine-glycine-aspartate consensus motif present in proteins of the extracellular matrix) that binds to $\alpha_v\beta_3$ integrin during angiogenesis. They demonstrated in vivo imaging of established teratomas in athymic mice after injection of ^{64}Cu-DOTA-RGD4 whereas teratomas could not be detected by PET using ^{18}F-FLT (binds to thymidine kinase) or ^{18}F-FDG (glucose analog) (Cao et al. 2009). As noted previously, however, long term tracking of cell fate or development of teratomas is unlikely with this modality due to the transient nature of the radionuclides.

While these imaging modalities and reporter gene constructs are being used to track teratoma formation and eliminate it, these modalities and methodologies could also be used to study the teratoma itself particularly in the areas of development and even tumorigenesis. Linking reporter genes with fluorescent proteins to genes of interest involved in specific pathways of developments or cancer would be extremely revealing.

3.2 Our experience with magnetic resonance microscopy

In our study of teratomas, the main imaging modality used has been high-resolution MRI (Castro et al.). Our goal in using this imaging technique has been to identify multiple tissue types in vivo and to correlate MRI intensities with histology. MRI performed at resolutions <100 μm per 2D voxel has been designated Magnetic Resonance Microscopy (MRM), and is used as a tool for virtual histological studies. Advancements in MRM have increased the resolution at which biological samples are imaged to as low as 10-16 μm in 2D images and 20-50 μm in 3D volume acquisitions. Many different "virtual tissue stains" can be imparted on MR histology sections by selecting imaging parameters that take advantage of MR contrast mechanisms. In our studies, teratomas have been imaged after removal from the euthanized mouse host and formalin fixation (see (Castro et al.) for details) using a 30 mm diameter 89 mm vertical-bore 11.7T Bruker AVANCE imaging system (Bruker BioSpin Corporation, Billerica, MA) housed at Carnegie Mellon University.

MRM datasets were roughly correlated with histology to determine what tissue types and structures were reliably identified using MR microscopy. Structures and tissues that could be identified from MRM images with little or no ambiguity were adipose tissue, cyst, cartilage, and epidermal lining. Successful registrations allowed homogeneous tissue patches in histology to be correlated with pixel intensity values in T_2-weighted images. Tissues with low intensity in MR images, defined as 3-22% of maximum (cyst) intensity, included adipose and necrotic tissues, mature bone/cartilage, and neuroectoderm. Tissues with medium intensity (22-52% of maximum intensity) included neuroectoderm, immature bone/cartilage, skin, muscle, and gastrointestinal. High-intensity tissues (53-100% of maximum intensity) were limited to cystic lining/fluid and gastrointestinal lining. As can be seen, considerable overlap of intensity and tissue types are present (Figure 1). The ability to transform high resolution MRM images to its corresponding histological appearance would be extremely valuable in the study of tissue development in teratomas in a longitudinal fashion. The ability to produce such "virtual histology" requires that MRM images are precisely aligned to their corresponding histological sections from the serially sectioned teratoma. Next the MRM image characteristics (intensity etc.) would have to be correlated with the images characteristics of the tissue in the histological sections. Several hurdles need to be addressed 1) Precise alignment of each histological section with the corresponding MRM images requires alignment in three-dimensions and 2) even with high-resolution scans the voxel dimensions are 42 um x 89um x 180 for MRM and 3.53 um x 3.53 um x 5 um for a typical histological section; a 1-2 orders of magnitude discrepancy. Single image registration between MRM and histology required tedious and vigilant sample alignment, but did allow for correlation between MR image intensity values and histological tissue types. An additional advantage of MRM is that it can deliver accurate volumetric measurements because true 3D images are obtained, and the sample is unaltered by histological processing.

These higher resolutions did not allow us to uniquely identify any additional tissue types directly from MR images, when compared with clinical imaging. However, MRM may allow direct, non-invasive identification of additional tissue types, because they did reveal texture patterns not visible in low-resolution MRI. Unidentified textures including radial patterns with alternating dark-bright spokes, faint stacks of filamentous parallel lines, dark finger-like projections, faint concentric lines against bright backgrounds, and dark patches with interior textures such as bright speckles or membrane-like lines. Correlating these textures and other regional patterns with the tissues they represent in histology, especially aided

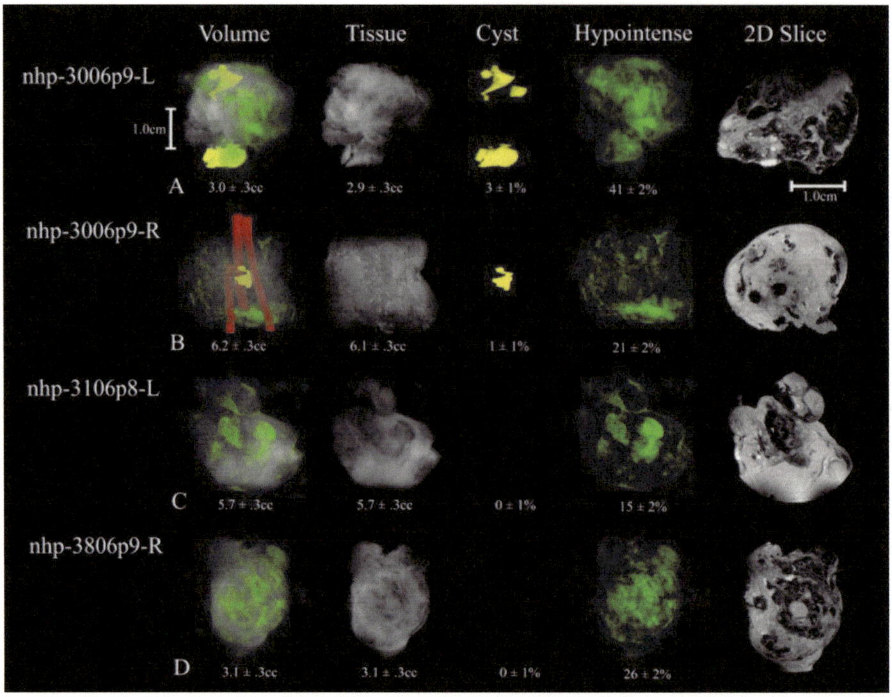

Fig. 1. High-resolution magnetic resonance imaging of teratomas. Teratomas derived from non-human primate ES cells were extracted, fixed and imaged. Hypointense (green; first and fourth columns) and cystic areas (yellow; first and third columns) are segregated and volumes assigned. In the second row, first column, the vertical rods are toothpicks placed into the teratoma at the time of imaging used as fiduciary markers for alignment studies. The far right colum shows a representative 2D slice.

by computational statistical methods, could identify additional tissue types with varying levels of confidence, directly from MRM of teratomas. Computational textural and shape analysis trained over many teratomas, as well as higher resolutions, may offer improved tissue identification by better resolving and considering these subtle texture variations.

4. Histopathology of teratomas

It seems that whenever we have inspected sections from teratomas under the microscope, we have found something, some structure that wasn't present (or seemingly wasn't present) on previous viewing. This is part of what makes them special and fascinating lesions to study. This emphasizes the numerous and varied developmental programs that are in motion simultaneously as these tumors grow. Few papers have carefully described and/or quantified the numerous specific tissue types present in most teratomas derived from ES cells. The description and quantification of tissue types will be critical if teratomas are to be used as models of development or subtle comparisons of teratomas derived under different experimental conditions are to be unraveled. We have identified the majority of tissue types from the hematoxylin and eosin (H&E) slides without the need for ancillary studies. Briefly, we emphasize two critical aspects of the histopathological analysis that allow for optimal examination of teratomas. One, we underscore the importance of having well-processed and well-stained H&E sections. This sounds easy, but in reality even though the histological equipment and techniques have been available for decades, the production of excellent slides for accurate tissue assessment, image production and analysis requires great experience within the laboratory. Well-processed tissue includes having the entire teratoma well-fixed (for several days preferably in paraformaldehyde or 10% phosphate buffered formalin for best penetration and antigenic preservation), serially sectioned at 2-3 mm using 130 mm tissue blades, and processed so that it is free of tissue drying and microtome cutting artifacts. This is particularly important since generally speaking, these tumors are large thus generating a tissue section with a large cross-sectional area compared to a small tissue biopsy. We recommend using a regressive H&E stain that provides for excellent nuclear detail and more consistent staining from batch to batch compared to a progressive staining technique (Carson 1997). Two, for detailed critical studies of teratomas, it is essential to have a versatile and experienced pathologist examine all of the tissue sections of the tumor carefully. The pathologist should be comfortable with the microscopic examination of a wide variety of organ systems including perinatal and/or pediatric specimens and experienced in identifying seemingly inconsequential structures within the milieu of tissues present. Perinatal and/or pediatric clinical pathology disciplines tend to have specimens that include fetal organs, organs with developmental arrest/neoplasia, extraembryonic tissues, and clinical teratomas.

Tissue components seen in our experimental teratomas and their germ layer derivation are shown in the table below.

ECTODERM	MESODERM	ENDODERM
Neuroepithelial	Mesenchyme (undefined immature connective tissue)	Pancreas*
Retinal*	Striated muscle	Liver*
Pigmented cells*	Smooth muscle	Respiratory epithelium
Mature neuroglial	Cartilage	Gastrointestinal epithelium
Immature neuroglial	Bone	Thyroid*
Skin	Adipose	Glandular (unspecified) tissue
Squamous epithelium	Kidney*	Yolk sac
	Skin dermis	
	Bone marrow	*-rarely or never seen in our teratomas

Table 2. Representative tissues seen in our teratomas and their germ layer origin

Only one previous study extensively examined in depth a series of teratomas derived from ES cells in the experimental setting. Gertow et al from the Karolinska Institute published their extensive analysis of teratomas derived from one hES cell line in 2004 (Gertow et al. 2004). This is really the first and only one of few papers that performed and documented extensive histopathological, immunohistochemistry, and FISH studies of teratomas derived from hES cell lines. They analyzed 5 teratomas including the application of 30 antibodies to specific antigens associated with various stages of tissue development from all three primordial germ layers. They found several interesting results: One, the vast majority of cells comprising the teratoma was derived from the stem cells with the exception of endothelial cells comprising feeding vessels and focal areas resembling choroid plexus. Two, they found no cells expressing markers of undifferentiated pluripotent cells. Three, most teratomas had multilineage tissues derived from all three germ layers meaning that different stages of development of the same general tissue type (i.e. mature and immature neuroglial) or different types of the same tissue were present (endochondral, intramembranous bone). Teratomas had extensive and predominant neural development with areas of cartilage, bone, and epithelial development usually within structures reminiscent of bronchi or intestine but derived from both ectoderm and endoderm. A minority of teratomas showed kidney development. Four, more immature areas had higher proliferation rates while mature areas had lower proliferation rates. Five, certain tissue types segregated with one another. Six, differentiation was higher in teratomas grown for longer periods. The latter three points seem to point to some organized pattern following a developmental program. Blum and Benvenisty from the Hebrew University in Jerusalem recently combined three different hES cell lines and produced teratomas that demonstrated equal contributions to tissues from all three lines (Blum and Benvenisty 2007). In addition, analysis of specific structures within the teratomas using laser capture and microsatellite and sex chromosome analysis demonstrated that some structures were polyclonal; that is derived from two cell lines indicating an inductive phenomenon rather than cell autonomous development.

Over the last several years, our group has focused on derivation and study of embryonic stem cells. The group has worked with mES cells, hES cells, and has derived several ES lines from non-human primates including several rhesus and baboon lines (Ben-Yehudah et al. ; Castro et al. ; Navara et al. 2007) (Figure 2).

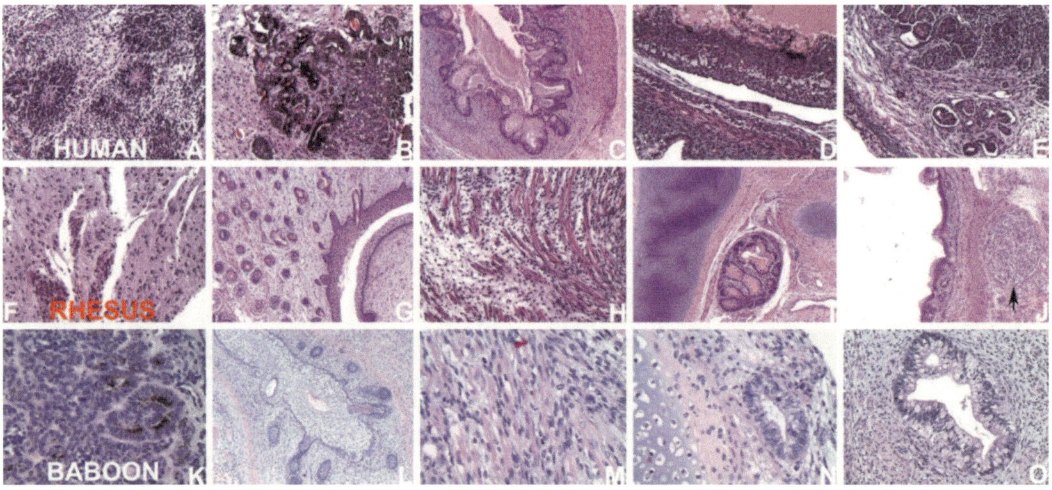

Fig. 2. Representative tissues seen in experimental teratomas derived from human, monkey, and baboon ES cell lines. Human teratomas (A-E): A-neuroepithelium, B-pigmented neural tissue (retinal), C-bowel, D-respiratory or bowel wall, E-primitive kidney. Rhesus teratomas (F-J): F-mature neuroglia, G-skin, H-immature striated muscle, I-immature bronchus, J-bowel wall with ganglion or myenteric plexus (arrow). Baboon teratomas (K-O): K-pigmented neural, L-skin, M-immature skeletal muscle, N-primitive gland adjacent to hyaline cartilage, O-primitive gland surrounded by primitive mesenchyme (H&E stained tissue sections).

More extensive histopathological analysis of teratomas derived from nhpES cells revealed at least 23 different and recognizable tissue entities. (The majority of tissue types present can be seen in Figure 1 of (Castro et al.)). All three germ layers were not only represented (by definition a teratoma) but represented by tissues of the same type at various stages of development and maturation. Neural tissue was represented by primitive neuroepithelium, immature neuroblastic areas, and mature neuroglial tissue resembling brain as well as mature myenteric plexus and ganglia. Skeletal muscle (mesoderm) showed immature myotubes, maturing striated fibrils, and mature striated muscle. Adipose tissue included immature fat containing lipoblasts and areas of mature adipose tissue. Endodermal derivatives included very primitive glandular structures lined by columnar cells with apical nuclei and basal clear cytoplasm (so called "piano key" epithelial cells of yolk sac), less primitive yet still undefined glandular epithelium surrounded by primitive mesenchyme, and well-defined pseudostratified ciliated (respiratory-type) and intestinal epithelium. Immature and mature types of the same tissue could be present in the same teratoma. Only one teratoma had a well-formed tooth organ present.

In summary, teratomas present a dazzling array of tissues at varying stages of development. Tissue recognition, quantification, and delineation of developmental stage are fundamental to their study as models of development, disease, and tumorigenesis.

5. Teratomas in the study of embryonic development

5.1 Quantification of tissues from experimental teratomas

A tumor has been produced in an immunodeficient mouse injected with cells, excised, plopped into formalin, sectioned, a few blocks produced which are sectioned onto slides and examined by the investigator or in some cases a pathologist to determine if tissues from all three germ layers are present. For the most part, if a derived cell line can produce a teratoma and thus prove that the cell line is pluripotent then the teratoma has done its job. The blocks are filed and that is the end. That fact that Muller et al wrote an article calling to standardize not only the reporting of teratoma production but also standardizing the assay speaks to the somewhat nonchalant attitude towards these lesions (Muller et al.). More recent literature has espoused a different look at the teratoma; as a lesion that models development or perhaps failed development since cells are taken out of their normal niche and neighborly contacts (Aleckovic and Simon 2008; Menendez, Bueno, and Wang 2006). Several reasons both practical and theoretical make teratomas an appealing strategy to study development. One, human embryos are inaccessible for studying early development. Embryos at very early stages of development are not available and ethical constraints are enormous. Two, mouse and human development differ significantly such that using mouse models to recapitulate human development is inaccurate. Three, in vitro differentiation of hES cells into embryoid bodies does not reach the complexity of embryonic development. The fact that isolated mouse ES cells can be injected back into a developing blastocyst and continue to contribute to normal embryo formation speaks volumes about how differences in the cellular and environmental milieu can have drastic changes in developmental potential. Some of these cellular and environmental cues in vivo may still be in play even after some degree of cell dissociation in the formation of the experimental teratoma.

We believe that identifying and quantifying the tissue milieu within teratomas both in vivo and ex vivo begin to answer fundamental descriptive questions regarding embryonic development both normal and aberrant. The manual quantification of the myriad of tissues present in most teratomas is a daunting and time consuming task considering the size of the tissue sections of most teratomas (150 mm^2), the number of blocks generated per teratoma (8-12), and the number of tissue types present (>23). A further daunting task is attempting to reconstruct the teratoma in three-dimensional space to visualize the spatial relationships of these tissues to one another. We have done this so far using a semiquantitative approach based on the microscopic assessment of each tissue section. For each section, the amount of each representative tissue belonging to a specific germ layer (i.e. ectoderm, mesoderm, and endoderm) is estimated from each slide of a serially sectioned teratoma using the following scale: 1-[1-20%], 2-[21-40%], 3-[41-60%], 4-[61-80%], and 5-[81-100%]. Estimated percentages are based on how much of the area of the tissue section is represented by that particular tissue. The median of the percentage range is taken for all blocks for each tissue type. For each germ layer, the medians representing the percentage range of the tissue types derived from that germ layer were summed and an overall percentage range assigned to that germ layer. Previously, we have performed semiquantification of tissue types in a limited number of teratomas derived from derived from mES, nhpES, and hES cells with the results in the following table:

	NHP (n=8)	MOUSE (n=3)	HUMAN (n=2)
Incubation (d)	77.8 (13.8)	68.3 (26.3)	70.5 (6.4)
Size (cm)	1.8 (0.3)*	1.9 (0.6)	2.6 (0.1)*
EC (median)	3	5	2
ME (median)	2	1	4
EN (median)	1	1	1

Table 3. Semiquantification of germ layer contributions in experimental teratomas from different species

No differences were seen for incubation days between teratomas derived from nhpES, mES, or hES cells. The human teratomas sampled were significantly larger than either nhp or mES cell derived lesions. Both nhpES and mES cell derived teratomas demonstrated higher median percentage of ectoderm derived tissue present in their teratomas compared to hES derived teratomas. However, hES cell derived teratomas demonstrated higher percentages of mesoderm derived tissues than nhpES or mES cell derived teratomas. No differences were seen for percentage of endoderm derived tissues between nhpES and mES cell derived teratomas. Significant differences at a p-value of 0.02 were seen between the percentage of endoderm derived tissues from nhpES and mES cells compared to hES cell derived teratomas (Ozolek et al. Presented at the 2007 International Society for Stem Cell Research Annual Meeting). While a limited analysis, some trends may exist. One, endoderm seems to be the least prevalent component within teratomas regardless of the species of ES cell from which they are derived. It may be that endodermal derivatives as a whole comprise a small percentage (by weight or volume) of eventual tissues and organs in these species or that the developmental program for certain endodermal derivatives is either not fully established or is aberrant. It is interesting that in our experience with examining numerous teratomas derived experimentally from ES cells, very few contain significant amounts of thyroid, liver, or pancreas—all of endodermal origin. Two, the amount of ectodermal derivatives seems to inversely correlate with the length of gestation (and by extension length of lifespan) for each species and correlates with the degree of nervous system maturity necessary for survival immediately after birth. The amount of mesodermal derivatives shows just the opposite. One could speculate that the mouse for example has a teleological need to have a more mature central nervous system relative to birth and would need ectodermal derivatives (nervous system, nerves, ganglia, etc.) to be the more abundant and more developed tissue. Our analysis of a large group of teratomas derived from two pedigreed "families" of nhpES cells (same male sperm donor and two different females; 3 ES cell lines in one family and 4 ES cell lines in the other family) demonstrated similar percentage ranges for tissues from each of the germ layers (ectoderm-2, mesoderm-3, and endoderm-1 by the above scoring system) for teratomas of one family compared to the other family (Castro and Ozolek, unpublished data). In all cases, injections of cells were done in both testes usually resulting in two teratomas. Most of these teratoma pairs showed a high concordance of similar tissue types and distributions between each other. Tissue by tissue comparison of teratomas from each of the two families showed no differences in the number of teratomas having a specific tissue with the exception of immature neuroglial tissue which was present in all teratomas from "family 2" and in only half of the teratomas from "family 1" (Castro and Ozolek, unpublished data). These teratomas show a slight predominance of tissue types derived from mesoderm mostly mesenchyme and muscle, followed by ectodermal derivatives (neuroglial and skin) followed by endodermal derivatives represented primarily by respiratory-type or intestinal epithelium (Figure 3). These results have several implications. One, the teratoma doesn't appear to develop in a single synchronous fashion, in other words; each specific tissue type is not synchronous in its development. Two, some tissue structure (organ formation) is apparent mostly exhibited by bronchial or intestinal development. In fact, in some instances the cross section of an intestinal segment observed in isolation from its surroundings could not be distinguished from that taken from a mature animal or human including the presence of myenteric neuroglial plexuses. Three, very few of these teratomas derived from nhpES cells that we have examined contain significant islands of developed pancreas, thyroid, liver (endoderm) or kidney (mesoderm); all observations also noted by Gertow et al in their teratoma analysis noted above (Gertow et al. 2004).

However, it should be mentioned that Hentze et al also performed similar semiquantification of germ layer percentages in teratomas generated from undifferentiated and differentiated ES cells (cardiomyocytes and beta cells). They found that mesoderm was the predominant tissue type in all teratomas followed by endoderm then ectoderm. Their brief description of the method used for semiquantifying germ layers was similar to ours in that they estimated tissue as a percentage of the tissue area of the slide. According to their methods, they first cut the teratomas into 0.5 gram pieces, then fixed and processed for histology (Hentze et al. 2009). Quantification of tissue from multiple small fragments could prove more challenging and time consuming if one is to examine the whole teratoma. Another source of discrepancy might revolve around how cysts were quantified. The majority of cysts present in a teratoma are going to be endodermal derivatives. In our methods we discounted the lumen as contributing to tissue area since it was an empty space filled with fluid and not actual tissue. The actual tissue element or mucosal lining would actually be a very small percentage of the total tissue present. One could argue the validity of either methodology.

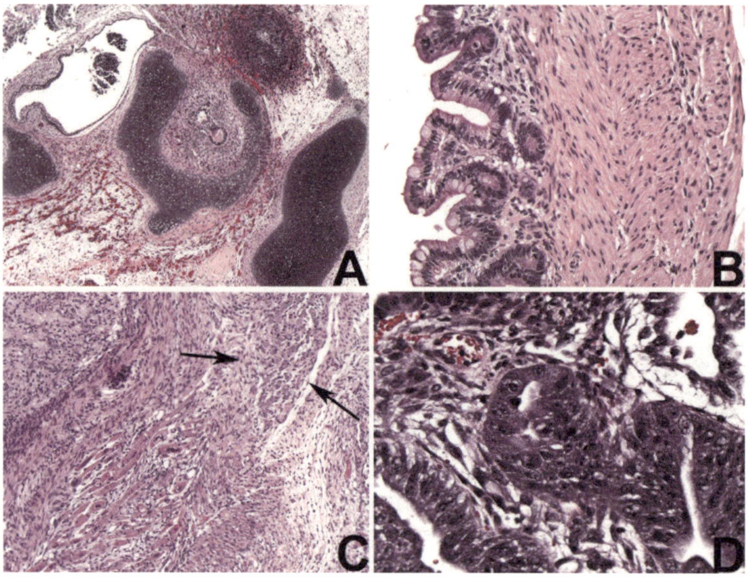

Fig. 3. Common histological themes and interesting observations in experimental teratomas. A-primitive bronchial development (center), B-bowel wall complete with mucosa, lamina propria, and two layers of smooth muscle layered in opposite directions (what would normally be found in mammalian bowel), C-ganglion/myenteric plexus amidst smooth muscle (between arrows), D-primitive glands composed of cells with hyperchromatic nuclei, high nuclear-to-cytoplasmic ratio and basal cytoplasmic clearing (upper right gland) reminiscent of yolk sac (H&E stained tissue sections).

5.2 Automated image analysis for detecting and quantifying tissue in experimental teratomas

As noted earlier, the prospect of manually quantifying tissue types within these teratomas is daunting. Our analysis of the teratomas derived from the nhpES cells took innumerable hours of careful microscopic examination, tabulation, and collation of data and this was only a semiquantitative analysis based on estimated percentages of tissues. Using more rigorous methods would have yielded slightly more accurate data, but would certainly have required two to three times the man hours to accomplish. The ability to accurately and quickly quantify tissues within numerous sections of a teratoma will be critical to identify consistent spatial relationships of tissues with each other, temporal heterogeneity of tissues in three-dimensions and going forward the ability to compare tissue types across multiple teratomas. These authors found only one study that has specifically addressed the issue of quantifying tissues within teratomas using automated imaging/image analysis technologies. Oh and colleagues from Singapore have developed a software platform called TeratomEye that employs three separate algorithms for identifying muscle, gut, and neural tissue with a user-friendly graphical interface (Oh et al. 2009). For muscle, color was used as the main discriminator. For gut, the presence of a lumen was critical in selecting candidates followed by comparison to a set of images of gut and non-gut epithelium. For neural tissue, the same process used for gut tissues was employed. Using these methodologies, they were able to identify muscle accurately 90% of the time with over 90% sensitivity and specificity, gut tissue 87.5% of the time with greater than 80% sensitivity and specificity, and for neural tissue 47.6%. It is unclear from the images provided in the paper whether all of the epithelium lining structures with lumens was indeed gut epithelium or perhaps early respiratory or other types of epithelium. The images of neural tissue demonstrated only neuroepithelium (with neuroepithelial rosettes) but not maturing or mature neuroglial tissues. A fruitful collaboration with Drs. Kovacevic, Rohde, and several graduate students at the Center for Bioimage Informatics of the Department of Biomedical Engineering at Carnegie Mellon University has resulted in several approaches to the problem of tissue quantification in teratomas. Our work has focused on the two critical components of this analysis: segmentation and identification (classification) of tissues. The ability to automatically identify the boundaries of tissue types and isolate tissues is crucial for both rapid identification and quantification of tissue types. This is of course easier said than done since the teratoma does not always follow our rules, but rather produces tissues closely intermingled with each other often without clear and well-defined boundaries (e.g. striated muscle, mesenchyme, adipose tissue, immature cartilage, immature neural). Segmentation is a critical step in many image analysis pipelines particularly if one is focused on the image analysis of one or several specific regions or areas of interest. In the initial phases of our collaborations, segmentation of tissue types was done manually by the pathologists who by doing so not only identified the region of interest but also provided the ground truth identification for each tissue type. The group at Carnegie Mellon University has made significant strides in developing a segmentation tool that can be applied across a variety of image platforms with minimal customization or "tweaking" necessary by the user based on the type of application (Chen et al. A pixel classification system for segmenting biomedical mages using intensity neighborhoods and dimension reduction. in press Proc IEEE Int Symp Biomed Imaging 2011 and Chen et al. A general system for automatic biomedical image segmentation using intensity neighborhoods. in press Int J Biomed Imaging, 2011) (Figure 4).

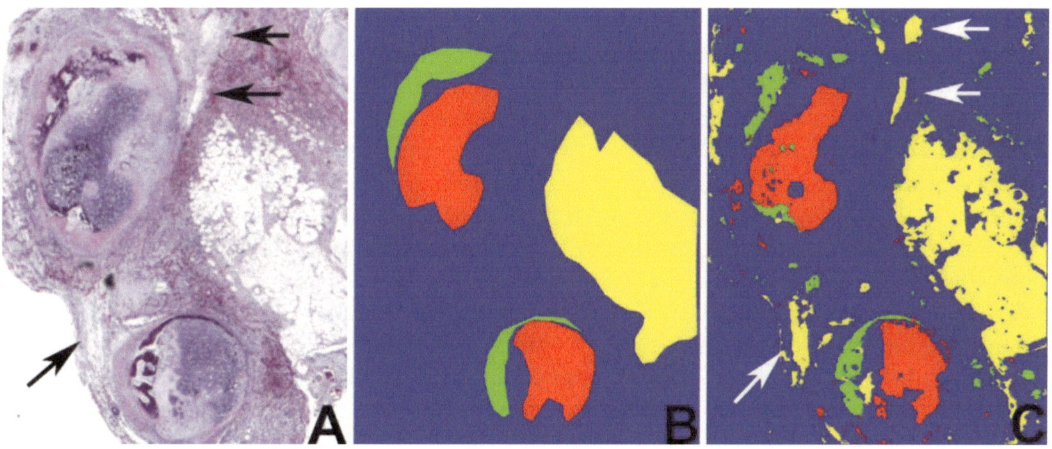

Fig. 4. Automated segmentation tool for distinguishing several common classes of tissues present in teratomas. A-Hematoxylin and eosin stained section of experimental teratoma. Image obtained using a Pathscan II whole slide scanner (Meyer Instruments, Inc. Houston, TX). Cartilage, bone, and adipose tissue (arrows) are readily discernible. B-Ground truth segmented by pathologist: cartilage (red), bone (green), and adipose (yellow). C-Automated segmentation of same image using methods described in text. Note that for adipose tissue in particular, the automated segmentation selected additional areas of adipose not initially segmented by hand (representative regions shown by white arrows).

From an image analysis perspective, tissue segmentation from routinely processed tissue stained with a conventional hematoxylin and eosin stain is challenging due to complex variation in texture, color, shape, and structure of the tissues of interest.
Briefly, this newer approach is achieved within a supervised learning strategy where pixels of a few segmented images are used to train a classifier capable of accurately determining the class (background, tissues of interest) of each pixel in unlabeled images of the same kind. These algorithms use pre-defined window sizes at a given pixel in which pixels inside the window are regarded as neighbors and intensity neighborhood vectors are constructed by reordering the pixels' intensities inside the window into a vector. Using these methods significantly better classification accuracy is achieved across the board for several prototype tissues seen in teratomas compared to other methodologies.

Statistical Measure	Bone	Cartilage	Fat	Background
Our Method	59.70%	73.18%	91.09%	88.93%
Color K-means	29.79%	51.06%	58.73%	55.20%

Table 4. Comparison of quantitative automated segmentation methods for tissue within teratomas

Other approaches aimed at tissue identification have been developed through this collaboration that include use of multiresolution classification, histogram/Earth Mover's Distance, pixel-level classification, and histopathology vocabulary. Multiresolution classifiers have been used since the images (standard H&E) possess many features that are localized in both space and frequency. For these analyses, the image is first converted to grayscale, decomposed into multiresolution subspaces, and features extracted (Haralick texture features and others). A neural network classifier and weighting algorithm are employed to produce a class. Using texture and nuclear features, classification accuracy for several classes of tissues within teratomas approached 88%. The addition of color features improved classification accuracy to over 90% (Bhagavatula et al.). Another approach using the histogram of pixels over the span of a given tissue of interest coupled with a K-nearest neighbor classifier and earth mover's distance produced an average accuracy of 92% for single tissue types (Castro and Bhagavatula. Multiresolution classification of tissue types in teratomas derived from human and non-human embryonic stem cells. Presented at the International Society for Stem Cell Research Annual Meeting 2009). The implementation of a tissue classifier based on the translation of words used to describe histopathological features has proven to be quite robust. In this classification scheme, the pathologist gives each tissue type a verbal description of visual descriptors in order of importance for definitively identifying that particular tissue. For instance in the case of neuroepithelial/neuroectoderm, the verbal descriptors ranked in order of importance to visually identifying it as neuroepithelial/neuroectoderm were the following:

1. Dark blue rim (pseudostratified nuclei) that appears several cells thick
   - Nuclei elongated, oval shape oriented perpendicular to base circumference
2. Round, oval, elliptical macro structure
3. Central lumen (may be light pink and/or white)

This effectively describes the Flexner-Wintersteiner rosettes of primitive neuroepithelium. This was done for 15-20 different tissue types. The engineer then translated those verbal descriptors into engineering synonyms. The pathologist then checked to see if these translated descriptions were still sufficient to classify the tissues. Based on this process a collection of vocabulary terms was assembled shown below:

Background/Fiber colors	Nuclear shape
Cytoplasm color	Nuclear organization
Clear areas (lumens)	Nuclear orientation
Nuclear color	Macro shape
Nuclear density	Background texture

Table 5. Vocabulary terms used for automated tissue classification

Using this approach an average classification accuracy of 97.75% was achieved over a 10-fold cross-validation for 4 tissue types (Bhagavatula et al Automated identification of tissues for digital pathology in press Transactions on Image Processing, 2011).

5.3 Role of immunohistochemistry in tissue identification and quantification

To this point little has been mentioned about ancillary histological techniques that may be of assistance in the classification and quantification of tissue types within teratomas. This has been for good reason since the goal of our image analysis collaborations has been to develop software platforms that can segment, identify, and quantify tissue from images derived from routinely processed tissue. This means the ability to segment, identify, and classify with high accuracy tissues from the same sections that pathologists' eyes have been using to classify tissues in teratomas and make diagnoses in surgical pathology for decades. The implications are that teratoma analyses can be done without more expensive ancillary histological techniques and increase specificity of tissue identification since as we will see using antibody mediated ancillary methods are not necessarily specific for one tissue type. Having said this, the use of ancillary techniques particularly immunohistochemistry can be very valuable not only for tissue identification and quantification but also key in elucidating developmental pathways. Hundreds of antibodies are available for these purposes with many suitable for formalin-fixed paraffin embedded (FFPE) tissue. The keys to using immunohistochemistry for applications such as tissue identification and quantification are the selection of the antibody, using proper immunohistochemistry techniques, and interpretation of staining. No one antibody stains all tissue derived from each germ layer. Thus, "pan-endodermal" or "pan-ectodermal" antibodies do not exist. In fact, the process is more difficult when considering the teratoma since tissues from all three primordial germ layers are not only present but are present in a developmentally heterogeneous fashion. This means that a single tissue type can be represented as it might appear histologically in early fetal life, late fetal life, and into childhood/adolescence/adulthood. Vimentin for instance is a commonly used antibody in both diagnostic and research arenas and is considered a marker of tissues derived from mesoderm or of mesenchymal origin and is used extensively in diagnostic soft tissue pathology. The antibody reacts to an intermediate filament that is present in most if not all fetal cells in early development and thus non-specific. NeuN is a marker of post-mitotic mature neurons, however, all mature neurons are not stained with NeuN (exceptions include cerebellar Purkinje cells, olfactory bulb mitral cells, and retinal photoreceptor cells) and immature neurons are not stained. The situation can be even more complicated when considering actin antibodies since 6 isoforms of actin are present, four of which are in muscle (alpha isoforms) and two in non-muscle (beta, gamma isoforms involved with the cytoskeleton and cell mobility). Therefore, a commonly used antibody such as muscle specific actin clone HHF35 will detect the alpha and gamma isoforms but not cells with the beta isoform. Smooth muscle actin, clone 1A4 detects not only smooth muscle but myofibroblasts and myoepithelial cells also. Antibodies are available that will detect actins in smooth, skeletal, and cardiac muscle. Detection of endodermal derivatives particularly yolk sac using IHC can be even more problematic. Antibodies to cytokeratins can be used to detect epithelial linings but are generally not specific for derivatives of endoderm (Dabbs 2006). Alpha-fetoprotein is produced by immature hepatocytes (liver), endoderm, and yolk sac elements but because of its solubility after alcohol-based processing can be variably present upon immunohistochemical staining. Glypican-3 is one of a family of cell surface heparan sulfate proteoglycans that is abundantly expressed in fetal life. In diagnostic pathology, this antigen has been shown to detect yolk sac elements (to the exclusion of embryonal carcinoma, germinoma and most cases of choriocarcinoma and teratoma elements) within human teratomas (Liu et al. ; Zynger et al. 2008). SALL4 also stains most yolk sac elements and is a nuclear stain but it is not as specific as it will also stain seminoma and embryonal carcinoma (Liu et al.). Taking into account the broad developmental spectrum of tissues present in teratomas, available antibodies (that have been tested true in FFPE tissues), and cost-to-benefit ratio (number of consecutive sections that must be done, staining, reagents, antibody cost, interpretation), the following table (Table 6) lists several antibodies that should highlight the majority of tissues present in a teratoma and provide delineation of germ layer derivation. Consecutive sections from each block of a serially sectioned teratoma could be stained with each antibody and reasonably accurate quantification of staining can then be performed using any number of image analysis software packages that can threshold by color/pixel intensity. We say "reasonably accurate" since antibodies only highlight the antigen they were raised against. One should be aware that the vast majority of antigens are typically present in one or at most two compartments of the cell thus an antibody to a particular antigen may only highlight the nucleus (in the case of most transcription factors) leaving the cytoplasm and membrane unstained or may highlight the membrane and not the cytoplasm or nucleus. Image analysis techniques that use threshold intensities will only segment the stained areas and not include the remainder of the cell thereby underestimating the number of pixels that contribute to that specific tissue. This point is illustrated in Figure 5.

Other tissue/developmental stage specific antibodies to consider that are also readily available for FFPE tissues would include myogenin or myoD1 (immature skeletal muscle; marker for rhabdomyosarcoma in diagnostic pathology), thyroid transcription factor (TTF-1; lung/thyroid), CD31 (PECAM-1; endothelial cells), NeuN (Neuronal Nuclei; mature post-mitotic neurons), Sox2 (neuroepithelium within neuroepithelial rosettes, primitive endoderm including yolk sac), S100 (early glia, schwannian stroma of myenteric plexuses, chondrocytes, adipocytes), NCAM (CD56; neural cell adhesion molecule, immature neuroglia), NFP (neurofilament protein, neurons), Ki-67 (proliferation marker for cells in any portion of the cell cycle). Antibodies to Oct4, Nanog, SSEA3, SSEA4, TRA-1-60, and TRA-1-81 are also available for detection of antigens associated with pluripotency, although these antigens are not typically present in teratomas (but can be seen in undifferentiated areas of teratocarcinomas).

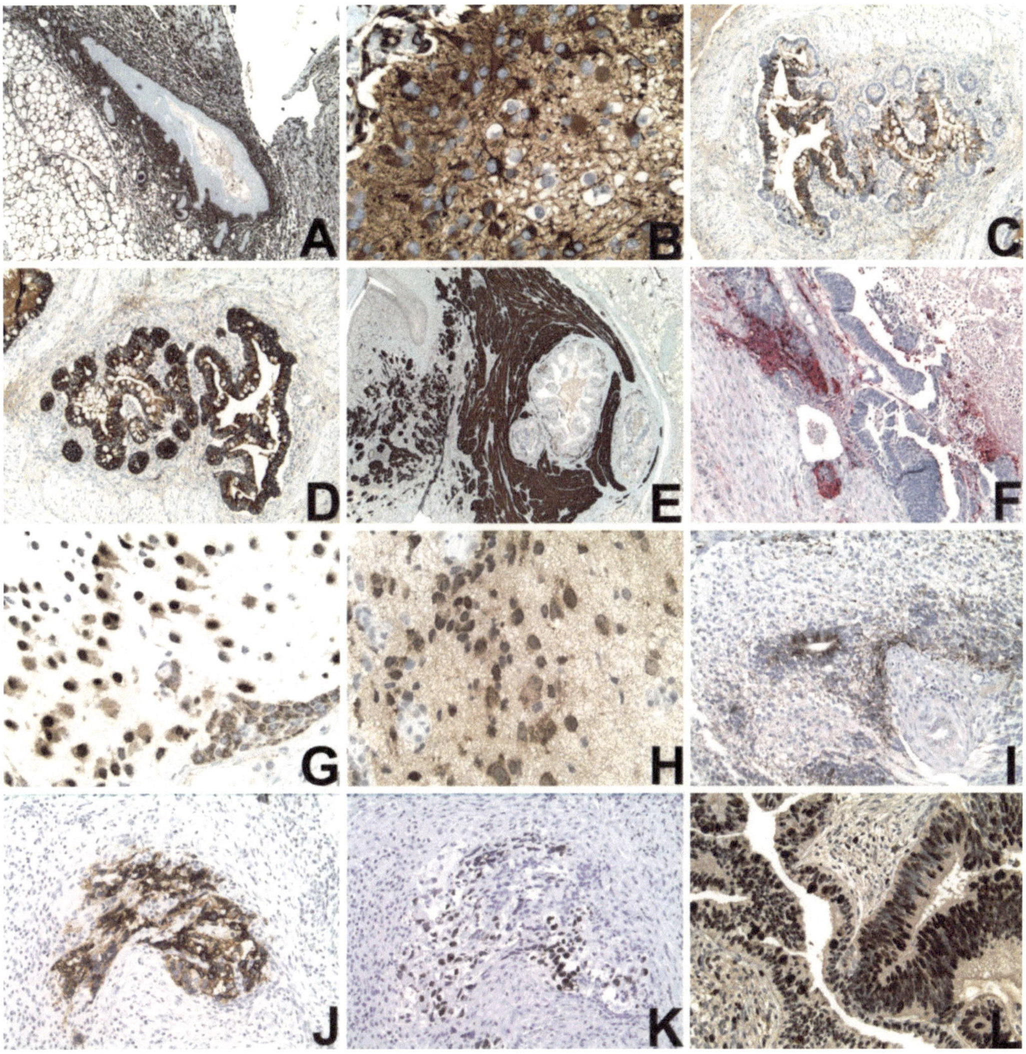

Fig. 5. Sample immunohistochemistry of experimental teratomas. A-Vimentin stained section of skin surrounded by adipose and connective tissue. Vimentin highlights most tissue present but excludes the epithelium (vimentin, 40X), B-Vimentin stained section of neuroglia. In this context vimentin highlights most of the fibrillary background and the cytoplasm of select cells some with cytoplasmic processes (darkly stained cells). These may represent developing astroglia or neural stem cells (vimentin, 200X), C-Cytokeratin 20 stain of epithelial structures. CK 20 is a low molecular weight Type I acidic cytokeratin that is present mainly in colon, pancreas, ovarian surface, and bladder epithelium. In this case the epithelium is focally highlighted (CK20, 100x), D-Cytokeratin 7 stain of same epithelium in (C). CK 7 is present in lung, breast, ovarian epithelium, mesothelium, and endometrium (CK7, 100x). E-All muscle actin stain surrounding mucinous epithelium (AMA, 40X), F-S100 stain of neuroglia highlights cellular and fibrillary staining of area adjacent to more primitive neuroepithelium (S100, red chromogen, 100X), G-NeuN stain of neuroglia showing crisp primarily nuclear staining of neurons (NeuN, 400X), H-PGP 9.5 stain of

neuroglia highlighting neurons and probable neuroblasts (PGP 9.5, 400x), I-glia fibrillary acidic protein (GFAP) stain of human teratoma highlighting some primitive neuroepithelium (GFAP, 200x), J-Glypican-3 stain of yolk sac tumor (taken from child with yolk sac tumor) (glypican-3, 200x), K-SALL4 stain of same focus of tumor seen in (J). Note nuclear staining pattern (SALL4, 200x), L-Ki-67 proliferation marker of neuroepithelium highlights nearly all cells in this focus (Ki-67, nuclear, 200x). Slides of yolk sac tumor in J and K kindly provided by Dr. S. Ranganathan, Children's Hospital of Pittsburgh.

In summary, the preceding sections have only scratched the surface of understanding specific developmental processes. Much of our effort has focused on being able to identify and quantify tissues within these tumors as a means of comparing tumors under experimental conditions. We believe this is a key analysis if teratomas are to be studied to better understand early developmental pathways and the impact of genetic and environmental perturbations on development. Automated image analysis may be crucial in providing the framework for which comparison of tissue composition and quantity can be made between experimental teratomas.

6. Teratomas as models of disease:

Since the isolation of hES cells in 1998, the application of hES cells to human health issues and their potential in this regard has grown exponentially. Directed differentiation of ES cells has produced specific tissue types that may be used for tissue engineering and tissue regeneration. ES cells may be helpful in drug testing, environmental mutagenesis, and toxicity of chemical/physical agents on organ systems and the developing embryo (Ahuja, Vijayalakshmi, and Polasa 2007). They could assist in the identification of molecules that facilitate differentiation into lineage-specific precursors. Applications for developmental biology, disease modeling, and cancer biology are all being investigated (Aleckovic and Simon 2008; Menendez, Bueno, and Wang 2006). In many of these applications, in vitro growth and manipulation of the stem cell population is sufficient and desirable. Early but restricted developmental potential is realized with embryoid body formation by hES cells in vitro. However, for some applications particularly when more prolonged growth or developmental potential is necessary, in vitro models may not be sufficient. Conclusions drawn from murine models of development/aberrant development, cancer, drug toxicity, and other human ailments will always spur some degree of doubt as to their relevance in the human.

The study of tissue development in teratomas could provide a platform for investigating the effects of ingested toxins on the early stages of embryonic development of many different tissue types derived from all three germ layers. This has great significance in understanding the effects of maternal exposure to agents and conditions such as alcohol, tobacco, illicit drugs, starvation, and poor nutrition on the developing fetus. The Barker hypothesis states that fetal and embryonic organ functions can be affected by an altered intrauterine environment and these alterations of organ function determine a setpoint of physiological and metabolic responses that predispose to diseases in adults (Gluckman et al. 2005; Lau and Rogers 2004; Miles, Hofman, and Cutfield 2005). Several epidemiological studies have documented increased deaths due to coronary artery disease and high blood pressure in adults who were born with low birth weights compared to those with birth weights in high percentiles. Intrauterine growth restriction has been linked to impaired glucose tolerance and diabetes in adulthood, reduced nephron count, hypertension, and thymic dysfunction. Indeed, small-for-gestational age infants show differences in growth and nutritional hormonal status with decreased levels of insulin-like growth factors, insulin-like growth factor binding proteins, leptin and others (Lau and Rogers 2004; Miles, Hofman, and Cutfield 2005). As a model of early human development, analysis of the teratoma derived from hES cells that have been subjected to various agents in culture (nicotine, alcohol) or treatment of the immunodeficient mouse host used for growing the teratoma could prove valuable in studying the molecular events of developing tissues in stressed or adverse intrauterine environments.

We have demonstrated accumulation of Tau protein in neurons within teratomas established from a mutant hES cell line produced by transfection with a mutant presenilin gene (unpublished observations) (Figure 6). This opens the door for teratomas to be used to study specific diseases including neurodegenerative diseases after genetic modification of hES cells.

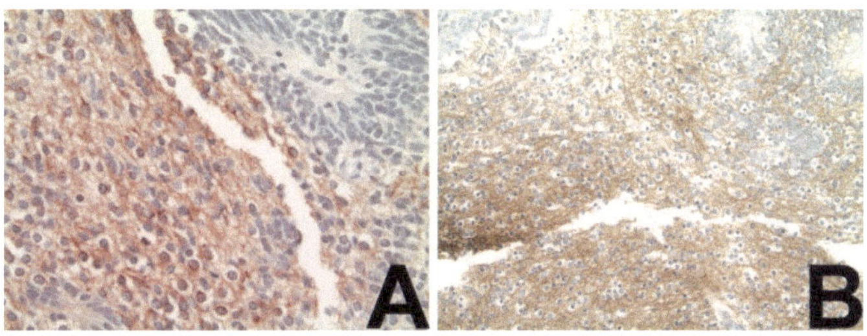

Fig. 6. Tau staining of hES cell line transfected with mutant presenilin gene. A-Staining predominantly in neuroglia adjacent to neuroepithelial rosette (upper right) (Tau, 400x), B-This same area stains with neural cell adhesion molecule signifying its immature state (NCAM, 200x).

Careful analysis of teratomas would begin to unlock the effects on developing tissues resulting from alterations of chromosomal number or specific gene(s). The above mentioned study by Prokhorova et al (Prokhorova et al. 2009) clearly illustrates that alterations of chromosomal number in hES cells can result in dramatic changes in teratoma gross phenotype.

Eiges et al established a hES cell line with full CGG expansion at the FMR1 locus from preimplanation embryos obtained by IVF from a premutation carrier female who had a brother with Fragile X syndrome. Using teratomas derived from this cell line, they were able to show that FMR1 inactivation (and subsequent decrease in Fragile X mental retardation protein (FMRP)) was dependent on differentiation of cells (Eiges et al. 2007). Mensah et al produced a trisomy 21 mES cell line and demonstrated that compared to the wild-type mES cells, teratomas derived from the trisomy 21 cells had significantly decreased amounts of neuroectodermal tissue, decreased mRNA for Tubb3 (neuron-specific gene) and gfap (glial gene), as well as reduced populations of both neuronal and glial cells (Mensah et al. 2007).

7. Teratomas in the study of tumorigenesis

7.1 The unifying theory of teratoma, germ cell tumor, and cancer in humans: More than speculation and theory?

We have already reviewed briefly the close link between ES cells and cancer cells. Mariusz Ratajczak and his group from Louisville have proposed a unifying theory of teratoma, germ cell tumor, and cancer development in humans. Actually, this idea is not new since several scientists in the 19th and early 20th century have proposed that some tumors result from embryonic rests or germ cells. More recent ideas have the origin of human teratomas and germ cell tumors from displaced visceral yolk sac elements (extraembryonic cell theory), primordial germ cells, and embryonic cells. Recent evidence from imprinting studies shows that primordial germ cells are likely candidates for the cell of origin of pediatric germ cell tumors (Gatcombe et al. 2004; Gonzalez-Crussi, Armed Forces Institute of Pathology (U.S.), and Universities Associated for Research and Education in Pathology. 1982; Gonzalez-Crussi, Winkler, and Mirkin 1978; Kristensen et al. 2008; Oosterhuis and Looijenga 2005; Oosterhuis et al. 2007; Schneider et al. 2001). What is new is that the purported cell to support these past theories may have been found. Ratajczak has isolated from murine bone marrow and organs a population of very small cells (5-6 um diameter) that have phenotypic features of embryonic stem cells including expression of Nanog and Oct4, high telomerase activity, lack of major histocompatibility antigen-1 expression, abundant euchromatin, and lack antigenic expression of hematopoietic or other stem cells. These cells had probably been excluded by flow cytometry in previous experiments trying to isolate stem cells from tissues due to their small size. These cells termed VSELs or very small embryonic/epiblast-like cells share imprinting patterns similar to those of primordial germ cells (PGCs) with upregulation of growth-repressive maternally imprinted genes (H19, p57, IgfR2) and downregulation of growth promoting paternally imprinted genes (Igf2, Rasgrf1). VSELs remain dormant within tissues and act as a reservoir for tissue specific stem cells that become activated upon tissue injury. The VSELs isolated by Ratajczak regain some of the somatic imprinting profile after about two weeks in culture but also lose expression of Oct4 and thus do not form teratomas in vivo ((Shin et al. 2009) and personal communication, Mariusz Ratajczak, M.D., Ph.D.). They have found certain culture conditions (C2C12 murine myoblast feeders) where VSELs will form embryoid bodies and stain for fetal alkaline phosphatase. They propose that transduction of VSELs with DNA methyltransferase may change imprinting to favor expression of growth promoting and proliferation genes (Ratajczak, Liu et al.) thus promoting experimental teratoma formation. A recent article by Parte et al demonstrates isolation of a similar population of very small embryonic-like pluripotent stem cells from ovarian surface epithelium (Parte et al.).

These authors promote that VSELs are the missing link between the germ line origin of cancer and the current stem cell theory of cancer. Teratomas/teratocarcinomas may form from Oct4+ VSELs that remain dormant in tissues (maintain somatic imprinting and may require additional mutations). Other germ cell tumors may develop from PGCs or VSELs with persistent genomic imprinting and additional mutations. Pediatric small round blue cell tumors (tumors with multipotential phenotypes such as hepatoblastoma, Wilms tumor) could be derived from mutated VSELs in tissues (epigenetic changes; i.e. loss of imprint in H19-Igf2 locus in Beckwith-Wiedemann). Solid tumors may arise perhaps due to chronic inflammation and/or tissue injury that recruit circulating and/or dormant VSELs into damaged tissues with possible heterokaryon formation and the formation of aneuploid cells. (Ratajczak, Shin, and Kucia 2009; Ratajczak, Shin et al. ; Ratajczak et al. 2008; Shin et al. 2009).

7.2 Teratomas as aids in the study of cancer

An interesting proposal by Tzukerman et al uses teratomas not necessarily as a model for tumorigenesis/cancer pathogenesis but as an environmental platform for studying tumor growth and behavior (Tzukerman et al. 2003). They question the validity of in vitro models to study tumor growth and invasive capability and current in vivo models that use immunocompromised mice. They postulate that murine in vivo models do not mimic what a tumor might do in a human host surrounded by human cells. Their study demonstrated that human ovarian cancer cells injected into teratomas derived from hES cells integrate into and invade surrounding human tissue (within the teratoma) and stimulate neoangiogenesis. They suggest that this type of analysis would be the intermediary step between in vitro and murine in vivo testing of anti-neoplastic agents and human clinical trials. Anti-neoplastic agents could be tested on tumors within a milieu of human cells derived from teratomas.

8. Conclusion

One of the obvious differences between teratomas and a developing human embryo is that we can look at a developing embryo and say (at some point) that the embryo will develop into a human being. We cannot say the same for the teratoma. We really don't know what a teratoma can or will develop into. The dissociation of the blastocyst in reality disarms the entire orderly developmental program forever and when blastocyst cells are collected, reaggregated, and rearranged in a foreign host they simply do not know which end should be up.

Human beings and all life for that matter are amazing in their construction and unconsciously coordinated physiological functions. Form and function are seamlessly woven into a biological tapestry. Teratomas derived from embryonic stem cells are likewise amazing tumors, a visual masterpiece under the microscope; modern or abstract tissue art if you will. Like many examples in the biomedical world, we know who they are but really know very little about what they are. At this point they are more like art in that they have been interpreted individually; "in the eye of the beholder" if you will for the purposes of delineating the presence of three germ layers. Teratomas may be able to help answer many questions. How much of each specific tissue type is present? Do the tissues present recapitulate normal development in three-dimensional space and time? How are the tissues present affected in type and quantity when derived from ES cells that have been manipulated genetically (induced chromosomal aberration, addition/deletion of specific genes) and environmentally (that is, by drugs or toxins, nutrients)? Do all ES cell lines have the equal potential of differentiating successfully into a specific cell type that could be potentially used to treat specific diseases such as Diabetes or Parkinson's disease? How does ES cell phenotype, cell number, site of injection, and host affect tissue formation in teratomas? What are the growth kinetics of teratomas? What influences success rate of teratoma formation? How do teratomas develop? What growth factors influence teratoma growth and tissue development? Do specific developmental programs (tissue or germ layer specific) always show specific developmental potentials? Why does a developmental spectrum (developmentally heterogeneous) of tissues exist within a single teratoma? Why do some "organs" develop more fully than others (i.e. intestine versus kidney)? Why are some tissues hardly ever seen (i.e. liver, thyroid)? Why do teratomas have difficulty forming blood vessels derived entirely from the injected ES cells? Do teratomas differ in their tissue composition between species? At the root of finding the answers to many of these questions will be the ability to carefully, accurately, quantitatively, and quickly delineate tissue types and three-dimensional tissue arrangements both in vivo and ex vivo. Thus techniques for imaging in vivo and ex vivo and image analysis of histological sections need to be investigated with full vigor and intensity. Targeting specific genes to track expression of known early embryological pathways within teratomas will be the key to unlocking developmental fate and for studying specific human chromosomal and genetic diseases.

9. Acknowledgments

The authors gratefully acknowledge our many collaborators and friends for their contributions to this work.

Gerald P. Schatten, Ph.D. and his group for their expertise in embryonic stem cell derivation and characterization and financial, emotional, and scientific support of this work.

Meena Sukhwani from Kyle Orwig's group for her expertise in embryonic stem cell injections and teratoma assay.

Eric T. Ahrens, Ph.D. and graduate student Parker H. Mills in the Department of Biological Sciences at Carnegie Mellon University for their brilliance and expertise in magnetic resonance microscopy.

Gustavo K. Rohde, Jelena Kovacevic, Matthew Fickus, Chen Cheng, Wei Wang, Ramu Bhagavatula, Amina Chebira, Melody Massar, Charles Jackson, Pablo Hennings, Garrett Jenkinson, Mukta Gore, Irina Khaimovich, and Shauna Ormon of the Center for Bioimage Informatics at Carnegie Mellon University for their amazing and outstanding work on automated tissue segmentation, identification, and quantification.

Sarangarajan Ranganathan, M.D. from the Division of Pediatric Pathology at the Children's Hospital of Pittsburgh for graciously contributing the glypican-3 and SALL4 stained sections of yolk sac tumor.

10. References

Ahuja, Y. R., V. Vijayalakshmi, and K. Polasa. 2007. Stem cell test: a practical tool in toxicogenomics. *Toxicology* 231 (1):1-10.

Aleckovic, M., and C. Simon. 2008. Is teratoma formation in stem cell research a characterization tool or a window to developmental biology? *Reprod Biomed Online* 17 (2):270-80.

Ben-Yehudah, A., C. A. th Easley, B. P. Hermann, C. Castro, C. Simerly, K. E. Orwig, S. Mitalipov, and G. Schatten. Systems biology discoveries using non-human primate pluripotent stem and germ cells: novel gene and genomic imprinting interactions as well as unique expression patterns. *Stem Cell Res Ther* 1 (3):24.

Bhagavatula, R., M. Fickus, W. Kelly, C. Guo, J. A. Ozolek, C. A. Castro, and J. Kovacevic. Automatic Identification and Delineation of Germ Layer Components in H&E Stained Images of Teratomas Derived from Human and Nonhuman Primate Embryonic Stem Cells. *Proc IEEE Int Symp Biomed Imaging* 2010:1041-1044.

Bloch, W., E. Forsberg, S. Lentini, C. Brakebusch, K. Martin, H. W. Krell, U. H. Weidle, K. Addicks, and R. Fassler. 1997. Beta 1 integrin is essential for teratoma growth and angiogenesis. *J Cell Biol* 139 (1):265-78.

Blum, B., and N. Benvenisty. 2007. Clonal analysis of human embryonic stem cell differentiation into teratomas. *Stem Cells* 25 (8):1924-30.

———. 2008. The tumorigenicity of human embryonic stem cells. *Adv Cancer Res* 100:133-58.

———. 2009. The tumorigenicity of diploid and aneuploid human pluripotent stem cells. *Cell Cycle* 8 (23):3822-30.

Bulic-Jakus, F., M. Ulamec, M. Vlahovic, N. Sincic, A. Katusic, G. Juric-Lekc, L. Serman, B. Kruslin, and M. Belicza. 2006. Of mice and men: teratomas and teratocarcinomas. *Coll Antropol* 30 (4):921-4.

Cao, F., Z. Li, A. Lee, Z. Liu, K. Chen, H. Wang, W. Cai, X. Chen, and J. C. Wu. 2009. Noninvasive de novo imaging of human embryonic stem cell-derived teratoma formation. *Cancer Res* 69 (7):2709-13.

Cao, F., K. E. van der Bogt, A. Sadrzadeh, X. Xie, A. Y. Sheikh, H. Wang, A. J. Connolly, R. C. Robbins, and J. C. Wu. 2007. Spatial and temporal kinetics of teratoma formation from murine embryonic stem cell transplantation. *Stem Cells Dev* 16 (6):883-91.

Carson, Freida L. 1997. *Histotechnology : a self-instructional text*. 2nd ed. Chicago: ASCP Press.

Castro, C. A., A. Ben-Yehudah, J. A. Ozolek, P. H. Mills, C. J. Redinger, J. D. Mich-Basso, D. A. McFarland, S. L. Oliver, E. T. Ahrens, and G. Schatten. Semiquantitative histopathology and 3D magnetic resonance microscopy as collaborative platforms for tissue identification and comparison within teratomas derived from pedigreed primate embryonic stem cells. *Stem Cell Res* 5 (3):201-11.

Chambers, I., and S. R. Tomlinson. 2009. The transcriptional foundation of pluripotency. *Development* 136 (14):2311-22.

Cohen, B., K. Ziv, V. Plaks, A. Harmelin, and M. Neeman. 2009. Ferritin nanoparticles as magnetic resonance reporter gene. *Wiley Interdiscip Rev Nanomed Nanobiotechnol* 1 (2):181-8.

Cooke, M. J., M. Stojkovic, and S. A. Przyborski. 2006. Growth of teratomas derived from human pluripotent stem cells is influenced by the graft site. *Stem Cells Dev* 15 (2):254-9.

Dabbs, David J. 2006. *Diagnostic immunohistochemistry*. 2nd ed. [Edinburgh]: Churchill Livingstone Elsevier.

Darr, H., Y. Mayshar, and N. Benvenisty. 2006. Overexpression of NANOG in human ES cells enables feeder-free growth while inducing primitive ectoderm features. *Development* 133 (6):1193-201.

Dolgin, E. Putting stem cells to the test. *Nat Med* 16 (12):1354-7.

Eiges, R., A. Urbach, M. Malcov, T. Frumkin, T. Schwartz, A. Amit, Y. Yaron, A. Eden, O. Yanuka, N. Benvenisty, and D. Ben-Yosef. 2007. Developmental study of fragile X syndrome using human embryonic stem cells derived from preimplantation genetically diagnosed embryos. *Cell Stem Cell* 1 (5):568-77.

Gatcombe, H. G., V. Assikis, D. Kooby, and P. A. Johnstone. 2004. Primary retroperitoneal teratomas: a review of the literature. *J Surg Oncol* 86 (2):107-13.

Gerecht-Nir, S., S. Osenberg, O. Nevo, A. Ziskind, R. Coleman, and J. Itskovitz-Eldor. 2004. Vascular development in early human embryos and in teratomas derived from human embryonic stem cells. *Biol Reprod* 71 (6):2029-36.

Gerecht-Nir, S., A. Ziskind, S. Cohen, and J. Itskovitz-Eldor. 2003. Human embryonic stem cells as an in vitro model for human vascular development and the induction of vascular differentiation. *Lab Invest* 83 (12):1811-20.

Gertow, K., J. Cedervall, C. Unger, K. Szoke, E. Blennow, M. P. Imreh, and L. Ahrlund-Richter. 2007. Trisomy 12 in HESC leads to no selective in vivo growth advantage in teratomas, but induces an increased abundance of renal development. *J Cell Biochem* 100 (6):1518-25.

Gertow, K., S. Wolbank, B. Rozell, R. Sugars, M. Andang, C. L. Parish, M. P. Imreh, M. Wendel, and L. Ahrlund-Richter. 2004. Organized development from human embryonic stem cells after injection into immunodeficient mice. *Stem Cells Dev* 13 (4):421-35.

Gilad, A. A., P. T. Winnard, Jr., P. C. van Zijl, and J. W. Bulte. 2007. Developing MR reporter genes: promises and pitfalls. *NMR Biomed* 20 (3):275-90.

Gluckman, P. D., W. Cutfield, P. Hofman, and M. A. Hanson. 2005. The fetal, neonatal, and infant environments-the long-term consequences for disease risk. *Early Hum Dev* 81 (1):51-9.

Goldberg Cohen, I., G. Beck, A. Ziskind, and J. Itskovitz-Eldor. 2006. Human embryonic stem cells for vascular development and repair. *Isr Med Assoc J* 8 (8):573-8.

Gonzalez-Crussi, F., Armed Forces Institute of Pathology (U.S.), and Universities Associated for Research and Education in Pathology. 1982. *Extragonadal teratomas*. Washington, D.C.: Published by the Armed Forces Institute of Pathology under the auspices of Universities Associated for Research and Education in Pathology.

Gonzalez-Crussi, F., R. F. Winkler, and D. L. Mirkin. 1978. Sacrococcygeal teratomas in infants and children: relationship of histology and prognosis in 40 cases. *Arch Pathol Lab Med* 102 (8):420-5.

Gutierrez-Aranda, I., V. Ramos-Mejia, C. Bueno, M. Munoz-Lopez, P. J. Real, A. Macia, L. Sanchez, G. Ligero, J. L. Garcia-Parez, and P. Menendez. Human induced pluripotent stem cells develop teratoma more efficiently and faster than human embryonic stem cells regardless the site of injection. *Stem Cells* 28 (9):1568-70.

Hentze, H., P. L. Soong, S. T. Wang, B. W. Phillips, T. C. Putti, and N. R. Dunn. 2009. Teratoma formation by human embryonic stem cells: evaluation of essential parameters for future safety studies. *Stem Cell Res* 2 (3):198-210.

Herszfeld, D., E. Wolvetang, E. Langton-Bunker, T. L. Chung, A. A. Filipczyk, S. Houssami, P. Jamshidi, K. Koh, A. L. Laslett, A. Michalska, L. Nguyen, B. E. Reubinoff, I. Tellis, J. M. Auerbach, C. J. Ording, L. H. Looijenga, and M. F. Pera. 2006. CD30 is a survival factor and a biomarker for transformed human pluripotent stem cells. *Nat Biotechnol* 24 (3):351-7.

Kishi, Y., Y. Tanaka, H. Shibata, S. Nakamura, K. Takeuchi, S. Masuda, T. Ikeda, S. Muramatsu, and Y. Hanazono. 2008. Variation in the incidence of teratomas after the transplantation of nonhuman primate ES cells into immunodeficient mice. *Cell Transplant* 17 (9):1095-102.

Knoepfler, P. S. 2009. Deconstructing stem cell tumorigenicity: a roadmap to safe regenerative medicine. *Stem Cells* 27 (5):1050-6.

Koch, C. A., C. E. Jordan, and J. L. Platt. 2006. Complement-dependent control of teratoma formation by embryonic stem cells. *J Immunol* 177 (7):4803-9.

Kooreman, N. G., and J. C. Wu. Tumorigenicity of pluripotent stem cells: biological insights from molecular imaging. *J R Soc Interface* 7 Suppl 6:S753-63.

Kristensen, D. M., S. B. Sonne, A. M. Ottesen, R. M. Perrett, J. E. Nielsen, K. Almstrup, N. E. Skakkebaek, H. Leffers, and E. R. Meyts. 2008. Origin of pluripotent germ cell tumours: the role of microenvironment during embryonic development. *Mol Cell Endocrinol* 288 (1-2):111-8.

Lanza, R. P. 2006. *Essentials of stem cell biology*. Amsterdam ; Boston: Elsevier/Academic Press.

Lau, C., and J. M. Rogers. 2004. Embryonic and fetal programming of physiological disorders in adulthood. *Birth Defects Res C Embryo Today* 72 (4):300-12.

Lee, A. S., C. Tang, F. Cao, X. Xie, K. van der Bogt, A. Hwang, A. J. Connolly, R. C. Robbins, and J. C. Wu. 2009. Effects of cell number on teratoma formation by human embryonic stem cells. *Cell Cycle* 8 (16):2608-12.

Legbo, J. N., W. E. Opara, and J. F. Legbo. 2008. Mature sacrococcygeal teratoma: case report. *Afr Health Sci* 8 (1):54-7.

Liu, A., L. Cheng, J. Du, Y. Peng, R. W. Allan, L. Wei, J. Li, and D. Cao. Diagnostic utility of novel stem cell markers SALL4, OCT4, NANOG, SOX2, UTF1, and TCL1 in primary mediastinal germ cell tumors. *Am J Surg Pathol* 34 (5):697-706.

Liu, J., E. C. Cheng, R. C. Long Jr, S. H. Yang, L. Wang, P. H. Cheng, J. J. Yang, D. Wu, H. Mao, and A. W. Chan. 2009. Noninvasive Monitoring of Embryonic Stem Cells in vivo with MRI Transgene Reporter. *Tissue Eng Part C Methods*.

Loh, Y. H., J. H. Ng, and H. H. Ng. 2008. Molecular framework underlying pluripotency. *Cell Cycle* 7 (7):885-91.

Marques, S. M., and J. C. Esteves da Silva. 2009. Firefly bioluminescence: a mechanistic approach of luciferase catalyzed reactions. *IUBMB Life* 61 (1):6-17.

Menendez, P., C. Bueno, and L. Wang. 2006. Human embryonic stem cells: A journey beyond cell replacement therapies. *Cytotherapy* 8 (6):530-41.

Mensah, A., C. Mulligan, J. Linehan, S. Ruf, A. O'Doherty, B. Grygalewicz, J. Shipley, J. Groet, V. Tybulewicz, E. Fisher, S. Brandner, and D. Nizetic. 2007. An additional human chromosome 21 causes suppression of neural fate of pluripotent mouse embryonic stem cells in a teratoma model. *BMC Dev Biol* 7:131.

Miles, H. L., P. L. Hofman, and W. S. Cutfield. 2005. Fetal origins of adult disease: a paediatric perspective. *Rev Endocr Metab Disord* 6 (4):261-8.

Muller, F. J., J. Goldmann, P. Loser, and J. F. Loring. A call to standardize teratoma assays used to define human pluripotent cell lines. *Cell Stem Cell* 6 (5):412-4.

Navara, C. S., J. D. Mich-Basso, C. J. Redinger, A. Ben-Yehudah, E. Jacoby, E. Kovkarova-Naumovski, M. Sukhwani, K. Orwig, N. Kaminski, C. A. Castro, C. R. Simerly, and G. Schatten. 2007. Pedigreed primate embryonic stem cells express homogeneous familial gene profiles. *Stem Cells* 25 (11):2695-704.

Oh, S. K., P. Chua, K. L. Foon, E. Ng, A. Chin, A. B. Choo, and R. Srinivasan. 2009. Quantitative identification of teratoma tissues formed by human embryonic stem cells with TeratomEye. *Biotechnol Lett* 31 (5):653-8.

Oosterhuis, J. W., and L. H. Looijenga. 2005. Testicular germ-cell tumours in a broader perspective. *Nat Rev Cancer* 5 (3):210-22.

Oosterhuis, J. W., H. Stoop, F. Honecker, and L. H. Looijenga. 2007. Why human extragonadal germ cell tumours occur in the midline of the body: old concepts, new perspectives. *Int J Androl* 30 (4):256-63; discussion 263-4.

Parte, S., D. Bhartiya, J. Telang, V. Daithankar, V. Salvi, K. Zaveri, and I. Hinduja. Detection, Characterization, and Spontaneous Differentiation In Vitro of Very Small Embryonic-Like Putative Stem Cells in Adult Mammalian Ovary. *Stem Cells Dev*.

Prelle, K., N. Zink, and E. Wolf. 2002. Pluripotent stem cells--model of embryonic development, tool for gene targeting, and basis of cell therapy. *Anat Histol Embryol* 31 (3):169-86.

Prokhorova, T. A., L. M. Harkness, U. Frandsen, N. Ditzel, H. D. Schroder, J. S. Burns, and M. Kassem. 2009. Teratoma formation by human embryonic stem cells is site dependent and enhanced by the presence of Matrigel. *Stem Cells Dev* 18 (1):47-54.

Ratajczak, M. Z., R. Liu, J. Ratajczak, M. Kucia, and D. M. Shin. The role of pluripotent embryonic-like stem cells residing in adult tissues in regeneration and longevity. *Differentiation* 81 (3):153-61.

Ratajczak, M. Z., D. M. Shin, and M. Kucia. 2009. Very small embryonic/epiblast-like stem cells: a missing link to support the germ line hypothesis of cancer development? *Am J Pathol* 174 (6):1985-92.

Ratajczak, M. Z., D. M. Shin, R. Liu, W. Marlicz, M. Tarnowski, J. Ratajczak, and M. Kucia. Epiblast/germ line hypothesis of cancer development revisited: lesson from the presence of Oct-4+ cells in adult tissues. *Stem Cell Rev* 6 (2):307-16.

Ratajczak, M. Z., E. K. Zuba-Surma, M. Wysoczynski, J. Ratajczak, and M. Kucia. 2008. Very small embryonic-like stem cells: characterization, developmental origin, and biological significance. *Exp Hematol* 36 (6):742-51.

Rosai, Juan, and Lauren Vedder Ackerman. 2004. *Rosai and Ackerman's surgical pathology*. 9th ed. 2 vols. Edinburgh New York: Mosby.

Schieke, S. M., M. Ma, L. Cao, J. P. McCoy, Jr., C. Liu, N. F. Hensel, A. J. Barrett, M. Boehm, and T. Finkel. 2008. Mitochondrial metabolism modulates differentiation and teratoma formation capacity in mouse embryonic stem cells. *J Biol Chem* 283 (42):28506-12.

Schneider, D. T., A. E. Schuster, M. K. Fritsch, J. Hu, T. Olson, S. Lauer, U. Gobel, and E. J. Perlman. 2001. Multipoint imprinting analysis indicates a common precursor cell for gonadal and nongonadal pediatric germ cell tumors. *Cancer Res* 61 (19):7268-76.

Seminatore, C., J. Polentes, D. Ellman, N. Kozubenko, V. Itier, S. Tine, L. Tritschler, M. Brenot, E. Guidou, J. Blondeau, M. Lhuillier, A. Bugi, L. Aubry, P. Jendelova, E. Sykova, A. L. Perrier, B. Finsen, and B. Onteniente. The postischemic environment differentially impacts teratoma or tumor formation after transplantation of human embryonic stem cell-derived neural progenitors. *Stroke* 41 (1):153-9.

Shin, D. M., E. K. Zuba-Surma, W. Wu, J. Ratajczak, M. Wysoczynski, M. Z. Ratajczak, and M. Kucia. 2009. Novel epigenetic mechanisms that control pluripotency and quiescence of adult bone marrow-derived Oct4(+) very small embryonic-like stem cells. *Leukemia* 23 (11):2042-51.

Tzukerman, M., T. Rosenberg, Y. Ravel, I. Reiter, R. Coleman, and K. Skorecki. 2003. An experimental platform for studying growth and invasiveness of tumor cells within teratomas derived from human embryonic stem cells. *Proc Natl Acad Sci U S A* 100 (23):13507-12.

Yu, J., and J. A. Thomson. 2008. Pluripotent stem cell lines. *Genes Dev* 22 (15):1987-97.

Zynger, D. L., M. J. Everton, N. D. Dimov, P. M. Chou, and X. J. Yang. 2008. Expression of glypican 3 in ovarian and extragonadal germ cell tumors. *Am J Clin Pathol* 130 (2):224-30.

Part 4

Pluripotency and Molecular Biology of Embryonic Stem Cells

14

Illuminating Hidden Features of Stem Cells

Gideon Grafi[1*], Rivka Ofir[3], Vered Chalifa-Caspi[2] and Inbar Plaschkes[2]
[1]French Associates Institute for Agriculture and Biotechnology of Drylands, Jacob Blaustein Institutes for Desert Research, Ben-Gurion University of the Negev,
[2]The National Institute for Biotechnology, Ben-Gurion University of the Negev,
[3]Ben Gurion University of the Negev and Dead Sea & Arava Science Center, Hazeva Branch,
Israel

1. Introduction

In recent years stem cells become a major topic both publicly and scientifically owing to their promise to cure diseases and restore organ functionality. Yet our understanding of the biology of stem cells and their inherent features is largely lagging behind the great promise of using these cells in transplantation therapies. In this chapter we highlight several aspects of the biology of stem cells/induced pluripotent stem (iPS) cells that are often overlooked. We define stem cells not only by their developmental capacities, namely, self-renewal and multi-lineage differentiation, but also by their inherent features. We present new bioinformatic data and draw the stem cell picture based on recent knowledge emerging from studying stem cells in plant and animal systems with emphasis on chromatin structure. We highlight some of the potentially hazardous pathways associated with cellular dedifferentiation (iPS cells) and with culturing stem cells. Notably, genomic modification associated with iPS cells is often discussed with respect to the methodology of introducing reprogramming genes into the host cells, namely, lentiviral/retroviral transduction, while ignoring the potential genomic modification naturally associated with dedifferentiation or with stress events. Based on our understanding of cellular processes accompanied stress response and cellular dedifferentiation we discuss strategies for improving the quantity and quality of iPS cells.

2. Inherent features of stem cells

2.1 What are stem cells?

One major, long recognized, pitfall in stem cell biology is that stem cells are commonly defined by their developmental potentialities, namely, self-renewal and multitype differentiation, rather than by their intrinsic features (Potten & Loeffler, 1990; McKay, 2000). This often leads to the erroneous assumption that reentry of stem cells to the cell cycle for the purpose of the so-called 'self-renewal' represents an inherent feature of stem cells. Consequently, the term stem cell culture has been established, leading biologists to incorrectly assume that stem cell features can be fully maintained under culture conditions (Grafi & Avivi, 2004). Considering that animal somatic cells are capable to become stem cells *via* dedifferentiation (Kurisaki et al., 2010), the traditional developmental capacity-based definition of stem cells may be useless in distinguishing between genuine stem cells and somatic cells that have the capacity to become stem cells. Obviously, defining stem cells by their inherent features or signature rather than their developmental capabilities is necessary. The signature of a given differentiated cell is commonly reflected in its gene expression profile. The attempts to uncover the 'stem cell signature' or the 'stemness genes' via transcriptome analysis of different stem cell culture lines were failed as these experiments yielded different 'signatures' and non-overlapping 'stemness genes' (Ramalho-Santos et al., 2002; Ivanova et al., 2002; Fortunel et al., 2003). This again highlighted the problem of defining stem cells by their developmental capacity, namely, self-renewal (cultured cells) (Grafi & Avivi, 2004). Contrary to the idea that stem cells represent a unique entity that is characterized by the expression of specific set of 'stemness' genes, it has been suggested that stem cells represent a unique transient state characterized by promiscuous expression of marker genes (Zipori, 2004). Consequently, several possibilities could describe the transcriptional landscapes of stem cells: (i) differentiation/lineage-specific genes are not expressed in stem cells, (ii) differentiation/lineage-specific genes are widely expressed in stem cells but at a very low level, and (iii) differentiation/lineage-specific genes are not expressed but assume a transcriptionally competent chromatin state (Zipori, 2004; Meshorer & Misteli, 2006; Efroni et al., 2008). The later is supported by the fact that stem cells, like dedifferentiating cells, acquire open, decondensed chromatin architecture, which is essential though not sufficient for initiating gene transcription (Williams et al., 2003; Grafi, 2004; Gaspar-Maia et al., 2011) as well as by the finding that some non-expressed genes or genes expressed at low levels in embryonic stem (ES) cells are primed but their transcription is attenuated by a unique combination of permissive and restrictive chromatin marks (Azuara et al., 2006; Bernstein et al., 2006). Apparently open chromatin configuration appears to be an intrinsic feature of stem cells that can be used to distinguish between actual stem cells and potential stem cells.

2.2 Chromatin structure and regulation in brief

The basic structural unit of chromatin is the nucleosome, which is composed of DNA wrapped around histone octamer made of two of each of core histone proteins, namely, H2A, H2B, H3 and H4. All core histone proteins share a common structural motif called the histone fold consisting of three alpha helices connected by short loops (Arents & Moudrianakis, 1995). The X-ray crystal structure of the nucleosome core particle showed that interactions between core histone proteins or between histone octamer and duplex DNA are largely dependent on the histone fold motif. The histone amino-terminal tail is unstructured and protruding outside the nucleosomal disk where it can contact with neighboring nucleosomes or with proteins or protein complexes that affect chromatin structure and function. The structure of chromatin is highly dynamic, facilitating the transition between permissive and repressive chromatin. This dynamic structure is controlled by multiple types of reversible chemical modifications that occur on the DNA (cytosine methylation) or on the DNA interacting core histone proteins. Most modifications of histone proteins occur on the N-terminal tails and include acetylation, methylation, phosphorylation, and ubiquitination. These chemical modifications can directly affect the interaction with DNA or generate binding sites for recruitment of proteins or protein complexes that affect the structure and function of chromatin and consequently differentiation and development. Gene promoters can be found in three fundamental states determined by their histone modification marks, namely, restrictive (e.g., methylated

H3K9/K27), permissive (e.g., methylated H3K4, acetylated H3K9) or both restrictive and permissive (e.g., methylated H3K27, methylated H3K4, also known as 'bivalent' state). For example, SET domain-containing histone methyltransferase proteins, such as Clr4 (Cryptic locus regulator) in *Schizosaccharomyces pombe* and SUV39H1 and SUV39H2 in humans are enzymes that methylate histone H3 specifically at lysine 9 (Rea et al., 2000). This methylation generates a binding site for Heterochromatin protein 1 (HP1) (Bannister et al., 2001; Lachner et al., 2001), which is required for the establishment of a condensed, repressive chromatin. Conversely, histone acetylation or methylation of histone H3 at lysine 4 are often associated with 'open' chromatin configuration and gene transcription (Eberharter & Becker, 2002; Lee & Workman, 2007; Eissenberg & Shilatifard, 2010). More recently it has been shown that some non-expressed genes or genes expressed at low levels in ES cells carry both permissive (H3K4me3; trimethylated H3K4) and restrictive (H3K27me3; trimethylated H3K27) chromatin marks (Azuara et al., 2006; Bernstein et al., 2006). This unique 'bivalent' chromatin state suggests a model in which many tissue-specific regulatory genes are 'primed' but their transcription is delayed until entry into a specific differentiation pathway that dictate either activation (e.g., recruitment of H3K27 demethylases) or silencing (e.g., recruitment of H3K4 demethylses) of the gene locus (Lan et al., 2008). Notably, bivalent domains were found to occur also in the rice genome in somatic cells where a large proportion of genes possessing the repressive mark H3K27me3 also contained permissive marks of H3K4me3 and H3K9ac (acetylated H3K9) (He et al., 2010).

2.3 Chromatin modifiers and the establishment of stem cell state

Stem cells as well as dedifferentiating cells have been shown to maintain their chromatin in an open, decondensed configuration relative to differentiated cells (reviewed in Gaspar-Maia et al., 2011). Accordingly, chromatin decondensation appears to be a hallmark of cells engaged in switching fate, a process, which requires the acquisition of a stem cell-like state, which is essential for normal development. For example, during fertilization, decondensation of sperm chromatin is essential for the formation of the male pronucleus, which is necessary for successful fertilization (Longo & Anderson, 1968). Interestingly, sperm chromatin decondensation can be mimicked in *Xenopus* egg extracts and requires nucleoplasmin, an acidic, thermostable, abundant nuclear protein capable of binding histone proteins (Laskey et al., 1978; Philpott et al., 1991; Katagiri & Ohsumi, 1994). Similarly, *Xenopus* egg extracts can induce chromatin decondensation of heterologous somatic nuclei. It has been shown that nuclei derived from chicken erythrocytes incubated in *Xenopus* egg extract were induced to replicate their DNA, a process preceded by two sequential phases of chromatin decondensation (Blank et al., 1992). Chromatin decondensation was reported in cells of regenerating tissues of the tubeworm *Owenia fusiformis* 12 h after amputation (Fontes et al., 1980), a stage, which might correlate with cellular dedifferentiation whereby somatic cells at the amputation site withdraw from their differentiated state and acquire stem cell-like state (Grafi, 2004). In plants, chromatin decondensation was found to be associated with cellular dedifferentiation induced following exposure to various stress conditions (e.g., protoplasting, exposure to dark) (Zhao et al., 2001; Tessadori et al., 2007; Damri et al., 2009, Grafi et al., 2011).

The evidence that stem cells may assume a more open chromatin conformation is dated about 40 years ago with the description of the morphological characteristics of the so-called stem cells in the bone marrow by Murphy et al. (1971). Using electron microscopy examination of 3-day post hypoxic marrow, they found that the large mononuclear cells are characterized by a fine nuclear chromatin (leptochromatic) with some peripheral condensation. Likewise, electron microscopy inspection of the erythropoietic cells of the chick showed that stem cells are characterized by large nuclei and homogenous euchromatin, while maturation is accompanied by an increase in nuclear condensation indicated by areas of heterochromatin (MacRae & Meetz, 1970). Also, ultrastructure of developing erythrocytes in the bone marrow of human adults showed that in the earliest stages, the nucleoplasm was chiefly composed of euchromatin, while during maturation heterochromatin was rapidly increased concomitantly with a reduction in nuclear size (Miura et al., 1974). Furthermore, using micrococcal nuclease, Weintraub (1978) showed that the nucleosome repeat length (NRL) increases from 190 to 212 basepairs during erythropoiesis in the chick, which was accompanied by a significant increase in the concentration of red cell specific histone H5, a linker histone necessary for stabilization of higher order chromatin structure (Robinson & Rhodes, 2006). This increase in NRL has recently been shown to be required for the formation of the 30 nm DNA fiber-induced chromatin compaction (Routh et al., 2008). Notably, similarly to animal stem cells, ultrastructural observations of the nuclei in the shoot apex (the plant stem cell niche) of the plant *Tradescantia paludosa* showed that a large proportion of the chromatin is organized as less condensed, diffused euchromatin fibrils (Booker & Dwivedi, 1973).

It has been demonstrated that chromatin decondensation of somatic nuclei in *Xenopus* egg extracts requires the chromatin-remodeling nucleosomal adenosine triphosphatase (ATPase) ISWI (Kikyo et al., 2000). More recently, it has been shown that embryonic stem cells are extensively engaged in global gene transcription, due in part to overrepresentation of chromatin-remodeling genes and the general transcription machinery; this activity is diminished upon differentiation (Efroni et al., 2008). Knockdown by RNAi of the SWI/SNF remodeling factor BRG1 and of the ISWI-related chromodomain helicase DNA binding protein 1-like (Chd1l) resulted in impairment of ES cell differentiation and proliferation (Efroni et al., 2008). Indeed, knockdown of Chd1 in ES cells established its importance for maintaining open chromatin, pluripotency of embryonic stem cells, as well as for reprogramming of somatic cell to pluripotent state (Gaspar-Maia et al., 2009). Furthermore, reprogramming of somatic cells by the expression of the four transcription factors, Oct4, Sox2, Klf4 and c-Myc is facilitated by combined expression of chromatin remodeling components of the BAF (Brg1/Brahma-associated factor) complex (Singhal et al., 2010).

2.4 Pattern of chromatin modifier gene expression in stem cells and dedifferentiating cells

Open chromatin configuration emerges as a fundamental theme in pluripotent stem cell biology, and should be primarily served as a major attribute for defining the stem cell state. This open chromatin conformation is crucial for maintaining the stemness state, which might confer stem cells with the capacity for rapid switching into the appropriate transcriptional program upon induction of differentiation and with the flexibility needed for differentiation into multiple cell types. Thus, in looking for molecular features defining stem cells we should highlight the molecular components regulating chromatin structure, namely chromatin modifier genes (CMGs). There are two possible ways for maintaining the unique chromatin state characteristic of stem cells, which we define as quantitative and qualitative approaches. The quantitative approach suggests that the flexibility needed for stem cells to differentiate into multiple cell types can be achieved by promiscuous expression of CMGs, while the qualitative approach suggests that the open chromatin conformation characteristic of stem cells is maintained by selective expression and repression of genes that promote the formation of permissive and restrictive chromatin, respectively.

The quantitative approach gains support from analysis of the gene expression profiles of the plant shoot apical meristem (SAM) stem cells (Yadav et al., 2009). By introducing fluorescent reporters into SAM-enriched *Arabidopsis* background line, Yadav et al. (2009) have isolated by FACS three distinct cell types of the SAM stem cell niche and analyzed the transcriptome profiles of these cells. One major attribute of SAM stem cells appears to be a flexible chromatin state demonstrated by the overrepresentation of genes involved in chromatin organization (Yadav et al., 2009). Remarkably, further analysis of the available microarray dataset of the SAM stem cells revealed that among the 445 CMGs, which were represented on the ATH1 array, stem cells showed the expression of 297 genes whose expression signal is higher than 256 ($>2^8$). This unusual expression of chromatin modifier genes in SAM stem cells might confer the flexibility required for maintaining the pluripotent state of stem cells.

Apparently, this widespread expression of CMGs observed in SAM stem cells is not mimicked in dedifferentiating plant cells. Transcriptome profiling of dedifferentiating protoplast cells (Damri et al., 2009) showed that among the 465 CMGs, which were represented on the ATH1 array only 95 genes displayed an expression signal higher than 256 ($>2^8$). However, close examination of these CMGs highlighted the qualitative approach revealing that many of the genes that were down-regulated in dedifferentiating protoplasts, such as histone deacetylase encoding genes SRT2, HDA14 and HDT4, as well as linker histone genes Hon1 and Hon2 are implicated in chromatin compaction, while many of the CMGs that were up-regulated, such as histone acetyltransferase encoding genes HAF1, HAC5 and HAG3 as well as histone demethylase genes JMJ21 and JMJ13 are implicated in chromatin decondensation. Thus the expression profile of CMGs favors the acquisition of an open chromatin conformation in dedifferentiating cells.

Since open chromatin configuration characterizes stem cells as well as dedifferentiated cells in plants and animals, the study of these cells in plants might have bearing to understanding inherent biological features of animal stem cells.

2.5 CMGs as a tool for monitoring pluripotency

We hypothesized that these features of quantitative and qualitative CMG expression of SAM stem cells and dedifferentiating cells, respectively, might hold also for human stem cells and dedifferentiating cells and could serve as a distinguishing feature between animal cells displaying various levels of pluripotency. To assess the qualitative approach, we first predefined a set of human CMG families, which were selected based on their predicted chromatin function, namely, histone deacetylases (HDACs) involved in heterochromatin formation, histone acetyltransferases (HATs), which are involved in euchromatin formation and the jumonji family of histone demethylases implicated in both euchromatin and heterochromatin formation. We then checked the expression level of these predefined CMGs (listed in Table 1) in a recently published meta-analysis of human gene expression datasets (Lukk et al., 2010). In this study, the authors collected and integrated high-quality data from 5,372 samples from 206 different studies, all generated using Affymetrix U133A arrays and deposited in public databases. Using text mining and curation, the authors binned the samples in 369 biological groups, each representing a particular cell or tissue type, disease state or cell line. Additionally, samples were grouped more crudely according to different criteria, one of which divided the samples into four meta-groups: Cell lines, Neoplasms,

Non-neoplastic diseases, and Normal. We have downloaded the entire gene expression matrix from the ArrayExpress repository (accession number E-MTAB-62), and reduced it to include only the 1033 samples of the "Normal" meta-group, and the 40 CMGs indicated in Table 1. Subsequently, probe sets which represented the same gene were consolidated as follows: (1) probe sets with signal intensity lower than 5 (in log2 scale) in all 1033 samples were removed, provided that at least one probe set for the same gene had signal intensity higher than 5 (in log2 scale) in at least one of the samples. (2) The signal intensities of the remaining probe sets were averaged to yield a single expression value per gene per sample. The resulting matrix was loaded into Partek Genomics Suite™. To allow for gene-wise comparison of expression profiles, expression values were standardized such that the mean and standard deviation of each gene's values were 0 and 1, respectively. The standardized dataset was subjected to bi-directional hierarchical clustering using Euclidean distance and complete linkage. Finally, sample annotations indicating 6 meta-groups (brain, cell line, hematopoietic system, incompletely differentiated, muscle, solid tissue) and operating laboratory were added. As shown in Fig. 1, based on their expression pattern, the CMGs were clearly clustered into three major clades. Most intriguing is clade 1 displaying

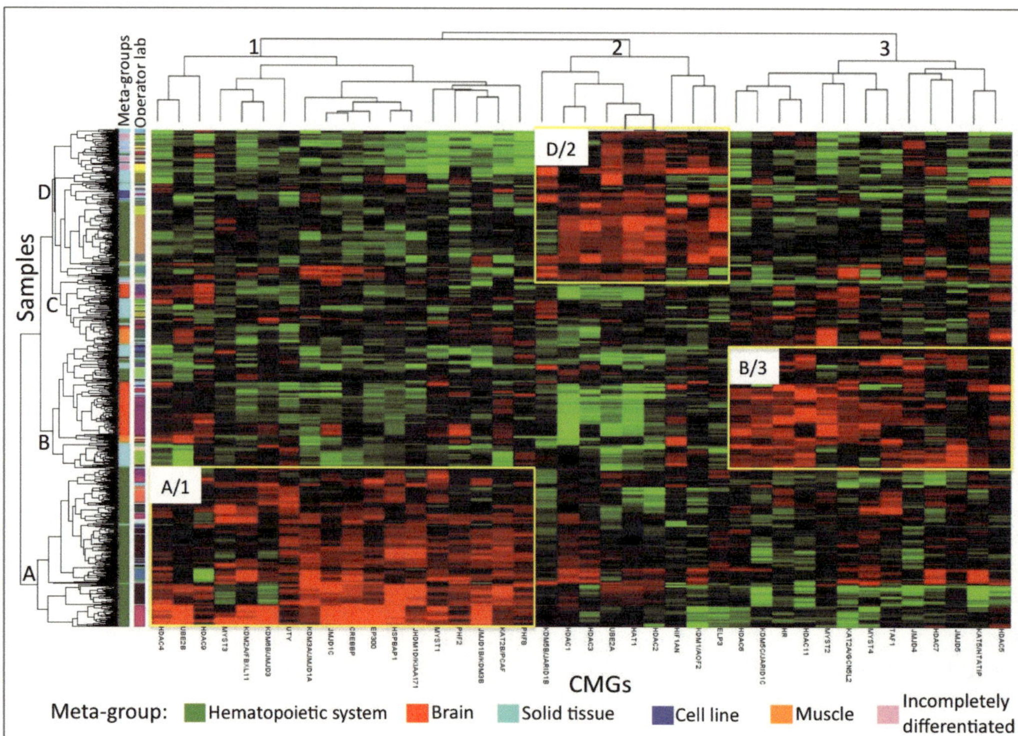

Fig. 1. Bi-directional hierarchical clustering of 40 CMGs expression in 1033 normal human tissues derived from public datasets compiled by Lukk et al. (2010). Gene-wise standardized expression level is shown as colored squares on a red-black-green scale (from up- to down-regulation, respectively). Samples are labeled by two color columns indicating their meta-group and operating laboratory. The three clustered CMG groups (1-3) displaying peculiar upregulation in a specific sample clade (A, B, D) are boxed yellow.

upregulation of 18 CMGs out of the 40 examined as they form a homogenous cluster of samples (clade A in Fig. 1) derived almost entirely from the hematopoietic system reported by various labs. Notably, 14 out of the 18 CMGs in clade 1 are implicated in euchromatin formation and gene activation (Table 1). Interestingly, 10 out of the 14 CMGs belong to the jumonji protein family of histone demethylases, most of which (8 genes) are implicated in removal of repressive methyl group and formation of euchromatin (Table 1). In clade 2, 9 CMGs are upregulated, most of which are implicated in heterochromatin formation (Table 1). This group defines a cluster (clade D in Fig. 1) composed of samples derived from the hematopoietic system (mostly differentiated lymphocytes, macrophages and granulocytes), cell line, solid tissue and incompletely differentiated cells. Clade 3 is characterized by the expression of 13 CMGs distributed almost equally between genes implicated in eu- and heterochromatin and defines a cluster (clade B in Fig. 1) composed of diverse samples derived mostly from the brain but also from hematopoietic system, muscle and solid tissue. Based on this analysis and the capacity for open chromatin configuration (Table 1) we predict that clade A represents a group of cells most of which from the hematopoietic system with the highest pluripotency level, clade B has moderate and clade D has the lowest pluripotency level. Clade C generates a cluster of cells derived from the brain, solid tissue and muscle with no clear clustering of the CMGs.

To evaluate the strength of our prediction we analyzed the expression pattern of the predefined CMGs in a microarray dataset (GSE18290) obtained for early stages of human embryos, which include one-, two-, four-, and eight-cell stage embryos, morula stage and the blastocyst stage (Xie et al., 2010). The raw CEL files of the respective Affymetrix Human Genome U133 Plus 2.0 Arrays were downloaded from NCBI GEO and further preprocessed using RMA in Partek Genomics Suite™. Expression values of 38 out of the 40 predefined CMGs were standardized and subjected to gene wise hierarchical clustering using Euclidean distance and complete linkage. The resulting heat map shown in Fig. 2 revealed a sharp transition in the expression of CMGs between four- and eight-cell embryos, which might reflect transition from high pluripotency level (totipotency) to a more reduced one. The finding that blastomers derived from two- and four-cell embryos retain totipotency (Van de Velde et al., 2008) may support this notion. The abrupt transition in pattern of CMG expression is consistent with the largest changes in transcriptome profile observed between four- and eight-cell stages of human embryos (Xie et al., 2010) as well as with the timing of zygote genome activation (ZGA) in humans, which commences between the four- and eight-cell stages of embryo development concomitantly with the degradation of maternal transcripts (Braude et al., 1988; Nothias et al., 1995). Based on the expression pattern of CMGs, cell samples could be divided into two major groups, namely group A, which is composed of one-to four-cell embryos displaying 18 (probably maternal) CMGs (clade 2), most of which (12 out of 18) are implicated in euchromatin formation and gene transcription. Group B is composed of eight-cell embryos, morula and blastocyst displaying upregulation of 20 CMGs (probably zygotic transcripts, clade 1) distributed equally between CMGs implicated in eu- and heterochromatin formation. Based on this analysis, it appears that overrepresentation of CMGs whose products are associated with open chromatin and gene transcription characterizes cells with high pluripotency level and thus supporting our hypothesis. We suggest that this attribute may faithfully serve as a tool for assessing the pluripotency level of the cell.

Gene name	Gene function	Predicted chromatin function	Group A/1	Group B/3	Group D/2
HDAC4	Histone deacetylase	**Hetero**	+	-	-
UBE2B	H2B ubiquitination	Eu	+	-	-
HDAC9	Histone deacetylase	**Hetero**	+	-	-
MYST3	HAT	Eu	+	-	-
KDM2A	H3K36 demethylase	**Hetero**	+	-	-
KDM6B	H3K27 demethylase	Eu	+	-	-
UTY	H3K27 demethylase	Eu	+	-	-
KDM3A	H3K9 demethylase	Eu	+	-	-
JMJD1C	H3K9 demethylase	Eu	+	-	-
CREBBP	HAT	Eu	+	-	-
EP300	HAT	Eu	+	-	-
HSPBAP1	Jumonji protein	unknown	+	-	-
JHDM1D	H3K9/K27 demethyl.	Eu	+	-	-
MYST1	HAT	Eu	+	-	-
PHF2	H3K9 demethylase	Eu	+	-	-
KAT2B	HAT	Eu	+	-	-
KDM3B	H3K9 demethylase	Eu	+	-	-
PHF8	H3K9/K27 demethyl.	Eu	+	-	-
KDM5B	H3K4 demethylase	**Hetero**	-	-	+
HDAC1	Histone deacetylase	**Hetero**	-	-	+
HDAC3	Histone deacetylase	**Hetero**	-	-	+
UBE2A	H2B ubiquitination	Eu	-	-	+
HAT1	Cytoplasmic HAT	unknown	-	-	+
HDAC2	Histone deacetylase	**Hetero**	-	-	+
HIF1AN	Jmj protein	**Hetero?**	-	-	+
KDM1	H3K4 demethylase	**Hetero**	-	-	+
ELP3	HAT	Eu	-	-	+
HDAC6	Histone deacetylase	**Hetero**	-	+	-
KDM5C	H3K4 demethylase	**Hetero**	-	+	-
HR	Jumonji protein	unknown	-	+	-
HDAC11	Histone deacetylase	**Hetero**	-	+	-
MYST2	HAT	Eu	-	+	-
KAT2A	HAT	Eu	-	+	-
MYST4	HAT	Eu	-	+	-
TAF1	HAT	Eu	-	+	-
JMJD4	unknown	unknown	-	+	-
HDAC7	Histone deacetylase	**Hetero**	-	+	-
KDM8	H3K36 demethylase	**Hetero**	-	+	-
KAT5	HAT	Eu	-	+	-
HDAC5	Histone deacetylase	**Hetero**	-	+	-

Table 1. Summary of the expression pattern of CMGs in the different clustered groups (shown in Fig. 1) and their predicted chromatin function. Eu, Euchromatin, HAT, histone acetyltransferase; Hetero, heterochromatin

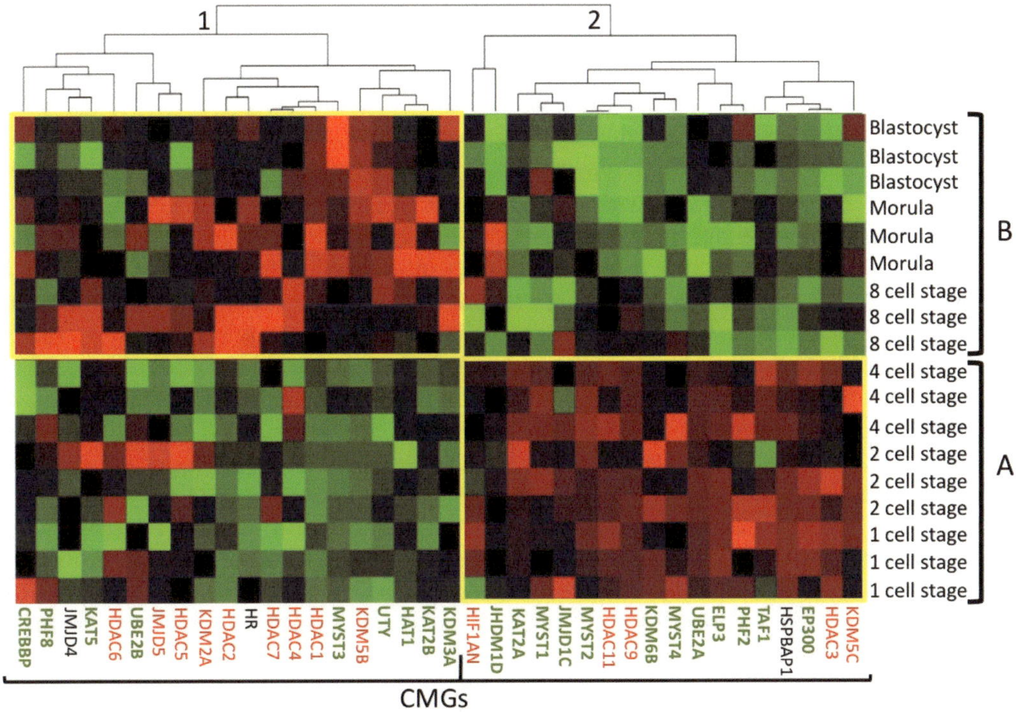

Fig. 2. Clustering of 38 CMGs expression in human embryos at various developmental stages derived from public datasets compiled by Xie et al. (2010, GSE18290). Gene-wise standardized expression level is shown as colored squares on a red-black-green scale (from up- to down-regulation, respectively). The two clustered CMG groups (upper dendrogram) displaying peculiar upregulation in a specific sample group (A and B) are boxed yellow. CMGs highlighted green and red are predicted to be involved in formation of euchromatin and heterochromatin, respectively.

3. Hazardous pathways associated with cellular dedifferentiation (iPS cells) and with culturing stem cells

It is now quite accepted that reprogramming of somatic cells in culture as well as culturing of stem cells are prone to hazardous genomic modifications and thus undermine their potential use in regenerative medicine. The complexity of hES cells is highlighted by the fact that injection of these cells (in their un-transformed state) into immunodeficient mice induced teratoma formation - a practice commonly used as a stringent assay to prove their existence. Indeed, animal embryonic stem cells often displayed genomic abnormalities in culture, which frequently resulted in malignant transformation (Lefort et al. 2009; Ben David & Benvenisty, 2011). Similarly, human iPS cells (Takahashi & Yamanaka, 2006), like human ES cells can acquire genetic abnormalities in the culture (Mayshar et al. 2010). These genetic aberrations are presumed to arise in part from culture adaptation while others are suspected to originate from the parent somatic cell. The later may be overcome by derivation of somatic cells from embryonic tissue (Ben-David et al., 2010). Yet, derivation of somatic cells from embryos, such as mouse embryonic fibroblasts (MEFs) did not improve significantly

reprogramming efficiency or reduced genetic and epigenetic abnormalities (Marión et al., 2009). Cell type of origin has a major effect on the properties of mouse iPS cells at early-passages, where iPS cells retain a transient epigenetic memory of their somatic cells of origin demonstrated by their differential gene expression and differentiation capacity, while at late passages these differences were attenuated (Polo et al., 2010).
Reprogramming of somatic cells to generate iPS cells is believed to commence with the introduction of the so-called 'pluripotent genes' into cultured somatic cells. Commonly, researchers are not aware that reprogramming has already initiated in somatic cells when they are removed from the *soma* and placed in tissue culture environments. Cell culturing of primary cells might impose an extreme stress over the cells resulting in reprogramming, dedifferentiation and acquisition of pluripotency prior to reentry of somatic cells [e.g., mouse embryonic fibroblasts (MEFs)] to the cell cycle (Fig. 3) - a stage, which is often overlooked. This aspect of cell culturing has been highlighted by Barbra McClintock (1984) in her Nobel article 'The significance of responses of the genome to challenge'. McClintock (1984) has recognized the potential for hazardous genetic variation that can be induced following exposure of cells to stress (e.g., cell culturing, virus infection) stating "Some responses to stress are especially significant for illustrating how a genome may modify itself when confronted with unfamiliar conditions. Changes induced in genomes when cells are removed from their normal locations and placed in tissue culture surroundings are outstanding examples of this. The establishment of a successful tissue culture from animal cells, such as those of rat or mouse, is accompanied by readily observed genomic restructuring." McClintock predicted that these aberrant genome responses to stress are likely to be induced by mobilization of transposable elements. Therefore, we should consider the possibility that the genotype(s) of somatic cells entering the cell cycle may not be identical to the genotype of the original somatic cells; a genotype(s) conferring increasing fitness for tissue culture conditions may prevail.

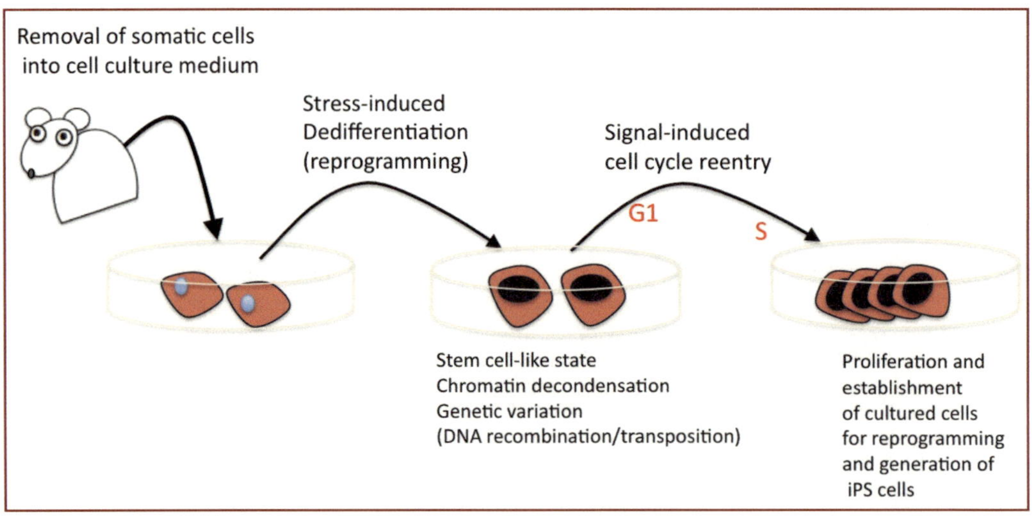

Fig. 3. Reprogramming of somatic cells. The removal of somatic cells from the body is stressful and might induce dedifferentiation (reprogramming) and acquisition of stem cell-like state prior to reentry into the cell cycle (G1-S transition). Note that due to genetic variation induced during dedifferentiation the genotype of cells entering the cell cycle may be different from the genotype of the original somatic cells.

In animals, somatic cell nuclear transfer (SCNT) as well as iPSCs, both are dedifferentiation-driven processes appear to be very limited (Tamada & Kikyo, 2004; William & Plath, 2008). This is probably due, at least in part, to a high frequency of p53-mediated DNA damage response and p53-dependent apoptosis, which are acting to ensure genomic integrity (Marión et al., 2009). Although, it is presumed that failure in reprogramming is selective for those cells with preexisting DNA damage, it cannot be excluded that DNA damage or 'irreversible genomic modifications' are induced in cells (e.g., protoplasts, primary MEFs) in the course of stress-induced cellular dedifferentiation (Fig. 3) or cell culturing via DNA recombination and DNA transposition (Pouteau et al., 1991; Hirochika, 1993; Grafi et al., 2007; Grafi, 2009; Grafi et al., 2011). Almost half of the human genome is composed of transposable elements, some of which still retain their capacity for transposition into genes, which could lead to genetic instability and human diseases (reviewed in Belancio et al. 2009). Indeed, Alu and Line-1 expression and retrotransposition have been reported in human ES cells and in human neural progenitor cells (Garcia-Perez et al., 2007; Coufal et al., 2009; Macia et al., 2011). Notably, some of these retroelements are often activated in various cell lines following exposure to stress (reviewed in Oliver & Greene, 2009).

Presently, it is not clear why the genome become vulnerable under certain stress conditions. One possibility is that following exposure to stress the genome is reacting by extensive and stochastic reorganization culminating in global chromatin decondensation and acquisition of dedifferentiated, stem cell-like state (reviewed in Grafi, 2009). Stochastic epigenetic modifications may release constraints over transposable elements resulting in their activation and transposition into other chromosomal sites. Stress-induced decrease in methylation of repetitive elements [long interspersed nucleotide element (LINE)-1 and Alu repetitive elements] was reported in blood samples derived from elderly individuals following exposure to traffic particles (Baccarelli et al., 2009). A whole-genome profiles of DNA methylation of several human iPS cell lines, showed aberrant reprogramming of DNA methylation; regions proximal to centromeres and telomeres display incomplete reprogramming of non-CG methylation, and differences in CG methylation and histone modifications (Lister et al., 2011).

4. Discussion

The capacity of somatic cells to dedifferentiate and acquire stem cell-like state is an important goal toward developing an efficient tool for use in regenerative medicine. Yet, the process of dedifferentiation is complex and unsafe resulting from increased incidents of DNA transposition/recombination-induced genetic variation and genome instability (Grafi, 2009). In addition, the findings that both embryonic stem cells (ESCs) and iPS cells show higher frequencies for genetic abnormalities relative to other cell lines limit their suitability for clinical use in regenerative medicine (Mayshar et al., 2010; Laurent et al., 2011). The generation of iPS cells from somatic cells by various means emerged as one reasonable approach for generating autologous iPS cells for clinical applications. This methodology solves the problem of transplant rejection and also moral concerns often raised regarding the use of ES cells. However, the process of generating iPS cells is very inefficient, has very low rate of success and may be subjected to hazardous genetic variation (Grafi, 2009) leading, at least partly, to p53-dependent cell death (Marión et al., 2009). The limited success in iPS cells is enigmatic inasmuch as dedifferentiation is an integral process of development both in plants and animals. It underlies the regenerative capacity of certain vertebrates

(Brockes & Kumar, 2008) as well as of plants (reviewed in Grafi, 2004) and, at least partly, the capacity for transdifferentiation, that is, the conversion of one cell type to another (Tosh & Slack, 2002).

We should consider the fact that the capacity for switching cell fate in animals suggests that the machinery needed for this transition is already exist in the framework of the cell including the pluripotent genes (e.g., OCT4, SOX2) whose ectopic expression in cultured cells may induce formation of pluripotent stem cells. Hence, why introducing these genes exogenously instead of activating the endogenous ones? Accumulating data suggest that reprogramming does not require the 'four factors' (OCT4, SOX2, KLF4, c-Myc) and can be carried out with only one factor OCT4 (Kim et al., 2009). This raises the question whether reprogramming is achieved due to ectopic expression of pluripotent genes or due to endogenous ones being activated as a consequence of the procedure itself (viral transduction), namely, virus-induced chromatin reorganization (Monier et al., 2000). Indeed, induction of pluripotency does not necessarily require exogenous factors and can be carried out by extracts derived from stem cells or undifferentiated human NCCIT carcinoma cells (Taranger et al., 2005). More recently it has been shown that pretreatment of somatic cells with chromatin modulators, namely, DNA methyltransferase and histone deacetylase inhibitors can improve reprogramming and the formation of hESC-like colonies by embryonic stem cell extracts (Han et al., 2010). Chromatin architecture is a fundamental theme in pluripotency and as such should be the primary means for activation of silent genes whose products involved in establishing of the pluripotent state. Support to this view are the open chromatin configuration characteristic of stem cells and the finding that chromatin remodeling factors play a critical role in maintaining open chromatin and reprogramming of somatic cells to pluripotent state (Efroni et al., 2008; Gaspar-Maia et al., 2009). Accordingly, it has recently been shown that chromatin-remodeling components of the BAF complex facilitate the reprogramming of somatic cells into pluripotent state (Singhal et al., 2010).

5. Conclusions

The use of CMGs for reprogramming of somatic cells has not been exploited sufficiently. We should consider manipulating of the activities of chromatin modifiers capable of facilitating the formation of open chromatin conformation. Emphasis should be given to those genes whose products actively remove repressive marks from histone tails, such as histone demethylases (Table 1), as a necessary step toward opening of otherwise closed chromatin. In this respect, together with existing tools such as the PluriTest (Muller et al., 2011), our bioinformatic data suggest that the transcription profile of CMGs can be formulated into a robust bioinformatic tool for assessing the pluripotency level of cells.

Obviously, the removal of cells (e.g., stem cells, somatic cells) from their normal location in the body and placing them under tissue culture conditions is hazardous and should be avoided, unless we find the way(s) to control the extent of genetic variation induced during dedifferentiation and cell culturing. An alternative approach has been suggested by Abramovich et al. (2008) in their article entitled 'Have we reached the point for in vivo rejuvenation?' The authors suggested to try and imitate natural rejuvenation processes and to test the possibility of inducing dedifferentiation and the pluripotent state in somatic cells *in vivo*. Future challenges will be to gain knowledge and find the appropriate means for inducing dedifferentiation at specific tissue or organ *in vivo* for efficient and safe regenerative medicine.

6. Acknowledgements

This chapter is dedicated to the memory of Dr. Amir Abramovich, a brilliant scientist who passed away untimely and who was so enthusiastic about the potential use of the process of dedifferentiation for the purpose of *in vivo* rejuvenation.

We thank Eitan Rubin for bioinformatics discussions and Vadim Fraifeld for critical reading of the chapter. This work was supported by The Israel Science Foundation (ISF) grant No. 476/09 to G.G. and V.C.-C.

7. References

Abramovich, A., Muradian, K.K. & Fraifeld, V.E. (2008). Have we reached the point for in vivo rejuvenation? *Rejuvenation Res* 11:489-492.

Arents, G. & Moudrianakis, E.N. (1995). The histone fold: a ubiquitous architectural motif utilized in DNA compaction and protein dimerization. *Proc Natl Acad Sci U S A.* 92:11170-11174.

Azuara, V., Perry, P., Sauer, S., Spivakov, M., Jørgensen, H.F., John, R.M., Gouti, M., Casanova, M., Warnes, G., Merkenschlager, M. & Fisher, A.G. (2006). Chromatin signatures of pluripotent cell lines. *Nat Cell Biol* 8:532-538.

Baccarelli, A., Wright, R.O., Bollati, V., Tarantini, L., Litonjua, A.A., Suh, H.H., Zanobetti, A., Sparrow, D., Vokonas, P.S. & Schwartz, J. (2009). Rapid DNA methylation changes after exposure to traffic particles. *Am J Respir Crit Care Med* 179:572-578.

Bannister, A.J., Zegerman, P., Partridge, J.F., Miska, E.A., Thomas, J.O., Allshire, R.C. & Kouzarides, T. (2001). Selective recognition of methylated lysine 9 on histone H3 by the HP1 chromo domain. *Nature* 410:120-124.

Belancio, V.P., Deininger, P.L. & Roy-Engel, A.M. (2009). LINE dancing in the human genome: transposable elements and disease. Genome Medicine 1:97.

Ben-David, U., Benvenisty, N. & Mayshar, Y. (2010). Genetic instability in human induced pluripotent stem cells: classification of causes and possible safeguards. *Cell Cycle* 9:4603-4604.

Ben-David, U. & Benvenisty, N. (2011). The tumorigenicity of human embryonic and induced pluripotent stem cells. *Nat Rev Cancer* 11:268-277.

Bernstein, B.E., Mikkelsen, T.S., Xie, X., Kamal, M., Huebert, D.J., Cuff, J., Fry, B., Meissner, A., Wernig, M., Plath, K., Jaenisch, R., Wagschal, A., Feil, R., Schreiber, S.L. & Lander, E.S. (2006). A bivalent chromatin structure marks key developmental genes in embryonic stem cells. *Cell* 125: 315-326.

Blank, T., Trendelenburg, M. & Kleinschmidt, J. (1992). Reactivation of DNA replication in erythrocyte nuclei by Xenopus egg extract involves energy-dependent chromatin decondensation and changes in histone phosphorylation. *Exp Cell Res* 202:224-232.

Booker, C.E. & Dwivedi, R.S. (1973). Ultrastructure of meristematic cells of dormant and released buds in Tradescantia paludosa. *Exp Cell Res* 82:255-261.

Brockes, J.P. & Kumar, A. (2008). Comparative aspects of animal regeneration. *Annu Rev Cell Dev Biol.* 24:525-249.

Braude, P., Bolton, V. & Moore, S. (1988). Human gene expression first occurs between the four-and eight-cell stages of preimplantation development. *Nature* 332:459–461.

Coufal, N.G., Garcia-Perez, J.L., Peng, G.E., Yeo, G.W., Mu, Y., Lovci, M.T., Morell, M., O'Shea, K.S., Moran, J.V. & Gage, F.H. (2009). L1 retrotransposition in human neural progenitor cells. *Nature* 460:1127-1131.

Damri, M., Ben-Meir, H., Avivi, Y., Caspi-Chalifa, V., Wolfson, M., Fraifeld, V. & Grafi, G. (2009). Senescing cells share common features with dedifferentiating cells. *Rejuvenation Research* 12:435-443.

Eberharter, A. & Becker, P.B. (2002). Histone acetylation: a switch between repressive and permissive chromatin. Second in review series on chromatin dynamics. *EMBO Rep* 3:224-229.

Efroni, S., Duttagupta, R., Cheng, J., Dehghani, H., Hoeppner, D.J., Dash, C., Bazett-Jones, D.P., Le Grice, S., McKay, R.D., Buetow, K.H., Gingeras, T.R., Misteli, T. & Meshorer, E. (2008). Global transcription in pluripotent embryonic stem cells. *Cell Stem Cell* 2:437-447.

Eissenberg, J.C. & Shilatifard, A. (2010). Histone H3 lysine 4 (H3K4) methylation in development and differentiation. *Dev Biol* 339: 240-249.

Fontes, M., Marilley, M., Le Parco, Y. & Thouveny, Y. (1980). Variations in accessibility of DNA during traumatic regeneration by Owenia fusiformis (polychaete annelid). *Cell Differ* 9:83-93

Fortunel, N.O., Out, H.H., Ng, H.H., Chen, J., Mu, X., Chevassut, T., Li, X., Joseph, M., Bailey, C., Hatzfeld, J.A., Hatzfeld, A., Usta, F., Vega, V.B., Long. P.M., Libermann, T.A. & Lim, B. (2003). Comment on " 'Stemness': transcriptional profiling of embryonic and adult stem cells" and "a stem cell molecular signature". *Science* 302:393.

Garcia-Perez, J.L., Marchetto, M.C., Muotri, A.R., Coufal, N.G., Gage, F.H., O'Shea, K.S. & Moran JV. (2007). LINE-1 retrotransposition in human embryonic stem cells. *Hum Mol Genet* 16:1569-1577.

Gaspar-Maia, A., Alajem, A., Meshorer, E. & Ramalho-Santos, M. (2011). Open chromatin in pluripotency and reprogramming. *Nat Rev Mol Cell Biol* 12:36-47.

Grafi, G. & Avivi, Y. (2004). Stem cells: a lesson from dedifferentiation. *Trends Biotech* 22:388-389.

Grafi, G. (2004). How cells dedifferentiated: a lesson from plants. *Dev Biol* 268:1-6

Grafi, G., Ben-Meir, H., Avivi, Y., Moshe, M., Dahan, Y. & Zemach, A. (2007). Histone methylation controls telomerase-independent telomere lengthening in cells undergoing dedifferentiation. *Dev Biol* 306:838-846.

Grafi, G. (2009). The complexity of cellular dedifferentiation: implications for regenerative medicine. *Trends Biotech* 27:329-332.

Grafi, G., Chalifa-Caspi, V., Nagar, T., Plaschkes, I., Barak, S. & Ransbotyn, V. (2011). Plant response to stress meets dedifferentiation. *Planta* 233:433–438.

Han, J., Sachdev, P.S. & Sidhu, K.S. (2010). A combined epigenetic and non-genetic approach for reprogramming human somatic cells. *PLoS One* 5:e12297.

He, G., Zhu, X., Elling, A.A., Chen, L., Wang, X., Guo, L., Liang, M., He, H., Zhang, H., Chen, F., Qi, Y., Chen, R. & Deng, X.W. (2010). Global epigenetic and transcriptional trends among two rice subspecies and their reciprocal hybrids. *Plant Cell* 22:17-33.

Hirochika, H. (1993). Activation of tobacco retrotransposons during tissue culture. *EMBO J.* 12:2521-2528.

Ivanova, N.B., Dimos, J.T., Schaniel, C., Hackney, J.A., Moore, K.A & Lemischka, I.R. (2002). A stem cell molecular signature. *Science* 298:601-604.

Katagiri, C. & Ohsumi, K. (1994). Remodeling of sperm chromatin induced in egg extracts of amphibians. *Int J Dev Biol* 38: 209-216.

Kikyo, N., Wade, P.A., Guschin, D., Ge, H. & Wolffe, A.P. (2000). Active remodeling of somatic nuclei in egg cytoplasm by the nucleosomal ATPase ISWI. *Science* 289:2360-2362.

Kim, J.B., Greber, B., Arau´zo-Bravo, M.J., Meyer, J., Park, K.I., Zaehres, H. & Schöler, H.R. (2009). Direct reprogramming of human neural stem cells by OCT4. *Nature* 461, 649–653.

Kurisaki, A., Ito, Y., Onuma, Y., Intoh, A. & Asashima, M. (2010). In vitro organogenesis using multipotent cells. *Hum Cell* 23:1-14.

Lachner, M., O'Carroll, D., Rea, S., Mechtler, K. & Jenuwein, T. (2001). Methylation of histone H3 lysine 9 creates a binding site for HP1 proteins. *Nature* 410:116-120.

Lan, F., Nottke, A.C. & Shi, Y. (2008). Mechanisms involved in the regulation of histone lysine demethylases. *Curr Opin Cell Biol* 20:316-325.

Laskey, R.A., Honda, B.M., Mills, A.D. & Finch, J.T. (1978). Nucleosomes are assembled by an acidic protein which binds histones and transfers them to DNA. *Nature* 275:416-420.

Laurent, L.C., Ulitsky, I., Slavin, I., Tran, H., Schork, A., Morey, R., Lynch, C., Harness, J.V., Lee, S., Barrero, M.J., Ku, S., Martynova, M., Semechkin, R., Galat, V., Gottesfeld, J., Izpisua Belmonte, J.C., Murry, C., Keirstead, H.S., Park, H.S., Schmidt, U., Laslett, A.L., Muller, F.J., Nievergelt, C.M., Shamir, R. & Loring, J.F. (2011). Dynamic Changes in the Copy Number of Pluripotency and Cell Proliferation Genes in Human ESCs and iPSCs during Reprogramming and Time in Culture. *Cell Stem Cell* 8:106-118.

Lee, K.K. & Workman J.L. (2007). Histone acetyltransferase complexes: one size doesn't fit all, *Nat. Rev. Mol. Cell Biol.* 8:284–295.

Lefort, N., Perrier, A.L., Laâbi, Y., Varela, C. & Peschanski, M. (2009). Human embryonic stem cells and genomic instability. *Regen Med* 4:899-909.

Lister, R., Pelizzola, M., Kida, Y.S., Hawkins, R.D., Nery, J.R., Hon, G., Antosiewicz-Bourget, J., O'Malley, R., Castanon, R., Klugman, S., Downes, M., Yu, R., Stewart, R., Ren, B., Thomson, J.A., Evans, R.M. & Ecker, J.R. (2011). Hotspots of aberrant epigenomic reprogramming in human induced pluripotent stem cells. *Nature* 471:68-73.

Longo, F.J. & Anderson, E. (1968). The fine structure of pronuclear development and fusion in the sea urchin, *Arbacia punctulata.* J *Cell Biol* 39:339-368.

Lukk, M., Kapushesky, M., Nikkilä, J., Parkinson, H., Goncalves, A., Huber, W., Ukkonen, E. & Brazma, A. (2010). A global map of human gene expression. *Nat Biotechnol* 28:322-324.

Macia, A., Muñoz-Lopez, M., Cortes, J.L., Hastings, R.K., Morell, S., Lucena-Aguilar, G., Marchal, J.A., Badge, R.M. & Garcia-Perez, J.L. (2011). Epigenetic control of retrotransposon expression in human embryonic stem cells. *Mol Cell Biol* 31:300-316.

MacRae, E.K. & Meetz, G.D. (1970). Electron microscopy of the ammoniacal silver reaction for histones in the erythropoietic cells of the chick. *J Cell Biol* 45:235-245.

Marión R.M., Strati K., Li H., Murga M., Blanco R., Ortega S., Fernandez-Capetillo O., Serrano M. & Blasco M.A. (2009). A p53-mediated DNA damage response limits reprogramming to ensure iPS cell genomic integrity. *Nature* 460:1149-1153.

Mayshar, Y., Ben-David, U., Lavon, N., Biancotti, J.C., Yakir, B., Clark, A.T., Plath, K., Lowry, W.E. & Benvenisty, N. (2010). Identification and classification of chromosomal aberrations in human induced pluripotent stem cells. *Cell Stem Cell* 7:521-531.

McClintock, B. (1984). The significance of responses of the genome to challenge. *Science* 226:792-801.

McKay, R. (2000). Stem cells—hype and hope. *Nature* 406:361–364.

Meshorer, E. & Misteli, T. (2006). Chromatin in pluripotent embryonic stem cells and differentiation. *Nat Rev Mol Cell Biol* 7:540-546.

Miura, A.B., Shibata, A., Akihama, T., Endo, Y. & Saito, Y. (1974). Ultrastructure of developing erythrocytes. *Tohoku J Exp Med* 112:299-313.

Monier, K., Armas, J.C., Etteldorf, S., Ghazal, P. & Sullivan, K.F. (2000). Annexation of the interchromosomal space during viral infection. *Nat Cell Biol* 2:661- 665.

Müller, F.J., Schuldt, B.M., Williams, R., Mason, D., Altun, G., Papapetrou, E.P., Danner, S., Goldmann, J.E., Herbst, A., Schmidt, N.O., Aldenhoff, J.B., Laurent, L.C. & Loring, J.F. (2011). A bioinformatic assay for pluripotency in human cells. *Nat Methods* 8:315-317.

Murphy, M.J. Jr., Bertles, J.F. & Gordon, A.S. (1971). Identifying characteristics of the haematopoietic precursor cell. *J Cell Sci* 9:23-47.

Nothias, J.Y., Majumder, S., Kaneko, K.J. & DePamphilis, M.L. (1995). Regulation of gene expression at the beginning of mammalian development. *J Biol Chem* 270:22077-22080.

Oliver, K.R. & Greene, W.K. (2009) Transposable elements: powerful facilitators of evolution. *BioEssays* 31:703-714.

Philpott, A., Leno, G.H. & Laskey, R.A. (1991). Sperm decondensation in Xenopus egg cytoplasm is mediated by nucleoplasmin. *Cell* 65:569-578.

Polo, J.M., Liu, S., Figueroa, M.E., Kulalert, W., Eminli, S., Tan, K.Y., Apostolou, E., Stadtfeld, M., Li, Y., Shioda. T., Natesan, S., Wagers, A.J., Melnick, A., Evans, T. & Hochedlinger, K. (2010). Cell type of origin influences the molecular and functional properties of mouse induced pluripotent stem cells. *Nat Biotechnol* 28:848-855.

Potten, C.S. & Loeffler, M. (1990). Stem cells: attributes, cycles, spirals, pitfalls and uncertainties. Lessons for and from the crypt. *Development* 110:1001-1020.

Pouteau, S., Huttnerl, E., Grandbastien, M.A. & Caboche, M. (1991). Specific expression of the tobacco Tntl retrotransposon
in protoplasts. *EMBO J.* 10:1911-1918.

Ramalho-Santos, M., Yoon, S., Matsuzaki, Y., Mulligan, R.C. & Melton, D.A. (2002). "Stemness": transcriptional profiling of embryonic and adult stem cells. *Science* 298:597-600.

Rea, S., Eisenhaber, F., O'Carroll, D., Strahl, B.D., Sun, Z.W., Schmid, M., Opravil, S., Mechtler, K., Ponting, C.P., Allis, C.D. & Jenuwein, T. (2000). Regulation of chromatin structure by site specific histone H3 methyltransferases. *Nature* 406:593-599.

Robinson, P.J. & Rhodes, D. (2006). Structure of the '30 nm' chromatin fibre: a key role for the linker histone. *Curr Opin Struct Biol* 16:336-343.

Routh, A., Sandin, S. & Rhodes, D. (2008). Nucleosome repeat length and linker histone stoichiometry determine chromatin fiber structure. *Proc Natl Acad Sci* U S A. 105:8872-8877.

Singhal, N., Graumann, J., Wu, G., Araúzo-Bravo, M.J., Han, D.W., Greber, B., Gentile, L., Mann, M. & Schöler, H.R. (2010). Chromatin-Remodeling Components of the BAF Complex Facilitate Reprogramming. *Cell* 141:943-955.

Takahashi, K. & Yamanaka, S. (2006). Induction of pluripotent stem cells from mouse embryonic and adult fibroblast cultures by defined factors. *Cell* 126:663-676.

Tamada, H. & Kikyo, N. (2004). Nuclear reprogramming in mammalian somatic cell nuclear cloning. *Cytogenet Genome Res* 105:285–291.

Taranger, C.K., Noer, A., Sørensen, A.L., Håkelien, A.M., Boquest, A.C. & Collas, P. (2005). Induction of dedifferentiation, genome wide transcriptional programming, and epigenetic reprogramming by extracts of carcinoma and embryonic stem cells. *Mol Biol Cell* 16:5719-5735.

Tessadori, F., Chupeau, M.C., Chupeau, Y., Knip, M., Germann, S., van Driel, R., Fransz, P. & Gaudin, V. (2007). Large-scale dissociation and sequential reassembly of pericentric heterochromatin in dedifferentiated Arabidopsis cells. *J Cell Sci* 120:1200-1208.

Tosh, D. & Slack, J.M. (2002). How cells change their phenotype. *Nat Rev Mol Cell Biol* 3:187-194.

Van de Velde, H., Cauffman, G., Tournaye, H., Devroey, P. & Liebaers, I. (2008). The four blastomeres of a 4-cell stage human embryo are able to develop individually into blastocysts with inner cell mass and trophectoderm. *Hum Reprod* 23:1742-1747.

Weintraub, H. (1978). The nucleosome repeat length increases during erythropoiesis in the chick. *Nucleic Acids Res* 5:1179-1188.

William, E.L. & Plath, K. (2008). The many ways to make an iPS cell. *Nature Biotechnol* 26:1246-1248.

Williams, L., Zhao, J., Morozova, N., Li, Y., Avivi, Y. & Grafi, G. (2003). Chromatin reorganization accompanying cellular dedifferentiation is associated with modifications of histone H3, redistribution of HP1, and activation of E2F-target genes. *Dev Dyn* 228:113-120.

Xie, D., Chen, C.C., Ptaszek, L.M., Xiao, S., Cao, X., Fang, F., Ng, H.H., Lewin, H.A., Cowan, C. & Zhong, S. (2010). Rewirable gene regulatory networks in the preimplantation embryonic development of three mammalian species. *Genome Res* 20:804-815.

Yadav, R.K., Girke, T., Pasala, S., Xie, M. & Reddy, G.V. (2009). Gene expression map of the Arabidopsis shoot apical meristem stem cell niche. *Proc Natl Acad Sci* U S A. 106:4941-4946.

Zhao, J., Morozova, N., Williams, L., Libs, L., Avivi, Y. & Grafi, G. (2001). Two phases of chromatin decondensation during dedifferentiation of plant cells: Distinction between competence for cell fate switch and a commitment for S phase. *J Biol Chem* 276:22772-22778.

Zipori, D. (2004). The nature of stem cells: state rather than entity. *Nat Rev Genet* 5:873-878.

15

Signaling Pathways in Mouse Embryo Stem Cell Self-Renewal

Leo Quinlan
Physiology, School of Medicine,
National University of Ireland Galway
Ireland

1. Introduction

At the pre-implantation blastocyst stage of development, the mammalian embryo is composed of a unique collection of cells of which three major populations predominate. The outermost layer the trophectoderm (TE) gives rise to the placenta, which acts to sustain the developing fetus connecting it to the mother host. The next is a cluster of cells known as the inner cell mass (ICM) these cells are said to be pluripotent (Fig. 1). A third group of cells known as the primitive endoderm, surrounds the ICM cells at the epiblast stage. As development proceeds the ICM cells rapidly divide and eventually begin to differentiate forming the three embryonic germ layers (ectoderm, mesoderm and endoderm). Effectively these pluripotent ICM cells are the precursors of all adult tissues. As these pluripotent cells commit to a specific cellular lineage, they lose their pluripotency. Embryonic stem (ES) cells are euploid pluripotent cell lines isolated directly from cultured preimplantation embryos. The first stable ES cell lines were isolated by immunosurgery from the ICM of implantation-delayed, mouse blastocysts (Martin, 1981; Evans and Kaufman, 1981). Mouse ES cells are very closely related to early ICM cells in terms of their developmental potential (Beddington and Robertson, 1989). This chapter will focus on mouse ES cells (mES) unless otherwise stated. Three features characterize mES cells;

1. They are isolated directly from the embryo (Robertson, 1987).
2. They can colonize the germ line when introduced to the embryo.
3. They possess unrestricted proliferative potential (Suda et al., 1987).

These features effectively mean that under appropriate conditions, a karyotype stable self-renewing, pluripotent population of cells can be propagated indefinitely *in vitro*. mES cells have other characteristics, which prove useful when comparing embryo derived stem cells to their differentiated progenies. mES cells have a euploid (2n) chromosome complement, a feature that allows their participation in germ cell development and the formation of chimeras (Bradley et al., 1984; Evans, 1994). The functional demonstration of mES cell developmental potential through chimera formation is the definitive proof of the pluripotent nature of the cell population in question. Biomarkers are often used as indicators of the stem cell state due to the time consuming and technically more difficult nature of getting functional proof of stemness. Many of the common markers are transcription factors expressed in the ICM and mES cells and have been shown to have functional roles in self-renewal and in the maintenance of pluripotency, in both isolated stem cells or the ICM. The surface markers expressed depend on species of origin, but common markers include members of the stage specific embryonic antigen (SSEA) family, alkaline phosphatase and Oct4 (Andrews, 2002; Pera et al., 2000; Shamblott et al., 1998). SSEA-1 is expressed in mouse preimplantation embryos from the eight-cell stage until the embryo differentiates into germ layers when it remains only in the ectodermal lineage (Solter and Knowles, 1979; Resnick et al., 1992; Pelton et al., 2002). Some of the best-characterized examples include the POU domain transcription factor Oct4, the homeodomain protein Nanog and the high-mobility group transcription factor Sox2. Of the three factors mentioned our understanding of Oct4 is best developed. Oct4 deficient embryos fail to initiate fetal development, indicating that Oct4 is essential for embryo development (Nichols et al., 1998). In mES cells there is an altered level of expression upon differentiation, a profile for down-regulation into TE and up-regulation in endoderm correlates with the Oct4 profile of expression *in vivo* embryos (Palmieri et al., 1994). mES cells can be maintained in an undifferentiated state *in vitro* with relative ease. They represent pluripotent embryonic cells, which are present only transiently *in vivo*. This enables their use as an *in vitro* model to elucidate the mechanisms of differentiation that these pluripotent cells undergo *in vivo*. The regulatory signaling and transcription networks that play a role in pluripotency and self-renewal it would seem have been conserved between mouse and human ES (hES) cells, however many differences also are found. As distinct from mES cells, hES cells do not appear to express high levels of SSEA1, but do have high levels of SSEA-3, −4, TRA-1–60 and TRA-1–81 (Brimble et al., 2007; Reubinoff et al., 2000). Other difference also exist, both cell types have a high ratio of nuclear to cytoplasmic volume, mES cells grow in nests and form three dimensional embryoid bodies while hES cells often grow in colonies as thin layers. Furthermore hES cells unlike mES can be maintained in a self-renewal cycle in the absence of the cytokine LIF or a feeder layer. Recently a population of cells with pluripotent capability was isolated from a post-implantation mouse epiblast at the later stages of development (Brons et al., 2007; Tesar et al., 2007). These epiblast stem cells (EpiS cells) do not require LIF support as ICM derived mES cells do (Rossant, J. 2007; Nagy et al., 2003). Morphologically EpiS cells are more like hES cells than mES cells as they tend to grow as thin flat layers. Considering these difference and others, the idea that mES and hES cells while both pluripotent, may in fact represent different and distinct stages in development.

Early studies with mES cells showed that the use of mitotically inactivated STO cells (Ware and Axelrad, 1972; Hooper, 1997) was essential in the maintenance of self-renewal and the pluripotent state. Later it was found that the requirement for the feeder layer could be circumvented by the addition of the cytokine LIF in the presence of serum. In the absence of feeder layers or LIF, mES cells differentiate into a variety of cell types (Doetschman et al., 1985) depending on the developmental cue or signaling pathway activated. The process of differentiation can be seen as a loss of pluripotency and mES like their in vivo counterparts are capable of multi-lineage differentiation. mES cells undergo a controlled pattern of differentiation when injected and reintegrated into a pre-implantation blastocyst. Under these circumstance mES cells respond as ICM cells to *in vivo* differentiation cues and fully participate in normal development. Furthermore they are capable of forming a wide range of normal cells including germ cells (Bradley *et al.*, 1984; Robertson *et al.*, 1986). *In vitro* mES cells can be induced to differentiate, by culturing cells in suspension or in a monolayer system under the effect of chemical inducers such as retinoic acid (Robertson, 1987). When mES cells are propagated under conditions that discourage their attachment to the substratum they form small aggregates, termed embryoid bodies (EBs). These three dimensional colonies organize in such a way that an endodermal layer develops on the outer surface and are now known as "simple embryoid bodies". The inner stock of cells remain undifferentiated not unlike the early events of embryogenesis, where the trophectoderm differentiates from the peripheral cells of the morula, while the inner cells remain pluripotent.

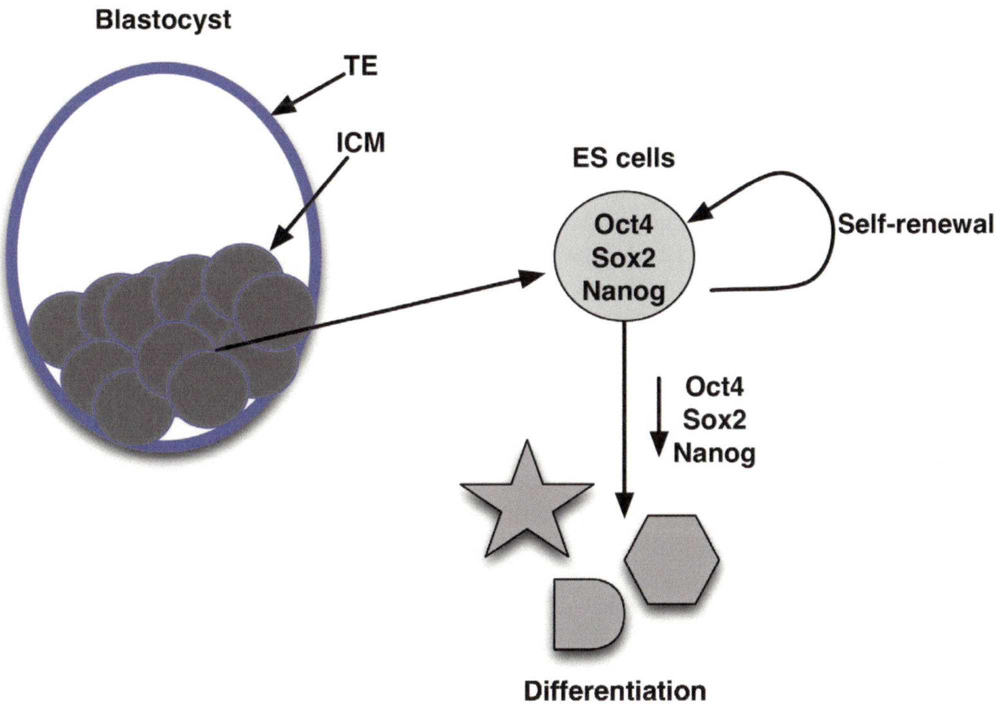

Fig. 1. Key transcription factors involved in self-renewal of ES cells.

Differentiated mES cells from EBs can give rise to a wide variety of cell types including neuronal (Bain *et al.*, 1995), hematopoietic (Suwabe *et al.*, 1998), endothelial (Yamashita *et al.*, 2000), cardiac (Maltsev *et al.*, 1993), smooth muscle (Yamashita *et al.*, 2000), chondrogenic (Kramer *et al.*, 2000) and osteoblastic cells (Buttery *et al.*, 2001). For stem cells there is a constant balancing act that must be maintained between self-renewal and the pluripotent phenotype versus cell lineage commitment and differentiation.

An understanding of the pathways and controlling factors involved in these fundamental cellular events is essential if we are to exploit the full potential of embryo derived stem cells for therapeutic uses in disease treatments and regenerative medicine in the future. This potential is real as it is clear that embryo derived stem cells are capable of unlimited self-renewal capacity and can differentiate into potentially any of over 200 cell types. The potency of these cells is maintained by a number of key regulatory factors, signaling pathways and extracellular signaling agents. The combination and interplay of these elements establishes a patterns of gene expression that sustains the pluripotent phenotype of ES cell. Some of the key regulators are transcription factors such as Oct4, Sox2 and Nanog, the signaling cascades involving phosphatidylinositol 3-kinase (PI3K) and the signaling molecules like LIF and Wnt proteins. Embryonic stem cells have great therapeutic potential however to fully realize the potential of these cells the signaling pathways that participate in controlling ES cell behavior must be better understood.

2. Pluripotency and Self-renewal

In the developing embryo, pluripotent cells first appear in the ICM, in the mouse blastocyst this is approximately at day 3.5 of embryonic development (E3.5). These cells persist as late as the pre-gastrulation stage. Thus *in vitro* mES cells represent a very transient population of cells existing for a very short period of time *in vivo*. Experimental extraction of these cells facilitates the expansion and maintenance of the pluripotent state *in vitro*. mES cells are pluripotent, which is defined as the ability to differentiate into all cell lineages that make up the adult organism (Buehr et al., 2003). Functional assessment of the pluripotency of mES cells can be determined by the capability of the cells to reintegrate, into the ICM of E3.5 blastocysts contributing to all cell lineages. Strict pluripotency has been shown only in the mouse, where mES cells completely integrate producing a high rate of chimerism in all tissues. In the presence of fetal serum, activation of STAT3 by LIF is sufficient to maintain mES cells in an undifferentiated state (Williams et al., 1988; Matsuda et al., 1999) however, this is not the case for hES cells (Dahéron et al., 2004). Potentially parallel pathways are at play in mouse and human ES cells through which ES cells achieve similar end points and sustain pluripotency. This characteristic is sustained by the ability of ES cells to self-renew. ES cells are capable of differentiation but also symmetrical division generating two identical undifferentiated pluripotent daughter cells. In real terms mES cells can be expanded indefinitely (years) in a self-renewal cycle, once culture conditions prevent differentiation. It appears that maintaining a pluripotent state during mES cell self-renewal is through active suppression of differentiation and the promotion of proliferation. The differentiation of mES cells can be induced by the ectopic expression of certain transcription factors. Forced expression of Gata6 in mES cells drives differentiation toward primitive endoderm lineages (Fujikura et al., 2002), while increasing Cdx2 results in trophectoderm formation (Niwa et al., 2005). Thus, the expression of genes promoting self-renewal, cell proliferation and suppressing cellular differentiation pathways must be stably maintained and passed on to each daughter cell. The regulation of self-renewal is of great interest and importance in developing our basic understanding but also for the development of regimes for cellular therapy. The ability to control and maintain the expansion of pluripotent cells is a cornerstone, if the true clinical potential of ES cell-derived therapies is to be realized for regenerative medicine. mES (Martin, 1981) and hES (Thomson et al., 1998) cells are similar in this regard although not absolutely identical, the molecular machinery and pathways involved is equally complex but involves a number of distinct players in each case (Sato et al., 2003). A good example of this is the vitamin A metabolite, all-trans-retinoic acid which has the effect of silencing self-renewal and driving a differentiation agenda for both mES and hES cells lines (Chen and Gudas, 1996; Mongan and Gudas, 2007). The self-renewal pathways in ES cells comprise complex networks of strategic actions of extracellular agents (including the presence or absence in culture of serum), intracellular signaling pathways and the control of key transcription factors. *In vitro*, LIF supports self-renewal and pluripotency of mES cells through activation of STAT3 (Smith et al., 1998), removal of LIF or suppression of STAT3 results in differentiation (Niwa et al., 1998). LIF receptor knockout mouse embryos are capable of passing the developmental stage required for mES cell derivation. Thus *in vivo* it is obvious that there is no magic bullet but other factors are also involved in this process (Nichols et al., 2001). The cytokine LIF was among one of the earliest molecules found to be associated with the maintenance of stem cell self-renewal *in vitro* and *in vivo*. More recently other factors have become known including the bone morphogenetic proteins (BMPs) 2 and 4 and glycogen synthase kinase-3. Under serum free culture conditions, mES cells require the presence of both LIF and BMPs to facilitate continued self-renewal (Ying et al., 2003). It appears that linked pathways are at play, LIF supporting self-renewal and proliferation, while BMP4 up-regulates members of the *Id* gene (inhibition of differentiation) family (Ying et al., 2003). Signaling pathway crosstalk with PI3K signaling has also been shown to play a part in self-renewal. The transcription factors Oct4, Sox2, and Nanog are now widely accepted as having a central role in promoting self-renewal and sustaining the undifferentiated phenotype (Ying et al., 2003). As if to emphasize the reliance of self-renewal on networked interactions, recent work has shown that a key set of promoter sequences bind Oct4, Sox2, and Nanog in ES cells (Chambers and Tomlinson, 2009; Avilion et al., 2003; Masui et al., 2007; Niwa et al., 2005). This however is not the total picture as further extensions of the co-dependence of the self-renewal network is coming to light all the time, interestingly some of these newer interactions are independent of the established tri-umbret of Oct4, Sox2, and Nanog (Ivanova et al., 2006). Other work is providing evidence to suggest that two independent pathways may be as play in ES cells. The established Oct4, Sox2, and Nanog networks may be acting to suppress differentiation and thus sustain pluripotency. While other transcription networks play a role in repression of specific cell lineage differentiation. Further extension of these interconnecting pathways includes the addition of the role of miRNA-encoding genes (Marson et al., 2008). In the following sections the role of specific transcription factors and signaling pathways will be expanded upon in the context of their role in the self-renewal of ICM derived mES cells.

2.1 Transcriptional networks

2.1.1 Oct4

Oct4 is a member of the POU (Pit-Oct-Unc) transcription factor family that regulates the expression of target genes by binding to a octameric sequence (Scholer et al., 1990). The key features of this family are the POU domain consisting of two sub-domains each of which bind to DNA. However the C-terminal is cell specific and may be essential for the expression of target gene in an orderly fashion as embryonic development proceeds. It is well established that the Oct4 gene (encoded by Pou5f1) is constitutively expressed in undifferentiated mES cells, in all pluripotent cells during mouse embryo development and is also an essential factor required in the generation of iPS cells (Niwa et al., 2000). Oct4 is also known as Oct3, Oct3/4, Otf3, and Otf4. In the mouse, Oct4 expression is up regulated beginning at the 4-cell stage and becomes localized to the pluripotent cell population (Yeom et al., 1996). The expression of Oct4 is common to human and mouse ES cells, and furthermore expression diminishes in both as cells differentiate. *In vivo* Oct4 knockout mouse embryos crash and do not develop beyond the blastocyst stage, they lack a pluripotent ICM cell population (Nichols et al., 1998), strongly suggesting a central role for Oct4 in maintaining pluripotency. In cells were Oct4 is repressed or in Oct4 knockouts, mES cells differentiate towards a trophoectodermal lineage. It has been reported that Oct4 inhibits trophectoderm lineage formation via an interaction with Cdx2 forming of an inhibitory complex (Niwa et al., 2005). Conversely up regulation or over-expression of Oct4 results in mES cell moving towards primitive endoderm (Niwa et al., 2000). These divergent effects of Oct4 suggest that it regulates the transcription of genes involved in coordination of multiple cellular functions and early cell fate decisions. Thus the actual level of Oct4 expression is important and a key level of expression is required to sustain pluripotency and self-renewal. In mES cells the expression of Oct4 is supported by the action of LIF and down regulated by the chemical inducer all-trans retinoic acid (Faherty et al., 2005; 2007). Suppression of STAT3 and accelerated expression of Oct4 also causes mES cells to differentiate (Niwa et al., 1998; Niwa et al., 2000). The role is Oct4 as a so called master regulator in sustaining pluripotency and self-renewal of mES cells is well known, however it is not a solo run (Nichols et al., 1998; Boyer et al., 2005). Oct4 alone without LIF, is not sufficient to sustain self-renewal and prevent mES cell differentiation, suggesting that additional factors also play a part. More recently it has been shown that Oct4 expression prevents stem cell differentiation by sustaining the expression of other pluripotency factors and inhibiting gene expression of lineage specific factors. Known targets for Oct4 include Fgf4, Rex1/Zfp42, and Sox2 (Zeng et al., 2004; Tomioka et al., 2002). Oct4 has been shown to act in concert with other factors and its DNA binding often occurs in conjunction with the HMG-family protein Sox2, an additional factor required for maintaining mES cell stemness (Chambers and Smith, 2004; Pesce and Schöler, 2001). Oct4 is a key regulator of ES cell fate, particularly in maintaining a pluripotent state. The requirement appears to be that Oct4 protein levels are constrained within the narrow band. It is clear that Oct4 has a critical role in sustaining pluripotency, however its control is unclear. The control of the level of expression appears somewhat auto-regulatory (Chew et al., 2005), but also depends on other factors including the transcription factor Nanog providing a feedback loop to sustain self-renewal (Pan et al., 2006)

2.1.2 Nanog

Nanog is an homeobox containing transcription factor of approximately 280 amino acids. In the developing mouse embryo Nanog plays a key role in determining the fate of the ICM cells, acting to sustain pluripotency and preventing differentiation (Chambers et al., 2003). Nanog was identified as a factor, which when over expressed, supported pluripotency even in the absence of a LIF based signal. In the embryo Nanog expression is first seen at the compacted morulae stage before becoming restricted to the ICM, post-implantation stage Nanog expression is drastically reduced. *In vitro*, Nanog expression is abundant in pluripotent cell types but absent from adult tissues (Chambers et al., 2003). As with their *in vivo* counterparts upon differentiation of mES cells the expression of Nanog is downregulated. Nanog-null embryos fail soon after implantation, stem cells derived from such embryos are pluripotent but are found to quickly differentiate (Chambers et al., 2003; Mitsui et al., 2003). Over-expression of Nanog without any other intervention is sufficient to sustain self-renewal even without LIF albeit the self-renewal capacity under these conditions is reduced. Under the same conditions the level of active STAT3 is not appreciably altered, furthermore increased STAT3 signaling does not appear to alter Nanog expression. These data would suggest that Nanog is neither a target for STAT3, or does it regulate STAT3 activity. However at one least report suggests Nanog is a direct downstream target for STAT3 in the maintenance of pluripotency (Suzuki et al., 2006). BMP signaling normally acts during embryonic development to induce mesoderm formation, but effects in mES cells can be quite different (Winnier et al., 1995). In mES cells low levels of BMPs in the absence of LIF promote mesoderm, while in the presence of LIF, mES cell pluripotency is sustained (Ying et al., 2003). BMP signaling is facilitated by downstream effectors including SMAD1, an effector which Nanog has been shown to interact with and leads to inhibition of BMP signaling (Suzuki et al., 2006). What is being proposed is that BMPs are at least initially pro-mesoderm lineage formation as evidenced by up-regulation of Brachyury a mesoderm marker. When activated STAT3 interacts with Brachyury and increases Nanog expression, the elevated levels of Nanog inhibit BMPs via SMAD1 interaction and thus maintain the undifferentiated pluripotent state. Nanog is now accepted as an important component in regulating the pluripotent phenotype however the mechanism of its own control and how it effects other genes is not entirely elucidated. It has been shown that Oct4 and Sox2 can bind to the Nanog promoter *in vitro* and *in vivo* (Rodda et al., 2005) suggesting that Oct4/Sox can act to up regulate Nanog expression. However other studies have shown that Nanog expression can be maintained in the absence of Oct4, thus other factors must contribute to the regulation. One such factor is a member of the forkhead family FoxD3, which is found in mES cells and the early embryo. FoxD3 knockout embryos have a similar fate as Nanog knockouts (Hanna et al., 2002). The exact mechanism by which Nanog is regulated and how it effects control of mES cell pluripotency is unknown, however it appears independent of STAT3 activation or the requirement for BMP4 at least when in serum free culture conditions. Nanog is capable of activating the Oct4 promoter forming a negative feedback loop upregulating Oct4 at times and suppressing Oct4 when levels are above normal (Pan et al., 2006). Furthermore it has been shown that Nanog is capable of activating Rex1 a target for Oct4 and Sox2 (Shi et al., 2006). Furthermore it has also been suggested that Nanog may interact with Wnt and BMP4 signaling independent of LIF which may help explain why the forced expression of Nanog in the absence of LIF sustains a level of self-renewal (Chambers et al., 2003; Mitsui et al., 2003). The regulation and comprehensive elucidation of the role of Nanog requires more experimental work before we can paint a true picture of its overall role in pluirpotency.

2.1.3 Sox2

Sox2 is a DNA-binding protein of the HMG family. In the mouse Sox2 is expressed predominatley at the blastocyst stage (Avilion et al., 2003). However unlike Oct4, Sox2 has a major role to play also later in development and in adult stem cells (Wood and Episkopou, 1999; Zappone et al., 2000). At early stages of development and in mES cells, Sox2 activates target genes through interaction with Oct4. Sox2 knockouts are lethal to mouse embryos and they fail to fully develop, furthermore ES cells derived from these embryos are unable to proliferate or self-renew (Avilion et al., 2003). As outlined for Oct4, a precise level of Sox2 appears to be key for pluripotency and to sustain self-renewal. Many studies have highlighted how Oct4 and Sox2 can in a direct way drive the expression of genes required for pluripotency including positive feedback on their own expression and that of Nanog (Chew et al., 2005; Tomioka et al., 2002). Together with the transcription factor Klf4, they activate the expression of Lefty1 (Nakatake et al., 2006). In mES cells a wide range of studies have focused on and delineated the functional role of Oct4, less is known about Sox2. Recent studies are beginning clear up the role of Sox2. As might have been anticipated mES cells deficient in Sox2 lose pluripotency and quickly differentiate supporting the perceived role of Sox2 in maintaining self-renewal. What is interesting is that in the Sox2 protein deficient system Oct-Sox enhancers are still active and up-regulation of Oct4 alone is sufficient to rescue these cells from differentiation. Thus it has been suggested that potentially other members of the Sox family may substitute for Sox2 in the co-activation process mediated in partnership with Oct4. What has become clear is that Sox2 plays a role in regulating many transcription factors that can affect Oct4 levels including Nanog. Furthermore in cellular reprogramming studies up-regulation of Oct4 in combination with Sox2 is sufficient to generate pluripotent cells (Takahashi and Yamanaka, 2006; Okita et al., 2007). Some fascinating studies looking at global protein phophorylation patterns in hES cells have revealed some interesting dynamics in the Oct4 and Sox2 pattern of activation (Burdon et al., 2002). Thus exploring phophorylation pathways from extracellular signals to gene transcription effects will be key to furthering our understanding of self-renewal, in this context, pathways like those involving LIF and PI3K will be key to disentangling the signaling and transcription circuits involved.

3. Signal transduction pathways

3.1 Leukemia inhibitory factor (LIF)

LIF is expressed in mouse preimplantation embryos from fertilization to the blastocyst stage but not in TE cells (Nichols *et al.*, 1996). LIF transcripts are also found in mES cells (Rathjen *et al.*, 1990) and endometrial glands of the mouse uterus which stops once implantation has occurred. Mouse LIF gene knockouts result in growth retardation and fertilized blastocysts fail to implant (Stewart *et al.*, 1992). Historically mES cells were derived and maintained on a feeder layer of embryonic fibroblast. Subsequently it was found that the use of conditioned media from these fibroblast cultures was sufficient to maintain mES cell self-renewal. It was then shown that a the active agent produced by the feeder layer capable of blocking mES cell differentiation was in fact a cytokine later identified as leukemia inhibitory factor (LIF) (Smith et al., 1988). LIF is the best-characterized effector of self-renewal in mES cells. It is a multifunctional cytokine, which has a wide variety of effects on various cell types (Hilton and Gough, 1991). The name LIF is based on initial observations that *in vitro* it is capable of inducing irreversible differentiation of the murine leukemia cell line M1 to macrophages. LIF is a highly glycosylated single chain polypeptide and a member of IL6 cytokine family (Taga and Kishimoto, 1997). LIF is known to be secreted by a number of cell types including fibroblasts, lymphocytes spleen and liver cells (Gough and Williams, 1989). LIF is a very potent agent, *in vitro* $10^{-9}g/L^{-1}$ (defined as $50U/mL^{-1}$) induces approximately 50% of a population of M1 cells (murine leukemia cell line) to differentiate. LIF induces a wide variety of effects on different cell types e.g. LIF has been shown to sustain the survival of murine primordial germ cells (DeFelici and Dolci, 1991) and stimulate the proliferation of myoblasts in culture. It has also been shown to affect bone growth and remodeling *in vitro* (Lorenzo et al., 1990; Wilson et al., 1992; Gearing et al., 1992; Ip et al., 1992; Taga and Kishimoto, 1997).

The cellular actions of LIF are effected via a specific cell membrane receptor. The LIF receptor is a heterodimeric complex composed of a glycoprotein subunit gp130 and the receptor subunit LIFR (also called LIFRβ) (Ernst and Jenkins, 2004). Studies in mES have shown that the gp130 subunit is the essential component in transmitting self-renewal signals (Nakamura *et al.*, 1998). Binding of LIF to the LIFR subunit induces dimerization with gp130, resulting in the formation of a high affinity receptor complex. The activated receptor switches on the constitutively bound tyrosine kinase Janus kinase (JAK). Activated JAK, phosphorylates both receptor subunits forming SH2 domain bind sites, which are capable of recruiting other signal transduction partners.

The SH2 domains facilitate the binding of signal transducers and activators of transcription (STAT) 1 and STAT3, which are phosphorylated by JAKs (Stahl et al., 1995). The activated STAT proteins form homodimers or heterodimers which then move to the nucleus, where they act as transcription factors (Auernhammer and Melmed, 2000). STAT3 is the principal STAT protein activated in mES cells stimulated with LIF (Niwa *et al.,* 1998). Activation of STAT3, has been shown to be critical for LIF/gp130 dependent self-renewal in mES cells (Niwa et al., 1998). Using a tamoxifen inducible form of STAT3 (fusion of STAT3 to estrogen receptor) it has been shown that activation of STAT3 is capable of sustaining self-renewal of mES in the presence of serum (Mastuda *et al.* 1999).

In the absence of fetal calf serum, in the presence of activated STAT3, BMP4 signaling maintains pluripotency. However, for hES cells LIF-STAT3 signaling cannot maintain pluripotency (Reubinoff et al., 2000) additional factors independent of LIF-STAT3 are required including basic fibroblast growth factor (bFGF) in the presence of Noggin which acts as a BMP pathway inhibitor. The exact mechanism of LIF-STAT3-dependent mES cell self-renewal is still not fully elucidated although models are arising (Fig. 2). A notable target for STAT3 is the transcription factor Myc (Cartwright et al., 2005) which along with others (Klf4, Oct4 and Sox2) has a role in cellular reprograming of somatic cells to a pluripotent state (Takahashi and Yamanaka, 2006). The forced up regulation of Myc supports self-renewal in the absence of LIF. Whereas cessation of LIF signaling results in a decrease in Myc expression presumably through a down-regulation of STAT3. Apart from the above-mentioned STATs a wide range of other downstream effector molecules can be activated through LIF receptor activation including extracellular regulated kinases (ERK), mitogen-activated protein kinases (MAPK) and phosphatydilinositol-3 kinase (PI3K). The network of interactions between intracellular pathways and extracellular ligands continues to develop a pace, with numerous overlaps being identified. In this context another kinase, glycogen synthase kinase 3 (GSK3) a key enzyme in the Wnt pathways is quickly activated resulting in Myc phosphorylation and its degradation. The activity of GSK3 may be controlled by PI3K either directly or indirectly due to LIF signaling. Another possible network connection is that between LIF, PI3K and the Wnt pathway in self-renewal comes from the data that shows improved results in the derivation of mES cells in the presence of the GSK3 inhibitor BIO. Thus from a signaling perspective multiple pathways may be involved in the maintenance of low levels of GSK3 activity to promote pluripotency and mES cell self-renewal. The array of signaling pathways and the level of crosstalk that exist between them and the LIF-STAT3 pathway in mES is slowly being deciphered giving us a clearer picture of the connections between LIF signaling and the transcriptional machinery controlling self-renewal.

3.2 PI3K Pathway

Phosphatidylinositol 3 kinases (PI3Ks) are recognized to modulate a wide range of cellular functions from growth, proliferation and self-renewal to simple metabolic control. They are a family of enzymes, which phosphorylate the 3'-OH position of the inositol ring of phosphoinositides. In 1987 (Whitman et al., 1987) identified two distinct phosphatidylinositol kinases (PIKs) isolated from fibroblasts. They further demonstrated that one of these enzymes associated with activated tyrosine kinase receptors. They called this kinase type I PIK. Subsequently the same group showed that the most abundant form of the previously identified enzymes, type II PIK, phosphorylates the D-4 position on the inositol ring and that type I PIK phosphorylated the inositol ring at the D-3 position.

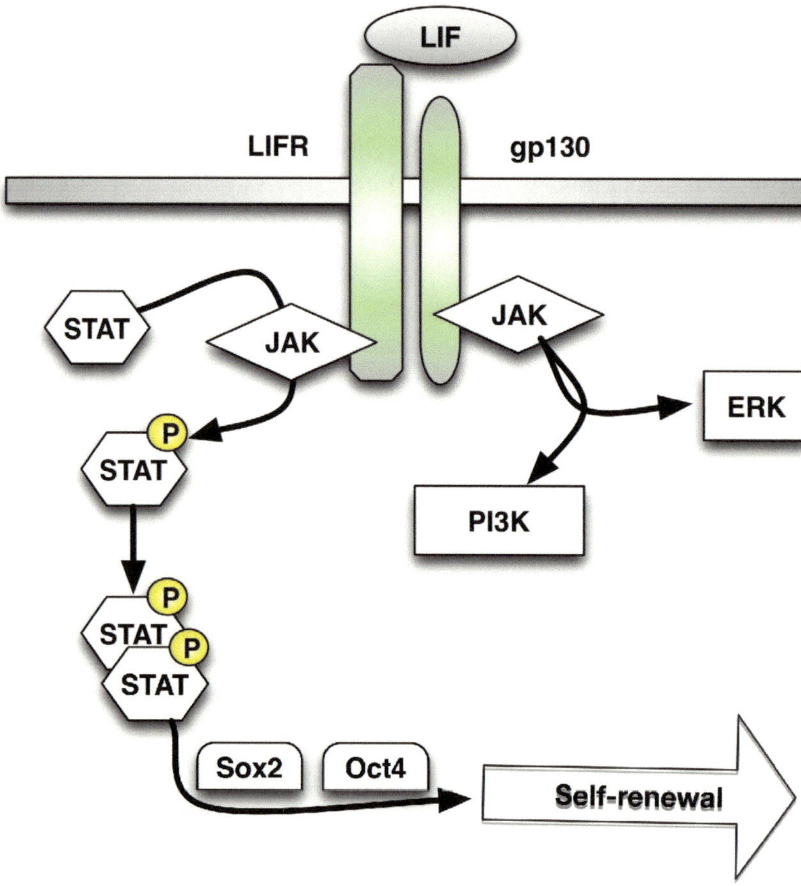

Fig. 2. LIF signal pathways and it integration into transcription machinery of self-renewal. Adapted from Niwa et al., 2009

Currently the family is divided into 3 classes based on structure and substrate preference (Wymann and Pirola, 1998; Vanhaesebroeck et al., 2001). Class I PI3Ks form heterodimers, consisting of a ~110 kDa catalytic subunit, and a regulatory subunit. The regulatory subunit comes in 4 main flavours (p85a, p55a, p50a, p85b, p55g) and a catalytic subunit in 3 major types (p110a, p110b, p110d) (Engelman et al., 2006). In vivo the primary substrate is phosphatidylinositol-4,5, bisphosphate (PtdIns(4,5)P2 or PIP2), which is converted to phosphatidylinositol-3,4,5, triphosphate (PtdIns(3,4,5)P3 or PIP3) (Cantley, 2002). This class of PI3Ks are activated by an array of plasma membrane receptors (for a review see Wymann et al., 2003). Class II PI3Ks produce PI(3)P and PI(3,4)P2 in vitro, but in vivo targets are less clear but the enzyme itself has been localized to the Golgi network. Class III PI3Ks produce only PI(3)P. Much of what we know about the functions of PI3K is because a potent and quite specific inhibitor is available. Wortmannin and LY294002 act as competitive ATP binders targeting the ATP-binding site of catalytic p110 subunit. The most interesting early finding was that wortmannin in the low nanomolar range blocked the respiratory burst of neutrophils (Baggiolini et al., 1987). Studies on purified enzymes have shown that the mammalian PI3K is the most sensitive to wortmannin (Yano et al., 1993). The use of these inhibitors has proved invaluable in the study of PI3K and its cellular effects (reviewed by Nakanishi et al., 1995). The best known product of PI3K action is PIP3 which has been shown to be an important second messenger capable of recruiting AKT and involved in numerous cellular pathways associated with growth, proliferation and survival (Cantley, 2002). The production of PIP3 facilitates the recruitment of pleckstrin homology (PH) domain containing proteins an important example of which is the protein kinase Akt which itself has multiple intracellular targets (Toker, 2002). Commonly in transformed cells the PI3K/Akt pathway is directly activated by the loss of PTEN, a negative regulator of PIP3 formation and an identified tumor suppressor. Maybe unsurprisingly in mES cells the role of PI3Ks was highlighted by the fact that in PTEN null mES cells, accelerated cell cycle progression was observed (Sun et al., 1999) which can be blocked by the PI3K inhibitor LY294002. However a role for PI3K signaling events has also been identified in the maintenance of pluripotency in mES cell derived for a number of species (Fig. 3) (Armstrong et al., 2006). Blocking PI3K signaling events results in elevated ERK/MAPK signaling (Paling et al., 2004) and there is evidence to suggest that ERK (Hamazaki et al., 2006) and Wnt (Sato et al., 2003) signaling are required to sustain pluripotency in both mouse and human ES cell lines. In the

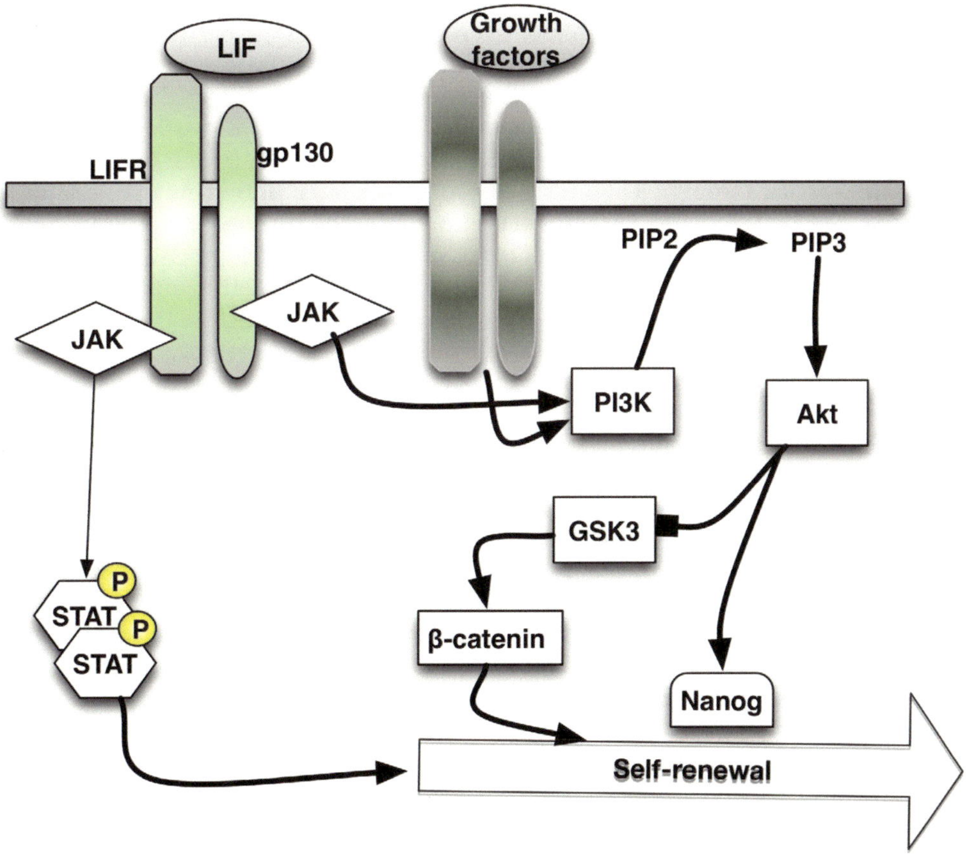

Fig. 3. Potential role for PI3K in self-renewal and LIF signaling in ES cells.

case of mES cells inhibition of PI3K pathways can induce differentiation in the presence or absence of LIF (Paling et al., 2004; Armstrong et al., 2006). However interestingly up regulation of Akt signaling is sufficient to maintain pluripotency of m ES cells (Watanabe et al., 2006). Another linkage for PI3K signaling and self-renewal comes from evidence that Nanog expression as well a number of Nanog target genes are modulated by PI3K signaling. Results have shown that the loss of pluripotent phenotype associated with PI3K blockage by LY294002 can be rescued by exogenous Nanog expression. Also regulation of GSK3 activity acting downstream of PI3Ks plays a role in Nanog expression. The evidence is clearly points out that PI3Ks play an important role in the signaling and maintenance of Nanog expression. PI3K effects are not limited directly to Nanog alone, inhibition of PI3K pathways results in the repression of rfx4, an identified Nanog target (Storm et al., 2007). However, interestingly, of the triad of master factors Oct4, Sox2, and Nanog, it appears that Nanog alone is sensitive to PI3K signaling pathways. However recently it has been shown that suppression of PI3K leads to a reduction in other self-renewal transcription factors including Klf4 (Storm et al., 2009), one of the targets in iPS generation. The role of PI3K in ES cells is complicated by the fact that self-renewal and cell proliferation are linked, and PI3Ks have been cast in major roles for both cellular processes.

3.3 Wnt pathway

The name "Wnt" comes from the fusion of the two names, int (based on the proto-oncogene integration-1 (Tanaka et al., 2002) and wg (based on wingless the segment polarity gene in Drosophila). The Wnt proteins are defined by amino acid sequence rather than by noted functional activities, but all Wnts share a number of common properties like numerous glycosylation sites and target sequences for secretion (Nusse and Varmus, 1992). Upon Wnt binding to its specific receptor, a signaling cascade is activated ultimately upregulating Wnt target genes. The Wnt signaling system is a highly conserved network controlling numerous other signaling transduction pathways from embryonic development to adult tissue homeostasis. Approximately 19 different WNT proteins have been identified acting on at least three different signaling pathways (Nusse and Varmus, 1992). The three pathways are the canonical Wnt pathway, acting via β-catenin and Tcf/Lef factors; the planar cell polarity (PCP) pathway; and the Wnt-Ca2þ pathway (Staal et al., 2008). This section will focus only on the canonical pathway. β-catenin is a well-known cytoplasmic protein and has a role in cell-cell adhesion acting to link membrane bound cadherins to the actin elements in the cytoskeleton. However it is now known to also act as a signaling molecule inside cells as part of the canonical Wnt signaling pathway (Reya and Clevers, 2005). In the absence of Wnt, β-catenin exists in a phosphorylated state in a complex marked for degradation by the ubiquitin-associated proteases. The β-catenin degradation complex includes the tumor suppressor proteins adenomatous polyposis coli gene (APC), Axin, and GSK3. Wnt signaling involves the Wnt ligand binding to the membrane receptor named Frizzled (Fz). Frizzled is a seven transmembrane receptor and the first receptor identified to bind the Wnt ligand (Bhanot et al., 1996). Activation of signal transduction by Wnt binding the Fz receptor requires a co-receptor attachment with a member of the low-density lipoprotein (LDL) family called Lrp5 and −6, this interaction is required for activation of the canonical Wnt signaling pathway (Li and Bu, 2005). Activation of Fz by Wnt results in the protection of β-catenin from proteosomal degradation. Thus the action of Wnt is to maintain the intracellular levels of β-catenin which then translocates to the nucleus where it forms a transcription complex with one of a number of transcription factors including Tcf1, Tcf3,

Tcf4, or Lef1 (Okamura et al., 1998). Tcf1 is found mainly in T lymphocytes, Tcf4 is widely expressed and found in stem cells of gut while Tcf3 is expressed in mES cells. In mES cells there is growing but often-conflicting evidence that Wnt signaling pathways are important components of mES cell self-renewal. Wnt pathways have been shown to sustain pluripotency but also are important for of adult progenitor cell proliferation. The focus on Wnt signaling and its role in pluripotency comes from studies using the GSK3 inhibitor 5-bromoindirubin-3-oxime (BIO) (Fig. 4) (Meijer et al., 2003; Sato et al., 2003).

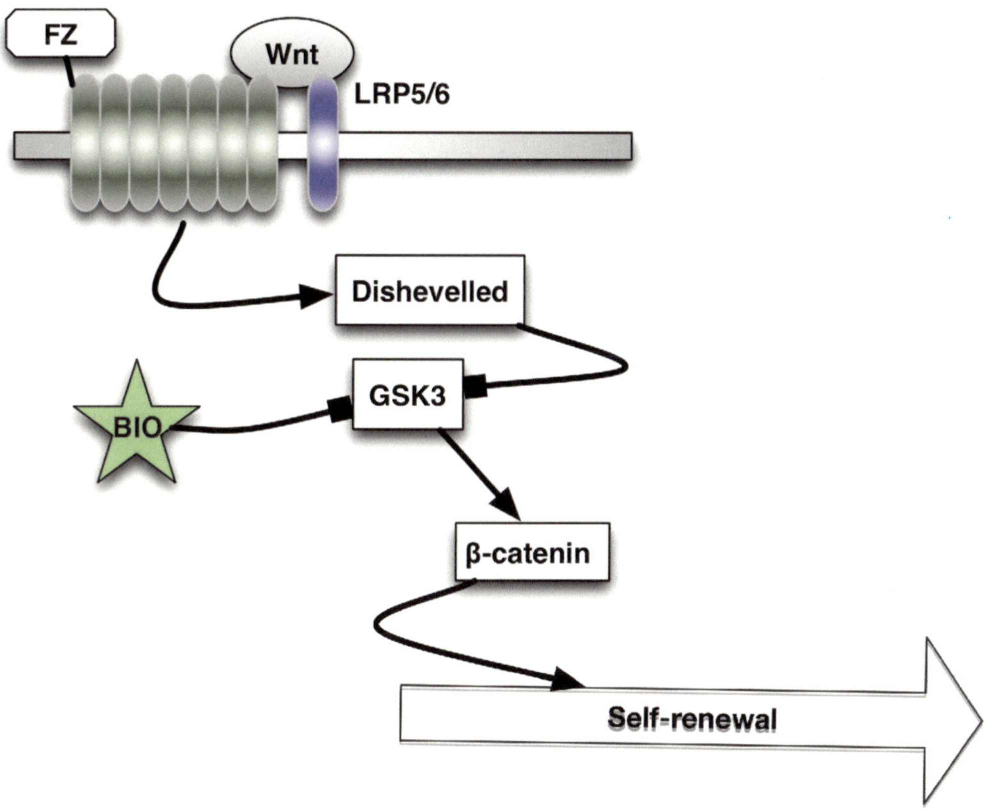

Fig. 4. Wnt signal pathway role in self-renewal.

Inhibition of GSK3 prolongs the existence of β-catenin, causing it to accumulate, increasing the pool, which can translocate to the nucleus and activate gene expression. BIO has been shown to be able to maintain pluripotency of mouse and human ES cells in the absence of LIF (Sato et al., 2003). In a similar vain activation of Wnt signaling indirectly by removing the inhibitory effect of APC sustains pluripotency, suggesting the Wnt signaling is required for self-renewal (Kielman et al., 2002). In addition, treatment with Wnt3a was found to stimulate hES cell proliferation (Singla et al., 2006). Oct4 over-expression increased β-catenin transcriptional activity in progenitor cells. The Wnt controlled transcription factor Tcf3 has been shown to repress Nanog and thus promote differentiation. More recent studies have shown that Lef1 acting along with β-catenin is able to up-regulate Oct4 expression and interact with Nanog and thus promote self-renewal. All these data suggest that Wnt/β-catenin signaling has some role in the mES cell self-renewal (Takao et al., 2007). In obvious contrast to LIF and BMP signaling in mouse and human ES cells there is no difference between the cell types with regard to Wnt/β-catenin signaling self-renewal (Hao et al., 2006). However, contrary to the above-portrayed role of Wnt in self-renewal, Wnt action has been shown to facilitate differentiation of mES cells into neural precursors and increases the expression of Brachyury a mesoderm marker (Yamaguchi et al., 1999). More work is required to elucidate the role of Wnt signaling in mES cell self-renewal and pluripotency and potential other effect for the non-canonical Wnt pathways.

4. Summary and conclusion

Even after prolonged periods and numerous expansions in culture ES cells retain the ability to respond to normal developmental signals and display no apparent bias for any one cell lineage when reintegrated to a developing embryo. Constructing a stable and coherent map of how ES cells achieve such a feat is a major challenge that must be met if the true potential of these cells is to be realized in a clinical setting. A fundamental breakthrough in this area came with the generation of a tetracycline-suppressible *Oct4* transgene in late 2000. This study clearly defined the effect of Oct4 loss on ES cell self-renewal (Niwa et al., 2000). Silencing *Oct4* resulted in ES cells differentiation into trophectoderm but most surprisingly an increase in Oct4 levels resulted in differentiation into mesoderm and endoderm cell linages (Niwa et al., 2000). *Sox2* null ES cells differentiate into trophoectodermal type lineage (Masui et al., 2007). In the absence of Sox2 expression Oct/Sox targets were sustained, leading to the suggestion that other Sox proteins may replace the role of Sox2. Somewhat like Oct4 but less clear cut it would seem from over-expression studies in mES cells that up-regulation of Sox2 results in formation of neural lineages (Kopp et al., 2008; Zhao et al., 2004). Nanog was initially identified as a molecule capable of supporting the pluripotent phenotype of ES cells in the absence of LIF (Chambers et al., 2003). Suppression of Nanog results in increase differentiation (Chambers et al., 2007; Ivanova et al., 2006). Sufficient evidence indicates that the levels of Oct4, Sox2, and Nanog, govern to a large extent the pluripotency of ES cells (Chambers and Smith, 2004). Interestingly the accumulated work on these factors also points to the fact that Oct4 and Sox2 may perhaps drive a process of differentiation countering the self-renewal process. Oct4 and Sox2 increase the production of fibroblast growth factor 4 (FGF4), a growth factor that pushes ES cells toward differentiation by making them more susceptible to leverage by specific lineage commitment factors. All the while Nanog works to resist ES cells differentiation. Artificial high levels of Nanog through constitutive expression systems prevent ES cell differentiation even when FGF signaling is active (Chambers et al., 2003; Ying et al., 2003). However the true role of Nanog and its functional relationship to Oct4 and Sox2 is increasingly more complex. The level of Nanog found in individual normal ES cells shows a high degree of heterogeneity and Nanog knockouts can sustain self-renewal and are pluripotent (Chambers et al., 2007). In general though it appears that cells that are deficient in Nanog have a high propensity to differentiate (Chambers et al., 2007; Mitsui et al., 2003). Our understanding of the extracellular signaling interactions with these transcriptional networks is to date very unclear. The closest we are coming to tying these elements together is looking at work done on LIF signaling in ES cells. ES cells traditionally maintained on feeder cell layers in the presence of serum or a defined serum replacement with the addition of LIF for mouse, and bFGF for human ES cells. These less than wholly defined conditions are entirely problematic for many reasons. A fundamental understanding of the pathways involved for both mouse and human stem cells is beginning to emerge. Niwa et al (2009) have shown that the LIF signaling is tightly linked into the transcriptional machinery of ES cell self-renewal. The Stat3 pathway activates Sox2 but not Nanog, while the PI3K–Akt pathway, effects predominately Nanog. Maintaining a pluripotent phenotype can be viewed as a getting the right balance between continued appropriate proliferation and inhibition of differentiation and/or cell death. Thus a key mechanism to sustaining ES cells in a pluripotent state may be to push and pull all at once i.e. push self-renewal factors and simultaneously block differentiation pathways. An example of this strategy is the effect of GSK3 inhibition in conjunction with inhibition of mitogen-activated protein kinase (MEK which facilitates long-term self-renewal of mouse ES cells with no requirement for cytokines). As we get closer to understanding the pathways in mouse ES cells in particular the issue of difference in comparison to human lines become all the more apparent. Human ES cells are significantly different from mouse ES cells in phenotype and signaling pathway profiles. There is the suggestion that the human cell lines may represent a later stage (epiblast stage, Epi Stage Cells (EpiSCs)) of development compared to the mouse lines. LIF for example does not support either human ES cells or EpiSCs, but does support mES cell self-renewal (Xu et al., 2005). Thus as we elucidate the pathways of pluripotency and self-renewal we will have to be mindful of the cells under study and the stage of development they represent. However these are not insurmountable tasks and growing volumes of data are beginning to delineate the signaling pathways and transcriptional networks controlling cell growth, proliferation and self-renewal. Only when we can clearly map the interactions of the many elements involved some competing and often conflicting signal pathways and key transcription regulators can we understand self-renewal and fully realise the potential of the pluripotent phenotype.

5. References

Andrews (2002). From teratocarcinomas to embryonic stem cells. Philos Trans R Soc Lond B Biol Sci *357*, 405-417.

Armstrong, L., Hughes, O., Yung, S., Hyslop, L., Stewart, R., Wappler, I., Peters, H., Walter, T., Stojkovic, P., et al. (2006). The role of PI3K/AKT, MAPK/ERK and NFkappabeta signalling in the maintenance of human embryonic stem cell pluripotency and viability highlighted by transcriptional profiling and functional analysis. Hum Mol Genet *15*, 1894-1913.

Auernhammer CJ, Melmed S. Leukemia-inhibitory factor neuroimmune modulator of endocrine function. Endocr Rev 2000; 21: 313-45.

Avilion, A.A., Nicolis, S.K., Pevny, L.H., Perez, L., Vivian, N., and Lovell-Badge, R. (2003). Multipotent cell lineages in early mouse development depend on SOX2 function. Genes Dev *17*, 126-140.

Baggiolini, M., Dewald, B., Schnyder, J., Ruch, W., Cooper, P.H., and Payne, T.G. (1987). Inhibition of the phagocytosis- induced respiratory burst by the fungal metabolite wortmannin and some analogues. Exp Cell Res *169*, 408-418.

Bain, G., Kitchens, D., Yao, M., Huettner, J.E. and Gottlieb, D.I. (1995) Embryonic stem cells express neuronal properties in vitro. *Developmental Biology*, 168, 342-57.

Beddington, R.S., and Robertson, E.J. (1989). An assessment of the developmental potential of embryonic stem cells in the midgestation mouse embryo. Development *105*, 733-37.

Bhanot, P., Brink, M., Samos, C.H., Hsieh, J.C., Wang, Y., Macke, J.P., Andrew, D., Nathans, J., and Nusse, R. (1996). A new member of the frizzled family from Drosophila functions as a Wingless receptor. Nature *382*, 225-230.

Boyer, L.A., Lee, T.I., Cole, M.F., Johnstone, S.E., Levine, S.S., Zucker, J.P., Guenther, M.G., Kumar, R.M., Murray, H.L., et al. (2005). Core transcriptional regulatory circuitry in human embryonic stem cells. Cell *122*, 947-956.

Bradley, A., Evans, M., Kaufman, M.H., and Robertson, E. (1984). Formation of germ-line chimaeras from embryo- derived teratocarcinoma cell lines. Nature *309*, 255-56.

Brimble SN, Sherrer ES, Uhl EW, Wang E, Kelly S, Merrill Jr AH et al. The cell surface glycosphingolipids SSEA-3 and SSEA-4 are not essential for human ESC pluripotency. Stem Cells 2007; 25: 54–62.

Brons IG, Smithers LE, Trotter MW, Rugg-Gunn P, Sun B, Chuva de Sousa Lopes SM et al. Derivation of pluripotent epiblast stem cells from mammalian embryos. Nature 2007; 448: 191–195.

Buehr, M., Nichols, J., Stenhouse, F., Mountford, P., Greenhalgh, C.J., Kantachuvesiri, S., Brooker, G., Mullins, J., and Smith, A.G. (2003). Rapid loss of Oct-4 and pluripotency in cultured rodent blastocysts and derivative cell lines. Biol Reprod *68*, 222-29.

Burdon, T., Smith, A., and Savatier, P. (2002). Signalling, cell cycle and pluripotency in embryonic stem cells. Trends Cell Biol *12*, 432-38.

Buttery, L.D., Bourne, S., Xynos, J.D., Wood, H., Hughes, F.J., Hughes, S.P., Episkopou, V. and Polak, J.M. (2001) Differentiation of osteoblasts and in vitro bone formation from murine embryonic stem cells. *Tissue Engineering*, *7*, 89-99

Cartwright P, McLean C, Sheppard A, Rivett D, Jones K, Dalton S. LIF/STAT3 controls ES cell self-renewal and pluripotency by a Myc-dependent mechanism. Development 2005; 132: 885-96.

Cantley, L.C. (2002). The phosphoinositide 3-kinase pathway. Science *296*, 1655-57.

Chambers, I., and Smith, A. (2004). Self-renewal of teratocarcinoma and embryonic stem cells. Oncogene *23*, 7150-160.

Chambers, I., and Tomlinson, S.R. (2009). The transcriptional foundation of pluripotency. Development *136*, 2311-322.

Chambers, I., Colby, D., Robertson, M., Nichols, J., Lee, S., Tweedie, S., and Smith, A. (2003). Functional expression cloning of Nanog, a pluripotency sustaining factor in embryonic stem cells. Cell *113*, 643-655.

Chambers, I., Silva, J., Colby, D., Nichols, J., Nijmeijer, B., Robertson, M., Vrana, J., Jones, K., Grotewold, L., and Smith, A. (2007). Nanog safeguards pluripotency and mediates germline development. Nature *450*, 1230-34.

Chen, A.C., and Gudas, L.J. (1996). An analysis of retinoic acid-induced gene expression and metabolism in AB1 embryonic stem cells. J Biol Chem *271*, 14971-980.

Chew, J.L., Loh, Y.H., Zhang, W., Chen, X., Tam, W.L., Yeap, L.S., Li, P., Ang, Y.S., Lim, B., et al. (2005). Reciprocal transcriptional regulation of Pou5f1 and Sox2 via the Oct4/Sox2 complex in embryonic stem cells. Mol Cell Biol *25*, 6031-046.

Dahéron, L., Opitz, S.L., Zaehres, H., Lensch, M.W., Lensch, W.M., Andrews, P.W., Itskovitz-Eldor, J., and Daley, G.Q. (2004). LIF/STAT3 signaling fails to maintain self-renewal of human embryonic stem cells. Stem Cells *22*, 770-78.

DeFelici, and Dolci (1991). Leukemia inhibitory factor sustains the survival of mouse primordial germ cells cultured on TM4 feeder layers. Dev. Biol. *147*, 281-84.

Doetschman, Eistetter, Katz, Schmidt, and Kemler (1985). The in vitro development of blastocyst-derived embryonic stem cell lines: formation of visceral yolk sac, blood islands and myocardium. J. Embryol. Exp. Morph. *87*, 27-45.

Engelman, J.A., Luo, J., and Cantley, L.C. (2006). The evolution of phosphatidylinositol 3-kinases as regulators of growth and metabolism. Nat Rev Genet *7*, 606-619.

Ernst M, Jenkins BJ. Acquiring signalling specificity from the cytokine receptor gp130. Trends Genet 2004; 20: 23-32.

Evans, M. (1994). Tissue culture of embryonic stem cells. Cell Biology: A Laboratory Handbook , 54-67.

Evans, M.J., and Kaufman, M.H. (1981). Establishment in culture of pluripotential cells from mouse embryos. Nature *292*, 154-56.

Faherty, S., Fitzgerald, A., Keohan, M., and Quinlan, L.R. (2007). Self-renewal and differentiation of mouse embryonic stem cells as measured by Oct4 expression: the role of the cAMP/PKA pathway. In Vitro Cell Dev Biol Anim *43*, 37-47.

Faherty, S., Kane, M.T., and Quinlan, L.R. (2005). SELF-RENEWAL AND DIFFERENTIATION OF MOUSE EMBRYONIC STEM CELLS AS MEASURED BY Oct4 GENE EXPRESSION: EFFECTS OF LIF, SERUM-FREE MEDIUM, RETINOIC ACID, AND dbcAMP. In Vitro Cell Dev Biol Anim *41*, 356-363.

Fujikura, J., Yamato, E., Yonemura, S., Hosoda, K., Masui, S., Nakao, K., Miyazaki, J.-i. and Niwa, H. (2002). Differentiation of embryonic stem cells is induced by GATA factors. Genes Dev. 16, 784-789.

Gearing, D.P., Comeau, M.R., Friend, D.J., Gimpel, S.D., Thut, C.J., McGourty, J., Brasher, K.K., King, J.A., Gillis, S., and Mosley, B. (1992). The IL-6 signal transducer, gp130: an oncostatin M receptor and affinity converter for the LIF receptor. Science *255*, 1434-37.

Gough, and Williams (1989). The pleiotropic actions of leukemia inhibitory factor. Cancer Cells *1*, 77-80.

Hamazaki, T., Kehoe, S.M., Nakano, T., and Terada, N. (2006). The Grb2/Mek pathway represses Nanog in murine embryonic stem cells. Mol Cell Biol *26*, 7539-549.

Hanna LA, Foreman RK, Tarasenko IA, *et al.* Requirement for Foxd3 in maintaining pluripotent cells of the early mouse embryo. Genes Dev 2002; 16:2650-2661.

Hao, J., Li, T.G., Qi, X., Zhao, D.F., and Zhao, G.Q. (2006). WNT/beta-catenin pathway up-regulates Stat3 and converges on LIF to prevent differentiation of mouse embryonic stem cells. Dev Biol *290*, 81-91.

Hilton, D.J., and Gough, N.M. (1991). Leukemia inhibitory factor: a biological perspective. J Cell Biochem *46*, 21-26.

Hooper, N.M. (1997). Glycosyl-phosphatidylinositol anchored membrane enzymes. Clin Chim Acta *266*, 3-12..

Ip, N.Y., Nye, S.H., Boulton, T.G., Davis, S., Taga, T., Li, Y., Birren, S.J., Yasukawa, K., Kishimoto, T., and Anderson, D.J. (1992). CNTF and LIF act on neuronal cells via shared signaling pathways that involve the IL-6 signal transducing receptor component gp130. Cell *69*, 1121-132.

Ivanova, N., Dobrin, R., Lu, R., Kotenko, I., Levorse, J., DeCoste, C., Schafer, X., Lun, Y., and Lemischka, I.R. (2006). Dissecting self-renewal in stem cells with RNA interference. Nature *442*, 533-38.

Kramer, J., Hegert, C., Guan, K., Wobus, A.M., Muller, P.K. and Rohwedel, J. (2000) Embryonic stem cell-derived chondrogenic differentiation in vitro: activation by BMP-2 and BMP-4. *Mechanisms of Development,* 92, 193-205.

Kielman, M.F., Rindapää, M., Gaspar, C., van Poppel, N., Breukel, C., van Leeuwen, S., Taketo, M.M., Roberts, S., Smits, R., and Fodde, R. (2002). Apc modulates embryonic stem-cell differentiation by controlling the dosage of beta- catenin signaling. Nat Genet *32*, 594-605.

Kopp, J.L., Ormsbee, B.D., Desler, M., and Rizzino, A. (2008). Small increases in the level of Sox2 trigger the differentiation of mouse embryonic stem cells. Stem Cells *26*, 903-911.

Li, Y., and Bu, G. (2005). LRP5/6 in Wnt signaling and tumorigenesis. Future Oncol *1*, 673-681.

Lorenzo, Sousa, and Leahy (1990). Leukemia inhibitory factor (LIF) inhibits basal bone resorption in fetal rat long bone cultures. Cytokine *2*, 266-271.

Maltsev, V.A., Rohwedel, J., Hescheler, J. and Wobus, A.M. (1993) Embryonic stem cells differentiate in vitro into cardiomyocytes representing sinusnodal, atrial and ventricular cell types. *Mechanisms of Development,* 44, 41-50.

Marson, A., Levine, S.S., Cole, M.F., Frampton, G.M., Brambrink, T., Johnstone, S., Guenther, M.G., Johnston, W.K., Wernig, M., et al. (2008). Connecting microRNA genes to the core transcriptional regulatory circuitry of embryonic stem cells. Cell *134*, 521-533.

Martin, G.R. (1981). Isolation of a pluripotent cell line from early mouse embryos cultured in medium conditioned by teratocarcinoma stem cells. Proc Natl Acad Sci U S A *78*, 7634-38.

Masui, S., Nakatake, Y., Toyooka, Y., Shimosato, D., Yagi, R., Takahashi, K., Okochi, H., Okuda, A., Matoba, R., et al. (2007). Pluripotency governed by Sox2 via regulation of Oct3/4 expression in mouse embryonic stem cells. Nat Cell Biol *9*, 625-635.

Matsuda, T., Nakamura, T., Nakao, K., Arai, T., Katsuki, M., Heike, T., and Yokota, T. (1999). STAT3 activation is sufficient to maintain an undifferentiated state of mouse embryonic stem cells. EMBO J *18*, 4261-69.

Meijer, L., Skaltsounis, A.L., Magiatis, P., Polychronopoulos, P., Knockaert, M., Leost, M., Ryan, X.P., Vonica, C.A., Brivanlou, A., et al. (2003). GSK-3-selective inhibitors derived from Tyrian purple indirubins. Chem Biol *10*, 1255-266.

Mitsui, K., Tokuzawa, Y., Itoh, H., Segawa, K., Murakami, M., Takahashi, K., Maruyama, M., Maeda, M., and Yamanaka, S. (2003). The homeoprotein Nanog is required for maintenance of pluripotency in mouse epiblast and ES cells. Cell *113*, 631-642.

Mongan, N.P., and Gudas, L.J. (2007). Diverse actions of retinoid receptors in cancer prevention and treatment. Differentiation *75*, 853-870.

Nakamura, T., Arai, T., Takagi, M., Sawada, T., Matsuda, T., Yokota, T. and Heike, T. (1998) A selective switch-on system for self-renewal of embryonic stem cells using chimeric cytokine receptors. *Biochemical and Biophysical Research Communications,* 248, 22-27.

Nakanishi, S., Yano, H., and Matsuda, Y. (1995). Novel functions of phosphatidylinositol 3-kinase in terminally differentiated cells. Cell Signal *7*, 545-557.

Nagy A, Gertsenstein M, Vintersten K, Behringer R. Manipulating the Mouse Embryo, 3rd edn. Cold Spring Harbor Laboratory Press, Chapter 2, 2003, pp 31–140.

Nakatake, Y., Fukui, N., Iwamatsu, Y., Masui, S., Takahashi, K., Yagi, R., Yagi, K., Miyazaki, J., Matoba, R., et al. (2006). Klf4 cooperates with Oct3/4 and Sox2 to activate the Lefty1 core promoter in embryonic stem cells. Mol Cell Biol *26*, 7772-782.

Nichols, J., Davidson, D., Taga, T., Yoshida, K., Chambers, I. and Smith, A.G. (1996) Complementary tissue-specific expression of LIF and LIF-receptor mRNAs in early mouse embryogenesis. *Mechanisms of Development,* 57, 123-131.

Nichols, J., Zevnik, B., Anastassiadis, K., Niwa, H., Klewe-Nebenius, D., Chambers, I., Schöler, H., and Smith, A. (1998). Formation of pluripotent stem cells in the mammalian embryo depends on the POU transcription factor Oct4. Cell *95*, 379-391.

Nichols J, Chambers I, Taga T, Smith A. Physiological rationale for responsiveness of mouse embryonic stem cells to gp130 cytokines. Development 2001; 128:2333-2339.

Niwa, H., Burdon, T., Chambers, I., and Smith, A. (1998). Self-renewal of pluripotent embryonic stem cells is mediated via activation of STAT3. Genes Dev *12*, 2048-060.

Niwa, H., Miyazaki, J., and Smith, A.G. (2000). Quantitative expression of Oct-3/4 defines differentiation, dedifferentiation or self-renewal of ES cells. Nat Genet *24*, 372-76.

Niwa, H., Toyooka, Y., Shimosato, D., Strumpf, D., Takahashi, K., Yagi, R., and Rossant, J. (2005). Interaction between Oct3/4 and Cdx2 determines trophectoderm differentiation. Cell *123*, 917-929.

Niwa, H., Ogawa, K., Shimosato, D., and Adachi, K. (2009). A parallel circuit of LIF signalling pathways maintains pluripotency of mouse ES cells. Nature *460*, 118-122.

Nusse, R., and Varmus, H.E. (1992). Wnt genes. Cell *69*, 1073-087.

Okamura, R.M., Sigvardsson, M., Galceran, J., Verbeek, S., Clevers, H., and Grosschedl, R. (1998). Redundant regulation of T cell differentiation and TCRalpha gene expression by the transcription factors LEF-1 and TCF-1. Immunity *8*, 11-20.

Okita, K., Ichisaka, T., and Yamanaka, S. (2007). Generation of germline-competent induced pluripotent stem cells. Nature *448*, 313-17.

Paling, N.R.D., Wheadon, H., Bone, H.K., and Welham, M.J. (2004). Regulation of embryonic stem cell self-renewal by phosphoinositide 3-kinase-dependent signaling. J Biol Chem *279*, 48063-070.

Palmieri, Peter, Hess, and Scholer (1994). Oct-4 transcription factor is differentially expressed in the mouse embryo during establishment of the first two extraembryonic cell lineages involved in implantation. Dev. Biol. *166*, 259-267.

Pan, G., Li, J., Zhou, Y., Zheng, H., and Pei, D. (2006). A negative feedback loop of transcription factors that controls stem cell pluripotency and self-renewal. FASEB J *20*, 1730-32.

Pelton, T.A., Sharma, S., Schulz, T.C., Rathjen, J., and Rathjen, P.D. (2002). Transient pluripotent cell populations during primitive ectoderm formation: correlation of in vivo and in vitro pluripotent cell development. J Cell Sci *115*, 329-339.

Pera, Reubinoff, and Trounson (2000). Human embryonic stem cells. J Cell Sci *113*, 5-10.

Pesce, M., and Schöler, H.R. (2001). Oct-4: gatekeeper in the beginnings of mammalian development. Stem Cells *19*, 271-78.

Rathjen, P.D., Nichols, J., Toth, S., Edwards, D.R., Heath, J.K. and Smith, A.G. (1990) Developmentally programmed induction of differentiation inhibiting activity and the control of stem cell populations. *Genes and Development,* 4, 2308-2318.

Resnick, Bixler, Cheng, and Donovan (1992). Long-term proliferation of mouse primordial germ cells in culture. Nature *359*, 550-51.

Reubinoff, B.E., Pera, M.F., Fong, C.Y., Trounson, A., and Bongso, A. (2000). Embryonic stem cell lines from human blastocysts: somatic differentiation in vitro. Nat Biotechnol *18*, 399-404.

Reya T, Clevers H. Wnt signalling in stem cells and cancer. Nature 2005; 434: 843-50.

Robertson (1987). *Teratocarcinomas and Embryonic Stem Cells. Embryonic Stem Cells: A Practical Approach.* Oxford: IRL Press

Sato, N., Meijer, L., Skaltsounis, L., Greengard, P., and Brivanlou, A.H. (2003). Maintenance of pluripotency in human and mouse embryonic stem cells through activation of Wnt signaling by a pharmacological GSK-3-specific inhibitor. Nature medicine *10*, 55-63.

Rodda DJ, Chew JL, Lim LH, *et al*. Transcriptional regulation of nanog by OCT4 and SOX2. J Biol Chem 2005; 280: 24731- 24737.

Rossant J. Stem cells and lineage development in the mammalian blastocyst. Reprod Fertil Dev 2007; 19: 111–118.

Scholer H, S Ruppert, N Suzuki, K Chowdhury, P Gruss. New type of POU domain in germ line-specific protein Oct4. Nature 1990; 344:435-9.

Shamblott, Axelman, Wang, Bugg, Littlefield, Donavan, Blumenthal, Huggins, and Gearhart (1998). Derivation of pluripotent stem cells from cultured human primordial germ cells. Proc Natl Acad Sci U S A *95*, 13726-731.

Shi, W., Wang, H., Pan, G., Geng, Y., Guo, Y., and Pei, D. (2006). Regulation of the pluripotency marker Rex-1 by Nanog and Sox2. J Biol Chem *281*, 23319-325.

Singla, D.K., Schneider, D.J., LeWinter, M.M., and Sobel, B.E. (2006). wnt3a but not wnt11 supports self-renewal of embryonic stem cells. Biochem Biophys Res Commun *345*, 789-795.

Smith, A.G., Heath, J.K., Donaldson, D.D., Wong, G.G., Moreau, J., Stahl, M., and Rogers, D. (1988). Inhibition of pluripotential embryonic stem cell differentiation by purified polypeptides. Nature *336*, 688-690.

Solter, D., and Knowles, B.B. (1979). Developmental stage-specific antigens during mouse embryogenesis. Curr Top Dev Biol *13 Pt 1*, 139-165.

Stahl N, Farruggella TJ, Boulton TG, Zhong Z, Darnell JE, Jr, Yancopoulos GD. Choice of STATs and other substrates specified by modular tyrosine-based motifs in cytokine receptors. Science 1995; 267: 1349-53.

Staal, F.J., Luis, T.C., and Tiemessen, M.M. (2008). WNT signalling in the immune system: WNT is spreading its wings. Nat Rev Immunol *8*, 581-593.

Storm, M.P., Bone, H.K., Beck, C.G., Bourillot, P.Y., Schreiber, V., Damiano, T., Nelson, A., Savatier, P., and Welham, M.J. (2007). Regulation of Nanog expression by phosphoinositide 3-kinase-dependent signaling in murine embryonic stem cells. J Biol Chem *282*, 6265-273.

Storm, M.P., Kumpfmueller, B., Thompson, B., Kolde, R., Vilo, J., Hummel, O., Schulz, H., and Welham, M.J. (2009). Characterization of the phosphoinositide 3-kinase-dependent transcriptome in murine embryonic stem cells: identification of novel regulators of pluripotency. Stem Cells *27*, 764-775.

Stewart, C.L., Kaspar, P., Brunet, L.J., Bhatt, H., Gadi, I., Kontgen, F. and Abbondanzo, S.J. (1992) Blastocyst implantation depends on maternal expression of leukaemia inhibitory factor. *Nature,* 359, 76-9.

Suda, Y., Suzuki, M., Ikawa, Y., and Aizawa, S. (1987). Mouse embryonic stem cells exhibit indefinite proliferative potential. J Cell Physiol *133*, 197-201.

Sun, H., Lesche, R., Li, D.M., Liliental, J., Zhang, H., Gao, J., Gavrilova, N., Mueller, B., Liu, X., and Wu, H. (1999). PTEN modulates cell cycle progression and cell survival by regulating phosphatidylinositol 3,4,5,-trisphosphate and Akt/protein kinase B signaling pathway. Proc Natl Acad Sci U S A *96*, 6199-6204.

Suwabe, N., Takahashi, S., Nakano, T. and Yamamoto, M. (1998) GATA-1 regulates growth and differentiation of definitive erythroid lineage cells during in vitro ES cell differentiation. *Blood,* 92, 4108-18.

Suzuki A, Raya A, Kawakami Y, et al. Nanog binds to Smad1
and blocks bone morphogenetic protein-induced differentiation of embryonic stem cells. Proc Natl Acad Sci USA 2006; 103:10294-10299.

Taga, T., and Kishimoto, T. (1997). Gp130 and the interleukin-6 family of cytokines. Annu Rev Immunol *15*, 797-819.

Takahashi, K., Tanabe, K., Ohnuki, M., Narita, M., Ichisaka, T., Tomoda, K., and Yamanaka, S. (2007). Induction of pluripotent stem cells from adult human fibroblasts by defined factors. Cell *131*, 861-872.

Takao, Y., Yokota, T., and Koide, H. (2007). Beta-catenin up-regulates Nanog expression through interaction with Oct-¾ in embryonic stem cells. Biochem Biophys Res Commun *353*, 699-705.

Tanaka, K., Kitagawa, Y., and Kadowaki, T. (2002). Drosophila segment polarity gene product porcupine stimulates the posttranslational N-glycosylation of wingless in the endoplasmic reticulum. J Biol Chem *277*, 12816-823.

Tesar PJ, Chenoweth JG, Brook FA, Davies TJ, Evans EP, Mack DL et al. New cell lines from mouse epiblast share defining features with human embryonic stem cells. Nature 2007; 448: 196–199.

Thomson, J.A., Itskovitz-Eldor, J., Shapiro, S.S., Waknitz, M.A., Swiergiel, J.J., Marshall, V.S., and Jones, J.M. (1998). Embryonic stem cell lines derived from human blastocysts. Science *282*, 1145.

Toker, A. (2002). Phosphoinositides and signal transduction. Cell Mol Life Sci *59*, 761-779.

Tomioka, M., Nishimoto, M., Miyagi, S., Katayanagi, T., Fukui, N., Niwa, H., Muramatsu, M., and Okuda, A. (2002). Identification of Sox-2 regulatory region which is under the control of Oct-3/4-Sox-2 complex. Nucleic Acids Res *30*, 3202-213.

Vanhaesebroeck, B., Leevers, S.J., Ahmadi, K., Timms, J., Katso, R., Driscoll, P.C., Woscholski, R., Parker, P.J., and Waterfield, M.D. (2001). Synthesis and function of 3-phosphorylated inositol lipids. Annu Rev Biochem *70*, 535-602.

Ware, and Axelrad (1972). Inheirited resisitance to N- and B-trophic murine leukemia viruses in vitro: Evidence that congenic mouse strains SIM and SIM.R differ at the Fv-I locus. Virology *50*, 339-348.

Watanabe, S., Umehara, H., Murayama, K., Okabe, M., Kimura, T., and Nakano, T. (2006). Activation of Akt signaling is sufficient to maintain pluripotency in mouse and primate embryonic stem cells. Oncogene *25*, 2697-2707.

Whitman, M., Kaplan, D., Roberts, T., and Cantley, L. (1987). Evidence for two distinct phosphatidylinositol kinases in fibroblasts. Implications for cellular regulation. Biochem J *247*, 165-174.

Williams, R.L., Hilton, D.J., Pease, S., Willson, T.A., Stewart, C.L., Gearing, D.P., Wagner, E.F., Metcalf, D., Nicola, N.A., and Gough, N.M. (1988). Myeloid leukaemia inhibitory factor maintains the developmental potential of embryonic stem cells. Nature *336*, 684-87.

Wilson, Metcalf, and Gough (1992). Cross-species comparsion of the sequence of the leukaemia inhibitory factor gene and its protein. Eur. J. Biochemistry *204*, 21-30.

Winnier G, Blessing M, Labosky PA, Hogan BL. Bone mor- phogenetic protein-4 is required for mesoderm formation and patterning in the mouse. Genes Dev 1995; 9:2105-2116 .

Wood, H.B., and Episkopou, V. (1999). Comparative expression of the mouse Sox1, Sox2 and Sox3 genes from pre-gastrulation to early somite stages. Mechanisms of development *86*, 197-201.

Wymann, M.P., and Pirola, L. (1998). Structure and function of phosphoinositide 3-kinases. Biochim Biophys Acta *1436*, 127-150.

Wymann, M.P., Zvelebil, M., and Laffargue, M. (2003). Phosphoinositide 3-kinase signalling--which way to target? Trends Pharmacol Sci *24*, 366-376.

Xu, R.H., Peck, R.M., Li, D.S., Feng, X., Ludwig, T., and Thomson, J.A. (2005). Basic FGF and suppression of BMP signaling sustain undifferentiated proliferation of human ES cells. Nat Methods *2*, 185-190.

Yamaguchi TP, Takada S, Yoshikawa Y, Wu N, McMahon AP. T (Brachyury) is a direct target of Wnt3a during paraxial mesoderm specification. Genes Dev 1999; 13: 3185-90.

Yamashita, J., Itoh, H., Hirashima, M., Ogawa, M., Nishikawa, S., Yurugi, T., Naito, M. and Nakao, K. (2000) Flk1-positive cells derived from embryonic stem cells serve as vascular progenitors. *Nature*, 408, 92-6.

Yano, H., Nakanishi, S., Kimura, K., Hanai, N., Saitoh, Y., Fukui, Y., Nonomura, Y., and Matsuda, Y. (1993). Inhibition of histamine secretion by wortmannin through the blockade of phosphatidylinositol 3-kinase in RBL-2H3 cells. J Biol Chem *268*, 25846-856.

Yeom, Y.I., Fuhrmann, G., Ovitt, C.E., Brehm, A., Ohbo, K., Gross, M., Hübner, K., and Schöler, H.R. (1996). Germline regulatory element of Oct-4 specific for the totipotent cycle of embryonal cells. Development *122*, 881-894.

Ying, Q.L., Nichols, J., Chambers, I., and Smith, A. (2003). BMP induction of Id proteins suppresses differentiation and sustains embryonic stem cell self-renewal in collaboration with STAT3. Cell *115*, 281-292.

Zappone, M.V., Galli, R., Catena, R., Meani, N., De Biasi, S., Mattei, E., Tiveron, C., Vescovi, A.L., Lovell-Badge, R., et al. (2000). Sox2 regulatory sequences direct expression of a (beta)-geo transgene to telencephalic neural stem cells and precursors of the mouse embryo, revealing regionalization of gene expression in CNS stem cells. Development *127*, 2367-382.

Zeng, X., Miura, T., Luo, Y., Bhattacharya, B., Condie, B., Chen, J., Ginis, I., Lyons, I., Mejido, J., et al. (2004). Properties of pluripotent human embryonic stem cells BG01 and BG02. Stem Cells *22*, 292-312.

Zeng, X., Tamai, K., Doble, B., Li, S., Huang, H., Habas, R., Okamura, H., Woodgett, J., and He, X. (2005). A dual-kinase mechanism for Wnt co-receptor phosphorylation and activation. Nature *438*, 873-77.

Zhao, S., Nichols, J., Smith, A.G., and Li, M. (2004). SoxB transcription factors specify neuroectodermal lineage choice in ES cells. Mol Cell Neurosci *27*, 332-342.

16

Building a Pluripotency Protein Interaction Network for Embryonic Stem Cells

Patricia Ng and Thomas Lufkin
Genome Institute of Singapore
Singapore

1. Introduction

Embryonic stem cells were isolated from the mouse in 1981. Two landmark papers (Evans & Kaufman, 1981; Martin, 1981) described the isolation from the blastocyst of a cell line, that grew rapidly, was maintained by passaging, had a normal karyotype unlike embryonal carcinoma cells and could be induced to differentiate into a wide variety of cell types by injecting them into the mouse or by culturing them in the absence of feeder cells *in vitro*. The most important property of these embryonic stem cells was the ability to differentiate into most cell types. This is termed "pluripotency". Understanding the molecular mechanisms of pluripotency will enable scientist to utilise stem cells more effectively, particularly in the field of tissue repair and regeneration (Murry & Keller, 2008). One approach towards the understanding of this molecular mechanism is to look at its protein interaction.

This chapter will explain what a protein interaction network is and why it is used for looking at pluripotency. It will cover methods used to build protein-interaction networks and the methods of validations for these protein interactions. The chapter will also present an integrated dataset that merges the current understanding into one protein interaction network. Base on this integrated network, we will discuss what constitutes key factors in pluripotency, how these key factors are connected in the network and the protein machineries that they recruit to set up the pluripotent state. Finally, the chapter will look at the future challenges in the completion and utilization of the protein interaction network for the manipulation of pluripotency.

2. What is the protein interaction network?

A protein interaction network comprises proteins as nodes and protein-protein interactions as undirected links between the nodes. Drawing networks allows researchers to manage and interpret large datasets. Interpretation of the dataset is done by adopting concepts from other fields such graph theory to describe network properties. Such interpretations can explain how the structure of the network is serving its biological function. For example, in the field of graph theory, several parameters can be computed for a network. These parameters describe the architecture of the network so that it can be compared to other networks. This can provide some insights into the behaviour of the network particularly if it is compared to a network that is similar and already better understood. The most fundamental parameter of a network is the number of links a node has. This is referred to as 'the degree of a node' and is a variable that is designated by "k". To describe all the nodes in the network, the number of nodes having different degree can be presented as a distribution curve (Figure 1). Depending on the pattern of the distribution curve, networks can be grouped into different classes (Barabasi & Oltvai, 2004). Random networks show a Poisson distribution (Figure 1A). Scale-free networks show a power law distribution (Figure 1B). The embryonic stem cell protein interaction network belongs to the class of scale-free networks.

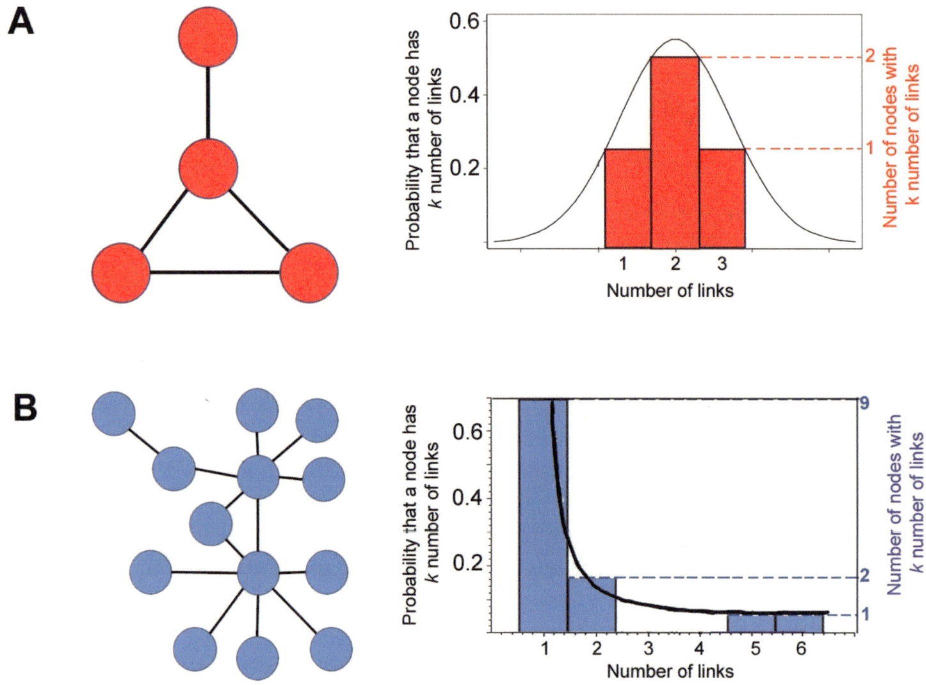

Fig. 1. Networks and their corresponding graphs. (A) Random network shows a normal distribution and (B) Scale-free network shows a power-law distribution when their number or fraction of nodes with different number of links are plotted.

In scale-free networks, most of the nodes will have very few links and only a few nodes will have many links (Figure 1B). Systems that are approximately scale-free include many biological networks like the yeast proteome, the prokaryote and eukaryote transcription network, all metabolic networks, and even the internet (Albert, 2005; Barabasi, 2009). Networks of this class show robustness against failure of single components. Besides degree distribution, some other network properties include the average number of neighbours, the average or characteristic path length, the network diameter and the clustering coefficient (Barabasi & Oltvai, 2004). Classifying networks by its degree distribution is one way graph theory can be used to associate universal laws or organizational principals to networks. Understanding network principals makes it easier to predict protein functions; to generate testable hypothesis; and to simulate manipulations of protein components to see if it gives desired consequences.

3. Building the protein interaction network

3.1 Methods used to build the network

Different methods can be used to build a protein interaction network. The simplest method is to build the network based on available information about protein interactions from the literature. A second method is to include interactions based on extrapolations of protein interactions in other organisms to the orthologs in the organism of interest. However, both of these methods are limited to known interactions. In addition, the second method may result in the inclusion of false interactions due to wrongly mapped orthologs or lost of conservation of interactions.

To discover protein-protein interactions without *a priori* knowledge and for network construction, a high throughput method is needed. For this, there are two tested approaches. There is the yeast 2-hybrid approach and the affinity purification-mass spectrometry approach. Interactors are identified by sequencing the plasmids encoding for the proteins (for the yeast 2-hyrbid approach) or by mass spectrometry (for the affinity purification approach). Both of these approaches have been tested in the construction of a protein interaction network for the yeast proteome (Gavin, et al., 2006; Gavin, et al., 2002; Ho, et al., 2002; Ito, et al., 2001; Uetz, et al., 2000). For the construction of a pluripotency-associated protein interaction network, the aim is to have only the components of pluripotency in this network. Hence this should be a subnetwork of the proteome. Oct3/4, a key factor for pluripotency is selected as the bait from which other proteins may be discovered. The network grows when proteins that interact with Oct3/4 are used as the next bait; and by iteration the protein interaction network for pluripotency can be completed. Caution should be taken during iteration of this process to avoid extending the network into the interactions of non pluripotency-associated proteins. Most of the datasets for embryonic stem cell pluripotency-associated network are generated by affinity purification-mass spectrometry. Use of the yeast 2-hybrid approach has been attempted on Oct3/4. But the number of interactors discovered via this approach (Li, et al., 2008) was significantly lower than that discovered via affinity purification-mass spectrometry. This suggests that Oct3/4 may recruit most of the subunits of macromolecules through only a few direct interactors or requires more interactors for contacts to be stabilized. Indirect interactors or cooperative interactors of Oct3/4 will not be discovered by the yeast 2-hybrid approach, which identifies only binary interactions.

3.2 Methods used to validate the network

A major concern with the use of affinity purification-mass spectrometry or yeast 2-hybrid approaches is the presence of false positives. In the yeast 2-hyrbid system, biologically irrelevant interactions can happen between two proteins inside the yeast nucleus to give a false-positive signal. In affinity purification-mass spectrometry, false positives are caused by background proteins that are not completely removed during affinity purification. Although mock purifications are included in experiments to allow identification of background proteins, there is a limitation in the mass spectrometer to capture all the peptides in a sample for identification. As such, sampling of the peptide population is not saturated. This causes estimations of relative abundance of proteins to be inaccurate and hence discrimination of noise from signal based on relative abundance of proteins in the actual versus the mock purifications also becomes inaccurate.

In view of these shortcomings of the approaches, validation of datasets becomes very important. The most direct method of validating a protein-protein interaction is by reciprocal co-precipitation. This can be done by expressing the two proteins in a cell culture system. However, some of these interactions are indirect and occur via a third protein, which if not present in the cell, would yield negative results in co-precipitation analysis. Even after direct association has been verified, it is important to further examine the functional significance of the interaction. Not all physical interactions have functional significance. For example, both Oct1 and Oct3/4 can interact with Sox2, but only the Oct3/4-Sox2 complex activates *Fgf4* expression (Yuan, et al., 1995). Hence, multiple validations are necessary to verify that a protein has a role in pluripotency. Validations that have been employed include: (i) evidence for the presence of the interactor in embryonic stem cells; (ii) evidence that the interactor and the bait co-exist in a common subcellular location; (iii) indication that the level of abundance of the interactor changes upon differentiation; (iv) indication that the interactor regulates the genes of known embryonic stem cell transcription factors or *vice versa;* (v) gain or loss of pluripotency of embryonic stem cells when the gene of the interactor is knocked-out, is suppressed by RNA interference, or is overexpressed. For this validation, pluripotency can be monitored by alkaline phosphatase staining, by embryonic stem cell morphology, by Oct3/4 or Nanog transcript level, by profiling of lineage markers and by the expression of stage-specific embryonic antigen (SSEA) 1, 3 and 4. Finally, validation can also be done by looking at the loss-of-function phenotypes in mice. For this, the gene of the interactors can be knocked-out, suppressed by RNA interference or overexpressed. Given that gene redundancy or functional redundancy is a phenomenon of pluripotency, validations that show no effect with a single gene knock-out could be further evaluated by double or triple knock-out of related genes.

4. Analysis of current datasets

Although the earliest protein-protein interaction network in embryonic stem cells was based on Nanog as that first bait protein, datasets of later work were mostly built upon Oct3/4 (Liang, et al., 2008; Pardo, et al., 2010; van den Berg, et al., 2010; Wang, et al., 2006). Other proteins that have been used as baits are Sall4, Tcfcp2l1, Dax1, Esrrb, Rex1, Nac1 and Zfp281 (van den Berg, et al., 2010). These proteins were used because they were found to interact with Oct3/4. To gain a more complete view of the pluripotency protein interaction network, datasets from the four published protein interaction networks of the embryonic stem cell was integrated as one (Figure 2). Integration of these datasets gives a network comprising 239 proteins. Of these, 131 proteins were directly associated to Oct3/4. As expected, the distribution of the nodes according to their degree of links follows a power law distribution curve (Figure 2). Theoretically, this would suggest that pluripotency is mediated by a highly robust mechanism that is insensitive to the loss of many of its individual components.
However, at this stage more work is required before such conclusion can become fully accepted. This is because the protein interaction network is currently incomplete. At this stage, the network structure can be strongly skewed by the methods used to generate the network (Futschik, et al., 2007). The observation that essential proteins like Oct3/4, tend to be more highly connected than nonessential proteins could be a true property or a consequence of their having been more thoroughly studied, or a combination of the two (Hakes, et al., 2008). As data accumulates, the power of system biology to catalogue and integrate data will become more meaningful (Pieroni, et al., 2008).

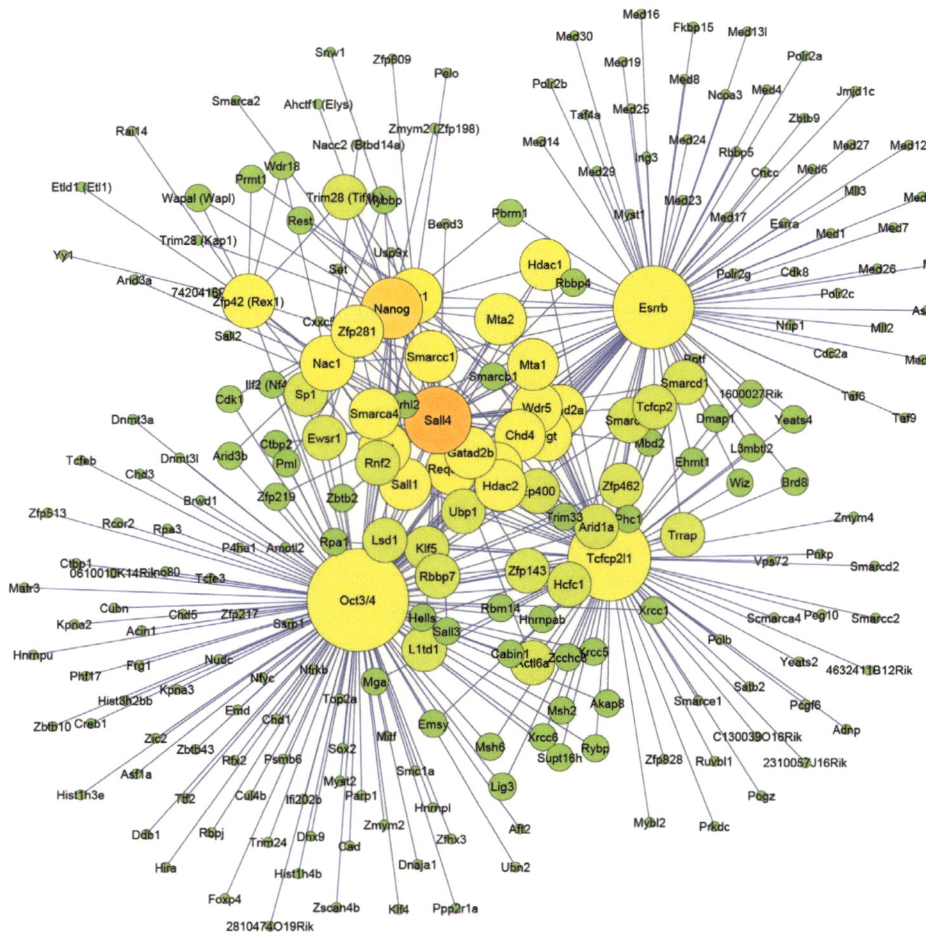

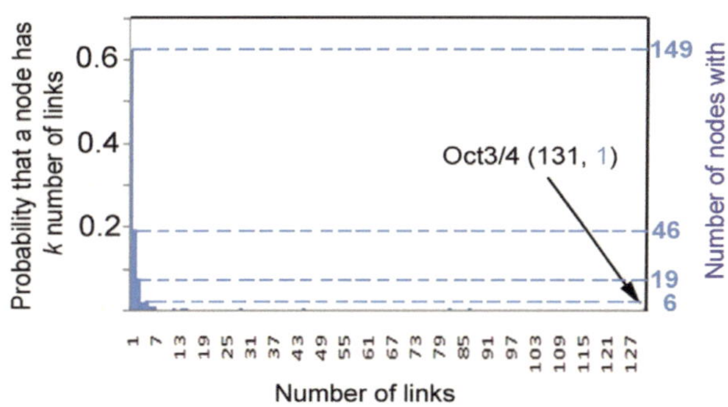

Number of nodes: 239
Connected components: 1
Average number of neighbors: 3.356
Characteristic path length : 2.497
Clustering coefficient : 0.310
Network diameter : 4

Fig. 2. The pluripotency-associated protein interaction network of mouse embryonic stem cells. The network is derived from a combined dataset from four publications (Liang, et al., 2008; Pardo, et al., 2010; van den Berg, et al., 2010; Wang, et al., 2006). Bigger and brighter colored nodes represents proteins with more links. Simple network parameters are indicated in the box. A graph showing that the nodes have a power law distribution for their number of links. The graph also indicated Oct3/4 as the node at the tail end of the distribution on the graph.

4.1 The key factors in pluripotency

From the literature, Oct3/4 is already known to be a key factor in pluripotency. In mice, loss of Oct3/4 results in embryos that fail to form a pluripotent inner cell mass (Nichols, et al., 1998). The inner cell mass in these embryos takes on a trophoblast lineage and subsequently fails to proliferate. In adult cells, provision of Oct3/4 together with various cocktails of transcription factors induces pluripotency (Nakagawa, et al., 2008; Takahashi & Yamanaka, 2006). While the other components of these cocktails can change, the inclusion of Oct3/4 is indispensible. The level of Oct3/4 is also important in the subsequent maintenance of pluripotency. While decrease of Oct3/4 to half its physiological level leads to conversion of embryonic stem cells to trophectoderm, an increase of Oct3/4 by less than two fold of its physiological dosage leads to conversion of embryonic stem cells into primitive endoderm and mesoderm (Niwa, et al., 2000). Finally, as the embryo develops, the level of Oct3/4 decreases in the cells that differentiate; but in germ cells where pluripotency is kept, Oct3/4 expression is maintained (Scholer, et al., 1990; Scholer, et al., 1989). Taken together, this is evidence for the role of Oct3/4 in inducing and in maintaining pluripotency. While the key role of Oct3/4 in pluripotency is obvious and does not need construction of the protein interaction network to point this out. The emergence of other protein hubs (nodes with high number of links) can suggest new key factors. Following Oct3/4, are two other proteins, Esrrb and Tcfcp2l1 that have 82 and 87 links respectively. The importance of Esrrb in pluripotency is corroborated by the observation that this protein can help in the induction of pluripotency in fibroblast (Feng, et al., 2009). Although there are no similar observations for Tcfcp2l1, its hub position in the network would suggest that this protein might be another important coordinator of pluripotency.

Recently, the use of RNA interference has offered a means to functionally screen the genome. This would be a complimentary approach to the protein interaction network to find key factors of pluripotency. To find genes that are needed for maintenance of pluripotency, individual genes are knock-down by RNA interference. Combining the datasets from several studies (Ivanova, et al., 2006; Zhang, et al., 2006), including two which were genome-wide screens (Ding, et al., 2009; Hu, et al., 2009), led to the identification of a total of 166 proteins. In concurrence with the identification of Esrrb as a hub protein in the protein interaction network, the same protein was found to be one of the 166 proteins that were important for the maintenance of pluripotency (Table 1). However, a total of only 15 genes, inclusive of Esrrb, from the list of 166 are in the protein interaction network. This suggests that there are other key components found via RNA interference that are not yet discovered by protein-protein interaction. On the reverse, there are 224 proteins in the protein interaction network that are not found by RNA interference. These proteins could be involved in the induction of pluripotency but not in maintenance of it. Alternatively, these proteins may not have been identified via RNA interference because there can be redundancy of function, which is one mechanism for the robustness of the network.

For human embryonic stem cell, no protein interaction network based on yeast 2-hybrid or affinity p4urification-mass spectrometry approaches have been generated. However, determinants of human embryonic stem cell pluripotency have been identified by a genome-wide RNA interference screen (Chia, et al., 2010). The screen identified a total of 566 genes and a protein interaction network base on these has been reported. Information regarding possible interactions between any of the 566 genes was mapped based on the online database STRING, which stores known interaction and includes transfers from orthologous

No.	Gene	Reference
1	Ehmt1	Hu
2	Esrrb	Ivanova
3	Hira	Hu
4	Mbd3	Hu
5	Mga	Hu
6	Nanog	Ivanova
7	Ncoa3	Hu
8	Oct3/4	Hu; Ivanova
9	Pcgf6	Hu
10	Rif1	Hu
11	Rnf2	Ding; Hu
12	Smc1a	Hu
13	Sox2	Hu; Ivanova
14	Yy1	Hu
15	Zfp219	Hu
16	1700067P10Rik	Hu
17	3110070M22Rik	Hu
18	5430407P10Rik	Hu
19	Acadsb	Ding
20	Acoxl	Hu
21	Adk	Hu
22	Aldoa	Hu
23	Amot	Hu
24	Apc	Ding; Hu
25	Ash2l	Zhang
26	Atg3	Hu
27	Atox1	Hu
28	BC018507	Ding
29	Bcl2l12	Hu
30	Bcorl1	Ding
31	Cbx1	Hu
32	Ccnb1ip1	Ivanova
33	Ccrn4l	Hu
34	Cdk9	Hu
35	Cdkn2aip	Hu
36	Cnih3	Hu
37	Cnot1	Ding
38	Cnot3	Hu
39	Coq3	Hu
40	Cpsf1	Hu
41	Cpsf2	Hu
42	Cpsf3	Ding; Hu
43	Ctr9	Ding; Hu
44	Cul3	Hu
45	Cxcl9	Hu
46	Cxxc1	Ding
47	D630039A03Rik	Hu
48	Dab2ip	Hu
49	Dazap1	Hu
50	Dppa4	Ivanova
51	Dppa5a (Dppa5)	Zhang
52	Ear11	Hu
53	Ecel1	Hu
54	Efr3b	Ding
55	Eif2s3x	Hu
56	Eif4a1	Hu
57	Eif4g2	Hu
58	Elof1	Hu
59	Eny2	Hu
60	Ep300	Hu
61	Epdr1	Hu
62	Eya1	Hu
63	Eya2	Hu
64	Fbxl8	Hu
65	Fip1l1	Ding; Hu
66	Fry	Hu
67	Gale	Hu
68	Ggh	Hu
69	Golga7	Hu
70	Grk6	Hu
71	Hao1	Hu
72	Hist1h3i	Hu
73	Hnrpul1	Hu
74	Hoxa7	Hu
75	Htatip2	Hu
76	Ift46	Hu
77	Il20	Hu
78	Il6st	Hu
79	Ing5	Hu
80	Ino80e (Ccdc95)	Hu
81	Iws1	Ding; Hu
82	Krtap16-7 (Krtap21-1)	Hu
83	Mapk14	Hu
84	Mcrs1	Ding; Hu
85	Med10	Hu
86	Metap2	Hu
87	Ms4a6b	Hu
88	Mtch2	Hu
89	MusD elements	Zhang
90	Ncapg2	Hu
91	Ncaph2	Hu
92	Ncl	Ding
93	Nedd8	Hu
94	Nfya	Ding
95	Nipbl	Hu
96	Nts	Hu
97	Nup188	Hu
98	Olfr114	Hu
99	Ostf1	Hu
100	P4ha3	Hu
101	Paf1	Hu
102	Pax7	Hu
103	Pcbp1	Hu
104	Pcid2	Hu
105	Pcna	Hu; Zhang
106	Peci	Hu
107	Piwil4	Hu
108	Plac1	Hu
109	Pole4	Hu
110	Ppp4c	Hu
111	Ptbp1	Ding
112	Rad21	Hu
113	Rbx1	Hu
114	Rexo1	Hu
115	Rfwd2	Hu
116	Rnf146	Hu
117	Rprd1b	Hu
118	Rtf1	Ding
119	Rutbc3	Hu
120	Samd11	Hu
121	Samd5	Hu
122	Sema4a	Hu
123	Setd1b	Hu
124	Sgsm3	Hu
125	Sh2bp1	Hu
126	Shfdg1	Ding
127	Slc16a11	Hu
128	Slc19a3	Hu
129	Smc1l1	Hu
130	Spesp1	Hu
131	Spire1	Hu
132	Sprr2i	Hu
133	Ssu72	Hu
134	Stambpl1	Hu
135	Syngr1	Hu
136	Syt13	Hu
137	Tbx3	Ivanova
138	Tcl1	Ivanova
139	Tekt1	Hu
140	Tgfb1	Hu
141	Thoc2	Ding
142	Thoc5 (Fmip)	Hu
143	Tle4	Zhang
144	Triap1	Hu
145	Trim16	Hu
146	Trim28	Hu
147	Trmt6	Hu
148	Tubd1	Hu
149	Uba1	Hu
150	Ube1x	Hu
151	Ube2m	Ding
152	Uble1b	Zhang
153	Uncx	Hu
154	Uqcr10	Hu
155	Vamp2	Hu
156	Wdr61	Ding; Hu
157	Wdr77	Zhang
158	Xpo7	Hu
159	Zadh2	Hu
160	Zfp13	Hu
161	Zfp42 (Rex1)	Zhang
162	Zfp628	Hu
163	Zfp759	Hu
164	Zfp771	Hu
165	Zfp786	Hu
166	Znhit4	Hu

Table 1. Pluripotency-associated genes found via RNA interference assay of mouse embryonic stem cells. Proteins that are also found in the protein interaction network are shaded grey.

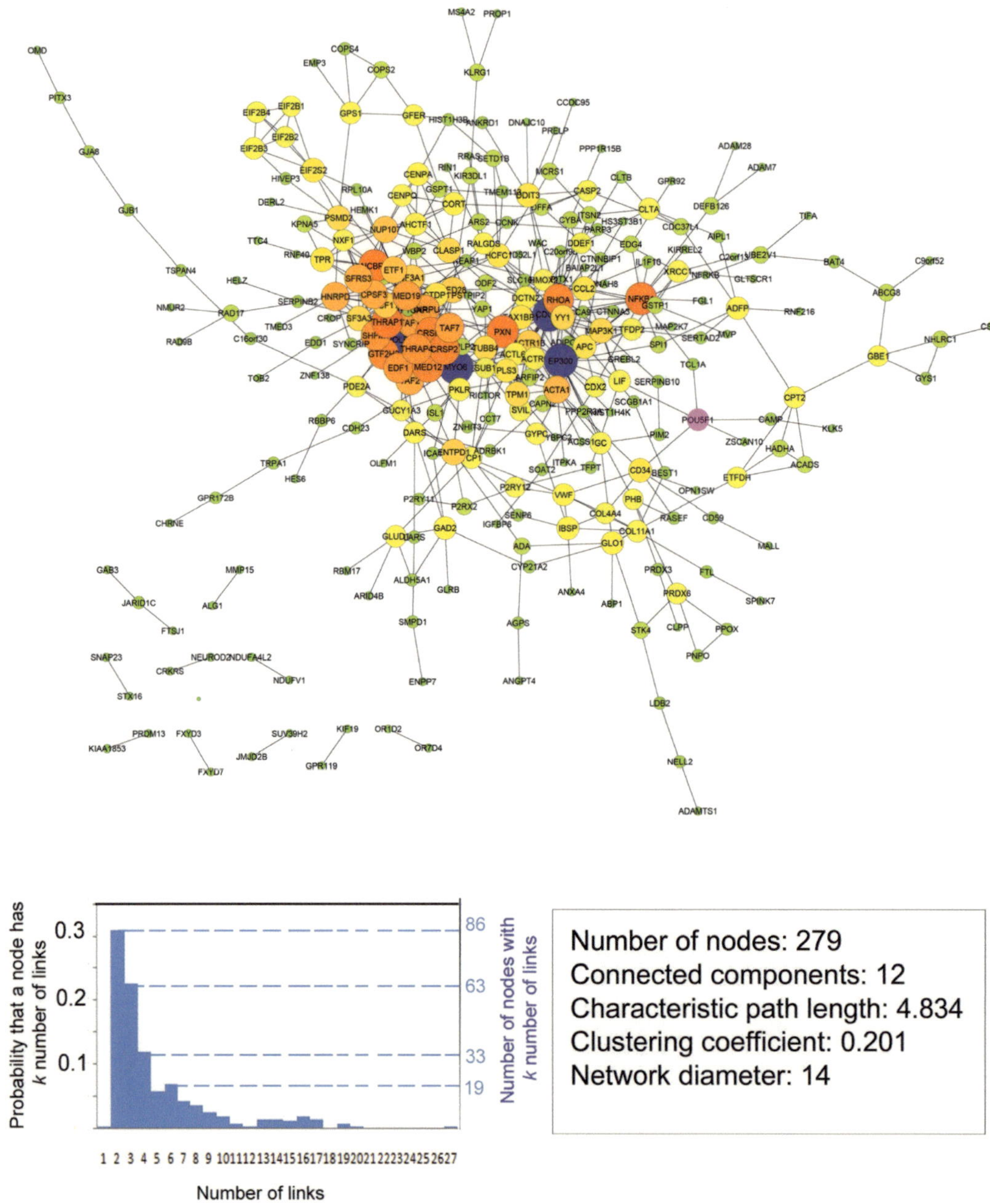

Fig. 3. The pluripotency-associated protein interaction network of human embryonic stem cell. Genes for pluripotency are discovered from RNA interference assay and interactions are based on the STRING database that also considers interactions from orthologs. Bigger and brighter colored nodes represents proteins with more links. The nodes with the top connectivity are colored blue. POU5F1 is colored purple for easy identification. The network has a power law distribution for their degree of links. Network parameters are indicated in the box.

interactions. Among the 566 genes, a total of 279 genes have some form of protein-protein interaction within the group and this network is shown in Figure 3.

The human network also shows a power law distribution (Figure 3). The hubs in the network are POLR2E with 26 links, MY06 with 19 links and EP300 and CDC42 both with 18 links. Notably human OCT4 is not one of the hubs. Again this is most probably an artifact of the incomplete network due to the lack of publications on OCT4 interactions. Although human OCT4 did not show up as a key factor, it is known to be important in pluripotency of human embryonic stem cell. Hence this emphasizes the need for more work in the construction of the network before reliable deductions can be made.

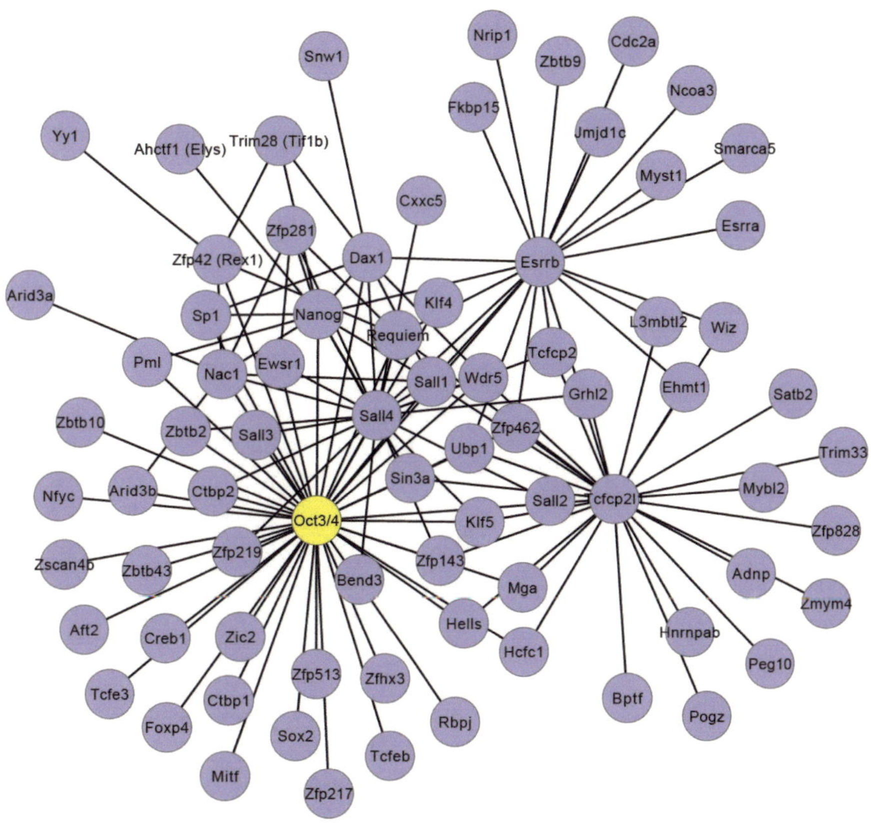

Fig. 4. The mouse embryonic stem cell protein interaction network for the transcription factors constructed based on the experimental datasets of four separate publications (Liang, et al., 2008; Pardo, et al., 2010; van den Berg, et al., 2010; Wang, et al., 2006).

4.2 How key factors network?

Proteins such as Oct3/4 and Esrrb are transcription factors and they appear to be key factors in pluripotency. On the genome, these transcription factors show clustering at embryonic stem cell-specific genes, supporting the notion that their collaborations forms codes for ensuring selective transcriptional activation (Chen, et al., 2008; Kim, et al., 2008). It remains to be confirmed if these clusterings require direct protein-protein interactions or simply are clustering at the same location. Protein-protein interaction between these transcription

factors could provide structural changes required for regulation of gene expression for pluripotency. It was suggested that collaborations involving more transcription factors would activate embryonic stem cell-specific genes. While transcription factors with little interactions would activate more general genes.
From the integrated dataset, proteins with the GO annotation "transcription factor" constitute a total of 78. Figure 4 shows a protein-interaction network of these transcription factors. Certainly, there are transcriptions factors that are important to pluripotency that do not cluster into the highly interactive zone because the network is incomplete. For example, Sox2 is important for regulating pluripotency genes but does not have many mapped collaborators probably because the Sox2-interactome has yet to be investigated by any lab. The current network therefore serves as a guide for future research directions.

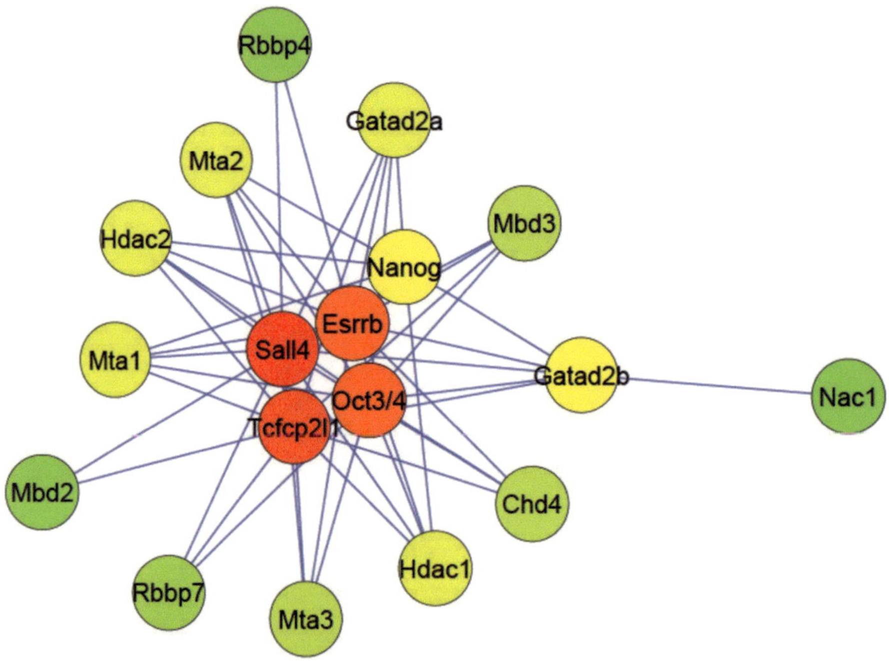

Fig. 5. The nucleosome remodeling deacetylase (NuRD) complex and its interactions with transcription factors in the pluripotency protein interaction network.

4.3 How key factors recruit protein machineries

The nucleosome remodeling histone deacetylase (NuRD) complex (Ahringer, 2000) was the most prominent complex identified in the embryonic stem cell protein-interaction network (Liang, et al., 2008; Pardo, et al., 2010; van den Berg, et al., 2010). All the components of this complex are in the network and each of the components interacts with one or more of the five transcription factors that was studied in greater detail (van den Berg, et al., 2010), namely Nanog, Esrrb, Oct4, Tcfcp2l1, and Sall4 which are themselves already tightly associated with one another (Figure 5).
This suggests that the transcription factors co-recruit the same machinery, NuRD for histone deacetylation as a gene repression mechanism to regulate pluripotency. Indeed case studies have shown that NuRD has specific developmental roles rather than being required for

general cellular functions (Ahringer, 2000; Ch'ng & Kenyon, 1999; Mannervik & Levine, 1999). Besides NuRD, other complexes have been reported in the study by Pardo *et al.* (Pardo, et al., 2010). Most of these are involved in chromosome remodeling. Confirmation of these findings would surely expand our knowledge of the extent to which each of these complexes contributes to pluripotency. This is because there is also converse evidence that chromosomal remodeling factors like the Polycomb Group and Polycomb Repressive Complex are not required for maintenance of pluripotency in embryonic stem cells (Azuara, et al., 2006; Boyer, et al., 2006; de Napoles, et al., 2004; Lee, et al., 2006; Montgomery, et al., 2005; Niwa, 2007; O'Carroll, et al., 2001). It is believed that the chromatin of the embryonic stem cell is "loose" so as to allow free accessibility to transcription factors, but at the same time repressors are there to serve to prevent spontaneous differentiation of the embryonic stem cells. Having the different chromatin modifiers inserted into the protein interaction network may help to clarify their role in pluripotency. Besides the chromatin modifiers, the basic transcriptional machinery was also found to be recruited to the protein interaction network by Esrrb (van den Berg, et al., 2010). However this mechanism appears not to be utilized by the other transcription factors in the network. It remains to be seen if this mechanism is directly related to the regulation of pluripotency.

5. Future challenges

Ironically, pluripotency is best demonstrated by its loss. A population of cells is pluripotent if it can differentiate into many cell types; but once that happens, pluripotency is lost. In the embryonic stem cell, molecules for pluripotency work to balance two opposing features: the readiness to differentiate and the prevention of differentiation. To understand the molecular mechanism of pluripotency, we need to keep in mind this concept of pluripotency. In simulations, pluripotency should demonstrate these two opposing forces. In the current protein interaction network both of these features of pluripotency are not distinguished. Furthermore, it is necessary to consider the multifunctionality of proteins. In this case, looking at proteins for the assignment of "jobs" may be more confusing than helpful. Alternatively, assignment of processes may be more meaningful if this was done to the edges of the network rather than the nodes. This approach of analysis can be illustrated by the following example. The interaction (edge) between Oct3/4 and Cdx2 serves the purpose of "gene repression"; and the interaction (edge) between Oct3/4 and Sox2 serves the purpose of "gene activation". Hence instead of annotating both functions to the Oct3/4 node, the annotations can be on the edges.

A protein interaction network by virtue of the protocols employed is a single snapshot of the protein-protein interactions of a population of cells at any given time. To understand how embryonic stem cells have the ability to differentiate into different cell types, further information will have to be integrated. The final protein interaction network should include information on protein subcellular location and protein concentration. All this information in the network will change as a function of time as the cell undergoes cell cycling and when the cell undergoes fate changes. A study on the system-level changes across the three mechanistic layers: epigenetic, transcriptional and translational during fate change in mouse embryonic stem cells show that changes in nuclear protein levels are not accompanied by concordant changes in the corresponding mRNA levels, suggesting that translational and post-translational mechanisms, rather than transcriptional regulation play important roles, during lost of pluripotency. (Lu, et al., 2009). For full understanding and successful simulation, information from the protein interaction network, the gene regulatory network and microRNA networks of ES cell should be fed back into one another. Integration of protein-protein interaction networks with transcriptional profiling networks has been done in yeast and has led to the discovery of new network features which are described as party hubs and date hubs (Vidal, et al., 2011). Party hubs are nodes that are coexpressed with its protein partners and date hubs are nodes that are not always transcribed at the same time and place as its partners. Biologically, party and date hubs may represent two kinds of protein-protein interactions. Transient protein-protein interactions that occur between transcription factors or between transcription factors and other protein complexes are date hubs. Static protein-protein interactions that occur between protein subunits of a stable protein complex are party hubs. The first type of interaction usually encodes instructions or messages while the second type of interaction functions mainly to execute the processes as a module. Identifying these interactions allow us to further understand how cell fate decisions are made and how these decisions are executed.

In view of the large number of proteins that have been associated with pluripotency. It is possible that there are also alternate means of achieving pluripotency. After all, pluripotency is a cellular state rather than a cellular composition. Proteins like Ronin (Dejosez, et al., 2008; Zwaka, 2008), which show strong associations with pluripotency, may operate via a separate network.

As data accumulation continues towards the point where the boundaries of the pluripotency-associated protein interaction network are felt, extra efforts will be needed towards looking for interactions amongst low concentration proteins and towards validation of this network. With a more complete embryonic stem cell protein interaction network, new hypothesis can be formulated. As more system biology data is generated from other fields, it will also become possible to compare between non-pluripotent and pluripotent networks. The embryonic stem cell protein interaction network, when ready, will serve as a point of comparison with other stem cells, with differentiated cells and with cancer cells. Such comparisons can potentially bring out unique features in each of these cellular conditions. Finally, in view of the differences between human and mice, the same work will have to be repeated with human embryonic stem cells. Knowledge gained from the challenges in mouse embryonic stem cell research ensues much faster progress with the human embryonic stem cell project.

6. Conclusion

Overall, we see great promise in getting answers and insights from a mature protein interaction network. Currently a total of 239 proteins form the mouse embryonic stem cell protein interaction network. More work is required in the construction of this network and this must be closely accompanied with attempts to annotate the purpose and nature of the interaction as discussed above. Another 151 proteins discovered to have a role in pluripotency by genome-wide RNA interference screening are yet to be connected to the protein interaction network. Multiple validations to confirm the involvement of these proteins in pluripotency are also necessary. In the network, the transcription factors show collaboration amongst themselves. A core group of transcription factors show recruitment of the same machinery, i.e. the NuRD. Some studies suggest that other chromatin modification machineries are also recruited. The role of these machineries remains to be investigated. When the network is reasonably saturated, system biology analysis should give greater insight into network properties. Inclusion of information on dynamic properties of the protein interaction network would also facilitate predictive capabilities

7. Acknowledgment

This work is supported by the Agency for Science, Technology and Research, Singapore.

8. References

Ahringer, J. (2000). NuRD and SIN3 histone deacetylase complexes in development. *Trends Genet*, Vol. 16, No. 8, (Aug), pp. (351-6),

Albert, R. (2005). Scale-free networks in cell biology. *J Cell Sci*, Vol. 118, No. Pt 21, (Nov 1), pp. (4947-57), 0021-9533 (Print)0021-9533 (Linking)

Azuara, V., Perry, P., Sauer, S., Spivakov, M., Jorgensen, H.F., John, R.M., Gouti, M., Casanova, M., Warnes, G., Merkenschlager, M. & Fisher, A.G. (2006). Chromatin signatures of pluripotent cell lines. *Nat Cell Biol*, Vol. 8, No. 5, (May), pp. (532-8),

Barabasi, A.L. (2009). Scale-free networks: a decade and beyond. *Science*, Vol. 325, No. 5939, (Jul 24), pp. (412-3), 1095-9203 (Electronic)0036-8075 (Linking)

Barabasi, A.L. & Oltvai, Z.N. (2004). Network biology: understanding the cell's functional organization. *Nat Rev Genet*, Vol. 5, No. 2, (Feb), pp. (101-13), 1471-0056 (Print)1471-0056 (Linking)

Boyer, L.A., Plath, K., Zeitlinger, J., Brambrink, T., Medeiros, L.A., Lee, T.I., Levine, S.S., Wernig, M., Tajonar, A., Ray, M.K., Bell, G.W., Otte, A.P., Vidal, M., Gifford, D.K., Young, R.A. & Jaenisch, R. (2006). Polycomb complexes repress developmental regulators in murine embryonic stem cells. *Nature*, Vol. 441, No. 7091, (May 18), pp. (349-53),

Ch'ng, Q. & Kenyon, C. (1999). egl-27 generates anteroposterior patterns of cell fusion in C. elegans by regulating Hox gene expression and Hox protein function. *Development*, Vol. 126, No. 15, (Aug), pp. (3303-12),

Chen, X., Xu, H., Yuan, P., Fang, F., Huss, M., Vega, V.B., Wong, E., Orlov, Y.L., Zhang, W., Jiang, J., Loh, Y.H., Yeo, H.C., Yeo, Z.X., Narang, V., Govindarajan, K.R., Leong, B., Shahab, A., Ruan, Y., Bourque, G., Sung, W.K., Clarke, N.D., Wei, C.L. & Ng, H.H. (2008). Integration of external signaling pathways with the core transcriptional network in embryonic stem cells. *Cell*, Vol. 133, No. 6, (Jun 13), pp. (1106-17),

Chia, N.Y., Chan, Y.S., Feng, B., Lu, X., Orlov, Y.L., Moreau, D., Kumar, P., Yang, L., Jiang, J., Lau, M.S., Huss, M., Soh, B.S., Kraus, P., Li, P., Lufkin, T., Lim, B., Clarke, N.D., Bard, F. & Ng, H.H. (2010). A genome-wide RNAi screen reveals determinants of human embryonic stem cell identity. *Nature*, Vol. 468, No. 7321, (Nov 11), pp. (316-20), 1476-4687 (Electronic)0028-0836 (Linking)

de Napoles, M., Mermoud, J.E., Wakao, R., Tang, Y.A., Endoh, M., Appanah, R., Nesterova, T.B., Silva, J., Otte, A.P., Vidal, M., Koseki, H. & Brockdorff, N. (2004). Polycomb group proteins Ring1A/B link ubiquitylation of histone H2A to heritable gene silencing and X inactivation. *Dev Cell*, Vol. 7, No. 5, (Nov), pp. (663-76),

Dejosez, M., Krumenacker, J.S., Zitur, L.J., Passeri, M., Chu, L.F., Songyang, Z., Thomson, J.A. & Zwaka, T.P. (2008). Ronin is essential for embryogenesis and the pluripotency of mouse embryonic stem cells. *Cell*, Vol. 133, No. 7, (Jun 27), pp. (1162-74),

Ding, L., Paszkowski-Rogacz, M., Nitzsche, A., Slabicki, M.M., Heninger, A.K., de Vries, I., Kittler, R., Junqueira, M., Shevchenko, A., Schulz, H., Hubner, N., Doss, M.X., Sachinidis, A., Hescheler, J., Iacone, R., Anastassiadis, K., Stewart, A.F., Pisabarro,

M.T., Caldarelli, A., Poser, I., Theis, M. & Buchholz, F. (2009). A genome-scale RNAi screen for Oct4 modulators defines a role of the Paf1 complex for embryonic stem cell identity. *Cell Stem Cell*, Vol. 4, No. 5, (May 8), pp. (403-15),

Evans, M.J. & Kaufman, M.H. (1981). Establishment in culture of pluripotential cells from mouse embryos. *Nature*, Vol. 292, No. 5819, (Jul 9), pp. (154-6),

Feng, B., Jiang, J., Kraus, P., Ng, J.H., Heng, J.C., Chan, Y.S., Yaw, L.P., Zhang, W., Loh, Y.H., Han, J., Vega, V.B., Cacheux-Rataboul, V., Lim, B., Lufkin, T. & Ng, H.H. (2009). Reprogramming of fibroblasts into induced pluripotent stem cells with orphan nuclear receptor Esrrb. *Nat Cell Biol*, Vol. 11, No. 2, (Feb), pp. (197-203), 1476-4679 (Electronic)1465-7392 (Linking)

Futschik, M.E., Tschaut, A., Chaurasia, G. & Herzel, H. (2007). Graph-theoretical comparison reveals structural divergence of human protein interaction networks. *Genome Inform*, Vol. 18, No. pp. (141-51),

Gavin, A.C., Aloy, P., Grandi, P., Krause, R., Boesche, M., Marzioch, M., Rau, C., Jensen, L.J., Bastuck, S., Dumpelfeld, B., Edelmann, A., Heurtier, M.A., Hoffman, V., Hoefert, C., Klein, K., Hudak, M., Michon, A.M., Schelder, M., Schirle, M., Remor, M., Rudi, T., Hooper, S., Bauer, A., Bouwmeester, T., Casari, G., Drewes, G., Neubauer, G., Rick, J.M., Kuster, B., Bork, P., Russell, R.B. & Superti-Furga, G. (2006). Proteome survey reveals modularity of the yeast cell machinery. *Nature*, Vol. 440, No. 7084, (Mar 30), pp. (631-6), 1476-4687 (Electronic)0028-0836 (Linking)

Gavin, A.C., Bosche, M., Krause, R., Grandi, P., Marzioch, M., Bauer, A., Schultz, J., Rick, J.M., Michon, A.M., Cruciat, C.M., Remor, M., Hofert, C., Schelder, M., Brajenovic, M., Ruffner, H., Merino, A., Klein, K., Hudak, M., Dickson, D., Rudi, T., Gnau, V., Bauch, A., Bastuck, S., Huhse, B., Leutwein, C., Heurtier, M.A., Copley, R.R., Edelmann, A., Querfurth, E., Rybin, V., Drewes, G., Raida, M., Bouwmeester, T., Bork, P., Seraphin, B., Kuster, B., Neubauer, G. & Superti-Furga, G. (2002). Functional organization of the yeast proteome by systematic analysis of protein complexes. *Nature*, Vol. 415, No. 6868, (Jan 10), pp. (141-7), 0028-0836 (Print)0028-0836 (Linking)

Hakes, L., Pinney, J.W., Robertson, D.L. & Lovell, S.C. (2008). Protein-protein interaction networks and biology--what's the connection? *Nat Biotechnol*, Vol. 26, No. 1, (Jan), pp. (69-72),

Ho, Y., Gruhler, A., Heilbut, A., Bader, G.D., Moore, L., Adams, S.L., Millar, A., Taylor, P., Bennett, K., Boutilier, K., Yang, L., Wolting, C., Donaldson, I., Schandorff, S., Shewnarane, J., Vo, M., Taggart, J., Goudreault, M., Muskat, B., Alfarano, C., Dewar, D., Lin, Z., Michalickova, K., Willems, A.R., Sassi, H., Nielsen, P.A., Rasmussen, K.J., Andersen, J.R., Johansen, L.E., Hansen, L.H., Jespersen, H., Podtelejnikov, A., Nielsen, E., Crawford, J., Poulsen, V., Sorensen, B.D., Matthiesen, J., Hendrickson, R.C., Gleeson, F., Pawson, T., Moran, M.F., Durocher, D., Mann, M., Hogue, C.W., Figeys, D. & Tyers, M. (2002). Systematic identification of protein complexes in Saccharomyces cerevisiae by mass spectrometry. *Nature*, Vol. 415, No. 6868, (Jan 10), pp. (180-3), 0028-0836 (Print) 0028-0836 (Linking)

Hu, G., Kim, J., Xu, Q., Leng, Y., Orkin, S.H. & Elledge, S.J. (2009). A genome-wide RNAi screen identifies a new transcriptional module required for self-renewal. *Genes Dev*, Vol. 23, No. 7, (Apr 1), pp. (837-48),

Ito, T., Chiba, T., Ozawa, R., Yoshida, M., Hattori, M. & Sakaki, Y. (2001). A comprehensive two-hybrid analysis to explore the yeast protein interactome. *Proc Natl Acad Sci U S A*, Vol. 98, No. 8, (Apr 10), pp. (4569-74), 0027-8424 (Print) 0027-8424 (Linking)

Ivanova, N., Dobrin, R., Lu, R., Kotenko, I., Levorse, J., DeCoste, C., Schafer, X., Lun, Y. & Lemischka, I.R. (2006). Dissecting self-renewal in stem cells with RNA interference. *Nature*, Vol. 442, No. 7102, (Aug 3), pp. (533-8),

Kim, J., Chu, J., Shen, X., Wang, J. & Orkin, S.H. (2008). An extended transcriptional network for pluripotency of embryonic stem cells. *Cell*, Vol. 132, No. 6, (Mar 21), pp. (1049-61),

Lee, T.I., Jenner, R.G., Boyer, L.A., Guenther, M.G., Levine, S.S., Kumar, R.M., Chevalier, B., Johnstone, S.E., Cole, M.F., Isono, K., Koseki, H., Fuchikami, T., Abe, K., Murray, H.L., Zucker, J.P., Yuan, B., Bell, G.W., Herbolsheimer, E., Hannett, N.M., Sun, K., Odom, D.T., Otte, A.P., Volkert, T.L., Bartel, D.P., Melton, D.A., Gifford, D.K., Jaenisch, R. & Young, R.A. (2006). Control of developmental regulators by Polycomb in human embryonic stem cells. *Cell*, Vol. 125, No. 2, (Apr 21), pp. (301-13),

Li, X., Sun, L. & Jin, Y. (2008). Identification of karyopherin-alpha 2 as an Oct4 associated protein. *J Genet Genomics*, Vol. 35, No. 12, (Dec), pp. (723-8), 1673-8527 (Print) 1673-8527 (Linking)

Liang, J., Wan, M., Zhang, Y., Gu, P., Xin, H., Jung, S.Y., Qin, J., Wong, J., Cooney, A.J., Liu, D. & Songyang, Z. (2008). Nanog and Oct4 associate with unique transcriptional repression complexes in embryonic stem cells. *Nat Cell Biol*, Vol. 10, No. 6, (Jun), pp. (731-9),

Lu, R., Markowetz, F., Unwin, R.D., Leek, J.T., Airoldi, E.M., MacArthur, B.D., Lachmann, A., Rozov, R., Ma'ayan, A., Boyer, L.A., Troyanskaya, O.G., Whetton, A.D. & Lemischka, I.R. (2009). Systems-level dynamic analyses of fate change in murine embryonic stem cells. *Nature*, Vol. 462, No. 7271, (Nov 19), pp. (358-62),

Mannervik, M. & Levine, M. (1999). The Rpd3 histone deacetylase is required for segmentation of the Drosophila embryo. *Proc Natl Acad Sci U S A*, Vol. 96, No. 12, (Jun 8), pp. (6797-801),

Martin, G.R. (1981). Isolation of a pluripotent cell line from early mouse embryos cultured in medium conditioned by teratocarcinoma stem cells. *Proc Natl Acad Sci U S A*, Vol. 78, No. 12, (Dec), pp. (7634-8),

Montgomery, N.D., Yee, D., Chen, A., Kalantry, S., Chamberlain, S.J., Otte, A.P. & Magnuson, T. (2005). The murine polycomb group protein Eed is required for global histone H3 lysine-27 methylation. *Curr Biol*, Vol. 15, No. 10, (May 24), pp. (942-7),

Murry, C.E. & Keller, G. (2008). Differentiation of embryonic stem cells to clinically relevant populations: lessons from embryonic development. *Cell*, Vol. 132, No. 4, (Feb 22), pp. (661-80),

Nakagawa, M., Koyanagi, M., Tanabe, K., Takahashi, K., Ichisaka, T., Aoi, T., Okita, K., Mochiduki, Y., Takizawa, N. & Yamanaka, S. (2008). Generation of induced pluripotent stem cells without Myc from mouse and human fibroblasts. *Nat Biotechnol*, Vol. 26, No. 1, (Jan), pp. (101-6),

Nichols, J., Zevnik, B., Anastassiadis, K., Niwa, H., Klewe-Nebenius, D., Chambers, I., Scholer, H. & Smith, A. (1998). Formation of pluripotent stem cells in the mammalian embryo depends on the POU transcription factor Oct4. *Cell*, Vol. 95, No. 3, (Oct 30), pp. (379-91),

Niwa, H. (2007). How is pluripotency determined and maintained? *Development*, Vol. 134, No. 4, (Feb), pp. (635-46),

Niwa, H., Miyazaki, J. & Smith, A.G. (2000). Quantitative expression of Oct-3/4 defines differentiation, dedifferentiation or self-renewal of ES cells. *Nat Genet*, Vol. 24, No. 4, (Apr), pp. (372-6),

O'Carroll, D., Erhardt, S., Pagani, M., Barton, S.C., Surani, M.A. & Jenuwein, T. (2001). The polycomb-group gene Ezh2 is required for early mouse development. *Mol Cell Biol*, Vol. 21, No. 13, (Jul), pp. (4330-6),

Pardo, M., Lang, B., Yu, L., Prosser, H., Bradley, A., Babu, M.M. & Choudhary, J. (2010). An expanded Oct4 interaction network: implications for stem cell biology, development, and disease. *Cell Stem Cell*, Vol. 6, No. 4, (Apr 2), pp. (382-95), 1875-9777 (Electronic)

Pieroni, E., de la Fuente van Bentem, S., Mancosu, G., Capobianco, E., Hirt, H. & de la Fuente, A. (2008). Protein networking: insights into global functional organization of proteomes. *Proteomics*, Vol. 8, No. 4, (Feb), pp. (799-816),

Scholer, H.R., Dressler, G.R., Balling, R., Rohdewohld, H. & Gruss, P. (1990). Oct-4: a germline-specific transcription factor mapping to the mouse t-complex. *Embo J*, Vol. 9, No. 7, (Jul), pp. (2185-95),

Scholer, H.R., Hatzopoulos, A.K., Balling, R., Suzuki, N. & Gruss, P. (1989). A family of octamer-specific proteins present during mouse embryogenesis: evidence for germline-specific expression of an Oct factor. *Embo J*, Vol. 8, No. 9, (Sep), pp. (2543-50),

Takahashi, K. & Yamanaka, S. (2006). Induction of pluripotent stem cells from mouse embryonic and adult fibroblast cultures by defined factors. *Cell*, Vol. 126, No. 4, (Aug 25), pp. (663-76),

Uetz, P., Giot, L., Cagney, G., Mansfield, T.A., Judson, R.S., Knight, J.R., Lockshon, D., Narayan, V., Srinivasan, M., Pochart, P., Qureshi-Emili, A., Li, Y., Godwin, B., Conover, D., Kalbfleisch, T., Vijayadamodar, G., Yang, M., Johnston, M., Fields, S. & Rothberg, J.M. (2000). A comprehensive analysis of protein-protein interactions in Saccharomyces cerevisiae. *Nature*, Vol. 403, No. 6770, (Feb 10), pp. (623-7), 0028-0836 (Print)0028-0836 (Linking)

van den Berg, D.L., Snoek, T., Mullin, N.P., Yates, A., Bezstarosti, K., Demmers, J., Chambers, I. & Poot, R.A. (2010). An Oct4-centered protein interaction network in embryonic stem cells. *Cell Stem Cell*, Vol. 6, No. 4, (Apr 2), pp. (369-81), 1875-9777 (Electronic)

Vidal, M., Cusick, M.E. & Barabasi, A.L. (2011). Interactome networks and human disease. *Cell*, Vol. 144, No. 6, (Mar 18), pp. (986-98), 1097-4172 (Electronic) 0092-8674 (Linking)

Wang, J., Rao, S., Chu, J., Shen, X., Levasseur, D.N., Theunissen, T.W. & Orkin, S.H. (2006). A protein interaction network for pluripotency of embryonic stem cells. *Nature*, Vol. 444, No. 7117, (Nov 16), pp. (364-8), 1476-4687 (Electronic) 0028-0836 (Linking)

Yuan, H., Corbi, N., Basilico, C. & Dailey, L. (1995). Developmental-specific activity of the FGF-4 enhancer requires the synergistic action of Sox2 and Oct-3. *Genes Dev*, Vol. 9, No. 21, (Nov 1), pp. (2635-45),

Zhang, J.Z., Gao, W., Yang, H.B., Zhang, B., Zhu, Z.Y. & Xue, Y.F. (2006). Screening for genes essential for mouse embryonic stem cell self-renewal using a subtractive RNA interference library. *Stem Cells*, Vol. 24, No. 12, (Dec), pp. (2661-8),

Zwaka, T.P. (2008). Ronin and caspases in embryonic stem cells: a new perspective on regulation of the pluripotent state. *Cold Spring Harb Symp Quant Biol*, Vol. 73, No. pp. (163-9),

17

Profile of Galanin in Embryonic Stem Cells and Tissues

Maria-Elena Lautatzis and Maria Vrontakis
University of Manitoba
Canada

1. Introduction

In the preimplantation mammalian embryo, cells of the inner cell mass can differentiate into any cell type present in the more mature embryo. As of 1981, in mice and 1998 in humans, it has been recognized that embryonic stem cells (ESCs) with a prolonged proliferative capacity can be derived from the inner cell mass in vitro (Evans and Kaufman 1981; Thomson, Itskovitz-Eldor et al. 1998). ESCs are pluripotent cells that can contribute to all tissues in vivo and to the three primary germ layers as well as extraembryonic tissues in vitro. Because pluripotency is maintained in these cells even after prolonged periods of culture, human ESCs have great therapeutic potential for tissue regeneration. Indeed, embryonic and adult stem cells (SCs) hold great promise for regenerative medicine, tissue repair, and gene therapy. Careful molecular characterization of embryonic pluripotency should help to optimize and scale up the in vitro production of ESCs for clinical applications.

The mechanisms regulating self-renewal and cell fate decisions in mammalian stem cells are poorly understood. As compared with differentiated cell types, stem cells express a significantly higher number of genes (represented by expressed sequence tags) of unknown function. The properties that distinguish stem cells from other cells are largely unknown, and the identification of signals that regulate stem cell differentiation remains fundamental to our understanding of cellular diversity.

Embryonic and adult stem cells have many similarities at the transcriptional level. The overlapping set of expressed gene products represents a molecular signature of stem cells (Bhattacharya, Miura et al. 2004; Assou, Le Carrour et al. 2007). A list of human and mouse genes involved in stemness has been generated (Assou, Le Carrour et al. 2007) and includes 92 stemness genes known to be expressed in mouse or human ESCs, e.g., OCT3/4, NANOG, Cripto/TDGFI, Cx43 and Galanin (Richards, Tan et al. 2004). Work in the field of embryogenesis has also contributed to our understanding of the function of these pluripotency-associated genes. The four most significantly overexpressed genes in undifferentiated embryonic tissues are Galanin, POU5FI, NANOG and DPPA4 (Zeng, Miura et al. 2004). In most studies, galanin has been highlighted as the most abundant transcript in ES culture as well as human and rodent embryonic tissues (Anisimov, Tarasov et al. 2002; Zeng, Miura et al. 2004). Both galanin and galanin receptors are expressed in ES cells, indicating a potential functional role for this protein (Tarasov, Tarasova et al. 2002). This chapter will be devoted to a description of the galanin expression profiles in embryonic tissues and stem cells as well as its possible functional role.

2. Galanin

Galanin was first identified from porcine intestinal extracts in 1978 by Professor Viktor Mutt and colleagues at the Karolinska Institute, Sweden, using a chemical assay technique that detects peptides according to their C-terminal alanine amide structures. Galanin is so-called because it contains an N-terminal glycine residue and a C-terminal alanine (Hokfelt and Tatemoto 2008). The structure of galanin was determined in 1983 by the same team (Tatemoto, Rokaeus et al. 1983), and galanin cDNA was first cloned from a rat anterior pituitary library in 1987 (Vrontakis, Peden et al. 1987). Galanin is a biologically active neuropeptide that is widely distributed in the central and peripheral nervous systems and the endocrine system. The N-terminus of galanin is highly conserved between species (almost 90% among vertebrates, with the first 15 amino acids being identical, indicating the likely importance of this molecule (Vrontakis 2002). Consistent with this sequence conservation, the first 15 amino acids of galanin are sufficient for agonistic receptor binding. Galanin is proteolytically processed from a 124-amino acid precursor peptide, preprogalanin, along with a 59- or 60-amino acid peptide known as galanin message associate peptide (GMAP) (Rokaeus and Brownstein 1986; Vrontakis, Peden et al. 1987; Evans and Shine 1991). Preprogalanin is encoded by a single-copy gene organized into 6 small exons (fig.1) spanning about 6 kb of genomic DNA (Kofler, Liu et al. 1996). The intron:exon organization of the galanin gene is conserved in all species studied thus far (Vrontakis 2002). Transcriptional studies of the galanin gene in multiple species concluded that the tissue-specific expression of this gene is achieved by enhancers as well as silencer

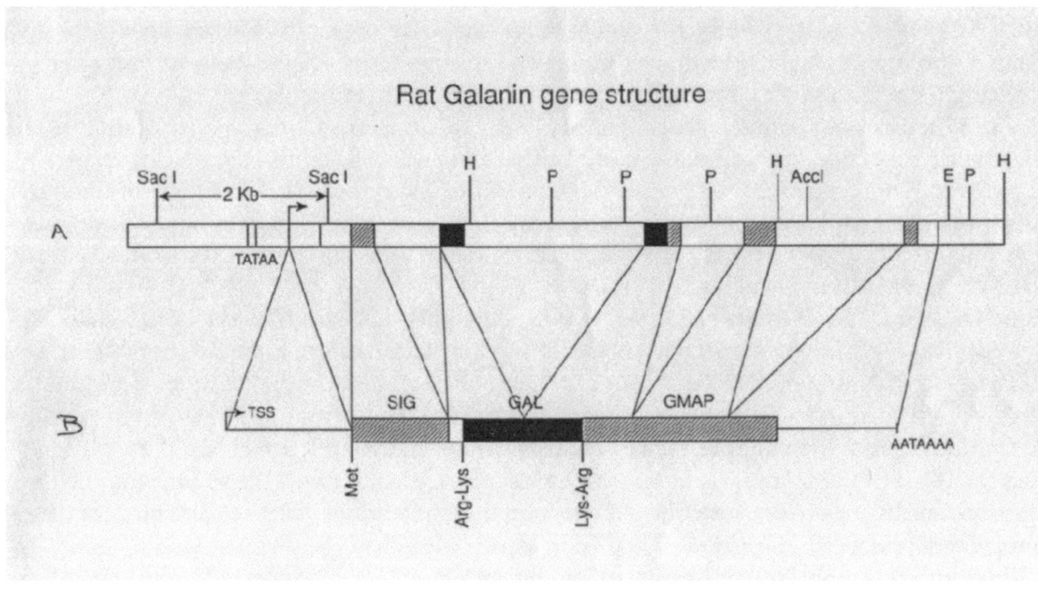

Fig. 1. Organization of the rat preprogalanin gene. A: Schematic representation of the rat preprogalanin gene. B: the position of the six exons with respect to the rat preprogalanin cDNA are shown. Abbreviations are as follows: ATG, translation initiation site; AATAAA, (Lang, Gundlach et al. 2007) the poly (A); TATA,TATA box;TSS, transcription start site; SIG, signal peptide; GAL, galanin; GAMP, galanin message associated peptide (Maria Vrontakis and Hong Zhang unpublished data)

sequences, which restrict expression to the appropriate cell type (Kofler, Evans et al. 1995; Corness, Burbach et al. 1997; Jiang, Spyrou et al. 1998; Rokaeus and Waschek 1998). We have sequence the 5′ flanking region of the rat galanin gene (Zhang, 1998) and have shown that the rat galanin promoter region contains some consensus sequences for known transcription factors. Up streamed of the modified TATA box, there is a conserved half-element (TGACG) for the protein CREB, which typically mediates gene expression by binding to the cyclic AMP response element (CRE). In the rat galanin promoter region, there are also several AP-1 binding sequences for the Jun/Fos protein families. Upstream of the CREB binding site there is a c-Ets element for the Ets factors. Furthermore, both negative and positive regulatory elements exist in the rat galanin gene. The negative regulatory elements appeared to be tissue specific since they are located differently in the different tissues. These negative transcription sites in the galanin promoter might be of importance for down regulating the gene during development.

The functional role of galanin remains largely unknown, as is the case for most other neuropeptides; however, Galanin has been implicated in many biologically diverse functions, including nociception, waking and sleep regulation, cognition, feeding, regulation of mood and regulation of blood pressure. It also has roles in development and can act as a trophic factor. Galanin has been linked to a number of diseases, including Alzheimer's disease, epilepsy, depression and eating disorders. Galanin appears to have neuroprotective activity, as its biosynthesis is increased 10- to 100-fold upon axotomy in the peripheral nervous system (whereas most neuropeptides are induced only 1.5- to 2-fold) or when seizure activity occurs in the brain. It may also promote neurogenesis (Mitsukawa, Lu et al. 2008). Galanin frequently co-localizes with classical neurotransmitters such as acetylcholine, serotonin and norepinephrine as well as with other neuromodulators such as Neuropeptide Y, Substance P and Vasoactive peptide (Lang, Gundlach et al. 2007). Expression of galanin at the mRNA and peptide levels is elevated following estrogen administration, neuronal activation, denervation and/or nerve injury as well as during development. The wide spectrum of galanin's activities indicates that galanin is an important messenger for intercellular communication within the nervous system and the neuroendocrine axis.

Galanin acts at specific membrane receptors to exert its effects. To date, three human and rodent galanin receptor subtypes have been cloned, namely, GalR1, GalR2 and GalR3 (Branchek, Smith et al. 2000). High conservation between species exists among receptors of a given subtype but not between subtypes in an individual species (Howard, Tan et al. 1997; Iismaa, Fathi et al. 1998; Kolakowski, O'Neill et al. 1998). All three galanin receptor subtypes are members of the G protein-coupled receptor superfamily, but the subtypes show substantial differences in their functional coupling and subsequent signaling activities, contributing to the diversity of the possible physiological effects of galanin (Fig. 2). GalR1, the most abundant receptor subtype in adult tissues, is associated with the Gi family, which mediates the inhibition of cAMP synthesis by adenylate cyclase. Furthermore, it opens G-protein-regulated inwardly rectifying potassium channels and stimulates mitogen-activated protein kinase (MAPK) activity. GalR2 acts through Gq/11 to regulate phospholipase C-mediated events. GalR3 couples to Gi/Go and mediates the opening of G protein-coupled inwardly rectifying potassium channels (Lang, Gundlach et al. 2007). Since the three galanin receptors exhibit distinct but overlapping patterns of expression in the central and peripheral nervous systems, a variety of ligands have been developed in an effort to elucidate the specific roles of each receptor (Langel and Bartfai 1998; Pooga, Jureus et al. 1998; Lu, Lundstrom et al. 2005). Galanin agonists have been shown to have therapeutic applications in the treatment of chronic pain. Conversely, galanin antagonists have therapeutic potential for the treatment of Alzheimer's disease, depression, and eating disorders.

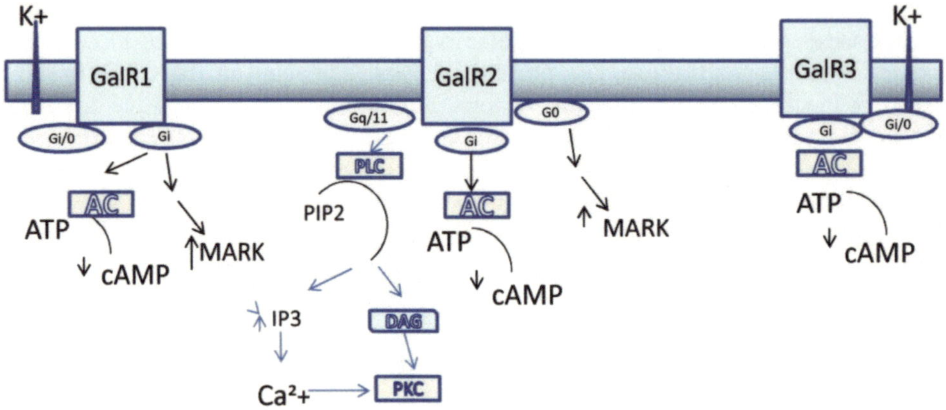

Fig. 2. Schematic illustration of the three galanin receptor subtypes and their intracellular transduction mechanisms. AC-adenyl cyclase , ATP-adenosine triphosphate, cAMP- cyclic adenosine monophosphate, DAG-diacyglycerol, IP3-inositol triphosphate, MAPK-mitogen activated protein kinase, PIP2-phosphatidyl 4,5-biposphate, PKC- protein kinase, PLC-phospholipase C.

2.1 Galanin in the early embryo

Galanin is one of the earliest neuropeptides to be expressed in the embryo. In the chicken embryo, galanin immunoreactive cells were first detected at E3.5 within the pharyngeal pouch region, the nodose ganglion, the primary sympathetic chain, the primitive splanchnic branches and the caudal portion of the Remark ganglion. These cells are derived from the neural crest. Indeed, galanin immunostaining appears at the same time as markers of neural crest cells. Transient galanin immunostaining was detected during the first week of development in cells displaying morphological features of migrating neuroblasts, but this expression domain had disappeared by E18 (Salvi, Vaccaro et al. 2001). At E4, galanin immunoreactivity was found in the spinal cord, medially in the motor column and in the intermediate zone. Neuroblasts appear coincident with galanin staining in the mesenchyme of the proventriculus/gizzard primordium (Salvi, Vaccaro et al. 1999; Salvi, Vaccaro et al. 2001). The precise role of galanin during chicken development remains unclear. The fact that in these experiments, galanin was present in undifferentiated or partially differentiated cells and the primitive sympathetic system well before these neurons reach their peripheral targets suggests that galanin has a developmental role in proliferation and migration.
Similar to the chicken, galanin-like immunoreactivity was detected in the mesenchyme and neural crest tissues of the early mouse embryo. At E10, we found that galanin-like immunoreactivity was readily detectable in the undifferentiated head and trunk mesenchyme (fig. 3) of mesenchymal or neural crest origin (Jones, Perumal et al. 2009), including the mesenchymal spiral ridges of the outflow tract of the heart and the endocardial cushions. The presence of galanin during these periods of morphogenesis

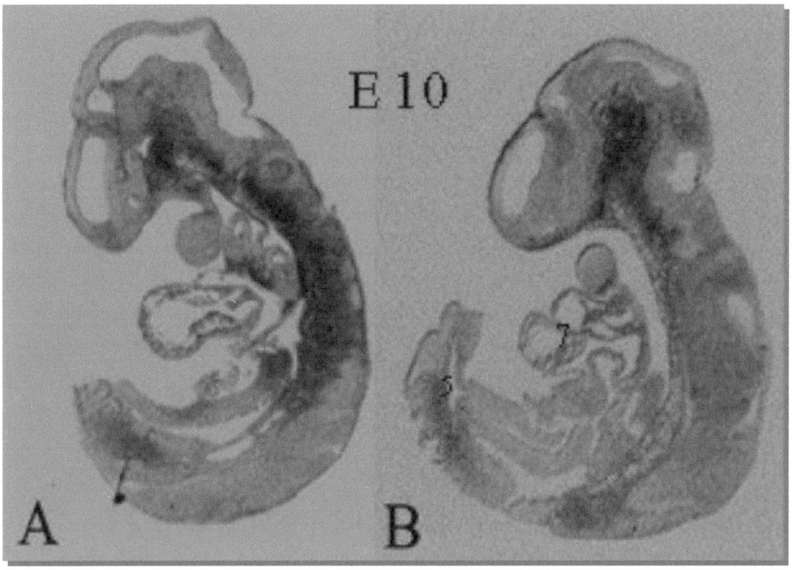

Fig. 3. Histochemistry profile of galanin like immunoreactivity in embryonic day 10 mouse embryo. A; sagittal and B; parasagittal section. Srong immonostaining for galanin is detected in the cephalic mesenchyme, trunk mesenchyme/somites, brachial arches, dorsal aorta and heart.

indicates a developmental role for this peptide in tissues of mesenchymal and neural crest origin in the early embryo. Galanin expression in mesenchymal cells during organogenesis was greater in tissues that depend on mesenchymal-epithelial interactions for their coordinated morphogenesis. Indeed, galanin staining is apparent during many instances of mesenchymal remodeling, e.g., during the formation of digits from limb buds, the formation of cartilage primordia in vertebrae and ribs, the formation of bones, the formation of the heart and in the mesenchyme of the kidney and genital organs (Jones, Perumal et al. 2009). It is surprising that at this early stage of development, galanin expression is largely outside the developing central nervous system. Thus, galanin might have different functions in the embryo and the adult. Although the functional significance of galanin expression in mesenchymal and neural crest cells is currently unclear, these data suggest a possible role for galanin in regulating stem/progenitor cell proliferation, migration and/or differentiation. This possibility is supported by our observation that galanin and its receptors are highly expressed in bone marrow mesenchymal stem cells (fig. 4) and facilitate cell migration both in vitro and in vivo (Louridas, Letourneau et al. 2009). Furthermore, the expression of galanin in neural crest cells may be relevant to our understanding of the molecular genetics of neuronal tumors. It has been shown that galanin and galanin receptors are expressed in cells of peripheral embryonic neuroectodermal tumors, such as glioblastomas and neuroblastomas (Berger, Tuechler et al. 2002; Berger, Santic et al. 2003; Berger, Santic et al. 2005). Perel et al. has suggested that galanin influences neuroblastoma development and tumor growth, counteracting differentiation as an autocrine/paracrine modulator (Perel, Amrein et al. 2002).

Galanin expression is also present in the mouse embryo at E7.5, during the late gastrulation stage. Here, galanin is abundantly expressed in the node (fig. 5) and primitive streak (Blum,

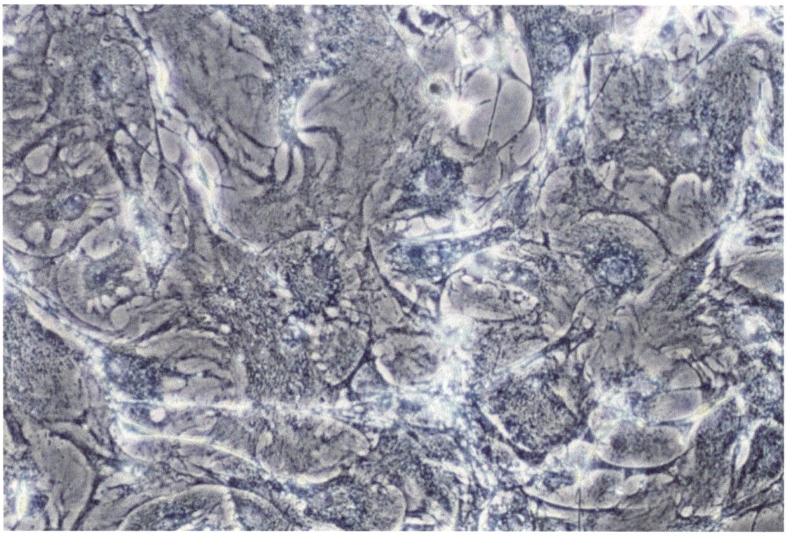

Fig. 4. Immunohistochemistry of bone marrow mesenchymal stem cells stained with a polyclonal galanin antibody. Strong staining is observed in both the cytoplasm and the nucleus of the cells.

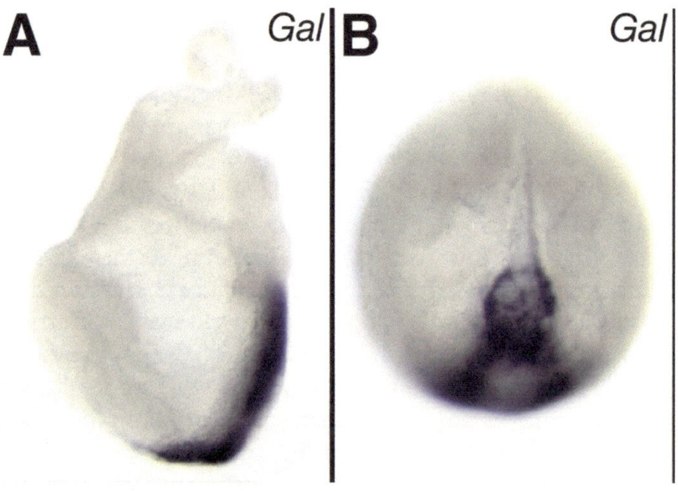

Fig. 5. E7.5 Galanin RNA in situ. A; is a lateral view of the embryo. B; is a distal view of the embryo. Copyright: This image is from Tamplin OJ, BMC Genomics 2008; 9(1):511, an open-access article, licensee BioMed Central Ltd

Andre et al. 2007) and thus represents a marker for the node and the notochord (Schweickert, Deissler et al. 2008; Tamplin, Kinzel et al. 2008). Shortly thereafter, at E8, expression in the primitive streak disappears. Nevertheless, the expression of a neuropeptide in the gastrula, that is, in the absence of any neural tissue, is quite surprising. In their studies, Tamplin et al. used Foxa2 mutant mice to identify novel marker genes for the node. Foxa2 is a forkhead transcription factor that is absolutely required for the formation of the node and the development of the three germ layers. Galanin expression was completely absent in the Foxa2 mutant embryos, indicating that galanin is a target of the Foxa2 gene as well as a regulatory factor involved in patterning.

There are also reports of galanin mRNA expression in preimplantation embryos (Kang, Yeo et al. 2003; Kimber, Sneddon et al. 2008). In the first report (Kang, Yeo et al. 2003), the galanin gene sequence was examined for methylation changes in bovine embryos derived by in vitro fertilization (IVF). The authors observed that the galanin sequence maintained an undermethylated status until the morula stage. By the blastocyst stage, certain CpG sites became specifically methylated, which may be an epigenetic sign for the galanin gene to initiate a differentiation program. Such changes in DNA methylation status are very unusual in pre-implantation mouse development. Shortly after fertilization, the paternal pronucleus is subjected to active demethylation (Mayer, Niveleau et al. 2000), whereas the maternal genome simultaneously undergoes de novo methylation. Afterward, a passive replication-coupled demethylation process occurs in successive cleavage stages up to the blastocyst stage (Dean, Santos et al. 2001). This methylation reprogramming process allows the mouse zygote to gain totipotency and commence the formation of a new individual. In mammals, there are several periods of genome-wide reprogramming of methylation patterns during in vivo development. Typically, a substantial part of the genome is demethylated and then, after some time, remethylated in a cell- or tissue-specific pattern. Thus, galanin methylation appears to play a critical role in cell fate determination and differentiation during development. The study of epigenetic mechanisms underlying the establishment and maintenance of the pluripotent state as well as the differentiation process is an area of intense investigation in ESC biology.

In the second study mentioned above (Kimber, Sneddon et al. 2008), Kimber et al. examined the expression of a number of genes known to be critical for early mouse development in human pre-implantation embryos. Developmental expression of a number of these genes (e.g., galanin, OCT3/4, CDX2, NANOG) was similar to that seen in murine embryos. Galanin mRNA was expressed in the cleavage stages (8-cell stage onward), suggesting a role for galanin in early cell fate decision in human embryos, which may have important implications for IVF treatment and the derivation of human ESCs (hESCs). Indeed, the same group reported that galanin mRNA and protein were both expressed in undifferentiated hESCs and human embryonal carcinoma cells but down regulated upon differentiation, shortly after the down regulation of OCT3/4, Nanog and FoxD3 (El-Bareg et al. 2007), implicating communication between these pluripotent genes in the pre-implantation human embryo and hESCs.

2.2 Galanin in ESCs

ESCs derived from the blastocysts of pre-implantation embryos are pluripotent and have the capability to generate all of the differentiated cell types present in the embryo. The mechanisms regulating self-renewal and cell fate decisions in mammalian stem cells are poorly understood. As compared with differentiated cells, stem cells express a significantly higher number of genes (represented by expressed sequence tags) of unknown function. The properties that distinguish stem cells from other cell types are largely unknown, and the identification of signals that regulate stem cell differentiation remains fundamental to our understanding of cellular diversity. Thus, an important step in the characterization of ESCs will involve the identification of a set of ESC-specific genes that function as markers or contribute to unique regulatory pathways. One approach to identify these signals is to generate stem cell gene expression profiles. Anisimov et al. used the genomic technique of serial analysis of gene expression (SAGE) to define the molecular bases of pluripotency and self-renewal (Anisimov, Tarasov et al. 2002). SAGE is a prominent technique for the quantitative and qualitative characterization of a cell's complete transcriptome (Velculescu, Madden et al. 1999). In their study, the authors performed SAGE on pluripotent mouse R1embryonic stem cells, sequencing a total of 140,313 SAGE tags. Because of the sensitivity of SAGE and the potential quantification of tags from contaminating cells, they cultivated ESCs without feeder layers in the presence of conditioned medium and leukemia inhibitory factor (LIF). After five passages, R1 ESCs maintained pluripotency and the ability to differentiate into cardiac myocytes, hematopoietic and neuron-like cells. One of the most abundant sequences in this SAGE catalogue was galanin. To determine whether the abundance of galanin was a characteristic of ES cells in general or possibly a feature limited to R1 ESCs cultivated under these defined conditions, they constructed other SAGE libraries from embryonal carcinoma (EC) P19 cells, embryonic germ (EG) cells and embryonic stem (ES) cells under different cultivation conditions. Galanin was highly expressed in each of these lines, indicating that high galanin expression is a distinguishing molecular feature of ESCs (Tarasov, Tarasova et al. 2002).

In addition to galanin, all three galanin receptors (GalR1, GalR2 and GalR3) are expressed in mouse R1 ESCs. Quantification of their relative abundances showed that GalR1 is barely detectable in R1 ESCs, while GalR2 and GalR3 are relatively abundant (GalR2 & GalR3 >> GalR1). Similarly, GalR1 is almost undetectable in P19 EC cells but highly abundant in fetal tissues (E16). GalR2 and GalR3 have similar levels of expression in P19 EC and R1 ESCs, and both receptors are widely distributed among fetal tissues (Tarasov, Tarasova et al. 2002). Unlike GalR1 and GalR3, the biological activity of GalR2 is exerted through activation of Gq and phospholipase C. It has also been suggested to play a prominent role during nervous system development (Burazin, Larm et al. 2000). Thus, the presence of galanin transcripts and the relative abundance of GalR2 and GalR3 in ES and EC cells suggest that galanin may be biologically active in ESCs.

Galanin function has been associated with LIF signaling. Addition of LIF into primary dorsal root ganglia (DRG) cultures significantly upregulated galanin expression (Ozturk and Tonge 2001). Similarly, LIF knockout mice have significantly lower levels of galanin (Sun and Zigmond 1996; Sun and Zigmond 1996). To determine whether the prominence of galanin in ESCs is mediated through an interaction with LIF, a series of further experiments were performed in which the medium containing LIF was substituted with non-conditioned maintenance medium without LIF. The absence of LIF actually increased galanin expression in R1 cells. Similarly, removing LIF had no effect on galanin expression in cultured hESCs (El-Bareg et al. 2007; Kimber, Sneddon et al. 2008), indicating that the abundance of galanin transcripts in ESCs is not regulated by LIF.

Several differences between human and mouse ESCs have been identified, including an inactive LIF pathway in human ESCs. Similar to the mouse, the transcriptome profile of hESCs was obtained using SAGE (Richards, Tan et al. 2004). A list of candidate marker genes responsible for stemness in human ESCs has also been created, with galanin highlighted as one of the most abundant genes (Richards, Tan et al. 2004). Transcription factors with a defined role in the maintenance of pluripotency and whose expression is downregulated upon differentiation, including POU5F1 (Oct3/4), SOX2, Galanin, REX1, NANOG, and FLJ10713, were previously identified in mouse ESCs (Anisimov, Tarasov et al. 2002; Ramalho-Santos, Yoon et al. 2002; Mitsui, Tokuzawa et al. 2003).

Using a large-scale oligonucleotide microarray, the profiles of 6 available human ESC lines were analyzed. The expression of defined genes was confirmed by reverse transcriptase polymerase chain reaction (RT-PCR), immunohistochemistry, focused microarrays and comparison to various databases maintained at the National Cancer Institute (Bhattacharya, Miura et al. 2004; (Zeng, Miura et al. 2004). A comparison of overexpressed genes identified 92 genes common to all six lines. These 92 genes constitute a molecular signature of "stemness" in human ESCs. Galanin was the most abundant, along with Oct3/4, Nanog, Sox2 and FOXD3. However, the exact molecular mechanisms involved in self-renewal and pluripotency are still not very clear.

In many respects, germ cell tumorigenesis resembles early embryogenesis. Embryonal carcinomas represent a histologic subgroup of testicular germ cell tumors, and EC cells may follow a differentiation trajectory in a manner similar to early embryogenesis. Using microarray analysis, the transcriptome of neoplastic tissues from the human testis was analyzed by Skotheim et al. (Skotheim, Lind et al. 2005). Selection for genes highly expressed in the undifferentiated, pluripotent embryonal carcinomas identified the major pluripotency markers, including Galanin, POU5F1(Oct3/4), NANOG, DPPA4. Again, Galanin was the most highly expressed gene. Galanin and POU5F1 were both up regulated at the protein level and thereby validated as diagnostic markers for undifferentiated tumor cells.

Preliminary data support the hypothesis that galanin exerts an effect on self-renewal and pluripotency of ESCs along with POU5FI, NANOG and DPPA4 because it is temporarily down regulated upon ESC differentiation and is also more abundant in undifferentiated embryonal carcinomas relative to differentiated carcinomas. Differential DNA methylation of specific sites in the galanin gene might represent an epigenetic signal for the galanin gene to initiate a differentiation program. This occurrence may explain why galanin continues to be expressed in somatic cells of neural crest and mesenchymal origin in the early embryo. Both de novo methylation and maintenance DNA methylation are critical for early development, but they are required for differentiation rather than maintenance of the undifferentiated state. Human ESCs have been shown to possess a unique DNA methylation signature as compared with differentiated cells and cancer cells (Bibikova, Chudin et al. 2006; Meissner, Mikkelsen et al. 2008; Amabile and Meissner 2009; Ball, Li et al. 2009; Meissner 2010), which supports the concept that a specific DNA methylation pattern may contribute to the pluripotent state. In particular, the pluripotency-associated genes Galanin, POU5F1(Oct3/4), NANOG and DPPA4 are largely unmethylated in ESCs and methylated in differentiated cells.

Understanding the epigenetic regulation of ESCs will help to shed light on the molecular basis of normal development as well as the abnormal processes that underlie cancer.

3. Conclusion

In conclusion, the neuroendocrine peptide galanin is one of the most highly expressed genes in both human and mouse ESCs and the embryonic tissues of many species. Galanin is thus considered a marker of "stemness" and pluripotency. All three galanin receptors are present in ESCs, suggesting that the peptide may be biologically active. There are enough indications to suggest a highly dynamic role of galanin in ESCs and in committing the fate of ES cells. The variety of cellular effect of galanin may depend on the environment surrounding the cells and possibly differential activation of its receptors. The switch from self-renewal to differentiation of ESCs might be triggered by a combination of other signals and coordinated changes in recruitment of epigenetic modulators and transcription factors to the promoter region. The strength of the intracellular signaling may affect the negative or positive regulatory elements of the galanin gene to use different intracellular pathways to mediate different cell function in ES cells.

4. References

Amabile, G. and A. Meissner (2009). "Induced pluripotent stem cells: current progress and potential for regenerative medicine." Trends Mol Med 15(2): 59-68.

Anisimov, S. V., K. V. Tarasov, et al. (2002). "SAGE identification of gene transcripts with profiles unique to pluripotent mouse R1 embryonic stem cells." Genomics 79(2): 169-76.

Assou, S., T. Le Carrour, et al. (2007). "A meta-analysis of human embryonic stem cells transcriptome integrated into a web-based expression atlas." Stem Cells 25(4): 961-73.

Ball, M. P., J. B. Li, et al. (2009). "Targeted and genome-scale strategies reveal gene-body methylation signatures in human cells." Nat Biotechnol 27(4): 361-8.

Berger, A., R. Santic, et al. (2003). "Galanin and galanin receptors in human gliomas." Acta Neuropathol 105(6): 555-60.

Berger, A., R. Santic, et al. (2005). "Galanin and galanin receptors in human cancers." Neuropeptides 39(3): 353-9.

Berger, A., C. Tuechler, et al. (2002). "Elevated expression of galanin receptors in childhood neuroblastic tumors." Neuroendocrinology 75(2): 130-8.

Bhattacharya, B., T. Miura, et al. (2004). "Gene expression in human embryonic stem cell lines: unique molecular signature." Blood 103(8): 2956-64.

Bibikova, M., E. Chudin, et al. (2006). "Human embryonic stem cells have a unique epigenetic signature." Genome Res 16(9): 1075-83.

Blum, M., P. Andre, et al. (2007). "Ciliation and gene expression distinguish between node and posterior notochord in the mammalian embryo." Differentiation 75(2): 133-46.

Branchek, T. A., K. E. Smith, et al. (2000). "Galanin receptor subtypes." Trends Pharmacol Sci 21(3): 109-17.

Burazin, T. C., J. A. Larm, et al. (2000). "Galanin-R1 and -R2 receptor mRNA expression during the development of rat brain suggests differential subtype involvement in synaptic transmission and plasticity." Eur J Neurosci 12(8): 2901-17.

Corness, J. D., J. P. Burbach, et al. (1997). "The rat galanin-gene promoter: response to members of the nuclear hormone receptor family, phorbol ester and forskolin." Brain Res Mol Brain Res 47(1-2): 11-23.

Dean, W., F. Santos, et al. (2001). "Conservation of methylation reprogramming in mammalian development: aberrant reprogramming in cloned embryos." Proc Natl Acad Sci U S A 98(24): 13734-8.

Evans, H. F. and J. Shine (1991). "Human galanin: molecular cloning reveals a unique structure." Endocrinology 129(3): 1682-4.

Evans, M. J. and M. H. Kaufman (1981). "Establishment in culture of pluripotential cells from mouse embryos." Nature 292(5819): 154-6.

Hokfelt, T. and K. Tatemoto (2008). "Galanin--25 years with a multitalented neuropeptide." Cell Mol Life Sci 65(12): 1793-5.

Howard, A. D., C. Tan, et al. (1997). "Molecular cloning and characterization of a new receptor for galanin." FEBS Lett 405(3): 285-90.
Iismaa, T. P., Z. Fathi, et al. (1998). "Structural organization and chromosomal localization of three human galanin receptor genes." Ann N Y Acad Sci 863: 56-63.
Jiang, K., G. Spyrou, et al. (1998). "Characterization of phorbolester-inducible human neuronal factors involved in trans-activation of the galanin gene." Biochem Biophys Res Commun 246(1): 192-8.
Jones, M., P. Perumal, et al. (2009). "Presence of galanin-like immunoreactivity in mesenchymal and neural crest origin tissues during embryonic development in the mouse." Anat Rec (Hoboken) 292(4): 481-7.
Kang, Y. K., S. Yeo, et al. (2003). "Precise recapitulation of methylation change in early cloned embryos." Mol Reprod Dev 66(1): 32-7.
Kimber, S. J., S. F. Sneddon, et al. (2008). "Expression of genes involved in early cell fate decisions in human embryos and their regulation by growth factors." Reproduction 135(5): 635-47.
Kofler, B., H. F. Evans, et al. (1995). "Characterization of the 5'-flanking region of the human preprogalanin gene." DNA Cell Biol 14(4): 321-9.
Kofler, B., M. L. Liu, et al. (1996). "Molecular cloning and characterisation of the mouse preprogalanin gene." Gene 182(1-2): 71-5.
Kolakowski, L. F., Jr., G. P. O'Neill, et al. (1998). "Molecular characterization and expression of cloned human galanin receptors GALR2 and GALR3." J Neurochem 71(6): 2239-51.
Lang, R., A. L. Gundlach, et al. (2007). "The galanin peptide family: receptor pharmacology, pleiotropic biological actions, and implications in health and disease." Pharmacol Ther 115(2): 177-207.
Langel, U. and T. Bartfai (1998). "Chemistry and molecular biology of galanin receptor ligands." Ann N Y Acad Sci 863: 86-93.
Louridas, M., S. Letourneau, et al. (2009). "Galanin is highly expressed in bone marrow mesenchymal stem cells and facilitates migration of cells both in vitro and in vivo." Biochem Biophys Res Commun 390(3): 867-71.
Lu, X., L. Lundstrom, et al. (2005). "Galanin receptor ligands." Neuropeptides 39(3): 143-6.
Mayer, W., A. Niveleau, et al. (2000). "Demethylation of the zygotic paternal genome." Nature 403(6769): 501-2.
Meissner, A. (2010). "Epigenetic modifications in pluripotent and differentiated cells." Nat Biotechnol 28(10): 1079-88.
Meissner, A., T. S. Mikkelsen, et al. (2008). "Genome-scale DNA methylation maps of pluripotent and differentiated cells." Nature 454(7205): 766-70.
Mitsui, K., Y. Tokuzawa, et al. (2003). "The homeoprotein Nanog is required for maintenance of pluripotency in mouse epiblast and ES cells." Cell 113(5): 631-42.
Mitsukawa, K., X. Lu, et al. (2008). "Galanin, galanin receptors and drug targets." Cell Mol Life Sci 65(12): 1796-805.
Ozturk, G. and D. A. Tonge (2001). "Effects of leukemia inhibitory factor on galanin expression and on axonal growth in adult dorsal root ganglion neurons in vitro." Exp Neurol 169(2): 376-85.
Perel, Y., L. Amrein, et al. (2002). "Galanin and galanin receptor expression in neuroblastic tumours: correlation with their differentiation status." Br J Cancer 86(1): 117-22.
Pooga, M., A. Jureus, et al. (1998). "Novel galanin receptor ligands." J Pept Res 51(1): 65-74.

Ramalho-Santos, M., S. Yoon, et al. (2002). ""Stemness": transcriptional profiling of embryonic and adult stem cells." Science 298(5593): 597-600.

Richards, M., S. P. Tan, et al. (2004). "The transcriptome profile of human embryonic stem cells as defined by SAGE." Stem Cells 22(1): 51-64.

Rokaeus, A. and M. J. Brownstein (1986). "Construction of a porcine adrenal medullary cDNA library and nucleotide sequence analysis of two clones encoding a galanin precursor." Proc Natl Acad Sci U S A 83(17): 6287-91.

Rokaeus, A. and J. A. Waschek (1998). "Tissue-specific enhancement and restriction of galanin gene expression in transgenic mice by 5' flanking sequences." Brain Res Mol Brain Res 60(2): 150-9.

Salvi, E., R. Vaccaro, et al. (1999). "Ontogeny of galanin-immunoreactive elements in the intrinsic nervous system of the chicken gut." Anat Rec 254(1): 28-38.

Salvi, E. P., R. Vaccaro, et al. (2001). "Ontogeny of galanin-immunoreactive elements in chicken embryo autonomic nervous system." Anat Rec 262(3): 266-78.

Schweickert, A., K. Deissler, et al. (2008). "Left-asymmetric expression of Galanin in the linear heart tube of the mouse embryo is independent of the nodal co-receptor gene cryptic." Dev Dyn 237(12): 3557-64.

Skotheim, R. I., G. E. Lind, et al. (2005). "Differentiation of human embryonal carcinomas in vitro and in vivo reveals expression profiles relevant to normal development." Cancer Res 65(13): 5588-98.

Sun, Y. and R. E. Zigmond (1996). "Involvement of leukemia inhibitory factor in the increases in galanin and vasoactive intestinal peptide mRNA and the decreases in neuropeptide Y and tyrosine hydroxylase mRNA in sympathetic neurons after axotomy." J Neurochem 67(4): 1751-60.

Sun, Y. and R. E. Zigmond (1996). "Leukaemia inhibitory factor induced in the sciatic nerve after axotomy is involved in the induction of galanin in sensory neurons." Eur J Neurosci 8(10): 2213-20.

Tamplin, O. J., D. Kinzel, et al. (2008). "Microarray analysis of Foxa2 mutant mouse embryos reveals novel gene expression and inductive roles for the gastrula organizer and its derivatives." BMC Genomics 9: 511.

Tarasov, K. V., Y. S. Tarasova, et al. (2002). "Galanin and galanin receptors in embryonic stem cells: accidental or essential?" Neuropeptides 36(4): 239-45.

Tatemoto, K., A. Rokaeus, et al. (1983). "Galanin - a novel biologically active peptide from porcine intestine." FEBS Lett 164(1): 124-8.

Thomson, J. A., J. Itskovitz-Eldor, et al. (1998). "Embryonic stem cell lines derived from human blastocysts." Science 282(5391): 1145-7.

Velculescu, V. E., S. L. Madden, et al. (1999). "Analysis of human transcriptomes." Nat Genet 23(4): 387-8.

Vrontakis, M. E. (2002). "Galanin: a biologically active peptide." Curr Drug Targets CNS Neurol Disord 1(6): 531-41.

Vrontakis, M. E., L. M. Peden, et al. (1987). "Isolation and characterization of a complementary DNA (galanin) clone from estrogen-induced pituitary tumor messenger RNA." J Biol Chem 262(35): 16755-8.

Zeng, X., T. Miura, et al. (2004). "Properties of pluripotent human embryonic stem cells BG01 and BG02." Stem Cells 22(3): 292-312.

Zhang, H, 1998 (1998). "Identification and characterization of the 5' flanking region of the rat galanin gene. Thesis University of Manitoba. http://hdl.handle.net/1993/1476

18

Rho-GTPases in Embryonic Stem Cells

Michael S. Samuel[1] and Michael F. Olson[2]
[1]*Centre for Cancer Biology, SA Pathology, Adelaide,*
[2]*The Beatson Institute for Cancer Research, Glasgow,*
[1]*Australia*
[2]*UK*

1. Introduction

The Rho family of small GTP-binding proteins is comprised of 22 members, including the most well characterized members RhoA, Rac1 and Cdc42 (Jaffe and Hall 2005). The Rho family proteins share a high degree of homology with the Ras proto-oncogene, and indeed were first identified as a result of this similarity (Ras homologue). Activity of these proteins is dependent upon their nucleotide binding state; inactive when associated with GDP but active following exchange of GDP for GTP, which induces conformational changes that promote association/activation of downstream effector proteins. The GDP/GTP cycle is regulated by GAPs that accelerate GTP hydrolysis by providing a critical catalytic amino acid leading to a return to the inactive state (Bernards and Settleman 2005), and GEFs that promote guanine nucleotide exchange and consequent Rho activation (Rossman et al. 2005). The number of GAPs and GEFs far exceeds the number of Rho proteins, and the roles of individual GAPs and GEFs in specific cell types and biological processes is currently an intensively studied field.

Although united by homology and function as regulators of the actin cytoskeleton, each of RhoA, Rac1 and Cdc42 has a distinct role in the organization of actin structures (Figure 1). RhoA is principally involved with the production of actin-myosin bundles and the generation of actomyosin contractile force. Rac1 contributes to the formation of actin meshworks that result in the emergence of large protrusive structures that lead to spreading or, if occurring in a polarized manner, will contribute to motility. Cdc42 promotes the formation of actin-rich filopodia. Together, coordinated programs of RhoA, Rac1 and Cdc42 activation/inactivation play prominent roles in processes such as endocytosis/exocytosis, adhesion and motility, which may subsequently impact upon proliferation and death/survival. Recent advances in the development of activation-state sensitive fluorescent probes have allowed temporal and spatial analysis of Rho protein activation, which has added significantly to our appreciation of Rho regulation and function (Hodgson et al. 2010). Much of the early research on Rho protein function relied upon over-expression of dominant-negative mutants that reduced affinity for GTP and constitutively-active mutants that reduced GTP hydrolysis; however, more refined analysis has become possible with the rise of RNAi and knockout methodologies (Heasman and Ridley 2008).

The study of Rho family proteins has historically focused on their roles as molecular switches acting downstream of cell surface receptors to regulate the actin cytoskeleton (Jaffe and Hall 2005). Significant effort has gone into classifying signaling from Rho proteins into

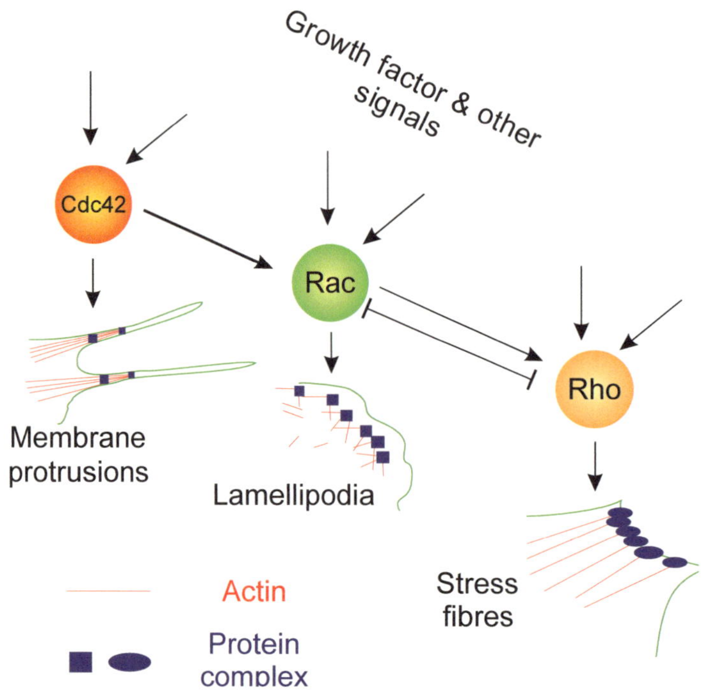

Fig. 1. Diagram of actin structures regulated by RhoA, Rac1 and Cdc42.

linear cascades, similarly to the classical Ras/Raf/MEK/ERK kinase cascade. However, recently a greater appreciation of the role of mechanical forces as fundamental influences in biology has emerged (Puceat et al. 2003). As central regulators of the actin-myosin cytoskeleton, an emerging concept is that many of the activities of Rho proteins may not be attributable to simple linear pathways, but instead are the product of modulating contraction and relaxation at the cellular and subcellular levels, with consequent effects on development and function at the tissue and organismal levels.

2. Embryonic Stem Cells

Pluripotent stem cells were first isolated from testicular teratocarcinoma (Pierce and Dixon 1959), a germ cell tumor type containing a population of pluripotent stem cells together with embryonic and extra-embryonic tissues that arise from these stem cells. Pluripotent stem cells of testicular teratocarcinomas are termed Embryonal Carcinoma (EC) cells and can give rise to collections of tumor cells having morphological characteristics of each of the three embryonic germ layers. In mice, EC cells have been demonstrated to be capable of contributing to every germ layer including the germ-line when injected into host blastocysts (Brinster 1974; Mintz and Illmensee 1975; Illmensee and Mintz 1976). Interestingly, under these conditions, EC cells are non-malignant, and chimeric mice containing tissues differentiated from EC cells are generally healthy. These observations formed the basis for the isolation of Embryonic Stem (ES) cells, which were derived from the pre-implantation embryo, arising when cells constituting the inner cell mass (ICM) of the pre-implantation blastocyst or the epiblast of the post-implantation blastocyst were placed in 2D-culture (Evans and Kaufman 1981; Martin 1981). Like EC cells, ES cells are pluripotent, being capable of giving rise to all tissues of the adult organism originating from the three germ layers, upon injection into a host blastocyst (Bradley et al. 1984). The great similarities observed between EC cells and ES cells led to an appreciation of the importance of the tissue microenvironment in informing cell behavior and fate.

A major attraction of murine ESC (mESC) research stemmed from the realization that mutations introduced into the mESC genome would be readily transmitted through the germ-line, enabling the establishment of strains of mice harboring specific genetic mutations (Capecchi 1989), thereby facilitating the elegant functional characterization of virtually any gene of interest. The first gene to be targeted and inactivated in mES cells was the X-linked gene *Hprt*, which encodes hypoxanthine guanine phosphoribosyltransferase, an enzyme involved in purine metabolism (Thomas and Capecchi 1987). In turn, an *Hprt*-deficient ES cell line was engineered to re-introduce the *Hprt* coding sequence and used to produce knock-in gene-targeted mice for the first time, which faithfully recapitulated the wild-type *Hprt* expression pattern (Thompson et al. 1989). Following on from these pioneering studies, techniques for establishing gene-targeted mice have been considerably improved and refined. Gene targeting in mES cells to generate loss of function or gain of function mutations with an exquisite degree of subtlety and control is now an established tool in biological research.

2.1 Maintenance of pluripotency

ES cells express markers of their undifferentiated state such as the octamer binding protein 4 (Oct4) (Rosner et al. 1990; Scholer et al. 1990), the SRY-related HMG-box gene 2 (Sox2) (Yuan et al. 1995), signal transducer and activator of transcription 3 (Stat3) (Niwa et al. 1998), the homeobox protein Nanog (Chambers et al. 2003; Mitsui et al. 2003) and alkaline phosphatase (AP) (Hahnel et al. 1990) that denote their capacity for both self-renewal and pluripotency. Of these, Oct4 and Sox2 have key roles in the maintenance of ES cell self-renewing capacity such that their expression is essential for the maintenance of pluripotency and their ectopic expression in somatic cells contributes to the generation of induced pluripotent (iPS) cells (Takahashi and Yamanaka 2006; Yu et al. 2007; Nakagawa et al. 2008).

Oct4 is a POU-domain transcription factor also termed POU5F1 and is indispensable for plutipotency. Oct4 deficient embryos develop to the morula stage, but are unable to form an ICM (Nichols et al. 1998) and in vitro culture of Oct4 deficient embryos failed to yield ES cells (Nichols et al. 1998). These observations are further elaborated by more recent work showing that selective deletion of the Oct4 gene in primordial germ cells (PGC) results in their death by apoptosis (Kehler et al. 2004). Oct4 expression is very tightly regulated and its transient increase and decrease during early stages of embryonic development have been termed the totipotent cycle (Yeom et al. 1996). While evidence for the absolute requirement for Oct4 in the maintenance of ES cells is very strong, there is controversy on whether it is required for the maintenance of adult stem cells. Although there are numerous reports of Oct4 expression in adult stem cells including in hematopoietic and mesenchymal stem cells and stem cells of epithelial tissues such as the pancreas, kidney, breast, uterus, lung and skin, a recent study in which its expression was systematically abrogated in several of these tissues has revealed that Oct4 is required for neither the maintenance of adult stem cells nor for wound healing (Lengner et al. 2007).

Sox2 is a HMG-box containing transcription factor closely related to the Y-chromosome located sex determining gene SRY. Its main role in the maintenance of pluripotency is thought to be closely related to the regulation of Oct4 transcription. Indeed Sox2 and Oct4 can jointly bind regulatory chromosomal regions associated with both the *Oct4* and *Sox2* genes (Chew et al. 2005; Masui et al. 2007) as well as regulating Nanog expression (Kuroda et al. 2005; Rodda et al. 2005).

2.2 Culturing ES cells

Since the initial isolation of ICM-derived mESCs in the early 1980s (Evans and Kaufman 1981; Martin 1981), conditions for the culture of ESCs have been developed and progressively refined. mESCs are propagated on a feeder layer of murine embryonic fibroblasts (MEFs) or in media containing leukemia inhibitory factor (LIF), under which conditions they maintain a pluripotent state (Williams et al. 1988). Withdrawal of LIF or culture in the absence of fibroblasts results in spontaneous differentiation of mESCs into a variety of lineages (Evans and Kaufman 1981; Martin 1981; Williams et al. 1988). The dependence of mES cells on LIF is thought to be related to LIF mediated activation of STAT3 signaling (Smith et al. 1988) which together with Oct4/Sox2, has a possible role in the regulation of Nanog expression.

Human ESCs (hESCs), which have been isolated from the epiblasts of human blastocysts (Thomson et al. 1998; Reubinoff et al. 2000) are also propagated on a feeder layer of MEFs, but LIF has no role in maintaining their pluripotency (Thomson et al. 1998; Reubinoff et al. 2000). Instead, a balance between Tgfβ/activin/nodal signaling and suppression of BMP signaling together with the FGF signaling pathway are important for self-renewal and the maintenance of pluripotency in this system (James et al. 2005; Vallier et al. 2005; Xu et al. 2005). However, as yet no reliable defined medium has been developed to enable the culture of hES cells in the absence of feeder cells. Like mESCs, hESCs spontaneously differentiate if cultured in the absence of a feeder layer, but unlike mESCs they undergo blebbing and apoptosis when maintained in a dissociated state (Watanabe et al. 2007).

hESCs are not only a valuable tool for the study of human development, but also have applications in regenerative medicine, toxicology and the development of new drugs to target human disease (Murry and Keller 2008). mESCs and hESCs are thus examples of the two major types of pluripotent stem cells, derived as they are from the ICM and the epiblast respectively.

3. Rho family GTPases in embryonic stem cells

One of the most interesting recent developments in ES research is the revelation that signaling through RhoA plays a key role in the survival of human embryonic stem cells. This was first appreciated in 2007, following a cell-based screen of biologically active compounds that promoted survival and proliferation of dissociated hESCs that identified Y27632, a selective inhibitor of the Rho-effector protein ROCK (Watanabe et al. 2007). The ROCK1 and ROCK2 serine/threonine kinases are central and critical regulators of actomyosin contractility (Coleman et al. 2001). Typically, these kinases are activated by association with active GTP-bound Rho proteins. Active ROCK promotes actomyosin contractility through a dual mechanism of simultaneously phosphorylating and activating the contractile force-generating regulatory myosin light chain (MLC) and the LIM kinases (Sugihara et al. 1998), which modulate filamentous actin stability. In contrast to hESC, mES cells do not require ROCK inhibition for survival even when disaggregated to a single cell suspension. Since that initial study, subsequent screens have identified additional ROCK selective inhibitors that promote the survival of hESC (Andrews et al. 2010; Pakzad et al. 2010) and neural stem cells (Xu et al.

2010), thereby independently validating the role of ROCK as a key regulator of ESC survival. The addition of Y27632 to the culture media is now standard practice and has greatly improved the reliability of hES cell survival (Olson 2008; Krawetz et al. 2009). The addition of Y-27632 can be directly to the cell culture medium or into the extracellular matrix upon which the hESCs are plated (Danovi et al. 2010). ROCK inhibitors have also been shown to improve recovery of cryopreserved ESC (Scott and Olson 2007; Wickman et al. 2010) and increase the efficiency of adenovirus-mediated gene transfer (Patwari and Lee 2008).

3.1 Rho signaling in ES cells

Recently, it has become clear that the actomyosin machinery downstream of Rho activation is essential for the blebbing and apoptosis that follow dissociation of hESCs (Martin 1981; Chen et al. 2010; Ohgushi et al. 2010), as inhibition of the myosin heavy chain ATPase with Blebbistatin, the use of actin disruption drugs or selective knock-down of ROCK1, ROCK2 or the myosin heavy and light chains all prolong survival of dissociated hESCs. Rho activation, coupled with Rac inhibition, was determined to be the driver of dissociation-induced hESC apoptosis via ROCK-mediated myosin light chain phosphorylation (Ohgushi et al. 2010). Activation of ROCK1 by caspase-mediated cleavage (Buecker et al. 2010) does not appear to contribute to apoptosis induced in this manner (Ohgushi et al. 2010). Overexpression of an active form of Ezrin, which strengthens the physical coupling between the plasma membrane and cortical actin cytoskeleton, was sufficient to block blebbing but not the dissociation-induced cell death, indicating that apoptosis was not caused by blebbing itself but the result of actomyosin contraction (Ohgushi et al. 2010). Although the dissociation-induced cell death was linked back to mitochondrial depolarization and cytochrome c release, further study will be required to determine how actomyosin contractility is coupled to the mitochondrial pathway of apoptosis (Ohgushi et al. 2010). It is also becoming clear that the particular sub-embryonic origin of the embryonic stem cell line determines whether Rho signaling is detrimental to survival on dissociation. While epiblast-derived hESCs are acutely sensitive to Rho signaling following dissociation, ICM-derived mESC have the capacity to survive dissociation without the need for inhibition of the actomyosin machinery (Ohgushi et al. 2010), a characteristic they share with human induced pluripotent stem cells (hiPSC), which display mESC-like morphological features (Evans and Kaufman 1981). On the other hand, epiblast-derived murine epiblast stem cells (mEpiSC) or mESCs differentiated into epiblast-like cells acquire a dependence on ROCK-inhibition in order to survive dissociation (Ohgushi et al. 2010). One theoretical possibility to account for these observations is that external pulling forces from adjacent cells in an epithelial sheet counteract the internal actomyosin contractile forces within individual cells such that the internal and external mechanical forces become balanced in all directions along the epithelial plane, thereby limiting their pro-apoptotic effects. Since mESCs are derived from the ICM prior to differentiation into epithelial-type cells and grow in disorganized three-dimensional cell collectives similar to the *bona fide* inner cell mass, they may not be dependent on external tension derived from cell-cell adhesions, such as those that occur in an epithelial sheet, for survival. In contrast, hESCs grow as tightly adherent two-dimensional sheets similar to the epiblast where pulling forces from adjacent cells would be sensed. In agreement with this model, when human induced pluriopotent stem cells (hiPSCs) were reprogrammed from fibroblasts through the expression of five reprogramming factors plus LIF, they acquired the ability to grow at low density or in suspension in parallel with changed in vitro growth characteristics to mESC-like disorganized three-dimensional structures (Tashiro et al. 2010). This exquisite sensitivity of epiblast and epiblast-like stem cells may reflect the critical importance of proper differentiation and spatial organization of the epiblast stage during embryonic development. If any individual cell in the epiblast layer were improperly positioned in the epithelial sheet, the potential consequences to the subsequent developmental stages and ultimately to the organism as a whole could be catastrophic.

3.2 Rac signaling in ES cells

The pro-apoptotic effect of Rho signaling in dissociated hESC is strongly counteracted by signaling through Rac. Indeed it has been shown that Rac1 is required for the survival of epiblast cells within the blastocyst during morphogenesis of the murine peri-implantation egg cylinder (He et al. 2010). During this process, the apoptosis mediated clearance of cells that are not in contact with the basement membrane (known as cavitation) is counteracted by signaling through Rac in those cells that remain apposed to the basement membrane (BM). In the absence of Rac1, cells in contact with the BM undergo apoptosis despite the survival signals that it normally provides (Kim et al. 2011). It is these BM-associated cells that give rise to the epiblast (He et al. 2010). Activation of Rac in the epiblast is mediated by the recruitment of the Crk adaptor protein and DOCK180 GEF (He et al. 2010). In turn, active Rac signals via PI3K and Akt to promote survival (He et al. 2010). Interestingly, a single dual-function protein, Abr, acts as Rho-GEF and Rac-GAP within dissociated hES cells in culture, simultaneously activating Rho and inactivating Rac upon cell dissociation, in a manner dependent on cell-cell interactions involving E-cadherin (Martin 1981; Ohgushi et al. 2010). The role of E-cadherin in hESC survival was also revealed in a chemical biology screen for small molecules that affected survival (Pakzad et al. 2010). One compound increased the survival of dissociated cells by reducing E-cadherin endocytosis, thus increasing the levels of cell-surface E-cadherin and consequently promoting cell-cell adhesions. In agreement with these observations, ectopic over-expression of E-cadherin was also sufficient to increase survival of dissociated hESCs (Rizzino 2010). However, when dissociated hESCs were grown on E-cadherin coated plates, they still underwent membrane blebbing and had significantly lower survival, indicating that homotypic E-cadherin interactions alone were not sufficient to promote survival (Ohgushi et al. 2010). These observations suggest the existence of a yet uncharacterized sensor that transmits a complementary signal derived from cell-cell adhesion that acts in concert with, or in parallel to, E-cadherin activation to repress actomyosin contractility and consequent cell death. Although mESCs are not sensitive to the same sort of dissociation-induced cell death, constitutive Rac1 deletion was found to induce membrane blebbing and eventual apoptosis of epiblast derived stem cells, possibly due to the lack of Rac1 activity to counter-balance the effect of RhoA activation (Kim et al. 2011). These Rac1 deleted cells also were defective in the formation of actin cytoskeleton structures such as lamellipodia and were significantly slower in migrating on collagen I coated dishes, revealing the critical role played by Rac1 in these biological activities. Similarly, Rac1 was found to be an important contributor to mESC migration on laminin (Li et al. 2010).

3.3 Cdc42 signaling in ES cells

Also implicated in murine peri-implantation development is the Cdc42 GTP-binding protein. Mouse embryoid bodies deficient for Cdc42 exhibited polarization defects characterized by aberrant adherens and tight cell-cell junction formation and failure of cavitation (Wu et al. 2007), in a process mediated by the atypical protein kinase C (aPKC) family of kinases. Despite the polarization defects, basement membrane formation, which requires polarized deposition and assembly of basement membrane components at the basal side of a cell layer, was unaffected by deletion of Cdc42 (Wu et al. 2007). Interestingly mES cells lacking Cdc42 had lower levels of active Rac1 although total Rac1 protein levels were unaffected (Wu et al. 2007), suggesting that some of the observed defects could be the result of reduced Rac1 activity. However, unlike Rac1 deficient mES cells that would undergo apoptosis while in contact with the basement membrane (Kim et al. 2011), deletion of Cdc42 still allowed survival of cells in contact with the BM (Wu et al. 2007). Additional defects in PIP_2-induced actin polymerization and cytoskeletal organization were likely to also contribute to defective adhesion and migration of mESC deleted of Cdc42 (Chambers et al. 2003; Wu et al. 2007). The motility of mESC plated on plated on laminin also were dependent on Cdc42 as revealed by siRNA-mediated knockdown (Li et al. 2010). These morphological, polarization and motility defects almost certainly contributed to early embryonic lethality in Cdc42 deficient mice (Chambers et al. 2003). These tantalizing observations point to complementary functions for Rho, Rac and Cdc42 during the processes of cavitation and the appearance of the epiblast, and underscore the importance of these proteins in appropriately mediating the survival or apoptotic clearance of cells during early morphogenesis. It therefore appears that the activity of the Rho family GTPases crucially determines the fate of pluripotent stem cells within the early developing embryo.

4. Additional functions of Rho proteins in ES cells

An interesting aspect of ESC is that under the right conditions, such as hanging drop suspension leading to the formation of embryoid bodies (Kurosawa 2007) , differentiation results in the production of cardiomyocytes that spontaneously contact and relax (beating) as they would in an intact heart (Wobus et al. 1991). Human ESC can also be differentiated into cardiomyocytes, which has generated considerable excitement in the field because of their value in examining the role of specific proteins in cardiac disease phenotypes, and also due to the eventual possibility that they might have therapeutic utility (Brinster 1974). To examine the role of Rac1 in the differentiation of mESC into cardiomyocytes, ectopic expression of constitutively-active Rac 1 deficient in GTPase activity (Rac1V12) or dominant-negative Rac1 with reduced affinity for GTP (Rac1N17 was used to elucidate the consequences of Rac1 gain-of-function and loss-of-function, respectively (Puceat et al. 2003). Expression of active Rac1V12 blocked the characteristic beating of embryoid bodies, due to a differentiation defect as indicated by reduced expression of cardiomyocyte differentiation markers such as MEF2C and ventricular myosin light chain 2 (MLCv2). In contrast, expression of a constitutively active form of RhoA did not block cardiomyocyte differentiation. Previous research had revealed that Rac1 regulates the activity of the NADPH oxidase that generates reactive oxygen species (ROS) (Di-Poi et al. 2001) , and when H_2O_2 was added to embryoid bodies for up to 7 days the effect on blocking cardiomyocyte differentiation by active Rac1V12 was mimicked, while the ROS scavenger catalase reduced the differentiation block induced by active Rac1V12 (Puceat et al. 2003). Consistent with this conclusion, expression of a point-mutant form of Rac1 that does not activate the NADPH oxidase (Rac1V12D38) did not block cardiomyocyte differentiation. Expression of the dominant-negative Rac1N17 to examine loss-of-function did not affect differentiation but did impair beating by interfering with the organization of sarcomeric units required for contraction (Puceat et al. 2003). In contrast to what occurred when Rac1 was expressed early, when the MLCv2 promoter was used to express active Rac1 in differentiated cardiomyocytes, increased beating was observed due to a facilitation of differentiation and prolonged proliferation (Puceat et al. 2003). Expression of dominant-negative Rac1N17 from the MLCv2 promoter had a similar effect as early expression on the organization of sarcomeric units. These results revealed that the role of Rac1 in cardiac differentiation is likely dependent on the developmental stage. Given the availability of mESC in which Rac1 can be conditionally deleted (Yuan et al. 1995), more refined analysis of the role of Rac1 in cardiac differentiation and disease should be possible.

5. Activating ROCK in mouse ICM-derived ES cells

Mechanical forces are increasingly appreciated as major influences in embryonic development. External mechanical forces can be produced by physical alterations to the microenvironment. These external forces are sensed by cells, leading to responses that allow the cell to adapt to the changed environmental circumstances. One way that cells respond to mechanical force is via integrin-mediated activation of Rho and ROCK resulting in increased cellular stiffness via increased actomyosin contracility, which is also known as reinforcement (Guilluy et al. 2011). There is considerable evidence that suppression of

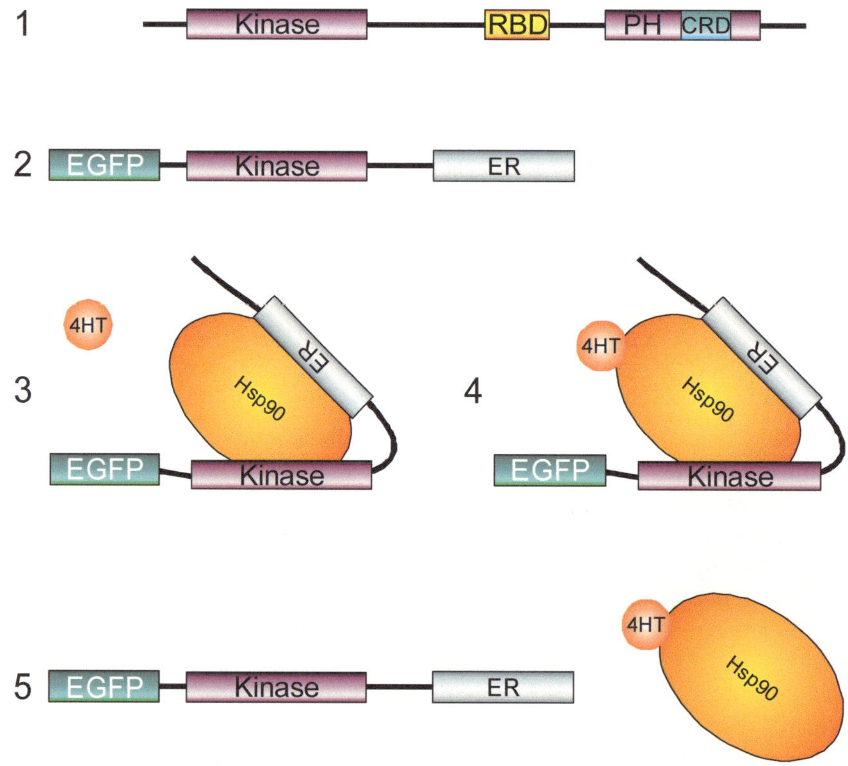

Fig. 2. Mechanism of conditional activation of ROCK. 1: Diagram of ROCK domains, RBD = Rho Binding Domain, PH = Pleckstrin Homology domain, CRD = Cysteine-Rich Domain. 2: Kinase domain of ROCK2 was fused to Enhanced Green Fluorescent Protein (EGFP) and the hormone-binding domain of Estrogen Receptor (ER) to create conditionally regulated ROCK:ER. 3: In the absence of ligand, Heat Shock Protein 90 (Hsp90) binds to the ER domain and represses catalytic activity. 4: Upon binding of estrogen analogues such as 4-hydroxytamoxifen (4HT), 5: Hsp90 is displaced thereby allowing for ROCK catalytic activity.

actomyosin contractility by inhibition of ROCK promotes the survival and continued proliferation of epiblast-derived hES cells. It is suggested, however, that this signaling axis is less important in ICM-derived mES cells. We therefore decided to take advantage of a system to conditionally activate ROCK within mES cells to determine whether ROCK activation and consequent actomyosin contractility had a role in their proliferation, survival and/or maintenance of pluripotency. Accordingly, we transduced G4 mES cells (George et al. 2007) with a pBabe-Puro retroviral vector (Morgenstern and Land 1990) encoding a conditionally-active version of ROCK fused to the hormone-binding domain of the estrogen receptor (Figure 2) (Croft and Olson 2006) to establish the pBabe-Puro-ROCK:ER mES cell line in which ROCK activity could be elicited by treatment with the estrogen analog 4-hydroxytamoxifen (4HT). As a negative control, cells were transduced with pBabe-Puro encoding a kinase-dead counterpart (KD:ER) to produce control pBabe-Puro-KD:ER mES cells that express of catalytically inactive control ROCK protein.
When maintained in 4HT, pBabe-Puro-ROCK:ER mES cells exhibited robust growth and a large number of colonies exhibiting a refractive colony morphology under transmitted light and fewer colonies exhibiting a differentiated morphology, consistent with a high degree of pluripotency (Figure 3). Consistent with this observation, 4HT treated pBabe-Puro-ROCK:ER mES cells express significantly higher levels of the pluripotency marker alkaline phosphatase (ALP) than 4HT treated pBabe-Puro-KD:ER mES cells or vehicle treated pBabe-Puro-ROCK:ER

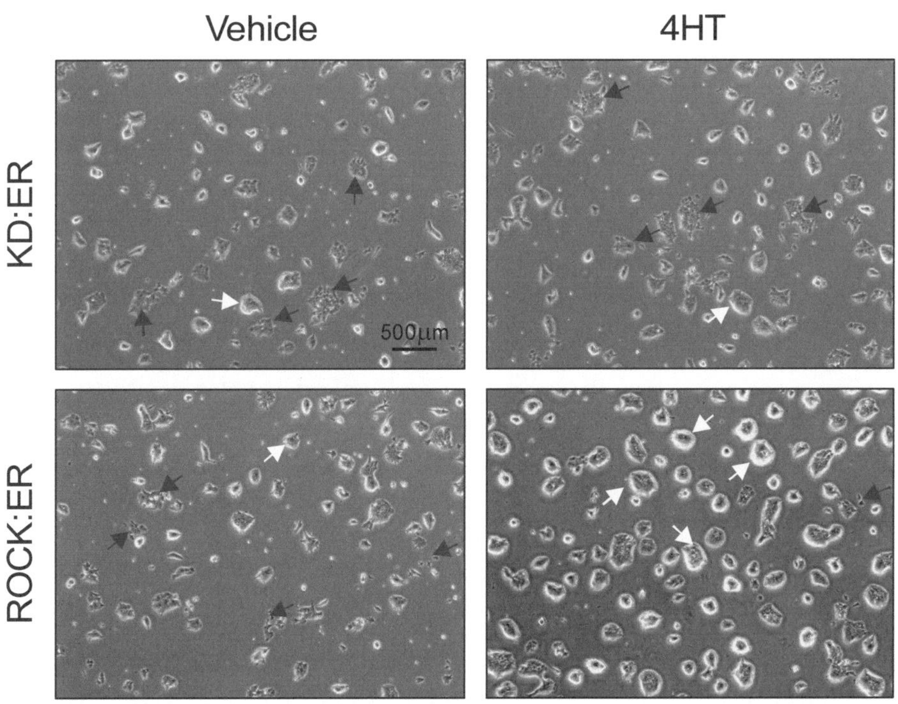

Fig. 3. Conditional ROCK activation in mES cells elicits a highly refractive colony morphology. Panels show brightfield images of pBabe-Puro-ROCK:ER and pBabe-Puro-KD:ER mES cells treated with Vehicle or 4HT. Flat colonies containing mainly differentiated cells (purple arrows) and raised colonies containing mainly undifferentiated cells (white arrows) are indicated. Scale bar denotes 500μm.

and pBabe-Puro-KD:ER mES cells (Figure 4A). To determine whether the increased ALP activity observed upon ROCK activation correlated with an increase in stemness, we then assessed the expression of two classical markers of pluripotency, Oct4 and Nanog. 4HT treated pBabe-Puro-ROCK:ER mES cells express significantly higher levels of Oct4 and Nanog than 4HT treated pBabe-Puro-KD:ER mES cells or vehicle treated pBabe-Puro-ROCK:ER and pBabe-Puro-KD:ER mES cells (Figure 4B). Consistent with this effect being mediated by the activity of ROCK, co-treatment of pBabe-Puro-ROCK:ER mES cells with 4HT and the selective ROCK inhibitor Y-27632 failed to induce Oct4 or Nanog expression (Figure 4B).

Taken together, these results strongly suggest that ROCK activation in mES cells promotes stemness and facilitates proliferation and survival. These observations are consistent with a previous report that inhibition of ROCK activity or silencing of ROCK expression in mESC causes a reduction in stem like properties including alkaline phosphatase activity and Oct3/4 expression, and increased expression of differentiation markers SOX-1, nestin and MAP2c when grown at high seeding densities (Chang et al. 2010). Interestingly, the effects of ROCK inhibition on morphology and colony formation were reversible if cells had been

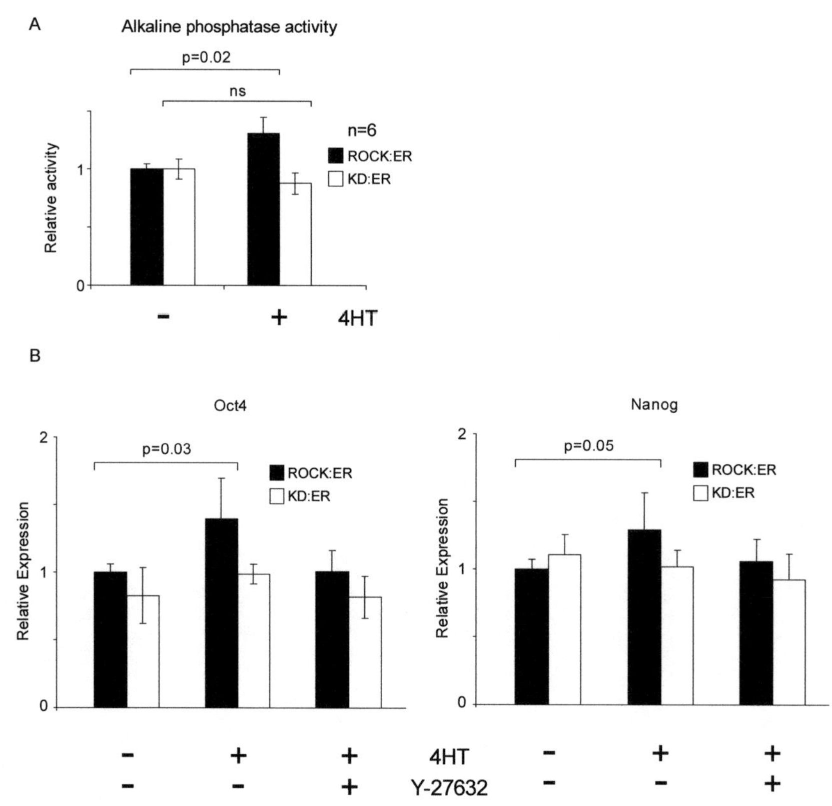

Fig. 4. Conditional ROCK activation in mES cells increases stemness. (A) Histogram shows alkaline phosphatase activity in pBabe-Puro-ROCK:ER and pBabe-Puro-KD:ER mES cells treated with Vehicle or 4HT. (B) Histograms show expression at the mRNA level of the stem cell markers Oct4 and Nanog in pBabe-Puro-ROCK:ER and pBabe-Puro-KD:ER mES cells treated with Vehicle or 4HT. All values are expressed as mean ± SD. P values were calculated using the Student's t-test.

grown at low initial densities where the reduction in stem like properties were not observed. However, at high cell densities where ROCK inhibition had repressed stem cell properties the effects were not reversible, suggesting that epigenetic reprogramming had occurred. It would be very interesting to determine whether the effects of ROCK activation on the maintenance of stemness would persist upon removal of tamoxifen and return of actomyosin contractility to basal levels.

6. Rho signalling in ES cell maintenance, proliferation, survival

There have been significant recent advances in our understanding of the requirement for specific Rho GTPases and downstream signaling pathways in ES cells from gene knockouts, RNAi and small molecule inhibitors. However, what has been missing is an understanding of where and when Rho proteins are activated and inactivated, for example during adhesion or differentiation. Activation-state sensitive fluorescent probes have been developed and used to characterize the temporal and spatial patterns of Rho activation during tumor cell migration and invasion (Vega et al. 2011). One exciting complementary area of research will be the determination of Rho protein activation with spatial and temporal resolution during ES cell growth and differentiation, ultimately through progressive developmental stages

7. References

Andrews, P. D. et al. (2010). "High-content screening of feeder-free human embryonic stem cells to identify pro-survival small molecules." Biochem J 432(1): 21-33.

Bernards, A. and J. Settleman (2005). "GAPs in growth factor signalling." Growth Factors 23(2): 143-149.

Bradley, A. et al. (1984). "Formation of germ-line chimaeras from embryo-derived teratocarcinoma cell lines." Nature 309(5965): 255-256.

Brinster, R. L. (1974). "The effect of cells transferred into the mouse blastocyst on subsequent development." J Exp Med 140(4): 1049-1056.

Buecker, C. et al. (2010). "A murine ESC-like state facilitates transgenesis and homologous recombination in human pluripotent stem cells." Cell Stem Cell 6(6): 535-546.

Capecchi, M. R. (1989). "Altering the genome by homologous recombination." Science 244(4910): 1288-1292.

Chambers, I. et al. (2003). "Functional expression cloning of Nanog, a pluripotency sustaining factor in embryonic stem cells." Cell 113(5): 643-655.

Chang, T. C. et al. (2010). "Rho kinases regulate the renewal and neural differentiation of embryonic stem cells in a cell plating density-dependent manner." PLoS One 5(2): e9187.

Chen, G. et al. (2010). "Actin-myosin contractility is reponsible for the reduced viability of dissociated human embryonic stem cells." Cell Stem Cell.

Chew, J. L. et al. (2005). "Reciprocal transcriptional regulation of Pou5f1 and Sox2 via the Oct4/Sox2 complex in embryonic stem cells." Mol Cell Biol 25(14): 6031-6046.

Coleman, M. L. et al. (2001). "Membrane blebbing during apoptosis results from caspase-mediated activation of ROCK I." Nat Cell Biol 3(4): 339-345.

Croft, D. R. and M. F. Olson (2006). "Conditional regulation of a ROCK-estrogen receptor fusion protein." Methods Enzymol 406: 541-553.

Danovi, D. et al. (2010). "Imaging-based chemical screens using normal and glioma-derived neural stem cells." Biochem Soc Trans 38(4): 1067-1071.

Di-Poi, N. et al. (2001). "Mechanism of NADPH oxidase activation by the Rac/Rho-GDI complex." Biochemistry 40(34): 10014-10022.

Evans, M. J. and M. H. Kaufman (1981). "Establishment in culture of pluripotential cells from mouse embryos." Nature 292(5819): 154-156.

George, S. H. et al. (2007). "Developmental and adult phenotyping directly from mutant embryonic stem cells." Proc Natl Acad Sci U S A 104(11): 4455-4460.

Guilluy, C. et al. (2011). "The Rho GEFs LARG and GEF-H1 regulate the mechanical response to force on integrins." Nat Cell Biol 13(6): 724-729.

Hahnel, A. C. et al. (1990). "Two alkaline phosphatase genes are expressed during early development in the mouse embryo." Development 110(2): 555-564.

He, X. et al. (2010). "Rac1 is essential for basement membrane-dependent epiblast survival." Mol Cell Biol 30(14): 3569-3581.

Heasman, S. J. and A. J. Ridley (2008). "Mammalian Rho GTPases: new insights into their functions from in vivo studies." Nat Rev Mol Cell Biol 9(9): 690-701.

Hodgson, L. et al. (2010). "Biosensors for characterizing the dynamics of rho family GTPases in living cells." Curr Protoc Cell Biol Chapter 14: Unit 14 11 11-26.

Illmensee, K. and B. Mintz (1976). "Totipotency and normal differentiation of single teratocarcinoma cells cloned by injection into blastocysts." Proc Natl Acad Sci U S A 73(2): 549-553.

Jaffe, A. B. and A. Hall (2005). "Rho GTPases: biochemistry and biology." Annu Rev Cell Dev Biol 21: 247-269.

James, D. et al. (2005). "TGFbeta/activin/nodal signaling is necessary for the maintenance of pluripotency in human embryonic stem cells." Development 132(6): 1273-1282.

Kehler, J. et al. (2004). "Oct4 is required for primordial germ cell survival." EMBO Rep 5(11): 1078-1083.

Kim, Y. Y. et al. (2011). "Cryopreservation of human embryonic stem cells derived-cardiomyocytes induced by BMP2 in serum-free condition." Reprod Sci 18(3): 252-260.

Krawetz, R. J. et al. (2009). "Human embryonic stem cells: caught between a ROCK inhibitor and a hard place." Bioessays 31(3): 336-343.

Kuroda, T. et al. (2005). "Octamer and Sox elements are required for transcriptional cis regulation of Nanog gene expression." Mol Cell Biol 25(6): 2475-2485.

Kurosawa, H. (2007). "Methods for inducing embryoid body formation: in vitro differentiation system of embryonic stem cells." Journal of bioscience and bioengineering 103(5): 389-398.

Lengner, C. J. et al. (2007). "Oct4 expression is not required for mouse somatic stem cell self-renewal." Cell Stem Cell 1(4): 403-415.

Li, L. et al. (2010). "Individual cell movement, asymmetric colony expansion, rho-associated kinase, and E-cadherin impact the clonogenicity of human embryonic stem cells." Biophys J 98(11): 2442-2451.

Martin, G. R. (1981). "Isolation of a pluripotent cell line from early mouse embryos cultured in medium conditioned by teratocarcinoma stem cells." Proc Natl Acad Sci U S A 78(12): 7634-7638.

Masui, S. et al. (2007). "Pluripotency governed by Sox2 via regulation of Oct3/4 expression in mouse embryonic stem cells." Nat Cell Biol 9(6): 625-635.
Mintz, B. and K. Illmensee (1975). "Normal genetically mosaic mice produced from malignant teratocarcinoma cells." Proc Natl Acad Sci U S A 72(9): 3585-3589.
Mitsui, K. et al. (2003). "The homeoprotein Nanog is required for maintenance of pluripotency in mouse epiblast and ES cells." Cell 113(5): 631-642.
Morgenstern, J. P. and H. Land (1990). "Advanced mammalian gene transfer: high titre retroviral vectors with multiple drug selection markers and a complementary helper-free packaging cell line." Nucleic Acids Res 18(12): 3587-3596.
Murry, C. E. and G. Keller (2008). "Differentiation of embryonic stem cells to clinically relevant populations: lessons from embryonic development." Cell 132(4): 661-680.
Nakagawa, M. et al. (2008). "Generation of induced pluripotent stem cells without Myc from mouse and human fibroblasts." Nat Biotechnol 26(1): 101-106.
Nichols, J. et al. (1998). "Formation of pluripotent stem cells in the mammalian embryo depends on the POU transcription factor Oct4." Cell 95(3): 379-391.
Niwa, H. et al. (1998). "Self-renewal of pluripotent embryonic stem cells is mediated via activation of STAT3." Genes Dev 12(13): 2048-2060.
Ohgushi, M. et al. (2010). "Molecular pathway and cell state responsible for dissociation-induced apoptosis in human pluripotent stem cells." Cell Stem Cell.
Olson, M. F. (2008). "Applications for ROCK kinase inhibition." Curr Opin Cell Biol 20(2): 242-248.
Pakzad, M. et al. (2010). "Presence of a ROCK inhibitor in extracellular matrix supports more undifferentiated growth of feeder-free human embryonic and induced pluripotent stem cells upon passaging." Stem Cell Rev 6(1): 96-107.
Patwari, P. and R. T. Lee (2008). "Mechanical control of tissue morphogenesis." Circ Res 103(3): 234-243.
Pierce, G. B. and F. J. Dixon, Jr. (1959). "Testicular teratomas. I. Demonstration of teratogenesis by metamorphosis of multipotential cells." Cancer 12(3): 573-583.
Puceat, M. et al. (2003). "A dual role of the GTPase Rac in cardiac differentiation of stem cells." Mol Biol Cell 14(7): 2781-2792.
Reubinoff, B. E. et al. (2000). "Embryonic stem cell lines from human blastocysts: somatic differentiation in vitro." Nat Biotechnol 18(4): 399-404.
Rizzino, A. (2010). "Stimulating progress in regenerative medicine: improving the cloning and recovery of cryopreserved human pluripotent stem cells with ROCK inhibitors." Regen Med 5(5): 799-807.
Rodda, D. J. et al. (2005). "Transcriptional regulation of nanog by OCT4 and SOX2." J Biol Chem 280(26): 24731-24737.
Rosner, M. H. et al. (1990). "A POU-domain transcription factor in early stem cells and germ cells of the mammalian embryo." Nature 345(6277): 686-692.
Rossman, K. L. et al. (2005). "GEF means go: turning on RHO GTPases with guanine nucleotide-exchange factors." Nat Rev Mol Cell Biol 6(2): 167-180.
Scholer, H. R. et al. (1990). "Oct-4: a germline-specific transcription factor mapping to the mouse t-complex." Embo J 9(7): 2185-2195.
Scott, R. W. and M. F. Olson (2007). "LIM kinases: function, regulation and association with human disease." J Mol Med 85(6): 555-568.

Smith, A. G. et al. (1988). "Inhibition of pluripotential embryonic stem cell differentiation by purified polypeptides." Nature 336(6200): 688-690.

Sugihara, K. et al. (1998). "Rac1 is required for the formation of three germ layers during gastrulation." Oncogene 17(26): 3427-3433.

Takahashi, K. and S. Yamanaka (2006). "Induction of pluripotent stem cells from mouse embryonic and adult fibroblast cultures by defined factors." Cell 126(4): 663-676.

Tashiro, K. et al. (2010). "Adenovirus vector-mediated efficient transduction into human embryonic and induced pluripotent stem cells." Cell Reprogram 12(5): 501-507.

Thomas, K. R. and M. R. Capecchi (1987). "Site-directed mutagenesis by gene targeting in mouse embryo-derived stem cells." Cell 51(3): 503-512.

Thompson, S. et al. (1989). "Germ line transmission and expression of a corrected HPRT gene produced by gene targeting in embryonic stem cells." Cell 56(2): 313-321.

Thomson, J. A. et al. (1998). "Embryonic stem cell lines derived from human blastocysts." Science 282(5391): 1145-1147.

Vallier, L. et al. (2005). "Activin/Nodal and FGF pathways cooperate to maintain pluripotency of human embryonic stem cells." J Cell Sci 118(Pt 19): 4495-4509.

Vega, F. M. et al. (2011). "RhoA and RhoC have distinct roles in migration and invasion by acting through different targets." J Cell Biol 193(4): 655-665.

Watanabe, K. et al. (2007). "A ROCK inhibitor permits survival of dissociated human embryonic stem cells." Nat Biotechnol 25(6): 681-686.

Wickman, G. R. et al. (2010). The Rho-Regulated ROCK Kinases in Cancer. The Rho GTPases in Cancer. K. L. van Golen, Springer New York: 163-192.

Williams, R. L. et al. (1988). "Myeloid leukaemia inhibitory factor maintains the developmental potential of embryonic stem cells." Nature 336(6200): 684-687.

Wobus, A. M. et al. (1991). "Pluripotent mouse embryonic stem cells are able to differentiate into cardiomyocytes expressing chronotropic responses to adrenergic and cholinergic agents and Ca2+ channel blockers." Differentiation; research in biological diversity 48(3): 173-182.

Wu, X. et al. (2007). "Cdc42 is crucial for the establishment of epithelial polarity during early mammalian development." Dev Dyn 236(10): 2767-2778.

Xu, R. H. et al. (2005). "Basic FGF and suppression of BMP signaling sustain undifferentiated proliferation of human ES cells." Nat Methods 2(3): 185-190.

Xu, Y. et al. (2010). "Revealing a core signaling regulatory mechanism for pluripotent stem cell survival and self-renewal by small molecules." Proc Natl Acad Sci U S A 107(18): 8129-8134.

Yeom, Y. I. et al. (1996). "Germline regulatory element of Oct-4 specific for the totipotent cycle of embryonal cells." Development 122(3): 881-894.

Yu, J. et al. (2007). "Induced pluripotent stem cell lines derived from human somatic cells." Science 318(5858): 1917-1920.

Yuan, H. et al. (1995). "Developmental-specific activity of the FGF-4 enhancer requires the synergistic action of Sox2 and Oct-3." Genes Dev 9(21): 2635-2645.

19

Cripto-1: At the Crossroads of Embryonic Stem Cells and Cancer

Nadia Pereira Castro, Maria Cristina Rangel, Tadahiro Nagaoka, Hideaki Karasawa, David S. Salomon and Caterina Bianco
Mammary Biology and Tumorigenesis Laboratory, National Cancer Institute, National Institutes of Health, Bethesda, USA

1. Introduction

Human Cripto-1 is a member of the Epidermal Growth Factor-Cripto-FRL-1-cryptic (EGF-CFC) family of peptides (Bianco et al., 2010; de Castro et al., 2010). During early vertebrate development, Cripto-1 functions as a co-receptor for transforming growth factor β (TGF-β) ligands, such as Nodal and growth and differentiation factor-1 and -3 (GDF-1 and GDF-3), through an heteromeric complex composed of Activin type II and type I (ALK4) serine threonine kinase receptors in the plasma membrane. Genetic studies in zebrafish and mice have demonstrated that Cripto-1/Nodal signaling is essential for the formation of the primitive streak, patterning of the anterior/posterior (A/P) axis, specification of mesoderm and endoderm and establishment of left/right (L/R) asymmetry (Bianco et al., 2010; de Castro et al., 2010). In adult tissues, Cripto-1 is expressed at low levels in all stages of mammary gland development and its expression increases during pregnancy and lactation. Overexpression of Cripto-1 in mouse mammary epithelial cells leads to their transformation *in vitro* and can enhance migration, invasion, branching morphogenesis and epithelial to mesenchymal transition (EMT) (Bianco et al., 2010; de Castro et al., 2010). Furthermore, transgenic mouse studies have shown that overexpression of a human Cripto-1 transgene in the mouse mammary gland under the transcriptional control of the mouse mammary tumor virus (MMTV) promoter results in mammary hyperplasias and papillary adenocarcinomas (Wechselberger et al., 2005). Cripto-1 is also expressed at high levels in approximately 50-80% of different types of human carcinomas, including breast, colon, pancreas, lung and ovary (Bianco et al., 2010; de Castro et al., 2010). Cripto-1 is therefore an example of an embryonic gene that plays an important role not only during embryogenesis, but is also implicated in malignant progression. Furthermore, Cripto-1 signaling might represent a common signaling pathway shared by embryonic stem cells and cancer cells. Recent studies have shown that Cripto-1 is an important component of critical core pathways that regulate stem cell self-renewal and differentiation (Hough et al., 2009). Chromatin immunoprecipitation (ChiP) analysis has revealed the presence of Oct-4 and Nanog binding sites in the Cripto-1 promoter region, suggesting that Cripto-1 is a direct transcriptional target of Oct-4 and Nanog transcription factors (Loh et al., 2006). More importantly, several signaling pathways that are critical for early embryonic development and regulate stem cell proliferation and differentiation have been shown to cross-talk with Cripto-1 (Bianco et al., 2010). Examples of embryonic signaling pathways that interact with Cripto-1 are the Wnt-β/catenin, the Notch, hypoxia and TGF-β signaling pathways. Interestingly, Cripto-1 and other embryonic genes that regulate stem cell function are also overexpressed in human tumors, confirming this connection between stem cells and cancer. Recently, Cripto-1 has been shown to be highly expressed in a small subset of stem-like cells in human embryonal carcinoma cells, in human malignant melanomas and in androgen-responsive and refractory human prostate cancer cells (Cocciadiferro et al., 2009; Strizzi et al., 2008; Watanabe et al., 2010). Therefore, Cripto-1 signaling pathway may represent an attractive target for treatment in cancer, because Cripto-1 targeting will eliminate not only differentiated cancer cells but also might target an undifferentiated subpopulation of tumor cells that exhibit stem-like characteristics, thereby leading to eradication of the tumor. In this review we will discuss the dual role of Cripto-1 as embryonic gene in the regulation of stem cell self-renewal and differentiation and as oncogene with re-expression in human cancers.

2. The EGF-CFC family

Human Cripto-1 is a member of the EGF-CFC family of peptides identified only in vertebrates, and plays an important role in embryonic development and in tumor progression. The EGF-CFC family includes monkey Cripto-1, mouse Cripto-1 (Cr-1=cfc2), chicken Cripto-1, zebrafish one-eyed pinhead (*oep*), *Xenopus* XCR1/FRL-1, XCR2 and XCR3, mouse cryptic (Cfc1) and human Cryptic (CFC1) (Bianco et al., 2005b). Structurally, these proteins contain an NH2-terminal signal peptide, a modified EGF-like domain, a CFC cysteine-rich domain and a short hydrophobic COOH-terminus containing short sequences for glycosylphosphatidylinositol (GPI) cleavage and attachment (Figure 1).

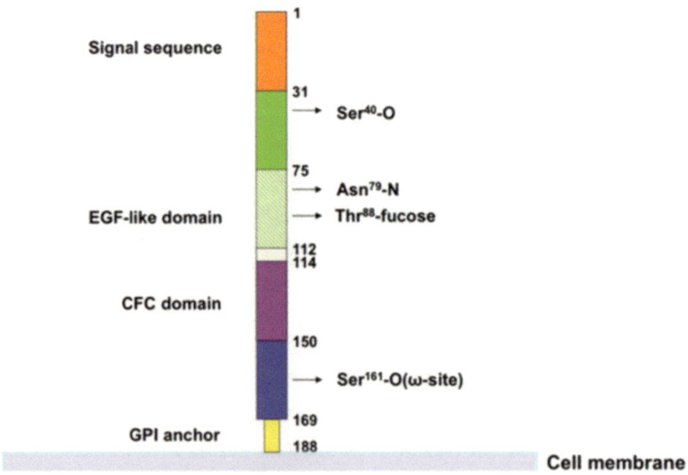

Fig. 1. Structure of the human Cripto-1 protein.

Cripto-1 contains several glycosylation sites and the residue Threonin 88 within the EGF-like domain modulates Cripto-1 ability to activate a Nodal-dependent signaling pathway. Cripto-1 can be cleaved from the cell membrane and can be released in the supernatant of cells by activity of the enzyme GPI-phospholipase D (GPI-PLD). EGF: epidermal growth factor, CFC: Cripto-FRL-1-Cryptic, GPI: glycosylphosphatidylinositol, Ser: Serine, Asn: Asparagine, Thr: Threonine.

GPI anchoring determines membrane localization of Cripto-1 within lipid rafts microdomains and is required for the paracrine activity of Cripto-1 as a Nodal co-receptor (Bianco et al., 2008; Watanabe et al., 2007b). Removal of the GPI anchor by GPI-PLD generates a soluble form of Cripto-1, which can therefore function both as a cell membrane anchored protein or as a soluble protein (Watanabe et al., 2007a; Watanabe et al., 2007b). In fact, soluble forms of Cripto-1 are biologically active in a number of different *in vitro* and *in vivo* assays (Minchiotti et al., 2001; Xu et al., 1999; Yan et al., 2002). Furthermore, Cripto-1 protein has several glycosyl modification sites, including O-linked glycosylation at Ser40 and Ser161 (ω-site for GPI-attachment), N-linked glycosylation at Asn79 and O-linked fucosylation at Thr88 (Figure 1) (Minchiotti et al., 2000; Schiffer et al., 2001; Shi et al., 2007; Watanabe et al., 2007a; Watanabe et al., 2007b). Among them, O-linked fucosylation of EGF-CFC proteins is required for their ability to function as co-receptors for the TGF-β-related protein Nodal (Schiffer et al., 2001). However, Cripto-1 O-fucosylation mutants are fully functional with regard to activation of Nodal independent signaling pathways (Bianco et al., 2008). Another study has demonstrated that is the Thr88 residue and not fucosylation of this residue that is necessary for Cripto-1 to function as a Nodal co-receptor (Schiffer et al., 2001; Shi et al., 2007).

3. Cripto-1 during embryonic development

In the embryonic development EGF-CFC proteins function as co-receptors for the TGF-β ligands Nodal, GDF-1 and GDF-3 (Bianco et al., 2010; de Castro et al., 2010 as cited in Yeo et al. 2001; Andersson at al. 2007). Genetic studies in zebrafish and mice have defined an essential role for Nodal that functions through *oep*/Cripto-1 in the formation of the primitive streak, patterning of the A/P axis, and specification of mesoderm and endoderm (mesoendoderm) (Bianco et al., 2010; de Castro et al., 2010 as cited in Chu et al. 2005). In later stages of embryonic development the EGF-CFC ortholog of Cripto-1, Cryptic, regulates the establishment of L/R axis, allowing asymmetric development of visceral organs (Bianco et al., 2010 as cited in Yan et al., 1999). Biochemical studies have demonstrated that EGF-CFC proteins bind directly to Nodal, GDF-1 or GDF-3 and the Activin type I receptor ALK4 (ActRIB), recruiting the Activin type II receptor and inducing activation and phosphorylation of Smad-2/Smad-3 signaling pathway (Bianco et al., 2010) (Figure 2). However, Cripto-1 can also regulate A/P axis specification independently of Nodal signaling (D'Andrea et al., 2008). In fact, Cripto-1$^{F78A/F78A}$ mouse embryos carrying the amino acid substitution F78A are unable to activate a Nodal signaling pathway but clearly establish an A/P axis and initiate germ layer formation and gastrulation movements (D'Andrea et al., 2008). During early mouse embryogenesis, Cripto-1 cannot be detected until prior to the gastrulation stage in the inner cell mass and in extra-embryonic trophoblast cells at day 4 of development. Increase in Cripto-1 expression is observed on day 6.5 of embryonic development when Cripto-1 is found in the primitive streak within epiblast cells undergoing EMT, which eventually give rise to mesoderm and endoderm (Bianco et al., 2010; de Castro et al., 2010). Cripto-1 is also detected in the ectoplacental cone at this stage. Cripto-1 expression then decreases on day 7 when it is detected mostly in the truncus arterious of the developing heart.

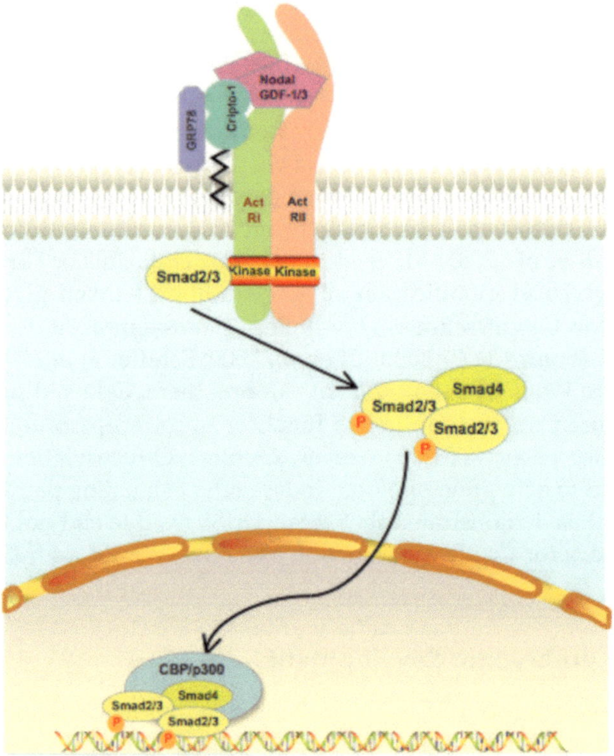

Fig. 2. Cripto-1/Nodal dependent signaling pathway.

Cripto-1 is a co-receptor for Nodal, GDF-1 and GDF-3, allowing them to interact with ALK4 (ActRIB). The Cripto-1/Nodal/ALK4/ActRII receptor complex triggers activation and phosphorylation of Smad-2 and Smad-3. Phosphorylated Smad-2 and Smad-3 form a complex with Smad-4 and they translocate into the nucleus. In the nucleus, Smad-2/-3/-4 complex interacts with CREB binding protein (CBP)/p300 and activates transcription of specific target genes. The heat shock protein GRP78 can also enhance Cripto-1/Nodal-dependent signaling pathway, as discussed later in the text. GDF-1/-3: growth and differentiation factor-1/3, GRP78: glucose-regulated protein 78.

With the exception of the developing heart, little if any expression of Cripto-1 can be detected in the embryo after day 8 (Bianco et al., 2010; de Castro et al., 2010 as cited in Dono et al., 1993; Minchiotti et al., 2000). Disruption of Cripto-1 in Cripto-1 $^{-/-}$ embryos is embryonically lethal and results in the formation of embryos that possess a head without a trunk, demonstrating that there is a severe deficiency in embryonic mesoderm and endoderm without a loss of anterior neuroectoderm formation (de Castro et al., 2010 as cited in Ding et al. 1998). Initiation of the primitive heart tube in Cripto-1 $^{-/-}$ mice is severely inhibited due to failure in the development of functional beating cardiomyocytes, as demonstrated by the absence of expression of terminal myocardial differentiation genes (de Castro et al., 2010 as cited in Xu et al., 1998). Cripto-1 $^{-/-}$ embryonic bodies (EB) derived from Cripto-1 $^{-/-}$ embryonic stem (ES) cells fail to form beating cardiomyocytes, while they differentiate into neuronal cells. Addition of a Cripto-1 recombinant protein to Cripto-1 $^{-/-}$

EBs during early time points (0–2 days) of differentiation is able to rescue cardiomyocyte differentiation. However, addition of Cripto-1 recombinant protein during later stages of differentiation fails to rescue cardiamyocyte differentiation, suggesting that Cripto-1 ability to promote cardiac differentiation of EB is restricted to an early window of this differentiation program (Minchiotti, 2005, as cited in Parisi et al., 2003). Interestingly, a microarray study revealed that Cripto-1 $^{-/-}$ ES cells had a reduced mRNA expression of the G protein coupled receptor APJ (also known as angiotensin type I-like receptor) and its ligand Apelin, as compared to control wild type ES cells. Gain of function experiments showed that APJ redirects the neural fate of Cripto-1 $^{-/-}$ ES cells and restores the cardiogenic program. Furthermore, comparison of Cripto-1, APJ and Apelin expression profile in mouse embryo by *in situ* hybridization revealed that expression of Apelin and APJ correlates with Cripto-1 expression. In fact, Apelin mRNA was clearly expressed in the developing primitive streak, whereas APJ was expressed in the primitive streak and adjacent mesoderm, resembling Cripto-1 expression pattern (D'Aniello et al., 2009). Therefore, APJ and Apelin are downstream targets of Cripto-1 signaling pathway and together with Cripto-1 they drive ES cells toward a cardiac lineage.

3.1 Cripto-1 in embryonic stem cells

Stem cells have the capacity to divide for an undetermined period of time and a potential to develop into many different cell types throughout early life and growth. Stem cells are distinguished from other cell types by two important characteristics. First, they possess the capability to differentiate into mature cells of any particular tissue (pluripotency). Second, they can undergo through numerous cell cycle divisions while maintaining their undifferentiated state (self-renewal). ES cells can be isolated from a 3- to 5-day-old embryo, called blastocyst, and have the potential to give rise to all specialized tissues and organs of a mature organism. Adult stem cells are found in various adult tissues, and function as a reservoir for cells that are lost during injury or disease (Bendall et al., 2008). Mouse embryonic stem cells (mES) or human embryonic stem cells (hES) have been very useful in the field of stem cell research. Comparison of gene expression profiles across species has shown that mouse and human ES cells share common highly conserved signaling pathways that regulate self-renewal and pluripotency, including the Cripto-1/Nodal signaling pathway. For example, in addition to Cripto-1, genes such as Oct-4, Lefty, Nodal, Sox-2, Utf-1 (undifferentiated embryonic cell transcription factor-1) and Tert (telomerase reverse transcriptase) are highly enriched in both mouse and human ES cells (Wei et al., 2005). Additionally, Cripto-1 has been identified as a pluripotency marker also in primate ES cells together with Oct-4, Nanog, Sox-2, Tert, LeftyA, and Rex-1 (Chang et al., 2010). In 2007 a comparative study of a large and diverse set of hES cell lines assessed the expression pattern of commonly used stem cell markers. All the hES cells analyzed exhibited similar expression profile for several stem cell markers, including the glycolipid stage specific embryonic antigens SSEA3 and SSEA4, the keratan sulfate antigens TRA-1-60, TRA-1-81, and the developmentally regulated genes including Nanog, Oct-4, Dnmt3b, Gabrb3, GDF-3 and Cripto-1 (Adewumi et al., 2007; Bianco et al., 2010; de Castro et al., 2010). Finally, Cripto-1 has been reported as a direct target gene of stem cell transcription factors (Bianco et al., 2010; de Castro et al., 2010; Hough et al., 2009). For instance, using the ChIP paired-end ditags method, Loh and collaborators mapped the binding sites of the transcription factors Oct-4 and Nanog in the mouse ES cell genome. Cripto-1 promoter was found to include Oct-4 and

Nanog binding sites, suggesting that key modulators of stem cell self-renewal and pluripotentiality directly regulate Cripto-1 expression in ES cells (Bianco et al., 2010; de Castro et al., 2010; Loh et al., 2006).

3.1.1 Cross-talk of Cripto-1 with stem cell signature pathways

Several signaling pathways regulate in a coordinate fashion early embryonic development and stem cell proliferation, maintenance and differentiation. Some of these signaling cascades have been shown to cross-talk with Cripto-1 signaling, suggesting a pivotal role played by Cripto-1 in stem cell self-renewal and maintenance. Among these signaling pathways are genes in the Wnt/β-catenin signaling pathway, TGF-β family members, the Notch pathway, and hypoxia inducible factor-1 alpha (HIF-1α) (Bianco et al., 2010; de Castro et al., 2010). A schematic diagram of the cross-talk of Cripto-1/Nodal signaling with other stem cell signature signaling pathways is shown in figure 3.

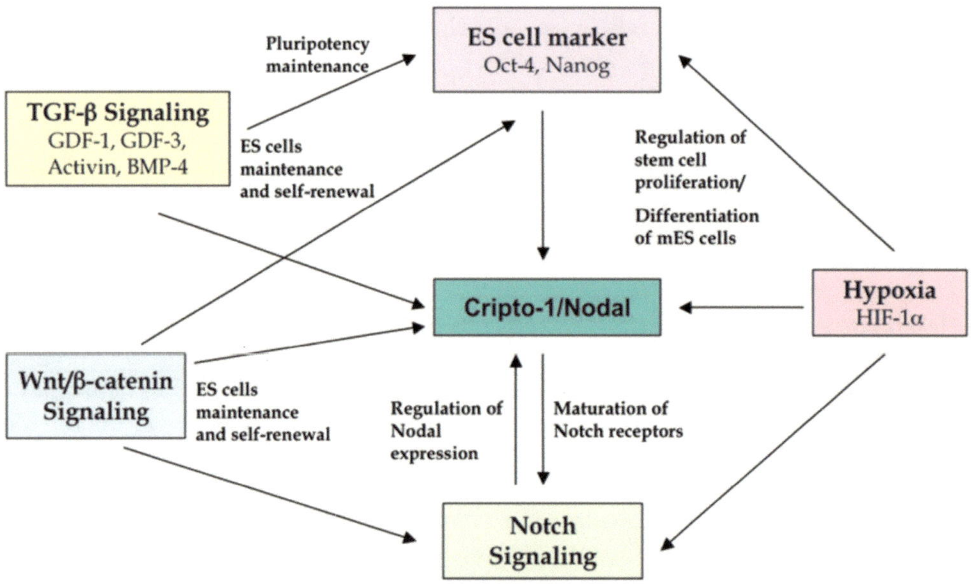

Fig. 3. Interplay of Cripto-1/Nodal signaling with other embryonic stem cell signaling pathways.

Cripto-1 signaling is a downstream target of Oct-4, Nanog, Wnt/β-catenin, Notch and Hypoxia/HIF-1α pathways. Cripto-1 also mediates signaling of TGF-β family members and enhances Notch signaling by facilitation of Notch receptor maturation. TGF-β: transforming growth factor-β, HIF-1α: hypoxia inducible factor-1α, GDF-1/-3: growth and differentiation factor-1/-3, BMP-4: bone morphogenetic protein-4, ES cells: embryonic stem cells, mES cells: mouse embryonic stem cells.

Activation of the canonical Wnt/β-catenin pathway has been reported to sustain the undifferentiated state of ES cells by triggering transcription of genes that regulate ES cell pluripotency and self-renewal, such as Oct-4, Nanog, Sox-2 and Cripto-1. For instance, Cripto-1 has been identified as a direct target gene in the canonical Wnt/β-catenin signaling pathway during embryonic development and in colon carcinoma cell lines and tissues

(Morkel et al., 2003). Remarkably, in β-catenin $^{-/-}$ or Wnt3 $^{-/-}$ mutant mouse embryos, Cripto-1 expression was found to be dramatically downregulated, suggesting that the Wnt/β-catenin signaling pathway is a positive regulator of Cripto-1 expression during embryonic development (Bianco et al., 2010). In addition, in *Xenopus*, the Cripto-1 ortholog XCR1/FRL-1 functions as a co-receptor for Wnt11 and together with frizzled receptor 7 and Glypican-4 mediates stabilization and activation of β-catenin (Bianco et al., 2010). Several studies have clearly implicated Wnt/β-catenin signaling pathway in initiation and maintenance of carcinomas of the skin, intestine, liver, and brain (Bianco et al., 2010; de Castro et al., 2010). Hence, the interaction between the Wnt/β-catenin and Cripto-1 signaling pathways might be significant also in cancer initiation and progression. Activation of TGF-β/Activin/Nodal signaling has also been reported to be required for the maintenance of undifferentiated state in human ES cells (James et al., 2005; Vallier et al., 2009). Some TGF-β family members, including Nodal, GDF-1 and GDF-3, require Cripto-1 for signaling and they are critical for ES cell maintenance and self-renewal (Bianco et al., 2010). Furthermore, Activin, TGF-β and bone morphogenetic protein (BMP)-4 can directly regulate Cripto-1 expression by binding to smad binding elements (SBE) within the Cripto-1 promoter in human embryonal carcinoma cells and in human colon cancer cells (Mancino et al., 2008). GDF-3 and Cripto-1 have also been identified as ES cell markers that are enriched in a population of ES cells which are uncommitted and have high self-renewal capacity (Hough et al., 2009). Finally, Activin/TGF-β signaling activity is required for Nanog expression in epiblast cells, suggesting a key role played by TGF-β family members in maintaining ES pluripotency and self-renewal (Shin et al., 2011; Wang et al., 2009). The Notch pathway is a key regulator of many developmental processes during fetal and adult differentiation. Studies have reported that inappropriate activation of Notch signaling cascade contributes to tumorigenesis (Wilson & Radtke, 2006; Wang et al., 2009). Interaction between Notch and Cripto-1/Nodal signaling pathways has also been described. Analysis of the transcriptional regulatory regions within the mouse Nodal promoter has identified the presence of CBF1 binding elements, suggesting a direct regulation of Nodal expression by Notch signaling (Bianco et al., 2010; as cited in Raya et al., 2003; Krebs et al 2003). In this regard, mouse embryo double mutants for Notch1 and Notch2 exhibit multiple defects in L/R asymmetry, which is regulated by Nodal. Moreover, a new insight into Notch and Cripto-1/Nodal signaling pathways has been reported. Watanabe and colleagues (2009) using a yeast two-hybrid system screened a mouse embryo or human colon cDNA prey library for potential Cripto-1 binding partners. Six candidate genes were isolated, including the mouse Notch3. By coimmunoprecipitation analysis Cripto-1 was found to directly bind to all four mammalian Notch receptors. Surprisingly, binding of Cripto-1 to Notch1 occurred mainly inside the cells in the endoplasmic reticulum/Golgi complex, where Cripto-1 enhanced cleavage of the Notch1 extracellular domain through a furin-like convertase. Enhanced cleavage of Notch1 receptor by Cripto-1 potentiates Notch signaling as demonstrated by enhanced ligand-induced activation of Notch signaling in Chinese hamster ovary cells expressing a CBF1-dependent Notch reporter assay (Watanabe et al., 2009). Hypoxia is an important regulator of stem cell proliferation and differentiation (Yeung et al., 2011). In fact, tissue stem cells reside within niches that are naturally hypoxic and low oxygen levels prevent their differentiation toward specific fates. Moreover, HIF-2α can directly regulate Oct-4 expression in ES cells by binding to hypoxia-responsive elements (HREs) within the promoter of mouse Oct-4. HIF-1α interacts with Notch signaling by binding to Notch intracellular domain and enhancing transcription of Notch downstream target genes Hes-1 and Hey-2. HIF-1α, in addition to regulating Notch1 signaling, can also regulate Cripto-1 expression by binding to HREs within the promoter region of mouse and human Cripto-1 gene in mouse ES cells and in human embryonal carcinoma cells (Bianco et al., 2009). In addition, hypoxic conditions enhanced differentiation of mES cells into beating cardiomyocytes when compared with mES cells grown under normoxic conditions. However, expression of Cripto-1 in mES was critical, because hypoxia was unable to induce differentiation of mES cells into cardiomyocytes in the absence of Cripto-1 expression.

4. Cripto-1 in cancer

Similarities between embryonic development and cellular transformation during oncogenesis have led to the identification of common signaling pathways, suggesting that reactivation of developmental signaling pathways might drive cell transformation and tumor progression in adult tissues (Bianco et al., 2010). Cripto-1 is a typical example of an embryonic gene that is re-expressed in human tumors, promoting cellular proliferation, migration, and tumor angiogenesis (Figure 4).

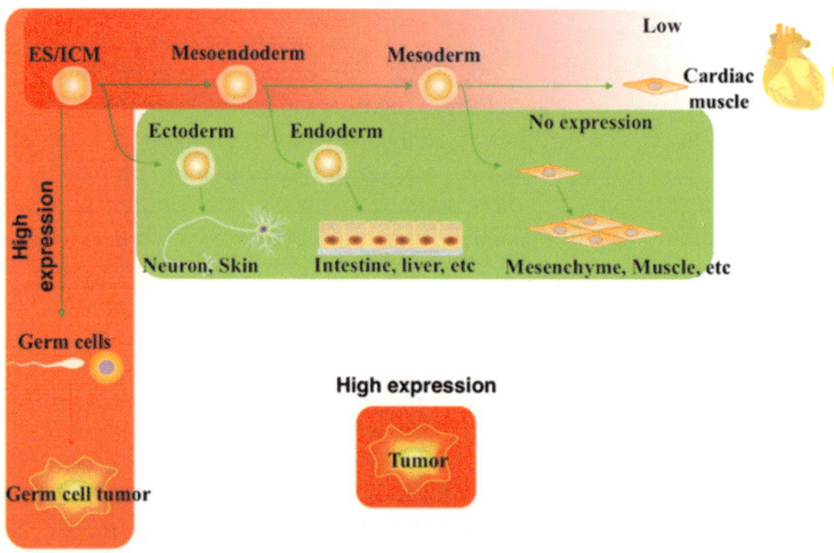

Fig. 4. Expression of Cripto-1 during embryogenesis and tumorigenesis.

Cripto-1 is highly expressed in undifferentiated embryonic stem cells and germ cells. Cripto-1 is important for mesoendoderm differentiation of ES cells and its expression is lost upon differentiation of ES cells toward the three germ cell layers. In the adult, Cripto-1 is re-expressed by tumor cells. Green color: no Cripto-1 expression, red color: Cripto-1 expression. ICM: Inner cell mass, ES: embryonic stem cells.

4.1 Cripto-1 oncogenic activities *in vitro* and *in vivo*

The first evidence of Cripto-1 oncogenic activity derives from studies demonstrating that Cripto-1 overexpression can induce *in vitro* transformation of a variety of cells, including NIH-3T3 fibroblasts and NOG-8 or CID9 mouse mammary epithelial cells (Ciardiello et al.,

1991; Ciccodicola et al., 1989; Niemeyer et al., 1998). Furthermore, Cripto-1 can enhance migration and invasion of a variety of normal mammary epithelial cells and human breast and cervical carcinoma cells, suggesting the involvement of Cripto-1 in tumor progression (Ebert et al., 2000; Normanno et al., 2004a; Wechselberger et al., 2001). Cripto-1 can also function as a potent angiogenic protein both *in vitro* and *in vivo*, enhancing the proliferation, migration and invasion of human umbilical vascular endothelial cells and stimulating their differentiation into vascular-like structures in matrigel. Likewise, overexpression of Cripto-1 promotes *in vivo* neovascularization of MCF-7 xenografts, supporting a role for Cripto-1 in modulating tumor angiogenesis (Bianco et al., 2005a). Since HIF-1α can enhance Cripto-1 expression by directly binding to the Cripto-1 promoter, it is possible that the hypoxic microenvironment within the growing tumor might enhance Cripto-1 expression, which in turn induces new vessel formation to sustain tumor growth (Bianco et al., 2009). Cripto-1 has also been directly implicated in mouse mammary tumor development (Sun et al., 2005; Wechselberger et al., 2005), as demonstrated by transgenic mouse models overexpressing the human Cripto-1 transgene in mouse mammary glands under the control of the MMTV or whey acidic protein (WAP) promoter (Sun et al., 2005; Wechselberger et al., 2005). The majority of nulliparous MMTV-Cripto-1 transgenic mice exhibited enhanced ductal branching, intraductal hyperplasias and hyperplastic alveolar nodules, and about 30-40% of aged multiparous female mice developed multifocal hyperplasias and papillary adenocarcinomas (Wechselberger et al., 2005). Interestingly, under a 2 years observation period, approximately 20% of female nulliparous or multiparous MMTV-Cripto-1 transgenic mice developed uterine leiomyosarcomas as compared to control mice. High levels of Cripto-1 transgene, phosphorylated forms of c-Src, AKT, glycogen synthase kinase-3β (GSK-3β) and dephosphorylated active-β-catenin were detected in the uterine tumors collected from MMTV-Cripto-1 transgenic mice (Strizzi et al., 2007), suggesting a role for Cripto-1 during uterine tumorigenesis either by activation of c-Src and Akt and/or via cross-talk with the canonical Wnt/β-catenin signaling pathway. Unlike the MMTV promoter, the WAP promoter is maximally active during mid-pregnancy and lactation. Approximately 50% of old nulliparous WAP-Cripto-1 mice developed multifocal intraductal hyperplasias, and more than 50% of multiparous WAP-Cripto-1 female mice developed multifocal mammary tumors of mixed histological subtypes (Sun et al., 2005). Histological analysis of the WAP-Cripto-1 mammary tumors revealed the presence of tumors containing glandular, papillary and undifferentiated carcinoma, as well as myoepithelioma and adeno-squamous carcinoma. Mammary tumors with mixed histology have also been described in MMTV-Wnt-1 transgenic mice, which have alterations in the canonical Wnt/β-catenin pathway. Interestingly, hyperactivation of a canonical Wnt/β-catenin pathway was detected in WAP-Cripto-1 mammary tumors, suggesting that the canonical Wnt/β-catenin pathway together with Cripto-1 might play an important role during mammary transformation *in vivo* (Miyoshi et al., 2002a; Miyoshi et al., 2002b; Sun et al., 2005). Finally, several studies have demonstrated that Cripto-1, in addition to function as an oncogene *in vitro* and in animal models, is overexpressed at the mRNA and protein level in a wide variety of human tumors, including colorectal, breast, gastric, pancreatic, ovarian, endometrium, nasopharynx and lung carcinomas, while very little or no expression is detected in normal adult tissues. Therefore, the selective expression of Cripto-1 in cancer cells suggests that it represents a promising target in cancer therapy. For an extensive review on Cripto-1 expression in human tumors see de Castro et al., 2010.

4.1.1 Intracellular signaling pathways activated by Cripto-1 during oncogenic transformation

While Cripto-1 functions mostly in a Nodal-dependent manner during embryogenesis, several studies have demonstrated that Cripto-1 induces cellular proliferation, motility, survival and EMT in a Nodal-independent fashion. Following binding to the GPI-linked heparan sulphate proteoglycan Glypican-1, Cripto-1 induces activation and phosphorylation of the cytoplasmic tyrosine kinase c-Src, which in turn activates mitogen-activated protein kinase (MAPK)/Phosphatydil inositol 3′ kinase (PI3K)/Akt signaling pathways (Figure 5) (Bianco et al., 2005b).

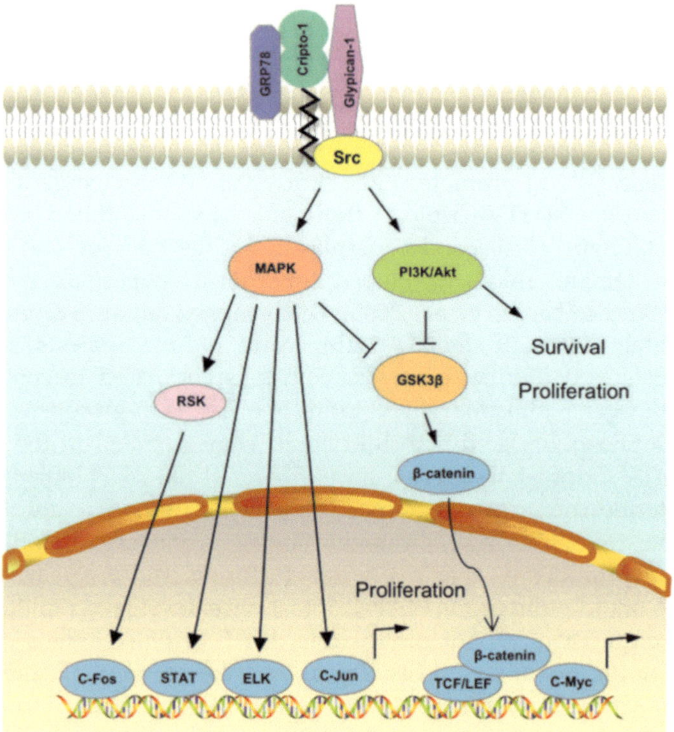

Fig. 5. Nodal-independent Cripto-1 signaling pathway.

Cripto-1 upon binding to Glypican-1 activates MAPK and Akt signaling pathways during tumor progression, enhancing cell proliferation and survival. MAPK and PI3K/Akt pathways can also inhibit GSK-3β leading to activation and stabilization of β-catenin. GRP78 can also enhance Cripto-1 activation of the MAPK/Akt signaling pathways. GRP78: glucose-regulated protein 78, MAPK: mitogen-activated protein kinase, PI3K: phosphatydil inositol 3′ kinase, GSK3β : glycogen synthase kinase 3β, TCF/LEF: T-cell factor/lymphoid enhancer factor, RSK: Ribosomal s6 kinase, STAT: signal transducers and activators of transcription.

Activation of the MAPK and PI3K/Akt signaling pathways by Cripto-1 is independent of Nodal and ALK4, since Cripto-1 can enhance phosphorylation of MAPK and Akt in cells lacking ALK4 and/or Nodal expression (Bianco et al., 2002). Furthermore, the tyrosine kinase c-Src is required by Cripto-1 to induce *in vitro* transformation of mouse mammary epithelial cells, indicating that inappropriate activation of c-Src by Cripto-1 in a Nodal- and

ALK4-independent manner might play a key role in Cripto-1 tumorigenic activity (Bianco et al., 2003). The interaction between Cripto-1 and Gypican-1 occurs within lipid raft microdomains, regions of the cell membrane that are enriched in GPI-linked proteins (Bianco et al., 2003). Cripto-1 mitogenic signaling is also dependent upon binding to GRP78, an endoplasmic reticulum (ER) heat shock chaperone protein that is expressed on the cell surface of tumor cells (Shani et al., 2008). Indeed, disruption of the Cripto-1/GRP78 complex at the cell membrane interferes with Cripto-1 oncogenic activity *in vitro*, preventing Cripto-1 activation of MAPK/Akt signaling pathway (Figure 5). Interestingly, unlike Glypican-1, GRP78 is also important for Nodal-dependent Cripto-1 signaling, enhancing activation of the Smad-2/Smad-3 signaling pathway (Kelber et al., 2009). Furthermore, GRP78 enhanced Cripto-1-dependent activation of the c-Src/MAPK/Akt signaling pathways, increased cellular proliferation and reduced cell adhesion, as evidenced by a decrease in E-cadherin expression in breast cancer cells (Kelber et al., 2009).

5. Cripto-1 and Cancer Stem Cells

Cancer stem cells (CSCs), also known as tumor initiating cells, share characteristics associated with embryonic stem cells, specifically the ability to give rise to all cell types within a particular tumor tissue. CSCs were first identified in the hematopoietic system and later they have also been reported in solid cancers, including cancers of the breast, lung, prostate, colon, brain, head and neck, and pancreas (Bianco et al., 2010; de Castro et al., 2010). CSCs represent a distinct population of cancer cells with innate chemo- and radio-resistance and therefore are responsible of tumor relapse (Bianco et al., 2010; de Castro et al., 2010 as cited in Huntly and Gilliland 2010). Moreover, CSCs are capable to self-renew and regenerate the original phenotype of the tumor when implanted into immunodeficient mice (Visvader & Lindeman, 2008). Similarities between embryonic development and cell transformation during oncogenesis have led to the identification of common contributing pathways, suggesting that reactivation of developmental signaling pathways might drive cell transformation and tumor progression in adult tissues (Bianco et al., 2010). Cripto-1 is a typical example of a common gene shared by embryonic cells and cancer cells contributing to early embryogenesis and cancer progression. More importantly, Cripto-1 is enriched in a subpopulation of cancer cells with stem-like characteristics. For instance, Watanabe and collaborators (2010) described a heterogeneous Cripto-1 expression pattern in human embryonal carcinoma (EC) with segregation of these cells into two distinct populations portraying high and low Cripto-1 expression. EC cells are pluripotent stem cells derived from germ cell teratocarcinomas and they represent the malignant counterparts of human ES cells. Interestingly, these two subpopulations showed different gene expression profiles and different *in vitro* and *in vivo* tumorigenic capability. The Cripto-1 high subpopulation of EC cells formed tumor spheres in a serum-free suspension culture with an efficiency significantly higher than the Cripto-1 low expressing EC cells. Furthermore, the Cripto-1 high expressing EC cells were able to generate tumors larger in size and with a shorter tumor latency period compared with tumors derived from Cripto-1 low expressing EC cells when injected subcutaneously into nude mice. Finally, regulators of pluripotent ES cells, such as Activin/Nodal signaling and the transcription factors Nanog and Oct-4, positively regulated Cripto-1 expression in Cripto-1 high expressing EC cells, suggesting the existence of a core transcriptional regulatory network of pluripotent stem cell transcription factors that cross-regulate each other expression (Loh et al., 2006). However, Cripto-1 expression is dispensable for pluripotent stem cell gene expression in EC cells, since siRNA-mediated knockdown of Cripto-1 has no effect on the expression of ES-related genes including Oct-4, Nanog, Sox-2 and Lefty. In another study, Cripto-1 has been exploited as a potential melanoma stem cells candidate marker in human malignant melanomas (Strizzi et al., 2008). Strizzi and colleagues, by FACS cell sorting based on Cripto-1 expression, isolated from a human melanoma cell line a subpopulation with stem-like characteristics. The Cripto-1 positive subpopulation of melanoma cells represented a relatively smaller size population with a more spindle-shaped morphology, and showed increased expression of the stem cell associated transcription factors, Oct-4 and Nanog, as compared to the parental melanoma cells. Finally, Cripto-1 and the stem cell markers Oct-4 and SUZ-12 were identified in a small subpopulation of stem-like cells in hormone-responsive and nonresponsive prostate tumor cell (Cocciadiferro et al., 2009). Altogether, these findings suggest that Cripto-1 might be useful in the identification of a subpopulation of cancer cells with stem like behavior that are resistant to conventional therapy and are therefore a major obstacle in the clinical treatment of cancer patients.

5.1 Cancer stem cells and EMT

During embryogenesis, tumor progression and metastasis, epithelial cells undergo dramatic morphological changes, acquiring mesenchymal properties in a process known as EMT. In the embryo, Cripto-1 is detected at high levels in epiblast cells undergoing EMT, which migrate through the primitive streak, eventually giving rise to the mesoderm and endoderm (Bianco et al., 2005b; Strizzi et al., 2005). In the tumor, the expression of EMT regulators at the tumor periphery is critical for tumor cells to acquire a mesenchymal phenotype that allow them to locally invade and escape from the primary tumor site, leading to the establishment of metastatic lesions (Micalizzi et al., 2010a). Cripto-1 is involved in tumor epithelial cells plasticity and may be an important EMT regulator together with Snail, Slug, Twist, and Six1 (Micalizzi et al., 2010b). It has been shown that mammary gland hyperplasias and tumors derived from MMTV-Cripto-1 transgenic mice express molecular markers and signaling molecules characteristics of EMT, suggesting that Cripto-1 might play an important role in facilitating migration and invasion of tumor cells (Strizzi et al., 2004). These findings might be significant since emerging evidence has suggested a link between stem cells and EMT (Hollier et al., 2009). In fact, EMT induction in immortalized human mammary epithelial cells resulted in the expression of stem cell markers and increased ability to form mammospheres *in vitro*, suggesting an important role for EMT in generating cancer stem-like cells in human breast tumors (Mani et al., 2008). Since Cripto-1 has been found to promote EMT in mouse mammary epithelial cells and mouse mammary tumors, contributing to a more aggressive mesenchymal phenotype, it is possible that Cripto-1 might support self-renewal, invasiveness and metastatic abilities of breast cancer stem-like cells through induction of an EMT program (Strizzi et al., 2004).

6. Cripto-1 as target for cancer therapy

High expression of Cripto-1 in human carcinomas and its enrichment in a stem-like cancer cell subpopulation strongly support Cripto-1 as a promising candidate for therapeutic intervention in cancer. Two different therapeutic approaches have been successfully used to impair Cripto-1 activity in cancer cells, including anti-Cripto-1 antisense (AS) oligonucleotides and neutralizing monoclonal antibodies (Adkins et al., 2003; Hu et al., 2007;

Normanno et al., 1996; Normanno et al., 2004b). For instance, a significant growth inhibition *in vitro* has been observed in ovarian, breast and colon cancer cells treated with anti-Cripto-1 AS oligonucleotides. A synergistic growth inhibitory effect was detected when anti-Cripto-1 AS oligonucleotides were combined together with oligonucleotides against TGF-α and amphiregulin (AREG), indicating that a variety of growth factors might cooperate in stimulating cancer cell proliferation (Casamassimi et al., 2000; De Luca et al., 2000; De Luca et al., 1999; Normanno et al., 2004b). Additionally, combination of anti-Cripto-1 AS oligonucleotides with anti-TGF-α and anti-AREG AS oligonucleotides was more effective *in vivo* as compared to single AS oligonucleotide administration, inhibiting the growth of colon cancer tumor xenografts in nude mice in a dose dependent manner (De Luca et al. 2000). Neutralizing antibodies that bind Cripto-1 have also been generated (Adkins et al., 2003; Hu & Xing, 2005). In particular, Adkins and colleagues of Biogen-Idec have generated a panel of monoclonal antibodies (mAbs) directed against a human recombinant Cripto-1 protein expressed as a human IgG1 Fc fusion protein (Adkins et al. 2003). The mAbs are capable to recognize different domains of the Cripto-1 protein, detecting its expression in human breast and colon tumor tissue samples and in human cancer cell lines. Among these anti-Cripto-1 mAbs, anti-CFC mAbs showed blocking activities *in vitro* and *in vivo* by interfering with the binding of Cripto-1 with ALK4 through the CFC domain. The anti-CFC blocking antibodies also blocked the interaction of Cripto-1 with Activin B (Adkins et al. 2003), restoring the growth inhibitory function of Activin B in tumor cells. Recently, among these anti-Cripto-1 mAbs, an N-terminal anti-Cripto-1 antibody has been selected to target Cripto-1 in human tumors, due to the higher binding affinity and range of reactivity of this antibody with tumor-derived Cripto-1 (Kelly et al., 2011). A humanized form of the anti-Cripto-1 N-terminal monoclonal antibody-conjugated to a DM4 toxin is currently been evaluated in phase I clinical trials for treatment of patients with refractory or relapsed Cripto-1 positive solid tumors (A phase I study of BIIB015 in relapsed/refractory solid tumors, protocol ID 207ST101/NCT00674947) (Bianco & Salomon, 2010). BIIB015 showed specificity for Cripto-1 *in vitro* and *in vivo* with a clinical response ranging between 50% of tumor inhibition to complete tumor regression in xenograft mouse models inoculated with lung (Calu-6), colon (CT-3), testicular (NCCIT) or breast (MDA-MB-231) human tumor cell lines (Kelly et al., 2011). In addition to mouse monoclonal antibodies against Cripto-1, anti-Cripto-1 rat monoclonal antibodies have also been generated. Rat anti-Cripto-1 monoclonal antibodies inhibited tumor cell growth and increased sensitivity of LS174T human colon cancer cells and MCF-7 human breast cancer cells to the cytotoxic effects of chemotherapeutic agents such as cisplatin, 5-Fluorouracil, carboplatin and epirubicin (Xing et al., 2004). Furthermore, a C13 rat monoclonal antibody inhibited cancer cell growth *in vivo* in LS174T tumor xenografts in SCID mice, inducing up to 80% growth inhibition of established tumors as compared to an untreated control groups. Interestingly, C4 and C13 rat monoclonal antibodies were also highly effective in reducing tumor growth of a human leukemia multidrug resistant (MDR) cell line (CEM/A7R) both *in vitro* and *in vivo* in an established xenograft tumor model (Hu et al., 2007).

6.1 Cripto-1 as a target for therapy in neurodegenerative and muscle degenerative diseases

Recent findings have demonstrated that Cripto-1 is a key player in the signaling pathway controlling neural induction in ES cells. Parisi and collaborators showed that Cripto-1 negatively regulated neuronal differentiation of ES cells (Parisi et al., 2003). Furthermore, disruption of Cripto-1 expression in mouse ES cells enhances neurogenesis and midbrain dopaminergic differentiation *in vitro* (Parish et al., 2005; Sonntag et al., 2005). These results suggest that blocking Cripto-1 expression and activity may provide a novel insight into the treatment of neurodegenerative diseases. Parkinson's disease is a progressive neurodegenerative disorder characterized by degeneration of dopaminergic neurons of the substantia nigra. Therefore methods that channel undifferentiated ES cells into dopaminergic neurons are under extensive evaluation, aiming to improve the safety of ES cells grafting that often results in uncontrolled proliferation and teratoma formation (Parish & Arenas, 2007). In a rat animal model of Parkinson's disease, mouse ES cells that are null for Cripto-1 expression (Cripto-1$^{-/-}$ ES cells), when grafted at low density into rats, differentiated into neuronal cells in the brain and were able to restore normal behavior without producing teratomas (Parish et al., 2005). Recently, a tetrameric tripeptide, which prevents Cripto-1/ALK4 interaction and interferes with Cripto-1 signaling was identified (Lonardo et al., 2010). The Cripto-1 blocking peptide was able to induce formation of neuroectoderm and increased midbrain dopaminergic neuron differentiation of mouse ES cells *in vitro* and *in vivo* (Lonardo et al., 2010). Moreover, mouse ES cells treated with Cripto-1 blocking peptide enhanced functional recovery and reduced tumor formation in Parkinsonian rats (Lonardo et al., 2010). Overexpression of Cripto-1 has also been observed in the cerebral cortical tissues of macaques that had been infected with a chimeric simian human immunodeficiency virus (SHIV) by cDNA array analysis (Stephens et al., 2006). SHIV enters the central nervous system (CNS) early after inoculation and cause encephalitis, characterized by transient meningitis and astrocytosis. Immunohistochemical analysis and RT-PCR confirmed widespread expression of Cripto-1 in the brains of SHIV-infected macaques. Although the functions of Cripto-1 in the neurons of the adult brain and the overexpression in the brains of SHIV-infected macaques are unclear, Cripto-1 may play a neuroinflammatory role and targeting Cripto-1 with various approaches, such as the tetrameric tripeptide described by Lonardo and colleagues, might inhibit progression of neurodegenerative disease in mammals. Cripto-1 may also be involved in degenerative muscle diseases, such as muscular dystrophy and spinal muscular atrophy. These diseases are characterized by skeletal muscle loss and few therapeutic approaches are available to restore the function of the lost muscle tissue (Tedesco et al., 2010). Cripto-1 is highly expressed in myoblast cells of regenerative muscles, whereas no expression of Cripto-1 is detected in normal muscle fibers. Furthermore, Cripto-1 induced a dose-dependent increase in proliferation and migration of myogenic precursor cells *in vitro* and enhanced skeletal muscle regeneration and angiogenesis after injury *in vivo* (Bianco & Salomon, 2010). These studies clearly suggest an involvement of Cripto-1 in neurodegenerative and muscle degenerative disease. Therefore, approaches that have been developed to target Cripto-1 in cancer might benefit also different diseases.

7. Conclusions

Critical signaling pathways are involved in modulating embryonic stem cell fate and behavior, maintaining a delicate balance between survival and self-renewal signals. Among these ES cell "signature" pathways, Cripto-1 is a critical gene that is used by ES cells. For instance, Cripto-1 is either a downstream target of ES transcription factors and/or signaling pathways or can modulate other ES cell signaling cascades (Fig. 3). Further, deregulation of stem cell self-renewal is probably a requirement for the initiation and formation of CSCs and therefore embryonic stem cell signature genes are also involved in cancer formation. Cripto-1 is indeed an embryonic gene that is re-expressed in an aberrant spatial and temporal manner in a variety of human tumors. Recent evidence has clearly demonstrated that Cripto-1 is expressed by a subset of cancer cells with stem-like characteristics (Watanabe et al., 2010). CSCs are considered to be a major obstacle in the complete eradication of tumors due their innate resistance to conventional therapy and therefore identification of surface markers that might discriminate CSCs from the bulk population of tumor cells is under active investigation. Therefore, Cripto-1 targeting in human tumors might have a major breakthrough in cancer therapy. Several approaches have been used to target Cripto-1 in cancer cells *in vitro* and *in vivo*, from antisense oligonucleotides to blocking monoclonal antibodies (Bianco & Salomon, 2010). Translation of Cripto-1 research to a clinical setting has been recently achieved. In fact, a humanized anti-Cripto-1 antibody conjugated with a potent cytotoxin is currently being evaluated in a Phase I clinical study in cancer patients. In conclusion, a complete understanding of Cripto-1 signaling in stem cell biology offers a great promise for improving stem cell mediated regenerative therapy as well as cancer therapy.

8. References

Adewumi, O., Aflatoonian, B., Ahrlund-Richter, L., Amit, M., Andrews, P.W., Beighton, G., Bello, P.A., Benvenisty, N., Berry, L.S., Bevan, S., Blum, B., Brooking, J., Chen, K.G., Choo, A.B., Churchill, G.A., Corbel, M., Damjanov, I., Draper, J.S., Dvorak, P., Emanuelsson, K., Fleck, R.A., Ford, A., Gertow, K., Gertsenstein, M., Gokhale, P.J., Hamilton, R.S., Hampl, A., Healy, L.E., Hovatta, O., Hyllner, J., Imreh, M.P., Itskovitz-Eldor, J., Jackson, J., Johnson, J.L., Jones, M., Kee, K., King, B.L., Knowles, B.B., Lako, M., Lebrin, F., Mallon, B.S., Manning, D., Mayshar, Y., McKay, R.D., Michalska, A.E., Mikkola, M., Mileikovsky, M., Minger, S.L., Moore, H.D., Mummery, C.L., Nagy, A., Nakatsuji, N., O'Brien, C.M., Oh, S.K., Olsson, C., Otonkoski, T., Park, K.Y., Passier, R., Patel, H., Patel, M., Pedersen, R., Pera, M.F., Piekarczyk, M.S., Pera, R.A., Reubinoff, B.E., Robins, A.J., Rossant, J., Rugg-Gunn, P., Schulz, T.C., Semb, H., Sherrer, E.S., Siemen, H., Stacey, G.N., Stojkovic, M., Suemori, H., Szatkiewicz, J., Turetsky, T., Tuuri, T., van den Brink, S., Vintersten, K., Vuoristo, S., Ward, D., Weaver, T.A., Young, L.A. & Zhang, W. (2007). Characterization of human embryonic stem cell lines by the International Stem Cell Initiative. *Nat Biotechnol*, Vol. 25, No.7, (Jul and 2007), pp. (803-816), 1087-0156 (Print)

Adkins, H.B., Bianco, C., Schiffer, S.G., Rayhorn, P., Zafari, M., Cheung, A.E., Orozco, O., Olson, D., De Luca, A., Chen, L.L., Miatkowski, K., Benjamin, C., Normanno, N., Williams, K.P., Jarpe, M., LePage, D., Salomon, D. & Sanicola, M. (2003). Antibody blockade of the Cripto CFC domain suppresses tumor cell growth in vivo. *J Clin Invest*, Vol. 112, No.4, (Aug and 2003), pp. (575-587)

Bendall, S.C., Stewart, M.H. & Bhatia, M. (2008). Human embryonic stem cells: lessons from stem cell niches in vivo. *Regen Med*, Vol. 3, No.3, (May and 2008), pp. (365-376), 1746-076X (Electronic)

Bianco, C., Adkins, H.B., Wechselberger, C., Seno, M., Normanno, N., De Luca, A., Sun, Y., Khan, N., Kenney, N., Ebert, A., Williams, K.P., Sanicola, M. & Salomon, D.S.

(2002). Cripto-1 activates nodal- and ALK4-dependent and -independent signaling pathways in mammary epithelial Cells. *Mol Cell Biol*, Vol. 22, No.8, (Apr and 2002), pp. (2586-2597)

Bianco, C., Cotten, C., Lonardo, E., Strizzi, L., Baraty, C., Mancino, M., Gonzales, M., Watanabe, K., Nagaoka, T., Berry, C., Arai, A.E., Minchiotti, G. & Salomon, D.S. (2009). Cripto-1 is required for hypoxia to induce cardiac differentiation of mouse embryonic stem cells. *Am J Pathol*, Vol. 175, No.5, (Nov and 2009), pp. (2146-2158)

Bianco, C., Mysliwiec, M., Watanabe, K., Mancino, M., Nagaoka, T., Gonzales, M. & Salomon, D.S. (2008). Activation of a Nodal-independent signaling pathway by Cripto-1 mutants with impaired activation of a Nodal-dependent signaling pathway. *FEBS Lett*, Vol. 582, No.29, (Dec 10 and 2008), pp. (3997-4002), 0014-5793 (Print)

Bianco, C., Rangel, M.C., Castro, N.P., Nagaoka, T., Rollman, K., Gonzales, M. & Salomon, D.S. Role of Cripto-1 in stem cell maintenance and malignant progression. *Am J Pathol*, Vol. 177, No.2, (Aug and pp. (532-540)

Bianco, C., Rangel, M.C., Castro, N.P., Nagaoka, T., Rollman, K., Gonzales, M. & Salomon, D.S. (2010). Role of Cripto-1 in stem cell maintenance and malignant progression. *Am J Pathol*, Vol. 177, No.2, (Aug and 2010), pp. (532-540), 1525-2191 (Electronic)

Bianco, C. & Salomon, D.S. (2010). Targeting the embryonic gene Cripto-1 in cancer and beyond. *Expert Opin Ther Pat*, Vol. 20, No.12, (Dec and 2010), pp. (1739-1749), 1744-7674 (Electronic)

Bianco, C., Strizzi, L., Ebert, A., Chang, C., Rehman, A., Normanno, N., Guedez, L., Salloum, R., Ginsburg, E., Sun, Y., Khan, N., Hirota, M., Wallace-Jones, B., Wechselberger, C., Vonderhaar, B.K., Tosato, G., Stetler-Stevenson, W.G., Sanicola, M. & Salomon, D.S. (2005a). Role of human cripto-1 in tumor angiogenesis. *J Natl Cancer Inst*, Vol. 97, No.2, (Jan 19 and 2005a), pp. (132-141)

Bianco, C., Strizzi, L., Normanno, N., Khan, N. & Salomon, D.S. (2005b). Cripto-1: an oncofetal gene with many faces. *Curr Top Dev Biol*, Vol. 67, and 2005b), pp. (85-133)

Bianco, C., Strizzi, L., Rehman, A., Normanno, N., Wechselberger, C., Sun, Y., Khan, N., Hirota, M., Adkins, H., Williams, K., Margolis, R.U., Sanicola, M. & Salomon, D.S. (2003). A Nodal- and ALK4-independent signaling pathway activated by Cripto-1 through Glypican-1 and c-Src. *Cancer Res*, Vol. 63, No.6, (Mar 15 and 2003), pp. (1192-1197)

Casamassimi, A., De Luca, A., Agrawal, S., Stromberg, K., Salomon, D.S. & Normanno, N. (2000). EGF-related antisense oligonucleotides inhibit the proliferation of human ovarian carcinoma cells. *Ann Oncol*, Vol. 11, No.3, (Mar and 2000), pp. (319-325), 0923-7534 (Print)

Chang, T.C., Liu, Y.G., Eddy, C.A., Jacoby, E.S., Binkley, P.A., Brzyski, R.G. & Schenken, R.S. (2010). Derivation and Characterization of Novel Nonhuman Primate Embryonic Stem Cell Lines from In Vitro-Fertilized Baboon Preimplantation Embryos. *Stem Cells Dev*, Vol. (Nov 4 and 2010), pp. 1557-8534 (Electronic)

Ciardiello, F., Dono, R., Kim, N., Persico, M.G. & Salomon, D.S. (1991). Expression of cripto, a novel gene of the epidermal growth factor gene family, leads to in vitro transformation of a normal mouse mammary epithelial cell line. *Cancer Res*, Vol. 51, No.3, (Feb 1 and 1991), pp. (1051-1054)

Ciccodicola, A., Dono, R., Obici, S., Simeone, A., Zollo, M. & Persico, M.G. (1989). Molecular characterization of a gene of the 'EGF family' expressed in undifferentiated human NTERA2 teratocarcinoma cells. *Embo J*, Vol. 8, No.7, (Jul and 1989), pp. (1987-1991)

Cocciadiferro, L., Miceli, V., Kang, K.S., Polito, L.M., Trosko, J.E. & Carruba, G. (2009). Profiling cancer stem cells in androgen-responsive and refractory human prostate tumor cell lines. *Ann N Y Acad Sci*, Vol. 1155, (Feb and 2009), pp. (257-262)

D'Andrea, D., Liguori, G.L., Le Good, J.A., Lonardo, E., Andersson, O., Constam, D.B., Persico, M.G. & Minchiotti, G. (2008). Cripto promotes A-P axis specification independently of its stimulatory effect on Nodal autoinduction. *J Cell Biol*, Vol. 180, No.3, (Feb 11 and 2008), pp. (597-605), 1540-8140 (Electronic)

D'Aniello, C., Lonardo, E., Iaconis, S., Guardiola, O., Liguoro, A.M., Liguori, G.L., Autiero, M., Carmeliet, P. & Minchiotti, G. (2009). G protein-coupled receptor APJ and its ligand apelin act downstream of Cripto to specify embryonic stem cells toward the cardiac lineage through extracellular signal-regulated kinase/p70S6 kinase signaling pathway. *Circ Res*, Vol. 105, No.3, (Jul 31 and 2009), pp. (231-238), 1524-4571 (Electronic)

de Castro, N.P., Rangel, M.C., Nagaoka, T., Salomon, D.S. & Bianco, C. (2010). Cripto-1: an embryonic gene that promotes tumorigenesis. *Future Oncol*, Vol. 6, No.7, (Jul and 2010), pp. (1127-1142), 1744-8301 (Electronic)

De Luca, A., Arra, C., D'Antonio, A., Casamassimi, A., Losito, S., Ferraro, P., Ciardiello, F., Salomon, D.S. & Normanno, N. (2000). Simultaneous blockage of different EGF-like growth factors results in efficient growth inhibition of human colon carcinoma xenografts. *Oncogene*, Vol. 19, No.51, (Nov 30 and 2000), pp. (5863-5871), 0950-9232 (Print)

De Luca, A., Casamassimi, A., Selvam, M.P., Losito, S., Ciardiello, F., Agrawal, S., Salomon, D.S. & Normanno, N. (1999). EGF-related peptides are involved in the proliferation and survival of MDA-MB-468 human breast carcinoma cells. *Int J Cancer*, Vol. 80, No.4, (Feb 9 and 1999), pp. (589-594)

Ebert, A.D., Wechselberger, C., Nees, M., Clair, T., Schaller, G., Martinez-Lacaci, I., Wallace-Jones, B., Bianco, C., Weitzel, H.K. & Salomon, D.S. (2000). Cripto-1-induced increase in vimentin expression is associated with enhanced migration of human Caski cervical carcinoma cells. *Exp Cell Res*, Vol. 257, No.1, (May 25 and 2000), pp. (223-229)

Hollier, B.G., Evans, K. & Mani, S.A. (2009). The epithelial-to-mesenchymal transition and cancer stem cells: a coalition against cancer therapies. *J Mammary Gland Biol Neoplasia*, Vol. 14, No.1, (Mar and 2009), pp. (29-43)

Hough, S.R., Laslett, A.L., Grimmond, S.B., Kolle, G. & Pera, M.F. (2009). A continuum of cell states spans pluripotency and lineage commitment in human embryonic stem cells. *PLoS One*, Vol. 4, No.11, and 2009), pp. (e7708)

Hu, X.F., Li, J., Yang, E., Vandervalk, S. & Xing, P.X. (2007). Anti-Cripto Mab inhibit tumour growth and overcome MDR in a human leukaemia MDR cell line by inhibition of Akt and activation of JNK/SAPK and bad death pathways. *Br J Cancer*, Vol. 96, No.6, (Mar 26 and 2007), pp. (918-927)

Hu, X.F. & Xing, P.X. (2005). Cripto monoclonal antibodies. *Drug News Perspect*, Vol. 18, No.5, (Jun and 2005), pp. (293-303), 0214-0934 (Print)

James, D., Levine, A.J., Besser, D. & Hemmati-Brivanlou, A. (2005). TGFbeta/activin/nodal signaling is necessary for the maintenance of pluripotency in human embryonic stem cells. *Development*, Vol. 132, No.6, (Mar and 2005), pp. (1273-1282), 0950-1991 (Print)

Kelber, J.A., Panopoulos, A.D., Shani, G., Booker, E.C., Belmonte, J.C., Vale, W.W. & Gray, P.C. (2009). Blockade of Cripto binding to cell surface GRP78 inhibits oncogenic Cripto signaling via MAPK/PI3K and Smad2/3 pathways. *Oncogene*, Vol. 28, No.24, (Jun 18 and 2009), pp. (2324-2336), 1476-5594 (Electronic)

Kelly, R.K., Olson, D.L., Sun, Y., Wen, D., Wortham, K.A., Antognetti, G., Cheung, A.E., Orozco, O.E., Yang, L., Bailly, V. & Sanicola, M. (2011). An antibody-cytotoxic conjugate, BIIB015, is a new targeted therapy for Cripto positive tumours. *Eur J Cancer*, Vol. (Mar 31 and 2011), pp. 1879-0852 (Electronic)

Loh, Y.H., Wu, Q., Chew, J.L., Vega, V.B., Zhang, W., Chen, X., Bourque, G., George, J., Leong, B., Liu, J., Wong, K.Y., Sung, K.W., Lee, C.W., Zhao, X.D., Chiu, K.P., Lipovich, L., Kuznetsov, V.A., Robson, P., Stanton, L.W., Wei, C.L., Ruan, Y., Lim, B. & Ng, H.H. (2006). The Oct4 and Nanog transcription network regulates pluripotency in mouse embryonic stem cells. *Nat Genet*, Vol. 38, No.4, (Apr and 2006), pp. (431-440), 1061-4036 (Print)

Lonardo, E., Parish, C.L., Ponticelli, S., Marasco, D., Ribeiro, D., Ruvo, M., De Falco, S., Arenas, E. & Minchiotti, G. (2010). A small synthetic cripto blocking Peptide improves neural induction, dopaminergic differentiation, and functional integration of mouse embryonic stem cells in a rat model of Parkinson's disease. *Stem Cells*, Vol. 28, No.8, (Aug and 2010), pp. (1326-1337), 1549-4918 (Electronic)

Mancino, M., Strizzi, L., Wechselberger, C., Watanabe, K., Gonzales, M., Hamada, S., Normanno, N., Salomon, D.S. & Bianco, C. (2008). Regulation of human Cripto-1 gene expression by TGF-beta1 and BMP-4 in embryonal and colon cancer cells. *J Cell Physiol*, Vol. 215, No.1, (Apr and 2008), pp. (192-203), 1097-4652 (Electronic)

Mani, S.A., Guo, W., Liao, M.J., Eaton, E.N., Ayyanan, A., Zhou, A.Y., Brooks, M., Reinhard, F., Zhang, C.C., Shipitsin, M., Campbell, L.L., Polyak, K., Brisken, C., Yang, J. & Weinberg, R.A. (2008). The epithelial-mesenchymal transition generates cells with properties of stem cells. *Cell*, Vol. 133, No.4, (May 16 and 2008), pp. (704-715)

Micalizzi, D.S., Farabaugh, S.M. & Ford, H.L. (2010a). Epithelial-mesenchymal transition in cancer: parallels between normal development and tumor progression. *J Mammary Gland Biol Neoplasia*, Vol. 15, No.2, (Jun and 2010a), pp. (117-134), 1573-7039 (Electronic)

Micalizzi, D.S., Wang, C.A., Farabaugh, S.M., Schiemann, W.P. & Ford, H.L. (2010b). Homeoprotein Six1 increases TGF-beta type I receptor and converts TGF-beta signaling from suppressive to supportive for tumor growth. *Cancer Res*, Vol. 70, No.24, (Dec 15 and 2010b), pp. (10371-10380), 1538-7445 (Electronic)

Minchiotti, G. (2005). Nodal-dependant Cripto signaling in ES cells: from stem cells to tumor biology. *Oncogene*, Vol. 24, No.37, (Aug 29 and 2005), pp. (5668-5675), 0950-9232 (Print)

Minchiotti, G., Manco, G., Parisi, S., Lago, C.T., Rosa, F. & Persico, M.G. (2001). Structure-function analysis of the EGF-CFC family member Cripto identifies residues essential for nodal signalling. *Development*, Vol. 128, No.22, (Nov and 2001), pp. (4501-4510), 0950-1991 (Print)

Minchiotti, G., Parisi, S., Liguori, G., Signore, M., Lania, G., Adamson, E.D., Lago, C.T. & Persico, M.G. (2000). Membrane-anchorage of Cripto protein by glycosylphosphatidylinositol and its distribution during early mouse development. *Mech Dev*, Vol. 90, No.2, (Feb and 2000), pp. (133-142), 0925-4773 (Print)

Miyoshi, K., Rosner, A., Nozawa, M., Byrd, C., Morgan, F., Landesman-Bollag, E., Xu, X., Seldin, D.C., Schmidt, E.V., Taketo, M.M., Robinson, G.W., Cardiff, R.D. & Hennighausen, L. (2002a). Activation of different Wnt/beta-catenin signaling components in mammary epithelium induces transdifferentiation and the formation of pilar tumors. *Oncogene*, Vol. 21, No.36, (Aug 15 and 2002a), pp. (5548-5556), 0950-9232 (Print)

Miyoshi, K., Shillingford, J.M., Le Provost, F., Gounari, F., Bronson, R., von Boehmer, H., Taketo, M.M., Cardiff, R.D., Hennighausen, L. & Khazaie, K. (2002b). Activation of beta -catenin signaling in differentiated mammary secretory cells induces transdifferentiation into epidermis and squamous metaplasias. *Proc Natl Acad Sci U S A*, Vol. 99, No.1, (Jan 8 and 2002b), pp. (219-224), 0027-8424 (Print)

Morkel, M., Huelsken, J., Wakamiya, M., Ding, J., van de Wetering, M., Clevers, H., Taketo, M.M., Behringer, R.R., Shen, M.M. & Birchmeier, W. (2003). Beta-catenin regulates Cripto- and Wnt3-dependent gene expression programs in mouse axis and mesoderm formation. *Development*, Vol. 130, No.25, (Dec and 2003), pp. (6283-6294)

Niemeyer, C.C., Persico, M.G. & Adamson, E.D. (1998). Cripto: roles in mammary cell growth, survival, differentiation and transformation. *Cell Death Differ*, Vol. 5, No.5, (May and 1998), pp. (440-449)

Normanno, N., Bianco, C., Damiano, V., de Angelis, E., Selvam, M.P., Grassi, M., Magliulo, G., Tortora, G., Bianco, A.R., Mendelsohn, J., Salomon, D.S. & Ciardiello, F. (1996). Growth inhibition of human colon carcinoma cells by combinations of anti-epidermal growth factor-related growth factor antisense oligonucleotides. *Clin Cancer Res*, Vol. 2, No.3, (Mar and 1996), pp. (601-609), 1078-0432 (Print)

Normanno, N., De Luca, A., Bianco, C., Maiello, M.R., Carriero, M.V., Rehman, A., Wechselberger, C., Arra, C., Strizzi, L., Sanicola, M. & Salomon, D.S. (2004a). Cripto-1 overexpression leads to enhanced invasiveness and resistance to anoikis in human MCF-7 breast cancer cells. *J Cell Physiol*, Vol. 198, No.1, (Jan and 2004a), pp. (31-39)

Normanno, N., De Luca, A., Maiello, M.R., Bianco, C., Mancino, M., Strizzi, L., Arra, C., Ciardiello, F., Agrawal, S. & Salomon, D.S. (2004b). CRIPTO-1: a novel target for therapeutic intervention in human carcinoma. *Int J Oncol*, Vol. 25, No.4, (Oct and 2004b), pp. (1013-1020)

Parish, C.L. & Arenas, E. (2007). Stem-cell-based strategies for the treatment of Parkinson's disease. *Neurodegener Dis*, Vol. 4, No.4, and 2007), pp. (339-347), 1660-2854 (Print)

Parish, C.L., Parisi, S., Persico, M.G., Arenas, E. & Minchiotti, G. (2005). Cripto as a target for improving embryonic stem cell-based therapy in Parkinson's disease. *Stem Cells*, Vol. 23, No.4, (Apr and 2005), pp. (471-476), 1066-5099 (Print)

Parisi, S., D'Andrea, D., Lago, C.T., Adamson, E.D., Persico, M.G. & Minchiotti, G. (2003). Nodal-dependent Cripto signaling promotes cardiomyogenesis and redirects the neural fate of embryonic stem cells. *J Cell Biol*, Vol. 163, No.2, (Oct 27 and 2003), pp. (303-314), 0021-9525 (Print)

Schiffer, S.G., Foley, S., Kaffashan, A., Hronowski, X., Zichittella, A.E., Yeo, C.Y., Miatkowski, K., Adkins, H.B., Damon, B., Whitman, M., Salomon, D., Sanicola, M. & Williams, K.P. (2001). Fucosylation of Cripto is required for its ability to facilitate nodal signaling. *J Biol Chem*, Vol. 276, No.41, (Oct 12 and 2001), pp. (37769-37778), 0021-9258 (Print)

Shani, G., Fischer, W.H., Justice, N.J., Kelber, J.A., Vale, W. & Gray, P.C. (2008). GRP78 and Cripto form a complex at the cell surface and collaborate to inhibit transforming growth factor beta signaling and enhance cell growth. *Mol Cell Biol*, Vol. 28, No.2, (Jan and 2008), pp. (666-677), 1098-5549 (Electronic)

0270-7306 (Linking)

Shi, S., Ge, C., Luo, Y., Hou, X., Haltiwanger, R.S. & Stanley, P. (2007). The threonine that carries fucose, but not fucose, is required for Cripto to facilitate Nodal signaling. *J Biol Chem*, Vol. 282, No.28, (Jul 13 and 2007), pp. (20133-20141), 0021-9258 (Print)

Shin, M., Alev, C., Wu, Y., Nagai, H. & Sheng, G. (2011). Activin/TGF-beta signaling regulates Nanog expression in the epiblast during gastrulation. *Mech Dev*, Vol. (Mar 22 and 2011), pp. 1872-6356 (Electronic)

Sonntag, K.C., Simantov, R., Bjorklund, L., Cooper, O., Pruszak, J., Kowalke, F., Gilmartin, J., Ding, J., Hu, Y.P., Shen, M.M. & Isacson, O. (2005). Context-dependent neuronal differentiation and germ layer induction of Smad4-/- and Cripto-/- embryonic stem cells. *Mol Cell Neurosci*, Vol. 28, No.3, (Mar and 2005), pp. (417-429), 1044-7431 (Print)

Stephens, E.B., Jackson, M., Cui, L., Pacyniak, E., Choudhuri, R., Liverman, C.S., Salomon, D.S. & Berman, N.E. (2006). Early dysregulation of cripto-1 and immunomodulatory genes in the cerebral cortex in a macaque model of neuroAIDS. *Neurosci Lett*, Vol. 410, No.2, (Dec 20 and 2006), pp. (94-99), 0304-3940 (Print)

Strizzi, L., Abbott, D.E., Salomon, D.S. & Hendrix, M.J. (2008). Potential for cripto-1 in defining stem cell-like characteristics in human malignant melanoma. *Cell Cycle*, Vol. 7, No.13, (Jul 1 and 2008), pp. (1931-1935)

Strizzi, L., Bianco, C., Hirota, M., Watanabe, K., Mancino, M., Hamada, S., Raafat, A., Lawson, S., Ebert, A., D'Antonio, A., Losito, S., Normanno, N. & Salomon, D. (2007). Development of leiomyosarcoma of the uterus in MMTV-CR-1 transgenic mice. *J Pathol*, Vol. 211, No.1, (Jan and 2007), pp. (36-44)

Strizzi, L., Bianco, C., Normanno, N. & Salomon, D. (2005). Cripto-1: a multifunctional modulator during embryogenesis and oncogenesis. *Oncogene*, Vol. 24, No.37, (Aug 29 and 2005), pp. (5731-5741)

Strizzi, L., Bianco, C., Normanno, N., Seno, M., Wechselberger, C., Wallace-Jones, B., Khan, N.I., Hirota, M., Sun, Y., Sanicola, M. & Salomon, D.S. (2004). Epithelial mesenchymal transition is a characteristic of hyperplasias and tumors in mammary gland from MMTV-Cripto-1 transgenic mice. *J Cell Physiol*, Vol. 201, No.2, (Nov and 2004), pp. (266-276)

Sun, Y., Strizzi, L., Raafat, A., Hirota, M., Bianco, C., Feigenbaum, L., Kenney, N., Wechselberger, C., Callahan, R. & Salomon, D.S. (2005). Overexpression of human Cripto-1 in transgenic mice delays mammary gland development and differentiation and induces mammary tumorigenesis. *Am J Pathol*, Vol. 167, No.2, (Aug and 2005), pp. (585-597)

Tedesco, F.S., Dellavalle, A., Diaz-Manera, J., Messina, G. & Cossu, G. (2010). Repairing skeletal muscle: regenerative potential of skeletal muscle stem cells. *J Clin Invest*, Vol. 120, No.1, (Jan 4 and 2010), pp. (11-19), 1558-8238 (Electronic)

Vallier, L., Mendjan, S., Brown, S., Chng, Z., Teo, A., Smithers, L.E., Trotter, M.W., Cho, C.H., Martinez, A., Rugg-Gunn, P., Brons, G. & Pedersen, R.A. (2009). Activin/Nodal signalling maintains pluripotency by controlling Nanog expression. *Development*, Vol. 136, No.8, (Apr and 2009), pp. (1339-1349)

Visvader, J.E. & Lindeman, G.J. (2008). Cancer stem cells in solid tumours: accumulating evidence and unresolved questions. *Nat Rev Cancer*, Vol. 8, No.10, (Oct and 2008), pp. (755-768)

Wang, Z., Li, Y., Banerjee, S. & Sarkar, F.H. (2009). Emerging role of Notch in stem cells and cancer. *Cancer Lett*, Vol. 279, No.1, (Jun 28 and 2009), pp. (8-12)

Watanabe, K., Bianco, C., Strizzi, L., Hamada, S., Mancino, M., Bailly, V., Mo, W., Wen, D., Miatkowski, K., Gonzales, M., Sanicola, M., Seno, M. & Salomon, D.S. (2007a). Growth factor induction of Cripto-1 shedding by glycosylphosphatidylinositol-phospholipase D and enhancement of endothelial cell migration. *J Biol Chem*, Vol. 282, No.43, (Oct 26 and 2007a), pp. (31643-31655), 0021-9258 (Print)

Watanabe, K., Hamada, S., Bianco, C., Mancino, M., Nagaoka, T., Gonzales, M., Bailly, V., Strizzi, L. & Salomon, D.S. (2007b). Requirement of glycosylphosphatidylinositol anchor of Cripto-1 for trans activity as a Nodal co-receptor. *J Biol Chem*, Vol. 282, No.49, (Dec 7 and 2007b), pp. (35772-35786), 0021-9258 (Print)

Watanabe, K., Meyer, M.J., Strizzi, L., Lee, J.M., Gonzales, M., Bianco, C., Nagaoka, T., Farid, S.S., Margaryan, N., Hendrix, M.J., Vonderhaar, B.K. & Salomon, D.S. (2010). Cripto-1 is a cell surface marker for a tumorigenic, undifferentiated subpopulation in human embryonal carcinoma cells. *Stem Cells*, Vol. 28, No.8, (Aug and 2010), pp. (1303-1314), 1549-4918 (Electronic)

Watanabe, K., Nagaoka, T., Lee, J.M., Bianco, C., Gonzales, M., Castro, N.P., Rangel, M.C., Sakamoto, K., Sun, Y., Callahan, R. & Salomon, D.S. (2009). Enhancement of Notch receptor maturation and signaling sensitivity by Cripto-1. *J Cell Biol*, Vol. 187, No.3, (Nov 2 and 2009), pp. (343-353)

Wechselberger, C., Ebert, A.D., Bianco, C., Khan, N.I., Sun, Y., Wallace-Jones, B., Montesano, R. & Salomon, D.S. (2001). Cripto-1 enhances migration and branching morphogenesis of mouse mammary epithelial cells. *Exp Cell Res*, Vol. 266, No.1, (May 15 and 2001), pp. (95-105)

Wechselberger, C., Strizzi, L., Kenney, N., Hirota, M., Sun, Y., Ebert, A., Orozco, O., Bianco, C., Khan, N.I., Wallace-Jones, B., Normanno, N., Adkins, H., Sanicola, M. & Salomon, D.S. (2005). Human Cripto-1 overexpression in the mouse mammary gland results in the development of hyperplasia and adenocarcinoma. *Oncogene*, Vol. 24, No.25, (Jun 9 and 2005), pp. (4094-4105)

Wei, C.L., Miura, T., Robson, P., Lim, S.K., Xu, X.Q., Lee, M.Y., Gupta, S., Stanton, L., Luo, Y., Schmitt, J., Thies, S., Wang, W., Khrebtukova, I., Zhou, D., Liu, E.T., Ruan, Y.J., Rao, M. & Lim, B. (2005). Transcriptome profiling of human and murine ESCs identifies divergent paths required to maintain the stem cell state. *Stem Cells*, Vol. 23, No.2, (Feb and 2005), pp. (166-185)

Wilson, A. & Radtke, F. (2006). Multiple functions of Notch signaling in self-renewing organs and cancer. *FEBS Lett*, Vol. 580, No.12, (May 22 and 2006), pp. (2860-2868), 0014-5793 (Print)

Xing, P.X., Hu, X.F., Pietersz, G.A., Hosick, H.L. & McKenzie, I.F. (2004). Cripto: a novel target for antibody-based cancer immunotherapy. *Cancer Res*, Vol. 64, No.11, (Jun 1 and 2004), pp. (4018-4023)

Xu, C., Liguori, G., Persico, M.G. & Adamson, E.D. (1999). Abrogation of the Cripto gene in mouse leads to failure of postgastrulation morphogenesis and lack of differentiation of cardiomyocytes. *Development*, Vol. 126, No.3, (Feb and 1999), pp. (483-494), 0950-1991 (Print)

Yan, Y.T., Liu, J.J., Luo, Y., E, C., Haltiwanger, R.S., Abate-Shen, C. & Shen, M.M. (2002). Dual roles of Cripto as a ligand and coreceptor in the nodal signaling pathway. *Mol Cell Biol*, Vol. 22, No.13, (Jul and 2002), pp. (4439-4449), 0270-7306 (Print)

Yeung, T.M., Gandhi, S.C. & Bodmer, W.F. (2011). Hypoxia and lineage specification of cell line-derived colorectal cancer stem cells. *Proc Natl Acad Sci U S A*, Vol. 108, No.11, (Mar 15 and 2011), pp. (4382-4387), 1091-6490 (Electronic)

20

Molecular Mechanisms Underlying Pluripotency and Lineage Commitment – The Role of GSK-3

Bradley W. Doble[1], Kevin F. Kelly[1] and James R. Woodgett[2]
[1]*McMaster University Stem Cell and Cancer Research Institute, Hamilton, Ontario*
[2]*Samuel Lunenfeld Research Institute, Mount Sinai Hospital, Toronto, Ontario*
Canada

1. Introduction

The highly related serine/threonine kinases GSK3α and GSK3β are transducers of Wnt/β-catenin, PI-3K, Notch and Hedgehog signalling pathways, placing them at the hub of key developmental and metabolic processes. There is accumulating evidence suggesting that GSK-3 inhibitors aid in the acquisition or sustenance of pluripotency in embryonic stem cells of mouse, rat and human origin. However, the mechanism through which GSK-3 inhibitors impart their effects is unclear due to the myriad cellular processes in which GSK-3 plays a role. Here, we review the studies that have examined the consequences of GSK-3 inhibition in pluripotent stem cells with a focus on key signalling pathways, which have been implicated in GSK-3 inhibitor-mediated effects.

2. Overview of GSK-3α and GSK-3β structure and function

Glycogen synthase kinase-3 is a serine/threonine protein kinase, which was named based on its ability to phosphorylate and inhibit glycogen synthase, the rate-limiting enzyme of glycogen synthesis (Embi et al., 1980; Woodgett et al., 1983). Mammals express two homologues of GSK-3, GSK-3α and GSK-3β, which are encoded by separate genes (Woodgett, 1990). Throughout this chapter we will use "GSK-3" to refer to both GSK-3α and GSK-3β. The two GSK-3 gene products share almost identical kinase domains, but differ substantially at their amino and carboxy termini. Notably, GSK-3α has a glycine-rich N-terminal extension, which is not present in GSK-3β. This is reflected in the relative molecular masses of GSK-3α and GSK-3β, with GSK-3α having a predicted MW of 51 kDa, and GSK-3β having a slightly lower MW of 47 kDa (see Fig. 1).

Both enzymes retain activity when their non-catalytic N-terminal extensions are deleted, whereas truncation mutants of their disparate non-catalytic C-terminal sequences display reduced activity, in part due to improper folding (Buescher and Phiel, 2010). There is a splice variant of GSK-3β (GSK-3β2), which has a 13 amino acid insert in the kinase domain (Mukai et al., 2002; Schaffer et al., 2003). The GSK-3β2 variant is less abundant than the GSK-3β1 variant. Even in the brain, where it is most enriched, GSK-3β2 protein levels are still approximately 2- to 3-fold less than those of GSK-3β1. GSK-3β2 displays reduced activity towards some substrates (e.g. Tau, CRMP2, CRMP4, phospho-GS2, Inhibitor-2) when compared to the GSK-3β1 variant, although some substrates (e.g. c-Myc and c-Jun) appear to be equivalently phosphorylated by either GSK-3β isoform (Mukai et al., 2002; Soutar et al., 2010).

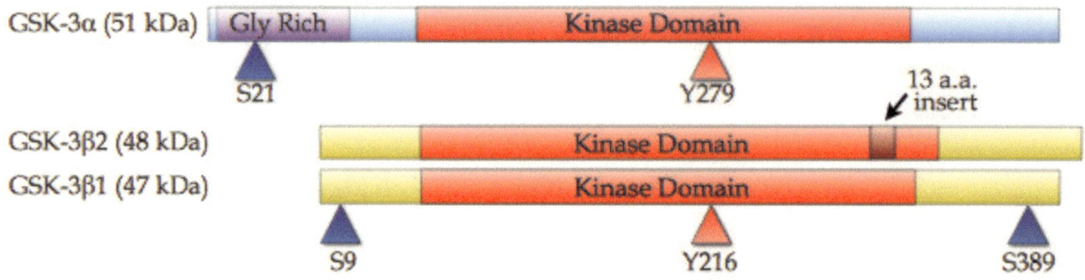

Fig. 1. Schematic depiction of GSK-3α and GSK-3β proteins. Red shaded triangles indicate the sites of tyrosine phosphorylation required for maximal catalytic activity. Blue shaded triangles point to sites of inhibitory phosphorylation.

GSK-3 is unusual in that it exhibits a preference for substrates that are pre-phosphorylated by another kinase at a position located four residues C-terminal to the GSK-3 target site (Fiol et al., 1987). Thus, the consensus sequence for a GSK-3 substrate is S/T-X-X-X-S-P/T-P, where the first serine or threonine (S/T) is the GSK-3 target site, X is any amino acid, and the final S/T residue is the site that is phosphorylated by a priming kinase. The GSK-3 substrates CRMP2 and CRMP4 deviate from the usual consensus sequence in that the number of residues between the priming site and the GSK-3 target site is four, instead of the usual three (Cole et al., 2004). Thus, the GSK-3 substrate consensus sequence is not invariant. Frequently, GSK-3 substrates contain several tandem target residues, spaced such that after an initial priming phosphorylation, GSK-3 will phosphorylate multiple consecutive target sites located N-terminal to the priming site, with GSK-3 acting as the priming kinase for each subsequent phosphorylation event. There are hundreds of potential GSK-3 substrates with properly spaced tandem GSK-3 target sites, based on bioinformatics analyses (Taelman et al., 2010), although less than 100 GSK-3 substrates have been reported in the literature and only a subset of these has been thoroughly validated.

Nascent GSK-3α and GSK-3β enzymes, as they fold into their final conformation, undergo a transient state in which they act as tyrosine kinases, serving to autophosphorylate themselves on a key tyrosine residue (Lochhead et al., 2006). The resultant phosphorylation on tyrosine 216 of GSK-3β or the equivalent tyrosine 279 of GSK-3α is required for full serine/threonine kinase activity of the fully mature enzymes. Subsequent kinase activity is specific for serine/threonine phosphorylation. In the absence of signals, that is, resting conditions, GSK-3 exhibits a high degree of activity. The primary mode of regulating GSK-3 is via inactivation, which can be accomplished through phosphorylation, protein-protein interactions or sequestration.

2.1 Regulation of GSK-3 activity through phosphorylation by upstream kinases.

2.1.1 Inhibitory N-terminal phosphorylation

When serine 9 of GSK-3β (or serine 21 of GSK-3α) is phosphorylated, it mimicks a primed, intramolecular pseudosubstrate that acts in cis as a competitive inhibitor of GSK-3 activity (Cross et al., 1995). The crystal structure of GSK-3β revealed that the negatively charged phosphate on primed substrates fits into a pocket of positively charged residues in the substrate binding cleft of the kinase, thereby placing the substrate into the proper orientation for GSK-3-mediated phosphorylation (Bax et al., 2001; Dajani et al., 2001; ter Haar et al., 2001). This pocket also binds the phosphorylated N-terminal peptide of GSK-3 when it is phosphorylated at S9/S21 - competing out exogenous substrates and hence, interfering with activity. There are several kinases capable of inhibiting GSK-3 through S9/S21 phosphorylation, including PKB/Akt, p70 ribosomal S6 kinase, p90 ribosomal S6 kinase, protein kinase A and certain protein kinase Cs (Cross et al., 1995; Fang et al., 2000; Goode et al., 1992; Sutherland and Cohen, 1994; Torres et al., 1999).

2.1.2 Inhibitory C-terminal phosphorylation (GSK-3β-specific)

Phosphorylation of serine 389 (S389), which is located near the carboxy terminus of mouse GSK-3β (threonine 390 is the equivalent residue in human GSK-3β), is also capable of generating a pseudosubstrate inhibitor of GSK-3β activity in a manner analogous to that obtained through phosphorylation of S9 (Thornton et al., 2008). There is no equivalent residue to S389 found in the sequence of GSK-3α, thus, S389 phosphorylation provides a mechanism through which GSK-3β activity may be differentially regulated from GSK3-α activity. The only kinase that has been linked to S389 phosphorylation to date is p38 MAPK, which appears to phosphorylate S389 in brain and thymocytes (Thornton et al., 2008).

2.2 Regulation through multi-protein complexes

GSK-3 can also be regulated through interactions with other proteins. This is most clearly demonstrated in the canonical Wnt/β-catenin signaling pathway where GSK-3 binds several proteins with distinct functional consequences. Details of this signalling pathway are discussed in section 4.1. In a multi-protein assembly known as the β-catenin destruction complex, a high-affinity interaction between GSK-3 and the scaffolding protein Axin or Axin2/Conductin is required for GSK-3 to efficiently phosphorylate the substrate β-catenin (Behrens et al., 1998; Hart et al., 1998; Ikeda et al., 1998a). GSK-3 interacts strongly with members of the GSK-3 binding protein (GBP)/FRAT family of proteins and these interactions were thought to compete with Axin/Axin2 interactions with GSK-3 as part of the mechanism of Wnt/β-catenin signalling [reviewed in: (van Amerongen and Berns, 2005)]. However, loss-of-function studies in mice indicate that the FRAT proteins are dispensable for Wnt/β-catenin signalling in mammals, given the lack of any obvious phenotype in triple knockout FRAT1,2,3-null mice (Van Amerongen, 2005). GSK-3 also interacts with the intracellular domain of LRP6, a co-receptor for Wnt/β-catenin signalling (Beagle et al., 2009; Mi et al., 2006; Piao et al., 2008; Wu et al., 2009). A model has been proposed in which Wnt ligand-induced phosphorylation of the intracellular domain of LRP6 results in GSK-3 binding and inactivation, thereby inhibiting GSK-3-mediated β-catenin phosphorylation (Wu et al., 2009).

2.3 Functional redundancy of GSK-3α and GSK-3β

As described above, the kinase domains of GSK-3α and GSK-3β are virtually identical, resulting in a high degree of functional overlap between the two proteins. For instance, with respect to the role of GSK-3 in Wnt/β-catenin signalling, there is clear functional redundancy between GSK-3α and GSK-3β, based on genetic studies in mouse embryonic stem cells (Doble et al., 2007). Still, there are striking differences in the phenotypes of mice lacking GSK-3α or GSK-3β. GSK-3α knockout mice are viable but insulin-sensitized

(MacAulay et al., 2007) and they display abnormal brain structure and behaviour (Kaidanovich-Beilin et al., 2009) as well as progressive cardiac hypertrophy and contractile dysfunction (Zhou et al., 2010). By contrast, GSK-3β knockout mice are inviable (Hoeflich et al., 2000). These animals die either prior to, or immediately after, birth and display cardiovascular developmental defects and liver degeneration (Hoeflich et al., 2000; Kerkela et al., 2008). It is currently unclear whether the different phenotypes arise due to different expression patterns of GSK-3α and GSK-3β (both isoforms are expressed in most tissues examined) or distinct subsets of substrates (although few, if any, substrates have been shown to be targeted by only one of the two isoforms).

3. Pathways and transcription factors that regulate pluripotency in mouse and human pluripotent stem cells

Stems cells are defined by their ability to self-renew (i.e. to make more stem cells) while retaining the ability to differentiate into one or more specialized cell types. Embryonic stem cells (ESCs) have the ability to differentiate into all cell types of the embryo proper (Evans and Kaufman, 1981; Rossant, 2001). In other words, they are *pluripotent*, a remarkable property that has obvious implications for tissue replacement therapies targetting diseased or damaged tissues. To retain their stem cell identity, both embryonic and adult stem cells (such as hematopoietic stem cells) have active signalling pathways that regulate their proliferation and pluripotency (reviewed in: (Molofsky et al., 2004)). In ESCs, self-renewal depends on the maintenance of a transcriptional program regulated, minimally, by a set of three transcription factors. These factors comprise the core transcriptional program responsible for the maintenance of mouse ES cell pluripotency and include; Oct-4, Sox-2 and Nanog (Boyer et al., 2005; Chew et al., 2005; Loh et al., 2006; Rodda et al., 2005).

The most commonly used media for maintaining mouse ESCs under serum-free conditions require supplementation with leukemia inhibitory factor (LIF) and bone morphogenetic protein-4 (BMP4), which initiate signalling through gp130/STAT- and Smad-mediated pathways, respectively (Niwa et al., 1998; Ying et al., 2003; Ying and Smith, 2003). Notably, these signalling molecules/pathways are incapable of maintaining undifferentiated human ESCs *in vitro* (Daheron et al., 2004; Humphrey et al., 2004; Xu et al., 2002). Feeder-free culture of human ESCs requires Activin/Nodal and FGF-2 supplementation instead (Beattie et al., 2005; James et al., 2005; Levenstein et al., 2006; Ludwig et al., 2006; Vallier et al., 2005). Despite a requirement for activation of ostensibly different signalling pathways, the transcriptional mediators of pluripotency appear to be similar in both human and mouse ESCs (Boyer et al., 2005; Chew et al., 2005; Loh et al., 2006; Rodda et al., 2005) with Nanog playing a central role (Pan and Thomson, 2007).

The difference in the medium requirements for mouse versus human ESCs (mESCs vs hESCs) likely reflects dissimilar developmental origins. There are two primary lines of evidence supporting the notion that hESCs represent a more advanced stage of development than the early blastocysts from which mESCs are obtained: i. Pluripotent mouse epiblast stem cells (EpiSCs), derived from an embryo at a later developmental stage than that typically used to obtain mESCs, have the same growth factor requirements (Activin/Nodal and FGF-2) as human "ESCs" (Brons et al., 2007; Tesar et al., 2007). ii. Human ESCs or induced pluripotent stem cells can attain a mESC-like, LIF-dependent state, if they are reprogrammed with pluripotency-inducing transcription factors in the presence of LIF (Buecker et al., 2010; Hanna et al., 2010). The pluripotent state of rodent

EpiSCs (and primate "ESCs") differs biochemically and epigenetically from that of true rodent ESCs; the two conditions have been termed *primed* and *naïve* pluripotent states, respectively (Nichols and Smith, 2009).
Of note, inactivation of GSK-3 through the use of small molecule inhibitors enhanced the maintenance of human pluripotent stem cells or pluripotent stem cells derived from the NOD strain of mouse in a naïve pluripotent state (Hanna et al., 2010; Hanna et al., 2009). Furthermore, the "ground state" of naïve pluripotency in mESCs can be maintained with fully defined serum-free medium by adding a combination of three small-molecule inhibitors (3i): SU5402 and PD184352, which serve to inhibit mitogen-activated protein kinase (MAPK) signaling; and CHIR99021, a high-specificity GSK-3 inhibitor (Ying et al., 2008). A proposed mechanism of the 3i cocktail is that inhibition of MAPK signalling blocks autoinductive FGF4 signalling required for mESC differentiation, while GSK-3 inhibition may play a role in allowing undifferentiated mESCs to survive and proliferate.

4. Effects of GSK-3 inhibition/ablation on pluripotent stem cells

There are now several reports in which inhibition of GSK-3 through the use of small molecules (e.g. CHIR99021 or BIO) has been shown to enhance the maintenance and derivation of embryonic stem cells obtained from mice, including the problematic C57BL/6 strain, from which *de novo* mESC cell line generation had been very difficult (Gertsenstein et al., 2010; Kiyonari et al., 2010; Sato et al., 2009). Furthermore, inhibition of GSK-3 has facilitated the isolation of rat ESCs, which had also been technically challenging (Buehr et al., 2008; Li et al., 2009). In the following sections, we provide an overview of the potential mechanisms through which GSK-3 inhibition imparts its effects on pluripotency.

4.1 The central role of GSK-3 in Wnt/β-catenin signalling

The Wnt family of secreted glycoproteins regulates cellular interactions during Wnt signalling and is involved in the genesis of a variety of human cancers, including those of the colon, liver and breast (Giles et al., 2003; Morin and Weeraratna, 2003). In the canonical Wnt/β-catenin pathway, the ultimate outcome of ligand-initiated signal transduction is largely dependent on the activation of genes regulated by T-cell factor (TCF) / lymphoid enhancer factor (LEF) transcription factors (Arce et al., 2006; Brantjes et al., 2002). TCF/LEF proteins, in the absence of a Wnt signal, repress transcription in conjunction with co-repressors such as members of the Groucho/TLE (Transducin-like Enhancer of split) family (Chen and Courey, 2000; Hurlstone and Clevers, 2002). In response to Wnt, newly accumulated β-catenin displaces Groucho/TLE from TCF/LEF and acts as a co-activator of TCF/LEF to initiate transcription (Brantjes et al., 2002; Chen and Courey, 2000; Daniels and Weis, 2005). There are four mammalian TCF/LEF family members: TCF1, LEF1, TCF3 and TCF4 (Arce et al., 2006). In mESCs, transcripts from all of the TCF/LEF genes are readily detected, but those encoding TCF3 predominate at steady state (Anton et al., 2007; Pereira et al., 2006). At the protein level, all TCF/LEF family members are readily detected by western blot analyses of mESC lysates (Kelly et al., 2011). TCF3 has been shown to negatively regulate the transcription of Nanog (Pereira et al., 2006; Yi et al., 2008) and to co-occupy promoters with Nanog and Oct-4 in mESCs (Cole et al., 2008).
GSK-3 plays a central role in the canonical Wnt/β-catenin signalling pathway (see Fig. 2), serving to keep steady state levels of the cytoplasmic, signalling pool of β-catenin low in the

absence of Wnt ligands (Dominguez et al., 1995; He et al., 1995; Peifer et al., 1994). A fraction of GSK-3 is complexed with the scaffolding proteins adenomatous polyposis coli (APC) and Axin1 (or Axin2), and efficiently phosphorylates β-catenin ["primed" by casein kinase-1 (CK1)] on a series of N-terminal domain residues earmarking it for polyubiquitination and proteasomal degradation (Doble and Woodgett, 2003). GSK-3 also phosphorylates APC and Axin, increasing their affinities for β-catenin (Hoeflich et al., 2000; Ikeda et al., 1998b; Jho et al., 1999; Yamamoto et al., 1999).

Activation of Wnt/β-catenin signalling requires two types of receptors. The first class belongs to a sub-family of single-pass transmembrane receptors, the low-density lipoprotein receptor related proteins (LRPs), LRP5 and LRP6 (He et al., 2004; Schweizer and Varmus, 2003; Tamai et al., 2000; Wehrli et al., 2000). The second type of receptor comprises a family of serpentine 7-pass transmembrane receptors known as Frizzleds (Fz)(Logan and Nusse, 2004; Orsulic and Peifer, 1996), which appear to function as G protein-coupled receptors (Katanaev et al., 2005; Liu et al., 2005; Schulte and Bryja, 2007). The binding of Wnt to Fz/LRP6 results in the recruitment of a trimolecular complex of Dishevelled, Axin and GSK-3 to the Fz/LRP6 heterodimer (Bilic et al., 2007; Zeng et al., 2008; Zeng et al., 2005). GSK-3 and CKI then phosphorylate residues of LRP6 to create an Axin docking site which enhances Axin's affinity for LRP6 and results in inhibition of the β-catenin destruction complex in an unclear manner, possibly involving direct GSK-3 inhibition through Axin-binding (Zeng et al., 2008; Zeng et al., 2005) and/or interactions with phospho-LRP6 (Wu et al., 2009). The reorganization of Axin upon LRP5/6 binding may also trigger sequestration of Axin-associated GSK-3 into multi-vesicular endosomes - preventing direct access to β-catenin (Taelman et al., 2010).

4.1.1 Wnt/β-catenin signalling reinforces the pluripotent state of mESCs

The first study to directly implicate GSK-3 and Wnt/β-catenin signalling in the regulation of ESC pluripotency suggested that human and mouse ESCs respond similarly to activation of Wnt/β-catenin signalling (Sato et al., 2004). By using a reporter construct harbouring TCF binding sites driving yellow fluorescent protein (YFP) expression, pathway activation was found to be highest in undifferentiated mESCs and was gradually lost as the cells differentiated, suggesting that endogenous β-catenin/TCF signalling supported the maintenance of pluripotency. Activation of the Wnt pathway in mESCs or hESCs, by treatment with either Wnt3a or the GSK-3 inhibitor BIO, prolonged the retention of pluripotency markers, although one caveat of this study is that these cells were not examined for more than 5-7 days, over multiple passages (Sato et al., 2004). Similar observations were made in mESCs treated with the GSK-3 inhibitor, LiCl (Anton et al., 2007), as well as Wnt-conditioned medium (Hao et al., 2006; Ogawa et al., 2006; Singla et al., 2006). As an aside, there is a fundamental difference between treatment with GSK-3 inhibitors and treatment with Wnt due to the inherent feedback control circuitry of Wnt signalling. Hence, the target genes of β-catenin include Axin2 and DKK1 which both act to reduce Wnt signalling. However, inhibition of GSK-3 short-circuits these controls leading to chronic activation of β-catenin signalling that is more reminiscent of tumours that harbour APC deletions or β-catenin mutations.

Some of the conclusions reached by Sato et al. conflict with an independent study investigating the effects of Wnt pathway manipulation in hESCs over multiple weeks (Dravid et al., 2005). In this study, addition of the Wnt antagonists Dkk1 or sFRP2 did not compromise the self-renewal ability of hESCs when cultured on a feeder layer, nor did the inclusion of recombinant Wnt3a support hESC propagation in the undifferentiated state in the absence of feeders. β-catenin/TCF activity increased significantly after BMP-4- or retinoic acid-induced hESC differentiation, supporting a pro-differentiation role for the Wnt pathway in hESCs (Dravid et al., 2005). Indeed, sustained Wnt pathway activation or GSK-3 inhibition in monolayer culture promotes hESC and mESC differentiation into mesoendodermal progenitors, which are capable of enhanced endothelial and cardiac specification (Bakre 2007). A bisindoylmaleimide GSK-3 inhibitor, designated 1m, which was shown to sustain pluripotency of mESCs (Bone et al., 2009), by contrast, has been shown to be effective in differentiating hESCs into definitive endoderm (Bone et al., 2011). These divergent effects of 1m-mediated GSK-3 inhibition are likely due to the caveats associated with the developmental origins of human and mouse ESCs, discussed in section 3 above.

4.1.2 Genetic Gain- and Loss-of-function studies of Wnt pathway regulators in ESCs

4.1.2.1 mESCS with mutant Adenomatous polyposis coli (Apc) display impaired differentiation

The scaffolding protein Adenomatous polyposis coli (Apc) is a key negative regulator of the Wnt/β-catenin pathway that is frequently mutated in human cancers, particularly those of the colon (Kwong and Dove, 2009). Kielman et al. used teratoma assays in syngeneic mice to evaluate the differentiation capacities of mESCs harbouring a spectrum of mutant forms of Apc (Kielman et al., 2002). These mutants exhibit varying abilities to regulate β-catenin levels, with some (i.e. $Apc^{MIN/MIN}$) being completely deficient in this regard, while others (i.e. $Apc^{1638N/1638N}$) retaining about 50% activity (Kielman et al., 2002). Interestingly, these cells exhibit a range of differentiation defects, with those harbouring the most disruptive Apc mutations having the most profound deficiencies. Although $Apc^{MIN/MIN}$ mESCs fail to form teratomas, $Apc^{1638N/1638N}$ teratomas fail to generate detectable neural tissue, cartilage and bone. These observations carried over to *in vitro* analyses: upon LIF withdrawal, only the most severe Apc mutant lines retained an undifferentiated morphology. Dysregulation of β-catenin levels was implicated as the mechanism underlying the differentiation blockade, since mESCs with a targeted deletion of β-catenin exon 3, which encodes the GSK-3 phosphorylation sites and so generates a stabilized form of the protein, also failed to differentiate (Kielman et al., 2002).

4.1.2.2 Ectopic expression of β-catenin reinforces pluripotency

A caveat of studies relying solely on GSK-3 inhibitors and/or Wnt stimulation is that such treatments can have pleiotropic effects that are not limited to β-catenin stabilization due to the many other targets of GSK-3. To directly examine whether increased β-catenin levels are sufficient to reinforce pluripotency, the effects of wild type and stabilized β-catenin overexpression were examined using an inducible system in mESCs (Ogawa et al., 2006). In this system, the expression of stabilized β-catenin, and to a lesser extent wild type β-catenin, prolonged the retention of an undifferentiated ESC morphology after LIF removal. This effect may be attributable, at least in part, to crosstalk between β-catenin and LIF/Stat3 signalling, since GSK-3 inhibition using BIO activated a Stat3 reporter construct (Ogawa et al., 2006). Takao et al. (2007) obtained similar results by expressing stabilized (S33A) β-catenin in mESCs, transiently selecting the cells for expression of puromycin resistance, and assaying for the retention of pluripotency markers after LIF withdrawal (Takao et al., 2007). Finally, our group demonstrated that mESCs stably expressing the S33A form of β-catenin exhibits enhanced self-renewal and delayed loss of pluripotency (Kelly et al., 2011). Intriguingly, similar results were observed using a deletion mutant of β-catenin lacking its C-terminal transactivation domain, which transactivates β-catenin/TCF target genes with ~10% the efficiency of the full length form, suggesting that sustenance of pluripotency by β-catenin occurs independently of TCF proteins.

One possible mechanism through which stabilized β-catenin could restrict exit from pluripotent state is through the direct modulation of the pluripotency regulators. In support of such a mechanism, the formation of β-catenin/Oct-4 complexes was implicated in the regulation of Nanog expression (Takao et al., 2007). Indeed, β-catenin/Oct-4 complexes were detected by co-immunoprecipitation using DKO mESC lysates, or wild type mESCs lysates after stimulation of the Wnt-β-catenin pathway using Wnt3a or GSK-3 inhibition. Furthermore, an established Oct reporter [PORE; (Botquin et al., 1998)] and specific Oct-4 target genes were induced after pathway stimulation, supporting a mechanism whereby β-catenin enhances Oct-4-mediated transcriptional activation (Kelly et al., 2011).

β-catenin also exerts some of its effects on ESCs through its function as a structural protein (see Fig. 2). The role of β-catenin in the cadherin/catenin complex has historically been a structural one, to "hardwire" the cadherin complex to the actin cytoskeleton (Nelson, 2008). More recently, this view has been called into question, owing to controversial studies proposing that a tripartite cadherin/β-catenin/actin complex does not form *in vivo* (Drees et al., 2005; Yamada et al., 2005). Nonetheless, analyses of β-catenin knockout mice strongly support its role as a key regulator of cadherin-mediated cell-cell adhesion (Haegel et al., 1995; Huelsken et al., 2000). Ablation of β-catenin causes embryonic lethality at ~d6.5, with the mutant embryos exhibiting defects in formation of embryonic ectoderm (Haegel et al., 1995), as well as anterior-posterior axis formation (Huelsken et al., 2000). Interestingly, expression levels and *adherens* junction localization of the β-catenin-related protein plakoglobin appear increased in β-catenin$^{(-/-)}$ embryos, suggesting that plakoglobin at least partly rescues adhesion defects caused by β-catenin deficiency (Huelsken et al., 2000). β-catenin$^{(-/-)}$ mESCs cultured in LIF-containing medium express markedly lower levels of Rex-1 transcript when compared to wild type mESCs, suggesting that these cells are prone to spontaneous differentiation (Anton et al., 2007). In support of this possibility, the expression of pluripotency markers is more rapidly lost in these cells after retinoic acid-induced differentiation (Wagner et al., 2010).

The dual roles of β-catenin have been re-examined recently by using genetic "rescue" experiments in Ctnnb1-deficient mESCs with a deletion of exons 3-6 (β-catenin$^{\Delta/\Delta}$) (Lyashenko et al., 2011). In mESCs lacking endogenous full-length β-catenin, plakoglobin partially compensates for the loss of β-catenin with respect to its cell adhesion functions in undifferentiated mESCs. Under standard mESC culture conditions, global gene transcript expression profiles of wild-type and β-catenin knockout mESCs were almost identical, with no observed differences in the transcript levels of the pluripotency factors: Nanog, Oct4, Sox2 or Rex1. Despite their normal appearance, β-catenin$^{\Delta/\Delta}$ mESCs were impaired in their ability to differentiate into mesendoderm and neuronal cell types. Stable re-expression of full-length or truncated β-catenin lacking its transactivation domain, in β-catenin$^{\Delta/\Delta}$ mESCs, was able to rescue their ability to differentiate into definitive endoderm and neuronal cells.

4.1.2.3 Generation and analysis of mouse embryonic stem cells lacking GSK-3

Although GSK-3 inhibition is strongly implicated in the sustenance of pluripotency through the use of small molecular inhibitors, some of these inhibitors are known to elicit off-target

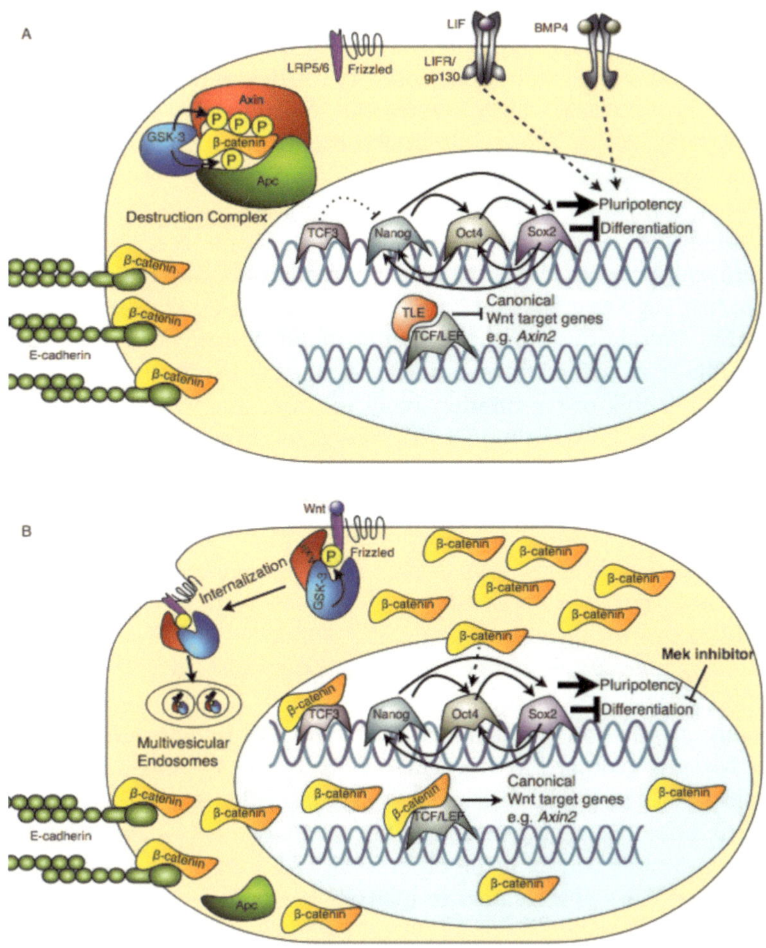

Fig. 2. Model illustrating the central role of β-catenin in mediating the pluripotency-enhancing effects of Wnt signaling (or GSK-3 inhibition/ablation) in mESCs. A. In the absence of Wnt, β-catenin is phosphorylated by GSK-3 in a "destruction complex", which tags it for proteosomal degradation. Levels of cytosolic and nuclear β-catenin are very low and canonical target genes are repressed via TCF/LEF factor interactions with TLE repressors. TCF3 exerts a weak repressive effect on *Nanog* gene expression but LIF and BMP supplementation are sufficient to prevent differentiation and ensure that the network of pluripotent transcription factors (Oct4, Sox2 and Nanog) supports pluripotency. B) In the presence of Wnt, β-catenin is no longer targeted for degradation as the destruction complex is disrupted and GSK-3 is sequestered in multivesicular endosomes (see text for details). β-catenin accumulates in the cytosol and translocates to the nucleus where it activates canonical Wnt/β-catenin target genes and associates with Oct-4 complexes. β-catenin also de-represses TCF3 targets such as *Nanog* through a mechanism that does not require its transactivation domain. Inhibition of Mek signal transduction, together with Wnt treatment (or GSK-3 inhibition) are sufficient to maintain mESCs in a pluripotent ground state.

effects, which could confound the interpretation of results (Bain et al., 2007). To circumvent such limitations, mouse embryonic stem cells lacking both alleles of GSK3α and GSK-3β (double knockout; DKO) were generated and characterized (Doble et al., 2007). The morphology of DKO mESCs is altered dramatically relative to the wild type line, as colonies closely resemble the compact, highly refractile morphology of mESCs treated with CHIR99021 (Kelly et al., 2011). These cells exhibit highly elevated levels of cytoplasmic and nuclear β-catenin and expression of several prototypical targets of the Wnt/β-catenin pathway (*axin2, brachyury and cdx1*) is strongly upregulated (Doble et al., 2007). One of the most striking consequences of GSK-3 ablation in mESCs is their profound inability to differentiate efficiently, particularly to the neurectoderm lineage. Using established differentiation assays such as embryoid bodies or teratomas, in striking contrast to wild type mESCs, DKO mESCs retain the expression of pluripotency markers (Oct-4, Nanog) and fail to express detectable amounts of neural markers, such as β-III-tubulin.

To determine whether this differentiation blockade was due to the hyperactivation of β-catenin/TCF target genes, the original DKO line was modified to allow for the site-specific integration of transgenes at the endogenous GSK-3β locus; thereafter, a dominant negative form of TCF4 (TCF4DN) was constitutively expressed from this locus under the control of the EF1α promoter (Kelly et al., 2011). Surprisingly, although the expression of TCF4DN efficiently attenuated the activation of β-catenin/TCF target genes, these cells were not rescued with regard to their impaired neurectoderm differentiation. However, after normalization of β-catenin levels through the stable expression of β-catenin-specific shRNAs, these cells were capable of neurectoderm differentiation in teratoma assays. These findings suggested that β-catenin, independent of its role as a transactivator of TCF target genes, regulates the acquisition and maintenance of the pluripotent state (Kelly et al., 2011).

Recent studies by two groups have revealed that TCF3 is regulated by β-catenin in a somewhat atypical manner (Wray et al., 2011; Yi et al., 2011). Of the four TCF/LEF family members, TCF3 is unusual in that there is little evidence that β-catenin directly activates TCF3 target genes. In most contexts, the other TCF/LEF family members, upon binding β-catenin, recruit transcriptional activators to increase target gene transcription. In mESCs, β-catenin appears to alleviate the repression of Tcf3 on genes such as Nanog, without recruiting transactivating factors (Wray et al., 2011). Derepression of TCF3 by β-catenin appears to be a significant part of the mechanism through which GSK-3 inhibition exerts its effects on mESC pluripotency (see Fig. 2).

4.2 PI3K signalling and GSK-3 inhibition

4.2.1 Signalling through PI3K and the sustenance of pluripotency

The PI3K pathway regulates various cellular processes, ranging from cell motility to growth, proliferation, survival and metabolism [reviewed in: (Cully et al., 2006; Yuan and Cantley, 2008)]. PI3Ks comprise a family of lipid kinases that phosphorylate the 3′ hydroxyl group of phosphoinositide lipids within membranes. After activation by upstream growth factors, cytokines and receptor tyrosine kinases (RTKs), PI3K phosphorylates PIP_2, to form PIP_3. This process is reversed by the tumour suppressor phosphatase and tensin homolog deleted on chromosome ten (PTEN), which dephosphorylates PIP_3 back to PIP_2, thus terminating the pathway. Upon generation, PIP_3 propagates the signal by binding to the pleckstrin homology (PH) domain of several downstream effector proteins, including the key effector protein kinase phosphoinositide-dependent kinase 1 (PDK1) which targets a variety of members of the AGC family of protein-serine kinases including Protein Kinase B (PKB; also known as Akt), resulting in the activation of PKB through its phosphorylation on T308. PKB elicits diverse effects through its phosphorylation of many signalling intermediates, including GSK-3, the Foxo transcriptional regulators and modulators of apoptotic signalling pathways. Aberrant activation of the PI3K pathway is thought to be a key event in the genesis and maintenance of several types of human cancers and various components of this system are frequently mutated in tumours (Luo et al., 2003; Cully et al., 2006; Yuan and Cantley, 2008).

The first indications that PI3K signalling may regulate self-renewal arose from the realization that signalling through the LIF receptor subunit, gp130, elicits pleiotropic effects through modulation of various signalling pathways, including the Ras/MAPK, JAK/STAT and PI3K pathways (Boeuf et al., 1997; Boeuf et al., 2001; Ernst et al., 1996; Takahashi-Tezuka et al., 1998). Indeed, direct stimulation of mESCs with LIF induces robust PI3K pathway activation (as assessed through the phosphorylation of Akt/PKB on S473, GSK3α/β on S21 and S9, respectively, and a downstream pathway target, ribosomal S6 kinase) (Paling et al., 2004). This effect is blocked by treatment with the PI3K inhibitor, LY294002, or after the expression of a dominant negative form of p85, the regulatory subunit of Class IA PI3Ks (Paling et al., 2004). Inhibition of PI3K activity by these means results in substantially reduced self-renewal of mESCs (Paling et al., 2004), by a mechanism that involves the regulation of Nanog expression (Kingham and Welham, 2009; Storm et al., 2007). Importantly, while treatment of mESCs with LY294002 blocks PI3K-mediated induction of Nanog expression, this effect is reversed by treatment with GSK-3 inhibitors, implicating GSK-3 in the regulation of self-renewal through the modulation of Nanog levels (Storm et al., 2007). More recently, by performing gene expression analyses on mESCs cultured in LIF-containing media with and without PI3K pathway inhibition, Storm et al. identified putative target genes of PI3K signalling (Storm et al., 2009). The majority of identified gene changers (~73%) were downregulated after PI3K inhibition, indicating that they are normally transcriptionally increased downstream of PI3K signalling. As was observed for Nanog (Storm et al., 2007), GSK-3 inhibition using BIO de-repressed the expression of some PI3K targets (Shp-1 and 1700061G19Rik), though many appeared to be regulated in a GSK-3-independent manner (Storm et al., 2009). Interestingly, knockdown of one of the GSK-3 targets, the tyrosine phosphatase Shp-1, compromised ESC self-renewal capacity, an effect that may involve its desphosphorylation of Stat3 (Storm et al., 2009). The identification of additional Shp-1 substrates will help clarify its mechanism of action in ESCs.

To determine whether constitutive activation of PI3K signalling was sufficient to support ESC self-renewal in the absence of LIF, the effects of stably expressing a constitutively active form of PKB (myr-PKB) in ESCs were examined (Watanabe et al., 2006). Both mouse and monkey ESCs expressing myr-PKB retained expression of pluripotency markers (Oct-4, Nanog, Rex-1) after LIF withdrawal, indicating that PI3K activation can reinforce pluripotency under conditions permissive of differentiation (Watanabe et al., 2006). We have confirmed these findings using stable ESC lines expressing activated alleles of PKB and PDK1 (Ling et al. unpublished).

4.2.2 Does PI3K signalling through GSK-3 modulate c-Myc stability to reinforce self-renewal?

PI3K induced phosphorylation and inhibition of GSK-3 does not lead to increased stabilization of β-catenin (likely due to insulation of the relatively small fraction of GSK-3 molecules associated with the Axin/APC destruction complex) (Ng et al., 2009). A possible mechanism through which PI3K/GSK-3 signalling may regulate ESC self-renewal is through the regulation of c-Myc stability. As a target of LIF/Stat3 signalling, maintenance of c-Myc expression reinforces the pluripotent state of mESCs by suppressing differentiation (Cartwright et al., 2005). Upon LIF withdrawal, c-Myc transcript and protein levels in mESCs are sharply reduced, through a mechanism involving both its transcriptional repression and increased targeting for proteasomal degradation. However, ectopic expression of c-Myc harbouring the T58A mutation, abrogating a target site for GSK-3 phosphorylation that promotes ubiquitation and turnover, sustains pluripotency in the absence of LIF (Cartwright et al., 2005). In a follow-up study from the same group, it was suggested that GSK-3's ability to phosphorylate c-Myc relies upon its nuclear translocation. Upon LIF withdrawal of mESCs, GSK-3 localization shifts from the cytoplasm to the nucleus, thereby gaining access to c-Myc, and destabilizing it through phosphorylation on T58. Moreover, expression of activated PKB (myr-PKB), triggering GSK-3 inactivation through its phosphorylation at S21/S9, blocks GSK-3 nuclear translocation and c-Myc T58 phosphorylation/degradation (Bechard and Dalton, 2009). Taken together, downstream LIF signalling involving GSK-3-mediated stabilization of c-Myc, may have a role in the regulation of ESC self-renewal.

5. Conclusions/Future studies

It is clear that in both human and mouse ESCs, inhibition of GSK-3 exerts strong effects on the biology of these unique cells. Whereas, in some contexts, GSK-3 inhibition promotes retention of naïve pluripotency, there are signalling environments in which GSK-3 inhibition promotes lineage-specific (mesendoderm) differentiation. Although we have focused on Wnt/β-catenin and PI3K signalling, GSK-3 also plays a role in other developmentally important signalling pathways, such as the Hedgehog pathway [reviewed in: (Jiang and Hui, 2008)]. How GSK-3-mediated regulation of substrates beyond those in the Wnt and PI-3K signalling pathways modulates ESC properties remains to be thoroughly investigated. Intriguingly, GSK-3 also appears to play a role in directing the expression of the de novo DNA methyltransferase Dnmt3a2, suggesting that GSK-3 may play a role in the regulation of the epigenome (Popkie et al., 2010). The role of GSK-3 in the epigenetic regulation of genes important for pluripotent stem cell self-renewal and differentiation is an exciting new direction for future studies.

6. Acknowledgments

We would like to acknowledge the Canadian Institutes of Health Research (JRW, BWD, KFK) and the Canada Research Chairs Program (BWD) for funding research from our laboratories that is cited in this chapter.

7. References

Van Amerongen, R., and Berns, A., (2005). Re-Evaluating the Role of Frat in Wnt-Signal Transduction. Cell Cycle 4, 1064-1072..

Anton, R., Kestler, H.A., and Kuhl, M. (2007). Beta-catenin signaling contributes to stemness and regulates early differentiation in murine embryonic stem cells. FEBS Lett *581*, 5247-5254.

Arce, L., Yokoyama, N.N., and Waterman, M.L. (2006). Diversity of LEF/TCF action in development and disease. Oncogene *25*, 7492-7504.

Bain, J., Plater, L., Elliott, M., Shpiro, N., Hastie, C.J., McLauchlan, H., Klevernic, I., Arthur, J.S.C., Alessi, D.R., and Cohen, P. (2007). The selectivity of protein kinase inhibitors: a further update. The Biochemical Journal *408*, 297-315.

Bax, B., Carter, P.S., Lewis, C., Guy, A.R., Bridges, A., Tanner, R., Pettman, G., Mannix, C., Culbert, A.A., Brown, M.J., *et al.* (2001). The structure of phosphorylated GSK-3beta complexed with a peptide, FRATtide, that inhibits beta-catenin phosphorylation. Structure *9*, 1143-1152.

Beagle, B., Mi, K., and Johnson, G.V.W. (2009). Phosphorylation of PPP(S/T)P motif of the free LRP6 intracellular domain is not required to activate the Wnt/beta-catenin pathway and attenuate GSK3beta activity. Journal of Cellular Biochemistry *108*, 886-895.

Beattie, G.M., Lopez, A.D., Bucay, N., Hinton, A., Firpo, M.T., King, C.C., and Hayek, A. (2005). Activin A maintains pluripotency of human embryonic stem cells in the absence of feeder layers. Stem Cells *23*, 489-495.

Bechard, M., and Dalton, S. (2009). Subcellular localization of GSK3beta controls embryonic stem cell self-renewal. Mol Cell Biol. 29, 2092-104.

Behrens, J., Jerchow, B.A., Würtele, M., Grimm, J., Asbrand, C., Wirtz, R., Kühl, M., Wedlich, D., and Birchmeier, W. (1998). Functional interaction of an axin homolog, conductin, with beta-catenin, APC, and GSK3beta. Science *280*, 596-599.

Bilic, J., Huang, Y.L., Davidson, G., Zimmermann, T., Cruciat, C.M., Bienz, M., and Niehrs, C. (2007). Wnt induces LRP6 signalosomes and promotes dishevelled-dependent LRP6 phosphorylation. Science *316*, 1619-1622.

Boeuf, H., Hauss, C., Graeve, F.D., Baran, N., and Kedinger, C. (1997). Leukemia inhibitory factor-dependent transcriptional activation in embryonic stem cells. J Cell Biol *138*, 1207-1217.

Boeuf, H., Merienne, K., Jacquot, S., Duval, D., Zeniou, M., Hauss, C., Reinhardt, B., Huss-Garcia, Y., Dierich, A., Frank, D.A., *et al.* (2001). The ribosomal S6 kinases, cAMP-responsive element-binding, and STAT3 proteins are regulated by different leukemia inhibitory factor signaling pathways in mouse embryonic stem cells. J Biol Chem *276*, 46204-46211.

Bone, H.K., Damiano, T., Bartlett, S., Perry, A., Letchford, J., Ripoll, Y.S., Nelson, A.S., and Welham, M.J. (2009). Involvement of GSK-3 in regulation of murine embryonic stem cell self-renewal revealed by a series of bisindolylmaleimides. Chem Biol *16*, 15-27.

Bone, H.K., Nelson, A.S., Goldring, C.E., Tosh, D., and Welham, M.J. (2011). A novel chemically directed route for the generation of definitive endoderm from human embryonic stem cells based on inhibition of GSK-3. Journal of Cell Science.124, 1992-2000.

Botquin, V., Hess, H., Fuhrmann, G., Anastassiadis, C., Gross, M.K., Vriend, G., and Schöler, H.R. (1998). New POU dimer configuration mediates antagonistic control of an osteopontin preimplantation enhancer by Oct-4 and Sox-2. Genes & Development *12*, 2073-2090.

Boyer, L.A., Lee, T.I., Cole, M.F., Johnstone, S.E., Levine, S.S., Zucker, J.P., Guenther, M.G., Kumar, R.M., Murray, H.L., Jenner, R.G., *et al.* (2005). Core transcriptional regulatory circuitry in human embryonic stem cells. Cell *122*, 947-956.

Brantjes, H., Barker, N., van Es, J., and Clevers, H. (2002). TCF: Lady Justice casting the final verdict on the outcome of Wnt signalling. Biological Chemistry *383*, 255-261.

Brons, I.G.M., Smithers, L.E., Trotter, M.W.B., Rugg-Gunn, P., Sun, B., Chuva de Sousa Lopes, S.M., Howlett, S.K., Clarkson, A., Ährlund-Richter, L., Pedersen, R.A., *et al.*

(2007). Derivation of pluripotent epiblast stem cells from mammalian embryos. Nature *448*, 191-195.

Buecker, C., Chen, H.-H., Polo, J.M., Daheron, L., Bu, L., Barakat, T.S., Okwieka, P., Porter, A., Gribnau, J., Hochedlinger, K., *et al.* (2010). A Murine ESC-like State Facilitates Transgenesis and Homologous Recombination in Human Pluripotent Stem Cells. Cell Stem Cell *6*, 535-546.

Buehr, M., Meek, S., Blair, K., Yang, J., Ure, J., Silva, J., Mclay, R., Hall, J., Ying, Q., and Smith, A.G. (2008). Capture of Authentic Embryonic Stem Cells from Rat Blastocysts. Cell *135*, 1287-1298.

Buescher, J.L., and Phiel, C.J. (2010). A noncatalytic domain of glycogen synthase kinase-3 (GSK-3) is essential for activity. J Biol Chem *285*, 7957-7963.

Cartwright, P., McLean, C., Sheppard, A., Rivett, D., Jones, K., and Dalton, S. (2005). LIF/STAT3 controls ES cell self-renewal and pluripotency by a Myc-dependent mechanism. Development *132*, 885-896.

Chen, G., and Courey, A.J. (2000). Groucho/TLE family proteins and transcriptional repression. Gene *249*, 1-16.

Chew, J.L., Loh, Y.H., Zhang, W., Chen, X., Tam, W.L., Yeap, L.S., Li, P., Ang, Y.S., Lim, B., Robson, P., *et al.* (2005). Reciprocal transcriptional regulation of Pou5f1 and Sox2 via the Oct4/Sox2 complex in embryonic stem cells. Mol Cell Biol *25*, 6031-6046.

Cole, A.R., Knebel, A., Morrice, N.A., Robertson, L.A., Irving, A.J., Connolly, C.N., and Sutherland, C. (2004). GSK-3 phosphorylation of the Alzheimer epitope within collapsin response mediator proteins regulates axon elongation in primary neurons. J Biol Chem *279*, 50176-50180.

Cole, M.F., Johnstone, S.E., Newman, J.J., Kagey, M.H., and Young, R.A. (2008). Tcf3 is an integral component of the core regulatory circuitry of embryonic stem cells. Genes Dev *22*, 746-755.

Cross, D.A., Alessi, D.R., Cohen, P., Andjelkovich, M., and Hemmings, B.A. (1995). Inhibition of glycogen synthase kinase-3 by insulin mediated by protein kinase B. Nature *378*, 785-789.

Cully, M., You, H., Levine, A.J., and Mak, T.W. (2006). Beyond PTEN mutations: the PI3K pathway as an integrator of multiple inputs during tumorigenesis. Nature Rev. Cancer *6*, 184-192.

Daheron, L., Opitz, S.L., Zaehres, H., Lensch, W.M., Andrews, P.W., Itskovitz-Eldor, J., and Daley, G.Q. (2004). LIF/STAT3 signaling fails to maintain self-renewal of human embryonic stem cells. Stem Cells *22*, 770-778.

Dajani, R., Fraser, E., Roe, S.M., Young, N., Good, V., Dale, T.C., and Pearl, L.H. (2001). Crystal structure of glycogen synthase kinase 3 beta: structural basis for phosphate-primed substrate specificity and autoinhibition. Cell *105*, 721-732.

Daniels, D.L., and Weis, W.I. (2005). Beta-catenin directly displaces Groucho/TLE repressors from Tcf/Lef in Wnt-mediated transcription activation. Nat Struct Mol Biol *12*, 364-371.

Doble, B.W., Patel, S., Wood, G.A., Kockeritz, L.K., and Woodgett, J.R. (2007). Functional redundancy of GSK-3alpha and GSK-3beta in Wnt/beta-catenin signaling shown by using an allelic series of embryonic stem cell lines. Dev Cell *12*, 957-971.

Doble, B.W., and Woodgett, J.R. (2003). GSK-3: tricks of the trade for a multi-tasking kinase. J Cell Sci *116*, 1175-1186.

Dominguez, I., Itoh, K., and Sokol, S.Y. (1995). Role of glycogen synthase kinase 3 beta as a negative regulator of dorsoventral axis formation in Xenopus embryos. Proc Natl Acad Sci U S A *92*, 8498-8502.

Dravid, G., Ye, Z., Hammond, H., Chen, G., Pyle, A., Donovan, P., Yu, X., and Cheng, L. (2005). Defining the role of Wnt/beta-catenin signaling in the survival, proliferation, and self-renewal of human embryonic stem cells. Stem Cells (Dayton, Ohio) *23*, 1489-1501.

Drees, F., Pokutta, S., Yamada, S., Nelson, W.J., and Weis, W.I. (2005). Alpha-catenin is a molecular switch that binds E-cadherin-beta-catenin and regulates actin-filament assembly. Cell *123*, 903-915.

Embi, N., Rylatt, D.B., and Cohen, P. (1980). Glycogen synthase kinase-3 from rabbit skeletal muscle. Separation from cyclic-AMP-dependent protein kinase and phosphorylase kinase. Eur J Biochem *107*, 519-527.

Ernst, M., Oates, A., and Dunn, A.R. (1996). Gp130-mediated signal transduction in embryonic stem cells involves activation of Jak and Ras/mitogen-activated protein kinase pathways. J Biol Chem *271*, 30136-30143.

Evans, M.J., and Kaufman, M.H. (1981). Establishment in culture of pluripotential cells from mouse embryos. Nature *292*, 154-156.

Fang, X., Yu, S.X., Lu, Y., Bast, R.C., Woodgett, J.R., and Mills, G.B. (2000). Phosphorylation and inactivation of glycogen synthase kinase 3 by protein kinase A. Proc Natl Acad Sci U S A *97*, 11960-11965.

Fiol, C.J., Mahrenholz, A.M., Wang, Y., Roeske, R.W., and Roach, P.J. (1987). Formation of protein kinase recognition sites by covalent modification of the substrate. Molecular mechanism for the synergistic action of casein kinase II and glycogen synthase kinase 3. J Biol Chem *262*, 14042-14048.

Gertsenstein, M., Nutter, L.M.J., Reid, T., Pereira, M., Stanford, W.L., Rossant, J., and Nagy, A. (2010). Efficient generation of germ line transmitting chimeras from C57BL/6N ES cells by aggregation with outbred host embryos. PLoS ONE *5*, e11260.

Giles, R.H., van Es, J.H., and Clevers, H. (2003). Caught up in a Wnt storm: Wnt signaling in cancer. Biochim Biophys Acta *1653*, 1-24.

Goode, N., Hughes, K., Woodgett, J.R., and Parker, P.J. (1992). Differential regulation of glycogen synthase kinase-3 beta by protein kinase C isotypes. J Biol Chem *267*, 16878-16882.

Haegel, H., Larue, L., Ohsugi, M., Fedorov, L., Herrenknecht, K., and Kemler, R. (1995). Lack of beta-catenin affects mouse development at gastrulation. Development *121*, 3529-3537.

Hanna, J., Cheng, A.W., Saha, K., Kim, J., Lengner, C.J., Soldner, F., Cassady, J.P., Muffat, J., Carey, B.W., and Jaenisch, R. (2010). Human embryonic stem cells with biological and epigenetic characteristics similar to those of mouse ESCs. Proc Natl Acad Sci U S A *107*, 9222-9227.

Hanna, J., Markoulaki, S., Mitalipova, M., Cheng, A.W., Cassady, J.P., Staerk, J., Carey, B.W., Lengner, C.J., Foreman, R., Love, J., *et al.* (2009). Metastable pluripotent states in NOD-mouse-derived ESCs. Cell Stem Cell *4*, 513-524.

Hao, J., Li, T.G., Qi, X., Zhao, D.F., and Zhao, G.Q. (2006). WNT/beta-catenin pathway up-regulates Stat3 and converges on LIF to prevent differentiation of mouse embryonic stem cells. Dev Biol *290*, 81-91.

Hart, M.J., de los Santos, R., Albert, I.N., Rubinfeld, B., and Polakis, P. (1998). Downregulation of beta-catenin by human Axin and its association with the APC tumor suppressor, beta-catenin and GSK3 beta. Curr Biol *8*, 573-581.

He, X., Saint-Jeannet, J.P., Woodgett, J.R., Varmus, H.E., and Dawid, I.B. (1995). Glycogen synthase kinase-3 and dorsoventral patterning in Xenopus embryos. Nature *374*, 617-622.

He, X., Semenov, M., Tamai, K., and Zeng, X. (2004). LDL receptor-related proteins 5 and 6 in Wnt/beta-catenin signaling: arrows point the way. Development *131*, 1663-1677.

Hoeflich, K.P., Luo, J., Rubie, E.A., Tsao, M.S., Jin, O., and Woodgett, J.R. (2000). Requirement for glycogen synthase kinase-3beta in cell survival and NF-kappaB activation. Nature *406*, 86-90.

Huelsken, J., Vogel, R., Brinkmann, V., Erdmann, B., Birchmeier, C., and Birchmeier, W. (2000). Requirement for beta-catenin in anterior-posterior axis formation in mice. J Cell Biol *148*, 567-578.

Humphrey, R.K., Beattie, G.M., Lopez, A.D., Bucay, N., King, C.C., Firpo, M.T., Rose-John, S., and Hayek, A. (2004). Maintenance of pluripotency in human embryonic stem cells is STAT3 independent. Stem Cells *22*, 522-530.

Hurlstone, A., and Clevers, H. (2002). T-cell factors: turn-ons and turn-offs. Embo J *21*, 2303-2311.

Ikeda, S., Kishida, S., Yamamoto, H., Murai, H., Koyama, S., and Kikuchi, A. (1998a). Axin, a negative regulator of the Wnt signaling pathway, forms a complex with GSK-3beta and beta-catenin and promotes GSK-3beta-dependent phosphorylation of beta-catenin. EMBO J *17*, 1371-1384.

Ikeda, S., Kishida, S., Yamamoto, H., Murai, H., Koyama, S., and Kikuchi, A. (1998b). Axin, a negative regulator of the Wnt signaling pathway, forms a complex with GSK-3beta and beta-catenin and promotes GSK-3beta-dependent phosphorylation of beta-catenin. EMBO J *17*, 1371-1384.

James, D., Levine, A.J., Besser, D., and Hemmati-Brivanlou, A. (2005). TGFbeta/activin/nodal signaling is necessary for the maintenance of pluripotency in human embryonic stem cells. Development (Cambridge, England) *132*, 1273-1282.

Jho, E., Lomvardas, S., and Costantini, F. (1999). A GSK3beta phosphorylation site in axin modulates interaction with beta-catenin and Tcf-mediated gene expression. Biochem Biophys Res Commun *266*, 28-35.

Jiang, J., and Hui, C.-C. (2008). Hedgehog signaling in development and cancer. Developmental Cell *15*, 801-812.

Kaidanovich-Beilin, O., Lipina, T.V., Takao, K., van Eede, M., Hattori, S., Laliberté, C., Khan, M., Okamoto, K., Chambers, J.W., Fletcher, P.J., *et al.* (2009). Abnormalities in brain structure and behavior in GSK-3alpha mutant mice. Mol Brain *2*, 35.

Katanaev, V.L., Ponzielli, R., Semeriva, M., and Tomlinson, A. (2005). Trimeric G protein-dependent frizzled signaling in Drosophila. Cell *120*, 111-122.

Kelly, K.F., Ng, D.Y., Jayakumaran, G., Wood, G.A., Koide, H., and Doble, B.W. (2011). β-catenin enhances Oct-4 activity and reinforces pluripotency through a TCF-independent mechanism. Cell Stem Cell *8*, 214-227.

Kerkela, R., Kockeritz, L., Macaulay, K., Zhou, J., Doble, B.W., Beahm, C., Greytak, S., Woulfe, K., Trivedi, C.M., Woodgett, J.R., *et al.* (2008). Deletion of GSK-3beta in mice leads to hypertrophic cardiomyopathy secondary to cardiomyoblast hyperproliferation. J Clin Invest *118*, 3609-3618.

Kielman, M.F., Rindapaa, M., Gaspar, C., van Poppel, N., Breukel, C., van Leeuwen, S., Taketo, M.M., Roberts, S., Smits, R., and Fodde, R. (2002). Apc modulates embryonic stem-cell differentiation by controlling the dosage of beta-catenin signaling. Nat Genet *32*, 594-605.

Kingham, E., and Welham, M.J. (2009). Distinct roles for isoforms of the catalytic subunit of class-IA PI3K in the regulation of behaviour of murine embryonic stem cells. J Cell Sci 122, 2311-21.

Kiyonari, H., Kaneko, M., Abe, S.-I., and Aizawa, S. (2010). Three inhibitors of FGF receptor, ERK, and GSK3 establishes germline-competent embryonic stem cells of C57BL/6N mouse strain with high efficiency and stability. Genesis *48*, 317-327.

Kwong, L.N., and Dove, W.F. (2009). APC and its modifiers in colon cancer. Advances in experimental medicine and biology *656*, 85-106.

Levenstein, M.E., Ludwig, T.E., Xu, R.H., Llanas, R.A., VanDenHeuvel-Kramer, K., Manning, D., and Thomson, J.A. (2006). Basic fibroblast growth factor support of human embryonic stem cell self-renewal. Stem Cells *24*, 568-574.

Li, W., Wei, W., Zhu, S., Zhu, J., Shi, Y., Lin, T., Hao, E., Hayek, A., Deng, H., and Ding, S. (2009). Generation of rat and human induced pluripotent stem cells by combining genetic reprogramming and chemical inhibitors. Cell Stem Cell *4*, 16-19.

Liu, X., Rubin, J.S., and Kimmel, A.R. (2005). Rapid, Wnt-induced changes in GSK3beta associations that regulate beta-catenin stabilization are mediated by Galpha proteins. Curr Biol *15*, 1989-1997.

Lochhead, P.A., Kinstrie, R., Sibbet, G., Rawjee, T., Morrice, N., and Cleghon, V. (2006). A chaperone-dependent GSK3beta transitional intermediate mediates activation-loop autophosphorylation. Molecular Cell *24*, 627-633.

Logan, C.Y., and Nusse, R. (2004). The Wnt signaling pathway in development and disease. Annual review of cell and developmental biology *20*, 781-810.

Loh, Y.H., Wu, Q., Chew, J.L., Vega, V.B., Zhang, W., Chen, X., Bourque, G., George, J., Leong, B., Liu, J., *et al.* (2006). The Oct4 and Nanog transcription network regulates pluripotency in mouse embryonic stem cells. Nat Genet *38*, 431-440.

Ludwig, T.E., Bergendahl, V., Levenstein, M.E., Yu, J., Probasco, M.D., and Thomson, J.A. (2006). Feeder-independent culture of human embryonic stem cells. Nat Methods *3*, 637-646.

Luo, J., Manning, B.D., and Cantley, L.C. (2003). Targeting the PI3K-Akt pathway in human cancer: rationale and promise. Cancer Cell *4*, 257-262.

Lyashenko, N., Winter, M., Migliorini, D., Biechele, T., Moon, R.T., and Hartmann, C. (2011). Differential requirement for the dual functions of beta-catenin in embryonic stem cell self-renewal and germ layer formation. Nat Cell Biol *13*, 1-10.

Macaulay, K., Doble, B.W., Patel, S., Hansotia, T., Sinclair, E.M., Drucker, D.J., Nagy, A., and Woodgett, J.R. (2007). Glycogen synthase kinase 3alpha-specific regulation of murine hepatic glycogen metabolism. Cell Metabolism *6*, 329-337.

Mi, K., Dolan, P.J., and Johnson, G.V.W. (2006). The low density lipoprotein receptor-related protein 6 interacts with glycogen synthase kinase 3 and attenuates activity. J Biol Chem *281*, 4787-4794.

Molofsky, A.V., Pardal, R., and Morrison, S.J. (2004). Diverse mechanisms regulate stem cell self-renewal. Current opinion in cell biology *16*, 700-707.

Morin, P.J., and Weeraratna, A.T. (2003). Wnt signaling in human cancer. Cancer Treat Res *115*, 169-187.

Mukai, F., Ishiguro, K., Sano, Y., and Fujita, S.C. (2002). Alternative splicing isoform of tau protein kinase I/glycogen synthase kinase 3beta. Journal of Neurochemistry *81*, 1073-1083.

Nelson, W.J. (2008). Regulation of cell-cell adhesion by the cadherin-catenin complex. Biochemical Society transactions *36*, 149-155.

Ng, S.S., Mahmoudi, T., Danenberg, E., Bejaoui, I., de Lau, W., Korswagen, H.C., Schutte, M., and Clevers, H. (2009). Phosphatidylinositol 3-kinase signaling does not activate the wnt cascade. Journal of Biological Chemistry *284*, 35308-35313.

Nichols, J., and Smith, A. (2009). Naive and primed pluripotent states. Cell Stem Cell *4*, 487-492.

Niwa, H., Burdon, T., Chambers, I., and Smith, A. (1998). Self-renewal of pluripotent embryonic stem cells is mediated via activation of STAT3. Genes Dev *12*, 2048-2060.

Ogawa, K., Nishinakamura, R., Iwamatsu, Y., Shimosato, D., and Niwa, H. (2006). Synergistic action of Wnt and LIF in maintaining pluripotency of mouse ES cells. Biochem Biophys Res Commun *343*, 159-166.

Orsulic, S., and Peifer, M. (1996). Cell-cell signalling: Wingless lands at last. Curr Biol *6*, 1363-1367.

Paling, N.R., Wheadon, H., Bone, H.K., and Welham, M.J. (2004). Regulation of embryonic stem cell self-renewal by phosphoinositide 3-kinase-dependent signaling. J Biol Chem *279*, 48063-48070.

Pan, G., and Thomson, J.A. (2007). Nanog and transcriptional networks in embryonic stem cell pluripotency. Cell Res *17*, 42-49.

Peifer, M., Pai, L.M., and Casey, M. (1994). Phosphorylation of the Drosophila adherens junction protein Armadillo: roles for wingless signal and zeste-white 3 kinase. Dev Biol *166*, 543-556.

Pereira, L., Yi, F., and Merrill, B.J. (2006). Repression of Nanog gene transcription by Tcf3 limits embryonic stem cell self-renewal. Mol Cell Biol *26*, 7479-7491.

Piao, S., Lee, S.-H., Kim, H., Yum, S., Stamos, J.L., Xu, Y., Lee, S.-J., Lee, J., Oh, S., Han, J.-K., *et al.* (2008). Direct inhibition of GSK3beta by the phosphorylated cytoplasmic domain of LRP6 in Wnt/beta-catenin signaling. PLoS ONE *3*, e4046.

Popkie, A.P., Zeidner, L.C., Albrecht, A.M., D'Ippolito, A., Eckardt, S., Newsom, D.E., Groden, J., Doble, B.W., Aronow, B., McLaughlin, K.J., *et al.* (2010). Phosphatidylinositol 3-kinase (PI3K) signaling via glycogen synthase kinase-3 (Gsk-3) regulates DNA methylation of imprinted loci. Journal of Biological Chemistry *285*, 41337-41347.

Rodda, D.J., Chew, J.L., Lim, L.H., Loh, Y.H., Wang, B., Ng, H.H., and Robson, P. (2005). Transcriptional regulation of nanog by OCT4 and SOX2. J Biol Chem *280*, 24731-24737.

Rossant, J. (2001). Stem cells from the Mammalian blastocyst. Stem Cells *19*, 477-482.

Sato, H., Amagai, K., Shimizukawa, R., and Tamai, Y. (2009). Stable generation of serum- and feeder-free embryonic stem cell-derived mice with full germline-competency by using a GSK3 specific inhibitor. Genesis 47, 414-22.

Sato, N., Meijer, L., Skaltsounis, L., Greengard, P., and Brivanlou, A.H. (2004). Maintenance of pluripotency in human and mouse embryonic stem cells through activation of Wnt signaling by a pharmacological GSK-3-specific inhibitor. Nat Med *10*, 55 - 63.

Schaffer, B., Wiedau-Pazos, M., and Geschwind, D.H. (2003). Gene structure and alternative splicing of glycogen synthase kinase 3 beta (GSK-3beta) in neural and non-neural tissues. Gene *302*, 73-81.

Schulte, G., and Bryja, V. (2007). The Frizzled family of unconventional G-protein-coupled receptors. Trends Pharmacol Sci *28*, 518-525.

Schweizer, L., and Varmus, H. (2003). Wnt/Wingless signaling through beta-catenin requires the function of both LRP/Arrow and frizzled classes of receptors. BMC Cell Biol *4*, 4.

Singla, D.K., Schneider, D.J., LeWinter, M.M., and Sobel, B.E. (2006). wnt3a but not wnt11 supports self-renewal of embryonic stem cells. Biochem Biophys Res Commun *345*, 789-795.

Soutar, M.P.M., Kim, W.-Y., Williamson, R., Peggie, M., Hastie, C.J., McLauchlan, H., Snider, W.D., Gordon-Weeks, P.R., and Sutherland, C. (2010). Evidence that glycogen synthase kinase-3 isoforms have distinct substrate preference in the brain. Journal of Neurochemistry *115*, 974-983.

Storm, M., Kumpfmueller, B., Thompson, B., Kolde, R., Vilo, J., Hummel, O., Schulz, H., and Welham, M.J. (2009). Characterization of the Phosphoinositide 3-Kinase-Dependent Transcriptome in Murine Embryonic Stem Cells: Identification of Novel Regulators of Pluripotency. Stem Cells *27*, 764-775.

Storm, M.P., Bone, H.K., Beck, C.G., Bourillot, P.Y., Schreiber, V., Damiano, T., Nelson, A., Savatier, P., and Welham, M.J. (2007). Regulation of Nanog expression by phosphoinositide 3-kinase-dependent signaling in murine embryonic stem cells. J Biol Chem *282*, 6265-6273.

Sutherland, C., and Cohen, P. (1994). The alpha-isoform of glycogen synthase kinase-3 from rabbit skeletal muscle is inactivated by p70 S6 kinase or MAP kinase-activated protein kinase-1 in vitro. FEBS Lett *338*, 37-42.

Taelman, V.F., Dobrowolski, R., Plouhinec, J.-L., Fuentealba, L.C., Vorwald, P.P., Gumper, I., Sabatini, D.D., and De Robertis, E.M. (2010). Wnt signaling requires sequestration of glycogen synthase kinase 3 inside multivesicular endosomes. Cell *143*, 1136-1148.

Takahashi-Tezuka, M., Yoshida, Y., Fukada, T., Ohtani, T., Yamanaka, Y., Nishida, K., Nakajima, K., Hibi, M., and Hirano, T. (1998). Gab1 acts as an adapter molecule linking the cytokine receptor gp130 to ERK mitogen-activated protein kinase. Mol Cell Biol *18*, 4109-4117.

Takao, Y., Yokota, T., and Koide, H. (2007). Beta-catenin up-regulates Nanog expression through interaction with Oct-3/4 in embryonic stem cells. Biochemical and biophysical research communications *353*, 699-705.

Tamai, K., Semenov, M., Kato, Y., Spokony, R., Liu, C., Katsuyama, Y., Hess, F., Saint-Jeannet, J.P., and He, X. (2000). LDL-receptor-related proteins in Wnt signal transduction. Nature *407*, 530-535.

ter Haar, E., Coll, J.T., Austen, D.A., Hsiao, H.M., Swenson, L., and Jain, J. (2001). Structure of GSK3beta reveals a primed phosphorylation mechanism. Nat Struct Biol *8*, 593-596.

Tesar, P.J., Chenoweth, J.G., Brook, F.A., Davies, T.J., Evans, E.P., Mack, D.L., Gardner, R.L., and Mckay, R.D.G. (2007). New cell lines from mouse epiblast share defining features with human embryonic stem cells. Nature *448*, 196-199.

Thornton, T.M., Pedraza-Alva, G., Deng, B., Wood, C.D., Aronshtam, A., Clements, J.L., Sabio, G., Davis, R.J., Matthews, D.E., Doble, B., *et al.* (2008). Phosphorylation by p38 MAPK as an alternative pathway for GSK3beta inactivation. Science *320*, 667-670.

Torres, M.A., Eldar-Finkelman, H., Krebs, E.G., and Moon, R.T. (1999). Regulation of ribosomal S6 protein kinase-p90(rsk), glycogen synthase kinase 3, and beta-catenin in early Xenopus development. Mol Cell Biol *19*, 1427-1437.

Vallier, L., Alexander, M., and Pedersen, R.A. (2005). Activin/Nodal and FGF pathways cooperate to maintain pluripotency of human embryonic stem cells. Journal of Cell Science *118*, 4495-4509.

Van Amerongen, R. (2005). Frat is dispensable for canonical Wnt signaling in mammals. Genes & Development *19*, 425-430.

Wagner, R.T., Xu, X., Yi, F., Merrill, B.J., and Cooney, A.J. (2010). Canonical Wnt/β-catenin regulation of liver receptor homolog-1 mediates pluripotency gene expression. Stem Cells *28*, 1794-1804.

Watanabe, S., Umehara, H., Murayama, K., Okabe, M., Kimura, T., and Nakano, T. (2006). Activation of Akt signaling is sufficient to maintain pluripotency in mouse and primate embryonic stem cells. Oncogene *25*, 2697-2707.

Wehrli, M., Dougan, S.T., Caldwell, K., O'Keefe, L., Schwartz, S., Vaizel-Ohayon, D., Schejter, E., Tomlinson, A., and DiNardo, S. (2000). arrow encodes an LDL-receptor-related protein essential for Wingless signalling. Nature *407*, 527-530.

Woodgett, J.R. (1990). Molecular cloning and expression of glycogen synthase kinase-3/factor A. EMBO J *9*, 2431-2438.

Woodgett, J.R., Davison, M.T., and Cohen, P. (1983). The calmodulin-dependent glycogen synthase kinase from rabbit skeletal muscle. Purification, subunit structure and substrate specificity. Eur J Biochem *136*, 481-487.

Wray, J., Kalkan, T., Gomez-Lopez, S., Eckardt, D., Cook, A., Kemler, R., and Smith, A. (2011). Inhibition of glycogen synthase kinase-3 alleviates Tcf3 repression of the pluripotency network and increases embryonic stem cell resistance to differentiation. Nature Cell Biology *13*, 1-9.

Wu, G., Huang, H., Garcia Abreu, J., and He, X.C. (2009). Inhibition of GSK3 phosphorylation of beta-catenin via phosphorylated PPPSPXS motifs of Wnt coreceptor LRP6. PLoS ONE *4*, e4926.

Xu, R.H., Chen, X., Li, D.S., Li, R., Addicks, G.C., Glennon, C., Zwaka, T.P., and Thomson, J.A. (2002). BMP4 initiates human embryonic stem cell differentiation to trophoblast. Nat Biotechnol *20*, 1261-1264.

Yamada, S., Pokutta, S., Drees, F., Weis, W.I., and Nelson, W.J. (2005). Deconstructing the cadherin-catenin-actin complex. Cell *123*, 889-901.

Yamamoto, H., Kishida, S., Kishida, M., Ikeda, S., Takada, S., and Kikuchi, A. (1999). Phosphorylation of axin, a Wnt signal negative regulator, by glycogen synthase kinase-3beta regulates its stability. J Biol Chem *274*, 10681-10684.

Yi, F., Pereira, L., Hoffman, J.A., Shy, B.R., Yuen, C.M., Liu, D.R., and Merrill, B.J. (2011). Opposing effects of Tcf3 and Tcf1 control Wnt stimulation of embryonic stem cell self-renewal. Nature Cell Biology *13*, 1-11.

Yi, F., Pereira, L., and Merrill, B.J. (2008). Tcf3 Functions as a Steady State Limiter of Transcriptional Programs of Mouse Embryonic Stem Cell Self Renewal. Stem Cells 26, 1951-60.

Ying, Q.L., Nichols, J., Chambers, I., and Smith, A. (2003). BMP induction of Id proteins suppresses differentiation and sustains embryonic stem cell self-renewal in collaboration with STAT3. Cell *115*, 281-292.

Ying, Q.L., and Smith, A.G. (2003). Defined conditions for neural commitment and differentiation. Methods Enzymol *365*, 327-341.

Ying, Q.L., Wray, J., Nichols, J., Batlle-Morera, L., Doble, B., Woodgett, J., Cohen, P., and Smith, A. (2008). The ground state of embryonic stem cell self-renewal. Nature *453*, 519-523.

Yuan, T.L., and Cantley, L.C. (2008). PI3K pathway alterations in cancer: variations on a theme. Oncogene *27*, 5497-5510.

Zeng, X., Huang, H., Tamai, K., Zhang, X., Harada, Y., Yokota, C., Almeida, K., Wang, J., Doble, B., Woodgett, J., *et al.* (2008). Initiation of Wnt signaling: control of Wnt coreceptor Lrp6 phosphorylation/activation via frizzled, dishevelled and axin functions. Development *135*, 367-375.

Zeng, X., Tamai, K., Doble, B., Li, S., Huang, H., Habas, R., Okamura, H., Woodgett, J., and He, X. (2005). A dual-kinase mechanism for Wnt co-receptor phosphorylation and activation. Nature *438*, 873-877.

Zhou, J., Lal, H., Chen, X., Shang, X., Song, J., Li, Y., Kerkelä, R., Doble, B.W., Macaulay, K., DeCaul, M., *et al.* (2010). GSK-3alpha directly regulates beta-adrenergic signaling and the response of the heart to hemodynamic stress in mice. J Clin Invest *120*, 2280-2291.

Part 5

Lessons from Development

21

Embryonic Stem Cells and the Germ Cell Lineage

Cyril Ramathal, Renee Reijo Pera and Paul Turek
Stanford University & The Turek Clinic
U.S.A.

1. Introduction

Stem cells possess the unique ability to either propagate by self-renewal or to differentiate to mature tissues under the influence of appropriate molecular cues. This remarkable feature, also termed "stem cell potency" has been the focus of both a historical and current research spotlight on stem cells, and particularly the potential for human stem cells in regenerative medicine and organ transplantation. There are several classes of stem cells, all varying in degrees of pluripotency. Depending on the developmental stage from where they originate, stem cell potencies range from totipotency (ability to transform into all cell types), pluripotency (most cell types), multipotency (many cell types), oligopotency (few cell types) to unipotency (one cell type) (Smith, 2001). Among these, embryonic stem cells (ESCs) represent the prototypical pluripotent stem cells.

Embryonic stem cells are localized to the inner cell mass of the developing mammalian blastocyst and can give rise to all future cell types and tissues of the organism. Additionally, a small population of embryonic stem cells is allocated to the germline (primordial germ cells). The evolution of these germline stem cells is different from the remaining cells which undergo gastrulation and give rise to the three germ layers (endoderm, mesoderm and ectoderm) during development (Ewen & Koopman, 2010). The survival, gonadal migration and proper epigenetic programming of primordial germ cells (PGCs) are major, distinct and early events that have an impact on future fertility and the successful transmission of genetic information from parent to offspring.

Finally, stem cells found in neonatal tissues, amniotic fluid, cord blood, and adult tissues are categorized as adult stem cells and are usually multipotent, oligopotent or unipotent. Adult stem cells are clinically important due to the absence of ethical and federal restrictions on their use. Furthermore, they are valued for their relative abundance and accessibility in somatic tissues. In contrast, human embryonic stem cells have been the focus of substantial research and discussion because of their unique potential to differentiate into almost all body cell types and tissues and the relative ease with which they are propagated in cell culture. The search for an alternative source of pluripotent stem cells other than from human embryos is a much sought after goal in the stem cell research community for several reasons. First, diversity in stem cells sources should be explored to better understand and evaluate their various clinical utilities. Second, immunological matching of stem cells to recipients is necessary to avoid rejection in stem cell-based regenerative therapies. Hence, the advent of induced pluripotency and nuclear reprogramming strategies have been well received as alternatives to derive human stem cells (Yamanaka & Blau, 2010). Finally, the derivation and use of stem cells require adherence to moral, ethical and political guidelines that are constantly changing. This chapter will review the formation of the human germline from embryonic stem cells and the subsequent derivation of gametes from mouse and human pluripotent stem cells, with an emphasis on the male germline.

1.1 Mouse and human embryonic stem cells

The first forays into understanding pluripotency in the mammalian embryo began with rather simple experiments in which pieces of mouse germ cell tumors known as teratocarcinomas were grown in culture and a variety of pluripotent cell types emerged (Solder, 2006). One cell type, termed Embryonal Carcinoma Cells (ECCs), produced all three germ layers as well as the original tumorigenic cells when isolated and transplanted into host mice. This established for the first time the pluripotent nature of embryonic cells. In the mouse pre-implantation embryo, embryonic stem cells (mESCs) first form within the inner cell mass (ICM). In studies by Evans and Kauffman and others, cells of the ICM were isolated from late blastocysts and grown in culture (Evans & Kaufman, 1981; Martin, 1981). These cells proliferated and demonstrated the capacity of self-renewal. The subsequent addition of Leukemia Inhibitory Factor (LIF) promoted the self-renewal of mESCs in culture by acting through the Stat3 pathway (Ying et al., 2008). Further studies showed that the ICM-derived cells could, if reintroduced into an embryo lacking an ICM, recolonize and form all three germ layers of the embryo, including the germline (Bradley et al., 1984). MESCs are characterized by the strong and persistent expression of important pluripotency genes such as *Oct4, Sox2, Nanog* and *Tcf,* among others. Thus, in mice, the pluripotent nature of cells derived from the inner cells mass was established relatively rapidly and easily than in the human.

In the human, embryonic stem cells (hESCs) are derived in a very similar manner to mouse ESCs from explants of the Inner Cell Mass (ICM) of *in vitro* fertilized blastocysts. Thomson *et al.* first isolated hESCs from pre-implantation human blastocysts and analyzed their morphology and expression of pluripotency genes (Thomson et al., 1998). Although they express high levels of pluripotency genes, hESCs share a morphology that is more consistent with that of early epiblast stem cells of the mouse (mEpiSCs). Specifically, they are "flattened." In addition, they depend on basic FGF and Activin signaling for self-renewal, possess large nuclear to cytoplasmic ratios, and respond less favorably to single cell clonability in contrast to mESCs (Vallier et al., 2005, 2009). While the TGFβ/Activin/Nodal signaling pathways are undoubtedly important for hESC pluripotency, the molecular mechanisms underlying these observations have yet to be explained. Meanwhile, the transcription factors Oct4, Nanog and Sox2 are clearly all involved in hESC pluripotency and newer molecules such as the WNTs (Wingless-Type MMTV Integration Site Family Members) have been implicated to drive hESC self-renewal through canonical signaling pathways (Sato et al., 2004). The functions of Oct4, Sox2 and Nanog are regarded as indispensable for hESC self-renewal as their expression typically reduces when hESCs differentiate. Finally, hESCs and mESCs are similar in that they both express a reportoire of cell surface proteins that distinguish them from other cell types, including SSEA-3, SSEA-4, Tra1-60 and Tra1-81, markers that are routinely used for characterization of ES and iPS cell lines.

Despite the fact that mESCs and hESCs are derived from seemingly equivalent epiblast cells in the blastocyst, there remains the question of whether hESCs possess as 'naïve' a state of pluripotency as do mESCs. This is because mESCs in culture behave more like the early mouse epiblast and demonstrate unrestricted potential for differentiation (Hanna & Jaenisch, 2010). Meanwhile, hESCs share some properties with mouse epiblast stem cells (pluripotency) but may be more restricted than mESCs (Vallier & Pedersen, 2009). This may be the case because current hESC derivation methods from ICM could actually select for cells in a different developmental stage than that from the mouse ICM (Nichols & Smith, 2011). Additionally, it has been observed that ESCS and embryonic germ cells (EGCs) share several biomarkers making it difficult to distinguish between the embryonic, pluripotent cells and primitive germ cells. The intimate relationship that exists between germline and embryonic stem cells in mice and humans is critically important to understand in the future, and will vastly improve current and future efforts to derive 'artificial' gametes.

1.2 Nuclear reprogramming and induced pluripotency

A set of landmark studies conducted by Yamanaka and others recently revealed that just four factors normally produced in the embryo are sufficient to reprogram differentiated, non-embryonic cells into pluripotent stem cells. This important finding bypasses many of the ethical issues associated with the use of human embryos, and also evades potential problems with immune rejection that might affect patients upon stem cells transplantation. The four Yamanaka reprogramming factors, Oct4, Sox2, Klf4 and c-Myc, when introduced into differentiated cells in excess amounts, reverses cell fate and by inducing these cells to revert back to a progenitor-like or ESC-like state. It is believed that this is a stochastic process whose mechanisms lie in the ability of these four transcription factors to induce epigenetic restructuring of the genome, thereby enabling expression of previously silenced genes (Hanna & Jaenisch, 2010). These artificially created pluripotent stem cells are termed induced pluripotent stem cells (iPSCs) and were first described by Takashaki and Yamanaka in rats in 2006 (Takahashi & Yamanaka, 2006). The same researchers also demonstrated that they could derive iPSCs from adult human cells, a finding subsequently confirmed by others (Nakagawa et al., 2008; Park et al., 2008; Takahashi et al., 2007; Yu et al., 2007).

A major concern with iPS cells is the level of genomic disruption that occurs with viral delivery of the reprogramming factors using retroviral or lentiviral vectors. Theoretically, these viral vectors, by integrating randomly into the host genome, may cause chromosomal translocations or transform reprogrammed cells into malignant cells. There are now several initiatives underway that seek to understand the degree of genome-wide mutations and changes in gene function that occur in iPSCs (Panopoulos et al., 2011). The art of generating pluripotent cells now involves making iPSCs with reprogramming proteins using techniques that both avoid the use of embryos and the downstream hazards of viral vectors. Additionally, it is now possible to employ just the reprogramming proteins of these factors without the use of viral vectors to create iPSCs (Zhou et al., 2009). Several gene delivery systems that have been described include employing excisable polycistronic cassettes (Sommer et al., 2010; Somers et al., 2010), *piggy*BAC introduction (Yusa et al., 2009), transfection and/or nucleofection of plasmids (Okita et al., 2008) and most recently, mRNA and miRNA delivery (Warren et al., 2010). There is also evidence that small chemicals can be used as agents to induce reprogramming in somatic cells (Zhu et al., 2010). Finally, direct reprogramming methods wherein cells from one lineage are directly trans-differentiated to another lineage (e.g. fibroblast to neural-like lineage) are gaining in popularity. Specifically, terminally differentiated cells like skin fibroblasts can be exposed to a cocktail of factors that simultaneously enable reversal of cell fate and induce differentiation to the second lineage

(Ieda et al., 2009; Uhlenhaut et al., 2009; Vierbuchen et al., 2010). The use of human, patient-specific iPSCs in regenerative medicine represents an exciting clinical avenue for the future, and will likely find application to human germ cell development.

1.3 Ethical and technical concerns in the use of stem cells

Among of the many ethical, technical, and safety concerns associated with the clinical application of stem cells, the primary concern surrounds the stem cell source. Human embryonic stem cells have generated the most controversy because the procedure of deriving a hESC line involves the sacrifice of a donated human embryo, even though the donated embryos most likely will be discarded (reviewed in Greely, 2006; Scott & Reijo Pera, 2008). Therefore, alternative sources of stem cells, such as iPSCs, which might achieve similar therapeutic goals, could be very valuable and bear fewer ethical issues than embryonic stem cells. Since one of the standard measures of a pluripotent stem cell is that it reliably forms teratomas when transplanted into animal models, careful long-term analysis and follow-up after transplantation will be needed in animal models before stem cells or directed differentiated cells are applied to humans. Yet, hESCs still represent the best model system with which to study early human development because of their differentiation potential and epigenetic programming and therefore remain the standard for studying stem cells properties. However, their suitability for transplantation studies is questionable mainly because of the risk of immune rejection from host human tissues (Swijnenburg et al., 2008). In contrast, patient-autologous iPSCs are considered more suitable for clinical transplantation due to better immunological suitability, but their differentiation profiles, epigenetic status, and potential for tumor formation are far less well characterized than are hESCs (Hyun et al., 2007). Therefore, it is imperative that studies with hESCs continue to allow better definition of the true potential of all other stem cell sources, including human iPSCs. Fortunately, there is currently considerable research devoted to deciphering the molecular mechanisms that underlie the cell fate reprogramming process. In parallel, many laboratories are now developing iPSC-based disease models for human degenerative diseases and cancers. These *in vitro* studies should provide the research community with insight into the mechanisms of disease progression and illuminate avenues of therapeutic potential. Consequently, hESCs will likely remain the "gold standard" stem cell type for developing basic scientific and clinical therapeutic protocols, and to which iPSCs and adult stem cells will be carefully compared for their therapeutic potential.

2. Mammalian germline formation

2.1 Primordial germ cell specification and candidate genes controlling germline establishment

The establishment of early germ cells and their successful maturation are complex processes, and require frequent changes in physiology, location and transcriptional profile of involved cells. Germline establishment in mammals occurs via "inductive" signaling, in contrast to lower organisms such as flies and worms where the germ cell identity is transmitted via the inheritance of germ "plasm" (McLaren, 1999, 2000; Saitou et al., 2002). In rodents and humans, the first glimpses of primordial germ cell (PGC) formation are observed in the embryo after implantation and gastrulation, when the epiblast, endoderm, mesoderm and ectoderm are first established (Matsui & Okamura, 2005). At this time, in response to molecular cues including Bone Morphogenetic Proteins (BMP4 and BMP8a) from the yolk sac, a population of pluripotent stem cells is segregated from the ICM and set physically apart from the extra-embryonic ectoderm or yolk sac of the embryo (**Figure 1**) (Lawson et al., 1999; Ying 2000, 2001). While BMPs provide the inductive signal to epiblast stem cells to supply PGCs, it is not clear what signals control the size of the PGC founder population or what molecules signal the termination of PGC specification. BMPs are crucial to specification of PGCs due to their activation of the ALK2 receptor and Smad-1/5 signaling pathways, as evidenced in mice (Hayashi et al., 2002). In the human embryo, the effects of BMPs on germline specification are unclear because there is very limited access to early embryonic samples. From work in our laboratory, it appears that similar BMP protein pathways are activated in the human embryo during gastrulation that direct PGC specification (Clark & Reijo Pera, 2006, Kee et al, 2006).

Additional mechanisms that may assist germline specification are the activation of pathways that either promote PGC survival and/or inhibit molecules that promote somatic differentiation of PGCs (Ewen & Koopman, 2010). As such, mouse and human PGCs retain expression of several biomarkers of pluripotency, including Oct4, Nanog and Sox2 which underlies their close resemblance to other pluripotent stem cells (Clark & Reijo Pera, 2006; Medrano et al., 2010; Nicholas et al., 2009). Interestingly, the transcription factor Sox2 is expressed on both mouse and human embryonic stem cells but unlike in the mouse, human Sox2 expression is diminished when PGCs migrate to the fetal gonad (Perrett et al., 2008). Coincident with the timing of germline specification in the mouse (E7.5), PGCs near the extraembryonic ectoderm begin to express germ-cell specific markers such as Blimp-1 (*Prdm1* in human), Stella (*Dppa3* in human), E-Cadherin, and Dazl and harbor alkaline phosphatase activity. Blimp-1/Prdm-1 is a transcriptional repressor whose activity is restricted to the germ lineage and appears to be critical for maintaining a PGC fate. There is strong evidence to suggest that in the mouse, Blimp-1 actively represses the somatic fate of PGCs by inhibiting expression of key somatic regulators during development (Ohinata et al., 2005; Hayashi et al., 2007). It is unclear whether Prdm-1 carries out a similar function in human germline formation. Once cell-cell communication has been established, the *Fragilis* (IFITIM-1 in human) and *Stella/ Dppa3* genes promote further development of PGCs and may do so in a similar fashion as Blimp-1 in mice (Saitou et al., 2002).

Migration of germ cells to the gonad begins at E8.5 in the mouse and during weeks 4 to 6 of human gestation (first trimester). At this stage, PGCs accumulating at the base of the allantois exit the extraembryonic ectoderm and begin migration to the developing gonads, also known as the genital ridge. During migration, PGCs also proliferate by undergoing mitosis and express a new set of biomarkers, including the CXCR4 receptor in mice and the proto-oncogene c-KIT and its ligand, KIT in both mice and humans (Molyneaux & Wylie, 2004; Gomperts et al., 1994). The *DAZ* gene homologue, DAZL is also expressed on migrating PGCs. In the mouse, migratory PGCs display pseudopodia that may assist in movement through the hindgut and it is plausible that human germ cells behave similarly since they are also observed in the hindgut during migration. Various somatic tissues interact with PGCs during the migratory path to the gonad. It is likely therefore, that these tissues express molecules and factors that guide or "cue" the PGCs and help maintain their survival. In the mouse, several candidate molecules have been identified, including receptors such as β-1 Integrin and extracellular matrix components such as Collagen I (Chuva de Sousa Lopes et al., 2005; De Felici et al., 2005). Although migration of PGCs is less well understood in human development, germ cells have been histologically observed during the late first trimester, when they undergo migration to the hindgut (Fujimoto et al., 1977; Gaskell et al., 2004; Goto et al., 2004). Male and female PGCs have been isolated from

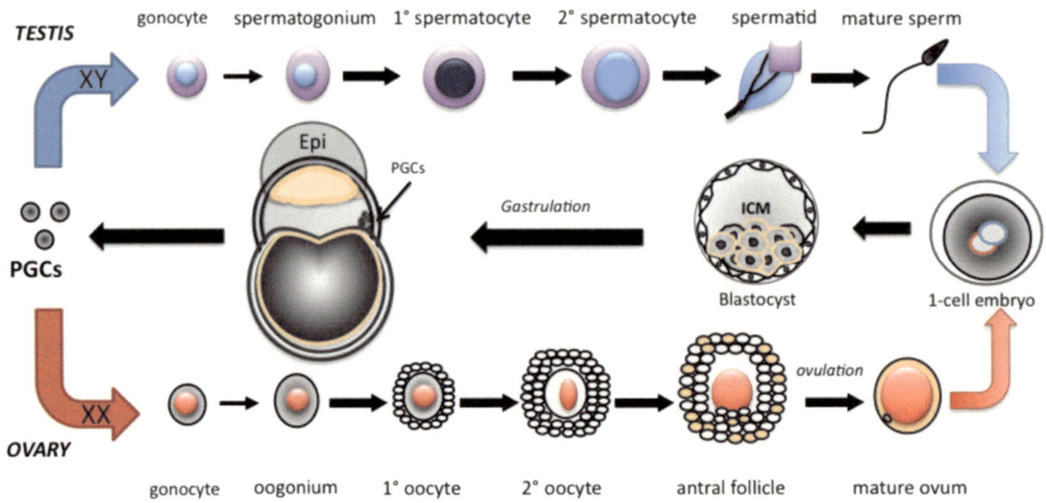

Fig. 1. Developmental Cycle of Mammalian Germ Cells.
Life cycle of the mouse and human embryo following fertilization, progressing through gastrulation and producing the germline. The germline develops in the gonads and transmits genetic information to the next generation, thus completing the cycle. Fertilization of oocytes by sperm promotes the formation of a 1-cell zygote that undergoes cell division and cleavage to form a blastocyst. The outer layer of blastocyst gives rise to the trophectoderm while the inner cell mass (ICM) contains embryonic stem cells (ESCs). During gastrulation (E7.5 in mouse; Day 15+ in human), the blastocyst cavitates and develops the three germ layers and the epiblast. The primordial germ cells (PGCs) are specified and localize near the extra-embryonic ectoderm, at the base of the allantois. Once PGCs are specified, they migrate to the fetal gonads and undergo sex-specific developmental to male and female gonocytes. Subsequently, male gonocytes undergo spermatogenesis while female gonocytes enter meiotic prophase I and begin oogenesis. Adapted from Schuh-Huerta et al., 2011.

10-week old fetuses and were observed to express Alkaline Phosphatase (AP), a marker of PGCs (Goto et al., 2004). The morphology of human PGCs also resemble the rounded shape of mouse PGCs. Female-specific germ cells have also been visualized at the ultrastructural level during gonadal development in human fetuses (Motta et al., 1997). Finally, recent studies by Kerr et al. with human fetal gonads (testis and ovary) provide a detailed analysis of pluripotency and germ cell-specific markers. The germ cells of the fetal testis are Oct4 +/Nanog +/ c-Kit+ from week 7 to 15 after which these cells become localized to the testis periphery. Meanwhile, the presumptive gonocytes in the week 15 testis show strong expression of Pumilio2 (PUM2), VASA and DAZL and express low to no pluripotency markers (Oct4, Nanog, c-Kit, Tra1-60, Tra1-81) (Kerr et al., 2007). In the human fetal ovary, as in the testis, the expression of pluripotency markers peaks by week 8 and then declines after week 9, as oocytes enter meiosis (Kerr et al., 2008). Interestingly, the cell surface markers SSEA-1 and SSEA-4 are co-expressed on the female germ cells from week 5 onwards although only SSEA-1 is restricted to the germ cell lineage.
Upon arrival at the genital ridge, germ cells express another germ-cell specific marker, VASA, a cytoplasmic protein that is implicated in translational regulation. The gene

encoding VASA expression, *DDX4* (*Mvh* in mouse) is highly conserved among species and is expressed exclusively in both male and female pre-meiotic germ cells (Gustafson & Wessel, 2010). This finding underlines the importance of the VASA protein in germline function and makes it an attractive candidate for further study. Along with VASA, other factors produced are germ cell nuclear antigen-1 (GCNA-1) and E-cadherin. The sex-specific character of the developing gonad is controlled by the chromosomal constitution of gonadal somatic cells. In particular, the *SRY* gene expressed on the Y chromosome in mammals is thought to be an essential regulator of various downstream targets including the *Sox9* gene that controls male gonadal development (McLaren, 1995, 2003). Once within the gonad, germ cells associate with Sertoli cells to form testis cords and this interaction induces the expression of VASA in post-migratory PGCs. VASA expression is induced in both male and female PGCs and persists until these cells enter meiosis and after which its levels diminish (Castrillon et al., 2000; Toyooka et al., 2000).

During germline development, an extensive remodeling of the epigenetic landscape occurs. This takes place during embryogenesis and during PGC migration to the gonad (Nicholas et al, 2009). The first wave of epigenetic remodeling occurs during implantation of the blastocyst and involves the erasure of all DNA methylation at CpG islands except those at imprinted gene loci. This transition is observed in all cells of the embryo, including the primitive germline. In female PGCs, there is another level of epigenetic change in the form of random X inactivation wherein one copy of the X chromosome pair is silenced. The second wave of epigenetic remodeling occurs when PGCs migrate to the primitive gonads and their paternal or maternal imprinted loci undergo a gradual process of erasure (Hajkova et al., 2002; Yamazaki et al., 2003). This phase occurs only in germ cells and may help to prime them for sex-specific DNA remethylation, when their developmental programs are established (Durcova-Hills et al., 2006). In the mouse, the re-establishment of imprints takes place prior to birth in the male prospermatogonia (E15) and only after birth in oocytes (Lucifero et al., 2002). In addition to DNA methylation changes, male and female germ cells also undergo post-translational histone modifications and RNA-mediated silencing (Reviewed in Tasler, 2009 & Nicolas et al., 2009).

The successful passage of germ cells through meiosis is a unique and highly rigorous process. However, between male and female embryos, the timing of meiosis is different. Male germ cells are restricted from entering meiosis while female germ cells enter meiosis within the embryo. Although the mechanisms for these contrasting behaviors are unclear, it appears that the gonadal cells provide the signal for (or against) meiotic entry (Brennan & Capel, 2004; Ewen & Koopman, 2010). One signal could be retinoic acid (RA) produced in the fetal ovary, which in turn induces *Stra8*, a key regulator of meiotic entry. In female germ cells destined to become oocytes, mitotic divisions cease and meiotic prophase begins with the correct stimuli (Borum, 1961); eventually oocytes arrest during Meiosis I prior to fetal birth and will only resume meiosis upon receiving hormonal signals during adulthood (Peters, 1970). Meanwhile, male germ cells transition from primordial status to the gonocyte stage, stop proliferating and remain quiescent in the fetal seminiferous tubules until after birth. The post-natal gonocytes then commit to a spermatogonial stem cell (SSC) fate and amplify through self-renewal or enter meiosis to initiate spermatogenesis. In both male and female germ cells, the synaptonemal complex proteins (SCPs) SCP-1, SCP-2 and SCP-3 are critical components of the meiotic machinery during chromosomal segregation (Chuma et al., 2001; Parra et al., 2004; Yuan et al., 2000). The completion of meiosis signals that germ cells have matured into haploid male and female gametes. At this stage, oocytes exclusively express GDF9 and spermatocytes express TEKT1. However, one feature that distinguishes human and mouse germline differentiation is the synchronization of meiotic entry, as in human fetal gonads, one can observe both pre-meiotic and meiotic germ cells in close proximity (Anderson, 2007).

Significant efforts have been made to culture mouse and human PGCs and gonocyes *in vitro.* In the case of mouse germ cells, the addition of endogenous factors known to affect germ cell development such as BMPs, RA, LIF and Forskolin have produced mixed results in maintaining PGCs in culture. For example, adding LIF enhanced PGC survival but the observations with the use of other factors not as clear. PGCs may also behave erratically in culture (showing low survival rates and non sex-specific behavior) because of the lack of a normal somatic environment (Childs et al., 2008). Studies of PGCs by Shamblott et al., Turnpenny et al. and Tu et al. with fetal human gonocytes resulted in a mixture of cellular phenotypes. Some gonocytes appeared rounded while others took on an elongated or 'spindly' appearance. In addition, they appeared to have different proliferation rates (Shamblott et al., 1998; Turnpenny et al., 2003; Tu et al., 2007). Currently, it is not at all clear that these cells resemble their germ cell counterparts *in vivo,* but improvements in culture conditions and the cellular microenvironment will certainly help in this regard.

2.2 Spermatogenesis and early oogenesis

As delineated earlier, mammalian germ cells populate the testis and ovary during development in a incredibly dynamic manner. During early prenatal mice and human embryo development, PGCs migrate to the primitive gonad (gonadal ridge) and associate with Sertoli cells to form primitive testicular cords (Brennan et al., 2004). Within the testicular cords, the primitive germline stem cells (now termed gonocytes) remain in the testis as the gonad differentiates. Eventually, Sertoli cells, peritubular myoid cells and gonocytes form more compact structures known as the seminiferous tubules. When the gonocytes migrate to the periphery of the tubules, they transform into prospermatogonia and then into spermatogonia (Gondos & Hobel, 1971). In the fetal testis, prospermatogonia enter mitotic arrest, a feature observed at E12.5 to E14.5 in the mouse. At the molecular level, during entry into meiosis, both male and female human gonocytes express DAZL proteins and *Vasa* transcripts and downregulate OCT3/4 expression (Anderson et al., 2007). Interestingly, it is the early migrating germ cells that share similar properties with embryonic stem cells and testicular germ cell tumors (Ezeh et al., 2005). From what is known, the development of gonocytes in the fetal ovary follows a similar path in that OCT3/4 expression is reduced while VASA, Germ Cell Nuclear Antigen (GCNA) and DAZL are expressed (McLaren, 2003). In contrast to the male gonocytes, the female gonocytes receive sex-specific signals from the fetal gonad to enter meiotic prophase. After initiating meiosis, the female gonocytes will develop into primordial follicles and subsequently into primary follicles at puberty. A key difference between mouse and human systems is the timing of primary follicle formation: the mouse achieves this stage at birth while in the human ovary, this occurs at puberty (Bukovsky et al., 2005). There is some speculation whether this difference in follicle development is due to autocrine signals produced from the oocyte itself or from the ovarian environment (Hutt & Albertini, 2007). A plausible hypothesis is that the immediate environment of early germ cells determines whether they are commit to spermatogenesis or oogenesis. The most obvious source of signals are the mesonephros, primitive Sertoli cells in the testis, and primitive Granulosa cells in the ovary.

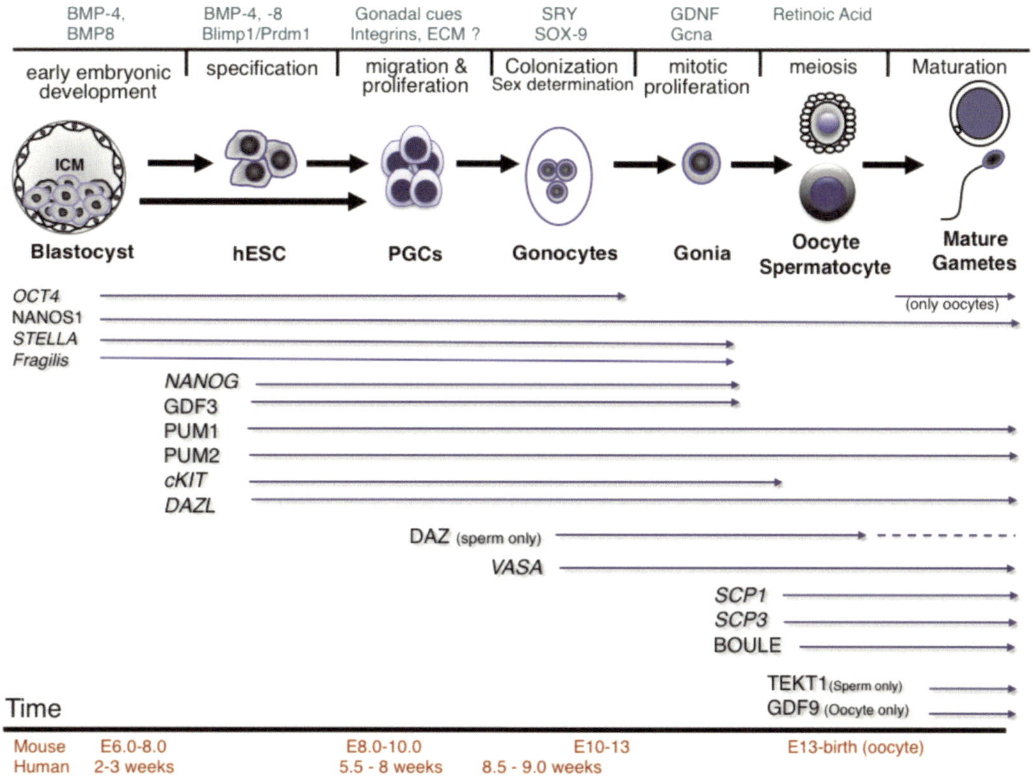

Fig. 2. Timeline of Germline Specification and Germ Cell Marker Expression. A temporal representation of the stages of human and mouse germline differentiation *in vivo*. At each cellular stage, important molecular and somatic signals controlling that stage are indicated above the diagram. Specific germ cell molecular markers are indicated on left with arrows depicting the duration of development during which they expression has been observed. Genes that are italicized are present in both mouse and human germ cells. At the bottom, an approximate timing of each stage during mouse or human germline development is indicated. Adapted from Schuh-Huerta & Reijo Pera, 2011.

The seminiferous tubule serves as the sperm production center, where approximately 123 x 10^6 spermatozoa are produced from germ cells daily, or about 1000 sperm/second (Amann et al., 1980; Rooij, 2009). Developing germ cells are arranged along the basement membrane in a highly ordered sequence and extend into the lumen of the tubule. At the most basal portion of the tubules lies the spermatogonial stem cell (SSC) population, closely associated with the adjacent Sertoli cells. Morphologic analysis of the various germ cells reveals at least 13 recognizable germ cell types in the human testis (Heller & Clermont, 1963, 1964). Each cell type is thought to represent a different step in spermatogenesis. From the least to the most differentiated, they have been named dark type A spermatogonia (Ad); pale type A spermatogonia (Ap); type B spermatogonia (B); preleptotene (R), leptotene (L), zygotene (z) and pachytene primary spermatocytes (p); secondary spermatocytes (II); and Sa, Sb, Sc, Sd_1, and Sd_2 spermatids (**Figure 3**). The early, type A spermatogonia are the most interesting germ cell type from a stem cell point of view (Rooij, 2009). In fact, time-course studies using GFP-based reporters with early type A, type Ad and type Ap spermatogonia in the mouse revealed that the early type A cell has the ability to divide, self-renew, and give rise to the Ad and Ap sun-populations (Nakagawa T. et al., 2007). These observations provide evidence for the existence of SSCs in the testis and their clonal behavior is protypical of other adult stem cells.

It is currently thought that pale type A (Ap) spermatogonia in the basal, stem cell niche of the seminiferous tubule divide at 16-day intervals and differentiate to type B spermatogonia, which then become spermatocytes (Clermont, 1972). The ability of SSCs within the testis stem cell niche to undergo stem cell renewal is governed by several known factors. The growth factor-receptor kit ligand/c-kit receptor system and the niche factor glial cell line-derived neurotrophic factor (GDNF) are important in this process (Oatley & Brinster, 2008). In fact, spermatogenesis in the rat is dependent on c-Kit receptor activity, whereas spermatogonial stem cell renewal may be c-kit independent (Dym, 1994). GDNF appears to provide a significant stimulus to self-renewal of SSCs through receptors for GDNF on SSCs such as c-Ret and GFR-1α(Meng et al., 2000). Despite this, our knowledge of other receptor-ligand systems that control human SSC renewal is limited at this point. During spermatogenesis, the cytoplasm between spermatogonial daughter cells remains conjoined after mitosis, forming cytoplasmic bridges between adjacent cells (Ewing et al., 1980). Cytoplasmic bridges are thought to be important for synchronized cellular proliferation, differentiation, and possibly regulation of gene expression. Thus, SSCs and early spermatogonia in the adult testis are critical for germ cell renewal and differentiation into sperm and raise important questions about the source of proliferative and differentiation signals for spermatogenesis. The majority of stages in mouse spermatogenesis delineated above are translatable to the human germ cell development pathway except for differences in the timing of each stage during development.

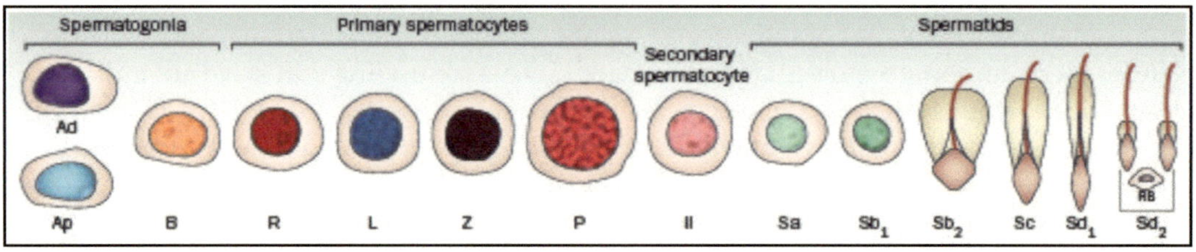

Fig. 3. Cells of the Human Germ Line Sequence. See text for details. With permission from Kee et al, 2009.

3. Germline stem cells

Under the premise that male and female gametes must be replenished, several investigators have queried the existence of elusive germline stem cells. Additionally, the pluripotency potential of embryonic germ cells and their similarity to mESCs has fueled the speculation about the existence of spermatogonial stem cells and their true differentiation potential both *in vivo* and *in vitro*. There is compelling data proving the existence and plasticity of testicular germline stem cells in the mouse and more recently the human testis. The exploration of germline stem cells in the human ovary has been confusing because of findings that challenged the long-held notion that females are endowed with a finite supply of oocytes. For instance, Johnson and colleagues observed that a small population of germline stem cells (GSCs) reside in the ovarian surface epithelium in the mouse and can undergo self renewal of the follicular pool within 24 hours (Johnson et al., 2004). However, these results have not been corroborated (Telfer et al., 2005). Johnson et al. reported that bone marrow stem cells enter the ovarian follicular pool through the vasculature and could provide a source for ovarian stem cells (Johnson et al., 2005). Despite the proof of successful transplantation in this study, they failed to demonstrate meiosis in the oocytes generated from ovarian stem cells or bone marrow-derived stem cells (Telfer et al., 2005). Furthermore, it was not clear whether these ovarian stem cell-derived oocytes could be artificially fertilized. Based on these findings, the existence of germline stem cells in the adult ovary remains unknown.

3.1 Testicular stem cells in animal models

Evidence for stem cells in the adult testis was first provided by Brinster and colleagues in 1994 when they transplanted a putative SSC population into mice and observed complete spermatogenesis with transplanted cells (Brinster RL & Zimmermann JW, 1994). Recent research has sought to characterize the true stem cell potential of early spermatogonia within the seminiferous tubule. Kanatsu-Shinohara et al first reported the generation of pluripotent stem cells from the mouse testis (Kanatsu-Shinohara et al., 2004). In their study, a population of cells termed germ line stem cells (GSCs) were isolated from neonatal mouse testis and propagated in culture (Kanatsu-Shinohara et al., 2003). Subsequently, they found that cell colonies with morphology quite distinct from GSCs, termed ESC-like cells, also appeared in culture. The formation of both GSCs and ESC-like cells depended on a GDNF signal and only ES-like cells expanded in regular mESC media. The expression profile and imprinting profiles indicated that the ES-like cells were more like ESC than GSCs. The definitive test of pluripotency or multipotency in stem cells is to transplant them and assess their ability to form all three germinal layers. In the case of spermatogonial stem cells or their transformed cell types, the ability to form germinal layers and sperm was tested following transplantation. Interestingly, transplantation of ESC-like cells into seminiferous tubules gave rise to teratomas while GSC transplants led to spermatogenesis without teratomas. When ESC-like cells were transplanted via subcutaneous delivery, teratomas containing all three germ layers formed. These ESC-like cells also had the potential to develop into various other cell types, including mature germ cells, showed chimera formation and germ line transmission. More recently, a study by Simon et al. revealed that when putative mouse SSCs were recombined with organs of fetal or neonatal origin, these cells had the ability to transdifferentiate into cells of all three germ layers as well as uterine, prostatic and skin epithelia (Simon et al., 2009). Thus, these experiments in rodents reveal that GSCs and ESC-like cells have very different stem cell characteristics. Unlike with neonatal testis, the authors were unable to generate ESC-like cells from adult mouse adult testis.

A major finding was reported in 2006 by Guan et al. wherein they successfully derived pluripotent stem cells from the adult mouse testis (Guan et al., 2006). With the introduction of a *Stra8*-eGFP reporter in which GFP expression was regulated by *Stra8* promoter activity, ES-like cells were separated from a heterogenous population of testicular cells. When these isolated ES-like cells were grown on mouse embryonic fibroblasts or cultured in medium containing LIF, they evolved into a subpopulation of cells that the authors termed multipotent adult germline stem cells (maGSCs). The pluripotency potential of maGSCs was verified by RT-PCR, *in vitro* differentiation to all three somatic germ layers, and by successful teratoma formation *in vivo*. Another transplantation study by Nakagawa et al. in 2007 employed the *Neurog3* (*Ngn3*) promoter to induce GFP expression and thereby enrich for two populations of undifferentiated spermatogonia (Nakagawa T. et al., 2007). The researchers took advantage of the expression of NGN3 on undifferentiated spermatogonia and, from their sorting studies, the first population of cells maintained the SSC progenitor pool through self-renewal while the second population revealed a pool of SSCs that diminished with age. More importantly, the study by Nakagawa and colleagues suggested that the population of SSCs in mice testes is ~2000 cells, substantially less than previous estimates of ~35000 cells (Nakagawa T. et al., 2007). A population of pluripotent stem cells in the adult mouse testis was again confirmed by Ko et al. in 2009. The authors showed that GSCs obtained from the adult mouse testis traverse through unipotency before conversion into multipotent ESC-like cells (Ko et al., 2009). They concluded that that the number of GSCs and culture duration are defining factors in deriving pluripotent stem cells from the germline: they termed these cells germline-derived pluripotent stem cells (gPSCs). In characteristic fashion of ESC-like cells, gPSCs demonstrated in vitro differentiation to all three somatic cell layers, teratoma formation and germ line transmission. Thus, several reports have documented the production of multi- or pluripotency germline stem cells in the mouse.

In parallel to these studies of mice multipotent germline stem cells, several groups have reported a collection of genes and biomarkers that are non-exclusively expressed by mouse SSCs (Reviewed in Caires et al., 2010). These include receptors and downstream facilitators of GDNF action such as GFR1α and RET proteins (Naughton et al., 2006). The gene *Ngn3* is also highly expressed on undifferentiated spermatogonial stem cells (Yoshida et al., 2004). The signaling activities of the PI3 Kinase family acting through AKT proteins are also important for SSC self-renewal because blocking these pathways suppresses GDNF activity and reduces spermatogenesis in mice (Oatley et al., 2006). Other molecules expressed on SSCs are *Oct4, Bclb6, Etv5, Lhx1* and *Nanos2* (Caires et al., 2010). In the future, we hope to employ these markers to more accurately isolate and definitively characterize putative SSC-like cells.

3.2 Testicular stem cells in the human

Based on the success in the adult rodent testes, several investigators have attempted the same types of studies in the human. In a landmark study, Conrad and colleagues devised a method by which they could harvest adult spermatogonial stem cells and subsequently reprogram them into pluripotent cells (Conrad et al., 2008). They employed the use of adult testis tissue from which they harvested a mixed cell population. The single cell suspension was then cultured *in vitro* in the presence of GDNF, as mouse studies have shown this growth factor to be critical for self-renewal of SSCs. They then purified SSCs by magnetic-activated cell separation (MACS) with CD49f bound beads. Further selection on collagen and laminin plates led to a highly enriched population of spermatogonial cells. When cultured in the same media used for hESCs and supplemented with LIF, the spermatogonial cells gradually arranged in multilayered, clustered colonies and displayed features consistent with pluripotent stem cells. The researchers termed these cells human adult germline stem cell (haGSC). Extensive characterization of haGSCs was then carried out by analyzing gene and protein expression and conducting microarray analysis for genome-wide changes in expression. Epigenetic reprogramming was assessed and various pluripotency markers were measured. Finally, the cells were karyotyped and a teratoma formation assay was successful in immunodeficient mice (Conrad et al., 2008). Overall, the authors observed significantly different expression profiles in haGCSs compared to the parent spermatogonial cells, and haGSCs acquired expression profiles similar to hESCs. Although a complete analysis of a novel cell population in the human testis, several caveats to this research exist. Entire testicles were used to derive spermatogonial cells, bringing into question the feasibility of this approach for clinical use (Kee et al., 2010). In addition, the global gene expression patterns of plurpotency markers obtained from these haGSCs have been challenged as being more closely associated with fibroblasts than hESCs (Ko et al, 2010). Nevertheless, the above experiments underline the concept that adult germline stem cells and by extension PGCs may share basic properties with hESCs.

To address the issue of limited availability of starting material, an alternative protocol for generating human pluripotent stem cells was reported by Kossack et al., 2009. Their approach required only a testis biopsy as starting substrate. The biopsies were enzymatically digested and testicular cells released in suspension were grown in basic media until small colonies of cells began to form. Selection for cells in colonies was performed by manually dissecting out colonies and then propagation on mouse embryonic fibroblasts (MEFs) in hESC culture conditions, supplemented with bFGF. Under these conditions, the spermatogonial cells acquired a hESC-like morphology and physical characteristics, and were termed human multipotent adult germline stem cells (hMGSCs). Careful analysis revealed that hMGSCs were karyotypically nomal, expressed both pluripotency and early germ cell markers and contained a generally hypomethylated DNA pattern consistent with hESCs rather than somatic cells. Albeit spontaneously differentiating to cells of endodermal, mesodermal, and ectodermal lineages, hMGSCs did not successfully form teratomas *in vivo*. Therefore, despite a simpler methodology using limited substrate, this study suggested that hMGSCs, unlike true hESCs, might be only multipotent and not fully pluripotent. A third study also used testicular tissue fragments as a cell source from which to separate spermatogonial stem cells (Golestaneh et al., 2009). Using 1g of testicular tissue, much larger than routine testis biopsy, they also derived pluripotent stem cells after isolated seminiferous tubules were enzymatically digested and grown in hESC media (supplemented with bFGF and TGF-β but not grown on MEF feeders). After several passages, they observed medium-sized colonies of ~500 cells which were termed ESC-like cells. As expected, these cells expressed pluripotency markers and they were able to differentiate to endoderm, mesoderm, and ectoderm. Similar to the finding of Kossack et al, only small teratomas formed from the transplanted ES-like cells, and only when large cell numbers (~$2x10^6$) were used. Thus, this study also suggests that the pluripotency of these ES-like cells may not be identical to true hESCs.

As eluded to above, a major limitation of deriving SSCs is the requirement for generous amounts of human testicular tissue as starting substrate. Protocols need to be refined to allow pluripotent stem cell generation from the smallest amount of testis tissue substrate to make this technology feasible for clinical use. Another limitation is the low efficiency of isolating SSCs from testis tissue and of generating pluripotent stem cells from SSCs. Optimization of culture conditions may reduce these inefficiencies in the future. Despite this, research over the last decade has uniquely highlighted the stem cell potential of SSCs derived from the human testis and their remarkable similarity to the differentiation potential of ESCs. A major drawback has been the inability of the SSCs to consistently produce large teratomas. This likely indicates that SSCs are only multipotent and not pluripotent (Kee et al., 2009), a distinction that has implications for the therapeutic potential of SSCs. Will they only be useful for obtaining sperm or will they have a broader potential for cell-based therapies in unrelated target tissues? Finally, the relatively invasive procedure for isolation of adult human germline stem cells from the testis compared to derivation of iPSCs may favor the latter approach as a preferred methodology for clinical use.

3.3 Azoospermia and genetic control of spermatogenesis

Human infertility is remarkably common, affecting 10-15% of couples. Most commonly, infertility is caused by defects in germ cell development (de Kretser, 1997; Skakkebaek et al., 1994). Microdeletions in the *AZF (Azoospermia Factor)* region of the human Y chromosome are the most common cause of human infertility and Non-Obstructive Azoospermia (NOA) in men. The associated clinical phenotypes are typically azoospermia or severe oligospermia, and on diagnostic testis biopsy reveal the complete absence of the germ cell lineage, to maturation arrest of spermatogenesis, to the production of a small number of germ cells (Gonzalves et al., 2005). The *AZF* region is highly repetitive and consists of numerous palindromic regions thought to have arisen out of repeated rounds of gene duplication and inversion (Kuroda-Kawaguchi et al., 2001; Skaletsky et al., 2003). The highly repetitive nature of the *AZF* region makes it particularly prone to homologous recombination between direct repeats and resulting in deletions. There is now a fairly comprehensive chromosomal map of the *AZF* regions with three specific regions currently identified, *AZFa, AZFb* and *AZFc,* that represent multi-gene segments on the Y chromosome (Kuroda-Kawaguchi et al., 2001). The most commonly deleted of these three is the *AZFc* region, resulting in phenotypes that range from complete absence of the germ cell lineage (termed Sertoli Cell Only [SCO] syndrome) to oligospermia (Reijo Pera, 1995, 1996). In spite of their deletion frequency, the function of genes that map to the Y chromosome *AZFc* are still poorly understood. This region harbors six multi-gene families including the *RBMY, PRY, VCY2, CDY1* and *DAZ* genes (Kuroda-Kawaguchi et al., 2001). Intriguingly, all of the protein-coding *AZF* genes are expressed exclusively in germ cells and some testicular cells, suggesting their importance in germ cell development (Navarro-Costa et al., 2010). Although precise functions have not been pinpointed, the predicted sequences and protein structures encoded by these genes provide a glimpse into their potential function. For instance, DAZ and RBMY genes are thought to be involved in RNA binding and transport, thereby positioning them as regulators of mRNA translation during spermatogenesis (Collier et al., 2005; Kee et al., 2009; Lee et al., 2006). Meanwhile, CDY and BPY2 genes are likely regulators of chromatin and genome organization (Caron et al., 2003; Lahn et al., 2002). Our overall understanding of infertility genetics will be greatly advanced by studying the molecular contribution of genes such as DAZ to meiotic recombination, DNA repair, epigenetic reprogramming and the development of cancer and other clinical aberrations.

4. From embryonic stem cells to germ cells

Our knowledge about mouse embryo and germline development from mESCs has allowed for *in vivo* and *in vitro* modeling of germline development. However, very limited access to early-stage human embryos and especially primordial human germ cells has greatly reduced the ability to conduct similar investigations into human germline development, let alone modelling it *in vitro*. This section reviews the highlights of experiments conducted during the last decade using *in vitro* model systems of mouse and human germline development from mESCs and hESCs (Table 1).

Source of cells	Germline Differentiation Strategy	Endpoint			Reference
		Cell Types Formed	*Gametes*	*Fertilization*	
mESC - XY, XX	Spontaneous differentiation in adherent culture and selection with *OCT4-GFP* reporter	Follicle & Oocyte-like cells, parthegenote-like cells	N	ND	Hubner *et al.*, 2003
mESC - XY	BMP-releasing EB formation, selection with *Mvh-GFP* reporter	*Mvh*-positive PGCs, formed sperm when transplanted *in vivo*	N	ND	Toyooka et al., *2003*
mESC - XY	EB formation with RA addition, selection of SSEA-1 positive cells	Haploid spermatids capable of fertilization	Y	Y	Geijsen *et al.*, 2004
mESC - XY	EB formation with testis-conditioned medium	Immature oocyte-like cells, oocyte gene expression	N	ND	Lacham-Kaplan *et al.*, 2006
mESC - XY	EB formation and Adherent culture differentiation	Follicle & Oocyte-like cells	N	ND	Novak *et al.*, 2006
mESC - XY	EB formation, selection with *Stra8-GFP* and subsequently with *Prm1-dsRed;* RA addition	Haploid, SSC-like cells which showed motility and form sperm	Y	Y, ICSI	Nayernia *et al.*, 2006
mESC - XY	EB formation with RA addition	Oocyte-like and Sperm-like cells	N	ND	Kerkis *et al.*, 2007
mESC	Eb formation and co-culture with ovarian granulosa cells	Oocyte-like cells, oocyte genes expressed	N	ND	Qing *et al.*, 2007
mESC - XX	EB formation and Adherent culture differentiation, selection with *Gdf9-GFP* reporter	Follicle-like structures, oocyte genes expressed	N	ND	Salvador *et al.*, 2008
mESC - XY	EB formation with RA and Testosterone addition	Putative male germ cells	Y	ND	Silva et al., 2008
hESC - XX, XY	EB formation	Immature male/female germ cells	N	ND	Clark *et al.*, 2004
hESC - XX	EB formation with BMP addition	VASA+ germ cells	N	ND	Kee *et al.*, 2006
hESC - XX	EB formation and Adherent culture differentiation	VASA+ Follicle-like structures	N	ND	Chen *et al.*, 2007
hESC - XY	Adherent culture on MEF feeders	Immature germ cells	N	ND	West *et al.*, 2008
hESC	Low-density adherent culture differentiation with Laminin substrate	CXCR4+ germ cells and sertoli cells	N	ND	Bucay *et al.*, 2008
hESC -XX	EB formation and Adherent culture differentiation, selection of SSEA-1 positive cells	Immature germ cells with PGC-like imprinting status	N	ND	Tilgner *et al.*, 2008
hESC - XX, XY hiPSC - XY	Co-culture of hESCs and hiPSCs with human fetal gonadal cells, selection with SSEA-1 and c-Kit.	Double-positive germ cells with PGC-like imprinting status (only with hESCs)	N	ND	Park *et al.*, 2009
hESC - XX, XY	Overexpression of DAZ, DAZL, BOULE and Adherent culture differentiation +/- BMPs	Haploid germ cells expressing germline markers and PGC-like imprinting	Y	ND	Kee *et al.*, 2009
hESC - XY	Adherent culture differentiation on KITL KO / WT feeders with addition of BMP4	Early germ-like cells expressing VASA and OCT4	N	ND	West *et al.*, 2010 a
hESC - XY	Adherent culture differentiation on MEF feeders	VASA+/OCT4+ clonally expanded early germ-like cells	Y	ND	West *et al.*, 2010 b
hESC - XX, XY hiPSC - XX, XY	*VASA-GFP* reporter or Overexpression of DAZ, DAZL, BOULE and Adherent culture differentiation + BMPs	Haploid, sperm-like germ cells expressing germline markers and PGC-like imprinting	Y	ND	Panula *et al.* 2010
ND = Not Determined			Y = Yes		
ICSI = Intracytoplasmic Sperm Injection			N = No		

Table 1. A Summary of the Recent Efforts to Derive Gametes from ESCs and iPSCs. Note: mESC – mouse embryonic stem cells; hESC – human embryonic stem cells; EB – embryoid body; BMP – bone morphogenetic protein; Mvh – mouse Vasa homolog; RA – retinoic acid; ICSI – intracytoplasmic sperm injection; MEF – mouse embryonic feeders.

4.1 Derivation of germ cells from mouse and human embryonic stem cells

The mouse has proven to be a robust and dependable system for analysis of germline development. Studies conducted *in vivo* during mouse early development have greatly informed *in vitro* investigations and have generated promising prospects for creating gametes from stem cells. The earliest *in vivo* evidence for embryonic stem cell-derived germ cells was suggested from experiments performed by Bradley *et al.* in 1984. They generated germline chimeras after injecting mESCs into blastocysts (Bradley et al., 1984). This work suggests that mESCs, if cultured under the right conditions, could be differentiated to PGCs. We now know that mESC share many developmental features and markers with PGCs which further supports the use of mESCs as a starting point. This is most convenient as mouse PGCs are difficult to grow and differentiate *in vitro* due to limited viability in culture (Resnick et al., 1992; Farini et al., 2005). Thus, we have learned a lot over the two decades about mouse embryonic and germ cell development to guide us as we more forward.

The first report on the creation of putative germ cells *in vitro* was reported in 2003. In this study, mESCs were transfected with a reporter Green Fluorescent Protein (GFP) under the control of an *Oct4* enhancer sequence (Hubner et al., 2003). The withdrawal of LIF, a factor that promotes self-renewal in mESCs, promoted PGC development and resulted in an increase in GFP-positive cells. More importantly, the GFP-positive cells assumed a follicle-like appearance and showed overall structural and molecular similarities to oocytes. Tooyaka et al. used a similar, reporter-based approach to derive male germ cells that resembled mouse spermatozoa (Tooyaka et al., 2003). They utilized the mouse VASA homologous gene promoter *Mvh* to drive GFP expression in mESC, which then spontaneously differentiated to embryoid bodies (EBs) in the absence of LIF. The putative PGC population when transplanted into a mouse neonatal testis, differentiated to give rise to spermatozoa. These two landmark studies set the stage for further investigation of deriving gametogenesis from mESC.

In 2004, Geijsen et al. derived male gametes from EBs after culture in the presence of retinoic acid (RA), a key paracrine factor produced in the testis (Geijsen et al., 2004). Additionally, by selecting cells that only expressed SSEA-1 and Oct4, the authors were able to enrich for mature, post-meiotic germ cells that were then capable of forming blastocytes after injection into oocytes. It was later shown by Kerkis et al. that the divergence of male and female germ cell programs during spontaneous differentiation of EBs *in vitro* was dependent on the length of culture time (Kerkis et al., 2007). Several attempts at oocyte or follicle derivation from mESC have been made using the EB-based approach with various culture and attachment conditions all of which yielded oocyte-like cells that lacked critical structural features such as the zona pellucida or expression of meiotic molecular machinery (Lacham-Kaplan et al., 2006; Novak et al., 2006). Other studies have employed slightly different approaches to derive putative oocytes from mESCs. Qing et al. obtained PGCs from cultured EBs and then co-cultured them on a layer of fetal ovarian granulosa cells and observed an increase in the meiotic marker SCP3 and upregulation of oocyte-specific genes such as *Gdf9*. The GDF9 positive cells probably resembled an immature oocyte as they lacked a zona pellucida and did not express mature oocyte proteins such as ZP4 (Qing et al., 2007). In an alternative approach developed by Salvador et al., a *Gdf9*-driven GFP reporter assay was introduced into mESC. The mESCs were then differentiated to EBs and GFP-positive cells were isolated from the EBs to isolate PGCs (Salvador et al., 2007). In summary, these studies may have derived primitive oocytes from mESCs, but failed to accomplish meiotic entry in the final cell products.

An alternative approach to deriving female germ cells was employed by Nicholas et al. by using a germ cell-specific reporter, *ΔPE:Oct4: GFP*. This reporting system was previously employed by Hubner et al. to separate GFP-expressing ESCs, PGCs and primary spermatogonia in the mouse (Nicholas et al., 2009). After verifying the expression of GFP on these subsets of cells, the authors spontaneously differentiated mESCs carrying this reporter to EBs, then FACS-sorted GFP positive cells, and observed a higher expression of germ cell and oocyte markers in the selected cells. They then assessed whether ESCs could give rise to mature oocytes *in vivo* by first forming aggregates of the GFP-positive cells with fetal ovarian tissue and then transplanting them to the kidney capsule. In doing so, they observed mature oocytes of confirmed ESC origin, some of which resembled primordial follicles with a surrounding layer of granulosa cells. An important insight from this work was that the timing of germline development, particularly with oocyte maturation, is a major parameter in deriving gametes. Furthermore, it highlights that the use of transplantation in combination with *in vitro* culturing may greatly improve the success of gamete production from mESCs.

The first study to show virtually complete gametogenesis *in vitro* was conducted by Nayernia et al. in 2006. A two-stage approach was employed to first derive pre-meiotic PGCs from mESCs with a *Stra8* driven GFP reporter induced with RA addition (Nayernia et al., 2006). GFP-positive cells were then separated by Fluorescence Activated Cell Sorting (FACS), grown in RA-supplemented medium and transfected with a second reporter, *Prm1-dsred*. The second reporter enriched for PGCs that had undergone meiosis and formed red-fluorescing, haploid cells. After further RA treatment, the haploid sperm-like cells were then injected into mouse oocytes by Intra-Cytoplasmic Sperm Injection (ICSI) to test the functionality of the derived germ cells. Indeed, blastocysts formed and viable offspring born that eventually succumbed to developmental complications and died prematurely. Potentially, the epigenetic reprogramming of the derived germ cells may not have been complete. Nevertheless, these efforts provided functional proof of gametogenesis *in vitro*.

Additional studies demonstrating *in vitro* spermatogenesis have employed organ culture systems as far back as 1960. Since then, there has been tremendous progress in the study of germ cell development, particularly in the mouse. This is best exemplified by a very recent study by Sato et al. in which functional mouse sperm were generated *in vitro* by employing a novel organ culture method. Specifically, they improved on existing organ culture methods by placing neonatal (mouse pup) testis fragments in FBS-soaked agar plugs in culture (Sato et al., 2011). With these plugs, they controlled the gas:liquid interface and also added exogenous factors. Their study conclusively showed through an analysis of reporter expression before and after meiosis, that functional, haploid spermatozoa were generated that could fertilize oocytes and give rise to viable, healthy pups. In addition, the generated pups were also shown to be naturally fertile as adults. In summary, attempts to generate germ cells from mESCs have yielded cells that resemble sperm or oocytes in gene expression, ploidy, epigenetic reprogramming and fertilization ability. However, these studies have also revealed inregularities in the duration of meiosis in culture, in the morphological features of oocytes, in the viability of offspring generated and in the germline transmission of exogenous genes to the offspring (Chuva de Sousa Lopes et al., 2010).

The evidence for germline differentiation from hESCs is more complex; as such literature is more preliminary than that from the mouse. Furthermore, doubts regarding the extent of pluripotency in hESCs vs. mESCs and the ethical constraints of testing the functionality of human germ cells are legitimate considerations with this research. Despite this, there is a tremendous need to understand human germline formation, the genes that control spermatogenesis and oogenesis and from a clinical standpoint, a need for *in vitro* derived functional gametes. Efforts to derive human germ cells began with a study by Clark et al. in 2004 in which spontaneously differentiated hESCs were used to create embryoid bodies (EBs) (Clark et al., 2004). Within the EB mixed cell population, they identified the levels of several known germ cell markers such as DAZL, STELLAR and c-KIT. Additionally, they observed that VASA levels were induced within a putative PGC population and other markers of meiosis were also upregulated. Collectively, their data indicated that PGCs had formed but had not entered meiosis. An important facet of their study was the identification of germline-associated genes such as DAZL and c-KIT as expressed in undifferentiated hESCs, which reminds us that hESCs may not be truly equivalent to cells of the ICM (Clark et al., 2006). The methods of Clark et al. were optimized in a follow-up study in which BMP-4, -7 and -8 were added exogenously during EB formation from hESCs and higher VASA expression was observed in the putative PGC population (Kee et al., 2006). Therefore, similar to the mouse model, the role of BMPs in human germline specification may be important. More recent studies using mESCs *in vitro* by Young et al. have illustrated how BMP2 and BMP4 but not BMP8b are important inductive agents for specifying the germ cell lineage in EBs (Young et al., 2010). These results help to clarify our understanding of BMP action in the germline as the goal of deriving haploid germ cells is pursued.

Chen et al. and Tilgner et al. also derived human PGCs from substrate EBs in independent studies. They observed that their EBs contained small structures that resembled ovarian follicles and that tested positive for c-KIT and low amounts of VASA protein (Chen et al., 2007). They also observed these structures in hESCs that were spontaneously differentiated in adherent cultures, suggesting that EBs are not essential for germ cell differentiation, at least in the female lineage. Meanwhile, Tilgner et al. used a more sophisticated strategy to separate germ cells from EBs and adherent cultures through FACs sorting for cells expressing the surface marker SSEA-1, a protein associated with very early germ cells (Tilgner et al., 2008). Despite evidence of complete epigenetic reprogramming in these early germ cells, the authors did not observe meiosis. In 2009, Park et al. built upon the findings of Tilgner et al. by isolating differentiated cells enriched in SSEA-1 as well as c-KIT expression. However, they used an interesting differentiation strategy and co-cultured hESC colonies with stromal cells from the human fetal gonad (Park et al., 2009). In this model, the paracrine factors produced from fetal stromal cells would ideally resemble those made *in vivo,* thereby theoretically inducing primitive germ cells to differentiate into PGCs. Their efforts were successful, yielding SSEA-1 /c-KIT positive PGCs that expressed numerous germ cell markers such as VASA, PRDM1 and DAZL and revealed evidence of partial epigenetic reprogramming.

Based on the findings of spontaneous differentiation of hESCs to PGCs, researchers have entertained the possibility that a small population of cells within a typical hESC population is pre-destined to become germ cells. Based on this hypothesis, several groups have explored whether simply optimizing culture conditions of hESCs can promote or induce germline differentiation. Stice and colleagues first reported that by the simple addition of basic FGF (bFGF) to hESCs growing on inactivated MEF feeder cells, they could produce approximately 69% VASA-positive cells with increases in expression of germline and meiotic markers (West et al., 2008). These data seem rather optimistic given that most other groups have reported much lower germ cell numbers under similar culture conditions. Concurrently, Bucay et al. modified their culture conditions by lowering the confluency of hESC colonies and observed a subpopulation of cells expressing CXCR4, a membrane receptor implicated in PGC migration (Bucay et al., 2009). Upon purification, the CXCR4-expressing cells could be grown on a laminin substrate and gave rise to putative PGCs and Sertoli cells. In recent studies, West et al. explore the effect of modulating KIT ligand (KIT-L) and BMP4 activities in culture of hESCs and observed that these are required for early germ-like cell (GLC) derivation (West et al., 2010a).

A major milestone in germline differentiation of both male and female human PGCs is the initiation and completion of meiosis. Deriving haploid gametes *in vitro* is a significant feat and one that researchers have struggled with for years in human germline differentiation. Kee and colleagues derived human haploid gametes for the first time in 2009 by introducing three major genes of the Deleted-in-Azoospermia (DAZ) family, DAZ, DAZL, and BOULE in hESCs and then differentiating the cells spontaneously in adherent culture conditions (Kee et al., 2009). Although a low number of haploid (1n) cells were observed (~2%) seven days after differentiation, they expressed numerous germ cell-specific markers and meiotic spreads indicated the presence of SCP3, a key member of the synaptonemal complex that forms during meiosis. These data also reinforce the importance of the DAZ family of genes for early (DAZL) and late (BOULE, DAZ) primordial germ cell formation. Since this work was published, another group has demonstrated meiosis in *in vitro* derived PGCs: West et al. had previously shown that they could isolate early VASA-positive germ-like cells (GLCs) and were able to subclone at least three populations of these cells from their mixed cultures and propagate them through at least 50 serial passages (West et al., 2010b). These cells were then be differentiated with FBS into a more homogenous population and the majority of cells entered meiosis (~71%) while also expressing germline and meiotic markers (West et al., 2010b). Taken together, these studies offer alternative and compelling strategies for derivation of PGCs from hESCs. While one (West et al., 2010b) suggests that GLCs are a natural clonal subpopulation of hESCs that can be propagated *in vitro*, the other (Kee et al., 2009) argues in favor of an active, gene-driven induction of germline differentiation.

4.2 Germline differentiation from human induced pluripotent stem cells

Ongoing advances in reprogramming of human somatic cell to pluripotent stem cells (iPSCs) or direct reprogramming of somatic cells to specific cell lineages, are a legitimate avenue for autologous, patient-derived stem cell therapies. By applying similar genetic approaches used for human ESCs, iPSCs can now be directed to give rise to cells of the germ cell lineage. We summarize here some of the recent efforts in this area. The use of human iPSCs (hiPSCs) to derive gametes was first shown by Park et al. in a study that also compared the derivation efficiency to hESCs (Park et al., 2009). In short, they grew hiPSCs on a feeder layer of fetal gonad stromal cells and subsequently isolated SSEA-1 + / c-KIT + cells which they characterized as a putative PGC population. These human induced PGCs (hiPGCs) expressed several germline markers but did not show evidence of imprinting erasure, a key feature of early PGCs. Further evidence for the creation of haploid hiPGCs from male and female hiPSCs was provided by Panula et al. recently wherein they used a GFP reporter under the control of the *Vasa* promoter to select for putative germ cells from spontaneously differentiated cultures (Panula et al., 2010). The *Vasa*-GFP transduced hiPSCs were differentiated adherently in feeder-free conditions in the presence of BMPs, giving rise to approximately 5% GFP-positive PGCs. These PGCs expressed a number of germ cell markers such as *Prdm1A, ACTC, Gata6, Pelota* and *IFITM1*. More importantly, the GFP-positive hiPGCs expressed the meiotic marker SCP3 in both punctate and elongated patterns, indicative of early and late meiosis, respectively. Interestingly, when compared to PGCs derived from both male or female hESCs, the efficiency of hiPGC derivation appears higher, suggesting that hiPSCs may be a better starting substrate for germline differentiation. In summary, these two investigations provide encouraging results about the ability to use hiPSCs to derive gametes *in vitro*. Although the differentiation potential of hiPSCs may vary greatly from one cell line to another, they appear to have strong potential as a genetic model system to study human germ cell development.

4.3 Clinical need and applications for *in vitro*-derived gametes

Currently many men suffer from non-obstructive azoospermia as described earlier that results in complete sterility. In addition, in the U.S. alone there are 55,000 childhood cancer survivors annually, many of whom were unable to bank sperm before receiving sterilizing treatments. Similarly, women have a finite supply of reproductively competent oocytes while others undergo premature ovarian failure, making childbearing difficult or impossible. Alternative options for parenthood in these cases are not always simple or inexpensive. Although sperm donation is economical, oocyte donation is quite expensive and the use of donated embryos is unusual (Nicholas et al., 2009). For men, recent strategies

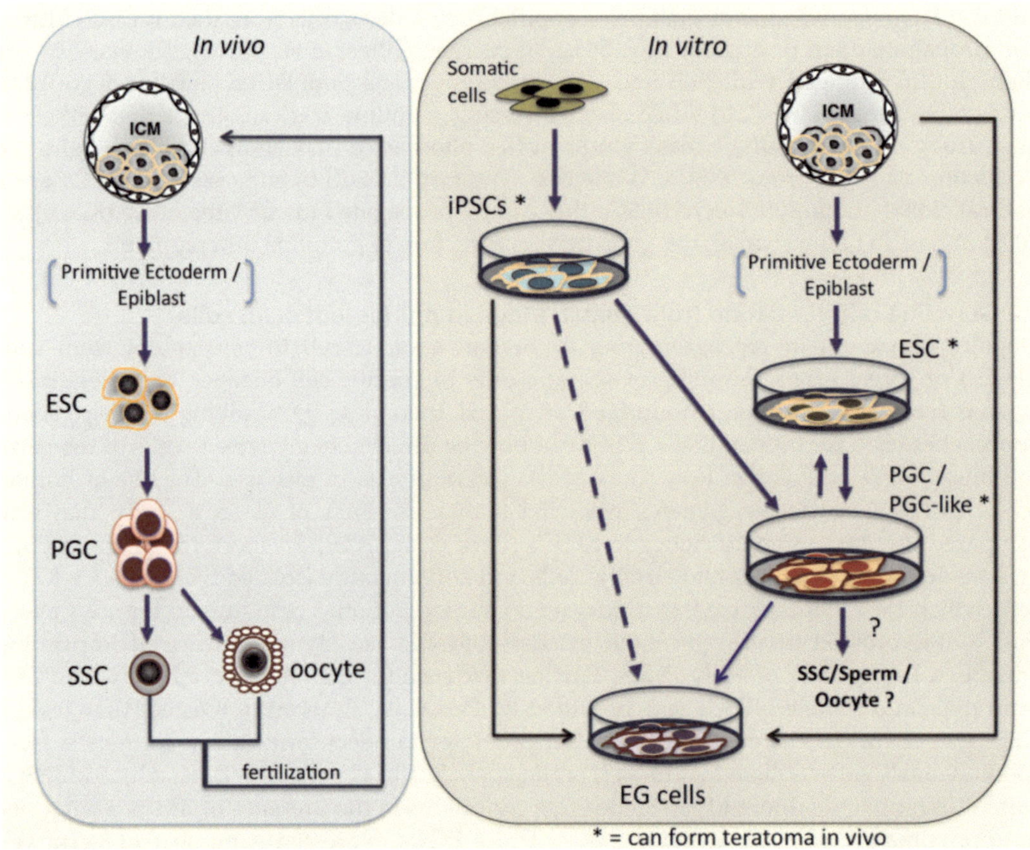

Fig. 4. Comparison of *in vivo* Germline Development and *in vitro* Germ Cell Derivation from ESCs and iPSCs.A depiction of *in vivo* development of germ cells and gametes from the mammalian embryo showing that embryonic stem cells (ESCs) originate from primitive ectoderm, by way of the inner cell mass (ICM). Primordial germ cells then migrate to gonads where they undergo self-renewal in the form of spermatogonial stem cells (SSCs) in the testis and oocytes in the ovary. On the right is a schematic of the cell lineages derived *in vitro*. ESCs cultured in dishes can be differentiated to PGCs via methods summarized in Table 1. PGCs cultured *in vitro,* at least in the mouse, can transform to embryonic germ (EG) cells. There is also evidence a subpopulation of hESCs cultured *in vitro* behave like EG cells. Finally, somatic cells can be reprogrammed with addition of Oct4, Sox2, Klf4 and c-Myc transcription factors to induce a stem cell-like fate. iPSCs can then be induced to form PGC-like cells or EG-like cells. Finally, ESCs, PGCs and iPSCs can all form teratomas when injected *in vivo* demonstrating their pluripotency.

such as extracting low numbers of testis sperm after prior localization with Fine Needle Aspiration (FNA) mapping, offer the hope of fatherhood for some candidates (Beliveau et al., 2011; Natali et al., 2011).

Although such strategies are gaining in popularity, the pregnancy success rates remains at about 30-40% using sophisticated in vitro fertilization (IVF) techniques (Centers for Disease Control [CDC], 2008). In addition, with these reproductive technologies, there remain important safety concerns about de novo sex chromosomal disorders, birth defects and imprinting-related disorders such as Beckwith-Wiedeman syndrome in offspring (Childs et al., 2008). Therefore, there is an obvious clinical need for high quality, *in vitro*-derived gametes. However, the prospect of employing current protocols and reagents to derive human gametes from hESCs appears unlikely at this time. On the other hand, the use of iPSCs and alternative stem cell sources such as adult stem cells, hold great promise in thisregard. In the foreseeable future, patients with genetic disorders affecting fertility could theorectially donate somatic cell biopsies that could then be reprogrammed into iPSCs and genetically modified to normal and subsequently utilized for gametogenesis *in vitro*.

However, there are numerous challenges to the application of iPSCs for clinical use (Hanna et al., 2010). For one, the exact level of authenticity and experimental proof required for the safe use of iPSC and hESC derived gametes in the clinic is entirely unknown. To address this issue, there are at least three major areas that the research community *can* and *should* address before taking *in vitro*-derived gametes to the clinic. Firstly, the chromosomal composition of the iPSCs must be identical to the original patient from where they were derived. There is now sufficient data detailing the gene mutations and copy number variations arising in iPSCs during reprogramming and clonal expansion. It would be wise to carefully screen for these variation in patient-derived samples (Reviewed in Panopoulos et al., 2011). Moreover, frequent chromosomal aberrations arise during meiotic recombination so it would be necessary to examine the fidelity of meiosis in gametes *in vitro*. Secondly, it is important to check for the proper erasure and subsequent re-establishment of maternal and paternal imprints in PGCs derived from both iPSCs and hESCs. There is published work with hESCs and iPSCs that has attempted to characterize imprinting status and DNA methylation patterns in these derived cells; the preliminary evidence seems to indicate that *in vitro* derived hiPGCs appear to undergo partial to full imprinting erasure (Kee et al., 2009; Panula et al., 2010). Whether the correct imprints are restored later on during germ cell development remains unknown. Thirdly, the most infallible method to test full functionality of *in vitro*-derived gametes is to test their ability to fertilize, form a blastocyst and viable offspring. In the mouse, this has been tested despite the premature death of offspring (Nayernia et al., 2006). One obstacle is how can we overcome ethical boundaries and challenges to do similar tests with human gametes? Perhaps for now, the primary medical application of human stem cell-derived gametes would be to use them as a reproductive toxicological screen for pharmaceutical compounds and other chemicals.

5. Conclusions

In the last decade, much progress has been made in our understanding of mouse and human embryology and embryonic stem cells. We have learned that complex genetic pathways that underlie the pluripotency aspects of hESCs and mESCs also underline our inability to easily control and direct these cells toward a specific lineage. In addition, the advent of diverse strategies to 'reprogram' somatic cells has opened up an almost limitless access to patient-specific stem cells that if used correctly, could provide very powerful cell-based therapies. Scientifically, the biology of germline development has always been an popular area of exploration for embryologists and stem cell biologists, especially because germ cells are unique in their ability to transmit genetic information from one generation to the next. If we can understand how germ cells develop from ESCs at the cellular and molecular level, we can apply this information to the derivation of male and female gametes. The goals of gamete formation include: 1) to address immediate health issues such as infertility in men and women due to sperm and egg dysfunction or absence; 2) to produce gametes that can be fertilized and produce an almost limitless source of mammalian embryos and ESCs for the study of specific diseases; 3) to devise methods with which to target hereditary and non-hereditary germline mutations or chromosomal abnormalities during the earliest stages of germ cell production.

The studies summarized in this review have shown that multipotent ES-like SSCs can develop into somatic cells of all three germ layers, but their ability to differentiate to the germline remains unclear or untested. In addition, it is believed that SSCs may only be multipotent and will be useful only for autologous transplantation and maybe for only a few tissue types. In this regard, one promising avenue for SSCs is differentiation down the spermatogenic pathway. Finally, there is a limited expansion potential of SSCs compared to iPSCs or hESCs and the imprinting status of these cells is less well defined than that of hESCs. A similar approach as used by Kee et al. in 2009 to overexpresss DAZ genes in hESCs could potentially be applied to differentiate SSC's to spermatids or sperm *in vitro*. Alternatively, one could envision that the research on *in vitro*-derived human oocytes could be similarly advanced and lead to a better understanding of the concept that there are renewable ovarian stem cells in humans. In conclusion, the potential of ESCs and iPSCs to be used for cell-based therapy is now being realized. Major challenges, both scientific and clinical, still exist in deriving germline cells that look and behave as reliably as their natural counterparts. With further advances in technology and the elucidation of new pathways in gerrmline function in the mouse and human systems, we predict that these challenges can be met and overcome in the future.

Keypoints:

- Mouse and human embryonic stem cells provide the ideal substrate for germline differentiation, both *in vivo* and *in vitro*.
- The mammalian germline is established as primordial germ cells immediately prior to gastrulation in mice and humans, and undergo similar physiological and molecular events.
- Induced pluripotent stem cells (iPSCs) and adult stem cells may provide alternative sources of pluripotent and multipotent stem cells with less potential for host rejection and decreased tumorigenicity.
- Testicular stem cells with multipotent and even pluripotent potential have been isolated in mouse models and, more recently, in humans.
- The *in vitro* derivation of early male and female germ cells is now achievable from mESCs, hESCs and human iPSCs with the application of various reporter systems and through embryoid body (EB) formation.

6. References

Amann, R. P., & S. S. Howards. (1980). Daily Spermatozoal Production and Epididymal Spermatozoal Reserves of the Human Male. *Journal of Urology*, Vol. 124, No. 2, (Aug 1980) pp. 211-215, 0022-5347

Anderson, R. A., N. Fulton, G. Cowan, S. Coutts, & P. T. Saunders. (2007). Conserved and Divergent Patterns of Expression of Dazl, Vasa and Oct4 in the Germ Cells of the Human Fetal Ovary and Testis. *BMC Developmental Biology*, Vol. 7, No.) pp. 136, 1471-213X

Beliveau, M. E., & P. J. Turek. (2011). The Value of Testicular 'Mapping' in Men with Non-Obstructive Azoospermia. *Asian Journal of Andrology*, Vol. 13, No. 2, (Mar) pp. 225-230, 1745-7262

Borum, K. (1961). Oogenesis in the Mouse. A Study of the Meiotic Prophase. *Experimental Cell Research*, Vol. 24, No., (Sep) pp. 495-507, 0014-4827

Bradley, A., M. Evans, M. H. Kaufman, & E. Robertson. (1984). Formation of Germ-Line Chimaeras from Embryo-Derived Teratocarcinoma Cell Lines. *Nature*, Vol. 309, No. 5965, (May 17-23) pp. 255-256, 0028-0836

Brennan, J., & B. Capel. (2004). One Tissue, Two Fates: Molecular Genetic Events That Underlie Testis Versus Ovary Development. *Nature Reviews Genetics*, Vol. 5, No. 7, (Jul) pp. 509-521, 1471-0056

Brinster, R. L., & J. W. Zimmermann. (1994). Spermatogenesis Following Male Germ-Cell Transplantation. *Proceedings of the National Academy of Sciences U S A*, Vol. 91, No. 24, (Nov 22) pp. 11298-11302, 0027-8424

Bucay, N., M. Yebra, V. Cirulli, I. Afrikanova, T. Kaido, A. Hayek, & A. M. Montgomery. (2009). A Novel Approach for the Derivation of Putative Primordial Germ Cells and Sertoli Cells from Human Embryonic Stem Cells. *Stem Cells*, Vol. 27, No. 1, (Jan) pp. 68-77, 1549-4918

Bukovsky, A., M. R. Caudle, M. Svetlikova, J. Wimalasena, M. E. Ayala, & R. Dominguez. (2005). Oogenesis in Adult Mammals, Including Humans: A Review. *Endocrine*, Vol. 26, No. 3, (Apr) pp. 301-316, 1355-008X

Caron, C., C. Pivot-Pajot, L. A. van Grunsven, E. Col, C. Lestrat, S. Rousseaux, & S. Khochbin. (2003). Cdyl: A New Transcriptional Co-Repressor. *The Journal of the European Molecular Biology Organization (EMBO)*, Vol. 4, No. 9, (Sep) pp. 877-882, 1469-221X

Castrillon, D. H., B. J. Quade, T. Y. Wang, C. Quigley, & C. P. Crum. (2000). The Human Vasa Gene Is Specifically Expressed in the Germ Cell Lineage. *Proceedings of the Nationall Acadym of Sciences U S A*, Vol. 97, No. 17, (Aug 15) pp. 9585-9590, 0027-8424

Chen, H. F., H. C. Kuo, C. L. Chien, C. T. Shun, Y. L. Yao, P. L. Ip, C. Y. Chuang, C. C. Wang, Y. S. Yang, & H. N. Ho. (2007). Derivation, Characterization and Differentiation of Human Embryonic Stem Cells: Comparing Serum-Containing Versus Serum-Free Media and Evidence of Germ Cell Differentiation. *Human Reproduction*, Vol. 22, No. 2, (Feb) pp. 567-577, 0268-1161

Childs, A. J., P. T. Saunders, & R. A. Anderson. (2008). Modelling Germ Cell Development in Vitro. *Molecular Human Reproduction*, Vol. 14, No. 9, (Sep) pp. 501-511, 1460-2407

Chuma, S., & N. Nakatsuji. (2001). Autonomous Transition into Meiosis of Mouse Fetal Germ Cells in Vitro and Its Inhibition by Gp130-Mediated Signaling. *Developmental Biology*, Vol. 229, No. 2, (Jan 15) pp. 468-479, 0012-1606

Chuva de Sousa Lopes, S. M., & B. A. Roelen. (2010). On the Formation of Germ Cells: The Good, the Bad and the Ugly. *Differentiation*, Vol. 79, No. 3, (Mar) pp. 131-140, 1432-0436

Chuva de Sousa Lopes, S. M., S. van den Driesche, R. L. Carvalho, J. Larsson, B. Eggen, M. A. Surani, & C. L. Mummery. (2005). Altered Primordial Germ Cell Migration in the Absence of Transforming Growth Factor Beta Signaling Via Alk5. *Developmental Biology*, Vol. 284, No. 1, (Aug 1) pp. 194-203, 0012-1606

Clark, A. T., M. S. Bodnar, M. Fox, R. T. Rodriquez, M. J. Abeyta, M. T. Firpo, & R. A. Pera. (2004). Spontaneous Differentiation of Germ Cells from Human Embryonic Stem Cells in Vitro. *Human Molecuar Genetics*, Vol. 13, No. 7, (Apr 1) pp. 727-739, 0964-6906

Clermont, Y. (1972). Kinetics of Spermatogenesis in Mammals: Seminiferous Epithelium Cycle and Spermatogonial Renewal. *Physiological Reviews*, Vol. 52, No. 1, (Jan) pp. 198-236, 0031-9333

Collier, B., B. Gorgoni, C. Loveridge, H. J. Cooke, & N. K. Gray. (2005). The Dazl Family Proteins Are Pabp-Binding Proteins That Regulate Translation in Germ Cells. *The Journal of the European Molecular Biology Organization (EMBO)*, Vol. 24, No. 14, (Jul 20) pp. 2656-2666, 0261-4189

Conrad, S., M. Renninger, J. Hennenlotter, T. Wiesner, L. Just, M. Bonin, W. Aicher, H. J. Buhring, U. Mattheus, A. Mack, H. J. Wagner, S. Minger, M. Matzkies, M. Reppel, J. Hescheler, K. D. Sievert, A. Stenzl, & T. Skutella. (2008). Generation of Pluripotent Stem Cells from Adult Human Testis. *Nature*, Vol. 456, No. 7220, (Nov 20) pp. 344-349, 1476-4687

de Kretser, D. M. (1997). Male Infertility. *Lancet*, Vol. 349, No. 9054, (Mar 15) pp. 787-790, 0140-6736

de Rooij, D. G. (2009). Spermatogonial Stem Cells, In: *Trends in Stem Cell Biology and Technology*, H. Baharvand, pp. 149-162, Humana Press, 1603279040, New York.

Durcova-Hills, G., P. Hajkova, S. Sullivan, S. Barton, M. A. Surani, & A. McLaren. (2006). Influence of Sex Chromosome Constitution on the Genomic Imprinting of Germ Cells. *Proceedings of the National Academy of Sciences U S A*, Vol. 103, No. 30, (Jul 25) pp. 11184-11188, 0027-8424

Dym, M. (1994). Spermatogonial Stem Cells of the Testis. *Proceedings of the National Academy of Sciences U S A*, Vol. 91, No. 24, (Nov 22) pp. 11287-11289, 0027-8424

Evans, M. J., & M. H. Kaufman. (1981). Establishment in Culture of Pluripotential Cells from Mouse Embryos. *Nature*, Vol. 292, No. 5819, (Jul 9) pp. 154-156, 0028-0836

Ewen, K. A., & P. Koopman. (2010). Mouse Germ Cell Development: From Specification to Sex Determination. *Molecular and Cellular Endocrinology*, Vol. 323, No. 1, (Jul 8) pp. 76-93, 1872-8057

Ewing, L. L., J. C. Davis, & B. R. Zirkin. (1980). Regulation of Testicular Function: A Spatial and Temporal View. *International Review of Physiology*, Vol. 22, No.) pp. 41-115, 0363-3918

Ezeh, U. I., P. J. Turek, R. A. Reijo, & A. T. Clark. (2005). Human Embryonic Stem Cell Genes Oct4, Nanog, Stellar, and Gdf3 Are Expressed in Both Seminoma and Breast Carcinoma. *Cancer*, Vol. 104, No. 10, (Nov 15) pp. 2255-2265, 0008-543X

Farini, D., M. L. Scaldaferri, S. Iona, G. La Sala, & M. De Felici. (2005). Growth Factors Sustain Primordial Germ Cell Survival, Proliferation and Entering into Meiosis in the Absence of Somatic Cells. *Developmental Biology*, Vol. 285, No. 1, (Sep 1) pp. 49-56, 0012-1606

Fujimoto, T., Y. Miyayama, & M. Fuyuta. (1977). The Origin, Migration and Fine Morphology of Human Primordial Germ Cells. *The Anatomical Record*, Vol. 188, No. 3, (Jul) pp. 315-330, 0003-276X

Gaskell, T. L., A. Esnal, L. L. Robinson, R. A. Anderson, & P. T. Saunders. (2004). Immunohistochemical Profiling of Germ Cells within the Human Fetal Testis: Identification of Three Subpopulations. *Biology of Reproduction*, Vol. 71, No. 6, (Dec) pp. 2012-2021, 0006-3363

Geijsen, N., M. Horoschak, K. Kim, J. Gribnau, K. Eggan, & G. Q. Daley. (2004). Derivation of Embryonic Germ Cells and Male Gametes from Embryonic Stem Cells. *Nature*, Vol. 427, No. 6970, (Jan 8) pp. 148-154, 1476-4687

Golestaneh, N., M. Kokkinaki, D. Pant, J. Jiang, D. DeStefano, C. Fernandez-Bueno, J. D. Rone, B. R. Haddad, G. I. Gallicano, & M. Dym. (2009). Pluripotent Stem Cells Derived from Adult Human Testes. *Stem Cells and Development*, Vol. 18, No. 8, (Oct) pp. 1115-1126, 1557-8534

Gomperts, M., M. Garcia-Castro, C. Wylie, & J. Heasman. (1994). Interactions between Primordial Germ Cells Play a Role in Their Migration in Mouse Embryos. *Development*, Vol. 120, No. 1, (Jan) pp. 135-141, 0950-1991

Gondos, B., & C. J. Hobel. (1971). Ultrastructure of Germ Cell Development in the Human Fetal Testis. *Z Zellforsch Mikrosk Anat*, Vol. 119, No. 1) pp. 1-20, 0340-0336

Gonsalves, J., Turek, P. J., Schlegel, P. N., Hopps, C. V., Weier, J. F., Pera, R. A. (2005). Recombination in men with Klinefelter syndrome. *Reproduction*, Vol 130, No. 2, (Aug) pp. 223-229,

Goto, T., J. Adjaye, C. H. Rodeck, & M. Monk. (1999). Identification of Genes Expressed in Human Primordial Germ Cells at the Time of Entry of the Female Germ Line into Meiosis. *Molecular and Human Reproduction*, Vol. 5, No. 9, (Sep) pp. 851-860, 1360-9947

Greely, H. T. (2006). Moving Human Embryonic Stem Cells from Legislature to Lab: Remaining Legal and Ethical Questions. *PLoS Medicine*, Vol. 3, No. 5, (Feb) pp. e143, 1549-1676

Guan, K., K. Nayernia, L. S. Maier, S. Wagner, R. Dressel, J. H. Lee, J. Nolte, F. Wolf, M. Li, W. Engel, & G. Hasenfuss. (2006). Pluripotency of Spermatogonial Stem Cells from Adult Mouse Testis. *Nature*, Vol. 440, No. 7088, (Apr 27) pp. 1199-1203, 1476-4687

Gustafson, E. A., & G. M. Wessel. (2010). Vasa Genes: Emerging Roles in the Germ Line and in Multipotent Cells. *Bioessays*, Vol. 32, No. 7, (Jul) pp. 626-637, 1521-1878

Hajkova, P., S. Erhardt, N. Lane, T. Haaf, O. El-Maarri, W. Reik, J. Walter, & M. A. Surani. (2002). Epigenetic Reprogramming in Mouse Primordial Germ Cells. *Mechanisms of Development*, Vol. 117, No. 1-2, (Sep) pp. 15-23, 0925-4773

Hanna, J. H., K. Saha, & R. Jaenisch. (2010). Pluripotency and Cellular Reprogramming: Facts, Hypotheses, Unresolved Issues. *Cell*, Vol. 143, No. 4, (Nov 12) pp. 508-525, 1097-4172

Hayashi, K., T. Kobayashi, T. Umino, R. Goitsuka, Y. Matsui, & D. Kitamura. (2002). Smad1 Signaling Is Critical for Initial Commitment of Germ Cell Lineage from Mouse Epiblast. *Mechanisms of Development*, Vol. 118, No. 1-2, (Oct) pp. 99-109, 0925-4773

Heller, C. G., & Y. Clermont. (1963). Spermatogenesis in Man: An Estimate of Its Duration. *Science*, Vol. 140, No., (Apr 12) pp. 184-186, 0036-8075

Heller, C. H., & Y. Clermont. (1964). Kinetics of the Germinal Epithelium in Man. *Recent Progress in Hormone Research*, Vol. 20, No.) pp. 545-575, 0079-9963

Hubner, K., G. Fuhrmann, L. K. Christenson, J. Kehler, R. Reinbold, R. De La Fuente, J. Wood, J. F. Strauss, 3rd, M. Boiani, & H. R. Scholer. (2003). Derivation of Oocytes from Mouse Embryonic Stem Cells. *Science*, Vol. 300, No. 5623, (May 23) pp. 1251-1256, 1095-9203

Ieda, M., J. D. Fu, P. Delgado-Olguin, V. Vedantham, Y. Hayashi, B. G. Bruneau, & D. Srivastava. (2010). Direct Reprogramming of Fibroblasts into Functional Cardiomyocytes by Defined Factors. *Cell*, Vol. 142, No. 3, (Aug 6) pp. 375-386, 1097-4172

Johnson, J., J. Bagley, M. Skaznik-Wikiel, H. J. Lee, G. B. Adams, Y. Niikura, K. S. Tschudy, J. C. Tilly, M. L. Cortes, R. Forkert, T. Spitzer, J. Iacomini, D. T. Scadden, & J. L. Tilly. (2005). Oocyte Generation in Adult Mammalian Ovaries by Putative Germ Cells in Bone Marrow and Peripheral Blood. *Cell*, Vol. 122, No. 2, (Jul 29) pp. 303-315, 0092-8674

Johnson, J., J. Canning, T. Kaneko, J. K. Pru, & J. L. Tilly. (2004). Germline Stem Cells and Follicular Renewal in the Postnatal Mammalian Ovary. *Nature*, Vol. 428, No. 6979, (Mar 11) pp. 145-150, 1476-4687

Kanatsu-Shinohara, M., K. Inoue, J. Lee, M. Yoshimoto, N. Ogonuki, H. Miki, S. Baba, T. Kato, Y. Kazuki, S. Toyokuni, M. Toyoshima, O. Niwa, M. Oshimura, T. Heike, T. Nakahata, F. Ishino, A. Ogura, & T. Shinohara. (2004). Generation of Pluripotent Stem Cells from Neonatal Mouse Testis. *Cell*, Vol. 119, No. 7, (Dec 29) pp. 1001-1012, 0092-8674

Kanatsu-Shinohara, M., N. Ogonuki, K. Inoue, H. Miki, A. Ogura, S. Toyokuni, & T. Shinohara. (2003). Long-Term Proliferation in Culture and Germline Transmission of Mouse Male Germline Stem Cells. *Biology of Reproduction*, Vol. 69, No. 2, (Aug) pp. 612-616, 0006-3363

Kee, K., V. T. Angeles, M. Flores, H. N. Nguyen, & R. A. Reijo Pera. (2009). Human Dazl, Daz and Boule Genes Modulate Primordial Germ-Cell and Haploid Gamete Formation. *Nature*, Vol. 462, No. 7270, (Nov 12) pp. 222-225, 1476-4687

Kee, K., J. M. Gonsalves, A. T. Clark, & R. A. Pera. (2006). Bone Morphogenetic Proteins Induce Germ Cell Differentiation from Human Embryonic Stem Cells. *Stem Cells and Development*, Vol. 15, No. 6, (Dec) pp. 831-837, 1547-3287

Kee, K., R. A. Pera, & P. J. Turek. (2010). Testicular Germline Stem Cells. *Nature Reviews Urology*, Vol. 7, No. 2, (Feb) pp. 94-100, 1759-4820

Kerkis, A., S. A. Fonseca, R. C. Serafim, T. M. Lavagnolli, S. Abdelmassih, R. Abdelmassih, & I. Kerkis. (2007). In Vitro Differentiation of Male Mouse Embryonic Stem Cells into

Both Presumptive Sperm Cells and Oocytes. *Cloning Stem Cells*, Vol. 9, No. 4, (Winter) pp. 535-548, 1536-2302

Kerr, C. L., C. M. Hill, P. D. Blumenthal, & J. D. Gearhart. (2008). Expression of Pluripotent Stem Cell Markers in the Human Fetal Ovary. *Human Reproduction*, Vol. 23, No. 3, (Mar) pp. 589-599, 1460-2350

Kerr, C. L., C. M. Hill, P. D. Blumenthal, & J. D. Gearhart. (2008). Expression of Pluripotent Stem Cell Markers in the Human Fetal Testis. *Stem Cells*, Vol. 26, No. 2, (Feb) pp. 412-421, 1549-4918

Kerr, C. L., M. J. Shamblott, & J. D. Gearhart. (2006). Pluripotent Stem Cells from Germ Cells. *Methods in Enzymology*, Vol. 419, No.) pp. 400-426, 0076-6879

Ko, K., N. Tapia, G. Wu, J. B. Kim, M. J. Bravo, P. Sasse, T. Glaser, D. Ruau, D. W. Han, B. Greber, K. Hausdorfer, V. Sebastiano, M. Stehling, B. K. Fleischmann, O. Brustle, M. Zenke, & H. R. Scholer. (2009). Induction of Pluripotency in Adult Unipotent Germline Stem Cells. *Cell Stem Cell*, Vol. 5, No. 1, (Jul 2) pp. 87-96, 1875-9777

Ko K, Arauzo-Bravo MJ, Tapia N, Kim J, Lin Q, Bernemann C, Han DW, Gentile L, Reinhardt P, Gerber B, Schneider RK, Kilesch S, Zenke M, Scholer HR (2010). Human Adult Germline Stem Cells in Question. *Nature*, Vol. 465, No. 7301, (June 24) E1, Discussion E3.

Kossack, N., J. Meneses, S. Shefi, H. N. Nguyen, S. Chavez, C. Nicholas, J. Gromoll, P. J. Turek, & R. A. Reijo-Pera. (2009). Isolation and Characterization of Pluripotent Human Spermatogonial Stem Cell-Derived Cells. *Stem Cells*, Vol. 27, No. 1, (Jan) pp. 138-149, 1549-4918

Kuroda-Kawaguchi, T., H. Skaletsky, L. G. Brown, P. J. Minx, H. S. Cordum, R. H. Waterston, R. K. Wilson, S. Silber, R. Oates, S. Rozen, & D. C. Page. (2001). The Azfc Region of the Y Chromosome Features Massive Palindromes and Uniform Recurrent Deletions in Infertile Men. *Nature Genetics*, Vol. 29, No. 3, (Nov) pp. 279-286, 1061-4036

Lacham-Kaplan, O., H. Chy, & A. Trounson. (2006). Testicular Cell Conditioned Medium Supports Differentiation of Embryonic Stem Cells into Ovarian Structures Containing Oocytes. *Stem Cells*, Vol. 24, No. 2, (Feb) pp. 266-273, 1066-5099

Lahn, B. T., Z. L. Tang, J. Zhou, R. J. Barndt, M. Parvinen, C. D. Allis, & D. C. Page. (2002). Previously Uncharacterized Histone Acetyltransferases Implicated in Mammalian Spermatogenesis. *Proceedings of the National Academy of Sciences U S A*, Vol. 99, No. 13, (Jun 25) pp. 8707-8712, 0027-8424

Lawson, K. A., N. R. Dunn, B. A. Roelen, L. M. Zeinstra, A. M. Davis, C. V. Wright, J. P. Korving, & B. L. Hogan. (1999). Bmp4 Is Required for the Generation of Primordial Germ Cells in the Mouse Embryo. *Genes and Development*, Vol. 13, No. 4, (Feb 15) pp. 424-436, 0890-9369

Lee, K. H., S. Lee, B. Kim, S. Chang, S. W. Kim, J. S. Paick, & K. Rhee. (2006). Dazl Can Bind to Dynein Motor Complex and May Play a Role in Transport of Specific Mrnas. *Journal of the European Molecular Biology Organization (EMBO)*, Vol. 25, No. 18, (Sep 20) pp. 4263-4270, 0261-4189

Lucifero, D., C. Mertineit, H. J. Clarke, T. H. Bestor, & J. M. Trasler. (2002). Methylation Dynamics of Imprinted Genes in Mouse Germ Cells. *Genomics*, Vol. 79, No. 4, (Apr) pp. 530-538, 0888-7543

Martin, G. R. (1981). Isolation of a Pluripotent Cell Line from Early Mouse Embryos Cultured in Medium Conditioned by Teratocarcinoma Stem Cells. *Proceedings of the National Academy of Sciences U S A*, Vol. 78, No. 12, (Dec) pp. 7634-7638, 0027-8424

Matsui, Y., & D. Okamura. (2005). Mechanisms of Germ-Cell Specification in Mouse Embryos. *Bioessays*, Vol. 27, No. 2, (Feb) pp. 136-143, 0265-9247

McLaren, A. (1995). Germ Cells and Germ Cell Sex. *Philosophical Transactions of the Royal Society of London. Series B, Biological Sciences*, Vol. 350, No. 1333, (Nov 29) pp. 229-233, 0962-8436

McLaren, A. (1999). Signaling for Germ Cells. *Genes and Development*, Vol. 13, No. 4, (Feb 15) pp. 373-376, 0890-9369

McLaren, A. (2000). Establishment of the Germ Cell Lineage in Mammals. *Journal of Cellular Physiology*, Vol. 182, No. 2, (Feb) pp. 141-143, 0021-9541

McLaren, A. (2003). Primordial Germ Cells in the Mouse. *Developmental Biology*, Vol. 262, No. 1, (Oct 1) pp. 1-15, 0012-1606

Medrano, J. V., A. I. Marques-Mari, & C. Simon. (2010). Germline Differentiation from Stem Cells, In: STEM CELLS: From Bench to Bedside, A. Bongso and E. H. Lee, pp. 557-582, World Scientific, http://ebooks.worldscinet.com/ISBN/9789814289399/9789814289399_0021.html.

Meng, X., M. Lindahl, M. E. Hyvonen, M. Parvinen, D. G. de Rooij, M. W. Hess, A. Raatikainen-Ahokas, K. Sainio, H. Rauvala, M. Lakso, J. G. Pichel, H. Westphal, M. Saarma, & H. Sariola. (2000). Regulation of Cell Fate Decision of Undifferentiated Spermatogonia by Gdnf. *Science*, Vol. 287, No. 5457, (Feb 25) pp. 1489-1493, 0036-8075

Molyneaux, K., & C. Wylie. (2004). Primordial Germ Cell Migration. *International Journal of Developmental Biology*, Vol. 48, No. 5-6) pp. 537-544, 0214-6282

Motta, P. M., S. Makabe, & S. A. Nottola. (1997). The Ultrastructure of Human Reproduction. I. The Natural History of the Female Germ Cell: Origin, Migration and Differentiation inside the Developing Ovary. *Human Reproduction Update*, Vol. 3, No. 3, (May-Jun) pp. 281-295, 1355-4786

Nakagawa, M., M. Koyanagi, K. Tanabe, K. Takahashi, T. Ichisaka, T. Aoi, K. Okita, Y. Mochiduki, N. Takizawa, & S. Yamanaka. (2008). Generation of Induced Pluripotent Stem Cells without Myc from Mouse and Human Fibroblasts. *Nature Biotechnology*, Vol. 26, No. 1, (Jan) pp. 101-106, 1546-1696

Nakagawa, T., Y. Nabeshima, & S. Yoshida. (2007). Functional Identification of the Actual and Potential Stem Cell Compartments in Mouse Spermatogenesis. *Developmental Cell*, Vol. 12, No. 2, (Feb) pp. 195-206, 1534-5807

Natali, A., & P. J. Turek. (2011). An Assessment of New Sperm Tests for Male Infertility. *Urology*, Vol., No., (Jan 20) pp., 1527-9995

Naughton, C. K., S. Jain, A. M. Strickland, A. Gupta, & J. Milbrandt. (2006). Glial Cell-Line Derived Neurotrophic Factor-Mediated Ret Signaling Regulates Spermatogonial Stem Cell Fate. *Biology of Reproduction*, Vol. 74, No. 2, (Feb) pp. 314-321, 0006-3363

Navarro-Costa, P., J. Goncalves, & C. E. Plancha. (2010). The Azfc Region of the Y Chromosome: At the Crossroads between Genetic Diversity and Male Infertility. *Human Reproduction Update*, Vol. 16, No. 5, (Sep-Oct) pp. 525-542, 1460-2369

Nayernia, K., J. Nolte, H. W. Michelmann, J. H. Lee, K. Rathsack, N. Drusenheimer, A. Dev, G. Wulf, I. E. Ehrmann, D. J. Elliott, V. Okpanyi, U. Zechner, T. Haaf, A. Meinhardt, & W. Engel. (2006). In Vitro-Differentiated Embryonic Stem Cells Give Rise to Male Gametes That Can Generate Offspring Mice. *Developmental Cell*, Vol. 11, No. 1, (Jul) pp. 125-132, 1534-5807

Nicholas, C. R., K. M. Haston, A. K. Grewall, T. A. Longacre, & R. A. Reijo Pera. (2009). Transplantation Directs Oocyte Maturation from Embryonic Stem Cells and Provides a Therapeutic Strategy for Female Infertility. *Human Molecuar Genetics*, Vol. 18, No. 22, (Nov 15) pp. 4376-4389, 1460-2083

Nichols, J., & A. Smith. (2011). The Origin and Identity of Embryonic Stem Cells. *Development*, Vol. 138, No. 1, (Jan) pp. 3-8, 1477-9129

Novak, I., D. A. Lightfoot, H. Wang, A. Eriksson, E. Mahdy, & C. Hoog. (2006). Mouse Embryonic Stem Cells Form Follicle-Like Ovarian Structures but Do Not Progress through Meiosis. *Stem Cells*, Vol. 24, No. 8, (Aug) pp. 1931-1936, 1066-5099

Oatley, J. M., M. R. Avarbock, A. I. Telaranta, D. T. Fearon, & R. L. Brinster. (2006). Identifying Genes Important for Spermatogonial Stem Cell Self-Renewal and Survival. *Proceedings of the National Academy of Sciences U S A*, Vol. 103, No. 25, (Jun 20) pp. 9524-9529, 0027-8424

Oatley, J. M., & R. L. Brinster. (2008). Regulation of Spermatogonial Stem Cell Self-Renewal in Mammals. *Annual Review of Cell and Developmental Biology*, Vol. 24, No.) pp. 263-286, 1081-0706

Ohinata, Y., B. Payer, D. O'Carroll, K. Ancelin, Y. Ono, M. Sano, S. C. Barton, T. Obukhanych, M. Nussenzweig, A. Tarakhovsky, M. Saitou, & M. A. Surani. (2005). Blimp1 Is a Critical Determinant of the Germ Cell Lineage in Mice. *Nature*, Vol. 436, No. 7048, (Jul 14) pp. 207-213, 1476-4687

Okita, K., M. Nakagawa, H. Hyenjong, T. Ichisaka, & S. Yamanaka. (2008). Generation of Mouse Induced Pluripotent Stem Cells without Viral Vectors. *Science*, Vol. 322, No. 5903, (Nov 7) pp. 949-953, 1095-9203

Page, J., A. Viera, M. T. Parra, R. de la Fuente, J. A. Suja, I. Prieto, J. L. Barbero, J. S. Rufas, S. Berrios, & R. Fernandez-Donoso. (2006). Involvement of Synaptonemal Complex Proteins in Sex Chromosome Segregation During Marsupial Male Meiosis. *PLoS Genetics*, Vol. 2, No. 8, (Aug 25) pp. e136, 1553-7404

Panopoulos, A. D., S. Ruiz, & J. C. Izpisua Belmonte. (2011). Ipscs: Induced Back to Controversy. *Cell Stem Cell*, Vol. 8, No. 4, (Apr 8) pp. 347-348, 1875-9777

Panula, S., J. V. Medrano, K. Kee, R. Bergstrom, H. N. Nguyen, B. Byers, K. D. Wilson, J. C. Wu, C. Simon, O. Hovatta, & R. A. Reijo Pera. (2011). Human Germ Cell Differentiation from Fetal- and Adult-Derived Induced Pluripotent Stem Cells. *Human Molecuar Genetics*, Vol. 20, No. 4, (Feb 15) pp. 752-762, 1460-2083

Park, I. H., R. Zhao, J. A. West, A. Yabuuchi, H. Huo, T. A. Ince, P. H. Lerou, M. W. Lensch, & G. Q. Daley. (2008). Reprogramming of

Human Somatic Cells to Pluripotency with Defined Factors. *Nature*, Vol. 451, No. 7175, (Jan 10) pp. 141-146, 1476-4687

Park, T. S., Z. Galic, A. E. Conway, A. Lindgren, B. J. van Handel, M. Magnusson, L. Richter, M. A. Teitell, H. K. Mikkola, W. E. Lowry, K. Plath, & A. T. Clark. (2009). Derivation of Primordial Germ Cells from Human Embryonic and Induced

Pluripotent Stem Cells Is Significantly Improved by Coculture with Human Fetal Gonadal Cells. *Stem Cells*, Vol. 27, No. 4, (Apr) pp. 783-795, 1549-4918

Perrett, R. M., L. Turnpenny, J. J. Eckert, M. O'Shea, S. B. Sonne, I. T. Cameron, D. I. Wilson, E. R. Meyts, & N. A. Hanley. (2008). The Early Human Germ Cell Lineage Does Not Express Sox2 During in Vivo Development or Upon in Vitro Culture. *Biology of Reproduction*, Vol. 78, No. 5, (May) pp. 852-858, 0006-3363

Peters, H. (1970). Migration of Gonocytes into the Mammalian Gonad and Their Differentiation. *Philosophical Transactions of the Royal Society of London. Series B, Biological Sciences*, Vol. 259, No. 828, (Aug 6) pp. 91-101, 0962-8436

Qing, T., Y. Shi, H. Qin, X. Ye, W. Wei, H. Liu, M. Ding, & H. Deng. (2007). Induction of Oocyte-Like Cells from Mouse Embryonic Stem Cells by Co-Culture with Ovarian Granulosa Cells. *Differentiation*, Vol. 75, No. 10, (Dec) pp. 902-911, 1432-0436

Reijo, R., R. K. Alagappan, P. Patrizio, & D. C. Page. (1996). Severe Oligozoospermia Resulting from Deletions of Azoospermia Factor Gene on Y Chromosome. *Lancet*, Vol. 347, No. 9011, (May 11) pp. 1290-1293, 0140-6736

Reijo, R., T. Y. Lee, P. Salo, R. Alagappan, L. G. Brown, M. Rosenberg, S. Rozen, T. Jaffe, D. Straus, O. Hovatta, & et al. (1995). Diverse Spermatogenic Defects in Humans Caused by Y Chromosome Deletions Encompassing a Novel Rna-Binding Protein Gene. *Nature Genetics*, Vol. 10, No. 4, (Aug) pp. 383-393, 1061-4036

Resnick, J. L., L. S. Bixler, L. Cheng, & P. J. Donovan. (1992). Long-Term Proliferation of Mouse Primordial Germ Cells in Culture. *Nature*, Vol. 359, No. 6395, (Oct 8) pp. 550-551, 0028-0836

Saitou, M., S. C. Barton, & M. A. Surani. (2002). A Molecular Programme for the Specification of Germ Cell Fate in Mice. *Nature*, Vol. 418, No. 6895, (Jul 18) pp. 293-300, 0028-0836

Salvador, L. M., C. P. Silva, I. Kostetskii, G. L. Radice, & J. F. Strauss, 3rd. (2008). The Promoter of the Oocyte-Specific Gene, Gdf9, Is Active in Population of Cultured Mouse Embryonic Stem Cells with an Oocyte-Like Phenotype. *Methods*, Vol. 45, No. 2, (Jun) pp. 172-181, 1095-9130

Sato, N., L. Meijer, L. Skaltsounis, P. Greengard, & A. H. Brivanlou. (2004). Maintenance of Pluripotency in Human and Mouse Embryonic Stem Cells through Activation of Wnt Signaling by a Pharmacological Gsk-3-Specific Inhibitor. *Nature Medicine*, Vol. 10, No. 1, (Jan) pp. 55-63, 1078-8956

Sato, T., K. Katagiri, A. Gohbara, K. Inoue, N. Ogonuki, A. Ogura, Y. Kubota, & T. Ogawa. (2011). In Vitro Production of Functional Sperm in Cultured Neonatal Mouse Testes. *Nature*, Vol. 471, No. 7339, (Mar 24) pp. 504-507, 1476-4687

Shamblott, M. J., J. Axelman, S. Wang, E. M. Bugg, J. W. Littlefield, P. J. Donovan, P. D. Blumenthal, G. R. Huggins, & J. D. Gearhart. (1998). Derivation of Pluripotent Stem Cells from Cultured Human Primordial Germ Cells. *Proceedings of the National Academy of Sciences U S A*, Vol. 95, No. 23, (Nov 10) pp. 13726-13731, 0027-8424

Schuh-Huerta SM & Reijo Pera RA (2011). Making germ cells from human embryonic stem cells, In: *Male Germline Stem Cells:*

Developmental and Regenerative Potential, K. Orwig & B. P. Hermann, pp. 49-86, Springer Humana Press, 978-1-61737-973-4, Pittsburgh, PA.

Silva, C., J. R. Wood, L. Salvador, Z. Zhang, I. Kostetskii, C. J. Williams, & J. F. Strauss, 3rd. (2009). Expression Profile of Male Germ Cell-Associated Genes in Mouse Embryonic Stem Cell Cultures Treated with All-Trans Retinoic Acid and Testosterone. *Molecular Reproduction and Development*, Vol. 76, No. 1, (Jan) pp. 11-21, 1098-2795

Simon, L., G. C. Ekman, N. Kostereva, Z. Zhang, R. A. Hess, M. C. Hofmann, & P. S. Cooke. (2009). Direct Transdifferentiation of Stem/Progenitor Spermatogonia into Reproductive and Nonreproductive Tissues of All Germ Layers. *Stem Cells*, Vol. 27, No. 7, (Jul) pp. 1666-1675, 1549-4918

Skakkebaek, N. E., A. Giwercman, & D. de Kretser. (1994). Pathogenesis and Management of Male Infertility. *Lancet*, Vol. 343, No. 8911, (Jun 11) pp. 1473-1479, 0140-6736

Skaletsky, H., T. Kuroda-Kawaguchi, P. J. Minx, H. S. Cordum, L. Hillier, L. G. Brown, S. Repping, T. Pyntikova, J. Ali, T. Bieri, A. Chinwalla, A. Delehaunty, K. Delehaunty, H. Du, G. Fewell, L. Fulton, R. Fulton, T. Graves, S. F. Hou, P. Latrielle, S. Leonard, E. Mardis, R. Maupin, J. McPherson, T. Miner, W. Nash, C. Nguyen, P. Ozersky, K. Pepin, S. Rock, T. Rohlfing, K. Scott, B. Schultz, C. Strong, A. Tin-Wollam, S. P. Yang, R. H. Waterston, R. K. Wilson, S. Rozen, & D. C. Page. (2003). The Male-Specific Region of the Human Y Chromosome Is a Mosaic of Discrete Sequence Classes. *Nature*, Vol. 423, No. 6942, (Jun 19) pp. 825-837, 0028-0836

Smith, A. G. (2001). Embryo-Derived Stem Cells: Of Mice and Men. *Annual Review of Cell and Developmental Biology*, Vol. 17, No.) pp. 435-462, 1081-0706

Solter, D. (2006). From Teratocarcinomas to Embryonic Stem Cells and Beyond: A History of Embryonic Stem Cell Research. *Nature Reviews Genetics*, Vol. 7, No. 4, (Apr) pp. 319-327, 1471-0056

Somers, A., J. C. Jean, C. A. Sommer, A. Omari, C. C. Ford, J. A. Mills, L. Ying, A. G. Sommer, J. M. Jean, B. W. Smith, R. Lafyatis, M. F. Demierre, D. J. Weiss, D. L. French, P. Gadue, G. J. Murphy, G. Mostoslavsky, & D. N. Kotton. (2010). Generation of Transgene-Free Lung Disease-Specific Human Induced Pluripotent Stem Cells Using a Single Excisable Lentiviral Stem Cell Cassette. *Stem Cells*, Vol. 28, No. 10, (Oct) pp. 1728-1740, 1549-4918

Sommer, C. A., A. G. Sommer, T. A. Longmire, C. Christodoulou, D. D. Thomas, M. Gostissa, F. W. Alt, G. J. Murphy, D. N. Kotton, & G. Mostoslavsky. (2010). Excision of Reprogramming Transgenes Improves the Differentiation Potential of Ips Cells Generated with a Single Excisable Vector. *Stem Cells*, Vol. 28, No. 1, (Jan) pp. 64-74, 1549-4918

Swijnenburg RJ, Schrepfer S, Govaert JA, Cao F, Ransohoff K, Sheikh AY, Haddad M, Connolly AJ, Davis MM, Robbins RC, Wu JC. (2008). Immunosuppressive therapy mitigates immunological rejection of human embryonic stem cell xenografts. *Proceedings of the National Academy of Sciences U S A.*, Vol. 105, No. 35, (Sep 2) pp. 12991-12996.

Takahashi, K., K. Tanabe, M. Ohnuki, M. Narita, T. Ichisaka, K. Tomoda, & S. Yamanaka. (2007). Induction of Pluripotent Stem Cells from Adult Human Fibroblasts by Defined Factors. *Cell*, Vol. 131, No. 5, (Nov 30) pp. 861-872, 0092-8674

Takahashi, K., & S. Yamanaka. (2006). Induction of Pluripotent Stem Cells from Mouse Embryonic and Adult Fibroblast Cultures by Defined Factors. *Cell*, Vol. 126, No. 4, (Aug 25) pp. 663-676, 0092-8674

Telfer, E. E., R. G. Gosden, A. G. Byskov, N. Spears, D. Albertini, C. Y. Andersen, R. Anderson, R. Braw-Tal, H. Clarke, A. Gougeon, E. McLaughlin, A. McLaren, K. McNatty, G. Schatten, S. Silber, & A. Tsafriri. (2005). On Regenerating the Ovary and Generating Controversy. *Cell*, Vol. 122, No. 6, (Sep 23) pp. 821-822, 0092-8674

Thomson, J. A., J. Itskovitz-Eldor, S. S. Shapiro, M. A. Waknitz, J. J. Swiergiel, V. S. Marshall, & J. M. Jones. (1998). Embryonic Stem Cell Lines Derived from Human Blastocysts. *Science*, Vol. 282, No. 5391, (Nov 6) pp. 1145-1147, 0036-8075

Tilgner, K., S. P. Atkinson, A. Golebiewska, M. Stojkovic, M. Lako, & L. Armstrong. (2008). Isolation of Primordial Germ Cells from Differentiating Human Embryonic Stem Cells. *Stem Cells*, Vol. 26, No. 12, (Dec) pp. 3075-3085, 1549-4918

Toyooka, Y., N. Tsunekawa, Y. Takahashi, Y. Matsui, M. Satoh, & T. Noce. (2000). Expression and Intracellular Localization of Mouse Vasa-Homologue Protein During Germ Cell Development. *Mechanisms of Development*, Vol. 93, No. 1-2, (May) pp. 139-149, 0925-4773

Tu, J., L. Fan, K. Tao, W. Zhu, J. Li, & G. Lu. (2007). Stem Cell Factor Affects Fate Determination of Human Gonocytes in Vitro. *Reproduction*, Vol. 134, No. 6, (Dec) pp. 757-765, 1470-1626

Turnpenny, L., S. Brickwood, C. M. Spalluto, K. Piper, I. T. Cameron, D. I. Wilson, & N. A. Hanley. (2003). Derivation of Human Embryonic Germ Cells: An Alternative Source of Pluripotent Stem Cells. *Stem Cells*, Vol. 21, No. 5) pp. 598-609, 1066-5099

Uhlenhaut, N. H., S. Jakob, K. Anlag, T. Eisenberger, R. Sekido, J. Kress, A. C. Treier, C. Klugmann, C. Klasen, N. I. Holter, D. Riethmacher, G. Schutz, A. J. Cooney, R. Lovell-Badge, & M. Treier. (2009). Somatic Sex Reprogramming of Adult Ovaries to Testes by Foxl2 Ablation. *Cell*, Vol. 139, No. 6, (Dec 11) pp. 1130-1142, 1097-4172

Vallier, L., & R. A. Pedersen. (2005). Human Embryonic Stem Cells: An in Vitro Model to Study Mechanisms Controlling Pluripotency in Early Mammalian Development. *Stem Cell Reviews*, Vol. 1, No. 2) pp. 119-130, 1550-8943

Vallier, L., T. Touboul, Z. Chng, M. Brimpari, N. Hannan, E. Millan, L. E. Smithers, M. Trotter, P. Rugg-Gunn, A. Weber, & R. A. Pedersen. (2009). Early Cell Fate Decisions of Human Embryonic Stem Cells and Mouse Epiblast Stem Cells Are Controlled by the Same Signalling Pathways. *PLoS One*, Vol. 4, No. 6) pp. e6082, 1932-6203

Vierbuchen, T., A. Ostermeier, Z. P. Pang, Y. Kokubu, T. C. Sudhof, & M. Wernig. (2010). Direct Conversion of Fibroblasts to Functional Neurons by Defined Factors. *Nature*, Vol. 463, No. 7284, (Feb 25) pp. 1035-1041, 1476-4687

Warren, L., P. D. Manos, T. Ahfeldt, Y. H. Loh, H. Li, F. Lau, W. Ebina, P. K. Mandal, Z. D. Smith, A. Meissner, G. Q. Daley, A. S. Brack, J. J. Collins, C. Cowan, T. M. Schlaeger, & D. J. Rossi. (2010). Highly Efficient Reprogramming to Pluripotency and Directed Differentiation of Human Cells with Synthetic Modified Mrna. *Cell Stem Cell*, Vol. 7, No. 5, (Nov 5) pp. 618-630, 1875-9777

West, F. D., D. W. Machacek, N. L. Boyd, K. Pandiyan, K. R. Robbins, & S. L. Stice. (2008). Enrichment and Differentiation of Human Germ-Like Cells Mediated by Feeder

Cells and Basic Fibroblast Growth Factor Signaling. *Stem Cells*, Vol. 26, No. 11, (Nov) pp. 2768-2776, 1549-4918

West, F. D., J. L. Mumaw, A. Gallegos-Cardenas, A. Young, & S. L. Stice. (2010). Human Haploid Cells Differentiated from Meiotic Competent Clonal Germ Cell Lines That Originated from Embryonic Stem Cells. *Stem Cells and Development*, Vol., No., (Dec 2) pp., 1557-8534

West, F. D., M. I. Roche-Rios, S. Abraham, R. R. Rao, M. S. Natrajan, M. Bacanamwo, & S. L. Stice. (2010). Kit Ligand and Bone Morphogenetic Protein Signaling Enhances Human Embryonic Stem Cell to Germ-Like Cell Differentiation. *Human Reproduction*, Vol. 25, No. 1, (Jan) pp. 168-178, 1460-2350

Yamazaki, Y., M. R. Mann, S. S. Lee, J. Marh, J. R. McCarrey, R. Yanagimachi, & M. S. Bartolomei. (2003). Reprogramming of Primordial Germ Cells Begins before Migration into the Genital Ridge, Making These Cells Inadequate Donors for Reproductive Cloning. *Proceedings of the National Academy of Sciences U S A*, Vol. 100, No. 21, (Oct 14) pp. 12207-12212, 0027-8424

Ying, Q. L., J. Wray, J. Nichols, L. Batlle-Morera, B. Doble, J. Woodgett, P. Cohen, & A. Smith. (2008). The Ground State of Embryonic Stem Cell Self-Renewal. *Nature*, Vol. 453, No. 7194, (May 22) pp. 519-523, 1476-4687

Ying, Y., X. M. Liu, A. Marble, K. A. Lawson, & G. Q. Zhao. (2000). Requirement of Bmp8b for the Generation of Primordial Germ Cells in the Mouse. *Molecular Endocrinology*, Vol. 14, No. 7, (Jul) pp. 1053-1063, 0888-8809

Ying, Y., X. Qi, & G. Q. Zhao. (2001). Induction of Primordial Germ Cells from Murine Epiblasts by Synergistic Action of Bmp4 and Bmp8b Signaling Pathways. *Proceedings of the National Academy of Sciences U S A*, Vol. 98, No. 14, (Jul 3) pp. 7858-7862, 0027-8424

Yoshida, S., A. Takakura, K. Ohbo, K. Abe, J. Wakabayashi, M. Yamamoto, T. Suda, & Y. Nabeshima. (2004). Neurogenin3 Delineates the Earliest Stages of Spermatogenesis in the Mouse Testis. *Developmental Biology*, Vol. 269, No. 2, (May 15) pp. 447-458, 0012-1606

Young, J. C., V. L. Dias, & K. L. Loveland. (2010). Defining the Window of Germline Genesis in Vitro from Murine Embryonic Stem Cells. *Biology of Reproduction*, Vol. 82, No. 2, (Feb) pp. 390-401, 1529-7268

Yu, J., M. A. Vodyanik, K. Smuga-Otto, J. Antosiewicz-Bourget, J. L. Frane, S. Tian, J. Nie, G. A. Jonsdottir, V. Ruotti, R. Stewart, Slukvin, II, & J. A. Thomson. (2007). Induced Pluripotent Stem Cell Lines Derived from Human Somatic Cells. *Science*, Vol. 318, No. 5858, (Dec 21) pp. 1917-1920, 1095-9203

Yuan, L., J. G. Liu, J. Zhao, E. Brundell, B. Daneholt, & C. Hoog. (2000). The Murine Scp3 Gene Is Required for Synaptonemal Complex Assembly, Chromosome Synapsis, and Male Fertility. *Molecular Cell*, Vol. 5, No. 1, (Jan) pp. 73-83, 1097-2765

Yusa, K., R. Rad, J. Takeda, & A. Bradley. (2009). Generation of Transgene-Free Induced Pluripotent Mouse Stem Cells by the Piggybac Transposon. *Nature Methods*, Vol. 6, No. 5, (May) pp. 363-369, 1548-7105

Zhou, H., S. Wu, J. Y. Joo, S. Zhu, D. W. Han, T. Lin, S. Trauger, G. Bien, S. Yao, Y. Zhu, G. Siuzdak, H. R. Scholer, L. Duan, & S. Ding. (2009). Generation of Induced

Pluripotent Stem Cells Using Recombinant Proteins. *Cell Stem Cell,* Vol. 4, No. 5, (May 8) pp. 381-384, 1875-9777

Zhu, S., W. Li, H. Zhou, W. Wei, R. Ambasudhan, T. Lin, J. Kim, K. Zhang, & S. Ding. (2010). Reprogramming of Human Primary Somatic Cells by Oct4 and Chemical Compounds. *Cell Stem Cell,* Vol. 7, No. 6, (Dec 3) pp. 651-655, 1875-9777

22

Techniques and Conditions for Embryonic Germ Cell Derivation and Culture

Maria P De Miguel[1], Candace L Kerr[2], Pilar López-Iglesias[1] and Yago Alcaina[1]
[1]*Cell Engineering Laboratory, La Paz Hospital Research Institute, Madrid*
[2]*Department of Gynecology and Obstetrics, Institute for Cellular Engineering, Johns Hopkins University School of Medicine, Baltimore, Maryland*
[1]*Spain*
[2]*USA*

1. Introduction

Embryonic germ cells (EGCs) are pluripotent stem cells derived from primordial germ cells (PGCs), which are unipotent cells that *in vivo* give rise to the gametes (McLaren, 2003).
Specification of PGCs takes place in the proximal posterior part of the epiblast shortly before gastrulation, when the epiblast is about to give rise to the three germ layers ectoderm, mesoderm, and endoderm. PGCs can be identified by expression of tissue non-specific alkaline phosphatase (TNAP) activity, various surface antigens stage specific embryonic antigen (SSEA1,3,4), mouse vasa homolog (Mvh) and intracellular proteins (Stella, Fragilis, Oct-4, Nanog and Blimp1 among others). From the proximal epiblast, PGCs migrate along the extraembryonic mesoderm at the base of the allantois, and then move into the epithelium of the hindgut. Later, PGCs start to move through the dorsal mesentery reaching the aorta-gonad-mesonephros (AGM) region and finish their migration at the developing genital ridges. When in gonads, PGC proliferate by mitosis until males enter in mitotic quiescence and in females enter into meiosis (De Felici, 2009).
Under special conditions, PGCs become pluripotent stem cells. In vivo PGCs can generate embryonal carcinoma cells (ECC), the pluripotent stem cells of testicular tumors (Stevens, 1967; Oosterhuis& Looijenga, 2005), while in vitro they can generate EGCs (Matsui et al., 1992; Resnick et al., 1992; Surani, 2007; De Felici et al., 2009; De Miguel et al., 2010). PGCs could be isolated and cultured as such during short periods (up to 10 days) maintaining their phenotype, until they undergo apoptosis. When exposed to a specific mixture of growth factors, PGCs generate EGC colonies. EGCs could be an important source of cells for germ cell or stem cell therapy and a valuable model for understanding development processes involved in reprogramming such as the acquisition of pluripotency.
EGCs were first derived in mice (Matsui et al., 1992; Resnick et al., 1992), and afterwards in a wide variety of mammals like cow (Cherny et al., 1994), goat (Jia et al., 2008), pig (Shim et al., 1997), and sheep (Ledda et al., 2010) among others. Importantly, in 1998 Shamblott et al. derived the first human EGC line, providing a potential source of pluripotent stem cells for therapy.
In this chapter, differences and similarities of EGC derivation and culture of different species are discussed, including species in which long term EGC lines derivation has not yet been achieved.

2. Mouse embryonic germ cells

PGCs in mouse arise at the epiblast adjacent to the extra-embryonic ectoderm around 6.5 days post coitum (dpc) (Lawson et al., 1999). At this stage mouse PGCs constitute a small cluster of TNAP positive cells which also express the POU domain transcription factor Oct-4, both of which are pluripotent markers of the inner cell mass (ICM) of the blastocyst and embryonic stem cells (ES). At 8-8.5 dpc, PGCs migrate and are found in the hindgut endoderm and at the base of the allantois. From there they migrate along the hindgut mesentery and begin to colonize the developing genital ridges at 10.5 dpc. At 13.5 dpc, an established population of 25,000-35,000 PGCs in the gonad will form the future gametes (Tam & Snow, 1981; Ginsburg et al., 1990).

PGCs from embryonic days 8.5 to 12.5 dpc are prevalently used for EGC derivation. They express the pluripotent markers Oct-3/4, Sox-2 and Nanog. However, in culture, PGCs proliferate for 7-10 days and then disappear either because they differentiate or die (De Miguel & Donovan, 2003). *In vitro* culture of mouse PGCs has identified many growth factors that affect their survival and/or proliferation. When PGCs in culture are exposed to leukemia inhibitory factor (LIF), basic fibroblast growth factor (bFGF), stem cell factor (SCF) and serum (Matsui et al., 1992; Resnick et al., 1992; Durcova-Hills et al., 2001) they become immortalized. In addition, unipotent PGCs turn into pluripotent cells called embryonic germ cells through a molecular reprogramming event. These cells are similar to embryonic stem (ES) cells derived from the early embryo or induced pluripotent stem cells (iPS) generated from differentiated tissue in terms of cell morphology and gene expression profile (Sharova et al., 2007; Durcova-Hills et al., 2008). Mouse EG cells also share many features that define pluripotency including the ability to differentiate into a variety of cell types *in vitro*, contribution to germline-competent chimaeras when introduced into blastocysts and the capacity to form spontaneous teratomas (Matsui et al., 1992; Labosky et al., 1994; Stewart et al., 1994).

However, EG cells differ from iPS and ES cells in some properties. EG cells have a different epigenetic state demonstrated by the erasure of genomic imprints and chromatin remodeling from certain imprinted genes like Igf2, Igf2rr, Dlk1 and H19, a process that occurs after specification of PGCs (Lee et al., 2002). Therefore, the pluripotency state from EG cells is somewhat distinct from the naïve pluripotency state of mouse ES cells (Gillich & Hayashi, 2011).

Although the findings in this respect are increasing, the molecular and epigenetic mechanisms that control the reprogramming conversion from PGCs into pluripotent EGCs are not fully understood. However, the signaling pathways and genes that control several processes like proliferation and specification of PGCs and the epigenetic restriction that separate PGCs from pluripotency have been identified. Several lines of evidence highlight key molecules involved in the pluripotency acquirement of PGCs. It has been proposed that downregulation of Prdm1/Blimp1 is a prerequisite for EG cell derivation (Durcova-Hills et al., 2008) and PGC-specific inactivation of the tumor suppressor PTEN enhances both EG cell production and testicular teratoma formation (Kimura et al., 2003). Also, an enforced activation of Akt, one of the major downstream effectors of PI3K, augments the efficiency of EG cell establishment and enables derivation of EG cells from late PGCs, such as from 14.5 dpc embryos (Kimura et al., 2008; Gillich & Hayashi, 2011).

2.1 Isolation and culture of mouse embryonic germ cells

Several protocols have been described for culturing mouse PGCs, their posterior conversion into EG cells as well as derivation of mouse EG cell lines. This part describes the basic protocol developed by the Donovan´s lab (Resnick et al., 1992; De Miguel & Donovan, 2003) for the derivation and establishment of mouse embryonic germ cells lines from mouse PGCs isolated from both 8.5 and 10.5 dpc embryos and some differences with protocols described by Matsui and Durcova-Hills & Surani (Matsui et al., 1992; Durcova-Hills & Surani, 2008).

2.1.1 Buffers and solutions

PBS: Phosphate buffered saline, without Ca^{+2} and Mg^{+2}, pH 7.0 (Gibco/Invitrogen).

Trypsin/EDTA: HBSS containing 0.05% trypsin and 0.53 mM EDTA (Sigma).

Fast Red/Naphtol phosphate solution (Alkaline phosphatase staining solution): a 1 mg/ml solution of Fast red TR salt (Sigma, stored at -20ºC) is made up in dH_2O. 40 µl/ml Naphtol AS-MX phosphate is then added (Sigma, stored at 4ºC). Used immediately.

2.1.2 Culture media

Basic culture medium for PGCs: DMEM (Dulbecco's modified Eagle's medium) high glucose, supplemented with 15% Fetal Bovine Serum (Gibco/Invitrogen) (we recommend to test different batches of Fetal Bovine Serum before purchasing since serum batch used for culture may be especially critical for the growth of PGCs and derivation of EG cells), 2 mM glutamine (Gibco/Invitrogen), 5 U/ml penicillin-streptomycin (Invitrogen), and 1 mM Na^+ pyruvate (Sigma).

Growth factors for EG cell derivation and culture: mSCF (R&D Systems) 10 ng/ml, mLIF (Millipore) 1000 U/ml, hbFGF (Gibco/Invitrogen) 1 ng/ml and Forskolin (Sigma) 100mM.

Basic culture medium for STO: (Sandoz Thioguanine- and Ouabain-resistant cell line, a transformed mouse embryonic fibroblast line): DMEM high glucose, supplemented with 10% Fetal Bovine Serum (Gibco/Invitrogen), 2 mM glutamine (Gibco/Invitrogen), 5 U/ml penicillin-streptomycin (Gibco/Invitrogen), 1 mM Na^+ pyruvate (Sigma). Protocol down below can be used with primary MEFs instead of STO cell line; in that case add 0.1mM non-essential amino acids (Gibco/ Invitrogen) to the medium.

Freezing medium: 9 ml of basic culture medium for PGCs (see above) plus 1 ml of dimethyl sulfoxide (DMSO). Mix thoroughly and keep on ice.

2.1.3 Fixatives

4% Paraformaldehyde: 4g paraformaldehyde in 100 ml PBS. To dissolve the paraformaldehyde, PBS is preheated at 90°C, and NaOH is added drop-wise until the solution turns clear. Solution is cooled down before use.

acid-ethanol: 1 ml acetic acid and 19 ml ethanol precooled at -20ºC.

2.1.4 PGCs Isolation and culture

8.5 days post coitum Embryo Dissection: C57BL/6J inbred mouse line is used to collect tissues containing PGCs (available from Jackson Labs). 8.5 day pregnant females are sacrificed by cervical dislocation. The day on which a vaginal plug is found is designated 0.5 dpc (coitus is assumed to take place at midnight). The abdomen is dissected using scissors and the uteri is removed. Using forceps, each implantation site is separated by cutting the uterus between them, very near to each embryo to allow the deciduum to project. The dissected implantation sites are placed in a Petri dish with ice cold PBS. The decidua is cut across with a pair of fine forceps (Dumont #55). Pressure is applied with fine forceps to the other side and base of the decidua to pop the embryo out of the slit in the deciduum. The extraembryonic membranes are dissected away and the posterior third of the embryo is removed, including the caudal end of the primitive streak and allantois. 50-100 PGCs are localized at the junction of the primitive streak and the base of allantois at this embryonic stage. The embryo fragments are collected in PBS and kept on ice until trypsinization.

10.5 days post coitum Embryo Dissection: At this stage most PGCs are in the developing genital ridges and the wall of the hindgut of the embryo. The embryos are removed from the implantation site and dissected out from the uterus and extraembryonic tissues. The embryo is cut in half, below the forelimbs. The caudal half of the embryo is kept and the anterior portion is discarded. A cut down is made at the midline of the embryo. The genital ridges lay on the dorsal body wall either side of the hindgut mesentery. The skin is opened out on either side of genital ridges and the forceps lifted upwards and towards the tail of the embryo. The two genital ridges, the dorsal aorta that is between them and hindgut are close to the dorsal body wall. The genital ridges are separated from the anterior end and pulled out and replaced in the PBS to squeeze the blood out from the aorta to reduce the number of contaminating cells. The genital ridge/aorta/dorsal mesentery PGC-containing portions are collected. The estimated number of PGCs per embryo is $1x10^3$ at 10.5 dpc.

12.5 days post coitum Embryo Dissection: The dissection of 12.5 dpc embryos is similar to that for the 10.5 dpc embryos. All the PGCs are in the genital ridge. At this stage in embryogenesis it is preferable to dissect the genital ridges away from the adhering tissue with fine forceps. The expected increase in embryo size is associated with a higher yield of PGCs isolation. The estimated number of PGCs per embryo is $10x10^3$ PGCs at 12.5 dpc. PGCs from older embryos than 10.5 dpc are not routinely isolated in our labs because the efficiency of EG cells derivation in culture is lower due to reduced cell proliferation in PGCs of male embryos and the entry into meiosis of PGCs in female embryos.

Enzymatic treatment: The PBS is eliminated and trypsin/EDTA is added to the tube. The embryo fragments are incubated in the solution for 8 min at 37°C in a water bath and the trypsin/EDTA solution is removed as much as possible being careful not to aspirate the tissue fragments. An appropriate volume of PGC basic culture medium is added to break up the tissue fragments by pipetting slowly up and down with a micropipette tip to obtain a single cell suspension. The cells are diluted into the required volume of basic culture medium for PGCs supplemented with growth factors.

PGC Culture: PGCs are cultured using a transformed mouse embryonic fibroblast line as feeder layer, the Sandoz Thioguanine- and Ouabain-resistant cell line (STO). STO cells are commercially available from the American Type Culture Collection (CRL-1503™). PGCs are plated onto confluent mitotically inactivated STO feeder cells and cultured in incubator at 37°C in an atmosphere of 95% air and 5% CO_2. Although soluble form of recombinant SCF is further added to the basic culture medium, both the transmembrane and the soluble form of SCF produced by STO cells is required for enhancing proliferation and long-term survival of cultured PGCs (Resnick et al., 1992). Notably, Sl^4m220 feeder cell line is also used by other investigators to culture PGCs and derive EG cell lines. Sl4m220 cells are derived from a homozygous null (*Sl/Sl*) murine embryo which contains a deletion of the gene coding SCF (*Sl* gene) and have been stably transfected with the membrane bound murine form of SCF (Matsui et al., 1991; Toksoz et al., 1992).

STO cells are maintained in basic culture medium for STO cells. For convenience these feeder cells should be prepared the day before the dissection. The STO cells are plated at a density of $10x10^4$ cells/cm^2 onto 0.1% w/v pregelatinized coated culture dishes. Next morning, STO cells are γ-irradiated with a dose of 50 Gy (5000 rads) to induce cell cycle arrest. After irradiation, the STO cells culture medium is removed and PGC basic culture medium supplemented with growth factors is added (STO cells can also be mitotically inactivated by treatment with mitomycin C (Sigma) at a concentration of 10μg/ml for 4 hours at 37°C). PGCs are plated on top of the STO feeder layer. The final dilution volume of basic culture PGC medium is calculated to plate approximately two 8.5-dpc embryos or 0.5 of one 11.5-dpc embryo per well of a 24-well culture plate. The number of PGCs isolated is estimated from the known numbers of PGCs present in each embryonic stage of the embryo. The medium is replaced every day by gentle aspiration of 2/3 of medium and fresh medium supplemented with growth factors is added.

2.1.5 Identification of PGCs and EG cells

Alkaline phosphatase staining: PGCs and EGCs colonies are distinguished from accompanying somatic and feeder cells by several techniques but the simplest method is alkaline phosphatase staining (Ginsburg et al., 1990). Note that this marker is not unique to germ cells. In mouse embryos, alkaline phosphatase activity is present also in the developing skeletal system (Kaufman, 1992), the developing gut (Merchant-Larios et al., 1985) and the neural tube (Kwong & Tam, 1984). Also, note that this marker stains dead cells, to stain live and dead cells use SSEA1 (see below).

The cultures are washed twice in PBS without Ca^{+2} and Mg^{+2}, and then fixed in 4% paraformaldehyde in PBS (see above) for 15 min at room temperature (RT). The cultures are washed three times in PBS, once in distilled water, and then are incubated in the dark in freshly made alkaline phosphatase staining solution for 20- 30 min at room temperature. After staining, cultures are washed in distilled water; the color reaction will stain PGCs red. The cells must be counted or photographed within a few days of staining, otherwise the cell morphology will deteriorate (Fig. 1).

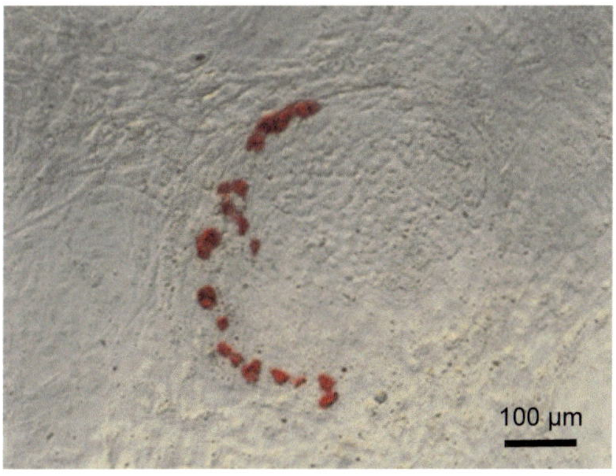

Fig. 1. Mouse primordial germ cells (8.5 dpc) cultured on STO feeder layer and stained for alkaline phosphatase activity in red.

Another way to distinguish between PGCs and EG cells is by their morphological characteristics and growth properties; PGCs grow as single cells, are mortal and will survive for only about 7-10 days in culture, whereas EG cells are immortal, form colonies and can be maintained indefinitely.

Immunofluorescence staining for SSEA-1 or Oct-3/4: PGCs and EG cells can be identified in culture using monoclonal antibodies that recognize pluripotency markers, such as SSEA-1 (anti-SSEA-1 monoclonal antibody can be obtained from the Developmental Studies Hybridoma Bank (http:// www.uiowa.edu/dshbwww/info.html) or commercially (R&D systems) (Donovan et al., 1986; Fenderson et al., 2006) and Oct-3/4 (BD Transduction Laboratories). Here we present a standard protocol for IF staining for SSEA-1 or Oct-3/4 on PGCs and EG cells (Durcova-Hills & Surani, 2008).

PGCs or EGCs are cultured on the appropriate feeder cells over 0.1% gelatin pre-coated microscope glass slides (Fisher) on 24-well culture plates. Cells are washed in PBS and the fixative solution is added for 15 minutes at RT. After that the fixative solution is removed and cells are washed with PBS. Cells are permeabilized and blocked with PBS solution containing 0.1% (w/v) Triton X-100 and bovine serum albumin (BSA; Sigma) at a concentration 1% (w/v) in PBS and incubated for 20 minutes at RT. Mouse anti SSEA-1 (1:200) or mouse anti Oct-3/4 (1:250) antibody diluted in antibody dilution buffer (PBS with 0.1% (w/v) BSA) are added to the fixed cells and incubated overnight at 4°C in a humidified chamber. Then samples are washed and the appropriate secondary antibody (anti-mouse IgM- FITC (1:100; Sigma) for SSEA-1 or Anti-mouse IgG-Alexa (red, 1:500; Molecular Probes) for Oct-3/4) diluted in antibody dilution buffer is added and incubated in a humidified dark chamber for 60 min at RT. Afterwards cells are washed with PBS and nuclei are counterstained with DAPI solution (Sigma) for 10 min at RT in a humidified chamber in the dark. The DAPI solution is aspirated and a drop of fluorescence mounting medium (Vectashield; Vector) is placed on an immunohistochemistry glass slide. The microscope glass slide with stained cells is placed onto the drop of mounting medium and any excess is removed with paper tissue. The samples should be examined under a fluorescence microscope with appropriate filters as soon as possible as the signal diminishes over time (Fig 2). Slides should be stored at 4°C (short term storage) or in a freezer at -20°C (for few days) in the dark.

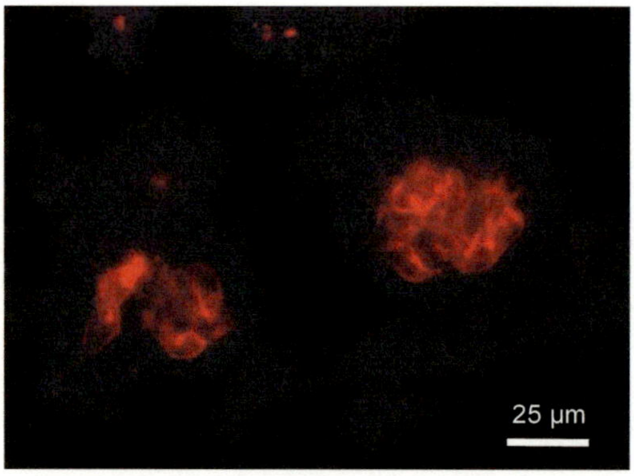

Fig. 2. Characterization of pluripotent EG cells. Detection of cell surface pluripotent marker SSEA-1 performed by immunofluorescence staining (red) in mouse EG cells.

2.1.6 Derivation of EG cells

For EG derivation, PGCs 7- 9 days are cultured in basic culture medium supplemented with leukemia inhibitory factor (LIF), basic fibroblast growth factor (bFGF) and stem cell factor (SCF). The pharmacological agent forskolin is a cAMP agonist, and one of the most potent

PGC mitogens (De Felici et al., 1993). It can also be added to PGC culture to increase derivation efficiency, however is not required for mouse EG derivation. By 7–10 days of primary culture, large colonies of EG cells resembling ES cell colonies should be growing on the STO feeder layer. If the generation of a EG cell line is desired, colonies of EG cells can then be subcultured by trypsinization onto fresh, mitotically inactive primary mouse embryonic fibroblasts (MEFs) or STO. The CF-1 mouse strain feeder layer MEFs isolated from 13.5 dpc embryos (also available at Millipore, PMEF-CFL) is used to subculture EG cells since in our hands EG cells grow better on this cell type. However, feeder layer Sl^4-m220 or STO have also been used to derive EG cell lines (Matsui et al., 1992). MEF feeder cells are plated the day prior to the EG colonies passage at a density 10 x 10^4 cells/cm^2 and irradiated (50 Gy) on the morning of use. EG tightly packed dome-shaped colonies can be visualized by fixing the cultures and staining for alkaline phosphatase or by observing living cultures under a microscope equipped with phase contrast or Hoffman modulation contrast optics.

To subculture EG cell colonies, primary cultures are washed twice with PBS and trypsin/EDTA solution is added, then the culture is incubated at 37°C in a humidified CO_2 incubator for 5–10 min. The trypsin/EDTA solution is removed and PGC basic culture medium added to neutralize the action of the trypsin. The solution is pipetted up and down to obtain a single cell suspension. Importantly, the extent of cell disruption of EGC clumps must be controlled to obtain 2 to 5-cell clumps as large clumps of EG colonies will differentiate if passaged. The cell suspension is transferred to a centrifuge tube and centrifuged 5 min at 400 *g* at RT, and then the pellet is resuspended in fresh basic culture medium supplemented with LIF. The cell suspension is plated onto mitotically inactivated MEF containing wells. To generate a EG cell line, cells can be grown in the same way as mouse ES cell lines and cultured on medium with only LIF, since the previously mentioned factors are no longer necessary. The medium is replaced every 24 hr by removing as much of the existing medium as possible and fresh medium supplemented with LIF. At early stages of EG cell derivation it may be difficult to see the small numbers of EG cells present. After 3–5 days the EG cell colonies should start growing and must be visible on the culture dish (Figs. 3&4).

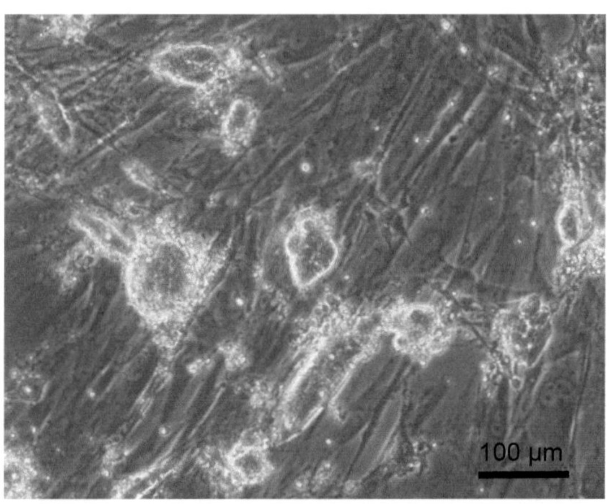

Fig. 3. Phase contrast image of mouse EGC colonies of a mouse EGC cell line cultured on MEFS.

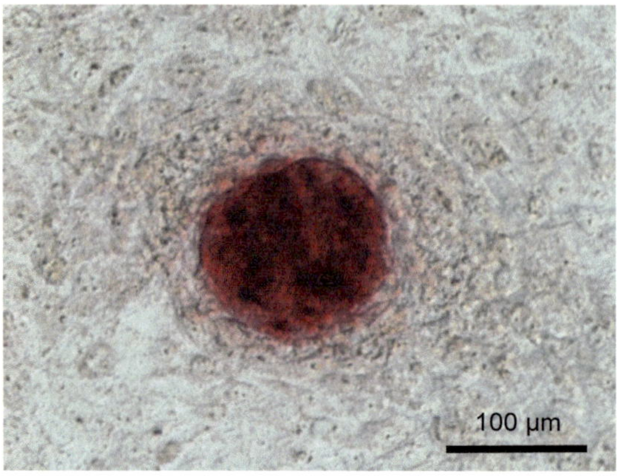

Fig. 4. Alkaline phosphatase staining of a mouse EGC colony.

Recently mouse EG cells have been derived from PGCs efficiently in similar conditions as those used for ES cell derivation from the ICM. By addition of SCF, bFGF, LIF and FBS to the primary culture of murine PGCs on Sl^4m220 feeders, an efficiency of EGC derivation of 1-2% has been reported (Leitch et al., 2010; Gillich & Hayashi, 2011). The addition of two inhibitors (MEK and GSK3 inhibitors) and LIF after two days of previous culture conditions yielded an efficiency of 2-10% (Leitch et al., 2010; Gillich & Hayashi, 2011).

2.1.7 Colony selection and Passaging EGC colonies

Single colonies of EG cells can be selected and expanded. Mitotically inactivated MEF feeder cells must be prepared 24 h before and irradiated on the morning of use. The medium from a culture dish of EG cells is removed and washed with PBS. Using an inverted or dissecting microscope, individual colonies are picked with a pipettor and transferd to a centrifuge conical tube containing a trypsin/EDTA solution. As many colonies as possible should be picked in a short period of time (so as to avoid over trypsinization of the colonies because it can affect the viability of cells) and the suspension placed into the incubator at 37°C for 5 to 10 min. The cell clumps are mixed by pipetting up and down until a single-cell suspension is obtained. To neutralize the effect of trypsin, complete medium is added and centrifuged for 5 min at 400 *g* at RT. The pellet is resuspended in complete medium and placed onto mitotically inactivated MEF feeder cells plate containing EGC complete medium with LIF. The medium with LIF factor should be replaced daily until the cells have expanded sufficiently that they can be passaged onto a larger culture dish plate.

EG colonies also can be picked using a pulled-glass pipette instead of using a pipettor to disaggregate cells in a trypsin-EDTA solution. Also picked EGCs colonies can be trypsinized in a microdrop under mineral oil before transferred onto feeder layer cells (Matsui et al., 1992; Durcova-Hills & Surani, 2008).

Once a EG cell line is derived as explained in the above paragraph, growing colonies can be picked and expanded on gelatin-coated plates without feeder layers albeit at a lower efficiency compared to the traditional method (Leitch et al., 2010).

3. Other mammalian EG cells

3.1. Rat embryonic germ cells

The first derivation of EG cells from rat has been recently reported (Leitch et al., 2010). Leitch and collaborators have efficiently derived and propagated rat EG cell lines with using culture conditions previously used for clonogenic expansion of rat pluripotent embryonic stem cells (Li et al., 2008). EG cell line generation was obtained adding a MEK inhibitor and a GSK3 inhibitor in combination with LIF without SCF or bFGF to the culture of PGCs at the first medium change. Rat EG cells express pluripotency markers similar to mouse EG cells (Nanog, Oct3/4, Sox2, AP, Klf4 and Rex1) and the EG cell lines derived were competent for multilineage colonization of chimaeras and embryoid body formation giving rise to cells of the three germ layers thus showing the ability to reprogram rat PGCs to the pluripotent state.

Rat EG cells were derived from rat embryos from embryonic day 10, equivalent to 8.5 dpc in the mouse. The protocol used for derivation of rat EG cell lines follows the same schedule as described above for mice (Leitch et al., 2010) but on day 3 of derivation (first medium change), the cultures were washed with PBS and the medium was replaced with serum-free N2B27 medium supplemented with 1μM PD0325901 MEK inhibitor, and 3 μM CHIR99021 GSK3 inhibitor and LIF. Cultures were then maintained continuously in these conditions. Picked colonies were expanded on gelatin-coated plates with neither feeders nor conditional medium. After colony picking, cells were expanded by dissociation with trypsin and replating every 2-4 days.

3.2 Buffalo embryonic germ cells

Recently, embryonic germ cell like-cells from 30-90 dpc fetuses of the Chinese buffalo have been derived. These EG-like cells were cultured for more than two weeks on buffalo embryonic fibroblast feeder cells in DMEM containing 20% FBS media and supplemented with LIF, bFGF and SCF. During isolation of EG-like cells, the mechanical method used for disaggregating cells was better than the trypsin digestion, enabling cells to reach a higher passage in culture. Buffalo EG-like cells grew in large densely packed colonies resembling mouse ES or EG cells colonies and were characterized by their expression of the pluripotency markers AP, SSEA-1, SSEA-3, SSEA-4 and Oct-4. EG-like cells were capable to differentiate into the three germ layers in vitro, although chimaera formation was not determined (Huang et al., 2007).

3.3 Pig EG cells

Pig EG cells have been derived from porcine primordial germ cells isolated from the gonadal ridges of fetuses on days 24-28 of gestation (approximate to human Carnegie Stages 18-20) (Piedrahita et al., 1998). Several breeds have been used: Hampshire X Yorkshire (Shim et al., 1997), Duroc and German Landrace (Mueller et al., 1999), and the Chinese mini pig (Tsung et al., 2003). Genital ridges are dissected from the embryos, washed with PBS and then underwent different enzymatic dissociation treatments depending on the author: 0.02% EDTA (Sigma) for 20 min at RT (Shim et al., 1997; Mueller et al., 1999), 0.25% trypsin + 0.02% EDTA or 1mg/ml dispase (Sigma) in PBS for 5 min or up to 15 min at 37ºC (Tsung et al., 2003). After incubation, PGCs are mechanically dissociated by gentle disruption of the tissues using fine forceps and then centrifuged at 800 x g for 5 min. Enzymatic treatment could be replaced by more roughly mechanically dissociation passing the disrupted tissues several times through a 20-gauge needle, then centrifuged for 3-5 min at 250 g to settle tissue fragments, and supernatant containing mostly single cells is collected and centrifuged at 1000 g for 5 min (Piedrahita et al., 1998). Colonies are disaggregated with 0.25% trypsin-EDTA for 10-15 min and passaged to fresh feeder layers at 4- to 10-day intervals (Shim et al., 1997; Piedrahita et al., 1998).

Similar to other species, feeder cells expressing membrane-bound SCF are required for survival and establishment of porcine EG cells (Lee & Piedrahita, 2000). STO cells, mitotically inactivated with 10 µg/ml mitomycin C (Sigma) for 2 h, are plated at a density of ~1.5-3 x 10^5 cells/cm^2 (Shim et al., 1997; Piedrahita et al., 1998).

The growth medium used to derive and maintain porcine EGCs is DMEM or DMEM:Ham´s F10 (1:1) (Piedrahita et al., 1998) containing 15% FBS, 1 mM L-glutamine, 0.1 mM or 0.01 mM nonessential amino acids (Gibco) (Piedrahita et al., 1998), 0.1 mM 2-mercaptoethanol (10 µM by Shim et al., 1997), penicillin (100 U/ml), and streptomycin (0.5 mg/ml) (Shim et al., 1997). FBS can be replaced with knockout serum replacement (KSR) (Petkov & Anderson, 2008) and 0.1 M Na^+ pyruvate (Gibco) has also added (Mueller et al., 1999). Cultures are maintained at 37-39°C in 5% CO_2, 95% air, and medium is changed every day. Growth factor supplementation is not necessary, but is common to use SCF (40 ng/ml), hbFGF (20-25 ng/ml), and LIF (10-20 ng/ml) (Piedrahita et al., 1998; Tsung et al., 2003). Lee & Piedrahita (2000) reported that even if it is not required for EGCs culture, supplementation with all three growth factors increases eight times the number of EGC colonies TNAP positive. In addition to AP, pig EGCs are Oct4, SSEA-1 and SSEA-4 and TRA-1-81 positive (Petkov & Anderson, 2008).

3.4 Goat EG cells

First attempts to generate EGCs from goat resulted in lines that survived briefly and after 3-4 passages differentiated (Kuhholzer et al., 2000; Lee & Piedrahita, 2000). However, the most successful attempt achieved to subculture goat EGC cells over 12 passages before spontaneous differentiation. Moreover, two chimeras out of 29 injected blastocysts were obtained: one aberrant chimera, showing cells representing the three germ layers and one viable chimera, showing chimerism in skin and blood. However, germ line transmission of the chimerism could not be confirmed (Jia et al., 2008).

Goat primordial germ cells were isolated from the gonadal ridges of fetuses from slaughterhouse. The fetuses age varied between groups, from 25 (Lee et al., 2000) to 32 dpc (Kuhholzer et al., 2000) or 28-42 dpc (Jia et al., 2008).

Tissues were washed with PBS with 0.02% EDTA, mechanically disaggregated and then incubated in 0.25% type IV collagenase (Sigma) for 30 min. To remove most somatic cells, cell suspension was filtered through a 100 mesh sterile gauze. After washing again in PBS and centrifugation at 1000 g for 5 min, cells were cultured on a goat embryonic fibroblast (GEF) feeder layer mitotically inactivated with mitomycin C at 100% density (Jia et al., 2008).

First attempts used STO cells as feeders and different media, such as DMEM: Ham's F10 medium (1:1), but colonies differentiated earlier in culture (Kuhholzer et al., 2000; Lee et al., 2000).

The growth medium used was DMEM supplemented with 15% KSR (Gibco), 0.1 mM nonessential amino acids, 0.1 mM 2-mercaptoethanol, 2 mM L-glutamine, 1 mM Na^+ pyruvate (Invitrogen), 100 U/ml penicillin and 0.1 mg/ml streptomycin (Sigma), 10 ng/ml recombinant human bFGF (Sigma), 10 ng/ml SCF and 1000 IU/ml recombinant murine LIF (Chemicon) at 37 °C in a humidified atmosphere with 5% CO_2 (Jia et al., 2008).

The EGC cells were subsequently passaged to a fresh MEF feeder layer in the same medium by picking up colonies after 10 to 12 days (Jia et al., 2008). Long term culture of EGCs was difficult because of spontaneous differentiation into epithelial-, neuronal-, and fibroblast-like cells after few passages (Kuhholzer et al., 2000; Lee et al., 2000). However, goat EGCs were positive for AP, SSEA-1, Nanog and c-kit, as their mouse counterparts (Jia et al., 2008).

4. Human EG cells

To date, six laboratories have reported successful derivation of human EGC lines (Shamblott et al., 1998; Turnpenny et al., 2003; Liu et al., 2004; Park et al., 2004; He et al., 2007; Hua et al., 2009). This very limited number of groups is due to the difficulties of acquiring human PGCs (therapeutic termination of pregnancy and ethical approval is needed), and the complexity of the derivation and culture of EGCs (Turnpenny et al., 2006; Kerr et al., 2006; Perret et al., 2008).

4.1 Growth medium

The methods used for human EGCs derivation are based in part in those described previously for mouse. The growth medium used to derive and maintain human EGCs is Dulbecco's modified Eagle's medium, DMEM-199 or KO-DMEM (Invitrogen) supplemented with 15% FBS or 20% Knockout serum (HyClone/Invitrogen), 0.1 mM nonessential amino acids (Invitrogen), 0.1 mM 2-mercaptoethanol (Sigma), 2mM glutamine (Invitrogen), 1mM Na^+ pyruvate (Invitrogen), 100 U/ml penicillin (Invitrogen), and 100 mg/ml streptomycin (Invitrogen). For human EGC derivation, 1000 U/ml human recombinant LIF (hLIF, Chemicon), 1–2 ng/ml human recombinant bFGF (Genzyme/R&D Systems), and 10 µM forskolin (Sigma) is added (Shamblott et al., 1998; Kerr et al., 2006). bFGF doses vary between groups, from 1 ng/ml (Shamblott et al., 1998) to 10 ng/ml (Hua et al., 2009). In contrast to mouse EGC derivation, supplementation with soluble SCF is not necessary for human EGCs. In agreement with mouse, transmembrane SCF expression of the feeder layer promoted human PGC growth (Kerr et al., 2006).

It is also possible to derive and maintain human EGCs without serum using 15-20% KSR (Invitrogen) (Turnpenny et al., 2003; Pan et al., 2005; Hua et al., 2009; Hiller et al., 2011). Whereas serum provides essential nutrients also provides some factors that promote differentiation of EGCs and make the analysis of the experiments difficult. Thus, it is reported that the derivation efficiency increases using KSR instead of FBS (Hua et al., 2009).

Recently it has been reported that the addition of 5-20 ng/ml recombinant BMP4 (R&D Systems) increases, in a dose-responsive manner, the efficiency of EGC derivation and maintenance. The results revealed an increase in EGC derivation of 5-fold with 20 ng/ml BMP4 and, after 3 weeks the survival efficiency increased by 50-fold in the presence of BMP4 (Hiller et al., 2011). Survival appeared to be in part related to the ability of BMP4 to inhibit spontaneous differentiation in these cultures, an inherent problem with long-term maintenance of human EGC cultures.

4.2 Feeder layer

Human EGCs have been mainly derived using the transformed mouse embryonic fibroblast line STO (Swelstad & Kerr, 2010). Although it is a clonal cell line, there are several phenotypic variations in STO cells from different isolates or even depending on the time in culture that affects the human EGC derivation (Kerr et al., 2006). Some groups have also successfully utilized primary MEFs (CF1) as feeder layers for EGC derivation (Liu et al., 2004; Hua et al., 2009) and even human embryonic fibroblast-like cells derived from gonadal ridges and dorsal mesenteries (see below, He et al., 2007).

The feeder layer can be mitotically inactivated with mitomycin C or by irradiation, either before or after plating. The most convenient method is to irradiate the STO cells after plating, although a large γ radiation unit is needed (Kerr et al., 2006). STO cells are cultured in the EGC growth medium without growth factor and are passaged for short periods, being disaggregated with 0.05% trypsin–EDTA. STO cells are plated at 5 x 10^4 cells per well in 96-well plates previously coated with 0.1% gelatin for 30 min. It is also possible to culture the STO cells in other plate types, maintaining a similar cell density of 1.5 x 10^5 cells/cm^2.

Feeder cells can be irradiated before plating, which has the advantage of the need of a smaller γ radiation unit, but the disadvantage of lower control of feeder cell density. Feeder cells are disaggregated as described, resuspended in growth medium and then exposed to 50 Gy of γ radiation or X rays. Finally, cells are cultured into gelatinized culture plates at ~1.5 x 10^5 cells/cm^2 and allowed to adhere overnight. Alternatively, feeder layers can be mitotically inactivated with 10 µg/ml mitomycin C (Sigma) (Park et al., 2004; Liu et al., 2004; Hua et al., 2009).

When MEFs are used instead of STO cells, these have to be seeded at 7.5 ×10^4 cells/cm^2, half the density of STO cells (Pan et al., 2005; Hua et al., 2009). There is not comparison reported between human EGCs derivation on MEFs versus STO cells or between irradiation against mitomycin C mitotic inactivation.

Co-culture of human EGCs with mouse feeder cells entails clinical restrictions due to the possible contamination by xenogenic proteins or pathogens. As a result, a new method has been developed for hEGCs culture using human embryonic fibroblast-like cells derived from gonadal ridges and dorsal mesenteries obtained from 5-10 week human embryos. The tissues are mechanical and enzymatically (0.25% trypsin 10 min) dissociated and then cultured in high glucose (HG)-DMEM, 10% FBS (Gibco), 1 mM Na^+ pyruvate, 100 U/ml penicillin, and 50 µg/ml streptomycin. These cells are plated at 3 x 10^5 cells in a 6-well plate (~3.2 x 10^4 cells/cm^2), and after 24 h treated with 12.5 mg/l of mitomycin C (Roche) (He et al., 2007).

4.3 PGC isolation

To derive human EGCs, PGCs are isolated from the fetal gonad between 5 and 11 weeks of gestation (Carnegie stage 15 onwards) at termination of pregnancy by using a drug protocol or surgically and with local research ethics committee approval with written informed consents (Kerr et al., 2006). Hua et al. (2009) used a wider margin with fetuses aged 4-13 weeks, but is not clear that EGC derivation is possible with PGCs of those younger or older ages since the authors did not specify the age of the PGCs that actually originated EGC colonies. Human EGC derivation is more difficult to standardize compared to mouse EGCs in part because of the age and genetic heterogeneity among the embryonic sources. The collection of mesenteries besides the gonadal ridges also varies between different groups. Most of them but two groups included mesenteries (Pan et al., 2005; Turnpenny et al., 2006). Here we are going to describe the detailed protocol reported by Kerr et al. (2006) pointing out the differences with other laboratories.

Dissected tissues are collected in 1 ml of ice cold growth medium, then soaked in Ca^{+2} and Mg^{+2} free Dulbecco's phosphate buffered saline (DPBS) for 5 min and then enzymatically disaggregated with 100µl of trypsin–EDTA solution (Invitrogen). The concentration of

trypsin and EDTA varies depending on the developmental stages, using 0.05% trypsin-0.5 mM EDTA solution at earlier stages, and 0.25% trypsin-0.5 mM EDTA at later stages.
Following that, the tissue is mechanically dissociated, using fine forceps and iris scissors, for 5-10 min at RT, and then placed at 37ºC for 5-10 min. Growth medium is added to the tube to stop the digestion and finally the cell suspension is pipetted 30-50 strokes with a 200 µl Pipetman (Gilson) (Kerr et al., 2006).
Slight variations in these protocols are used by other groups: A previous 0.01% EDTA treatment for 10 minutes (Turnpenny et al., 2003), different enzymatic dissociation treatments between using only trypsin (without EDTA) (Liu et al., 2004), collagenase IV + DNase I (Turnpenny et al., 2003; Park et al., 2004), and even without mechanically dissociation and just incubation with 0.125% trypsin and 0.02% EDTA for 10-20 min or with 0.125% collagenase for 20-40 min at 37°C (Hua et al., 2009). No comparisons between different PGC isolation protocols have been reported, so it is unknown which one is the more efficient.
The cell suspension is plated on the previously prepared feeder layer, culturing each gonad on 4-10 wells of a 96 well plate, keeping in mind that derivation efficiency could be affected by the plating density (Kerr et al., 2006).
To get rid of accompanying somatic cells, PGCs could be sorted before plating (Kerr et al., 2008; Hiller et al., 2011). PGCs are isolated using magnetic cell sorting technology and an indirect labeling of cells with magnetically tagged goat anti-mouse IgM antibodies towards a mouse-anti-SSEA1 antibody (Miltenyi Biotech). After tissue dissociation, cells are incubated with SSEA1 antibody (1:5 dilution) for 15 min on ice, then secondary antibody is applied at 1:100 dilution for another 30 min on ice and sorted on magnetic columns. Culturing a pure population of PGCs allows ruling out possible effects of the somatic cells of the gonads in EGC derivation. When PGCs are counted after sorting, approximately 50 PGCs are seeded in each well of a 96-well plate (Hiller et al., 2011).
The plate is incubated at 37ºC in 5% (or 8% (Shamblott et al., 1998) CO_2 with 95% humidity for 7 days. Growth medium is changed every day, removing 90% of the old medium (Kerr et al., 2006).

4.4 Passage of EGC cultures

No EGC colonies are normally seen in most human EGC cultures during the first 7 days and only solitary PGCs could be observed when staining for TNAP activity. Passages are performed after 7 days, subculturing the PGCs onto fresh feeder cells. Compared to other pluripotent stem cells, hEGCs are also challenging to maintain due to the difficulty in disaggregating colonies, so the passage is a critical step in the EGC culture (Kerr et al., 2006).
Medium has to be removed and the wells washed with Ca^{+2} and Mg^{+2} free DPBS. To disaggregate the cells, 40 µl of freshly thawed trypsin-EDTA solution (0.05%-0.25% trypsin and 0.5 mM EDTA) is added to each well and incubated for 5 min at 37ºC. Instead of trypsin, 0.1% type IV collagenase could be used (He et al., 2007). Then a Pipetman and 200 µl tip is used to scrape the bottom of the wells and gently pipetting the culture 20-30 times. To stop digestion fresh growth medium is added and pipetting another 10-30 times. Cell suspension is placed into twice the number of former wells on feeder cells (Kerr et al., 2006).
Subsequent passages are repeated every 7 days and after 2-3 weeks EGC colonies can be seen in some of the wells (Kerr et al., 2006).
Although initially 50% of the wells on average produce EGC colonies (Kerr et al., 2006), after 2 to 3 weeks, large and recognizable EGC colonies are seen only at approximately 10 to 20% of the wells (Swelstad & Kerr, 2010) (Figs. 5-7).

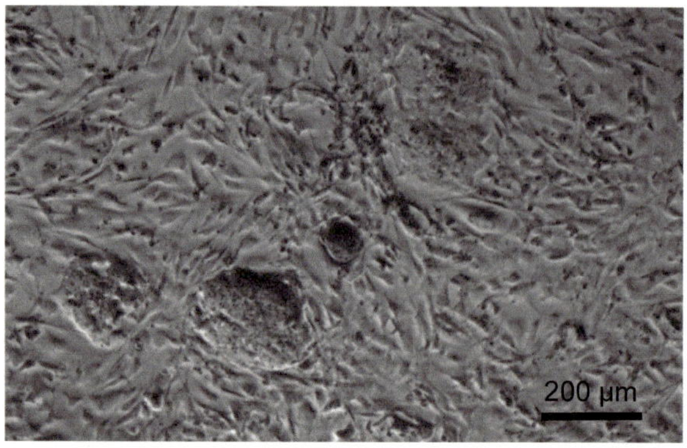

Fig. 5. Phase contrast image of human EGC colonies after derivation on STO feeder cells.

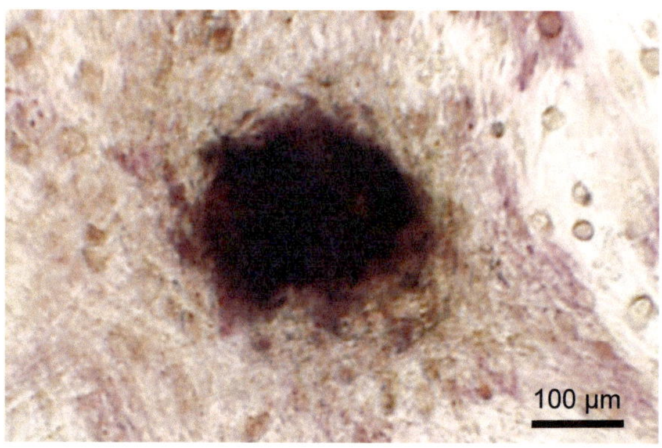

Fig. 6. Human EGC colony stained for alkaline phosphatase activity in red.

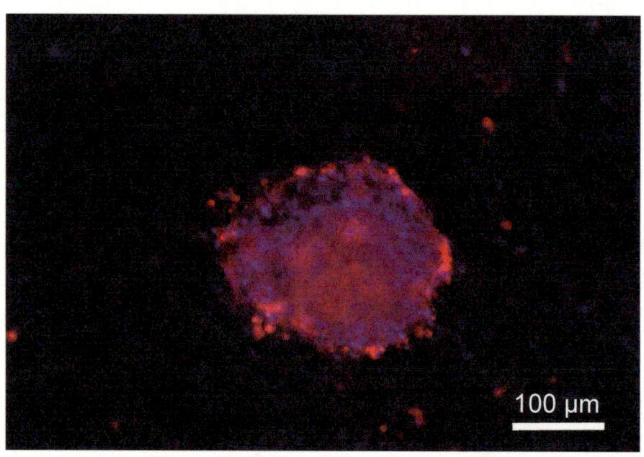

Fig. 7. Characterization of pluripotent human EG cells. Indirect immunofluorescent detection of pluripotent marker, Tra-1-81 (red). Dapi (blue) stains nuclei.

A relationship between the conversion efficiency and the embryonic stage or sex of the embryo has been reported in both male and female gonads to be associated with an increase in Oct-4 expressing PGCs which peak around 10 weeks of age for both sexes (Kerr et al., 2008a,b).

5. Attempts in challenging species

Finally, several attempts to derive EG cells from species are ongoing, and the challenges of EG derivation of these particular species are examined.

5.1 Rabbit EG cells

No stable EGCs lines have been derived yet from rabbit primordial germ cells. Kakegawa et al. (2008) reported the obtaining of EGC-like colonies that can form embryonic bodies (EBs) and differentiate in several cell types in vitro but did not develop teratomas after 20 days of injection of 3 x 10^6 cells. In addition, EGC-like colonies were lost in culture after 4 passages. These cells expressed Oct3/4, Sox2, and SSEA-1 and were AP positive as their mouse counterparts.

Rabbit PGCs were isolated from New Zealand White rabbit's embryos of 9.5-11.5 days of gestation, which resemble 8.5-11.5 dpc mouse embryos. Genital ridges were dissected and the dorsal mesentery was removed. Tissues were then washed with PBS and incubated in 0.04% trypsin, 0.25% EDTA (Sigma) for 5 minutes at RT. After the enzymatic digestion, PGCs were dissociated using a glass capillary tube and seeded on mitotically inactivated MEF feeder layer. Culture medium was composed of knockout-DMEM (Invitrogen) with 20 % KRS (Invitrogen), L-glutamine, Na^+ pyruvate, MEM nonessential amino acids and 2-mercaptoethanol. No growth factor addition was necessary for EG-like colonies formation, but the number of colonies increased with the addition of LIF (Chemicon), bFGF (Upstate), and forskolin (Sigma) at different concentrations. The greatest improvement was seen with 1000 IU/ml LIF, 20 ng/ml bFGF and 20 µM forskolin. For passaging, cultures were dissociated with 0.04% trypsin, 0.25% EDTA, and centrifuged onto fresh feeder cells at 4- to 7-day intervals (Kakegawa et al., 2008).

5.2 Sheep EG cells

Sheep EGC colonies have been derived and maintained over 14 passages. Their pluripotency has been proved by in vitro differentiation into a range of tissue types and by embryonic body formation. However, attempts to produce teratoma formation or chimaeric generation were attempted but without success (Ledda et al., 2010). Sheep EG expressed AP, Oct3/4, Sox2, Nanog and SSEA-1 (Ledda et al., 2010).

Primordial germ cells from sheep were isolated from fetuses of 20–28 days (similar to human Carnegie stages 13-19) following protocols similar to those previously described for mouse. PGCs were cultured on mitotically inactivated MEF feeder layers at near 100% of density on DMEM supplemented with 20% serum, L-glutamine, Na^+ pyruvate, MEM nonessential aminoacids, 2-mercaptoethanol and 1000 IU/ml LIF. Supplementation with 10-30 ng/ml bFGF and forskolin improved in vitro proliferation and long term survival of PGCs (Ledda et al., 2010).

5.3 Cow EG cells

Bovine EG-like cells were derived from 29-35 days of gestation PGCs. Genital ridges from the fetuses obtained from slaughterhouses were dissected without mesonephros. Tissues were enzymatically disrupted with trypsin and EDTA and then seeded on a feeder layer of neomycin-resistant LIF-producing STO cells which express recombinant human LIF (McMahon & Bradley, 1990), or on bovine primary embryonic fibroblasts. Passages were performed picking the colonies from the feeder layers and subculturing (Cherny et al., 1994). As with sheep and rabbit EGCs, bovine EG-like cells also demonstrated the ability to generate embryoid bodies and differentiate to various cell types. Bovine EGCs were demonstrated positive for Oct-4, and Hes-1. Pluripotency in vivo was not demonstrated, as no teratomas or chimaeras were developed and EGCs only could be maintained in culture for up to 10 passages (7 weeks) (Cherny et al., 1994).

5.4 Non-human primate EG cells

Non-human primates pose an interesting alternative source for studying the highmarks of EGC derivation relevant to human biology and preclinical applications. While mouse has been the most extended animal model for EGCs research, there remain significant differences between mouse and human EGCs in terms of growth characteristics, marker expression, signaling patterns and imprinting status. Furthermore, human EGC derivation is limited to a diverse genetic background and human embryo donations are anonymous it is not possible to analyze epigenetic patterns relating them to parental lines. Thus, it seems reasonable to suggest that questions regarding the process of reprogramming and differentiation via genomic imprints could potentially be more easily and appropriately addressed using non-human primate EGCs.

To date, there are no reports on the isolation of stable long term EGC lines from non-human primate PGCs, but short term EGCs have been developed from the groups of Schatten and Simerly (Simerly et al., 2010).

Baboon embryos were collected between days 28-35 post coitum (similar to human Carnegie Stages 12-16), dissecting the lower half of the embryo and excising the genital ridge, dorsal mesentery and part of the gut. Tissues were dissociated with 0.05% trypsin treatment and cells cultured on irradiated STO feeder cells in 96-well plates at a density of ~10×10^4 cells/cm^2.

The growth medium used was DMEM high glucose with 15% FBS (Hyclone, Fisher), 2mM glutamine, 5 U/ml penicillin-streptomycin and 1mM Na^+ pyruvate, supplemented with various growth factor combinations, similar to both mouse and human protocols. These included different concentrations of hbFGF, 20 µM retinoic acid, 1000 U/ml hLIF, 10 ng/ml hSCF (Cell Signaling), 100 µM forskolin (Sigma), 20 ng/ml stromal cell-derived factors α & β (α,β-SDF), and 50 ng/ml BMP-4. BMP-4 was previously used to improve human EGC culture (Hiller et al., 2011) and retinoic acid has been shown to replace bFGF for mEGC derivation (Koshimizu et al., 1996). Three different protocols have been tried: 10 ng/ml hbFGF, LIF, hSCF and forskolin; 20 ng/ml hbFGF, LIF, hSCF, α,β-SDF and BMP-4; and retinoic acid, LIF, hSCF, α,β-SDF and BMP-4. Medium was removed and replenished with fresh growth medium daily and passaging onto fresh STO feeders was performed weekly. EGC colonies were established using all three protocols, but after 2-3 passages start to differentiate.

6. Conclusions

In summary, EGC lines have been derived from a variety of mammals including human and are being attempted in many others. Comparison of differences in methodology used as well as in pluripotency markers, achieved time of culturing and functional pluripotency demonstration, are summarized in Tables 1 and 2.

	Age (dpc)	Feeders	Medium	GFs	Passages	Pluripotency
Mouse (Matsui et al., 1992; De Miguel & Donovan 2003)	8.5-12.5	STO MEFs (P1 onwards)	DMEM + 15% FBS	LIF, bFGF, SCF, Forskolin	> 1 year	Chimaeras teratomas
Rat (Leitch et al., 2010)	10	Sl4-m220 MEFs (P1)	N2B27 serum free	LIF, MSKi, GSK3i	35	Chimaeras
Buffalo (Huang et al., 2007)	30-90	BuffaloEFs	DMEM + 20% FBS	LIF, bFGF, SCF	8	In vitro diff
Pig (Piedrahita et al., 1998)	24-28	STO	DMEM + 15% FBS	LIF, bFGF, SCF	> 10 months	Chimaeras
Goat (Jia et al., 2008)	25-42	GoatEFs MEFs	DMEM + 15% KSR	LIF, bFGF, SCF	12	Chimaeras
Human (Shamblott et al. 1998; Kerr et al., 2006)	5-11 week	STO	DMEM + 15% FBS	LIF, bFGF	>20	In vitro diff
Rabbit (Kakegawa et al., 2008)	9.5-11.5	MEFs	KO-DMEM + 20% KSR	LIF, bFGF, Forskolin	4	In vitro diff EBs
Sheep (Ledda et al., 2010)	20-28	MEFs	DMEM + 20% serum	LIF; bFGF + Forskolin	14	In vitro diff EBs
Cow (Cherny et al., 1994)	29-35	SNL BovineEF			>10	In vitro diff EBs
Baboon (Simerly et al., 2010)	28-35	STO	DMEM + 15% FBS	LIF, bFGF, SCF	3	ND

Table 1. Comparison of different methods of EGC lines derivation from different species. GF: growth factors. EBs: in vitro embryoid bodies. dpc: days post coitum. FBS: fetal bovine serum. KSR: knockout serum replacement. MEFs: mouse embryo fibroblasts. EFs: embryonic fibroblasts. KO: knockout. SNL: STO Neomycin-resistant LIF-producing cells. ND: not determined.

	Mouse	Rat	Buffalo	Pig	Goat	Human	Rabbit	Sheep	Cow	Baboon
TNAP	+ CC RT (De Miguel & Donovan, 2003)	+ CC (Leitch et al., 2010)	+ CC (Huang et al., 2007)	+ CC (Petkov & Anderson, 2008)	+ CC (Jia et al., 2008)	+ CC (Shamblott et al., 1998)	+ CC (Kakegawa et al., 2008)	+ CC (Ledda et al., 2010)	+ CC (Cherny et al., 1994)	**ND**
Oct4	+ IF; RT (Durcova-Hills et al., 2008)	+ IF; RT (Leitch et al., 2010)	+ IF RT (Huang et al., 2007)	+ ICC (Petkov & Anderson, 2008)	**ND**	+ ICC (Liu et al., 2004)	+ ICC (Kakegawa et al., 2008)	+ RT (Ledda et al., 2010)	+ RT (Cherny et al., 1994)	**ND**
Sox2	+ IF; RT (Durcova-Hills et al., 2008)	+ IF; RT (Leitch et al., 2010)	+ RT (Huang et al., 2007)	**ND**	**ND**	- RT (Perrett, et al., 2008)	+ ICC (Kakegawa et al., 2008)	+ RT (Ledda et al., 2010)	**ND**	**ND**
Nanog	+ RT (Durcova-Hills et al., 2008)	+ IF; RT (Leitch et al., 2010)	+ RT (Huang et al., 2007)	**ND**	+ RT (Jia et al., 2008)	+ RT (Turpenny et al., 2006)	**ND**	+ RT (Ledda et al., 2010)	**ND**	**ND**
cMyc	+ IF; RT (Durcova-Hills et al., 2008)	**ND**	**ND**	**ND**	**ND**	**ND**	**ND**	**ND**	**ND**	**ND**
Klf4	+ IF; RT (Durcova-Hills et al., 2008)	+ RT (Leitch et al., 2010)	**ND**	**ND**	**ND**	**ND**	**ND**	**ND**	**ND**	**ND**
Rex1	+ RT (Durcova-Hills et al., 2008)	+ RT (Leitch et al., 2010)	**ND**	**ND**	**ND**	**ND**	**ND**	**ND**	**ND**	**ND**
SSEA1	+ IF (Durcova-Hills et al., 2008)	**ND**	+ IF (Huang et al., 2007)	+ ICC (Petkov & Anderson, 2008)	+ ICC (Jia et al., 2008)	+ ICC (Shamblott et al., 1998)	+ ICC (Kakegawa et al., 2008)	+ ICC (Ledda et al., 2010)	**ND**	**ND**
SSEA3	- µarry (Sharova et al., 2007)	**ND**	+ IF (Huang et al., 2007)	+ IF (Petkov & Anderson, 2008)	**ND**	+ ICC (Shamblott et al., 1998)	**ND**	**ND**	**ND**	**ND**

	Mouse	Rat	Buffalo	Pig	Goat	Human	Rabbit	Sheep	Cow	Baboon
SSEA4	- µarry (Sharova et al., 2007)	**ND**	+ IF (Huang et al., 2007)	+ ICC (Petkov & Anderson, 2008)	**ND**	+ ICC (Shamblott et al., 1998)	**ND**	**ND**	**ND**	**ND**
Mvh	**ND**	**ND**	**ND**	**ND**	**ND**	**ND**	- RT (Kakegawa et al., 2008)	**ND**	**ND**	+ IF (Simerly et al., 2010)
TRA1-81	- µarry (Sharova et al., 2007)	**ND**	**ND**	+ ICC (Petkov & Anderson, 2008)	**ND**	+ ICC (Shamblott et al., 1998)	**ND**	**ND**	**ND**	**ND**
TRA1-60	- µarry (Sharova et al., 2007)	**ND**	**ND**	**ND**	**ND**	+ ICC (Shamblott et al., 1998)	**ND**	**ND**	**ND**	**ND**
TERT	**ND**	**ND**	**ND**	**ND**	**ND**	+ ICC (Turpenny et al., 2003)	**ND**	**ND**	**ND**	**ND**
Others	**Stella+** IF; RT (Durcova-Hills et al., 2008)	**ND**	**gp130+** RT (Huang et al., 2007)	**ND**	**ckit+** ICC (Jia et al., 2008)	**ckit+** IF (Hua et. al., 2009) **EMA1+** ICC (Turpenny et al., 2003)	**ND**	**ND**	**Oct6+** RT **Hes1+** RT (Cherny et al., 1994)	**ND**

Table 2. Comparison of differences in expression of markers in EGCs of different species. CC: Cytochemistry. IF: immunofluorescence. ICC: immunocytochemistry. RT: RT-PCR. µarry: microarray. ND: Not determined.

7. Acknowledgements

This work was supported in part by grant SAF2010-19230 from the Ministry of Science and Innovation, Spain and the BioMedical Foundation Mutua Madrileña, Spain.

8. References

Cherny, RA., Stokes, TM., Merei, J., Lom, L., Brandon, MR. & Williams RL. (1994). Strategies for the isolation and characterization of bovine embryonic stem cells.*Reproduction, Fertility and Development* Vol.6(No.5): 569-75.

De Felici, M. (2009). Primordial germ cell biology at the beginning of the XXI century, *International Journal of Developmental Biology* Vol.53(No.7): 891-4.

De Felici, M., Dolci, S. & Pesce, M. (1993). Proliferation of mouse primordial germ cells in vitro: a key role for cAMP, *Developmental Biology* Vol.157(No.1): 277-80.

De Felici, M., Farini, D. & Dolci, S. (2009). In or out stemness: comparing growth factor signalling in mouse embryonic stem cells and primordial germ cells, *Current Stem Cell Research&Therapy* Vol.4(No.2): 87-97.

De Miguel, MP. & Donovan, PJ. (2003). Isolation and culture of embryonic germ cells. *Methods in Enzymology* Vol.365: 353-63.

De Miguel, MP., Fuentes-Julian, S. & Alcaina, Y. (2010). Pluripotent stem cells: Origin, maintenance and induction, *Stem Cells Reviews and Reports* Vol.6: 633-49.

Donovan, PJ., Stott, D., Cairns, LA., Heasman, J. & Wylie, CC. (1986). Migratory and postmigratory mouse primordial germ cells behave differently in culture, *Cell* Vol.44: 831–8.

Durcova-Hills, G., Ainscough, J. & McLaren, A. (2001). Pluripotential stem cells derived from migrating primordial germ cells, *Differentiation* Vol.68(No.4-5): 220-6.

Durcova-Hills, G., Tang, F., Doody, G., Tooze, R. &Surani, MA. (2008). Reprogramming primordial germ cells into pluripotent stem cells, *PLoS One* Vol.3(No.10): e3531.

Durcova-Hills, G. & Surani, A. (2008). Reprogramming primordial germ cells (PGC) to embryonic germ (EG) cells, *Current Protocols in Stem Cell Biology* Vol.1:5:1A.3.1-1A.3.20.

Fenderson, BA., De Miguel, MP., Pyle, AD. & Donovan, PJ. (2006). Staining embryonic stem cells using monoclonal antibodies to stage-specific embryonic antigens, *Methods in Molecular Biology* Vol.325: 207-24.

Gillich, A. & Hayashi, K. (2011). Switching stem cell state through programmed germ cell reprogramming. *Differentiation*. [Epub ahead of print].

Ginsburg, M., Snow, MH. & McLaren, A. (1990). Primordial germ cells in the mouse embryo during gastrulation. *Development* Vol.110(No.2): 521-8.

He, J., Wang, Y. & Li, YL. (2007). Fibroblast-like cells derived from the gonadal ridges and dorsal mesenteries of human embryos as feeder cells for the culture of human embryonic germ cells, *Journal of Biomedical Science* Vol.14(No.5): 617-28.

Hiller, M., Liu, C., Blumenthal, PD., Gearhart, JD. & Kerr, CL. (2011). Bone morphogenetic protein 4 mediates human embryonic germ cell derivation, *Stem Cells and Development* Vol.20(No.2): 351-61.

Hua, J., Yu, H., Liu, S., Dou, Z., Sun, Y., Jing, X., Yang, C., Lei, A., Wang, H. & Gao, Z. (2009). Derivation and characterization of human embryonic germ cells: serum-free culture and differentiation potential, *Reproductive Biomedicine Online* Vol.19(No.2): 238-49.

Huang, B., Xie, TS., Shi, DS., Li, T., Wang, XL., Mo, Y., Wang, ZQ. & Li, MM. (2007). Isolation and characterization of EG-like cells from Chinese swamp buffalo (Bubalus bubalis), *Cell Biology International* Vol.31(No.10): 1079-88.

Jia, W., Yang, W., Lei, A., Gao, Z., Yang, C., Hua, J., Huang, W., Ma, X., Wang, H. & Dou, Z. (2008). A caprine chimera produced by injection of embryonic germ cells into a blastocyst, *Theriogenology* Vol.69(No.3): 340-8.

Kakegawa, R., Teramura, T., Takehara, T., Anzai, M., Mitani, T., Matsumoto, K., Saeki, K., Sagawa, N., Fukuda, K. & Hosoi, Y. (2008). Isolation and culture of rabbit primordial germ cells, *Journal of Reproduction and Development* Vol.54(No.5): 352-7.

Kaufman, MH. (1992). *The Atlas of Mouse Development*, Academic Press, San Diego.

Kemper, CH. & Peters, PW. (1987). Migration and proliferation of primordial germ cells in the rat. *Teratology* Vol.36(No.1): 117-24.

Kerr, CL., Shamblott, MJ. & Gearhart, JD. (2006). Pluripotent stem cells from germ cells, *Methods in Enzymology* Vol.419: 400-26.

Kerr, CL., Hill, CM., Blumenthal, PD. &Gearhart, JD. (2008). Expression of pluripotent stem cell markers in the human fetal ovary, *Human Reproduntion* Vol.23(No.3): 589-99

Kerr, CL., Hill, CM., Blumenthal, PD. & Gearhart, JD. (2008). Expression of pluripotent stem cell markers in the human fetal testis, *Stem Cells* Vol.26(No.2): 412-21.

Kimura, T., Suzuki, A., Fujita, Y., Yomogida, K., Lomeli, H., Asada, N., Ikeuchi, M., Nagy, A., Mak, T. W. & Nakano, T. (2003). Conditional loss of PTEN leads to testicular teratoma and enhances embryonic germ cell production, *Development* Vol.130: 1691-700.

Kimura, T., Tomooka, M., Yamano, N., Murayama, K., Matoba, S., Umehara, H., Kanai, Y. & Nakano, T. (2008). AKT signaling promotes derivation of embryonic germ cells from primordial germ cells. *Development* Vol.135(No.5): 869-79.

Koshimizu, U., Taga, T., Watanabe, M., Saito, M., Shirayoshi, Y., Kishimoto, T. & Nakatsuji, N. (1996). Functional requirement of gp130-mediated signaling for growth and survival of mouse primordial germ cells in vitro and derivation of embryonic germ (EG) cells. *Development* Vol.122(No.4): 1235-42.

Kühholzer, B., Baguisi, A. & Overström, EW. (2000). Long-term culture and characterization of goat primordial germ cells, *Theriogenology* Vol.53(No.5): 1071-9.

Kwong, WH. & Tam, PPL. (1984). The pattern of alkaline phosphatase activity in the developing mouse spinal cord, *Journal of Embryology & Experimental Morphology* Vol.82: 241-51

Labosky, PA., Barlow, DP. & Hogan, BL. (1994). Mouse embryonic germ (EG) cell lines: transmission through the germline and differences in the methylation imprint of insulin-like growth factor 2 receptor (Igf2r) gene compared with embryonic stem (ES) cell lines, *Development* Vol.120(No.11): 3197-204.

Labosky, PA.& Hogan, BL. (2008). Mouse primordial germ cells: isolation and in vitro culture *Methods in Molecular Biology* Vol.461: 187-99.

Lawson, KA., Dunn, NR., Roelen, BA., Zeinstra, LM., Davis, AM., Wright, CV., Korving, JP. & Hogan, BL. (1999). Bmp4 is required for the generation of primordial germ cells in the mouse embryo, *Genes & Development* Vol.13(No.4): 424-36.

Ledda, S., Bogliolo, L., Bebbere, D., Ariu, F. & Pirino, S. (2010). Characterization, isolation and culture of primordial germ cells in domestic animals: recent progress and insights from the ovine species. *Theriogenology* Vol.74(No.4): 534-43.

Lee, CK., Moore, K., Scales, N., Westhusin, M., Newton, G., Im, KS. & Piedrahita, JA. (2000). Isolation and genetic transformation of primordial germ cell (PGC)-derived cells from cattle, goats, rabbits and rats, *Asian - Australasian Journal of Animal Sciences* Vol.13: 587-94.

Lee, CK. & Piedrahita, JA. (2000). Effects of growth factors and feeder cells on porcine primordial germ cells in vitro. *Cloning* Vol.2(No.4): 197-205.

Lee, J., Inoue, K., Ono, R., Ogonuki, N., Kohda, T., Kaneko-Ishino, T., Ogura, A. & Ishino, F. (2002). Erasing genomic imprinting memory in mouse clone embryos produced from day 11.5 primordial germ cells, *Development* Vol.129(No.8): 1807-17.

Leitch, HG., Blair, K., Mansfield, W., Ayetey, H., Humphreys, P., Nichols, J., Surani, MA. & Smith, A. (2010). Embryonic germ cells from mice and rats exhibit properties consistent with a generic pluripotent ground state. *Development* Vol.137(No.14): 2279-87.

Li, P., Tong, C., Mehrian-Shai, R., Jia, L., Wu, N., Yan, Y., Maxson, RE., Schulze, EN., Song, H., Hsieh, CL., Pera, MF. & Ying, QL. (2008). Germline competent embryonic stem cells derived from rat blastocysts, *Cell* Vol.135(No.7): 1299-310.

Liu, S., Liu, H., Pan, Y., Tang, S., Xiong, J., Hui, N., Wang, S., Qi, Z. & Li, L. (2004). Human embryonic germ cells isolation from early stages of post-implantation embryos, *Cell and Tissue Research* Vol.318(No.3): 525-31.

Matsui, Y., Toksoz, D., Nishikawa, S., Nishikawa, S., Williams, D., Zsebo, K. & Hogan, BL. (1991). Effect of Steel factor and leukaemia inhibitory factor on murine primordial germ cells in culture, *Nature* Vol.353: 750-2.

Matsui, Y., Zsebo, K. & Hogan, BL. (1992). Derivation of pluripotential embryonic stem cells from murine primordial germ cells in culture, *Cell* Vol.70(No.5): 841-7.

McLaren, A. (2003). Primordial germ cells in the mouse, *Developmental Biology* Vol.262: 1-15.

McMahon, AP. & Bradley, A. (1990). The Wnt-1 (int-1) proto-oncogene is required for development of a large region of the mouse brain, *Cell* Vol.62(No.6): 1073-85.

Merchant-Larios, H., Mendlovic, F. & Alvarez-Buyulla, A. (1985). Characterisation of alkaline phosphatase from primordial germ cells and ontogenesis of this enzyme in the mouse, *Differentiation* Vol.29: 145-51.

Mueller, S., Prelle, K., Rieger, N., Petznek, H., Lassnig, C., Luksch, U., Aigner, B., Baetscher, M., Wolf, E., Mueller, M. & Brem, G. (1999). Chimeric pigs following blastocyst injection of transgenic porcine primordial germ cells, *Molecular Reproduction and Development* Vol.54(No.3): 244-54.

Oosterhuis, JW. & Looijenga, LH. (2005). Testicular germ-cell tumours in a broader perspective, *Nature Reviews Cancer* Vol.5(No.3): 210-22.

Pan, Y., Chen, X., Wang, S., Yang, S., Bai, X., Chi, X., Li, K., Liu, B. & Li, L. (2005). In vitro neuronal differentiation of cultured human embryonic germ cells, *Biochemical and Biophysical Research Communications* Vol.327(No.2): 548-56.

Park, JH., Kim, SJ., Lee, JB., Song, JM., Kim, CG., Roh, S. & Yoon, HS. (2004). Establishment of a human embryonic germ cell line and comparison with mouse and human embryonic stem cells, *Molecules and Cells* Vol.17(No.2): 309-15.

Perrett, R M., Turnpenny, L., Eckert, JJ., O'Shea, M., Sonne, SB., Cameron, IT., Wilson, DI., Meyts, ER. & Hanley, NA. (2008). The Early Human Germ Cell Lineage Does Not Express SOX2 During In Vivo Development or upon In Vitro Culture, Biology of Reproduction Vol.78 (No.5): 852-8.

Pesce, M., De Felici, M. (1995). Purification of Mouse Primordial Germ Cells by MiniMACS Magnetic Separation System, *Developmental Biology* Vol.170(No.2): 722-5.

Petkov, SG. & Anderson, GB. (2008). Culture of porcine embryonic germ cells in serum-supplemented and serum-free conditions: the effects of serum and growth factors on primary and long-term culture, *Cloning Stem Cells* Vol.10(No.2): 263-76.

Piedrahita, JA., Moore, K., Oetama, B., Lee, CK., Scales, N., Ramsoondar, J., Bazer, FW. & Ott, T. (1998). Generation of transgenic porcine chimeras using primordial germ cell-derived colonies, *Biology of Reproduction* Vol.58(No.5): 1321-9.

Resnick, JL., Bixler, LS., Cheng, L. & Donovan, PJ. (1992). Long-term proliferation of mouse primordial germ cells in culture, *Nature* Vol.359(No.6395): 550-1.

Shamblott, MJ., Axelman, J., Wang, S., Bugg, EM., Littlefield, JW., Donovan, PJ., Blumenthal, PD., Huggins, GR. & Gearhart, JD. (1998). Derivation of pluripotent stem cells from cultured human primordial germ cells, *Proceedings of the National Academy of Sciences* Vol.95(No.23): 13726-31.

Sharova, LV., Sharov, AA., Piao, Y., Shaik, N., Sullivan, T., Stewart, CL., Hogan, BL. & Ko, MS. (2007). Global gene expression profiling reveals similarities and differences among mouse pluripotent stem cells of different origins and strains, *Developmental Biology* Vol.307(No.2): 446-59.

Shim, H. & Anderson, GB. (1998). In vitro survival and proliferation of porcine primordial germ cells, *Theriogenology* Vol.49(No.3): 521-8.

Shim, H., Gutiérrez-Adán, A., Chen, LR., BonDurant, RH., Behboodi, E. & Anderson, GB. (1997). Isolation of pluripotent stem cells from cultured porcine primordial germ cells, *Biology of Reproduction* Vol.57(No.5): 1089-95.

Simerly, C., Castro, C., Gori, M., Turpin, J., Champagne, J., Grund, K., Bauer, C., De Miguel, MP., Donovan, P. & Schatten, G. (2010). Towards Deriving Baboon Pluripotent Embryonic Germ Cells. *8thAnnual Meeting of the International Society for Stem Cell Research*. San Francisco, CA., mes-2010.

Stevens, LC. (1967). Origin of testicular teratomas from primordialgerm cells in mice, *Journal of the National Cancer Institute* Vol.38(No.4): 549-52.

Stewart, CL., Gadi, I. & Bhatt, H. (1994). Stem cells from primordial germ cells can reenter the germ line, *Developmental Biology* Vol.161(No.2): 626-8.

Surani, MA. (2007).Germ cells: the eternal link between generations,*Comptes rendus biologies* Vol.330(No.6-7): 474-8.

Swelstad, BB. & Kerr, L. (2010). Current protocols in the generation of pluripotent stem cells: theoretical, methodological and clinical considerations, *Stem Cells and Cloning: Advances and Applications* Vol. 3: 13–27.

Tam, PP., Snow, MH. (1981). Proliferation and migration of primordial germ cells during compensatory growth in mouse embryos, *Journal of Embryology & Experimental Morphology* Vol.64: 133-47.

Toksoz, D., Zsebo, KM., Smith, KA., Hu, S., Brankow, D., Suggs, SV., Martin, FH. & Williams, DA. (1992). Support of human hematopoiesis in long-term bone marrow cultures by murine stromal cells selectively expressing the membrane-bound and secreted forms of the human homolog of the steel gene product, stem cell factor, *Proceedings of the National Academy of Sciences* Vol.89: 7350–4.

Tsung, HC., Du, ZW., Rui, R., Li, XL., Bao, LP., Wu, J., Bao, SM. & Yao Z. (2003). The culture and establishment of embryonic germ (EG) cell lines from Chinese mini swine, *Cell Research* Vol.13(No.3): 195-202.

Turnpenny, L., Brickwood, S., Spalluto, CM., Piper, K., Cameron, IT., Wilson, DI. & Hanley, NA. (2003). Derivation of human embryonic germ cells: an alternative source of pluripotent stem cells, *Stem Cells* Vol.21(No.5): 598-609.

Turnpenny, L., Spalluto, CM., Perrett, RM., O'Shea, M., Hanley, KP., Cameron, IT., Wilson, DI. & Hanley, NA. (2006). Evaluating human embryonic germ cells: concord and conflict as pluripotent stem cells, *Stem Cells* Vol.24(No.2): 212-20.

23

Pluripotent Gametogenic Stem Cells of Asexually Reproducing Invertebrates

Valeria V. Isaeva
A N. Severtsov Institute of Ecology and Evolution of the Russian Academy of Sciences, A.V. Zhirmunsky Institute of Marine Biology of the Far Eastern Branch of the Russian Academy of Sciences
Russia

1. Introduction

In this review, morphological and some functional properties of stem cells in different representatives of animals with asexual reproduction (sponges, hydroids, planaria, colonial rhizocephalan crustaceans and colonial ascidia) are considered in comparison with metazoan germline cells and *in vitro* mammalian embryonic stem cells.

Stem cells are an essential and defining feature developed during evolution of all multicellular organisms (Lohman, 2008; Batygina, 2010; Funayama et al., 2010). The study of mammalian embryonic stem cells has become a hot, intensely developing field in biology, biotechnology, and biomedicine. However, comparative studies of stem cells in various multicellular organisms are required to understand molecular mechanisms of maintaining pluri/totipotency, "stemness", which still remain far from clear, as well as mechanisms that regulate gametogenesis, reproduction, development and regeneration.

According to the generally accepted view, stem cells are cells of embryos or adult organisms capable of self-renewing by mitotic reproduction and differentiation into specialized cell types (Weissman et al., 2001; Cogle et al 2003; Müller, 2006; Lohman 2008; Rinkevich et al., 2009; Sköld et al., 2009; Funayama et al., 2010).

Two main types of stem cells should be considered: the cells of the germline and the somatic stem cell lineages (Hogan, 2001; Rinkevich, 2009; Srouji & Extavour, 2011). Germ cells are the only cell type capable of generating a new whole organism in animals; everlasting germline cycle continues from one generation to the next, thus the germline escapes the mortality that all somatic cells of an organism ultimately confront (the common view from Weismann, 1892, to Cinalli et al., 2008). The true innovation in the evolution was the generation of gametogenic germ line, the loss of gametogenic potential from the majority of cells of the organism and protection of the germ line throughout development (Extavour, 2008; Strouji & Extavour, 2011). During development, germ cells are set aside from all somatic cells of the embryo (Cinalli et al., 2008).

E. Davidson and colleagues have developed the "set-aside" hypothesis (Davidson et al., 1995; Cameron et al., 1998; Jenner, 2000; Collins & Valentine, 2001). Cameron et al. (1998) called "set-aside cells" the patches of larval cells that give rise to the adult body plan in animals with indirect development while most of the larval cells have a dead-end fate. These cells are developmentally set aside from the embryo-larva differentiation process, have an essentially unlimited division capacity, and they produce new populations of cells that are organized into the parts of the adult body plan (Cameron et al., 1998). So many adult organs are not derived from cells within larval organs, but from pluripotent cells sequestered, set aside during larval life as primordia from which adult structures form, such as imaginal discs of insects (Collins & Valentine, 2001).

Generally, we can regard all stem cells as "set-aside cells", reserve cells.

The reproductive strategy of multicellular organisms can include sexual and asexual reproduction. In organisms with exclusively sexual reproduction, primary germ cells become segregated during embryogenesis. Two main types of germline cell specification were defined as preformation (early specification of primary germ cells by means of asymmetric distribution of maternal cytoplasmic determinants) and epigenesis, i.e. later specification of germ lineage cells by inductive signals (Extavour & Akam, 2003; Travis, 2007; Frank et al., 2009). Recently, Strouji and Extavour (2011) refer to former "preformation" as "inheritance mode", and former "epigenesis" as "inductive mode". Besides, a third variant of germ cell specification, along with preformation and epigenesis, is recognized as somatic embryogenesis, in asexually reproducing animals, which have stem cells able to differentiate into germ and somatic cells throughout the life of an individual or a colony (Buss, 1987, 1999; Blackstone & Jasker, 2003; Frank et al., 2009; Rinckevich et al., 2009). Pluri/totipotent stem cells of these animals provide a cell source for gametogenesis, asexual reproduction and regeneration (Isaeva et al., 2008b, 2009; Frank et al., 2009; Sköld et al., 2009). Stem cells that have the potential to become both somatic and primary germ cells and are morphologically indistinguishable from the latter were defined as "primary stem cells" (Sköld et al., 2009).

In animals with asexual reproduction by somatic embryogenesis, the germ lineage remains non-segregated even in the adult organism, the gametes of which differentiate from stem cells (Blackstone & Jasker, 2003; Rinkevich, 2009; Sköld et al., 2009). Examples of such reserve stem cells, capable of differentiating both into germ and somatic cells, include sponge archaeocytes, cnidarian interstitial cells, planarian neoblasts, and stem cells of colonial ascidians (reviews: Isaeva et al., 2008b, 2009; Frank et al., 2009; Rinkevich et al., 2009; Sköld et al., 2009; Isaeva, 2010; Srouji, Extavour, 2011). These self-renewing stem cells maintains continuously throughout the life of an individual or a colony, being predecessors of germ cells and all the types (or a wide spectrum) of somatic cells, so ensuring both sexual and asexual reproduction. Thus, in invertebrates with asexual reproduction, no early segregation of a germ cell lineage takes place, and a self-renewing reserve of stem cells with broad or unlimited morphogenetic potential is maintained over the entire lifespan.

Similarly in plants, floral stem cells arise from stem cells of shoot apical meristema (Verdeil et al., 2007; Lohman, 2008; Batygina, 2010). Somatic embryogenesis is viewed as a condition characteristic to the lower metazoans; in the course of evolution, metazoans switch from somatic embryogenesis to preformation and epigenesis, and a subsequent return to somatic embryogenesis is a rare event (Blackstone & Jasker, 2003). If epigenesis was used by Urbilateria to specify the germ line, then preformation must have evolved convergently several times during the bilaterian radiation (Extavour, 2008).

In organisms with asexual, agamous development, the organism that has developed from the zygote is capable of natural cloning and forming numerous genetically identical individuals or modular units of a colony. Clonal morphogenesis, called somatic embryogenesis, as applied to animals, was usually termed blastogenesis (Berrill, 1961; Ivanova-Kazas, 1996) while in plants it was termed somatic embryogenesis or embryoidogenesis (see Verdeil et al., 2007; Batygina, 2010). In the life cycle of colonial animals, one generation of oozooid (individual that has developed from an egg) alternates with numerous generations of blastozooids, with respectively alternating morphogenetic processes: embryogenesis and blastogenesis (Ivanova-Kazas, 1996). In animal kingdom, natural cloning is a widespread phenomenon that includes polyembryony, budding, fission (architomy, paratomy, autotomy) etc.

Both striking similarities and considerable differences in stem cell systems have been observed between plants and animals (Lohman, 2008; Batygina, 2010; Sablowski, 2010). Lohman (2008) considers as the most important differences the capacity of plants to maintain totipotent stem cells throughout their entire lives and the dramatic developmental plasticity of plant cells; besides, plant cells are unable to move within the organism by active migration. Similarities in the stem cell pools of plants and animals suggest that there might have been strong evolutionary constraints that shaped the path for the development of stem cell systems (Lohman, 2008; Sablowski, 2010).

Our study points towards elucidation of the evolutionary conservative cellular, sub-cellular and molecular bases of "stemness", focusing on the comparative investigation of pluripotent stem cells in reproducing asexually representatives of five metazoan types: archaeocytes in the sponge *Oscarella malakhovi* (Porifera), interstitial cells in the colonial hydroids *Obelia longissima* and *Ectopleura crocea* (Cnidaria), neoblasts in the planarian *Girardia (Dugesia) tigrina* (Plathelmintes), stem cells in the colonial parasitic rhizocephalan crustaceans *Peltogasterella gracilis* and *Polyascus polygenea* (Arthropoda), the colonial ascidian *Botryllus tuberatus* (Chordata), and also mammalian embryonic stem cells as a benchmark, "standard reference", using *in vitro* culture, electron microscopic, histological, some histochemical, immunochemical, and molecular methods. Our hypothesis is the evolutionary conserved structural and functional organization of pluri/totipotent cells with gametogenic potentiality and of the molecular mechanisms maintaining it, in embryonic stem, reserve "primary" stem, and germline cells of different metazoan animals. Germplasm granules (or nuage) were studied as an ultrastructural marker and key organelle of germline cells and pluripotent stem cells of invertebrates. We have chosen as a molecular marker the evolutionarily conserved members of DEAD-box family (*vasa* and *pl10* related genes) whose expression was revealed by other researchers in germinal granules of germline cells in various metazoan taxa (see below). The reaction revealing the activity of alkaline phosphatase, formerly used for the identification of embryonic stem and primary germ cells of mammals and other vertebrate animals, was applied as a cytochemical marker of the pluripotent cells of invertebrates. The proliferating cell nuclear antigen, PCNA, was used as a marker of the proliferation of self-renewing stem cells.

Sponge archaeocytes, cnidarian interstitial cells, and planarian neoblasts are classical, long-explored stem cells. Rhizocephalan crustaceans (Arthropoda: Crustacea: Cirripedia: Rhizocephala) parasitizing free-living crustaceans, mainly decapods, until now are often considered as incapable of asexual reproduction and coloniality (for example, Blackstone & Jasker, 2003; Sköld et al., 2009). However, at the parasitic stage of the life cycle, many species of rhizocephalan have asexual reproduction without separation of blastozooids, resulting in the emergence of colonial organization (Høeg, 1992; Høeg & Lützen, 1995; Isaeva et al., 2003, 2004; Shukalyuk et al., 2005, 2007, 2011) that is unique among crustaceans, all arthropods, and all Ecdysozoa. Direct evidence of colonial organization at the parasitic stage of life cycle has been obtained only for a few rhizocephalan species. We visualized asexual reproduction, revealed and studied stem cells in the stolons and buds of the colonial

rhizocephalans *Polyascus polygenea* and *Peltogasterella gracilis* (Isaeva et al., 2003, 2004; Shukalyuk et al., 2005, 2007). Hemoblasts of the colonial ascidians *Botryllus schlosseri*, *Botrylloides leachi* and other species are considered as putative totipotent or pluripotent stem cells (Pancer et al., 1995; Burighel & Cloney, 1997; Stoner et al., 1999; Cima et al., 2001; Laird et al., 2005; Sunanaga et al., 2006), although stem cells have not been morphologically identified in the genus *Botryllus*. We have found stem cells, morphologically similar to hemoblasts in *B. leachi* (Cima et al., 2001) and other studied ascidians (Burighel & Cloney, 1997), in the early buds and vascular system of the colonial ascidian *Botryllus tuberatus* (Akhmadieva et al., 2007).
The data of our team showed that studied pluripotent reserve stem cells serve as the predecessors of germ and somatic cells and display some evolutionarily conserved features of the morphological and functional organization typical also for cells of the germ line and embryonic stem cells (Isaeva et al., 2003, 2004, 2005, 2008b, 2009, 2011; Akhmadieva et al., 2007; Shukalyuk, Isaeva, 2005; Shukalyuk et al., 2005, 2007, 2011; Isaeva, 2010; Isaeva, Akhmadieva, 2011).

2. Stem cells in asexually reproducing animals share common features with germ cells and embryonic stem cells

2.1 Pluri/totipotency

In animals with asexual reproduction, the differentiation of "primary" stem cells into germ and somatic cells is delayed, the germ lineage in these animals is not segregated (Tuzet, 1964; Buss, 1987, 1999; Sköld et al., 2009); these stem cells serve as the cellular source of asexual and sexual reproduction as well as of regeneration (Isaeva et al., 2008b, 2009; Frank et al., 2009). Pluripotent stem cells can differentiate into a very wide spectrum of somatic cells in adult organisms.
Depending on the width of the potential range of cell differentiation, totipotent, pluripotent, multipotent, oligopotent, and unipotent stem cells are distinguished, but the usage of this terminology is not unified (Müller, 2006; Newton, 2006; Isaeva et al., 2008b, 2009; Rinkevich et al., 2009; Sköld et al., 2009). Totipotent cells can give rise to all cell types of a developing organism; the zygote and the cells of the early mammalian embryo are totipotent. Stem cells of invertebrates reproducing asexually are traditionally often but not always referred to as totipotent, if their ability to differentiate into gametes and all somatic cells of the organism is shown (reviews: Isaeva et al., 2008b, 2009; Rinkevich et al., 2009; Sköld et al., 2009). For instance, archaeocytes of sponges are regarded as totipotent (Simpson, 1984; Müller, 2006) or pluripotent (Funayama, 2008; Funayama et al., 2010). Funayama believes that both archaeocytes and choanocytes of sponges are pluripotent. Neoblasts of planarians are considered totipotent (Shibata et al., 1999; Gschwentner et al., 2001; Peter et al., 2001; Sköld et al., 2009) or pluripotent (Shibata et al., 2010). Similarly, stem cells of colonial rhizocephalan crustaceans can also be considered totipotent (Isaeva et al., 2004, 2008b, 2009; Shukalyuk et al., 2005, 2007) or pluripotent. The stem cells in colonial ascidians are named both totipotent and pluripotent cells (Stoner et al., 1999; Weissman, 2000; Laird & Weissman, 2004; Laird et al., 2005; Sunanaga et al., 2006). Stem cells giving rise to germline and many but not all somatic cell types are referred to as multipotent. Cnidarian interstitial cells are usually believed to be multipotent, especially in the genus *Hydra* (Bode, 1996; Mochizuki et al., 2001; Frank et al., 2009), but the interstitial cells of *Hydractinia echinata* are recognized as totipotent (Frank et al., 2009).

Estimations of the potentiality of female germ line cells are also contradictory: such cells can be qualified as unipotent, since they produce only one type of differentiated cells, and totipotent, taking into account their potential of developing into a whole organism (Hogan, 2001; Seydoux & Braun, 2006; Strome & Lehman, 2007). Nussbaum (1880) recognized the germ line cells as totipotent and principally different from somatic cells with their limited potency. Accepting the concept of the maintenance of totipotency by cells of the female germ lineage, the author believes that the ability of the stem cells in asexually reproducing invertebrates to differentiate into female gametes and potentially to develop into a whole organism gives grounds for considering them as totipotent independently of the width of their somatic derivates (Isaeva et al., 2008b, 2009).

If we understand cell totipotency as the ability of a single cell to produce a whole organism, only the zygote and the blastomeres of the early embryo of mammals and other animals with regulative type of development are totipotent, having the potential to form an entire living organism. Mammalian embryonic stem cells derived from inner cell mass of the blastocyst have pluripotency: they are able to differentiate into tissues of all three germ layers but cannot produce a whole embryo (Cogle et al 2003). On the other hand, mammalian embryonic stem cells are able to give oogenic cells and oocytes entering meiosis and parthenogenetically producing blastocyst-like cell masses (Hübner et al., 2003; Daley, 2007; Kerkis et al., 2007).

In plants, a new individual can develop from one totipotent somatic cell (Verdeil et al., 2007; Lohman, 2008; Batygina, 2010; Sablowski, 2010). In the asexual reproduction of plants, for instance in polyembryony, a new individual develops from one stem cell, and the pattern of the cell divisions is similar to the cleavage of the zygote (Batygina, 2010). Thus, a single totipotent stem cell of plants may be similar to zygote. The ability of plants to maintain totipotent stem cells over the entire life span of the organism is considered to be their fundamental difference from animal stem cells (Lohman, 2008; Verdeil et al., 2007).

As for the stem cells of invertebrates with asexual reproduction, traditionally named totipotent, it is usually not one stem cell, but some kind of a complex, an aggregate of the stem cells gives rise to the new organism or zooid in asexual reproduction (Blackstone & Jasker, 2003; Rinkevich et al., 2009). It was experimentally shown that one stem cell of a trypsinized cysticercus of the parasitic cestode *Taenia crassiceps* injected into the host organism may form a whole cysticercus in the host organism (Toledo et al., 1997), but it is a rare exception from the common rule. In rhizocephalan crustaceans, the endoparasitic organism develops from a few cells introduced into the host organism by the larva (Høeg & Lützen, 1995). The number of stem cells in the earliest bud of the blastozooid of the colonial rhizocephalans *Peltogasterella gracilis* and *Polyascus polygenea*, according to the author's data, is about 10–15 cells (Isaeva, 2010). The number of stem cells capable of producing the new individual in cnidarians, free living flat worms and colonial ascidians is estimated as lying between 100 and 300 (Rinkevich et al., 2009). A similar number of archaeocytes is required for producing a whole organism of the freshwater sponge *Ephydatia fluviatilis*, as experimentally determined by Nikitin (1977), who showed the necessity of a "critical mass" of cell aggregates for the development of a sponge organism. This "mass effect" is explained by the formation of sufficient concentrations of required metabolites in cell aggregates. The "critical mass" of cells is probably necessary also for the formation of an individual or a blastozooid in invertebrates. So, the ability of one cell or a small number of stem cells to develop into a whole organism is determined by an adequate environment.

To avoid confusion and misunderstanding, the term "pluripotent" is using here according to the recent reviews (Seydoux & Braun, 2006; Funayama et al., 2010; Strouji & Extavour, 2011). The problem of cell line having unlimited morphogenetic potential stems from A. Weismann' "germ plasm" theory. August Weismann (1834–1914) was the first to discover and describe metazoan stem cells (*Stammzellen*) and primary germ cells (*UrKeimzellen*) in the course of his detailed study of colonial hydroids (Weismann, 1883; see also Frank et al., 2009). According to Weismann (1892, 1893), stem cells, retaining the "germ plasm" (*Keimplasma*, the nuclear hereditary substance containing the determinants of germ cells), are capable of differentiating into gametes, providing for the continuity of the immortal germ line, the germline way (*Keimbahn*) and the continuation of life in the sequence of generations. It was Weismann who determined that the germ cells of hydroids originate from the embryonic reserve of undifferentiated cells, and he did not associate the concept of germ cell line with the early isolation of this line (Weismann, 1892). Weismann has shown that the germ cells of hydroids differentiate not during embryonic development, but much later, in generations formed by budding (*Knospen-Generationen*: Weismann, 1883). The idea of early germ cell segregation and of the continuity of these cells in the sequence of generations was first of all formulated by Nussbaum (1880), who believed that interstitial cells of *Hydra* maintain the germline way. It was later shown that the interstitial cells of hydroids, stem cells continuously undergoing the mitotic cycle, produce both germ cells and some types of somatic cells (see Bode, 1996; Bosch, 2008; Frank et al., 2009). Thus, hydroids, with their late specification of germline cells, differentiating from the interstitial cells during the entire lifespan of the colony and producing also somatic derivates, paradoxically became the main object of the studies that resulted in the emergence of the idea of early segregation of the totipotent germ cell line from somatic cells (Nussbaum, 1880), and to Weismann's "germ plasm" theory (Weismann, 1883, 1892, 1893). Weismann supposed that germ cells preserve all the factors of inheritance, whereas each somatic cell loses, in the course of differentiation, part of the germ plasm and of the initial potential of the egg (Weissman, 1892, 1893); Weismann's concept has been criticized in the light of modern biological data (see, e.g., Frank et al., 2009). But Weismann's views were not as rigid as the views of some his followers: "Weismann was not a weismannist" (Winter, 2001, p. 518).

2.2 Self-renewal of stem cells

The term "self-renewal" denotes the ability of stem cells to reproduce mitotically during a long period, and in the case of the stem cells of adult organisms, during entire life span of the organism (Weissman et al., 2001; Lohman 2008; Rinkevich et al., 2009; Sköld et al., 2009). Particularly, pluripotent stem, primary germ, and gonial cells have a common property - self-renewal through mitotic reproduction over long periods or throughout life span of the organism (Houston & King, 2000; Hogan, 2000; Sköld et al., 2009).
For instance, sponge archaeocytes are self-renewing, mitotically active, telomerase-positive and bromodeoxyuridine incorporating cells (Müller, 2006; Funayama et al., 2010). Interstitial cells in hydra and other cnidarians can produce both germline cells and some but not all somatic cell types, since epidermal and gastrodermal cells are also capable of mitotic reproduction; so the stem cell system in hydroids includes interstitial, epidermal and gastrodermal stem cells continuously undergoing the mitotic cycle (Campbell, 1974; Thomas & Edwards, 1991; Bode, 1996). For the identification of stem cells capable of mitotic reproduction, bromodeoxyuridine, a thymidine analogue, was successfully used to reveal DNA synthesis in interstitial cells in hydra (Teragawa & Bode, 1990), and neoblasts in

flatworms (Gschwentner et al., 2001; Peter et al., 2001). Ethynyl deoxyurudine, another thymidine analogue, was employed for the same purpose in the ctenophore *Pleurobrachia pileus* (Alié et al., 2011). High activity of telomerase was shown in cells of the embryos, gonads, and early buds in colonial ascidian *Botryllus schlosseri* (Laird & Weissman, 2004).

PCNA (proliferating cell nuclear antigen) assay is used to reveal cells that do not cease to divide mitotically (Hall & Woods, 1990). The PCNA assay was also used to identify neoblasts in planarian; it has been shown that such a test is a reliable tool for neoblast identification in *Girardia japonica* (Orii et al., 2005). We used the immunochemical test PCNA to reveal proliferating cells in the colonial rhizocephalan *Peltogasterella gracilis* and the colonial hydroid *Obelia longissima*. Stem cells in budding stolons proved to be the only PCNA-positive cells in the *P. gracilis* interna while differentiated epithelial cells of the stolon and cells of the trophic system of the interna were PCNA-negative (Shukalyuk et al., 2005). Such test in *O. longissima* has revealed a more intense staining of interstitial and gonial cells, although reproducing epidermal and gastrodermal cells were also positive; differentiated somatic cells manifested poor or negative reaction (Isaeva et al., 2011).

Mammalian embryonic stem cells express telomerase, the protein associated with a pluripotent and immortal phenotype (Cogle et al 2003).

2.3 Gametogenic potentiality

The pluripotent stem cells ensure both sexual and asexual reproduction, being predecessors of the germ and all the somatic cells. The ability to differentiate into gametogenic and somatic cells was shown for archaeocytes and choanocytes in sponges (Simpson, 1984; Müller, 2006; Funayama, 2008; Funayama et al., 2010), interstitial cells in hydra and other cnidarians (Thomas & Edwards, 1991; Bode, 1996; Isaeva et al., 2011), turbellarian neoblasts (Shibata et al., 1999; Peter et al., 2001; Isaeva et al., 2005), stem cells of colonial rhizocephalans (Isaeva et al., 2004; Shukalyuk et al., 2005), ascidian stem cells (Pancer et al., 1995; Stoner & Weissman, 1996; Stoner et al., 1999; Weissman, 2000). So, pluripotent stem cells in invertebrates with asexual reproduction are potentially gametogenic cells.

Particularly, sponges have no permanent germline; archaeocytes and choanocytes are gametogenic cells (Tuzet, 1964; Blackstone & Jasker, 2003; Sköld et al., 2009). Sponge archaeocytes are considered to be the main cell source in sexual and asexual reproduction (Simpson, 1984; Müller, 2006; Funayama, 2008). Probably, the stem system of sponges includes two types of pluripotent stem cells: archaeocytes and choanocytes; both cell types are able to differentiate into germ and somatic cells; choanocytes can transform to archaeocytes, which later produce other cell types (Funayama, 2008; Funayama et al., 2010).

Cnidarian interstitial cells can produce germline cells (Campbell, 1974; Thomas & Edwards, 1991; Bode, 1996). The gonial cells and early oocytes developing from interstitial cells are distinct from them only by a greater size (Thomas & Edwards, 1991).

Gametogenic potentiality was most convincingly displayed using planarians (Baguña et al., 1989; Peter et al., 2001; Shibata et al., 1999). Neoblasts of asexually reproducing planarians and other Turbellaria are able to differentiate into germ and somatic cells of all types (Agata & Watanabe, 1999; Shibata et al., 1999; Peter et al., 2001; Orii et al., 2005). According to our data, neoblasts can become gonial cells: among individuals in asexual race of *Girardia tigrina* that reproduced exclusively by architomy during 40 years, one planaria laid cocoons after spontaneous sexualization; histological and ultrastructural study of this planaria demonstrated the presence of gonads, gonial cells, and oocytes (Isaeva et al., 2005).

Pluripotent stem cells of colonial rhizocephalans are also the predecessors of somatic and germ cells, so ensuring the reproductive strategy with alternation of asexual and sexual reproduction. In the colonial rhizocephalans *Polyascus polygenea* and *Peltogasterella gracilis* the stem cells migrated into the developing ovary becoming oogonial cells (Isaeva et al., 2004; Shukalyuk et al., 2005).

In colonial ascidians, germline as well as somatic cells differentiate from circulating hemoblasts (Pancer et al., 1995; Stoner & Weissman, 1996; Stoner et al., 1999). The differentiation of *Botryllus primigenus* hemoblasts into female germline cells was observed, and the primary germ cells are morphologically indistinguishable from hemoblasts (Sunanaga et al., 2006). Fusion of the vascular systems of colonies in ascidian *B. schlosseri* can lead to the replacement of germ and somatic cells in one colony with those from another one (Pancer et al., 1995; Stoner & Weissman, 1996; Stoner et al., 1999; Weissman, 2000; Laird et al., 2005). It remains unclear if the germline and somatic lineages are segregated or they descend from the same initial stem cell population in botryllids (Stoner et al., 1999).

Pluripotent embryonic stem cells of mammals have capability to differentiate *in vitro* into female and male germ cells (Hübner et al., 2003; Geijsen et al., 2003; Toyooka et al., 2003; Clark et al., 2004; Daley, 2007; Kerkis et al., 2007).

2.4 Morphological features of pluripotent stem cells

Pluripotent stem cells of asexually reproducing invertebrates and germline cells of all studied metazoan animals share common morphological and functional features: a high nuclear/cytoplasmic ratio, a large rounded nucleus with diffuse chromatin and a prominent nucleolus, thin rim of undifferentiated basophilic cytoplasm, including specific electron-dense cytoplasmic granules or nuage, and a set of specific regulatory molecules (Shukalyuk & Isaeva, 2005; Isaeva et al., 2008b, 2009, 2011; Extavour, 2008; Rinkevich et al., 2009; Isaeva, 2010; Strouji & Extavour, 2011; Shukalyuk et al., 2011). Germ cells can almost always be unambiguously distinguished from somatic cells by the same characteristic morphology (Extavour, 2008).

The morphological organization of pluripotent stem and gonial cells in the studied representatives of Porifera, Cnidaria, Platyhelminthes, Arthropoda, and Chordata shares common features typical for germline cells in other studied Metazoa (Isaeva et al., 2003, 2004, 2005, 2008b, 2009; Akhmadieva et al., 2007; Shukalyuk & Isaeva, 2005; Shukalyuk et al., 2005, 2007, 2011; Isaeva et al., 2011). For example, archaeocytes of the sponge *Oscarella malakhovi*, interstitial and gonial cells of the colonial hydroids *Obelia longissima* and *Ectopleura crocea* as well as neoblasts of the planarian *Girardia tigrina* have an ultrastructural morphology typical for stem and germ cells of all metazoan animals (Isaeva et al., 2005, 2011; Isaeva & Akhmadieva, 2011). Stem cells in the rhizocephalans *Polyascus polygenea* (Isaeva et al., 2004) and *Peltogasterella gracilis* (Shukalyuk et al., 2005) similarly demonstrate all morphological properties that are common to embryonic, pluripotent stem and germ cells of other metazoan animals.

2.4.1 Germinal granules (nuage)

Germline cells

The cells of the germ line can be identified and retraced during development of an organism owing to the availability of the "germ plasm" as cytoplasmic markers presented by granular or fibrillar material not surrounded by a membrane. "Germ plasm", Weismann's famous term (Weismann, 1892, 1893) now is understood metaphorically. According to this modern understanding, the "germ plasm" contains electron-dense, RNA-enriched material structured as compact germinal granules or a more dispersed "nuage", a specific ultrastructural marker of metazoan germline cells (see Matova & Cooley, 2001; Seydoux & Braun, 2006; Strome & Lehman, 2007). The germinal granules or nuage are considered key organelles of germline cells (Ikenishi, 1998; Amikura et al., 2001; Matova & Cooley, 2001; Chuma et al., 2006; Seydoux & Braun, 2006; Lim & Kai, 2007; Strome & Lehman, 2007).

The specific electron-dense material of germinal granules was denoted in the early XX century by the German terms *Keimbahnchromidien* and *Keimbahnplasma*, and this terminology was directly related to Weismann's; later the English terms *germ plasm, germ cell determinants, polar, perinuclear, chromatoid, germinal, germ granules (bodies), dense bodies, ectosomes*, the French term *nuage*, and many others were introduced (see Mahowald, 1971, 2001; Beams & Kessel, 1974; Eddy, 1975; Ikenishi, 1998; Houston & King, 2000; Isaeva & Reunov, 2001; Matova & Cooley, 2001; Seydoux & Braun, 2006; Lim and Kai, 2007; Strome & Lehnam, 2007; Isaeva et al., 2008b, 2009; Frank et al., 2009; Isaeva, 2010). Complex structures, including electron-dense material and mitochondria in vitellogenic oocytes, are called also the *Balbiani body, vitelline (yolk) body, mitochondrial cloud, intermitochondrial cement, yolk nucleus*: "the chaos in the nomenclature" complicates the comparison of these structures in different species (Kloc et al., 2004).

The presence of germinal (perinuclear) granules is an evolutionary conserved feature of germline cells in multicellular animals. These specific organelles have been found in more than 80 species of seven animal types (Eddy, 1975). The structure of these organelles is similar, but they can be represented in cells of different organisms and at different life cycle stages as either a few large granules (bodies) or as a cloud (nuage) of fine-dispersed material. In oogenesis, the germinal bodies transform morphologically but do not disappear in female germ cells throughout the life cycle: for instance, the polar granules are gradually replaced with nuage during polar cell migration in *Drosophila* (Mahowald, 1971). The continuity of the maternally inherited germ plasm material throughout the life cycle has been demonstrated in *Drosophila, Xenopus*, and nematode (Mahowald, 1971, 2001; Ikenishi, 1998).

In the plant kingdom, oogonial cells of the brown alga *Undaria pinnatifida* include electron-dense bodies similar to the germinal granules of metazoan oogonial cells and oocytes; similar bodies were found in the cytoplasm of some other species of higher and lower plants (Alexandrova & Reunov, 2008), giving evidence of the common pattern of the morphofunctional organization of plant and animal reproductive cells. This similarity indicates a very conserved features of reproductive cells in Metazoa and Metaphyta.

Germlinal granules in pluripotent stem cells

In some asexually reproducing invertebrates, stem cells capable to differentiate into germ and somatic cells can be also identified by the presence of specific electron-dense cytoplasmic structures, morphologically similar or identical to germinal granules or nuage in germline cells. The "germ plasm," containing germinal granules or dispersed "nuage" material, becomes acceptable as a specific ultrastructural marker and a key organelle in pluripotent, potentially gametogenic stem cells of asexually reproducing invertebrates (Shibata et al., 1999; Mochizuki et al., 2001; Isaeva et al., 2008b, 2009, 2011; Frank et al., 2009; Strouji & Extavour, 2011), although the data concerning the structural and molecular organization of germ determinants in the cells of various metazoan taxa are rather fragmentary.

The germinal granules in stem cells of invertebrates, whose life cycle includes asexual reproduction, were earlier have been revealed in the interstitial cells of *Pelmatohydra robusta* (Noda, Kanai, 1977) and the planarian neoblasts (Hori, 1982; Rieger et al., 1991; Auladell et al., 1993; Agata & Watanabe, 1999; Shibata et al., 1999). Noda and Kanai (1977) shown that not only germline cells but also interstitial cells contain germ plasm with "dense bodies" associated with nuclear pores and mitochondria. The number and size of these dense bodies in *P. robusta* decreases as somatic cells (cnidoblasts) differentiate from interstitial cells and increases during early oogenesis (Noda & Kanai, 1977). In planarian, the germinal granules (chromatoid bodies) were observed both in the germline cells and in the neoblasts, where they likewise lie near the nuclear envelope in contact with the mitochondria (Hori, 1982; Auladell et al., 1993; Shibata et al., 1999). Chromatoid bodies disappear during neoblast differentiation into somatic cells, while in oogenic cells they are present throughout the life cycle (Hori, 1982; Auladell et al., 1993; Shibata et al., 1999). These observations led to a suggestion that chromatoid bodies have the function related to pluri/totipotency maintenance (Shibata et al., 1999).

Typical electron-dense germinal granules have not been previously described in the archaeocytes or any other cells of sponges. Dense fibrillar bodies were found in the oogonia and oocytes of different sponges (see Tuzet, 1964; Isaeva, Akhmadieva, 2011). In the cytoplasm of archaeocytes in the sponge *Oscarella malakhovi* we have found germinal granules of a typical morphology located near the nuclear envelope (Isaeva & Akhmadieva, 2011). The electron-dense germinal bodies revealed in interstitial cells of the colonial hydroids *Obelia longissima* and *Ectopleura crocea* (Akhmadieva et al., 2005; Isaeva et al., 2011) were similar in their ultrastructure to the "dense bodies" in interstitial and germ cells of *Pelmatohydra robusta* (Noda & Kanai, 1977) as well as in oocytes of *Hydra carnea* (Honegger et al., 1989) and other cnidarians (Thomas & Edwards, 1991). Such germinal granules are common for interstitial cells and oocytes of *O. longissima.* The germinal granules (usually termed chromatoid bodies) have been found near the nuclear envelope (often in contact with nuclear pores) surrounded by mitochondria in neoblasts and gonial cells of the planarian *Girardia tigrina* (Isaeva et al., 2005).

The cytoplasm of embryonic and stem cells in the studied rhizocephalans *Peltogasterella gracilis* and *Polyascus polygenea* contains germinal granules morphologically similar to those in germ cells. In particular, stem cells in rhizocephalan *P. gracilis* feature the presence of the germinal bodies with a typical ultrastructural morphology; all or most blastomeres of cleaving *Peltogasterella gracilis* embryos contain prominent germinal granules (Shukalyuk et al., 2005, 2007, 2011).

In the colonial ascidian *Botryllus tuberatus,* we observed in the cytoplasm of some stem cells in early buds small electron-dense bodies (Akhmadieva et al., 2007), which are similar to the nuage material often found in vertebrates.

So, pluripotent gametogenic stem cells in studied asexually reproducing sponges, cnidarians, turbellarians, arthropods, and chordates feature the presence of the germinal granules. So germinal granules (or more dispersed nuage material) can be used as a specific ultrastructural marker and a key organelle of pluripotent stem cells of asexually reproducing invertebrates (Shibata et al., 1999; Mochizuki et al., 2001; Isaeva et al., 2008b, 2009, 2011; Frank et al., 2009; Isaeva, 2010; Isaeva & Akhmadieva, 2011).

Electron-dense granular structures were observed also in embryonic stem cells of mouse (Shukalyuk et al., 2011). Thus, germinal granules were found not only in cells of the germ line but also in pluripotent stem cells of asexually reproducing invertebrates (sponges, hydroids, turbellarians, colonial rhizocephalan crustaceans and ascidians) and pluripotent mESC *in vitro*.

In some somatic metazoan cells, processing bodies (P-bodies) have been found; their function is translation and they are considered as a structural and functional analog of the germinal granules (see Seydoux & Braun, 2006; Kotaja et al., 2006).

2.5 Molecular markers of stem cells

The ultrastructural and molecular organization of germinal granules of germline cells is evolutionarily conserved in all studied representatives of the animal kingdom from sponges to mammals (Ding & Lipshitz, 1993; Ikenishi, 1998; Houston & King, 2000; Matova & Cooley, 2001; Mochizuki et al., 2001; Extavour & Akam, 2003; Juliano et al., 2006; Seydoux & Braun, 2006; Strome & Lehman, 2007; Extavour, 2008; Ewen-Campen et al. 2010; Srouji & Extavour, 2011). It has been shown that some molecules localized in germinal granules are involved in specification of germline cells, and some genes encoding them are highly conserved evolutionary in all studied metazoans (Mahowald, 2001; Matova & Cooley, 2001; Mochizuki et al., 2001; Sato et al., 2001; Seydoux & Braun, 2006; Strome & Lehman, 2007). Germinal granules components include proteins, mRNAs, and noncoding RNAs; as far as is known, RNA-binding proteins are involved in mRNA localization, protection, and translation control (Extavour & Akam, 2003; Leatherman & Jongens, 2003; Chuma et al., 2006; Seydoux & Braun, 2006; Hayashi et al., 2007; Lim & Kai, 2007; Strome & Lehman, 2007; Ewen-Camden et al., 2010). The germinal granules are thought to function as a specific cytoplasmic regulatory center preventing the expression of somatic differentiation genes, maintaining the totipotency in germline cells, necessary for the conception of a new organism, preventing somatic gene expression and protecting the cells from somatic differentiation, that confirmed by data on the transcription "silence" of germline cells (Leatherman & Jongens, 2003; Chuma et al., 2006; Seydoux & Braun, 2006; Strome & Lehman, 2007; Cinalli et al., 2008; Extavour, 2008).

Germline cells can be distinguished from somatic cells by localization of mRNA or protein products of germ-cell-specific genes, notably the *vasa* and *nanos* gene family products (Extavour, 2008). Genes representing the core of the germline program, as *vasa, piwi, nanos* show striking evolutionary conservation (Extavour & Akam, 2003).

Several conserved molecules are expressed in both germ and pluripotent stem cells; these include Piwi family proteins, Tudor family proteins, and *PL10* gene products, *vasa* family members, and possibly *nanos* (Extavour, 2008; Ewen-Camden et al., 2010; Srouji & Extavour, 2011). Like germline cells, stem cells of asexually reproducing invertebrates are also characterized by the expression of protein products of genes related to *vasa, piwi, nanos,* and some other genes (reviews: Rinkevich et al., 2009; Sköld et al., 2009; Funayama et al., 2010). Nanos, a CCHC zinc finger RNA-binding protein expressed in germline stem cells in planarians (Sato et al., 2006), and in hydra (Mochizuki et al., 2000).

The piRNA-binding proteins of Argonaute subfamily, coding by *piwi*-related genes plays a central role in RNA silencing in a small RNA-mediated manner or via translational regulation; small regulatory RNAs include small interfering RNAs and Piwi-interacting RNAs mediating the epigenetic regulation of gene expression; these small RNAs can exert regulation at the transcriptional level, by affecting chromatin structure, or post-transcriptionally, by affecting mRNA stability or translation (Ambrose & Chen 2007). In metazoan germ cells, *piwi*-related genes express, presumably involving in germline and stem cell maintenance; Piwi proteins as well as several microRNAs are highly evolutionary conserved within plant and animal kingdoms (Filipowicz et al., 2005; Ambrose & Chen 2007; Funayama et al., 2010; Shibata et al. 2010; Alié et al., 2011). Gametogenic pluripotent stem cells in sponges (Funayama et al., 2010), cnidarians (Seipel et al., 2004) and flatworms (Shibata et al., 2010) express *piwi* homologs.

So stem cells of invertebrates with asexual reproduction, as well as cells of the germ lineage, also display the expression of proteins related to Piwi, Nanos, and some others (reviews: Rinkevich et al., 2009; Sköld et al., 2009; Srouji & Extavour, 2011). In stem cells of the sponge *Ephydatia fluviatilis*, the activity of a gene related to *piwi*, which is expressed in germ and stem cells of other animals and plants and whose function is to maintain the totipotency (pluripotency) of these cells, was detected (Funayama, 2008; Funayama et al., 2010).

Mammalian embryonic stem cells express gene *Oct4* coding transcription factor associated with a pluripotent and immortal phenotype (Cogle et al., 2003), however, this gene unique to deuterostomes (see Srouji & Extavour, 2011). In undifferentiated mouse embryonic and induced pluripotent stem cells, which, along with invertebrate stem cells, are potentially gametogenic cells retaining pluripotency, expression of Miwi/Piwi, Nanog/Nanos and Oct4 was found (Shukalyuk et al., 2011). Thus, embryonic stem, germ and pluripotent stem cells of various metazoan animals share *piwi* gene expression.

2.5.1 *Vasa* and other members of DEAD-box family

The first identified component of the granules of germ plasm was the protein product of the *Drosophila vasa* gene (Hay et al., 1988), RNA helicase, belonging to the family of proteins containing conserved DEAD-box sequences (Luking et al., 1998; Raz, 2000; Extavour & Akam, 2003; Rosak & Linder, 2004). Proteins of the DEAD family are found in all eukaryotes (from yeast up to plants and animals) and are involved in splicing, editing, processing, nuclear-cytoplasmic traffic, initiation of translation, and degradation of RNA (Raz, 2000; Rosak & Linder, 2004). The Vasa family is thought to have evolved from the PL10 family of helicases, which share significant structural similarity with *vasa* gene products (Mochizuki et al., 2001). PL10 products are usually localized to both germ cells and pluripotent cell types (Extavour, 2008). Germinal granules of germline cells in various animals, from sponges to vertebrates, were found to contain a protein product (RNA-helicase) of the *vasa* gene or related genes, a key determinant and a universal marker of germline cells in metazoans, which is necessary for the formation and maintenance of the structural organization of germinal granules and, presumably, for the maintenance of pluri/totipotency of cells; in animals that reproduce only sexually, the expression of *vasa*-related genes is always exclusively confined to the germcell line during the entire course of development, from early embryo up to gametogenesis (Ding & Lipshitz, 1993; Ikenishi, 1998; Shibata et al., 1999; Castrillon et al., 2000; Raz, 2000; Matova & Cooley, 2001; Mochizuki et al., 2001; Extavour & Akam, 2003; Seydoux & Braun, 2006; Sunanaga et al., 2006; Strome & Lehman, 2007; Ewen-Campen et al. 2010). Antibodies against Vasa protein react with both polar granules and nuage in *Drosophila* (Hay et al., 1988), which confirms the functional identity of these structures (Mahowald, 2001). The specifics of the structural organization of the cytoplasm and of the functional activity of germline cells has resulted from evolutionarily conserved mechanisms common to all studied metazoan animals, and the presence of Vasa-like proteins in the germ plasm of different animals indicates the conservation of molecular mechanisms underlying the formation and maintenance of the germ plasm (Raz, 2000; Extavour & Akam, 2003; Juliano et al., 2006; Extavour, 2008; Ewen-Campen et al., 2010).

In the polyembryonic wasp *Copidosoma floridanum*, the secondary embryos develop either into normal larvae and then into fertile insects or into soldier larvae with the defense function. During development, at the stage of four blastomeres, one of them is Vasa-positive, and it gives rise to primary germline cells; embryos containing the Vasa-positive blastomere become fertile imagoes while the caste of soldiers without germline cells develops from the embryos depleted of the Vasa-positive blastomere, that confirms the involvement of Vasa protein in germline and caste determination in *C. floridanum* (Donnell et al., 2004; Corley et al., 2005).

The presence of a Vasa-like protein was demonstrated not only in germline cells but also in large interstitial cells of hydra *Pelmatohydra robusta* (Mochizuki et al., 2001) and neoblasts of planarians (Shibata et al., 1999). Sequences of Vasa homologs have been found in the sponge *Ephydatia fluviatilis* (Mochizuki et al. 2000, 2001).

We have revealed the evolutionarily conserved sites of genes of the DEAD family, particularly *vasa* and *pl10* related genes, in DNA of the rhizocephalan crustaceans *Polyascus polygenea, Peltogasterella gracilis* and *Clistosaccus paguri* (Shukalyuk et al., 2007). Based on deduced sequencing of the protein products of these genes and on data from the genetic bank, a phylogenetic tree showing the close relationships of rhizocephalans to other arthropods was constructed. Selective expression of RNA of the *vasa*-related gene in stem and germline cells and its localization in the germinal granules of embryonic cells of *P. polygenea* were revealed. So selective expression of the *vasa*-related genes, the evolutionarily conserved markers and determinants of germline cells formerly revealed in germ cells of various metazoan animals and in stem cells of cnidarians and planarians, was also observed in embryonic, stem and germ cells of rhizocephalans (Shukalyuk et al., 2007). Thus, the expression of a gene related to *vasa* is specific not only to germline cells, but also to pluripotent gametogenic stem cells in asexually reproducing invertebrates, and the products of *vasa*-related genes involved in the determination of germ cells and in maintenance of cellular pluri/totipotency can be a useful molecular marker of pluripotent stem cells (Shibata et al., 1999; Mochizuki et al., 2001; Shukalyuk et al., 2007, 2009; Rinkevich et al., 2009; Sköld et al., 2009; Srouji & Extavour, 2011). The maintenance of the stem cell morphofunctional organization involves evolutionary conserved developmental mechanisms common for studied asexually reproducing multicellular animals.

Recently, Alié et al. (2011) found the expression of *piwi*, *vasa* and *PL10* in the male and female germlines and within pluri/multipotent stem cells in *Pleurobrachia pileus* (Ctenophora) without asexual reproduction. The authors suggest that *piwi*, *vasa* and *PL10* belong to a gene network ancestrally acting in two contexts: the germline and stem cells, whatever the nature of their progeny (Alié et al., 2011).

The data on *vasa-*, *PL10-* and *piwi*-related genes in pluripotent stem cells of invertebrates are presented in Table 1.

2.5.2 Mitochondrial components of germinal granules

The material of germinal granules (nuage) includes products of the nuclear genome; besides, there is evidence for the mitochondrial origin of some molecular components of germinal granules. The contact with mitochondria is a typical property of structured germinal granules in diverse multicellular animals (Isaeva & Reunov, 2001; Matova & Cooley, 2001; Carré et al., 2002). Ribosomal RNAs of mitochondrial origin and several other products of

	vasa	*PL10*	*piwi*
Porifera			Funayama et al., 2010
Cnidaria	Mochizuki et al., 2001; Rebscher et al., 2008	Mochizuki et al., 2001	Seipel et al., 2004
Ctenophora	Alié et al., 2011	Alié et al., 2011	Alié et al., 2011
Plathelminthes	Agata et al., 2006; Pfister et al., 2008	Shibata et al., 1999	Shibata et al. 2010
Annelida	Rebscher et al., 2007	Rebscher et al., 2007	Rebscher et al., 2007
Arthropoda (Rhizocephala)	Shukalyuk et al., 2007, 2011	Shukalyuk et al., 2007, 2011	
Echinodermata	Juliano & Wessel, 2009		Juliano et al., 2006
Chordata (Tunicata)	Rosner et al., 2009	Rosner et al., 2005	Brown et al., 2009

Table 1. Gene expression in pluripotent stem cells of invertebrates

the mitochondrial genome were revealed in the germinal granules of *Drosophila*, planarian and the frog *Xenopus* (Ding & Lipshitz, 1993; Kobayashi et al., 1998, 2005; Kashikawa et al., 1999; Amikura et al., 2001; Mahowald, 2001; Matova & Cooley, 2001; Leatherman & Jongens, 2003; Seydoux & Braun, 2006). The germinal granules are commonly surrounded with polysomes (Mahowald, 2001); polysomes in *Drosophila* embryos were shown to contain ribosomes similar to mitochondrial ones by size properties (Amikura et al., 2001; Kobayashi et al., 2005). In *Xenopus* 16S rRNA can be found outside of mitochondria only in the germ plasm granules (Kobayashi et al., 1998). Mitochondrial (both large and small) ribosomal RNA has also been detected in the chromatoid bodies in turbellarian neoblasts (Sato et al., 2001).

The presence of mitochondrial rRNAs outside of the mitochondria in association with germinal granules has been generally accepted; it becomes apparent that mitochondrial rRNAs and other products of the mitochondrial genome are involved in the formation of germline cells in diverse multicellular animals (Ikenishi, 1998; Kloc et al., 2000; Mahowald, 2001; Amikura et al., 2001; Matova & Cooley, 2001; Leatherman & Jongens, 2003; Seydoux & Braun, 2006). It has been suggested that products of both the nuclear and mitochondrial genomes are essential for the structural organization and functioning of the germinal granules of germ plasm (Kobayashi et al., 1998, 2005; Ding & Lipshitz, 1993; Isaeva & Reunov, 2001; Isaeva et al., 2005, 2011).

The export of mitochondrial rRNA from mitochondria to the polar granules in *Drosophila* depends on the activity of nuclear genes *oskar, vasa,* and *tudor* (Amikura et al., 2001; Matova & Cooley, 2001). Vasa protein or its homolog, a component of the germinal granules in different animals, has been found in the mitochondrial matrix of germline cells in *Xenopus* embryos (Watanabe et al., 1992). Similarly, the protein encoded by the nuclear gene *tudor* is present both in the polar granules and inside mitochondria in the early *Drosophila* embryos (see Ding & Lipshitz, 1993). In addition to these proteins, the germ determinants include many more components encoded by the nuclear genome.

The export of the ribosomal RNAs from mitochondria to the germinal granules is no longer questioned, but the mechanism underlying a transport is considered unprecedented and enigmatic (Kashikawa et al., 1999; Ding & Lipshitz, 1993; Amikura et al., 2001). Our ultrastructural data indicating the disruption of the outer mitochondrial membrane and the transformation of the mitochondrial matrix with inner membrane cristae into material of germinal granules in representatives of various animal taxa may clarify the mechanism of the export of mitochondrial components into germinal granules (Reunov et al., 2000; Reunov et al., 2004; Isaeva et al., 2005, 2011; Isaeva & Akhmadieva, 2011). This phenomenon enables us to suppose the participation of mitochondria in the biogenesis of the germinal granules (Isaeva & Reunov, 2001; Isaeva et al., 2005, 2011).

Destruction of the outer mitochondrial membrane and transformation of the mitochondrial matrix to the material of germinal granules or nuage have been revealed in the gonial cells of echinoderms and vertebrates (Reunov et al., 2000, 2004), planarian (Isaeva et al., 2005), sponge (Isaeva & Akhmadieva, 2011) and hydroids (Isaeva et al., 2011). The ultrastructural evidence of mitochondrial origin of the germinal granules (chromatoid bodies) in gonial cells and neoblasts of planarian *Girardia tigrina* has been obtained: the transformation of mitochondrial matrix with inner membrane cristae into the germinal bodies was observed (Isaeva et al., 2005). The mitochondrial derivatives devoid of the outer membrane but still containing flattened vesicles as remnants of inner membrane cristae is a usual picture observed in the germ granules or nuage of stem and gonial cells in the studied representatives of diverse taxa including the sponge *Oscarella malakhovi* and the hydroids *Obelia longissima* and *Ectopleura crocea* (Isaeva & Akhmadieva, 2011; Isaeva et al., 2011). We suggest that the structural frame of germinal granules derives from mitochondria and is filled with transcription products of the nuclear genome. The release of the mitochondrial matrix material in the germ plasm is the way to incorporate the mitochondrial derivatives into germinal granules, mediating the involvement of mitochondrial genome products in the biogenesis of the macromolecular complex of germinal determinants (Isaeva & Reunov, 2001; Isaeva et al., 2005, 2011). We propose also flows of molecular information connecting the nucleus, mitochondria, and germinal granules and involving in pluri/totipotency maintenance (Isaeva et al., 2005). The maintenance of the preexisted structural and functional organization of the germinal granules as cytoplasmic regulatory centers is likely controlled by ancient conserved mechanisms common for all multicellular animals (Isaeva & Reunov, 2001; Isaeva et al., 2005, 2011).

2.5.3 Alkaline phosphatase activity in stem cells

The histochemically detectable high level of alkaline phosphatase activity has become an empirical marker of mammalian primary germ and embryonic stem cells *in vivo* and *in vitro* (Chiquoine, 1954; Mintz, 1959; Talbot et al., 1993; Thompson et al., 1998; Lacham-Kaplan, 2004). High activity of alkaline phosphatase has been determined in the cultured embryonic stem cells, not only of mammals, but also of other vertebrates (for examples, Hong et al., 1998). Germ cells can be distinguished from somatic cells by high levels of alkaline phosphatase activity in vertebrate germline cells (Extavour, 2008).

No similar research has been carried out on invertebrates until quite recently. We applied cytochemical methods to show alkaline phosphatase activity for stem cell identification in the rhizocephalans *Peltogasterella gracilis* and *Polyascus polygenea* at the colonial parasitic stage of their life cycle to reveal the common feature of pluripotent stem cells of vertebrate and invertebrate animals. A high level of alkaline phosphatase activity, comparable to that of mouse embryonic stem cells *in vitro*, has been revealed in the cytoplasm of stem cells in the studied colonial rhizocephalans (Isaeva et al., 2003; Shukalyuk et al., 2005). The stem cells were identified due to their high alkaline phosphatase activity, in contrast with the differentiated somatic cells of the endoparasitic interna characterized by a poor nonspecific staining of yellow color (Isaeva et al., 2003; Shukalyuk et al., 2005). In the blastomeres of dividing embryos of *P. polygenea* the alkaline phosphatase activity was confined to germinal granules located near the nucleus (Shukalyuk et al., 2005).

High alkaline phosphatase activity has also been recorded in interstitial and gonial cells of the colonial hydroid *O. longissima* (Isaeva et al., 2011). Histochemical assay for alkaline phosphatase revealed intense staining of some hemocytes (apparently, hemoblasts as stem cells) and cells of early buds in ascidian *Botryllus tuberatus* distinguished the stem cells from differentiated somatic cells (Akhmadieva et al., 2007).

Thus, the stem and gonial cells in rhizocephalans, hydroids and ascidians selectively express alkaline phosphatase activity. Specific brick red staining of stem cells in the studied representatives of colonial cnidarians, arthropods, and chordates was similar in color and intensity to that of cultured mouse embryonic stem cells used as "standard reference" (Isaeva et al., 2003; Shukalyuk et al., 2005). Our data is the evidence of the common functional characteristic of stem cells in such distant taxa as chordates, arthropods and cnidarians. We applied this classical histochemical method developed on the mammalian embryonic germ and stem cells to identify invertebrate stem cells, that reveals an opportunity for the application of this cytochemical reaction to the specific marking of stem cells of invertebrates in other taxonomic groups.

So classical reaction revealing the activity of alkaline phosphatase, earlier used for the identification of primary germ cells and embryonic stem cells in vertebrates, became applicable as a cytochemical marker of both gametogenic and pluri/totipotent stem cells of invertebrates (Isaeva et al., 2003; Laird et al., 2005; Shukalyuk et al., 2005; Akhmadieva et al., 2007; Rinkevich et al., 2009; Sköld et al., 2009). Among plants, a high alkaline phosphatase activity was found in the early gametangia of the brown alga *Undaria pinnatifida* (Alexandrova & Reunov, 2008).

2.6 Amoeboid cell motility of stem cells

Archaeocytes of sponges are characterized by amoeboid motility and active migration (Simpson, 1984; Müller, 2006; Funayama, 2008; Funayama et al., 2010). Archaeocytes are defined as large amoeboid cells actively migrating within the mesohyl (Funayama, 2008; Funayama et al., 2010). According to our data, migrating archaeocytes morphologically similar to those described previously in other sponge species participate in *Oscarella malakhovi* budding (Isaeva, Akhmadieva, 2011).

In hydra and other cnidarians, interstitial cells are capable of active migration (Campbell, 1974; Thomas & Edwards, 1991; Bode, 1996). Migration of numerous interstitial stem and oogonial cells inside the stolon and their participation in the formation of medusoid generation was also observed in *Obelia longissima* and *Ectopleura crocea* (Isaeva et al., 2011).

Turbellarian neoblasts can migrate to the injured surface and sites of gonad formation (Rieger et al., 1991; Auladell et al., 1993; Agata & Watanabe, 1999; Shibata et al., 1999); amoeboid neoblasts and gonial cells in planarian *Girardia tigrina* demonstrated the migratory possibility (Isaeva et al., 2005).

Undifferentiated rhizocephalan stem cells have been found inside each early stolon bud; similar cells migrate within the stolons in *Peltogasterella gracilis* and *Polyascus polygenea* (Isaeva et al., 2004; Shukalyuk et al., 2005).

The primary germline cells are known to emerge outside of the future gonad and later traverse through several developing somatic tissues on their journey to the emerging gonad using both amoeboid motility and passive morphogenetic movements (Matova & Cooley, 2001; Kunwar & Lehmann, 2003; Travis, 2007; Cinalli et al., 2008).

Thus, pluripotent stem cells of asexually reproducing invertebrates are similar to primary germ cells in their ability to amoeboid movement and extensive migrations within the organism, directed to asexual reproduction sites, to the wound surface resulting from fission or damage, or to the gonads, respectively (Isaeva et al., 2009; Isaeva, 2010). In contrast, plant stem cells, with the rigid cellulose wall, are unable to migrate within the organism, and only passively moving together with the tissue, due to cell proliferation and expansion (Lohman, 2008).

2.7 Plasticity of stem cells in morphogenesis

Comparison of normal morphogenesis with its experimental changes helps us understand the plasticity of embryogenesis and blastogenesis. The plasticity of early animal embryogenesis is clearly shown in experiments with dissociated cells *in vitro* demonstrating the remarkable potential of cell self-organization (Isaeva et al., 2008a; Presnov et al., 2010). More than 100 years ago E. Wilson (1907) performed his famous experiments with dissociated sponge cells, reaggregated and developed into small sponges. Later Nikitin (1974) showed in experiments with dissociated cells of the sponge *Ephydatia fluviatilis* that homogeneous cell aggregates formed from stem cells (nucleolar amoebocytes, i.e. archaeocytes) are able to develop to a whole organism, while aggregates of other cell types inevitably died. Similar experimental studies were later carried out on sea urchin embryos: it was shown that reaggregates of dissociated embryonic cells *in vitro* formed "embryoids" (Giudice, 1962; Spiegel & Spiegel, 1986) which were able to develop into more or less normal larvae (Giudice, 1962) and, after metamorphosis, became fertile sea urchins (Hinegardner, 1975). Such experiments demonstrate the remarkable self-organization potential of embryonic stem cell *in vitro* (Isaeva et al., 2008a). Chimerical reaggregates of embryonic stem cells of sea urchins can form secondary blastula- or gastrula-like embryoids (Isaeva et al., 2008a; Presnov et al., 2010). This phenomenon is similar to the natural larval cloning by budding or fragmentation producing secondary larvae in starfish and other echinoderms (Jaeckle, 1994; Vickery & McClintock, 2000; Rinkevich et al., 2009).

So changes in the initial conditions of morphogenesis *in vitro* lead to changed self-organization of an embryonic cell system; similar modifications of embryogenesis may occur also in the course of evolution under considerably changed conditions of early development, for example, in endoparasitism. The endoparasitic rhizocephalan interna with stem cells is "culturing" in the host organism and using host hemolymph as a nutritive medium; these extremely favorable conditions lead to stem cell proliferation and expansion of cells and tissues. In parasitoid insects with polyembryony, a system similar to cell culture is formed. Polyembryony, the development of a whole embryo from one of the early blastomeres, i.e. asexual reproduction at an early embryonic stage, is known at least in six animal phyla (Craig et al., 1997; Sköld et al., 2009). For instance, among insects, polyembryony has been described in some parasitoid members of Hymenoptera and Strepsiptera (Johannsen & Butt, 1941; Hagan, 1951). Polyembryony has been studied in detail in the parasitoid wasp *Copidosoma floridanum*: the zygote forms a morula, consistsing of about 200 cells; repeatedly dividing mitotically active embryonic cells produce more than a thousand of secondary morulae, which form a polymorula, or polygerm (Donnell et al., 2004; Corley et al., 2005). Polyembryony is similar to budding (Perez, 1931; Ghiselin, 1987), and the latter is quite widespread in the animal kingdom as a way of asexual reproduction. In polyembryony, the stage of cleavage, cell reproduction, becomes longer than in common embryonic development.

The embryonic stem cells of mammals *in vitro* are similar to the pluripotent stem cells of rhizocephalans or other asexually reproducing animals. Stem cells in free-living asexually reproducing invertebrates use the own parental organism as a nutritive medium. Germ and pluripotent stem cells are "privileged", "predatory" cells inclined to "parasitism" as has been displayed for colonial ascidians (Buss, 1999; Pancer et al., 1995; Rinkevich, 2009); these cells can survive starvation through "cannibalism" (Kerszberg & Wolpert, 1998).

Sexual reproduction and early stages of embryogenesis are relatively conservative in all the animal kingdom due to the monophyly of metazoans (Sköld et al., 2009). Since asexual reproduction emerged in the course of the evolution of different metazoan lineages repeatedly and independently, asexual reproduction is more variable and less conservative than embryogenesis. The stage of cleavage is missing in blastogenesis, and stem cells can be likened to the embryonic cells of the morula stage. The integration of blastogenesis in the process of early embryogenesis in animals with polyembryony disrupts the conservatism of embryonic development (Isaeva, 2010). Polyembryony and the breaking of the conservatism of embryogenesis are known also in plants (Batygina, 2010).

The data on the asexual reproduction in some arthropods and chordates contradicts the dogma that asexual reproduction is common exclusively among the lower animals. In particular, the statement that vertebrates are incapable of natural cloning (Blackstone & Jasker, 2003) is disproved by known facts about facultative polyembryony in mammals, which has become obligate in some armadillo species, e.g. in *Dasypus novemcinctus* (Loughry et al., 1998).

The self-renewing pool of totipotent stem cells in colonial invertebrates provides the cellular basis for realization of the reproductive strategy including both asexual and sexual reproduction. The principal difference between the reproductive strategy that includes asexual reproduction and the strategy with exclusively sexual reproduction concerns the maintenance of the pluri/totipotent stem cell lineage with gametogenic potential during the entire life span of an asexually reproducing organism; a self-renewing reserve of pluripotent stem cells is the cellular source ensuring the reproductive strategy that includes sexual and asexual reproduction.

The problem of cells dedifferentiation in asexual reproduction and regeneration is less obvious, and the solution of this problem requires special markers. The notion of the high plasticity of the development and fate of cells in colonial animals (Frank et al., 2009; Rinkevich et al., 2009; Sköld et al., 2009), similar to that found in plants (Skold et al., 2009), appears sufficiently justified. In plants, however, differentiated cells retain the ability to dedifferentiate and become totipotent stem cells (Lohman, 2008; Batygina, 2010); animal cells at the stage of terminal differentiation usually have no such ability.

2.8 Evolutionary transition from preformation to epigenesis in colonial Rhizacephala

Extavour (2008) considered the transition from epigenesis to preformation as the repeated evolutionary event, but she thinks that examples of epigenesis in phyla where preformation is plesiomorphic never observed. However, the blastogenesis in colonial species of rhizocephalan crustaceans (Arthropoda: Crustacea: Cirripedia: Rhizocephala) involves a deep reorganization of development; we observe evolutionary secondary transition from preformation to epigenesis. We found germinal bodies in all or most blastomeres of cleaving embryos of *Polyascus polygenea* at stage of 16-32 blastomeres. In each germinal body selective expression of mRNA transcript of *vasa*-like gene and also high selective activity of alkaline phosphatase were revealed (Shukalyuk et al., 2005, 2011). These data indicates the evolutionary secondary rearrangement of the developmental mode and transition from ancestral preformation with mosaic cleavage and early segregation of the germ line (that is plesiomorphic feature in Crustacea and all Arthropoda) to epigenesis with equipotential blastomeres containing germinal granules and asexual reproduction resulting in colonial organization (Isaeva, 2010). Thus, the evolutionary transition from preformation to epigenesis is possible as well as more frequent evolutionary transition from epigenesis to preformation. In Rhizocephala, the radical transformation of the ancestral reproductive strategy involved all the levels of organization, from molecular and subcellular to species-specific (Kasyanov, 2001). The cellular basis for the reproductive strategy of Rhizocephala, including both sexual and asexual reproduction, is self-renewing pool of pluri/totipotent stem cells (Isaeva et al., 2008b, 2009).

In recent reviews (Blackstone & Jasker, 2003; Sköld et al., 2009) and in modern textbooks, crustaceans as well as all other arthropods and the entire Ecdysozoa clade, are considered as acolonial and aclonal, though in some rhizocephalan crustaceans the colonial organization as result of asexual cloning has already been described (Høeg & Lützen, 1995; Isaeva et al., 2003, 2004, 2008; Glenner et al., 2003; Shukalyuk et al., 2005). We have shown budding of a stolon filled with stem cells in colonial rhizocephalans *P. polygenea* and *P. gracilis* (Isaeva et al., 2003, 2004, Shukalyuk et al., 2005, 2007). Stem cells in rhizocephalans demonstrate all morphological properties shared by stem cells in other studied animals with asexual reproduction. The colonial nature of Rhizocephala has probably been denied because their colonial interna was not clearly visualized until quite recently. We have visualized the process of asexual reproduction by budding with maintaining colonial unity and connections between the developing blastozooids and common stolon in *P. polygenea* and *P. gracilis*, leaving no doubt about the colonial organization of these crustaceans at the endoparasitic stage of their life cycle (Isaeva & Shukalyuk, 2007; Isaeva et al., 2008b; Isaeva et al., 2004; Shukalyuk et al., 2005, 2007).

3. Conclusion

On the morphological and gene expression levels, germ cells and stem cells are very similar (Shukalyuk & Isaeva, 2005; Extavour, 2008; Isaeva et al., 2008b, 2009, 2011; Rinkevich et al., 2009; Sköld et al., 2009; Isaeva, 2010; Srouji & Extavour, 2011). Pluri/totipotent gametogenic stem cells are similar to germ and embryonic stem cells; evidence of the evolutionary conserved morphological and functional characteristics of pluripotent stem cells typical also to cells of the germ line have been obtained in representatives of such various metazoan phyla as Porifera, Cnidaria, Plathelminthes, Arthropoda and Chordata (Isaeva et al., 2003, 2004, 2005, 2008b, 2009, 2011; Shukalyuk et al., 2005, 2007, 2011; Akhmadieva et al., 2007; Isaeva & Akhmadieva, 2011).

The data supported our hypothesis that pluripotent, potentially gametogenic stem cells display evolutionarily conserved features of the morphological and functional organization typical for cells of the germ line and embryonic stem cells. In asexually reproducing invertebrates, from sponges and hydroids to some arthropods and chordates, stem cells share with cells of early embryos evolutionary conserved features presumably involved in maintenance of pluri/totipotency, including the gametogenic program. Such invertebrate cells capable of both gametogenesis and asexual reproduction (blastogenesis) are similar in their potential to mammalian embryonic stem cells. We propose that evolutionary and ontogenetically related cells of early embryos and pluripotent stem cells belong to populations of cells that retain a wide or unlimited morphogenetic potential.

Our results along with literature data allow suggest the existence of evolutionary conservative, common for all studied metazoan representatives, from sponges to chordates, cellular, sub-cellular and molecular bases of pluripotency and "stemness" of stem and germ cells.

Many authors called pluripotent stem cells of animals with asexual reproduction *somatic* cells (Blackstone & Jasker, 2003; Extavour & Akam, 2003; Extavour, 2008; Rinkevich, 2009; Sköld et al., 2009; Funayama et al., 2010). The term *somatic embryogenesis* (Buss, 1987; Blackstone & Jasker, 2003) clearly shows that stem cells providing for asexual reproduction are considered as somatic cells. However, pluri/totipotent stem cells in asexually reproducing animals, as well as primary germ cells in sexual reproduction, are not belonging to any germ layer, tissue, or population of specialized somatic cells (Isaeva et al., 2008b, 2009; Isaeva, 2010). The population of these pluripotent stem cells is a diaspora of amoeboid cells, dispersed in the organism; these stem cells do not display contact inhibition of cell reproduction and movement. The morphological and functional organization of the stem cells capable for both gametogenesis (and subsequent embryogenesis) and asexual reproduction (blastogenesis) and germ-line cells shares common properties. Our notion of the evolutionary and ontogenetic similarity between the stem cells of asexually reproducing animals, cells of the germ line and embryonic stem cells (Isaeva et al., 2008b, 2009) brings us back to the concept of Weismann (1883), who wrote about the indistinguishability of gametogenic cells and undifferentiated cells that preserve the "germ plasm." It was later proposed that pluripotent gametogenic stem cells and getmline cells could be evolutionarily and ontogenetically related (Weissman, 2000; Extavour & Akam, 2003; Travis, 2007; Extavour, 2008; Strouji & Extavour, 2011). Primary germ cells and pluri/totipotent stem cells share many morphological characters and rely on the activity of related genes (Extavour & Akam, 2003; Hayashi et al., 2007; Travis, 2007; Sköld et al., 2009; Srouji & Extavour, 2011). Pluripotent stem cells of animals with asexual reproduction were termed "primary stem cells" (Skøld et al., 2009); such pluripotent stem cells are predecessors of primary germ cells (Blackstone & Jasker 2003; Sköld et al., 2009). Evolutionary conserved mechanism ensures germline specification remaining cell pluripotency (Strouji & Extavour, 2011). Thus, the pluripotent stem cells are not identical to somatic cells. Besides, morphogenesis in animal asexual reproduction does not completely recapitulate embryogenesis. So, the term *somatic embryogenesis* (Buss, 1987; Blackstone & Jasker, 2003) is not completely correct, the term *blastogenesis* seems preferable (Berrill, 1961).

Stem cells of animals with asexual reproduction, as well as cells of the germ lineage, probably originate in the early embryogenesis either from the early totipotent blastomeres or from their derivates that retain pluri/totipotency. The author believes that the evolutionarily and ontogenetically related cells of early embryos, primary stem and primary germ cells belong to cell populations capable of realizing the developmental program, including gametogenesis (and, potentially, subsequent embryogenesis) and blastogenesis (Isaeva et al., 2008b, 2009; Isaeva, 2010).

Thus, published and original data indicate evolutionary conservation and similarity of the studied morphofunctional properties of stem cells in metazoans with asexual reproduction (from sponges and cnidarians to chordates), germline and embryonic stem cells. In invertebrates with asexual reproduction, stem cells can differentiate into both germline and somatic cells; these pluri/totipotent stem cells represent a source of cells for the life strategy realization including sexual and asexual reproduction. Further research on the stem cells of various metazoan animals may reveal the evolutionary conserved basis of cellular totipotency and potential immortality.

4. Acknowledgment

This study was supported by a grant from the Russian Federation for Basic Research (no. 09-04-00019).
I am most grateful always to my collegues Andrey Shukalyuk, Anna Akhmadieva, Yana Alexandrova, Alexey Chernyshev and Arkady Reunov for our collective work.

5. References

Agata, K., Nakajima, E., Funayama, N., Shibata, N., Saito, Y., Unesono, Y. (2006). Two Different Evolutionary Origin of Stem Cell Systems and their Molecular Basis, *Seminar in Cell and Developmental Biology*, Vol.17, pp. 503-509, ISSN 1084-9521

Agata, K. & Watanabe, K. (1999). Molecular and Cellular Aspects of Planarian Regeneration, *Seminars in Cell and Developmental Biology*, Vol.10, pp. 377–383, ISSN: 1084-9521

Akhmadieva, A.V., Shukalyuk, A.I., Aleksandrova, Ya.N. & Isaeva, V.V. (2007). Stem Cells in Asexual Reproduction of Colonial Ascidian *Botryllus tuberatus, Russian Journal of Marine Biology*, Vol.33, No.2, pp. 134–137, ISSN 1063-0740

Alexandrova,Ya.N. & Reunov, A.A. (2008). The Oogonia of Macroalga *Undaria pinnatifida* are Alkaline-Phosphatase Positive and Contain Germinal Body-like Structure, *Journal of Phycology*, Vol.44, pp. 712-715, ISSN: 1529-8817

Alié, A., Leclère, L., Jager, M., Dauraud, C., Chang, P., Guyader, H., Quéinnec, E., & Manuel, M. Somatic Stem Cells Express *Piwi* and *Vasa* Genes in an Adult Ctenophore: Ancient Association of "Germline Genes" with Stemness, Developmental Biology, Vol.350, pp. 183-197, ISSN 0012-1606

Ambrose, V. & Chen, X. (2007). The Regulation of Genes and Genomes by Small RNA, *Development*, Vol.134, pp. 1635-1641, ISSN 1011-6370

Amikura, R., Kashikawa, M., Nakamura, A., & Kobayashi, S. (2001). Presence of Mitochondria-type Ribosomes Outside Mitochondria in Germ plasm of *Drosophila* Embryos, *The Proceedings of the National Academy of Sciences USA*, Vol. 98, pp. 9133-9138, ISSN 1091-6490

Auladell, C., Garcia-Valero, J., & Baguna, J. (1993). Ultrastructural Localization of RNA in the Chromatoid Bodies of Undifferentiated Cells (Neoblasts) in Planarians by the RNAse-Gold Complex Technique, *Journal of Morphology*, Vol. 216, pp. 319–326, ISSN 1097-4687

Baguña, J., Salo, E., & Auladell, C. (1989). Regeneration and Pattern Formation in Planarians. III. Evidence that Neoblasts are Totipotent Stem Cells and the Source of Blastema Cells, *Development*, Vol. 107, pp. 77–86, ISSN 1011-6370

Batygina, T.B. Stem Cells and Morphogenetic Developmental Programs in Plants. (2010). In: *Dauther Cells: Properties, Characteristics, and Stem Cells.* Eds. Hitomi, A. & Katoaka, V., Nova Science Publishers, New York, pp. 51-128, ISBN 978-1-60741-638-8.

Beams, H.W. & Kessel, R.G. (1974). The Problem of Germ Cell Determinants, *International Review of Cytology*, Vol. 39, pp. 413–479, ISSN:0074-7696

Berrill, N.J. (1961). *Growth, Development, and Pattern*, Freeman and Company: San Francisco, London, 556 p.

Blackstone, N.W. & Jasker, B.D. (2003). Phylogenetic Consideration of Clonality, Coloniality, and Mode of Germline Development in Animals, *Journal of Experimental Zoology*, Vol. 287, pp. 35-47, ISSN: 15525015, 15525007

Bode, H.R. (1996). The Interstitial Cell Lineage of Hydra: a Stem Cell System that Arose Early in Evolution, *Journal of Cell Science*, Vol. 109, pp. 1155-1164, ISSN 0021-9533

Bosch, T.C.G. (2008). Stem Cells in Immortal *Hydra,* In: *Stem Cells From Hydra to Man,* Ed. Bosch, T.C.G., Springer, pp. 37-57, e-ISBN: 978-1-4020-8274-0

Brown, F.D., Keeling, E.L., Le, A.D., Swalla, B.J. (2009). Whole Body Regeneration in a Colonial Ascidian, *Botrylloides violaceus, Journal of Experimental Zoology, B Molecular and Developmental Evolution,* Vol.312, pp.885-900, ISSN 1552-499X

Burighel, P. & Cloney, R.A. (1997). Urochordata: Ascidiacea, In: *Microscopic Anatomy of Invertebrates,* N.Y.: Wiley-Liss.,Vol. 15, pp. 221–347, ISBN 9780471561224

Buss, L.W. (1987). *The Evolution of Individuality,* Princeton University Press, Princeton, NJ, ISBN 978-0691084688

Buss, L.W. (1999). Slime Molds, Ascidians and the Utility of Evolutionary Theory, *The Proceedings of the National Academy of Sciences USA,* Vol. 96, P. 8801–8803, ISSN-0027-8424

Cameron, R.A., Peterson, K.J. & Davidson, E.H. (1998). Developmental Gene Regulation and the Evolution of Large Animal Body Plan, *American Zoologist,* Vol. 38, No 4, pp. 609-620, ISSN 0003-1569

Campbell, R.D. (1974). Cnidaria, In: *Reproduction of Marine Invertebrates,* New York; London: Academic Press, Vol. 1, pp. 133–199, ISBN 0-12-2825-01 (-02,-03)-9

Carré, D., Djediat, C., & Sardet, C. (2002). Formation of a Large Vasa-Positive Germ Granules and Its Inheritance by Germ Cells in the Enigmatic Chaetognaths, *Development,* Vol. 129, pp. 661–670, ISSN 1011-6370

Castrillon, D.H., Quade, B.J., Wang, T.Y., et al. (2000). The Human VASA Gene is Specifically Expressed in the Germ Cell Lineage, *The Proceedings of the National Academy of Sciences USA,* Vol. 97, pp. 9585–9590, ISSN-0027-8424

Cima, F., Perin, A., Burighel, P., & Ballarin, L. (2001). Morpho-functional Characterization of Haemocytes of the Compound Ascidian *Botrylloides leachi* (Tunicata, Ascidiacea), *Acta Zoologica,* Vol. 82, pp. 261–274, ISSN 1463-6395

Chiquoine, A.D. (1954). The Identification, Origin and Migration of the Primordial Germ Cells of the Mouse Embryo, *The Anatomical Record,* Vol. 118, pp. 135- 146, ISSN: 1932-8486

Chuma, S., Hosokawa, M., Kitamura, K., Kasai, S., Fujioka, M., Hiyoshi, M., Takamune, K., Noce, T., & Nakatsuji N. (2006). Tdrd1/Mtr-1, a tudor-related Gene, is Essential for Male Germ-cell Differentiation and Nuage/Germinal Granule Formation in Mice. *The Proceedings of the National Academy of Sciences USA.* October, Vol. 103, pp. 15894–15899, ISSN 1091-6490 www.pnas.org

Cinalli, R. M., Rangan,P. & Lehmann, R. (2008). Germ Cells Are Forever. *Cell,* Vol. 132, pp 559-562, ISSN 0092-8674

Cogle, C. R., Guthrie, S.M., Sanders, R. C., Allen, W. L., Scott, E. W. & Petersen, B.E. (2003). An Overview of Stem Cell Research and Regulatory Issues *Mayo Foundation for Medical Education and Research,* Vol. 8, pp. 993-1003, ISSN: 1942-5546.

Collins, A.G. & Valentine, J.W. (2001). Defining Phyla: Evolutionary Pathways to Metazoan Body Plans, *Evolution and Development,* Vol. 3, No 6, pp. 432-442, ISSN 1520541X, 1525142X

Corley, L.S., White, M.A. & Strand, M.R. (2005). Both Endogenous and Environmental Factors Affect Embryo Proliferation in the Polyembryonic Wasp *Copidosoma floridanum, Evolution and Development,* Vol. 7, pp. 115-121, ISSN 1520541X, 1525142X

Clark, A.T., Bodnar, M.S., Fox, M., Rodriquez, R.T., Abeyta, M.J., Firpo, M.T. & Pera, R.A.R. (2004). Spontaneous Differentiation of Germ Cells from Human Embryonic Stem Cells in vitro, *Human Molecular Genetics,* Vol. 13, pp. 727-739, ISSN 0964-6906

Daley, G.Q. (2007). Gametes from Embryonic Stem Cells: a Cup Half Empty or Half Full? *Science*, Vol. 316, pp. 409-410, ISSN 0036-8075

Davidson, E.H., Peterson, K.J. & Cameron, R.A. (1995). Origin of Bilaterian Body Plans: Evolution of Developmental Regulatory Mechanisms, Science, Vol.270, pp. 1319-1325, ISSN 0036-8075

Ding, D. & Lipshitz, H.D. (1993). A Molecular Screen for Polar-localized Maternal RNAs in the Early Embryo of *Drosophila, Zygote*, Vol. 1, pp. 257–271, ISSN 0967-1994

Donnell, D.M., Corley, L.S., Chen, G. & Strand, M.R. (2004). Caste Determination in a Polyembric Wasp Involves Inheritance of Germ Cells, *The Proceedings of the National Academy of Sciences USA*, Vol. 101, pp. 10095-10100, ISSN 0027-8424

Eddy, E.M. (1975). Germ Plasm and the Differentiation of the Germ Cell Line, *International Review of Cytology*, Vol. 43, pp. 229–280, ISSN:0074-7696

Extavour, C.G.M. (2008). Urbisexuality: the Evolution of bBlaterian Germ Cell Specification and Reproductive Systems, Key Themes in Evolutionary Developmental Biology In: *Evolving Pathways*, Eds. Minelli, A. & Fusco, G. Cambridge, New York, Melbourne, Madrid, Cape Town, Singapore, San Paulo: Cambridge University Press, pp. 321-342, ISBN-13 978-0-511-37860-7

Extavour, C.G. & Akam, M. (2003.) Mechanisms of Germ Cell Specification across the Metazoans: Epigenesis and Preformation, *Development*, Vol. 130., pp. 5869-5884, ISSN 1011-6370

Ewen-Kampen, B., Schwager, E.E. & Extavour, C.G.M. (2010). The Molecular Machinery of Germ Line Specification, *Molecular Reproduction and Development*, Vol. 77, pp. 3-18, ISSN 1098-2795

Filipowicz, W., Jaskiewicz, L., Kolb, F.A. & Pillai, R.S. (2005). Post-transcription Gene Silencing by siRNAs and miRNAs, *Current Opinion in Structural Biology*, Vol. 15, pp. 331-341, ISSN 0959-440X

Frank, U., Plickert, G. & Müller, W.A. (2009). Cnidarian Interstitial Cells: the Dawn of Stem Cell Research, In: *Stem Cells in Marine Organisms*, Eds. Rinkevich B., & Matranga V., Springer: Dordrecht, Heidelberg, London, New York, pp. 33-59. e-ISBN 978-90-481-2767-2

Funayama, N. (2008). Stem cells of sponge, In: *Stem Cells. From Hydra to Man*, Ed. Bosch Th.C.G., Springer, pp. 17-36, e-ISBN: 978-1-4020-8274-0

Funayama, N., Nakatsukasa, M., Mohri, K., Masuda, Y., & Agata, K. (2010). Piwi Expression in Archeocytes and Choanocytes in Demosponges: Insights into the Stem Cell System in Demosponges, *Evolution and Development*, Vol. 12, No 3, pp. 275–287, ISSN 1520541X, 1525142X

Geijsen, N., Horoschak, M., Kim, K., Gribnau, J., Eggan, K. & Daley, G.Q. (2003). Derivation of Embryonic Germ Cells and Male Gametes from Embryonic Stem Cells, *Nature*, Vol. 427, pp. 148-154, ISSN 0028-0836

Ghiselin, M.T. (1987). Evolutionary Aspects of Marine Invertebrate Reproduction, In: *Reproduction of Marine Invertebrates*, Vol. 9, General Aspects: Seeking Unity in Diversity, Eds. Giese A.C., Pearse J.C. & Pearse, V.B. Palo Alto: Blackwell Scientific Publications and Pacific Grove: The Boxwood Press, pp. 609-665, ISBN 0792370899

Giudice, G. (1962) Restitution of Whole Larvae from Disaggregated Cells of Sea Urchin Embryo, *Developmental Biology*, Vol. 5, pp. 402-411, ISSN 0012-1606.

Glenner, H., Lützen, J. & Takahashi, T. (2003). Molecular and Morphological Evidence for a Monophyletic Clade of Asexually Reproducing Rhizocephala: *Polyascus*, New Genus (Cirripedia), *Journal of Crustacean Research*, Vol. 23, pp. 548–557, ISSN 0011-216X.

Gschwentner, R., Ladurner, P., Nimeth, K. & Rieger, R. (2001). Stem Cells in a Basal Bilaterian. S-phase and Mitotic Cells in *Convolutriloba longifissura*, *Cell and Tissue Research*, Vol. 304, pp. 401-408, ISSN 0302-766X

Hagan, H.R. (1951). *Embryology of the Viviparous Insects*, New York: The Ronald Press Company, 472 p.

Hall, P.A. & Woods, A.L. (1990). Immunohistochemical Markers of Cellular Proliferation: Ahievements, Problems and Prospects, *Cell and Tissue Kinetics*, Vol. 23, pp. 23–55, ISSN 0008-8730

Hay, B., Jan, L.Y., & Jan, Y.N. (1988). A Protein Component of *Drosophila* Polar Granules is Encoded by *vasa* and Has Extensive Sequence Similarity to ATP-dependent Helicases, *Cell*, Vol. 55, pp. 577–587, ISSN 1535-9786

Hayashi, K., De Sousa Lopez, S.M.C. & Surani, M.A. (2007). Germ Cell Specification in Mice, *Science*, Vol. 316, pp. 394-396 ISSN 1095-9203

Hinegardner, R.T. (1975). Morphology and Genetics of Sea Urchin Development, *American Zoologist*, Vol. 15, pp. 679-689, ISSN 0003-1569

Hogan, B. (2001). Primordial Germ Cells as Stem Cells, In: *Stem Cell Biology*, Eds. Marshak, D.R., Gardner, R.L. & Gottlieb, D, New York: Cold Spring Harbor Laboratory Press, pp. 189-204, ISBN 0613923901

Høeg, J.T. (1992). Rhizocephala, In: *Microscopic Anatomy of Invertebrates, Vol. 9: Crustacea*, New York: Wiley-Liss, pp. 313–345, ISBN 9780471561149

Høeg, J.T. & Lützen, J. (1995). Life Cycle and Reproduction in the Cirripedia Rhizocephala, *Oceanography and Marine Biology: an Annual Reviw*, Vol. 33, pp. 427-485, ISBN-13 9781439853641

Honegger, T.G., Zurrer, D. & Tardent, P. (1989). Oogenesis in *Hydra carnea*: A New Model Based on Light and Electron Microscopic Analyses of Oocyte and Nurse Cell Differentiation, *Tissue and Cell*, Vol. 21, pp. 381–393, ISSN 0040-8166.

Hong, Y., Winkler, C., & Schartl, M. (1998). Production of Medakafish Chimeras from a Stable Embryonic Stem Cell Line, *The Proceedings of the National Academy of Sciences USA*, Vol. 95, pp. 3679–3684, ISSN 0027-8424

Hori, I. (1982). An Ultrastructural Study of the Chromatoid Body in Planarian Regenerative Cells, *Journal of Electron Microscopy*, Vol. 31, pp. 63–72, ISSN 1477-9986

Houston, D.W. & King, M.L. (2000). Germ Plasm and Molecular Determinants of Germ Cell Fate, *Current Topics in Developmental Bioogy*, Vol. 50, pp. 155–181, ISSN:0070-2153

Hübner, K., Fuhrmann, G., Kohler J., Reinhold, R., De La Fuente., R., Wood, J., Strauss, J., Boiani, M. & Scholer, H. (2003). Derivation of Oocytes From Mouse Embryonic Stem Cells, *Science*, Vol. 300, pp. 1251-1256, ISSN 0036-8075

Ikenishi, K. (1998). Germ Plasm in *Caenorabditis elegans, Drosophila* and *Xenopus, Development, Growth and Differentiation*, Vol. 40, pp. 1–10, ISSN: 1440-169X

Isaeva, V.V. (2010). The Diversity of Ontogeny in Animals with Asexual Reproduction, and Plasticity of Early Development, *Russian Journal of Developmental Biology*, Vol. 41, No. 5, pp. 285–295, ISSN 0062-3604

Isaeva, V.V. & Akhmadieva, A.V. (2011). Germinal Granules in Archaeocytes of the Sponge *Oscarella malakhovi* Ereskovsky, 2006, *Russian Journal of Marine Biology*, Vol.37, pp. 209-216, ISSN 1063-0740.

Isaeva, V.V. Akhmadieva, A.V., Alexandrova, Ya. Chernushev, A.V. & Shukalyuk, A.I. (2011). Germinal Granules in Interstitial Cells of Colonial Hydroids *Obelia longissima* Pallas, 1766 and *Ectopleura crocea* Agassiz, 1862, *Russian Journal of Marine Biology*, Vol. 37, No. 4, pp. 303–310, ISSN 1063-0740

Isaeva, V., Alexandrova, Ya. & Reunov, A. (2005). Interaction between Chromatoid Bodies and Mitochondria in Neoblasts and Gonial Cells of the Asexual and Spontaneously Sexualized Planarian *Girardia (Dugesia) tigrina, Invertebrate Reproduction and Development,* Vol. 48 (1-3), pp. 119-128, ISSN 0792-4259

Isaeva, V., Akhmadieva, A.V., Alexandrova, Ya.N. & Shukalyuk, A.I. (2009). Morphofunctional Organization of Reserve Stem Cells Providing for Asexual and Sexual Reproduction of Invertebrates, *Russian Journal of Developmental Biology,* Vol. 40, No. 2, pp. 57–68, ISSN 0062-3604

Isaeva, V.V., Kasyanov, N.V. & Presnov E.V. (2008a). *Analysis Situs* of Spatial-Temporal Architecture in Biological Morphogenesis, In: *Progress in Mathematical Biology Research,* Ed. J.T. Kelly, New York: Nova Science Publishers, pp. 141-189, ISBN 978-1-60456-171-5

Isaeva, V.V. & Reunov, A.A. (2001). Germ Plasm and Germ Cell Line Determination: Role of Mitochondria, *Russian Journal of Marine Biology,* Vol. 27, No. 4, pp. 231–237, ISSN 1063-0740

Isaeva, V.V., Shukalyuk, A.I. & Akhmadieva, A.V. (2008b). Stem Cells in Reproductive Strategy of Asexually Reproducing Invertebrates, *Russian Journal of Marine Biology,* Vol. 34, No. 1, pp. 1–8, ISSN 1063-0740

Isaeva, V.V., Shukaljuk, A.I. & Kizilova, E.A. (2003). Revealing of Stem Cells in Colonial Interna of the Rhizocephalans *Peltogasterella gracilis* and *Sacculina polygenea* at Parasitic Stage of their Life Cycle, *Tsitologiya,* Vol. 45, No. 8, pp. 758–763, ISSN 0041-3771

Isaeva, V.V., Shukalyuk, A.I., Korn, O.M. & Rybakov, A.V. (2004). Development of Primordial Externae in the Colonial Interna of *Polyascus polygenea* (Crustacea: Cirripedia: Rhizocephala), *Crustacean Research,* No 33,pp. 61-71, ISSN 0287-3478

Ivanova-Kazas, O.M. (1996). Blastogenesis, Cormogenesis, and Evolution, *Russian Journal of Marine Biology,* Vol. 22, No. 5, pp. 285–294, ISSN 1063-0740

Jaeckle, W.B. (1994). Multiple Modes of Asexual Reproduction by Tropical and Subtropical Sea Star Larvae: an Unusual Adaptation for Genet Dspersal and Srvival, *Biological Bulletin,* Vol. 186, pp. 62-71, ISSN:0006-3185

Jenner, R.A. (2000). Evolution of Animal Body Plan: the Role of Metazoan Phylogeny at the Interface between Pattern and Process, *Evolution and Development,* Vol. 2, No 4, pp. 208-221, ISSN 1520-541X.

Johannsen O.A., Butt F.H. (1941). *Embryology of Insects and Myriapods. The developmental History of Insects, Centipedes, and Millipedes from Egg Deposition to Hatching,* New York, London: McGraw-Hill Book Company, 462 p.

Juliano, C.E., Voronina, E., Stack, C., Aldrich, M., Cameron, A.R. & Wessel, G.M. (2006). Germ Line Determinants Are not Localized Early in Sea Urchin Development, but do Accumulate in the Small Micromere Lineage, *Developmental Biology,* Vol. 300, pp. 406–415, ISSN 0012-1606

Juliano, C.E., & Wessel, G.M. (2009). An Evolutionary transition of Vasa regulation in Echinoderms, *Evolution and Development,* Vol.11, pp. 560-573, ISSN 1520541X, 1525142X

Kashikawa, M., Amikura, R., Nakamura, A. & Kobayashi, S. (1999). Mitochondrial Small Ribosomal RNA is Present on Polar Granules in Early Cleavage Embryos of *Drosophila melanogaster, Development, Growth and Differentiation,* vol. 41, pp. 495–502, ISSN 1440-169X

Kerkis, A., Fonseca, S.A.S., Serafim, R.C., Lavagnolli, T.M.C., Abdelmassih, S., Abdelmassih, R. & Kerkis, I. (2007). *In vitro* Differentiation of Male Mouse Embryonic Stem Cells into Both Presumptive Sperm Cells and Oocytes, *Cloning Stem Cells*, Vol. 9, No. 4, pp. 535–548, ISSN 1536-2302

Kerszberg, M. & Wolpert, L. (1998). The Origin of Metazoa and The Egg: a Role for Cell Death, *Journal of Theoretical Biology*, Vol. 193, pp. 535–537, ISSN: 0022-5193

Kloc, M., Bilinski, S. & Etkin, L.D. (2004). The Balbiani Body and Germ Cell Determinants: 150 Years Later, *Current Topics in Developmental Biology*, Vol. 59, pp. 1-36, ISSN:0070-2153

Kobayashi, S., Amikura, R., & Okada, M. (1998). Mitochondrial Large Ribosomal RNA in Germ Plasm of *Xenopus* Embryos, *Current Biology*, Vol. 8, pp. 1117–1120, ISSN 0960-9822

Kobayashi, S., Sato, K., & Hayashi, Y. (2005). The Role of Mitochondrial rRNAs and Nanos Protein in Germline Formation in *Drosophila* Embryos, *Zoological Science*, Vol. 22, pp. 943-954, ISSN: 0289-0003

Kotaja, N., Bhattacharyya, S.N., Jaskiewicz, L., Kimmins, S., Parvinen, M., Witold Filipowicz, W., & Sassone-Corsi P. (2006). The Chromatoid Bodies of Male Germ Cells: Similarity with Processing Bodies and Presence of Dicer and microRNA Pathway Components, *The Proceedings of the National Academy of Sciences USA*, vol. 103, pp. 2647–2652, ISSN 0027-8424

Lacham-Kaplan, O. (2004). *In Vivo* and *In Vitro* Differentiation of Male Germ Cells in the Mouse, *Reproduction*, Vol. 128, pp. 147–152, ISSN 1470-1626

Laird, D.J., De Tomaso, A.W. & Weissman, I.L. (2005). Stem Cells are Units of Natural Selection in a Colonial Ascidian, *Cell*, Vol. 123, pp. 1351-1360, ISSN 0092-8674

Laird, D.J. & Weissman, I.L. (2004). Telomerase Maintained in Self-renewing Tissues during Serial Regeneration of the Urochordate *Botryllus schlosseri*, *Developmental Biology*, Vol.273, pp. 185-194, ISSN 1062-3604

Leatherman, J.L. & Jongens, T.A. (2003). Transcriptional Silencing and Translational Control: Key Features of Early Germline Development, *BioEssays*, Vol. 25, Ppp 326-335, ISSN 1579-4377

Lim, A.K. & Kai, T. (2007). Unique Germ-line Organelle, Nuage, Functions to Repress Selfish Genetic Elements in *Drosophila melanogaster*, *The Proceedings of the National Academy of Sciences USA*, Vol. 104, pp. 6714-6719, ISSN 1091-6490

Lohman, J.U. (2008). Plant Stem Cells: Divide et Impera. In: *Stem Cells. From Hydra to Man*, Ed. Bosch, Th.C.G. Springer Science, pp. 1-15, e-ISBN 978-1-4020-8274-0

Loughry, W.C., Prodöhl, P.A., McDonough, C.M. & Avise, J.C. (1998). Polyembryony in Armadillos, *American Scientist*, Vol. 86, pp. 274-279, ISSN 0003-0996

Luking, A., Stahl, U. & Schmidt, U. (1998). The Protein Family of RNA Helicases, *Critical Review in Biochemistry and Molecular Biology*, Vol. 33, pp. 259–292, ISSN 1040-9238

Mahowald, A.P. (1971). Polar Granules in *Drosophila*. III. The Continuityof Polar Granules during the Life Cycle of *Drosophila*, *Journal of Experimental Zoology*, Vol. 170, pp. 551–563, ISSN 1552-499X

Mahowald, A.P. (2001). Assembly of the *Drosophila* Germ Plasm, *International Review of Cytology*, Vol. 203, pp. 187-213, ISSN 0074-7696

Marshak, D.R., Gottlieb, D. & Gardner, R.L. (2001). Introduction: Stem Cell Biology, In: *Stem Cell Biology*, Eds. Marshak, D.R., Gardner, R.L. & Gottlieb, D. Cold Spring Harbor Labiratory Press. Cold Spring Harbor, New York, pp. 1-16, ISBN 0-87969-673-7

Matova, N. & Cooley, L. (2001). Comparative Aspects of Animal Oogenesis, *Developmental Biology*, Vol. 16, pp. 1–30, ISSN 1062-3604

Mintz, B. (1959). Continuity of the Female Germ Cell Line from Embryo to Adult, *Archiv of Microscopic Morphology*, Vol. 48, pp. 155–172

Mochizuki, K., Nishimiya-Fujisawa, C. & Fujisawa, T. (2000). Expression and Evolutionary Conservation of nanos-related Genes in Hydra, *Development Genes and Evolution*, Vol. 210, pp. 591–602, ISSN 0949-944X

Mochizuki, K., Nishimiya-Fujisawa, C. & Fujisawa, T. (2001). Universal Occurrence of the *vasa*-related Genes among Metazoans and their Germline Expression in Hydra, *Development, Genes and Evolution*, Vol. 211, pp. 299-308, ISSN 0949-944X

Müller, W.E.G. (2006). The Stem Cell Concept in Sponges (Porifera): Metazoan traits, *Seminars in Cell and Developmental Biology*, Vol. 17, pp. 481-491, ISSN 1084-9521

Newton, D.E. (2006). *Stem Cell Research*. New York: Infobase Publishing. 284 p., ISBN 10-0-8160-6576-4

Noda, K. & Kanai, C. (1977) An Ultrastructural Observation of *Pelmatohydra robusta* at Sexual and Asexual Stages, with a Special Reference to "Germinal Plasm," *Journal of Ultrastructural Research*, vol.61, pp. 284-294, ISSN 0022-5320

Nikitin, N.S. (1977). Experimental Morphological Study of Morphogenetic Potential of Homogeneous Conglomerates of Different Cell Types from the Freshwater Sponge *Ephydatia fluviatilis, The Soviet Journal of Developmental Biology*, Vol. 8, pp. 460-467, ISSN 0049-1748

Nussbaum, M. Zur Differenzirung des Geschlechts im Thierreich, *Archiv für Mikroskopy und Anatomy*, 1880. Band 18, S. 1-121

Orii, H., Sakurai, T., & Watanabe, K. (200)., Distribution of the Stem Cells (Neoblasts) in the Planarian *Dugesia japonica, Development, Genes and Evolution*, Vol. 215, pp. 143–157, ISSN 0949-944X

Pancer, Z., Gershon, H. & Rinkevich, B. (1995). Coexistence and Possible Parasitism of Somatic and Germ Cell Lines in Chimeras of the Colonial Urochordate *Botryllus schlosseri, Biological Bulletin*, Vol. 189, pp. 106-112, ISSN 0006-3185

Perez, C. (1931). Cycle évolutif des Rhizocéphales du genre Chlorogaster, *Archivio Zoologico Italiano*, Vol. 26 (Atti Dell' XI Congreso Internazionale di Zoologia, Padova 1930), pp. 1319-1328

Laird, D.J. & Weissman, I.L. (2004). Telomerase Maintained in Self-renewing Tissues During Serial Regeneration of the Urochordate *Botryllus schlosseri, Developental Biology*, Vol. 273, pp. 185-194, ISSN 0012-1606

Peter, R., Ladurner, P. & Rieger, R.M. (2001). The Role of Stem Cell Strategies in Coping with Environmental Stress and Choosing between Alternative Reproductive Modes: Turbellaria Rely on a Single Cell Type to Maintain Individual Life and Propagate Species, *Marine Ecology*, Vol. 22, pp. 35-51, ISSN 1439-0485

Pfister, D., De Mulder, K., Hartenstein, V., Kuales, G., Borgonie, G., Marx, F., Morris, J., & Ladurner, P. (2008). Flatworm Stem Cells and the Germ Line, *Developmental Biology*, Vol.319, pp. 146-159, ISSN 1062-3604

Presnov, E., Isaeva, V., & Kasyanov, N. (2010). Topological Determination of Early Morphogenesis in Metazoa, *Theory in Bioscience*, Vol. 129, pp, 259-270, ISSN:1431-7613

Raz, E. (2000). The Function and Regulation of the *vasa*-like Genes in Germ-Cell Development, *Genome Biology*, Vol. 3, pp. 1017.1-1017.6, ISSN 1474-7596

Rebscher, N., Zelada-González, F., Banish, T., Raible, F. & Arendt, D. (2007). Vasa Unveils a Common Origin of Germ Cells and of Somatic Stem Cells from the Posterior Growth Zone in the Polychaete *Platynereis dumerlii, Developmental Biology*, Vol.306, pp. 599-611, ISSN 1062-3604

Rebscher, N., Volk, C., Teo, R., & Plickert, G. (2008). The Germ Plasm Component vasa Allowstracing of the Interstitial Stem Cells in the Cnidarian *Hydractinia echinata. Developmental Dynamics*, Vol.237, pp. 1736–1745.

Reunov, A., Isaeva, V., Au, D., & Wu, R. (2000). Is It Possible the Arising of Nuage Constituents from Mitochondria? *Development, Growth and Differentiation*, Vol. 42, pp. 129–143, ISSN 1440-169X

Reunov, A.A., Neznanova, S.Yu., Aleksandrova, Ya.N., & Isaeva, V.V. (2004). An Ultrastructural Study of the Interaction of Germinal Granules and Mitochondria in *Apostichopus japonicus* (Echinodermata: Holothuroidea) and *Pleuronectes asper* (Teleostei: Pleuronectidae), *Russian Journal of Marine Biology*, Vol. 30, No. 3, pp. 244–246, ISSN 1063-0740

Rieger, R.M., Tyler, S., Smith, J.P.S. III & Rieger, G.E. (1991). Plathelminthes: Turbellaria, In: *Microscopic Anatomy of Invertebrates, vol. 3. Plathelminthes and Nemertea*, Ed.Harrison, F.W., New York: Wiley-Liss, ISBN 0521853761

Rinkevich, B. (2009). Stem Cells: Autonomy Interactors that Emerge as Causal Agents and Legitimate Units of Selection, In: *Stem Cells in Marine Organisms*, Eds. Rinkevich, B. & Matranga, V., Springer: Dordrecht, Heidelberg, London, New York, pp. 1-20, e-ISBN 978-90-481-2767-2

Rinkevich, Y., Matranga, V. & Rinkevich, B. (2009). Stem Cells in Aquatic Invertebrates: Common Premises and Emerging Unique Themes, In: *Stem Cells in Marine Organisms*, Eds. Rinkevich, B. & Matranga, V., Springer: Dordrecht, Heidelberg, London, New York, pp. 61-104, e-ISBN 978-90-481-2767-2

Rosak, S. & Linder, P. (2004). DEAD-box Proteins: the Driving Forces Behind RNA Methabolism, *Review Molecular and Cell Biology*, Vol. 5, pp. 232–240, ISSN 1471-0072

Rosner, A., Moiseeva, E., Rinkevich, Y., Lapidot, Z., Rinkevich, B. (2009). Vasa and the Germ Line Lineage in a Colonial Urochordate, *Developmental Biology*, Vol.331, pp. 113-128, ISSN 1062-3604

Sablowski, R. (2010). Stem Cells in Plants and Animals, *Nature Education*. Vol. 3, pp. 4-9.

Sato, K., Sugita, T., Kobayashi, K., Fujita, K., Fuji, T., Matsumoto, Y., Mikami, T., Nishizuka, N., Nishizuka, S., Shojima, K., Suda, M., Takahashi, G., Himeno, H., Muto, A. & Ishida, S. (2001). Localization of Mitochondrial Ribosomal RNA on the Chromatoid Bodies of Marine Planarian Polyclad Embryos, *Development, Growth and Differentiation*, Vol. 43, pp.107–114, ISSN 1440-169X

Seipel K., Yanze, N. & Schmid, V. (2004). The Germ Line and Somatic Stem Cell Gene Cniwi in the Jellyfish Podocoryne carnea, *The International Journal of Deveopmental Biology*, Vol. 48, pp. 1–7, ISSN 0214- 6282

Seydoux, G. & Braun, R.E. (2006). Pathway to Totipotecy: Lessons from Germ Cells, *Cell*, Vol. 127, pp. 891-904, ISSN 1943-4456

Seto, A. G., Kingston, R. E., & Lau, N. C. (2007). The Coming of Age for Piwi Proteins, *Molecular Cell Research*, Vol. 26, pp. 603–609, ISSN 1097-2765

Shibata, N., Rauhana, L., & Agata, K. (2010). Cellular and Molecular Dissection of Pluripotent Adult Somatic Stem Cells in Planarians, *Development, Growth and Differentiation*, Vol. 52, pp. 27–41, ISSN 1440-169X

Shibata, N., Umesono, Y., Orii, H., Sakurai, T., Watanabe, K. & Agata, K. (1999). Expression of *vasa (vas)*-related Genes in Germline Cells and Totipotent Somatic Stem Cells of Planarians, *Developmental Biology*, Vol. 206, pp. 73–87, ISSN 1608-3326

Shukalyuk, A., Golovnina, K., Baiborodin, S., Gunbin, K., Blinov, A.G. & Isaeva V.V. (2007). *vasa*-related Genes and their Expression in Stem Cells of Colonial Parasitic

Rhizocephalan Barnacle *Polyascus polygenea* (Arthropoda: Crustacea: Cirripedia: Rhizocephala), *Cell Biology International*. Vol. 31, No 2,pp. 97-108, ISSN 1065-6995.

Shukalyuk, A.I., Isaeva, V.V. (2005). Stem and Germ Cells in Asexually Reproducing Animals: a Comparative Study, The American Society for Cell Biology. 45th Annual Meeting (San Francisco, December, 2005), *Molecular Biology of the Cell*, Vol. 16, No 11 (Suppl.), pp. 752a.

Shukalyuk, A.I., Isaeva, V.V., Akhmadieva A.V., & Alexandrova Ya.N. (2011). Stem Cells in Asexually Reproducing Invertebrates, Embryonic Stem Cells and Germline Cells Share Common, Evolutionary Conserved Features, International Society for Stem Cell Research, 9th Annual Meeting, June 15-18, 2011, Friday Poster Abstracts, Toronto, Canada, p. 77, www.isscr.org.

Shukalyuk, A., Isaeva, V., Kizilova, E. & Baiborodin, S. (2005). Stem cells in Reproductive Strategy of Colonial Rhizocephalan Crustaceans (Crustacea: Cirripedia: Rhizocephala), Invertebrate Reproduction and Development, Vol. 48, pp. 41-53, ISSN: 0792-5077

Simpson, T.L. (1984). *The Cell Biology of Sponges*, Springer Verlag: Berlin, Heidelberg, New York. 662 p.

Smith, A. Embryonic stem cells, (2001). In: *Stem cell biology*, Eds. Marshak, D.R., Gardner, R.L. & Gottlieb, D., New York: Cold Spring Harbor Laboratory Press, pp. 205-230, ISBN 0-87969-673-7

Sköld, H.N., Obst, M., Sköld, M. & Åkesson, B. (2009). Stem Cells in Asexual Reproduction of Marine Invertebrates, In: *Stem Cells in Marine Organisms*, Eds. Rinkevich, B. & Matranga, V. Springer: Dordrecht, Heidelberg, London, New York, pp. 105-137, e-ISBN 978-90-481-2767-2

Spiegel, E. & Spiegel, M. (1986). Cell-Cell Interactions During Sea Urchin Morphogenesis, In: *Developmental Biology: A Comprehensive Synthesis*. Vol. 2, New York, London: Plenum Press, pp. 195-240, ISBN 978-0-306-42166-2

Spiegel, M., Spiegel, E. (1975). The Reaggregation of Dissociated Embryonic Sea Urchin Cells, *American Zoologist*. Vol. 15, pp. 583-606, ISSN 0003-1569

Srouji, J. & Extavour, C. (2011). Redefining Stem Cells and Assembling Germ Plasm: Key Transition in the Evolution of the Germ Plasm, In: *Key transition in animal evolution*, Eds. DeSalle, R. & Schierwater, B. Science Publishers New York, USA Abingdon, UK, pp. 360-397, ISBN 978-1-57808-695-5

Stoner, D.S., Rinkevich, B., & Weissman, I.L. (1999). Heritable Germ and Somatic Cell Lineage Competitions in Chimeric Colonial Protochordates, *The Proceedings of the National Academy of Sciences USA*, Vol. 96, pp. 9148–9153, ISSN-0027-8424

Stoner, D.S., & Weissman, I.L. (1996). Somatic and Germ Cell Parasitism in a Colonial Ascidian: Possible Role for a Highly Polymorphic Allorecognition System, *The Proceedings of the National Academy of Sciences USA*, Vol. 93, pp. 15254-15259, ISSN-0027-8424

Strome, S. & Lehman, R. (2007). Germ versus Soma Decisions: Lessons from Flies and Worms, *Science*, Vol. 316, pp. 392-393, ISSN 1095-9203 (online)

Sunanaga, N., Saito, Y. & Kawamura, K. (2006). Postembryonic Epigenesis of Vasa-positive Germ Cells from Aggregated Hemoblasts in the Colonial Ascidian, *Botryllus primigenus*, *Development Growth and Differentiation*, Vol. 48, pp. 87-100, ISSN 0012-1592

Talbot, N.C., Rexroad, C.E. Jr., Pursel, V.G. & Powell, A.M. (1993). Alkaline Phosphatase of Pig and Sheep Epiblast Cells in Culture, *Molecular Reproduction and Development*, Vol. 36, pp. 139–147, ISSN 1098-2795

Teragawa, C.K. & Bode, H.R. (1990). Spatial and Temporal Patterns of Interstitial Cell Migration in *Hydra vulgaris, Developmental Biology*, Vol. 138, pp. 63–81, ISSN: 0012-1606

Thomas, M.B. & Edwards, N.C. (1991). Cnidaria: Hydrozoa, In: *Microscopic Anatomy of Invertebrates*, Vol. 2, pp. 91–183. ISBN 0521853761

Thompson, J.A., Itskovitz-Eldon, J., Shapiro, S.S., Waknitz, M.A., Swiergiel, J.J., Marshall, V.S. & Jones, J.M. Embryonic Stem Cells Lines Derived from Human Blastocysts, *Science*, 1998, Vol. 282, pp. 1145−1147, ISSN 0036-8075

Toledo, A., Cruz, C., Fragoso, G., Laclette, J.P., Merchant, M.T., Hernandez, M. & Sciutto, E. (1997). *In vitro* Culture of *Taenia crassiceps* Larval Cells and Cyst Regeneration after Injection into Mice, *Journal of Parasitology*, Vol. 83, pp. 189–193, ISSN:0022-3395

Toyooka, Y., Tsunekawa, N., Akasu, R., & Noce, N. (2003). Embryonic Stem Cells Can Form Germ Gells in vitro, *The Proceedings of the National Academy of Sciences USA*, Vol. 100, pp. 11457–11462, ISSN 1091-6490

Tuzet, O. (1964). L'Origine de la Lignée Germinale et la Gametogénèze chez les Spongiares, In: *L'origine de la Lignée Germinale chez les Vertébrés et chez Quelques Groupes d'Invertébrés*. Ed. Wolff, E., Herman: Paris, pp. 79-111

Travis, J. (2007). A Close Look at Urbisexuality, *Science*, Vol. 316, pp. 390-391, ISSN 0036-8075

Verdeil, J.-L., Alemanno, L., Niemenak, N. & Trabarger, T.J. (2007). Pluripotent versus Totipotent Plant Stem Cells: Dependence versus Autonomy? *Trends in Plant Science*, Vol. 12, No.6, pp. 245-252, ISSN 1360-1385

Vickery, M.S. & McClintock, J.B. (2000). Effects of Food Concentration and Availability on the Incidence of Cloning in Planktotrophic Larvae of the Sea Star *Pisaster ochraceus, Biological Bulletin*. Vol. 199, pp. 298-304, ISSN 0006-3185

Weismann, A. (1883). *Die Entstehung der Sexualzellen bei den Hydromedusen: Zugleich ein Beitrag zur Kenntniss des Baues und der Lebenserscheinungen dieser Gruppe*, Fischer, Jena, 295 S.

Weismann, A. (1892). *Das Keimplasma. Eine Theorie der Vererbung*, Jena: Verlag von Gustav Fisher, 682 S.

Weismann. A. (1893). *The Germ Plasm. A Theory of Heredity*, New York: Charles Scriber's Sons, 468 p.

Weissman I.L. (2000). Stem Cells: Units of Development, Units of Regeneration, and Units in Evolution, *Cell*, Vol. 100, pp. 157-168, ISSN 0092-8674

Weissman, I. L., Anderson, D. J. & Gage, F. (2001). Stem and Progenitor Cells: Origins, Phenotypes, Lineage Commitments, and Transdifferentiations, *Annual Review of Cell Developmental Biology*, Vol. 17, pp. 387–403, ISSN 1081-0706

Wilson, H.W. (1907). On Some Phenomena of Coalescence and Regeneration in Sponges, *Journal of Experimental Zoology*, Vol. 5, pp. 250-252

Winter, R. G. (2001). August Weismann on Germ-plasm Variation, *Journal of the History of Biology*, Vol. 34, pp. 517–555, ISSN 0022-5010

Wylie, C. (1999). Germ Cells, *Cell*, Vol. 96, pp. 165–174, ISSN 0092-8674